T0189986

Lecture Notes in Computer Science   9772

Commenced Publication in 1973
Founding and Former Series Editors:
Gerhard Goos, Juris Hartmanis, and Jan van Leeuwen

More information about this series at http://www.springer.com/series/7409

De-Shuang Huang · Kang-Hyun Jo (Eds.)

Intelligent Computing Theories and Application

12th International Conference, ICIC 2016
Lanzhou, China, August 2–5, 2016
Proceedings, Part II

 Springer

Editors
De-Shuang Huang
Tongji University
Shanghai
China

Kang-Hyun Jo
University of Ulsan
Ulsan
Korea (Republic of)

ISSN 0302-9743   ISSN 1611-3349   (electronic)
Lecture Notes in Computer Science
ISBN 978-3-319-42293-0   ISBN 978-3-319-42294-7   (eBook)
DOI 10.1007/978-3-319-42294-7

Library of Congress Control Number: 2016943868

LNCS Sublibrary: SL3 – Information Systems and Applications, incl. Internet/Web, and HCI

Printed on acid-free paper

This Springer imprint is published by Springer Nature
The registered company is Springer International Publishing AG Switzerland

Preface

The International Conference on Intelligent Computing (ICIC) was started to provide an annual forum dedicated to the emerging and challenging topics in artificial intelligence, machine learning, pattern recognition, bioinformatics, and computational biology. It aims to bring together researchers and practitioners from both academia and industry to share ideas, problems, and solutions related to the multifaceted aspects of intelligent computing.

ICIC 2016, held in Lanzhou, China, August 2–5, 2016, constituted the 12th International Conference on Intelligent Computing. It built upon the success of ICIC 2015, ICIC 2014, ICIC 2013, ICIC 2012, ICIC 2011, ICIC 2010, ICIC 2009, ICIC 2008, ICIC 2007, ICIC 2006, and ICIC 2005 that were held in Fuzhou, Taiyuan, Nanning, Huangshan, Zhengzhou, Changsha, China, Ulsan, Korea, Shanghai, Qingdao, Kunming, and Hefei, China, respectively.

This year, the conference concentrated mainly on the theories and methodologies as well as the emerging applications of intelligent computing. Its aim was to unify the picture of contemporary intelligent computing techniques as an integral concept that highlights the trends in advanced computational intelligence and bridges theoretical research with applications. Therefore, the theme for this conference was "Advanced Intelligent Computing Technology and Applications." Papers focused on this theme were solicited, addressing theories, methodologies, and applications in science and technology.

ICIC 2016 received 639 submissions from 22 countries and regions. All papers went through a rigorous peer-review procedure and each paper received at least three review reports. Based on the review reports, the Program Committee finally selected 236 high-quality papers for presentation at ICIC 2016, included in three volumes of proceedings published by Springer: two volumes of *Lecture Notes in Computer Science* (LNCS), and one volume of *Lecture Notes in Artificial Intelligence* (LNAI).

This volume of *Lecture Notes in Computer Science* (LNCS) includes 82 papers.

The organizers of ICIC 2016, including Tongji University and Lanzhou University of Technology, China, made an enormous effort to ensure the success of the conference. We hereby would like to thank the members of the Program Committee and the referees for their collective effort in reviewing and soliciting the papers. We would like to thank Alfred Hofmann, of Springer, for his frank and helpful advice and guidance throughout and for his continuous support in publishing the proceedings. Moreover, we would like to thank all the authors in particular for contributing their papers. Without the high-quality submissions from the authors, the success of the conference would not have been possible. Finally, we are especially grateful to the IEEE Computational Intelligence Society, the International Neural Network Society, and the National Science Foundation of China for their sponsorship.

May 2016
De-Shuang Huang
Kang-Hyun Jo

Preface

Organization

General Co-chairs

De-Shuang Huang	China
Cesare Alippi	Italy
Jie Cao	China

Program Committee Co-chairs

Kang-Hyun Jo	Korea
Vitoantonio Bevilacqua	Italy
Jinyan Li	Australia

Organizing Committee Co-chairs

Aihua Zhang	China
Ce Li	China

Organizing Committee Members

Weirong Liu	China
Erchao Li	China
Xiaolei Chen	China
Hui Chen	China
Suping Deng	China
Lin Zhu	China
Gang Wang	China

Award Committee Chair

Kyungsook Han	Korea

Tutorial Co-chairs

Laurent Heutte	France
Abir Hussain	UK

Publication Co-chairs

M. Michael Gromiha	India
Valeriya Gribova	Russia
Juan Carlos Figueroa	Colombia

Workshop/Special Session Chair

Ling Wang   China

Special Issue Co-chairs

Henry Han   USA
Phalguni Gupta   India

International Liaison Chair

Prashan Premaratne   Australia

Publicity Co-chairs

Evi Syukur   Australia
Chun-Hou Zheng   China
Jair Cervantes Canales   Mexico

Exhibition Chair

Lin Zhu   China

Program Committee

Andrea F. Abate	Fengfeng Zhou	Ming Jiang
Akhil Garg	Francesco Pappalardo	Jijun Tang
Vangalur Alagar	Shan Gao	Joaquin Torres
Angel Sappa	Liang Gao	Jun Zhang
Angelo Ciaramella	Kayhan Gulez	Kang Li
Bingqiang Liu	Hongei He	Ka-Chun Wong
Shuhui Bi	Huiyu Zhou	Seeja K.R
Bin Liu	Fei Han	Kui Liu
Cheen Sean Oon	Huanhuan Chen	Min Li
Chen Chen	Mohd Helmy Abd Wahab	Jianhua Liu
Wen-Sheng Chen	Hongjie Wu	Juan Liu
Michal Choras	Indrajit Saha	Yunxia Liu
Xiyuan Chen	Ivan Vladimir Meza Ruiz	Haiying Ma
Chunmei Liu	John Goulermas	Maurizio Fiasche
Costin Badica	Jianbo Fan	Marzio Pennisi
Dah-Jing Jwo	Jiancheng Zhong	Peter Hung
Daming Zhu	Junfeng Xia	Qiaotian Li
Dongbin Zhao	Jiangning Song	Qinmin Hu
Ben Niu	Jian Yu	Robin He
Dunwei Gong	Jim Jing-Yan Wang	Wei-Chiang Hong

Emanuele Lindo Secco
Shuigeng Zhou
Shuai Li
Shihong Yue
Saiful Islam
Jiatao Song
Shuo Liu
Shunren Xia
Surya Prakash
Shaoyuan Li
Tingwen Huang
Vasily Aristarkhov
Fei Wang
Xuesong Wang
Weihua Sun

Weidong Chen
Wei Wei
Zhi Wei
Shih-Hsin Chen
Wu Chen Su
Shitong Wang
Xiufen Zou
Xiandong Meng
Xiaoguang Zhao
Minzhu Xie
Xin Yin
Xinjian Chen
Xiaoju Dong
Xingsheng Gu
Xiwei Liu

Yingqin Luo
Yongquan Zhou
Yun Xiong
Yong Wang
Yuexian Hou
Chenghui Zhang
Weiming Zeng
Zhigang Luo
Fa Zhang
Liang Zhao
Zhenyu Xuan
Shanfeng Zhu
Quan Zou
Zhenran Jiang

Additional Reviewers

Yong Chen
Peng Xie
Yunfei Wang
Selin Ozcira
Stephen Tang
Badr Abdullah
Xuefeng Cui
Lumin Zhang
Chunye Wang
Qian Chen
Kan Qiao
Yingji Zhong
Wei Gao
Tao Yi
Liuhua Chen
Faliu Yi
Xiaoming Liu
Sheng Ding
Xin Xu
Zhebin Zhang
Shankai Yan
Yueming Lyu
Giulia Russo
Marzio Pennisi
Zhile Yang
Enting Gao
Min Sun

Lingjiao Pan
Ying Bi
Chao Jin
Shiwei Sun
Mohd Shamrie Sainin
Xing He
Xue Zhang
Junqing Li
Chen Chen
Wei-Shi Zheng
Chao Wu
Tingli Cheng
Francesco Pappalardo
Neil Buckley
Bolin Chen
Pengbo Wen
Long Wen
Bogdan Czejdo
Jing Wu
Weiwei Shen
Ximo Torres
Lan Huang
Jingchuan Wang
Savannah Bell
Alexandria Spradlin
Christina Spradlin
Li Liu

Geethan Mendiz
Jingsong Shi
Lun Li
Cheng Lian
Jin-Xing Liu
Obinna Anya
Lai Wei
Yan Cui
Peng Xiaoqing
Vivek Kanhangad
Yong Xu
Morihiro Hayashida
Yaqiang Yao
Chang Li
Jiang Bingbing
Haitao Li
Wei Peng
Jerico Revote
Xiaoyu Shi
Jia Meng
Jiawei Wang
Jing Jin
Yong Zhang
Biao Xu
Vangalur Alagar
Kaiyu Wan
Surya Prakash

Yongjin Li
Changning Liu
Xionghui Zhou
Hong Wang
Gongjing Chen
Yuntao Wei
Fangfang Zhang
Jia Liu
Jing Liu
Jnanendra Sarkar
Sayantan Singha Roy
Puneet Gupta
Shaohua Li
Zhicheng Liao
Adrian Lancucki
Julian Zubek
Srinka Basu
Xu Huang
Liangxu Liu
Qingfeng Li
Cristina Oyarzun Laura
Rina Su
Xiaojing Gu
Peng Zhou
Zewen Sun
Xin Liu
Yansheng Wang
Xiaoguang Zhao
Qing Lei
Yang Li
Wentao Fan
Hongbo Zhang
Minghai Xin
Yijun Bian
Yao Yu
Vasily Aristarkhov
Qi Liu
Vibha Patel
Jun Fan
Bojun Xie
Jie Zhu
Long Lan
Phan Cong Vinh
Zhichen Gong
Jingbin Wang
Akhil Garg

Wei Liao
Tian Tian
Xiangjuan Yao
Chenyan Bai
Guohui Li
Zheheng Jiang
Li Hailin
Huiyu Zhou
Baohua Wang
Kasi Periyasamy
Li Nie
Zhurong Wang
Ella Pereira
Danilo Caceres
Meng Lei
Changbin Du
Shaojun Gan
Yuan Xu
Chen Jianfeng
Chuanye Tang
Bo Liu
Bin Qian
Xuefen Zhu
Haoqian Huang
Fei Guo
Jiayin Zhou
Raul Montoliu
Oscar Belmonte
Farid Garcia-Lamont
Alfonso Zarco
Yi Gu
Ning Zhang
Jingli Wu
Xing Wei
Shenshen Liang
Nooraini Yusoff
Yanhui Guo
Nureize Arbaiy
Wan Hussain Wan Ishak
Yizhang Jiang
Pengjiang Qian
Si Liu
Chen Aiguo
Yunfei Yi
Rui Wang
Jiefang Liu

Aijia Ouyang
Hongjie Wu
Andrei Velichko
Wenlong Hang
Lijun Quan
Min Jiang
Tomasz Andrysiak
Faguang Wang
Liangxiu Han
Leonid Fedorischev
Wei Dai
Yifan Zhao
Xiaoyan Sun
Yiping Liu
Hui Li
Yinglei Song
Elisa Capecci
Tinting Mu
Francesco Giovanni Sisca
Austin Brockmeier
Cheng Wang
Juntao Liu
Mingyuan Xin
Chuang Ma
Marco Gianfico
Davide Nardone
Francesco Camastra
Antonino Staiano
Antonio Maratea
Pavan Kumar Gorthi
Antonio Brunetti
Fabio Cassano
Xin Chen
Fei Wang
Chen Xu
Gianpaolo Francesco
  Trotta
Alberto Cano
Xiuyang Zhao
Zhenxiang Chen
Lizhi Peng
Nagarajan Raju
e Wang
Yehu Shen
Liya Ding
Tiantai Guo

Zhengjun Xi
Lvzhou Li
Yuan Feng
Fabio Narducci
Silvio Barra
Jie Guo
Dongliang Xu
Murillo Carneiro
Junlin Chang
Shilei Qiao
Chao Lu
Guohui Zhang
Hongyan Sang
Mi Xiao
Andrei Mocanu
Qian Zhang
Eduyn Lopez
Joao Bertini
nuele Secco
Changxing Ding
Kuangyu Wang
George Caridakis
Zhenhuan Zhu

Andras Kupcsik
Yi Gu
Gabriela Ramirez
Parul Agarwal
Junming Zhang
Angelo Ciaramella
Xiannian Fan
Zhixuan Wei
Xiaoling Wang
Zhe Liu
Zheng Tian
Bei Ye
Jiqian Li
Biranchi Panda
Yan Qi
Shiming Yang
Bing Zhang
Fengzhu Xiong
Qinqin Zhang
Amelia Badica
Juning Gao
Yi Xiong
Haya Alaskar

Ala Al Kafri
Wen Shen
Yu Chen
Wei Zhang
Hongjun Su
Wenrui Zhao
Xihao Hu
Wenxi Zhang
Lihui Shi
Peng Li
Zhenhua Lai
Jingang Wang
Haitao Zhu
Mohammed Khalaf
Zhanjun Wang
Zhengmao Zou
Weihua Wang
Ahmed J. Aljaaf
Li Tang
Jun Yin
Tameem Ahmad
Haiying Ma

Contents – Part II

Nature Inspired Computing and Optimization

Genetic Algorithms

Signal Processing

Differential Evolution

Memetic Algorithms

Swarm Intelligence and Optimization

Soft Computing

Protein Structure and Function Prediction

Advances in Swarm Intelligence: Algorithms and Applications

Evolutionary Computation
and Learning

A Hybrid Scatter Search Algorithm to Solve the Capacitated Arc Routing Problem with Refill Points

Eduyn Ramiro López-Santana[1(✉)],
Germán Andrés Méndez-Giraldo[1],
and Carlos Alberto Franco-Franco[2]

[1] Universidad Distrital Francisco José de Caldas, Bogotá, Colombia
{erlopezs, gmendez}@udistrital.edu.co
[2] Universidad de la Sabana, Chía, Cundinamarca, Colombia
carlosfrfr@unisabana.edu.co

Abstract. This paper presents a hybrid scatter search algorithm to solve the capacitated arc routing problem with refill points (CARP-RP). The vehicle servicing arcs must be refilled on the spot by using a second vehicle. This problem is addressed in real-world applications in many services systems. The problem consists on simultaneously determining the vehicles routes that minimize the total cost. In the literature is proposed an integer linear programming model to solve the problem. We propose a hybrid algorithm based on Scatter Search, Simulated Annealing and Iterated Local Search. Our method is tested with instances from the literature. We found best results in the objective function for the majority instances.

Keywords: Scatter search · Arc routing · Hybrid algorithms · Refill points · Simulated annealing · Iterated local search

1 Introduction

The arc routing problem (ARP) consists in find an optimal schedule to traversal of a given subset of edges or arcs, which are contained in a graph. This scheduling problem has several applications in real-world services operations as garbage collection, snow clearing, road network maintenance, mail delivery, milk delivery, among others. Some of this problems where mentioned by Assad and Golden [1], also they present some algorithms. The basic problem consists in serve a subset of arcs or edges in a graph by a set of vehicles, subject to a set of constraints. The goal is minimize the total cost of traversed edges by a subset of feasible vehicle routes.

An extension of the ARP is the capacitated arc routing problem (CARP), this problem consist in satisfied the required demands over the edges or arcs at minimum cost while the capacity of vehicles is ensure [2]. There are several variations of classical CARP. The classical CARP is defined on an undirected graph [3], but several real life applications of the problem must take into account the existence of one-way streets and streets where the two sides must be serviced in parallel. These variations are referred as

© Springer International Publishing Switzerland 2016
D.-S. Huang and K.-H. Jo (Eds.): ICIC 2016, Part II, LNCS 9772, pp. 3–15, 2016.
DOI: 10.1007/978-3-319-42294-7_1

CARP on Directed Graphs (DCARP) or Mixed Graphs (MCARP) and are describe in detail by [4].

Others variations consist in CARP with alternative objective functions different to classical objective of minimizing the total routing cost. These could be minimizing the total number of vehicles used, equalizing the load of the tours, or minimizing the length of the longest tour [3]. For instance, the Multi Objective CARP defined as the classical CARP where the objective is to minimize the total routing cost, and the makespan, i.e., the length of the longest tour is presented by [5, 6].

The CARP also include the time windows as service requirement in an edge. This is named CARP with Time Windows (CARPTW). A greedy randomized adaptive search procedure with path relinking is presented by [7] for the CARPTW. Their results indicate that the proposed metaheuristic is competitive with the best-published algorithms. However, contrary to the vehicle routing problem with time windows, this problem received little attention.

In cases where each vehicle is located in one of several depots from which it must start and end its tour, it is named as Multi Depot CARP (MD-CARP). Others variations consists in CARP with periodic visits [8] named the Periodic CARP (PCARP) and the CARP with stochastic information [9].

A particular extension of CARP is with Mobile Depots. In [10], is considered a version of the CARP with two different types of servicing vehicles, where only one of them unloads at the depot. The other type of vehicle unloads onto the first type. This problem has real life application in collection problems, where satellite vehicles with small capacity unload into one of several large vehicles, which in turn are the only ones to unload at the depot. The CARP with refill points (CARP-RP) was stated by [11]. It is a particular CARP with Multi depot where the depots are traveling to the vehicles to refill. In this problem, two types of vehicles are given. The usual service vehicles that service the edges by traversing them, and the refilling vehicles that can meet the service vehicle at any point in the graph for refilling purposes. A mathematical model is given for the CARP-RP and a Cutting Plane algorithm for solving the problem is suggested by [11]. In [12], the authors describe a solution procedure for a capacitated arc routing problem with refill points and multiple loads based in route first-cluster second heuristic procedure. A genetic algorithm to solve this problem is proposed by [13], the show has their algorithm uses the split algorithm to calculate the fitness of individuals and to determine feasible vehicle routes and replenishment locations, and adopts local search as the mutation to expand search space.

The CARP is NP-hard and exact methods or algorithms are useful only on small instances [3]. The exact algorithms consist in different methods. For instance, it use tour based on rules [14] or valid inequalities and use cutting planes [15]. Also, it is transformed the problem into equivalent node-routing problems and use existing exact algorithms [16], formulated the problem as a capacitated vehicle routing problem and use a branch and cut an price algorithm for solving the problem [17], and transform the problem into a node routing problem using a column generation scheme [18].

For solving large-scale problems, due to the complexity of the problem, there have been several types of heuristic and metaheuristic algorithms. A metaheuristic algorithm was developed for solving a routing winter gritting vehicles, for this problem is proposed a two phase solution, the first one obtains an optimal solution for the Chinese postman problem and the second phase consist in using simulated annealing for the constrained problem [19]. Another algorithm consists in using variable neighborhood search decent comparing the method with tabu search [20], there is also proposed a tabu search heuristic [21]. There is also another combined methods like scatter search [8], memetic algorithms [22–24] and guide local search [25].

Scatter Search (SS) is an evolutionary method proposed by Glover [26] based on the issue of capturing relevant information of an individual constraints and integrating it into new surrogate constraints in order to generate composite decision rules and new trial solutions. SS as evolutionary method maintains a population of solutions that evolves in successive generations. Rego and Leao [27] provide a general scatter search design for problems dealing with the optimization of cycles (or circuits) on graphs. The authors shows a tutorial of SS applied to solve a vehicle routing problem. Another application of SS to routing problems is presented in [28], the authors provide a SS framework to solve the vehicle routing problem with time windows (VRPTW). Their approach provided a means to combine solutions, diversify, and intensify the meta-heuristic search process. However, the literature related with the CARP and SS is scarce. Chu et al. [8] present an application of the SS to PCARP. Their results show that the SS strongly improves its initial solutions and clearly outperforms the greedy heuristic.

In this paper, we attempt to use a hybrid scatter search to solve the CARP-RP. We use an evolutionary algorithm based in a scatter search (SS) procedure and involves an Iterated local search (ILS) procedure and a simulated annealing (SA) procedure.

The remainder of this paper is organized as follows: Sect. 2 introduces the problem statement according with [11]. Section 3 describes our hybrid scatter search approach. Section 4 shows some numerical results from a set of instances of literature and compare the proposed method with a local search and simulated annealing algorithms. Finally, Sect. 5 concludes this work and provides possible research directions.

2 Problem Statement

The CARP-RP problem consists in determining a set of vehicles that satisfied the demand over a subset of arcs or edges; the arcs that has to be traversed must be visited in one direction but can be traversed several times. The vehicles that are in charge of satisfying the demand are refilled by a second set of vehicles. The vehicles used to service the arcs are called servicing vehicles which has a finite capacity and the second type of vehicles are called refilling vehicle and are in charge of refill the servicing vehicles in any arc or edge of the network. The objective function is minimize the total cost of satisfying the demand and refill the vehicles.

This problem has applications in the road network maintenance services, the problem consist in painting or repainting the road markings, these roads are painted by special vehicles, and there is a fleet of trucks that replenish the painting vehicles. The mathematical formulation as mixed integer linear programming model for this problem is described in [11]. We consider two types of vehicles. The first one is service vehicle (SV) with a cost incurred on traversing an arc. The second one is a refill vehicle

(RV) with a cost incurred on traversing an arc. Our objective functions consist in the sum of SV cost and RV cost in the optimal solution.

3 Solution Approach

We propose a hybrid algorithm to solve the CARP-RP. Our proposal method is inspired in a scatter search (SS) procedure and involves an Iterated local search (ILS) procedure and a simulated annealing (SA) procedure. SS consists in two phases. The first phase creates an initial solution set. The second phase is an iteratively procedure that operates on a set of reference solutions to generate new solutions by weighted linear combinations of structured subsets of solutions.

SS was proposed by Glover [26] in 1977, is a population-based metaheuristic that operates on a set of feasible solutions and it is described in detail by [29]. We applied the framework proposed in [27] to solve the CARP-RP. SS operates on a set of reference solutions (RS) to generate new solutions by weighted linear combinations of structured subsets of solutions [27]. The RS is required to be made up of high quality and diverse solutions and the goal is to produce weighted centers of selected sub-regions that project these centers into regions of the solution space to be explored by auxiliary heuristic procedures. We use ILS and SA procedures as auxiliary heuristic procedures in order to improve the solutions. Table 1 shows the pseudo code of proposed method to solve the CARP-RP. In the next sections, we describe in detail our procedures.

Table 1. Pseudocode of proposal method based on SS to CARP-RP

Create Initial Solution Set→ ISS
Improvement initial(ISS) → S
Reference set update (S) → RS
While (the stopping criterion does not satisfied)
Subset generation (RS) → S'
Solution combinations(S') → S''
Repair Solution (S'') → S'''
Improvement solutions (S''')→ S''''
Reference set update (S'''') → RS
End while

3.1 Initial Solution Set

SS starts with an initial solutions set composed by different elements, i.e., diverse solutions. Thus, a systematic procedure to generate these solutions is appropriate. We take the representation of a solution proposed by [27]. The authors use a permutation

P as a seed to generate subsequent permutations. We define the subsequence $P(h:s)$ where s is a positive integer between 1 and h, so that $P(h:s) = (s, s+h, s+2h, \ldots, s+rh)$, where r is the largest nonnegative integer such that $s + rh \leq n$. Finally, we define the permutation $P(h)$ for $h < n$, to be $P(h) = (P(h:h), P(h:h-1), \ldots, P(h:1))$.

The representation is two vectors with n elements each one. In the first, the order indicates the sequence that the mandatory arcs are traversed. The second indicates if the capacity is fulfilled, then a refill is done. This is called refill point (RF) and is denoted by 1. The mandatory arcs are labeled from 1 to n. An example with eight mandatory arcs is shown in (1). Note that the arc number 8 is not mandatory and the refill point is after the arc 5, i.e., in the path between arcs 5 and 3.

$$\begin{pmatrix} 1, 2, 4, 5, 3, 7, 6, 9 \\ 0, 0, 0, 1, 0, 0, 0, 0 \end{pmatrix} \tag{1}$$

In addition, each arc has a start node and end node, and each arc has a direction. For example, for the arc 1, the starting node is v_1 and the ending node is v_3, and for the arc 2, the starting node is v_2 and the ending node is v_4. Then, in the solution (1) the vehicle must to go from node v_3 to v_2. We assume that the connection between nodes v_3 and v_2 is the shortest path. This is computed with the Floyd-Warshall Algorithm [30].

For the refill points, we take the shortest path between the ending node of the last arc visited and the depot node. Moreover, the shortest path between depot node and the starting node of the first arc in the path immediately after the refill point. These shortest paths are computed with the Floyd-Warshall Algorithm.

3.2 Improvement Initial Method

This procedure looks to transform an initial solution into one or more enhanced solutions. We take the arcs of the solution between the refill is executed and then we consider the standard 2-opt procedure to find a 2-optimal tour for each individual segment of tour. For instance, in (1) we present an example with eight mandatory arcs. The arc to exchange are randomly selected.

$$\text{Before } 2 - \text{opt} \, (1, 2, 4, 5, refill, 3, 7, 6, 9) \tag{2}$$

$$\text{After } 2 - \text{opt} \, (4, 2, 1, 5, refill, 7, 3, 9, 6) \tag{3}$$

3.3 Reference Set Update Method

This method creates and maintains a set of reference solutions. It is a typical method used in any evolutionary technique. This consists in a set of solutions that containing high evaluation combinations of attributes replaces the less promising solutions at each iteration (generation) of the method in order to enhance the quality of the population.

Our reference set (RS) is defined by two distinct subsets B and D, representing respectively the subsets of high quality and diverse solutions. The size of RS is b, then de size of B and D are $b/2$. First, we delete the repeated solutions in the set of solution S. Then, we select the best $b/2$ solutions from S. Now, to create the set D of diverse solutions we generate it by successively selecting the solution, which mostly differs from the ones currently belonging to RS.

We use the diversity measure proposed by [27]. The authors define $d_{ij} = (S_i \cup S_j) \setminus (S_i \cap S_j)$. as the distance between solutions S_i and S_j, which gives the number of edges by which the two solutions differ from each other. The solutions are included in RS according with the *maxmin* criterion, which maximizes the minimum distance of each candidate solution among all solutions currently in the reference set. The diversity measure is calculated repeatedly until the RS is completed.

3.4   Subset Generation Method

This method is the first step in the scatter search phase. The method generates subsets of the reference set as a basis for creating combined solutions. Rego and Leao [27] states that this method is typically designed to organize subsets of solutions to cover different promising regions of the solution space. We adopt the three types of subsets suggested by [27]. It can be summarized as:

- subsets containing only solutions in B,
- subsets with only solutions in D, and
- subsets mixing in solutions in B and D in different proportions.

3.5   Solution Combination Method

To generate new solution, we make different sets as:

- All 2-element subsets,
- 3-element subsets derived from two element subsets by augmenting each 2-element subset to include the best solution (as measured by the objective function value) not in this subset,
- 4-element subsets derived from the 3-elelement subsets by augmenting each 3-element subset to include the best solution (as measured by the objective function value) not in this subset,
- the subsets consisting of the best b elements (as measured by the objective function value), for $b = 5, \ldots, B$.

We build new solutions by weighted linear combinations on the subsets defined. To restrict the number of solutions we generated only one solution in each subset by a convex linear combination defined as follows. Let E be a subset defined in RS, and let $H(E)$ denote the convex-hull of E. We generate solutions $S \in H(E)$ represented as

$$S = \sum_{t=1}^{r} \lambda_t S_t \tag{4}$$

$$\sum_{t=1}^{r} \lambda_t = 1 \tag{5}$$

$$\lambda_t \geq 0, t = 1, .., r, \tag{6}$$

where the multiplier λ_t represents the weight assigned to solution S_t. We compute these multipliers by

$$\lambda_t = \frac{\frac{1}{C(S_t)}}{\sum_{t=1}^{r} \frac{1}{C(S_t)}}, \tag{7}$$

so that the better (lower cost) solutions receive higher weight than less attractive (higher cost) solutions. Then, we calculate the score of each variable x_{ij} relative to the solutions in E by computing:

$$score(x_{ij}) = \sum_{t=1}^{r} \lambda_t x_{ij}^t, \tag{8}$$

where x_{ij} means that is an edge in the solution S. Finally, as variables are required to be binary, the value is obtained by rounding its score to give $x_{ij} = \lfloor score(x_{ij}) + 0.5 \rfloor$. The new solution is made with variables with a value of one. If the solution is infeasible, a repair method is necessary.

3.6   Repair Solution Method

The feasibility criterion is reviewed in this step is to verify that all required arcs are in the solution after performing the combination of the solutions of the subsets. If all arcs are in the solution any procedures does not made otherwise it verifies that mandatory arcs are introduced in the solution by a greedy method. This method add an arc according to the minimum value for the shortest path between tail of the last arc in the solution and the head of the arcs that have not yet been incorporated into this.

3.7   Improvement Solution Method

With a set of solutions, we use a combined method in order to improve its quality. We use ILS with a SA embedded procedure. For this, we create a cycle for each solution generated in the reconstruction method. Table 2 shows the pseudocode of ILS procedure. For each generated solution is always saved the best at all times of the iteration, in addition to the solution that is selected refill points are generated by that node and low for this distance is chosen.

Table 3 presents the pseudocode for SA procedure. The decrease procedure of temperature parameter is geometric, i.e., in each iteration the new temperature is computed as at, where a is setting as 0.85.

Table 2. Pseudocode of ILS

Set b as number of iterations
For k → 1 to ns
SA(S_k) →current
For i→1 to b
Disturb(current) → s
SA(s) → s^*
If s^* is accepted
s →current
End if
End-for
End-for

Table 3. Pseudocode of SA

t→ temperature
s →Best
Repeat
disturb(s) →R
If Cost(R)<Cost(S) or random<$e^{\frac{Cost(S)-Cost(R)}{t}}$
R → s
Decrease(t)
If Cost(S)<Cost(Best)
s →Best
End if
End-If
Until t<Tmin
Return Best

4 Numerical Results

In order to test our algorithm, we use the set of instances proposed in [31]. We transform the set of CARP into CARP-RP using the procedure described in [11]. The instances contain two types of costs: servicing cost and traversing cost, the servicing cost are the cost of traversing and servicing an arc, while the traversing cost denotes the cost of traversing the edge without service it. The instances also contain the number of edges and the demand over the required edges; those with zero demand are no required edges. Finally, the instances contain the number of vehicles and the capacity.

We run 20 experiments for each instance. The stopping criterion was 10 iterations of SS procedure. All tests in this work were run using Visual Basic 2010 on a Windows 8 64-bit machine, with an Intel i5 3337 processor (2 × 1.8 GHz) and 6 GB of RAM.

Table 4 provides the results of our algorithm. We test our algorithm on 33 instances, for all them we run six iterations. This table is organized as follows: column one and five show the instance's name, the second and sixth columns provide the best objective function found by our algorithm. The third and seventh columns contain the average objective function, and finally in the fourth and last one column indicate the total amount of time spend by our algorithm in seconds.

We can observe in Table 4 that our algorithm find good solutions in multiple runs, in all the instances the average solutions are very closer to the best solution found by the algorithm. In fact, for nine instances, the average solution is equal to the best solution found and on average; there is a gap of 1.8 % to the best solution. Due to we use the same number of iterations in all the runs, the algorithm spends 122.33 s on average for solving each instance.

Table 4. Results of proposed method

Instance	Best objective function	Average objective function	Time (s)	Instance	Best objective function	Average objective function	Time (s)
Val10A	473.0	474.3	152.7	Val5B	458.0	465.7	125.0
Val10B	436.0	440.0	149.3	Val5C	436.0	436.0	124.3
Val10C	437.0	446.3	154.7	Val5D	463.0	463.0	127.0
Val10D	454.0	454.0	160.3	Val4A	384.0	398.7	128.3
Val9A	384.0	386.7	151.7	Val4B	431.0	442.3	127.3
Val9B	386.0	391.0	148.3	Val4C	439.0	442.0	128.0
Val9C	383.0	389.7	151.3	Val4D	396.0	412.7	129.7
Val8A	409.0	409.0	124.0	Val3A	111.0	117.0	123.0
Val8B	440.0	448.0	124.3	Val3B	104.0	106.0	124.7
Val8C	406.0	410.7	125.3	Val3C	104.0	104.0	124.0
Val7A	325.0	325.0	134.7	Val2A	284.0	289.0	123.3
Val7B	338.0	340.0	135.7	Val2B	257.0	262.7	124.3
Val7C	326.0	326.0	140.3	Val2C	268.0	268.0	124.0
Val6A	231.0	231.0	123.0	Val1A	210.0	228.0	124.0
Val6B	246.0	255.7	124.7	Val1B	189.0	194.3	122.3
Val6C	231.0	236.7	123.7	Val1C	185.0	192.0	123.7

The computational time obtained in Table 4 is high, thus we compare the results with other metaheuristics applied to the same problem, in order to know its performance. We did not compare our method with an optimization model because the Benavent's instances problems are harder to solve. Table 5 presents the results of a Local Search and Simulated Annealing procedures developed to the same problem. The first column show the instance's name. Columns two and six provide the objective function (O.F.) value. Computational time in seconds are presented in columns three and seven. Columns four and eight shows the improvement in O.F. (Imp. O.F.) regarding to the results of Table 4 for each instance. The minus sign represents a decrease and positive sign an increment. Column five and nine deals with the improvement in computational time (Imp. Time) regarding to the results of Table 4 for each instance. The interpretation is the same of O.F.

Regarding with O.F. obtained with local search procedure, for all instances, our procedure improved the values with an average of 56 %. While, for the computational only in 13 instances the time was decreased with an average of 13 %. For the others 20 instances, the computational times was increased in average of 36 %. Likewise, with the simulated annealing our procedure improved 32 of the instances, achieving a reduction of 26 %. Only for instance Val3A our procedure did not improve the O.F. As to computational time, our procedure reduced three instances with an average of 30 %, while the increment was of 383 %. To sum up, our procedure improved the O.F. but the computational time is greater than other metaheuristics, thus we need to accelerate our proposed method.

Table 5. Comparison of proposed method with local search and simulated annelaing algorithms

Instance	Local search				Simulated annealing			
	O.F.	Time (s)	Imp. O.F. (−)	Imp. Time (−)	O.F.	Time (s)	Imp. O.F. (−)	Imp. Time (−)
Val10A	1107.0	183.2	−57 %	−17 %	570.5	144.2	−17 %	6 %
Val10B	1118.0	188.1	−61 %	−21 %	585.5	146.2	−26 %	2 %
Val10C	1146.0	184.8	−62 %	−16 %	605.9	143.9	−28 %	7 %
Val10D	1240.0	182.9	−63 %	−12 %	721.9	143.1	−37 %	12 %
Val9A	878.0	179.0	−56 %	−15 %	469.8	214.6	−18 %	−29 %
Val9B	867.0	184.4	−55 %	−20 %	466.4	214.7	−17 %	−31 %
Val9C	880.0	184.8	−56 %	−18 %	458.4	218.6	−16 %	−31 %
Val8A	886.0	118.4	−54 %	5 %	564.6	52.2	−28 %	138 %
Val8B	882.0	117.6	−50 %	6 %	538.9	51.8	−18 %	140 %
Val8C	1065.0	122.4	−62 %	2 %	722.0	51.9	−44 %	142 %
Val7A	728.0	128.5	−55 %	5 %	415.0	100.9	−22 %	33 %
Val7B	735.0	128.2	−54 %	6 %	414.5	102.4	−18 %	33 %
Val7C	848.0	129.6	−62 %	8 %	488.0	98.7	−33 %	42 %
Val6A	533.0	95.4	−57 %	29 %	296.8	30.3	−22 %	306 %
Val6B	564.0	96.5	−56 %	29 %	334.9	27.0	−27 %	361 %
Val6C	641.0	102.9	−64 %	20 %	408.0	27.3	−43 %	353 %
Val5A	985.0	138.3	−55 %	−4 %	568.3	40.0	−22 %	233 %
Val5B	1005.0	136.2	−54 %	−8 %	581.4	40.1	−21 %	212 %
Val5C	1045.0	138.8	−58 %	−10 %	606.5	40.3	−28 %	209 %
Val5D	1178.0	124.7	−61 %	2 %	729.1	40.0	−36 %	218 %
Val4A	944.0	126.2	−59 %	2 %	574.0	69.8	−33 %	84 %
Val4B	956.0	131.7	−55 %	−3 %	558.3	70.1	−23 %	82 %
Val4C	981.0	143.8	−55 %	−11 %	570.6	66.3	−23 %	93 %

(*Continued*)

Table 5. (*Continued*)

Instance	Local search				Simulated annealing			
	O.F.	Time (s)	Imp. O.F. (−)	Imp. Time (−)	O.F.	Time (s)	Imp. O.F. (−)	Imp. Time (−)
Val4D	1096.0	150.7	−64 %	−14 %	674.8	65.9	−41 %	97 %
Val3A	183.0	78.7	−39 %	56 %	109.0	11.3	2 %	988 %
Val3B	194.0	71.1	−46 %	75 %	110.5	11.3	−6 %	1003 %
Val3C	251.0	71.8	−59 %	73 %	157.4	11.3	−34 %	997 %
Val2A	500.0	66.5	−43 %	86 %	319.3	11.5	−11 %	973 %
Val2B	534.0	67.1	−52 %	85 %	321.4	11.3	−20 %	1005 %
Val2C	763.0	75.2	−65 %	65 %	506.4	11.5	−47 %	980 %
Val1A	355.0	76.0	−41 %	63 %	215.2	12.1	−2 %	923 %
Val1B	378.0	77.8	−50 %	57 %	242.5	12.2	−22 %	899 %
Val1C	471.0	79.9	−61 %	55 %	296.6	12.2	−38 %	913 %

5 Conclusions

We developed a hybrid algorithm for the capacitated arc routing problem with refill points. This is an interesting scheduling problem with several real-world services applications as road network maintenance and pumping stations networks, scheduling of technicians, waste collection, among others.

Our approach combined an iterated local search and simulated annealing with a scatter search procedure. We tested our proposed method on a set of instances where it was observed that the computational time was very consistent between them with each one representing a different size. Our method allows a comparison point to possible developments or applications about this problem.

Future work should focus in to improve the decision-aid tool to allow speeding up the method. In addition, to decrease the computational time is possible to combine other techniques such as path relinking, variable neighborhood search, among others. Another real world constraints and features can be explored as heterogeneous vehicles, stochastic travel times, other objective functions, among others.

Acknowledgements. This work was supported in part by the Centro de Investigaciones y Desarrollo Científico at Universidad Distrital Francisco José de Caldas (Colombia) under Grant No. 2-602-468-14.

References

1. Assad, A.A., Golden, B.L.: Arc routing methods and applications. In: Handbooks in Operations Research and Management Science, pp. 375–483. Elsevier (1995)
2. Golden, B.L., Wong, R.T.: Capacitated arc routing problems. Networks **11**, 305–315 (1981)

3. Wøhlk, S.: A decade of capacitated arc routing. In: Golden, B., Raghavan, S., Wasil, E. (eds.) The Vehicle Routing Problem: Latest Advances and New Challenges. Operations Research/Computer Science Interfaces Series, vol. 43, pp. 29–48. Springer, Heidelberg (2008)

4. Belenguer, J.-M., Benavent, E., Lacomme, P., Prins, C.: Lower and upper bounds for the mixed capacitated arc routing problem. Comput. Oper. Res. **33**, 3363–3383 (2006)

5. Lacomme, P., Prins, C., Sevaux, M.: Multiobjective capacitated arc routing problem. In: Fonseca, C.M., Fleming, P.J., Zitzler, E., Deb, K., Thiele, L. (eds.) EMO 2003. LNCS, vol. 2632, pp. 550–564. Springer, Heidelberg (2003)

6. Lacomme, P., Prins, C., Sevaux, M.: A genetic algorithm for a bi-objective capacitated arc routing problem. Comput. Oper. Res. **33**, 3473–3493 (2006)

7. Reghioui, M., Prins, C., Labadi, N.: GRASP with path relinking for the capacitated arc routing problem with time windows. In: Giacobini, M. (ed.) EvoWorkshops 2007. LNCS, vol. 4448, pp. 722–731. Springer, Heidelberg (2007)

8. Chu, F., Labadi, N., Prins, C.: A scatter search for the periodic capacitated arc routing problem. Eur. J. Oper. Res. **169**, 586–605 (2006)

9. Fleury, G., Lacomme, P., Prins, C.: Stochastic Capacitated Arc Routing Problem (2005)

10. Pia, A., Filippi, C.: A variable neighborhood descent algorithm for a real waste collection problem with mobile depots. Int. Trans. Oper. Res. **13**, 125–141 (2006)

11. Amaya, A., Langevin, A., Trépanier, M.: The capacitated arc routing problem with refill points. Oper. Res. Lett. **35**, 45–53 (2007)

12. Amaya, C.-A., Langevin, A., Trépanier, M.: A heuristic method for the capacitated arc routing problem with refill points and multiple loads. J. Oper. Res. Soc. **61**, 1095–1103 (2010)

13. Shan, H.U., Dan, L.I.N.: Algorithms for solving CARP-RP-ML problem. Comput. Eng. **38**, 168–170 (2012)

14. Hirabayashi, R., Saruwatari, Y., Nishida, N.: Tour construction algorithm for the capacitated arc routing problem. Asia Pac. J. Oper. Res. **9**, 155–175 (1992)

15. Belenguer, J.M., Benavent, E.: A cutting plane algorithm for the capacitated arc routing problem. Comput. Oper. Res. **30**, 705–728 (2003)

16. Baldacci, R., Maniezzo, V.: Exact methods based on node-routing formulations for undirected arc-routing problems. Networks **47**, 52–60 (2006)

17. Longo, H., de Aragão, P.M., Uchoa, E.: Solving capacitated arc routing problems using a transformation to the CVRP. Comput. Oper. Res. **33**, 1823–1837 (2006)

18. Tagmouti, M., Gendreau, M., Potvin, J.-Y.: Arc routing problems with time-dependent service costs. Eur. J. Oper. Res. **181**, 30–39 (2007)

19. Eglese, R.W.: Routeing winter gritting vehicles. Discret. Appl. Math. **48**, 231–244 (1994)

20. Hertz, A., Mittaz, M.: A variable neighborhood descent algorithm for the undirected capacitated arc routing problem. Transp. Sci. **35**, 425–434 (2001)

21. Hertz, A., Laporte, G., Mittaz, M.: A tabu search heuristic for the capacitated arc routing problem. Oper. Res. **48**, 129–135 (2000)

22. Fung, R.Y.K., Liu, R., Jiang, Z.: A memetic algorithm for the open capacitated arc routing problem. Transp. Res. Part E: Logist. Transp. Rev. **50**, 53–67 (2013)

23. Liu, M., Singh, H.K., Ray, T.: Application specific instance generator and a memetic algorithm for capacitated arc routing problems. Transp. Res. Part C: Emerg. Technol. **43**, 249–266 (2014)

24. Wang, Z., Jin, H., Tian, M.: Rank-based memetic algorithm for capacitated arc routing problems. Appl. Soft Comput. **37**, 572–584 (2015)

25. Beullens, P., Muyldermans, L., Cattrysse, D., Van Oudheusden, D.: A guided local search heuristic for the capacitated arc routing problem. Eur. J. Oper. Res. **147**, 629–643 (2003)

26. Glover, F.: Heuristics for integer programming using surrogate constraints. Decis. Sci. **8**, 156–166 (1977)
27. Rego, C., Leão, P.: A scatter search tutorial for graph-based permutation problems. In: Sharda, R., Voß, S., Rego, C., Alidaee, B. (eds.) Metaheuristic Optimization via Memory and Evolution. Operations Research/Computer Science Interfaces Series, vol. 30, pp. 1–24. Kluwer Academic Publishers, Boston (2005)
28. Russell, R.A., Chiang, W.-C.: Scatter search for the vehicle routing problem with time windows. Eur. J. Oper. Res. **169**, 606–622 (2006)
29. Laguna, M., Martí, R., Martí, R.C.: Scatter Search: Methodology and Implementations in C. Operations Research/Computer Science Interfaces Series, vol. 1. Springer Science & Business Media, New York (2003)
30. Ahuja, R.K., Magnanti, T.L., Orlin, J.B.: Network Flows: Theory, Algorithms, and Applications. Prentice Hall, Upper Saddle River (1993)
31. Benavent, E., Campos, V., Corberan, A., Mota, E.: The capacitated arc routing problem: lower bounds. Networks **22**, 669–690 (1992)

A Novel Fitness Function Based on Decomposition for Multi-objective Optimization Problems

Cai Dai[1], Xiujuan Lei[1(✉)], and Xiaofang Guo[2]

[1] College of Computer Science, Shaanxi Normal University,
Xi'an 710062, China
{cdai0320,xjlei}@snnu.edu.cn
[2] School of Science, Xi'an Technological University, Xi'an, Shaanxi, China
gxfang1981@126.com

Abstract. Research on multi-objective optimization problems (MOPs) becomes one of the hottest topics of intelligent computation. The diversity of obtained solutions is of great importance for multi-objective evolutionary algorithms. To this end, in this paper, a novel fitness function based on decomposition is proposed to help solutions converge toward to the Pareto optimal solutions and maintain the diversity of solutions. First, the objective space is decomposed in a set of sub-regions based on a set of direction vectors and obtained solutions are classified. Then, for an obtained solution, the size of the class which contains the solution and an aggregation function value of the solution are used to calculate the fitness value of the solution. Aggregation function which decides whether the target space is divided evenly plays a very important role in the fitness function. A hyperellipsoidal function is designed for any-objective problems. The proposed algorithm has been compared with NSGAII and MOEA/D on various continuous test problems. Experimental results show that the proposed algorithm can find more accurate Pareto front with better diversity in most problems, and the hyperellipsoidal function works better than the weighted Tchebycheff.

Keywords: Evolutionary algorithm · Multi-objective optimization · Fitness function · Aggregation function · Elliptic function

1 Introduction

Since there are many problems with several optimization objectives or criteria in the real world, multi-objective optimization has become a hot research topic. Unlike the single-objective optimization problem, the multi-objective problems (MOPs) are typically characterized by conflicting objectives. In this scenario, the Pareto optimal solutions, also known as the non-dominated solutions, are used [1], and, the multi-objective optimization algorithms for MOPs should be able to: (1) discover solutions as close to the Pareto optimal solutions as possible; (2) find solutions as diversely as possible in the obtained non-dominated front; (3) find solutions distributed in the non-dominated front as evenly as possible. However, achieving these three goals simultaneously is still a challenging task for multi-objective optimization algorithms. Among various

© Springer International Publishing Switzerland 2016
D.-S. Huang and K.-H. Jo (Eds.): ICIC 2016, Part II, LNCS 9772, pp. 16–25, 2016.
DOI: 10.1007/978-3-319-42294-7_2

multi-objective optimization algorithms, multi-objective evolutionary algorithms (MOEAs), which make use of the population evolution to get the optimal solutions, are a kind of effective methods for solving MOPs. Nowadays, there exist many MOEAs [2–16], such as multi-objective genetic algorithms [3, 4], multi-objective particle swarm optimization algorithms [5, 6], multi-objective differential evolution algorithms [7, 8], multi-objective immune clone algorithms [9, 10], and group search optimizer [11, 12]. To enhance the performance of MOEAs, some mathematical improvements, such as orthogonal experimental design, are often applied [13, 14].

An effective MOEA should well maintain population diversity. For this purpose, often used techniques can be classified into three main categories: (1) the first category uses an external elitist archive which stores the non-dominated solutions found so far. Some such archive strategies, such as Crowding distance [3], adaptive grid [17], are widely used in the existing MOEAs (e.g., [18–20]); (2) the second category is referred to as IBEAs (indicator-based evolutionary algorithms), which utilize an indicator function, such as the hypervolume [21–23], as the fitness function, and the strong search ability of IBEAs has been demonstrated in the literature [24]; (3) the third category makes use of the decomposition to maintain the diversity of solutions, and a representative of this category is MOEA/D [25] (multi-objective evolutionary algorithm based on decomposition). MOEA/D works well on a wide range of multi-objective problems with many objectives, discrete decision variables and complicated Pareto sets [26–28], and it has also been used as a basic element in some hybrid algorithms [29–31]. Also, MOEA/D and its variants have been successfully applied to a number of application areas [32, 33].

In this paper, we propose a novel fitness function-based evolutionary algorithm (NFEA). Decomposition is applied to the fitness function. The target space is evenly decomposed into a number of subtarget space. In other words, the multi-objective optimization problem is decomposed into a number of scalar optimization subproblems and optimizes them simultaneously. In MOEA/D, each subproblem is optimized by using information from its several neighboring subproblems. Each subproblem is only affected by its several neighboring subproblems. This is said that only its several neighboring subproblems can effectively improve its optimization. In NFEA, each subproblem can improve its optimization by all other subproblems. This improves the performance of NFEA and NFEA has lower computational complexity at each generation than MOEA/D and NSGA-II.

Aggregation function plays a very important role in the algorithm. It is well-known that the weighted sum cannot appropriately handle multi-objective problems with non-convex Pareto fronts whereas the weighted Tchebycheff can handle them. Moreover, the weighted Tchebycheff has nice theoretical properties [34]. However, for non-convex problems, aggregation function based on weighted Tchebycheff is theoretically shown to be the only function to capture the whole Pareto front [34]. The weighted Tchebycheff is not good at finding Pareto optimal solutions around the edges of the Pareto front of three or more objectives problem. Aggregation function based on elliptic function and ellipsoidal function has a good performance in capturing non-convex fronts [35]. In order to fit four or more objectives problems, a hyperellipsoidal function which is an extension of elliptic function and ellipsoidal function is designed.

This paper is organized as follows. Section 2 introduces four decomposition approaches for multi-objective optimization problems. In Sect. 3, we describe in detail the NFEA algorithm. Section 4 compares NFEA with MOEA/D and NSGA-II, experimental results and analysis are then given in this section. Section 5 concludes this paper.

2 Hyperellipsoidal Function Approach

An m-objective minimization problem can be written as:

$$\begin{cases} \min F(x) = (f_1(x), f_2(x), \cdots, f_m(x)) \\ s.t. g_i(x) \le 0, i = 1, 2, \cdots, q \\ h_j(x) = 0, j = 1, 2, \cdots p \end{cases} \tag{1}$$

Where $x = (x_1, \cdots, x_n) \in X \subset R^n$ is called decision variable and X is an n-dimensional decision space. $f_i(x)(i = 1, \cdots, m)$ is the i-th objective to be minimized, $g_i(x)(i = 1, 2 \cdots, q)$ defines the i-th inequality constraint and $h_j(x)(j = 1, 2, \cdots, p)$ defines the j-th equality constraint. Furthermore, all the constraints determine the set of feasible solutions which is denoted by Ω, and $Y = \{F(x)|x \in \Omega\}$ is denoted as the objective space.

Elliptic function and ellipsoidal function have a good performance in capturing non-convex fronts for two or three objectives problems. The ideas may be extended for a many-objective problem. A hyperellipsoidal function is presented in this paper. In this approach, the scalar optimization problem is in the form:

$$min\, g(F(x), a, \gamma, Z^*) = \left(\sum_{i=1}^{m} \lambda_i (f_i(x) - Z_i^*) \right)^2 / a^2 + \sum_{i=2}^{m} (f_i'(x))^2$$

$$where\, f_i'(x) = \begin{cases} \lambda_{i-1} \left(\sum_{j=i}^{m} \lambda_j (f_j(x) - Z_j^*) \right) - f_{i-1}(x) \sum_{j=i}^{m} \lambda_j^2, if\, i = 2, \cdots m - 1 \\ \lambda_m (f_{m-1}(x) - Z_{m-1}^*) - \lambda_{m-1} (f_m(x) - Z_m^*), if\, i = m \end{cases} \tag{2}$$

Where, $Z^* = (Z_1^*, \cdots, Z_m^*)$ is the hyperellipsoidal centered and $\lambda = (\lambda_1, \cdots, \lambda_m)$ is the semi major axis, $a = \sqrt{e^2 + 1}$, and e is the eccentricity of the ellipse. In this paper, $Z^* = (Z_1^*, \cdots, Z_m^*)$ is the reference point, and the transformation matrix is an unit orthogonal matrix. When m = 2, the hyperellipsoidal function is an elliptic function; and when m = 3, the hyperellipsoidal function is an ellipsoidal function.

3 A New Algorithm: NFEA

In this section, we present NFEA, a new fitness function and decomposition algorithm, in which decomposition technique is applied to the fitness function. The main difference between our approach and other multi-objective evolutionary algorithms lies in the strategy that the algorithm firstly let solutions maintain diversity and while force the solutions converge to the Pareto front. The algorithm emphasizes solutions located in less crowded regions found during the evolutionary process and aggregation function increases the convergence pressure.

3.1 The New Fitness Function

A critical and difficult aspect of multi-objective evolutionary algorithms is finding a suitable fitness function. The fitness value of each individual should indicate the capability of various features of this member to survive. We propose a new fitness function which can let solutions maintain diversity. In order to maintain the diversity of the solutions, we first classify the individuals. A mathematical model is introduced to classify the individuals. This mathematical model is as follows:

$$Z^i = \left\{ x \middle| x. \in POP, \Delta\big(F(x), \lambda^i\big) = \max_{1 \le j \le N} \left\{ \Delta\big(F(x), \lambda^j\big) \right\} \right\}, i = 1, \cdots M \quad (3)$$

where $POP = \{x^1, x^2, \cdots, x^{K_1}\}$ is the current population with K_1 solutions, $W = \{\lambda^1, \lambda^2, \cdots, \lambda^N\}$ is a set of weight vectors, and $\Delta\big(F(x), \lambda^i\big)$ is a specific aggregation function which will be given in the following. In this paper, the weighted Tchebycheff approach and the hyperellipsoidal function approach are used as aggregate functions (aggregation functions). For convenience, the algorithm with the weighted Tchebycheff approach is called NFEA-TW and with the hyperellipsoidal function approach is called NFEA-HE.

If a set $Z^i (i = 1, \cdots m)$ is not empty, the fitness function value of each solution x in Z^i is calculated by the following formula:

$$fitness(x) = |Z^i| + \sigma + \theta\Delta\big(F(x), \lambda^i\big) \quad (4)$$

where $x \in Z^i$; if x is a non-dominated solution of Z^i, σ is equal to 0, otherwise, σ is equal to 1; $|Z^i|$ indicates the number of solutions in set Z^i; θ is a small positive constant to make $\theta\Delta\big(F(x), \lambda^i\big) < 1$. In this work, our main objectives are to improve the convergence and maintain the diversity by minimizing the fitness value of solutions. In particular, minimizing $|Z^i|$ to make each sub-objective space have more than one solution can improve the diversity of the solutions of POP in objective space; minimizing σ to keep the non-dominated solutions of Z^i can improve the convergence. Moreover, minimizing $\Delta\big(F(x), \lambda^i\big)$ to make the angle between the objective vector of the solution x and the weight vector λ^i close to 0 can make the solutions in POP distribute more evenly in objective space.

3.2 Replacement

NFEA is an elitist algorithm. The update strategy uses the $(\lambda + \mu)$ type of deterministic replacement where λ indicates the size of the parent population and μ indicates the size of the descendant population. In this replacement strategy, the parent and descendant populations are combined, then we select λ best individuals which are kept to the next generation. This means that μ individuals will be removed from the parent and descendant populations. In this work, λ and μ are equal to M. Delete rules are as follows:

Step 1. let $k = 1$; if $k > \mu$, stop; otherwise go to Step 2.

Step 2. Find a set Z^i whose size is the maximum among $\{Z^i (i = 1, \cdots, M)\}$. Find a solution $x \in Z^i$ whose $\Delta(F(x), \lambda^i)$ is the maximum among Z^i. Delete the solution x from the set Z^i, then let $Z^i = Z^i \backslash x$ and $k = k + 1$. Go to step 1.

3.3 The NFEA Algorithm

The steps of the algorithm NFEA are as follows:

Step 1. (Initialization) Given population size N and set of vectors size M. In this paper, we have $M = N$. Randomly generate an initial population POP(0) and the set of vectors W. Let $k = 0$.

Step 2. (Fitness) Calculate the fitness value of each individual x in POP(k) by formula (4). Then some better individuals which are put into the parent poplation POP are selected from the population POP(k). In this paper, binary tournament selection with replacement is used.

Step 3. (Crossover and Mutation) Apply crossover operator and mutation operator to the parent population to generate offspring. The set of all these offspring is denoted as O.

Step 4. (Update) Select N best individuals among POP(k) $\cup$ O to put into POP(k + 1). Let $k = k + 1$.

Step 5. (Termination) If stop condition is satisfied, stop; otherwise, go to Step 2.

4 Experimental Study

In this section, we compare NFEA with the hyperellipsoidal function (NFEA-HE) with MOEA/D [27], NSGA-II [3] and NFEA with the weighted Tchebycheff (NFEA-TW) through computation experiments. These three algorithms are compared with NFEA-HE, because (1) NSGA-II is a well-known multi-objective evolutionary algorithm based on fitness function and NFEA is also a multi-objective evolutionary algorithm based on fitness function; (2) MOEA/D is the representative of the multi-objective evolutionary algorithms based on decomposition and NFEA uses the decomposition technology; (3) NFEA-HE and NFEA-TW are compared to show that aggregation function plays a very important role in the fitness function.

4.1 Test Instances

The performance of NFEA is analyzed using four DTLZ [25] test problems (DTLZ1-DTLZ4 for three-objective and four-objective) and four two-objective test problems (F1-F4) which are introduced by Zhang [39].

4.2   Parameter Settings

Algorithms are implemented on a personal computer (Intel Xeon CPU 2.53 GHz RAM). For comparisons fair, all of the algorithms use the same operating. The solutions are coded as real vectors. The literatures [36] have suggested that simulated binary crossover (SBX) [37] is more suitable for DTLZ test instances and differential evolution (DE) [38] for problems with complicated PS shapes. Therefore, crossover (SBX for DTLZ1-DTLZ4 and DE for F1-F4) and mutation (polynomial mutation [37]) operators are applied directly to real parameter values in these algorithms. In this paper, the parameters are set as follows: crossover rate $CR = 0.5$ and scaling factor $F = 0.5$ in DE operator, crossover probability $P_c = 0.9$ and distribution $\eta_c = 20$ in SBX operator; distribution index $\eta_m = 20$ and mutation probability $P_m = 0.1$ in mutation operator; eccentricity of ellipse in Hyperelliposidal function $a^2 = 1000$; population size and weight vectors in MOEA/D, NFEA-TW and NFEA-HY: population size N and weight vectors in these three algorithms are controlled by an integer H; the integer H is set 104 for m = 2, 13 for m = 3 and 8 for m = 4 in the three algorithms; the population sizes N is 105 for 2-objective and 3-objective and 165 for 4-objective in all four algorithms; number of the weight vectors in the neighborhood in MOEA/D: T = 20; number of decision variables: it is set to be 30 for F1-F4, and 12 for all the three-objective and four-objective test instances; number of independent runs: we run each algorithm for each test instance 30 times independently; four algorithms stop after 500 generations. Moreover, the inverted generational distance (*IGD*) metric [39] is used in assessing the performance of the algorithms in our experimental studies.

4.3   Experimental Results and Analysis

Table 1 shows the mean and standard deviation of IGD metric values of the final solutions obtained by each algorithm in each test instance. In the form of *IGD* metric, Table 1 shows that the means of *IGD* metric values obtained by NFEA-HE are the lowest among the four algorithms in each test problem and all the standard deviations also are the smallest. This indicates that for these test problems NFEA-HE has the absolute advantage in finding a better uniform set of solutions and better convergence near the true Pareto front than the other three algorithms and has a strong stability. For four 2-objective problems, the means of *IGD* metric values obtained by NSGA-II are lower than NFEA-TW. However, for 3-objective and 4-objective problems, the means of *IGD* metric values obtained by NFEA-TW is lower than NSGA-II and is about the same with MOEA/D. This indicates that for 3-objective and 4-objective problems NFEA-TW is able to find better diverse sets of solutions than NSGA-II.

From the results of the previous analysis can be drawn that the performance of NFEA-TW is better than NSGA-II and is about the same with MOEA/D for 3-objective and 4-objective test problems, and NFEA-TW can obtain the better diversity of solutions on the Pareto front than MOEA/D. The aggregation function NFEA-TW and MOEA/D use are the the weighted Tchebycheff. These indicate that the framework of NFEA is feasible and effective, and the computation complexity is lower than MOEA/D. The performance of NFEA-HE is better than other three algorithms for these

Table 1. Mean and standard deviation of *IGD* metric values

Instance	NFEA-HE		NFEA-TW		NSGA-II		MOEA/D	
	Mean	Std	Mean	Std	Mean	Std	Mean	Std
F1	0.0058	0.0001	0.0082	0.0026	0.0073	0.0001	0.0063	0.0001
F2	0.0475	0.0150	0.1286	0.0563	0.1274	0.0496	0.0580	0.0346
F3	0.0296	0.0051	0.0803	0.0147	0.0394	0.0042	0.0833	0.0625
F4	0.0271	0.0056	0.0817	0.0196	0.0783	0.0046	0.0625	0.0512
DTLZ1-3D	0.0179	0.0001	0.0245	0.0020	0.0249	0.0011	0.0241	0.0004
DTLZ2-3D	0.0492	0.0001	0.0609	0.0008	0.0649	0.0012	0.0625	0.0006
DTLZ3-3D	0.0562	0.0004	0.0693	0.0018	0.0704	0.0027	0.0681	0.0014
DTLZ4-3D	0.0577	0.0013	0.0640	0.0021	0.0641	0.0026	0.0726	0.0028
DTLZ1-4D	0.0313	0.0001	0.0839	0.0030	0.1501	0.0055	0.0638	0.0009
DTLZ2-4D	0.0943	0.0001	0.1690	0.0083	0.1901	0.0138	0.1683	0.0015
DTLZ3-4D	0.0947	0.0002	0.1685	0.0075	6.7886	2.3962	0.1689	0.0023
DTLZ4-4D	0.0948	0.0003	0.1976	0.0252	0.2121	0.0137	0.1852	0.0057

all these problems. This shows that the hyperellipsoidal function works better and has a better performance in capturing non-convex fronts than the weighted Tchebycheff. It also shows that the target space than the weighted Tchebycheff divided by the hyperellipsoidal function is more uniform.

5 Conclusion

The multi-objective evolutionary algorithm based on a new fitness function is proposed and a hyperellipsoidal function is proposed. We test the performance of the new algorithm with the hyperellipsoidal function on a variety of continuous problems taken from the literature. These test problems have special complexities that a multi-objective evolutionary algorithm has to deal with. We compare NFEA-HE with NSGA-II and MOEA/D. The experiments indicate that NFEA-HE performs well and the hyperellipsoidal function works better than the weighted Tchebycheff. In addition, we also compare NFEA-HE with NFEA-TW. The experiments indicate that the hyperellipsoidal function has a better performance in capturing non-convex fronts than the weighted Tchebycheff. The performance of NFEA-HE on solving multi-objective optimization problems with rules Pareto front is good. How NFEA-HE can deal well multi-objective optimization problems with no rules or discontinuous Pareto front and many-objective optimization problems are our further work and such work is currently under way.

Acknowledgments. This work was supported by National Natural Science Foundation of China (no. 61502290), China Postdoctoral Science Foundation (no. 2015M582606), Industrial Research Project of Science and Technology in Shaanxi Province (no. 2015GY016), Fundamental Research Funds for the Central Universities (no. GK201603094) and Natural Science Basic Research Plan in Shaanxi Province of China (no. 2016JQ6045).

References

1. Coello, C.A.C., Van Veldhuizen, D.A., Gary, B.L.: Evolutionary Algorithms for Solving Multiobjective Problems. Kluwer, New York (2002)
2. Zhou, A.M., Qu, B.Y., Li, H., Zhao, S.Z., Suganthan, P.N., Zhang, Q.F.: Multiobjective evolutionary algorithms: a survey of the state of the art. Swarm Evol. Comput. **1**(1), 32–49 (2011)
3. Deb, K., Pratap, S.A., Meyarivan, T.: A fast and elitist multiobjective genetic algorithm: NSGA–II. IEEE Trans. Evol. Comput. **6**(2), 182–197 (2002)
4. Kumphon, B.: Genetic algorithms for multi-objective optimization: application to a multi-reservoir system in the chi river basin Thailand. Water Resour. Manage **27**(12), 4369–4378 (2013)
5. Pires, E.J.S., Machado, J.A.T., Oliveira, P.B.D.: Entropy diversity in multi-objective particle swarm optimization. Entropy **15**(12), 5475–5491 (2013)
6. Xue, B., Zhang, M.J., Browne, W.N.: Particle swarm optimization for feature selection in classification: a multi-objective approach. IEEE Trans. Cybern. **43**(6), 1656–1671 (2013)
7. Qu, B.Y., Suganthan, P.N.: Multi-objective differential evolution with diversity enhancement. J. Zhejian Univ. Sci. C-Comput. Electron. **11**(7), 538–543 (2010)
8. Baatar, N., Jeong, K.Y., Koh, C.S.: Adaptive parameter controlling non-dominated ranking differential evolution for multi-objective optimization of electromagnetic problems. IEEE Trans. Magn. **50**(2), 709–712 (2014)
9. Gong, M.G., Jiao, L.C., Du, H.F., Bo, L.F.: Multiobjective immune algorithm with nondiminated neighbor-based selection. Evol. Comput. **16**(2), 225–255 (2008)
10. Shang, R.H., Jiao, L.C., Liu, F., Ma, W.P.: A novel immun clonal algorithm for MO problems. IEEE Trans. Evol. Comput. **16**(1), 35–50 (2012)
11. Wang, L., Zhong, X., Liu, M.: A novel group search optimizer for multi-objective optimization. Expert Syst. Appl. **39**(3), 2939–2946 (2012)
12. Zhan, Z.H., Li, J.J., Cao, J.N., Zhang, J., Chung, H.H., Shi, Y.H.: Multiple populations for multiple objectives: a coevolutionary technique for solving multiobjective optimization problems. IEEE Trans. Cybern. **43**(2), 445–463 (2013)
13. Sanchez, M.S., Ortiz, M.C., Sarabia, L.A.: Selection of nearly orthogonal blocks in 'ad-hoc' experimental designs. In: 8th Colloquium on Chemiometricum Mediterraneum (CCM), vol. 133, pp. 109–120 (2014)
14. Liu, S.H., Ye, W.H., Lou, P.H., Tang, D.B.: Structural dynamic optimization for carriage of gantry machining center using orthogonal experimental design and response surface method. J. Chin. Soc. Mech. Eng. **33**(3), 211–219 (2012)
15. Li, H., Zhang, Q.F.: Multiobjective optimization problems with complicated Pareto sets, MOEA/D and NSGA-II. IEEE Trans. Evol. Comput. **13**(2), 284–302 (2009)
16. Zhao, S.Z., Suganthan, P.N., Zhang, Q.F.: Decomposition-based multiobjective evolutionary algorithm with an ensemble of neighborhood sizes. IEEE Trans. Evol. Comput. **16**(3), 442–446 (2012)
17. Knowles, J.D., Corne, D.W.: Approximating the nondominated front using the Pareto archived evolution strategy. Evol. Comput. **8**, 149–172 (2000)
18. Coello, C.A.C., Lechuga, M.S.: MOPSO: a proposal for multiple objective particle swarm optimization. In: Proceedings of Congress Evolutionary Computation, pp. 1051–1056 (2002)
19. Coello, C.A.C., Pulido, G.T., Lechuga, M.S.: Handling multiple objectives with particle swarm optimization. IEEE Trans. Evol. Comput. **8**(3), 256–279 (2004)

20. Sierra, M.S., Coello C.A.C.: Improving PSO-based multiobjective optimization using crowding, mutation and ε-Dominance. In: Proceedings of Evolutionary Multi-Criterion Optimization, pp. 505–519 (2005)
21. Friedrich, T., Horoba, C., Neumann, F.: Multiplicative approximations and the hypervolume indicator. In: Proceedings of 2009 Genetic and Evolutionary Computation Conference, pp. 571–578 (2009)
22. Zitzler, E., Thiele, L.: Multiobjective evolutionary algorithms: a comparative case study and the strength pareto approach. IEEE Trans. Evol. Comput. 3(4), 257–271 (1999)
23. Asadzadeh, M., Tolson, B.: Pareto archived dynamically dimensioned search with hypervolume-based selection for multi-objective optimization. Eng. Optim. 45(12), 1489–1509 (2013)
24. Wagner, T., Beume, N., Naujoks, B.: Pareto-, aggregation-, and indicator-based methods in many-objective optimization. In: Obayashi, S., Deb, K., Poloni, C., Hiroyasu, T., Murata, T. (eds.) EMO 2007. LNCS, vol. 4403, pp. 742–756. Springer, Heidelberg (2007)
25. Zhang, Q.F., Li, H.: MOEA/D: a multiobjective evolutionary algorithm based on decomposition. IEEE Trans. Evol. Comput. 11(6), 712–731 (2007)
26. Nebro, A.J., Durillo, J.J.: A study of the parallelization of the multi-objective metaheuristic MOEA/D. In: Blum, C., Battiti, R. (eds.) LION 4. LNCS, vol. 6073, pp. 303–317. Springer, Heidelberg (2010)
27. Li, H., Zhang, Q.F.: Multiobjective optimization problems with complicated Pareto sets, MOEA/D and NSGA-II. IEEE Trans. Evol. Comput. 13(2), 284–302 (2009)
28. Zhao, S.Z., Suganthan, P.N., Zhang, Q.F.: Decomposition-based multiobjective evolutionary algorithm with an ensemble of neighborhood sizes. IEEE Trans. Evol. Comput. 16(3), 442–446 (2012)
29. Sindhya, K., Miettinen, K., Deb, K.: A hybrid framework for evolutionary multi-objective optimization. IEEE Trans. Evol. Comput. 17(4), 495–511 (2012)
30. Tan, Y.Y., Jiao, Y.C., Li, H., Wang, X.K.: MOEA/D plus uniform design: a new version of MOEA/D for optimization problems with many objectives. Comput. Oper. Res. 40(6), 1648–1660 (2013)
31. Jan, M.A., Khanum, R.A.: A study of two penalty-parameterless constraint handling techniques in the framework of MOEA/D. Appl. Soft Comput. 13(1), 128–148 (2013)
32. Chang, P.C., Chen, S.H., Zhang, Q.F., Lin, J.L.: MOEA/D for flowshop scheduling problems. In: IEEE Congress on Evolutionary Computation, CEC 2008, pp. 1433–1438 (2008)
33. Konstantinidism, A., Charalambous, C., Zhou, A., Zhang, Q.F.: Multi-objective mobile agent-based sensor network routing using MOEA/D. In: IEEE Congress on Evolutionary Computation, CEC 2010, pp. 1–8 (2010)
34. Miettinen, K.: Nonlinear Multiobjective Optimization. Kluwer Academic Publishers, Boston (1998)
35. Tekinalp, O., Karsli, G.: A new multiobjective simulated annealing algorithm. J. Global Optim. 39(1), 49–77 (2007)
36. Zhang, Q., Liu, W., Tsang, E., Virginas, B.: Expensive multiobjective optimization by MOEA/D with Gaussian process model. IEEE Trans. Evol. Comput. 14(3), 456–474 (2010)
37. Deb, K.: Multi-objective Optimization Using Evolutionary Algorithms. Wiley, New York (2001)

38. Price, K., Storn, R.M., Lampinen, J.A.: Differential Evolution: A Practical Approach to Global Optimization. Natural Computing Series. Springer, Heidelberg (2005)
39. Huband, S., Hingston, P., Barone, L., While, L.: A review of multiobjective test problems and a scalable test problem toolkit. IEEE Trans. Evol. Comput. **10**(5), 477–506 (2006)

MREP: Multi-Reference Expression Programming

Qingke Zhang[1], Xiangxu Meng[1], Bo Yang[2], and Weiguo Liu[1(✉)]

[1] School of Computer Science and Technology,
Engineering Research Center of Digital Media Technology,
Ministry of Education, Shandong University, Jinan 250101, China
qingke.zhang@mail.sdu.edu.cn,
{mxx,weiguo.liu}@sdu.edu.cn
[2] Shandong Provincial Key Laboratory of Network Based Intelligent
Computing, University of Jinan, Jinan 250022, China
yangbo@ujn.edu.cn

Abstract. MEP is a variant of genetic program applied to solve the symbol regression and classification problems. It can encode multiple solutions of a problem in a single chromosome. However, when the ratio of genes reuse is low, it may not get a high accuracy result within limited iterations and may fall into the trap of local optimum. Therefore, we proposed a novel genetic evolutionary algorithm named MREP (multi-reference expression programming). The MREP chromosome is encoded in a two-dimensional structure and each gene in one chromosome can refer other sub-layer's gene randomly. The main contribution can be described as follows: Firstly, a novel chromosome encoding scheme is proposed based on a two-dimensional structure. Secondly, two different cross-layer reference strategies are designed to enhance the code reuse of genes located at different layers in one chromosome. Two groups experiments were conducted on eight symbol regression functions. The statistical results reveal that the MREP performs better than the compared algorithms and can solve the symbol regression functions problem efficiently.

Keywords: Multi-expression programming · Cross-layer-reference · Two dimensional operators · Genetic programming · Symbol regression

1 Introduction

Evolutionary computation is defined as randomized search procedures inspired by the working mechanism of genetics and natural selection [1]. It is an effective method to solve optimization, search and learning problems. The classical evolutionary computing algorithms is genetic algorithms (GA), genetic programming (GP), evolution strategies (ES) and evolutionary programming (EP). Among them, GP is a symbolic optimization technique for nonlinear system modeling. Since the introduction of standard GP by Koza (2010) [2], numerous GP paradigms and variants have been proposed. These GP variants can be classified into two streams: standard GP and Linear structure GP (LGP) [3]. In standard GP, a genetic program is represented in a tree structure. In linear GP, a genetic program is represented in a linear list of machine code

© Springer International Publishing Switzerland 2016
D.-S. Huang and K.-H. Jo (Eds.): ICIC 2016, Part II, LNCS 9772, pp. 26–38, 2016.
DOI: 10.1007/978-3-319-42294-7_3

instructions or high-level language statements. Recently, several linear variants of LGP have been proposed. Some of them are multi-expression programming (MEP), grammatical evolution (GE) [4], gene expression programming (GEP) [5], cartesian genetic programming (CGP) [6] and Genetic Algorithm for Deriving Software (GADS) [7]. Generally, a GP chromosome encodes a single expression (computer program). This is also the case for GEP, GE and LGP chromosomes. By contrast, MEP has a special ability to encode several expressions in a single chromosome. Furthermore, when solving symbolic regression problems, the MEP chromosome decoding process has the same complexity as other techniques such as GE and GEP. In [8], a systematic comparison of several linear genetic programming algorithms has been carried out on several numerical experiments. The results obtained show that MEP has the best overall behavior for some problems such as symbolic regression and even-parity. Now multi expression programming has been widely used in many areas, such as classification problems [9], stock market forecast [10], TSP [11], and data prediction [12–16].

Although MEP has some advantages over the single expression chromosome encoding algorithms, if code-reuse ability is not fully utilized, the number of symbols in chromosome is increasing sharply than the number of symbols in a GEP or GE chromosome [8]. Additionally, the diversity of decoding expressions can't be effectively guaranteed. Therefore, we proposed a novel algorithm named multi-reference expression programming (MREP). In MREP, a chromosome is encoded into a two-dimensional format which contains several layers. The gene can arbitrarily reuse other layer's genes located in its previous position in a 2-D chromosome. The novel gene encoding mechanism is designed to increase the ratio of code-reuse for each gene in a two-dimensional (2-D) chromosome rather than in a conventional one-dimensional (1-D) chromosome. The encoding methods can bring more solutions (expressions) and enrich the diversity of population within the limited number of genes.

The rest of the paper is structured as follows. The background of standard MEP is described in Sect. 2. The presentation of the proposed MREP algorithm is described in Sect. 3. Numerical experimental results and analysis are listed in Sect. 4. Conclusion are shown in last section.

2 MEP Algorithm

The standard MEP algorithm adopts a steady state as its underlying mechanism. The MEP algorithm starts by creating a random population of individuals. The following steps are repeated until a stop condition is satisfied. Firstly, two parents are selected using a tournament method selection procedure. Then the parents are recombined by crossover operation to generate two offspring. The offspring are considered for mutation. Finally, the best offspring replaces the worst individual in the current population if the offspring performs better than the worst individual. The algorithm returns the best expression evolved along a fixed number of generations.

MEP genes are (represented by) substrings of a variable length. The number of genes per chromosome is constant and equals to the chromosome length. Each gene encodes a terminal (an element in the terminal set T) or a function symbol (an element in the function set F). A gene that encodes a function includes pointers towards the function arguments. Function arguments always have indices of lower values than the position of the function itself in the chromosome. Therefore, the first symbol of each chromosome should be encoded into a terminal symbol. In this way only syntactically correct programs are obtained. For example, assuming that the set of functions $F \rightarrow \{+, -, *, /\}$ and terminals $T \rightarrow \{x0, x1, c\}$, where c represents a constant. x0 and x1 represent the variables. A MEP chromosome with six genes is shown in Fig. 1.

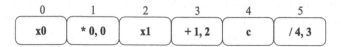

Fig. 1. MEP chromosome.

The MEP chromosomes are read from left-right way starting with the first position x0. A terminal symbol denotes a simple expression, while a function symbol represents a complex expression (made up by linking the operands specified by the argument positions with the current function symbol). For instance, genes 0, 2, and 4 in Fig. 1 encode simple expressions formed by a single terminal symbol from terminal set T. These expressions associated with genes 0, 2, and 4 are: $E_0 = x0$, $E_2 = x1$, $E_4 = c$. Gene 1 indicates the operation $*$ on the operands located at positions 0 and 0 of the chromosome. Therefore, Gene 1 encodes the expression: $E_1 = x0 * x0$. Gene 3 indicates the operation + on the operands located at positions 0 and 2 of the chromosome. Therefore, Gene 3 encodes the expression: $E_3 = x0 * x0 + x1$. Gene 5 indicates the operation / on the operands located at positions 4 and 3 of the chromosome. Therefore, Gene 5 encodes the expression: $E_5 = c/((x0 * x0) + x1)$. MEP encodes a number of expressions in single chromosome, therefore, it is required to choose one of the expressions $(E_0–E_5)$ to present the chromosome according to its fitness value. Only the expression with best fitness can be selected to represent current chromosome.

3   MREP Algorithm

Multi-Reference Expression Programming algorithm is a variant of MEP. The chromosome of MREP use the linear encoding structure not a tree structure which is similar to MEP. However, MREP individual encodes its chromosome into a two-dimensional structure rather than one-dimensional structure. The evolutionary process starts by creating a random population of individuals, each of which is encoded into a 2-D structure. Then the following steps are conducted iteratively until a stop condition is reached. Firstly, two individuals are randomly selected as parents by using a tournament procedure. Then, the selected parents are recombined by 2-D crossover to generate two offspring. The offspring are considered to conduct the mutation operation. Finally, the better offspring is used to replace the worst individual among the population. The details of MREP are described as follows.

3.1   Encoding Principle

MREP chromosome is encoded into a 2-D grid, which contains R rows and C columns. For the chromosome, each row represents a layer (or a sub-chromosome) and each layer owns the same number of columns (or genes). Each element in the grid denotes a gene. Therefore, the number of genes in one chromosome is R * C. The chromosome length is constant and equal to the columns in that chromosome. Figure 2 shows an example of the MREP chromosome which contains 5 layers and 6 genes in each layer (R = 5, C = 6). A genes position in the 2-D chromosome is represented by its coordinate as G (layer, index), where layer $\in$ [0, R), index $\in$ [0, C).

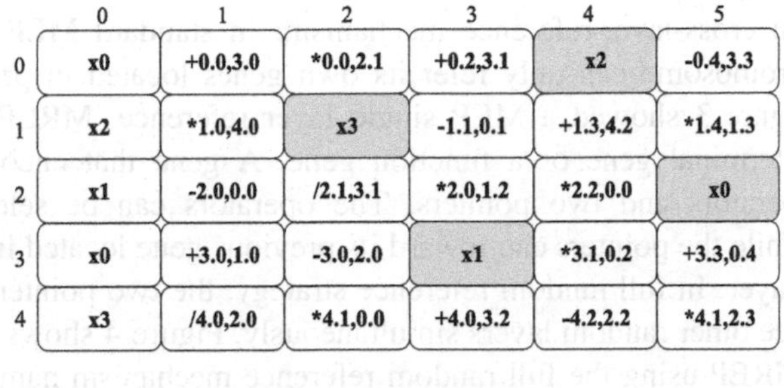

	0	1	2	3	4	5
0	x0	+0.0,3.0	*0.0,2.1	+0.2,3.1	x2	-0.4,3.3
1	x2	*1.0,4.0	x3	-1.1,0.1	+1.3,4.2	*1.4,1.3
2	x1	-2.0,0.0	/2.1,3.1	*2.0,1.2	*2.2,0.0	x0
3	x0	+3.0,1.0	-3.0,2.0	x1	*3.1,0.2	+3.3,0.4
4	x3	/4.0,2.0	*4.1,0.0	+4.0,3.2	-4.2,2.2	*4.1,2.3

Fig. 2. MREP chromosome

MREP genes can be classified into two types: function gene (an element in the function set *F*) and terminal gene (an element in the terminal set *T*). The first column of the 2-D chromosome must contain terminals which are randomly selected from the set of terminals *T*. All other genes are determined by randomly selecting a terminal from the set of functions. A gene that encodes a function contains three parts: one is the operators selected from the function sets, other two parts are two pointers toward the other genes which located in current gene's previous arbitrary positions in a random layer. The code principle ensured that no cycle arises while the chromosome is decoded. For example, Suppose the function set $F = \{+, -, *, /\}$ and the terminal set $T = \{x0, x1, x2, x3\}$, in Fig. 2, a gene located at G (2, 3) was encoded into a function as {*2.0, 1.2}. The first pointer towards a gene located at G (2, 0), the second pointer towards a gene located at G (1, 2). Therefore, the gene G (2, 3) can be decoded as a function: $F_{G\ (2,3)} = F_{G\ (2,0)} * F_{G\ (1,2)}$. Among these R * C genes, the number of symbols for each gene are strings of a variable length, the maximum number of symbols in a MREP chromosome can be computed as follows in (1):

$$Max\_Number\_of\_Symbols = n + 1 * (R * C - 1) + R \qquad (1)$$

where n is the number of arguments of the function with the greatest number of arguments. For instance, if the function symbol is a unary operator, then $n = 1$, if the function symbol is a binary operator, then $n = 2$. The maximum number of effective symbols is achieved when all the genes are encoded with a function gene. In MREP, each chromosome could involve more information and expressions with other layer's genes. The encoding methods can bring more solutions (expressions) and enrich the diversity of population within the limited number of genes.

3.2   Cross Layer Reference Mechanism

In MREP, the proposed cross-layer encoding scheme enables the gene in different layers can communicate with each other. The reference ratio of each gene in some degree can dominate the diversity of the population. We analysis the structure of MREP and designed two different random cross-layer reference strategy. One is the full random cross layer reference; another is single random cross layer reference. Both of them are described in detail as follows.

Fully random cross-layer-reference mechanism: In standard MEP algorithm, the gene in each chromosome can only refer its own genes located in previous gene of current gene. Figure 3 showed a MEP single layer reference. MREP genes can be encoded into a terminal gene or a function gene. A gene that encodes a function contains one operators and two pointers. The operators can be selected from the function set F, while the pointers can toward its previous gene located in other layer or its current own layer. In full random reference strategy, the two pointers of a function gene can reference other random layers simultaneously. Figure 4 shows a fully random reference. The MREP using the full random reference mechanism named MREP_FR.

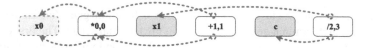

Fig. 3. MEP single-layer reference (All the dashed lines point to genes located at the same layer.)

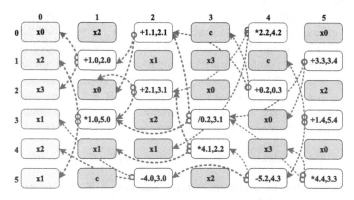

Fig. 4. Fully random cross-layer reference (All the dashed lines point to genes located at a random layer.)

Single random cross-layer-reference mechanism: The single random reference integrated the feature of the same layer's reference by MEP and the random cross-layer reference by MREP. Both reference strategy is used in the function gene. As the general MREP chromosome, a gene that encodes a function contains one operator and two pointers. The operator can be selected from the function set F, while the pointers can toward its previous gene located in other layer or its current own layer. For the single random cross-layer reference, one pointer towards its previous gene which located in current gene's layer, while another pointer towards its previous gene located in a random layer. Figure 5 shows a single cross-layer random reference strategy. The MREP using the single random reference mechanism named MREP_SR. It is the default reference mechanism used by MREP. The performance and efficiency of the two different cross-layer reference strategies (MREP_FR, MREP_SR) will be tested in experimental parts.

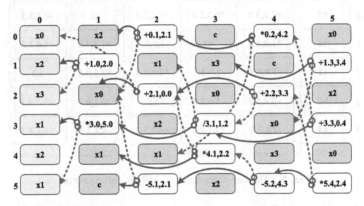

Fig. 5. Single random cross-layer reference (The solid line points to a gene located at the same layer, while the dashed line points to a gene located at a random layer)

3.3 Decoding and Fitness Evaluation

In the evaluation process, the first step is to decode all genes and compute each gene's fitness for the chromosome. When the genes in MREP individuals are translated into computer programs (expressions), they are decoded starting with the first position G (0, 0) and then follows the direction from top-down and then left-right.

The decoding process was similar to MEP. For instance, in Fig. 5, a gene located at G (4, 3) was encoded into a function as {*4.1, 2.2}. The first pointer towards a gene located at G (4, 1), the second pointer towards a gene located at G (2, 2). Therefore, the expression of G (4, 3) can be decoded as a function: G (4, 3) = G (4, 1) *G (2, 2). The gene G (4, 1) is encoded as a terminal symbol as G (4, 1) = x1. G (2, 2) is a function gene expressed by G (2, 2) = G (2, 1) + G (0, 0) = x0 + x0. Therefore, the gene located at G (4, 3) can be decoded as an expression: G (4, 3) = G (4, 1) * G (2, 2) = x1* (x0 + x0). Other genes in MREP chromosome share the same decoding principle similar to the gene G (4, 3).

The decoding order follows the direction from top-down and then left-right for each chromosome. Figure 6 shows the order of the decoding process. The principle of the decoding order can guarantee that a gene can refer its previous genes located in a random layer no matter it's upper layer or down layer arbitrarily. Otherwise, if the decoding order was inversed, a gene referring one previous gene which located under the current layer can't be decode validly, just as the referred gene has not been decoded primarily.

MREP evaluate each gene expression according to the decoding order showed in Fig. 6. The fitness of each expression encoded in a chromosome is computed during the individual evaluation. This evaluation is performed by reading the chromosome only once and storing partial results in a matrix.

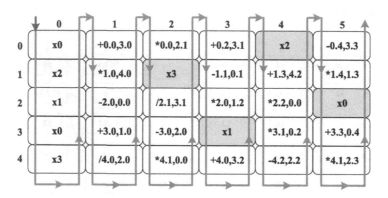

Fig. 6. The decoding order of MREP chromosome

For example, in Fig. 2, the gene G(3, 4) use $*$ operator to multiply gene G(3, 1) and gene G(0, 2). Therefore, the fitness of gene G(3, 4) can be calculated immediately through a multiply operation as the formula by fitness$_{G(3, 4)}$ = fitness$_{G(3, 1)}$ $*$ fitness$_{G(0, 2)}$. The gene with the best fitness among all genes can be chosen to represent the chromosome. The fitness of a MREP chromosome C can be calculated by the formulas (2) and (3).

$$fitness(C) = \min_{i \in 0, R, j \in 0, C} fitness(G(i,j)) \qquad (2)$$

$$fitness(G(i,j)) = \sum_{k=0}^{N} |E_{i,j}^k - O^k| \qquad (3)$$

where N is the number of samples. G (i, j) denotes a gene located at the ith layer and j^{th} position in chromosome C. $E_{i,j}^k$ represents the experimental value calculated by the gene G (i, j) on the kth sample data; O^k denotes the target value of the kth sample data. R and C denote the rows and columns of the chromosome, respectively.

3.4   Genetic Operators

The evolution operators used within MREP algorithm are selection, crossover and mutation. Since the genetic operators preserve the chromosome structure, all the off-spring are syntactically correct expressions. In order to narrow down the searching scope and improve the evolution speed, we introduced some new function operators, such as *sqrt, pow, log, exp, sin* et al., in function set F as the prior knowledge. Besides, for some real number problems, it is not effective for the general MEP in retrieving the complicated target function, as the gene was not assigned in a high probability with real number parameters. Therefore, some real constant was added to the terminal set T. More details about the genetic operators are described in the following.

- **Crossover:** Two parents are selected randomly and recombined by crossover. In MREP, four types of crossover operators namely, one-point crossover, two-point crossover, random crossover and interchange operator are designed based on a 2-D chromosome. The interchange operator was designed for a single chromosome by interchange two random selected sub-layers in one chromosome.
- **Mutation:** Terminal symbol, function symbol and gene index may be used to mutate the gene. In order to prevent the chromosome structure from damaging, the first column genes in MREP chromosome must be encoded to a terminal symbol. If the current gene changes into a function symbol, the function arguments must be function pointers to the previous genes. These different types of genes can mutate to each other under a given probability.

4   Experiment Results and Analysis

4.1   Test Functions and Test Datasets

In the following experiments, we compared MREP, the standard MEP (marked by MEP_S), and an improved MEP with multi-population (marked by MEP_M) algorithm on the following symbol regression functions listed in Table 1. Some of the test functions are selected from [8]. The multi-expression programming and multi reference expression programming algorithms are implemented in C language.

Table 1. Test symbol regression functions

Functions	Symbol regression functions	Accuracy
f_1	$f_1(x) = x^4 + x^3 + x^2 + x$	1.0E−6
f_2	$f_2(x) = x^5 - 2x^3 + x$	1.0E−6
f_3	$f_3(x) = x^6 - 2x^4 + x^2$	1.0E−6
f_4	$f_4(x) = y^{-2} * (x^2 + x)$	1.0E−6
f_5	$f_5(x) = x^{-2} + \sqrt{x}$	1.0E−6
f_6	$f_6(x) = \log(x^3 + x^2)$	1.0E−6
f_7	$f_7(x) = e^x + \cos(x^3)$	1.0E−6
f_8	$f_8(x) = xy * e^{-x^2 - y^2}$	1.0E−6

The experimental test data is composed by the values of variables and the targets. A sampling of the numerical variables is generated in the real interval $(-10, +10)$ with 16 valid bits over 50 randomly points. The corresponding target data is generated by the exact calculation of the eight test functions. The experiment parameters are showed in Table 2. The mean value (μ) and standard deviation (σ) of the test results are used as the measurement. They are calculated as follows in (4).

$$\mu = \frac{1}{N} \cdot \sum_{i=1}^{N} f_i, \quad \sigma = \sqrt{\frac{1}{N} \cdot \sum_{i=1}^{N} (f_i - \mu)^2} \tag{4}$$

where N denotes the repeat testing times, f_i denotes the best fitness obtained in each test, and μ denotes the average fitness among the N test results.

Table 2. Parameter settings.

Parameters	MREP	MEP_M	MEP_S
Generations	1000	1000	1000
Population size	40	40	240
Population Group	1	6	1
Length	40	40	40
Layers	6	1	1
Operators	+, −,*, ÷, 1/(), *pow, sqrt,*		
	exp, log, sin, cos		
Mutation Prob.	0.4	0.4	0.4
Crossover Prob.	0.6	0.6	0.6
Function Prob.	0.5	0.5	0.5
Terminal Prob.	0.4	0.4	0.4
Constant Prob.	0.1	0.1	0.1

4.2 Experiment-I: Performance Comparisons

In this part, three multiple expression programming algorithms namely MREP, standard MEP (MEP_S), an improved MEP with multiple populations (MEP_M) are compared on the eight test functions. All benchmark functions were tested 100 times independently aiming to reduce statistical errors. The mean fitness (Mean), standard deviation (Std.), best fitness (Best), worst fitness (Worst), and success rate (Rate) are used to measure the performance of compared algorithms. The best results (minimum result) of the mean fitness of the test algorithms are shown in bold.

Tables 3 and 4 showed the statistical results on the eight benchmark functions from f_1 to f_8. According to the fitness calculation formula in (3), the mean fitness reflects the mean absolute error between the experimental value and the target value. Therefore, the smaller of the mean value the better of the performance. The statistical results in Tables 3 and 4 revealed that MREP performs better than other two algorithms on function f_1, f_2, f_4, f_5, f_6, f_7, and f_8. Besides, Table 5 shows the mean iteration of each algorithm on those successfully discovered functions on f_1, f_2, f_3, f_4, f_5 and f_6, MREP can retrieve those symbol regression functions successfully with fewer average number of iterations compared with other algorithms. The MEP with multiple sub-populations outperform the standard MEP in function accuracy and the mean evolution generations.

Compared with other two algorithms, the better performance of MREP can attribute to its encoding scheme which enhanced the ratio of gene reuse by cross-layer-reference mechanism. Genes in MEP can only reuse the genes in the same chromosome, while the MREP genes can refer more genes located at different layers (sub-chromosome) which can enrich the diversity of the genes among the whole population.

Table 3. Results of the compared algorithm-1

Algorithm	Item	f_1	f_2	f_3	f_4
MREP	Mean	**1.22E−12**	**2.73E−12**	1.24E−11	**1.81E−12**
	Std.	6.50E−14	2.44E−13	1.50E−12	1.04E−14
	Best	1.15E−12	2.49E−12	1.50E−12	1.04E−14
	Worst	1.31E−12	3.10E−12	1.32E−11	1.81E−12
	Rate	100.00 %	100.00 %	100.00 %	100.00 %
MEP_M	Mean	1.61E−12	3.93E−12	**1.20E−11**	**1.81E−12**
	Std.	2.44E−13	8.50E−13	1.03E−12	1.79E−12
	Best	9.96E−13	2.71E−12	1.09E−11	1.79E−12
	Worst	1.92E−12	4.77E−12	1.32E−11	1.82E−12
	Rate	100.00 %	100.00 %	100.00 %	100.00 %
MEP_S	Mean	1.72E−12	4.14E−12	1.30E−11	1.90E−12
	Std.	5.50E−12	9.09E−13	1.03E−11	4.00E−14
	Best	9.97E−13	2.54E−12	1.14E−11	1.79E−12
	Worst	5.64E−11	6.40E−12	1.15E−10	1.93E−12
	Rate	100.00 %	100.00 %	100.00 %	100.00 %

Table 4. Results of the compared algorithm-2

Algorithm	Item	f_5	f_6	f_7	f_8
MREP	Mean	**6.37E−13**	**1.69E−14**	**1.46E−14**	**1.44E−13**
	Std.	1.02E−13	5.33E−16	0.00E+00	0.00E+00
	Best	5.06E−13	1.66E−14	1.46E−14	1.44E−13
	Worst	7.70E−13	1.79E−14	1.46E−14	1.44E−13
	Rate	100.00 %	100.00 %	100.00 %	100.00 %
MEP_M	Mean	7.06E−13	1.70E−14	9.28E−01	1.64E−13
	Std.	6.13E−14	5.44E−16	1.86E + 00	3.40E−14
	Best	6.41E−13	1.66E−14	1.46E−14	1.44E−13
	Worst	7.90E−13	1.77E−14	4.64E+00	2.56E−13
	Rate	100.00 %	100.00 %	90.00 %	100.00 %

(Continued)

Table 4. (*Continued*)

Algorithm	Item	f_5	f_6	f_7	f_8
MEP_S	Mean	6.83E−13	1.76E−14	3.75E−01	7.78E−01
	Std.	9.95E−14	1.02E−16	1.04E+00	3.39E+00
	Best	5.42E−13	1.75E−14	1.46E−14	1.56E−13
	Worst	8.06E−13	1.77E−14	3.83E+00	1.56E+01
	Rate	100.00 %	100.00 %	88.00 %	95.00 %

Table 5. The average number of iterations on the successfully discovered functions

Algorithm	f_1	f_2	f_3	f_4	f_5	f_6
MREP	5.80	4.00	1.00	30.0	18.2	4.80
MEP_M	7.50	11.6	1.00	89.8	19.6	13.0
MEP_S	30.95	32.08	1.73	154.6	42.8	80.4

4.3   Experimental-II: Analysis of the Cross Layer Reference

In MREP, the proposed cross-layer reference encoding scheme was designed to enable the genes in different layers communicating with each other. The reference ratio of each gene in some degree can dominate the diversity of the population. In Sect. 3, two different random cross-layer reference strategy in MREP chromosome are proposed, one is the full random cross-layer reference (marked by MREP_FR), and another is single random cross-layer reference (marked by MREP_SR). Both of them are evaluated on the eight symbol regression functions and shared the same parameter settings as shown in Table 2. Table 6 shows the statistical average results generated by 30 tests. The average number of successful evolution iterations (Avg.) is used to compare the efficiency of both the reference strategies.

Table 6. Comparasion between the two random cross-layer reference strategies

f	MREP_FR			MREP_SR		
	Mean ± Std.	Rate	Avg.	Mean ± Std.	Rate	Avg.
$f1$	1.18E−12 ± 9.26E−14	**100 %**	26.8	2.51E−12 ± 3.53E−13	**100 %**	**8.30**
$f2$	2.96E−12 ± 8.35E−13	**100 %**	10.0	2.51E−12 ± 5.39E−13	**100 %**	**4.20**
$f3$	1.32E−11 ± 6.53E−15	**100 %**	1.00	1.24E−11 ± 6.51E−13	**100 %**	**1.00**
$f4$	1.74E−12 ± 4.45E−14	**100 %**	28.1	1.75E−12 ± 8.09E−14	**100 %**	**14.9**
$f5$	6.82E−13 ± 1.10E−13	**100 %**	25.1	7.04E−13 ± 9.42E−14	**100 %**	**11.6**
$f6$	1.71E−14 ± 9.05E−16	**100 %**	11.2	1.74E−14 ± 8.76E−16	**100 %**	**5.80**
$f7$	1.46E−14 ± 0.00E+00	**100 %**	46.0	1.53E−14 ± 1.54E−15	**100 %**	**36.6**
$f8$	1.44E−13 ± 0.00E+00	**100 %**	22.2	1.45E−13 ± 3.20E−15	**100 %**	**18.3**

The best result of the average number of iterations (Avg.) computed by algorithms are shown in bold. It can be seen that MREP_SR converges faster than MREP_FR with fewer iterations. And either way, the result reveals that the gene adopted the hybrid reference mechanism (by mixed the same layer's reference with cross-layer's reference) can enhance the ratio of gene-reuse and generate better results than general MEP algorithms. In some degree, the experiments proved that the ratio of gene's reuse is crucial to influence the performance of multi-expression programming.

5 Conclusions

This paper presents a novel genetic evolutionary algorithm named MREP. MREP individual encodes its chromosome into a two-dimensional structure rather than one-dimensional structure utilized by MEP. Furthermore, a novel reference model was created by cross-layer reference mechanism to enrich the diversity of gene and enhance the code-reuse. Two groups experiments were conducted on eight symbol regression functions. The statistical result reveals that the MREP performs better than other compared algorithms. As the chromosome of MREP is encoded into a 2-D structure and the chromosome genes in one column has no refer relationships, thus the genes evaluated process can be easily parallelized by using the multi-threading technology.

Acknowledgement. This work was supported by National Natural Science Foundation of China under Grant No. 61572230, No. 61173078, No. 61573166 and Shandong Provincial Natural Science Foundation under Grant ZR2015JL025. The authors would like to thank the anonymous reviewers for providing comments to help us improve the contents of this paper.

References

1. Kallel, L., Bart, N., Alex, R.: Theoretical aspects of evolutionary computing. Springer Science & Business Media, New York (2013)
2. Koza, B.J.: Evolving caching algorithms in C by GP. In: Genetic Programming. MIT Press (2010)
3. Banzhaf, W., Nordin, P., Keller, R.E., Francone, F.D.: Genetic Programming: An Introduction: On the Automatic Evolution of Computer Programs and Its Applications, December 1998
4. Ryan, C., Neill, M.O.: Grammatical evolution: a steady state approach. Late Breaking Papers Genetic Programming, pp. 180–185 (1998)
5. Ferreira, C.: Gene expression programming: a new adaptive algorithm for solving problems. Eprint Arxiv Cs (2), 87–129 (2001)
6. Miller, J.F.: Gecco 2013 tutorial: Cartesian genetic programming. In: Conference Companion on Genetic and Evolutionary Computation, pp. 715–740 (2013)
7. Paterson, N.R.: Genetic programming with context sensitive grammars. In: Proc. Eurogp Lncs 63(84), 113117 (2002)
8. Oltean, M., Groan, C.: A comparison of several linear genetic programming techniques. Complex Syst. **4**, 285–313 (2003)
9. Baykasolu, A., Ozbakir, L.: Mepar-miner: multi-expression programming for classification rule mining. Eur. J. Oper. Res. **183**(2), 767–784 (2007)

10. Groan, C., Abraham, A., Ramos, V., Han, S.Y.: Stock market prediction using multi expression programming. In: Portuguese Conference on Artificial Intelligence, EPIA 2005, pp. 73–78 (2006)
11. Oltean, M., Dumitrescu, D.: Evolving TSP heuristics using multi expression programming. In: Bubak, M., Albada, G.D., Sloot, P.M., Dongarra, J. (eds.) ICCS 2004. LNCS, vol. 3037, pp. 670–673. Springer, Heidelberg (2004)
12. Alavi, A.H., Gandomi, A.H., Modaresnezhad, M., Mousavi, M.: New ground-motion prediction equations using multi expression programing. J. Earthquake Eng. 15(4), 511–536 (2011)
13. Cattani, P.T., Johnson, C.G.: ME-CGP: multi expression Cartesian genetic programming. In: IEEE Congress on Evolutionary Computation, CEC 2010, Barcelona, Spain, 18–23 July, pp. 1–6 (2010)
14. Garg, A., Garg, A., Lam, J.S.L.: Evolving functional expression of permeability of fly ash by a new evolutionary approach. Transport Porous Media 107(2), 555–571 (2015)
15. Yang, B., Zhang, Q., Wang, L., Li, Y.: Inference of differential equations by MMEP for cement hydration modeling. In: IEEE International Conference on Computer Supported Cooperative Work in Design, pp. 4–10 (2013)
16. Zhang, Q., Yang, B., Wang, L., Jiang, J.: An improved multi-expression programming algorithm applied in function discovery and data prediction. Int. J. Inf. Commun. Technol. 5 (5), 218–233 (2013)

Independent Component Analysis

Extraction of Independent Components from Sparse Mixture

Jian-Xun Mi[1,2(✉)], Cong Li[1,2], and Chao Li[1,2]

[1] College of Computer Science and Technology, Chongqing University of Posts
and Telecommunications, Chongqing, China
mijianxun@gmail.com
[2] Chongqing Key Laboratory of Computational Intelligence,
Chongqing University of Posts and Telecommunications, Chongqing, China

Abstract. In this paper we study extraction of independent components from
the instantaneous sparse mixture with additive Gaussian noise. We model the
problem as a dictionary-learning-like objective function which tries to discover
independent atoms and corresponding sparse mixing matrix. The objective
function involves fidelity term, L1 normalization term and Negentropy term
which respectively limits noise, maximizes the sparseness of mixing matrix and
non-Gaussianity of each atom. An alternative iteration algorithm is proposed to
solve the optimization. According to our simulation, the proposed method
outperforms FastICA and K-SVD.

Keywords: Independent component analysis · Sparse mixture · Instantaneous
mixture · Negentropy

1 Introduction

Independent component analysis (ICA) is a very popular method to discover the
independent blind sources of data [1, 2]. The method assumes that blind sources are
independent to each other and their mixture is instantaneous. ICA is the most widely
applied blind source separation (BSS) method for its simplicity and good performance.

We assume that there is a n-variate random vector $\mathbf{x} = (x_1, x_2, \cdots, x_n)^{\mathrm{T}}$ with zero
mean follows the ICA model, and that is, each components of $\mathbf{x}$ is a linear mixture of l
mutually independent components in the vector $\mathbf{s} = (s_1, s_2, \cdots, s_l)^{\mathrm{T}}$. Usually the
mixing matrix $\mathbf{A}$ is assumed to be full-rank and square so that $l = n$ and $\mathbf{A} \in R^{n \times n}$. The
common ICA model can be written as:

$$\mathbf{x} = \mathbf{As}, \tag{1}$$

where we only consider noise-free cases. Independent components (ICs) $\mathbf{s}$ are known as
a type of blind sources and the blindness refers to both original sources and the mixing
process. In ICA, one uses an unmixing matrix $\mathbf{W} = [\mathbf{w}_1; \cdots; \mathbf{w}_l](\mathbf{w}_i \in R^{1 \times l})$ to recover
ICs from the mixture:

© Springer International Publishing Switzerland 2016
D.-S. Huang and K.-H. Jo (Eds.): ICIC 2016, Part II, LNCS 9772, pp. 41–51, 2016.
DOI: 10.1007/978-3-319-42294-7_4

$$\mathbf{y} = \mathbf{W}\mathbf{x} \qquad (2)$$

where $\mathbf{y} = (y_1, y_2, \cdots, y_l)^{\mathrm{T}}$ is an estimation of elements in **s**. The estimation of underlying sources by ICA losses the information on energy and permutation of original ICs. However ICA is still attractive especially in medical system where the wave of a source is more meaningful than its energy. For example, ICA can identify artifacts and signals of interest from brain imaging signals, such as magnetoencephalography (MEG) [3] or electroenchaphalography (EEG), and functional magnetic resonance imaging (fMRI) [4]. And ICA can also be used to extract the fetal electrocardiography (FECG) from the electrocardiography (ECG) recordings measured on the mother's skin [5].

Many ICA algorithms have been proposed to estimate the unmixing matrix. Early effort by Bell and Sejnowski used infomax principle to develop ICA algorithm. Amari et al. proposed natural gradient based algorithm which is essentially connected to maximum likelihood estimation. The "nonlinear PCA" approach was introduced by Hyvarinen et al. From the perspective of information theory, the objective of main stream ICA algorithms could be seen as minimization of mutual information between the extracted components. It is proved that maximization of the negentropy of ICs is equal to minimization of their mutual information. The negentropy is known as a measure of non-Gaussianity of extracted components which is not easy to be exactly estimated. Hyvarinen et al. gave a high efficient estimation of negentropy and proposed a fixed-point algorithm for ICA referred to as FastICA. In practical applications, noises are unavoidable and combined into the mixture in the process of data acquisition [6]. Some ICA algorithms can still work up to some extent if the impact of noise on interesting ICs is not severe. However to broaden the application of ICA, many studies focus on the noisy IC extraction problem. Here we assume the noises are additive and the noisy ICA model is defined by

$$\mathbf{x} = \mathbf{A}\mathbf{s} + \mathbf{e}, \qquad (3)$$

where **e** is additive noise term which is assumed to be Gaussian distributed in the paper. To reduce the impact of noises, objective function for noise-free case needs redesign. For example Gaussian moments could be used to replace forth order moments in classical ICA [7].

In this paper, we consider that case that each noisy observation is a sparse mixture of ICs. That is to say, the mixing vector $\mathbf{a}_i$, a row in mixing matrix **A**, corresponding to observation $\mathbf{x}_i$ only has a few non-zero entries. Hence we still study the noisy ICA problem as Eq. (3) as well as consider the sparse priori on **A**. We have to emphasize that the early study of sparse priori in ICA [8] only considered the sparseness of **W**, which is the inverse of **A**. And a sparse W is not necessary to result in a sparse **A**. Actually it is hard to directly restrict the sparsity of mixing vectors in the traditional ICA model. In recent years, dictionary learning (DL) is a hot research topic in machine learning area. For example, Aharon et al. proposed K-SVD algorithm to generate an over-complete dictionary from a training data set [9]. A link is established between ICA and DL in this paper due to we treat IC extraction as a dictionary learning problem. The atoms in the learned dictionary are mutually independent with each.

The paper is organized as follows: we introduce DL problem and ICA in Sect. 2. Then we propose our DL-like IC extraction model in Sect. 3 along with an alternative iteration algorithm derived to solve the proposed objective function. We show simulations of proposed method in Sect. 4 and then conclude the paper in Sect. 5.

2 Extracting ICs Using DL-like Model

2.1 Dictionary Learning

For a typical DL study, the observation data is assumed to be decomposed into two parts, i.e., dictionary matrix and coding matrix. Actually, DL can be expressed as the same model as (3) (random vector $\mathbf{s}$ and $\mathbf{x}$ are rewritten to sample matrix form using capital $\tilde{\mathbf{S}} = [\tilde{\mathbf{s}}_1; \cdots; \tilde{\mathbf{s}}_l] \in R^{l \times p}$ and $\tilde{\mathbf{X}} = [\tilde{\mathbf{x}}_1; \cdots; \tilde{\mathbf{x}}_n] \in R^{n \times p}$ respectively where p denotes the length of an atom). Now $\tilde{\mathbf{S}}$ is seen as the dictionary matrix which produces every observed sample $\mathbf{x}_i$ with the relation

$$\tilde{\mathbf{x}}_i = \mathbf{a}_i \tilde{\mathbf{S}} \tag{4}$$

where every row of $\tilde{\mathbf{S}}$ is an atom vector and $\mathbf{a}_i$ is a row in $\mathbf{A}$. From the respective of DL, one wants to learn very rich representation of observations so that the dictionary $\tilde{\mathbf{S}}$ needs to be over-complete, i.e., the number of atoms is greater than the length of atom. Hence each observed sample $\mathbf{x}_i$ is a sparse representation on all atoms, satisfying

$$\underset{\mathbf{a}_i}{\arg\min} \|\mathbf{a}_i\|_1 \text{ subject to } \|\tilde{\mathbf{x}}_i - \mathbf{a}_i \tilde{\mathbf{S}}\|_F^2 \leq \varepsilon \tag{5}$$

where ε is a small positive constant and $\|\cdot\|_1$ denotes L1-norm of a vector which can be used as an effective approximation of L0-norm to constrain the number of non-zero entries in $\mathbf{a}_i$. To merge all sample representations in to a unified formula, the objective function of DL is rewritten as

$$\underset{\mathbf{A}, \tilde{\mathbf{S}}}{\arg\min} \|\mathbf{A}\|_1 \text{ subject to } \|\tilde{\mathbf{X}} - \mathbf{A}\tilde{\mathbf{S}}\|_F^2 \leq \varepsilon \tag{6}$$

where $\|\cdot\|_F$ denotes the Frobenius norm of a matrix. Above objective function can be further transformed to an equivalent form

$$\underset{\mathbf{A}, \tilde{\mathbf{S}}}{\arg\min} \|\tilde{\mathbf{X}} - \mathbf{A}\tilde{\mathbf{S}}\|_F^2 + \lambda \|\mathbf{A}\|_1 \tag{7}$$

where λ is used to balance the fidelity term and the sparse constraint. To solve Eq. (7), one can alternately update $\mathbf{A}$ and $\tilde{\mathbf{S}}$ by K-means algorithm or K-SVD algorithm. There are some collections between DL and ICA that the latter extracts a complete dictionary consisting of ICs. In terms of noisy measurement of data, DL is more similar to noisy ICA than classical ICA which does not consider the noise. The main difference between these two is that DL cannot produce independent atoms.

2.2 Negentropy-Based ICA

Unlike DL, the main aim of ICA is to produce the underlying mutually independent components by minimize the mutual information between the components

$$\min I(\mathbf{y}) = \sum_{i=1}^{n} H(y_i) - H(\mathbf{y}) \tag{8}$$

where $H(\cdot)$ represents entropy function. According to the central limit theorem, the above objective function of ICA is equal to maximize the non-Gaussianity of each component y_i which is measured by Negentropy. Since it is hard to directly calculate the Negentropy of a component, a simple approximation is given by

$$J(y_i) \propto [E\{H(y_i)\} - E\{H(v)\}]^2 \tag{9}$$

where v is a Gaussian variable having zero mean and with variance equal to the variance of y_i, and H is any non-quadratic function. Many ICA algorithms are Negentropy-based and the FastICA is the most famous one. A typical objective function of Negentropy-based ICA is

$$\arg\min_{w_i\,i=1,\cdots,n} \sum_{i=1}^{n} J(\mathbf{y}_i) = \sum_{i=1}^{n} [E\{H(\mathbf{y}_i)\} - E\{H(v)\}]^2 \text{ and } y_i = \mathbf{w}_i\mathbf{x}$$

$$\text{subject to } E(y_iy_j) = \begin{cases} 1 & i = j \\ 0 & i \neq j \end{cases} \tag{10}$$

where the constraint condition is to avoid identical solution and to restrict the output source to have standard variance. FastICA uses a fix-point optimization method to solve the objective function with very high efficiency but its convergence behavior is not easy to analyze [10, 11]. Equation (10) can also be solved under Constrained ICA (cICA) [12, 13] framework by a Newton-like optimization and cICA framework can be further extended to incorporate inequality constrains.

Main stream ICA studies do not consider noise measurement since ICA algorithms work well unless noise magnitude is too high. To deal with strong noise case, noisy ICA algorithm is considered as a variation of ICA which uses a different G, which is also a non-quadratic function, to estimate ICs despite the noises brought by measurement [7].

3   DL-like Model for Extracting ICs

For some applications, each observed data $\mathbf{x}_i$ is a linear combination of only a few ICs. For example, facial features extracted by ICA tend to be localized in space and facial image can be expressed as a sparse representation of these local features, which is more robust to noises than globe features such as eigenface. Early study on sparsity of mixing matrix by Hyvarinen and Karthikesh [8] handled the problem using a sparsity conjugate priors in ordinary ICA. However, the input data are assumed to be whitened and the sparsity is imposed on the unmixing matrix not the mixing matrix directly. Since a sparse matrix does not mean its inverse matrix is sparse, their method is not capable of directly producing a sparse mixing matrix. Besides the research has not considered the noisy measurement.

However DL discovers the underlying sources along with corresponding coding matrix which actually is the mixing matrix for ICA. In the objective function of DL, the solution of Eq. (7) directly outputs the mixing matrix $\mathbf{A}$ whose sparsity is constrained by L1-norm. However the atoms learned by DL are not mutually independent. So that we propose a new DL-like model to learn independent underlying sources which is given as follows

$$\underset{\mathbf{A}, \tilde{\mathbf{S}}}{\arg\min} \; \Psi(\mathbf{A}, \tilde{\mathbf{S}}) = \frac{1}{2} G_1(\mathbf{A}, \tilde{\mathbf{S}}) + \lambda_1 G_2(\mathbf{A}) + \lambda_2 G_3(\tilde{\mathbf{S}})$$

$$\text{subject to } G_4(\tilde{\mathbf{S}}) = 0 \tag{11}$$

where $G_1(\mathbf{A}, \tilde{\mathbf{S}}) = \left\| \tilde{\mathbf{X}} - \mathbf{A}\tilde{\mathbf{S}} \right\|_F^2$ is to ensure the fidelity; $G_2(\mathbf{A}) = \|\mathbf{A}\|_1$ is to enforce the sparseness of the mixing matrix; $G_3(\tilde{\mathbf{S}}) = \sum_{i=1}^{l} -\tilde{J}(\tilde{\mathbf{s}}_i)$ is used to make atoms mutually independent ($\tilde{J}(\cdot)$ estimates the "sample negentropy" of each atom) which is calculated by

$$G_3(\tilde{\mathbf{S}}) = \sum_{i=1}^{l} -\tilde{J}(\tilde{\mathbf{s}}_i) = \sum_{i=1}^{l} -\left(\frac{1}{p} \sum_{j=1}^{P} H(\tilde{s}_{i,j}) - c_i \right)^2 \tag{12}$$

where $c_i = E\{H(v)\}$ and v is a Gaussian variable having zero mean and with variance equal to the sample variance of $\tilde{\mathbf{s}}_i$, and we use function $H(t) = \log(\cosh(t))$ for the negentropy estimation; $G_4(\tilde{\mathbf{S}}) = \left\| \tilde{\mathbf{S}}\tilde{\mathbf{S}}^T / (p-1) - \mathbf{I} \right\|_F^2$ is the constraint term to ensure the learned atoms are not identical, where $\mathbf{I}$ is an unit matrix, λ_1 and λ_2 are parameters to balance the fidelity term, sparsity term, and negentropy term.

To solve Eq. (11), alternative iteration optimization is used to update $\mathbf{A}$ and $\tilde{\mathbf{S}}$ alternatively. First, we fix $\tilde{\mathbf{S}}$ to update the mixing matrix $\mathbf{A}$ by following objective function:

$$\underset{\mathbf{A}}{\arg\min} \; \Psi(\mathbf{A}) = \frac{1}{2} G_1(\mathbf{A}) + \lambda_1 G_2(\mathbf{A}) \tag{13}$$

Optimization of Eq. (13) can be decomposed into l independent sparse representation problems as

$$\Psi(\mathbf{A}) = \frac{1}{2}\left\|\tilde{\mathbf{X}} - \mathbf{A}\tilde{\mathbf{S}}\right\|_F^2 + \lambda_1 \|\mathbf{A}\|_1$$

$$= \sum_{i=1}^{l} \frac{1}{2}\left\|\tilde{\mathbf{x}}_i - \mathbf{a}_i\tilde{\mathbf{S}}\right\|_2^2 + \lambda_1 \sum_{i=1}^{l} \|\mathbf{a}_i\|_1 . \qquad (14)$$

$$= \sum_{i=1}^{l} \left(\frac{1}{2}\left\|\tilde{\mathbf{x}}_i - \mathbf{a}_i\tilde{\mathbf{S}}\right\|_2^2 + \lambda_1 \|\mathbf{a}_i\|_1 \right)$$

In this phase, the sparseness of $\mathbf{A}$ is increased and the sparse representation optimization issues can be solve by many public available codes such as *l1_ls* toolbox. Next we fix $\mathbf{A}$ to update $\tilde{\mathbf{S}}$ by solving:

$$\arg\min_{\tilde{\mathbf{s}}} \ \Psi(\tilde{\mathbf{S}}) = \frac{1}{2}G_1(\tilde{\mathbf{S}}) + \lambda_2 G_3(\tilde{\mathbf{S}})$$

$$\text{subject to } G_4(\tilde{\mathbf{S}}) = 0. \qquad (15)$$

The gradient descent algorithm is employed and we calculate the gradients by

$$\frac{1}{2}\frac{\partial}{\partial\tilde{\mathbf{S}}}G_1(\tilde{\mathbf{S}}) = \frac{1}{2}\frac{\partial}{\partial\tilde{\mathbf{S}}}\left\|\tilde{\mathbf{X}} - \mathbf{A}\tilde{\mathbf{S}}\right\|_F^2$$

$$= \frac{1}{2}\frac{\partial}{\partial\tilde{\mathbf{S}}}tr\left[\left(\tilde{\mathbf{X}} - \mathbf{A}\tilde{\mathbf{S}}\right)^T\left(\tilde{\mathbf{X}} - \mathbf{A}\tilde{\mathbf{S}}\right)\right] \qquad (16)$$

$$= -\mathbf{A}^T\tilde{\mathbf{X}} + \mathbf{A}^T\mathbf{A}\tilde{\mathbf{S}}$$

and

$$\frac{\partial}{\partial\tilde{\mathbf{S}}}G_3(\tilde{\mathbf{S}}) = -\begin{pmatrix} \frac{\partial\tilde{J}(\tilde{s}_{1,1})}{\partial\tilde{s}_{1,1}} & \cdots & \frac{\partial\tilde{J}(\tilde{s}_{1,p})}{\partial\tilde{s}_{1,p}} \\ \vdots & \ddots & \vdots \\ \frac{\partial\tilde{J}(\tilde{s}_{l,1})}{\partial\tilde{s}_{l,1}} & \cdots & \frac{\partial\tilde{J}(\tilde{s}_{l,p})}{\partial\tilde{s}_{l,p}} \end{pmatrix} \qquad (17)$$

where $\frac{\partial\tilde{J}(\tilde{s}_{i,j})}{\partial\tilde{s}_{i,j}} = \left(\frac{1}{p}\sum_{j=1}^{P} H(\tilde{s}_{i,j}) - c_i\right)\left(\frac{1}{p}H'(\tilde{s}_{i,j})\right)$ and $H'(\tilde{s}_{i,j}) = \tanh(\tilde{s}_{i,j})$. $\tilde{\mathbf{S}}$ is updated by

$$\tilde{\mathbf{S}} \leftarrow \tilde{\mathbf{S}} - \gamma\frac{\partial\frac{1}{2}G_1(\tilde{\mathbf{S}}) + \lambda_2 G_3(\tilde{\mathbf{S}})}{\partial\tilde{\mathbf{S}}} \qquad (18)$$

The equality constraint is to prevent producing identical atoms and to restrict the variance, which is fulfilled by symmetric orthogonalization. That is, we first do iterative step of atoms by Eq. (18), and afterwards orthogonalize all atoms by

$$\tilde{\mathbf{S}} = \frac{(\sqrt{p}-1)\tilde{\mathbf{S}}}{(\tilde{\mathbf{S}}\tilde{\mathbf{S}}^T)^{\frac{1}{2}}}. \tag{19}$$

The orthogonalization needs to calculate an inverse matrix which is not time-consuming if only a small number of independent atoms are extracted. We summarize above alternative iteration optimization in Fig. 1.

Task: Find a dictionary $\tilde{\mathbf{S}}$ consisting of independent atoms to represent observed data $\tilde{\mathbf{X}}$ by a sparse mixing matrix $\mathbf{A}$ as follows:

$$\min \left\| \tilde{\mathbf{X}} - \mathbf{A}\tilde{\mathbf{S}} \right\|_F^2 + \lambda \|\mathbf{A}\|_1$$

Initialization: Set $\tilde{\mathbf{S}}^{(0)}$ with the output of FastICA conducting on $\tilde{\mathbf{X}}$.

Repeat until convergence:

1. Fix $\tilde{\mathbf{S}}$ to update $\mathbf{A}$. Use any L1-norm regulation based linear regression solver to calculate the following SR problem one by one,

$$\arg\min_{\mathbf{a}_i} \frac{1}{2} \left\| \tilde{\mathbf{x}}_i - \mathbf{a}_i \tilde{\mathbf{S}} \right\|_2^2 + \lambda_1 \sum_{i=1}^l \|\mathbf{a}_i\|_1, \ i = 1, \cdots, l.$$

2. Fix $\mathbf{A}$ to update $\tilde{\mathbf{S}}$.

   Repeat until convergence

   (1) Conduct one-step iteration of following gradient descent

   $$\tilde{\mathbf{S}} \leftarrow \tilde{\mathbf{S}} - \gamma \frac{\partial \frac{1}{2} G_1(\tilde{\mathbf{S}}) + \lambda_2 G_3(\tilde{\mathbf{S}})}{\partial \tilde{\mathbf{S}}};$$

   (2) Orthogonalize all atoms by

   $$\tilde{\mathbf{S}} = \frac{(\sqrt{p}-1)\tilde{\mathbf{S}}}{(\tilde{\mathbf{S}}\tilde{\mathbf{S}}^T)^{\frac{1}{2}}}.$$

Fig. 1. The DL-like IC extraction algorithm

Unlike dictionary learning by K-SVD algorithm and K-means algorithm which attempt to ensure the sparseness of mixing matrix, we simultaneously change all atoms in dictionary update stage to primarily guarantee the maximization of total negentropy. Also in sparse representation stage, our algorithm differs from the previous DL algorithms in terms of aims. Our algorithm imposes sparsity constraint on mixing matrix to obtain the original genuine sparse mixture whose sparseness is reduced due to noisy measurement. For DL algorithm, the aim of sparse representation is that a data sample is sparsely represented by the learned dictionary leads to the fact that the dictionary should consist of very rich atoms.

Comparing to ICA algorithms, our method is capable of incorporating sparse prior on mixing matrix into discovering of ICs. The mixing matrix learned by ICA is no longer sparse if the observation data is affected by noise.

A full description of our DL-like IC extraction algorithm is summarize in Fig. 1.

4   Simulations

The proposed algorithm is tested on a synthetic dataset and compared with FastICA algorithm and K-SVD dictionary learning algorithm. Six synthetic independent sources are used as shown in Fig. 2. The observed data is a sparse mixture of these independent sources, which is done by using a 6×6 mixing matrix $\mathbf{A}^*$ which has no more than three nonzero entries in each row. We add Gaussian noise with variance equal to 0.4 to each observation channel. An example of observation data is shown in Fig. 3. The simulation is implemented on a PC with Matlab 2014b version.

To evaluate the results, a performance index is used, defined as

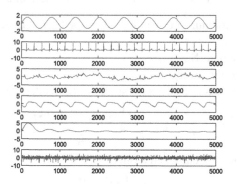

Fig. 2.   Six used independent sources.

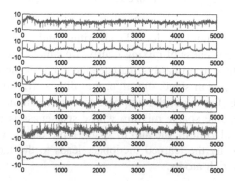

Fig. 3.   Observed mixture data.

$$E_p = \sum_{i=1}^{m} \left(\sum_{j=1}^{m} \frac{|p_{ij}|}{\max_k |p_{ik}|} - 1 \right) + \sum_{j=1}^{m} \left(\sum_{i=1}^{m} \frac{|p_{ij}|}{\max_k |p_{kj}|} - 1 \right). \tag{20}$$

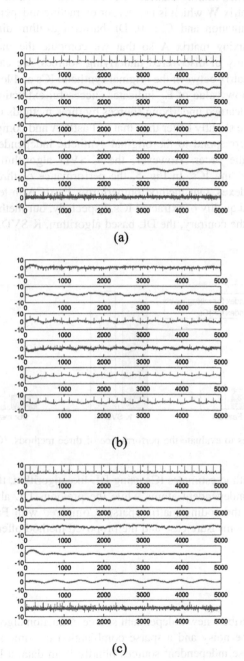

Fig. 4. Extracted underlying sources by three methods. (a) ICs estimated by FastICA; (b) K-SVD learned atoms; (c) Independent atoms learned by our method.

The index is to measure whether a matrix $\mathbf{P}$ is a permutation matrix. If $\mathbf{P}$ is an ideal permutation matrix, this index is equal to zero. We use this index to measure the accuracy of both learned mixing matrices and extracted independent sources. FastICA outputs unmixing matrix $\mathbf{W}$ which is perfect for extracting independent sources only if $\mathbf{P} = \mathbf{WA}^*$ is a permutation and $E_p = 0$. DL-based algorithm directly calculates the approximation of mixing matrix $\mathbf{A}$ so that we compute the index as $\mathbf{P} = \mathbf{A}^{-1}\mathbf{A}^*$. To evaluate the quality of discovered underlying sources, we calculate index $\mathbf{E}_{\tilde{s}}$ of sample covariance matrix between the original synthetic ICs and learned sources by the three algorithms. A low value of $\mathbf{E}_{\tilde{s}}$ indicates successful extraction of the real ICs.

We illustrate the learned underlying sources by three methods in Fig. 4. The results output our method are slightly better than that of FastICA and components extracted by both methods can correspond to the original synthetic independent sources approximately except for the order. However, the K-SVD algorithm is not capable of extracting real underlying ICs. In Fig. 5, the performance of three methods are also evaluated by two indexes. Our method outperforms FastICA in terms of both mixing matrix estimation and quality of extracted ICs. Especially, our method extracts ICs with better accuracy. On the contrary, the DL based algorithm, K-SVD, works much worse than other two.

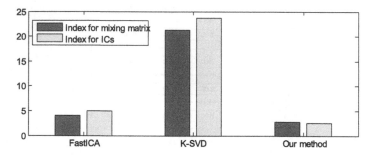

Fig. 5. Two indexes to evaluate the performance of three methods. (Color figure online)

Although our method estimates ICs using DL-like algorithm, the method is proved to be able to learn independent atoms while the previous DL algorithm only learns atoms which makes the coding matrix sparse. Compared with FastICA, our method works better for noisy measure case and the proposed method offers a new perspective to conduct ICA.

5 Conclusion

In this paper we describe a new independent source extraction algorithm. The observed data is assumed to be noisy and a sparse combination of some independent components. To extract these independent sources blindly from data, a DL-like algorithm is proposed to estimate the independent atoms. We employ the alternative iteration optimization scheme to alternatively update mixing matrix and dictionary. The update of mixing matrix is done by solve a number of sparse representation problems. And dictionary update is done by gradient descent approach. Our method differs from common ICA algorithms which extract ICs by calculating projection matrix while we directly maximize the negentropy of all atoms. The simulation results show that our algorithm outperforms FastICA and K-SVD based DL algorithm in terms of both the quality of extracted ICs and the accuracy of estimated mixing matrix. Our future work is to extend our model to extract rich independent atoms, and to this end we need to develop new algorithms for large scale data.

Acknowledgements. This work was supported partially by the National Nature Science Foundation of China under Grant Nos. 61403053. And partially this research is funded by Chongqing Natural Science Foundation (the project No. is No. cstc2014jcyjA40018,cstc2014j-cyjA40022, and cstc2014kjrcqnrc40002) and by Chongqing education committee under Grant No. KJ1500402 and KJ1500417.

References

1. Besic, N., Vasile, G., Chanussot, J., Stankovic, S.: Polarimetric incoherent target decomposition by means of independent component analysis. IEEE Trans. Geosci. Remote Sens. **53**, 1236–1247 (2015)
2. Mi, J.-X.: A novel algorithm for independent component analysis with reference and methods for its applications. PLoS ONE **9**, e93984 (2014)
3. Hämäläinen, M., Hari, R., Ilmoniemi, R.J., Knuutila, J., Lounasmaa, O.V.: Magnetoencephalography—theory, instrumentation, and applications to noninvasive studies of the working human brain. Rev. Mod. Phys. **65**, 413 (1993)
4. Huettel, S.A., Song, A.W., McCarthy, G.: Functional Magnetic Resonance Imaging. Sinauer Associates, Sunderland (2004)
5. Hyvärinen, A.: Independent component analysis: recent advances. Philos. Trans. Math. Phys. Eng. Sci. **371**, 20110534 (2013)
6. Tian, X., Cai, L., Chen, S.: Noise-resistant joint diagonalization independent component analysis based process fault detection. Neurocomput. B **149**, 652–666 (2015)
7. Zhao, Y., He, H., Mi, J.: Noisy component extraction with reference. Front. Comput. Sci. **7**, 135–144 (2013)
8. Hyvärinen, A., Raju, K.: Imposing sparsity on the mixing matrix in independent component analysis. Neurocomputing **49**, 151–162 (2002)
9. Aharon, M., Elad, M., Bruckstein, A.: k-SVD: an algorithm for designing overcomplete dictionaries for sparse representation. IEEE Trans. Sign. Process. **54**, 4311–4322 (2006)
10. Oja, E., Zhijian, Y.: The FastICA algorithm revisited: convergence analysis. Neural Netw. IEEE Trans. **17**, 1370–1381 (2006)
11. Wei, T.: A convergence and asymptotic analysis of the generalized symmetric FastICA algorithm. IEEE Trans. Sign. Process. **63**, 6445–6458 (2015)
12. De-Shuang, H., Jian-Xun, M.: A new constrained independent component analysis method. Neural Netw. IEEE Trans. **18**, 1532–1535 (2007)
13. Lu, W., Rajapakse, J.C.: Approach and applications of constrained ICA. IEEE Trans. Neural Netw. **16**, 203–212 (2005)

Compressed Sensing, Sparse Coding

Leaf Clustering Based on Sparse Subspace Clustering

Yun Ding, Qing Yan, Jing-Jing Zhang, Li-Na Xun,
and Chun-Hou Zheng[✉]

School of Electrical Engineering and Automation,
Anhui University, Hefei, China
zhengch99@126.com

Abstract. Leaf recognition is one of the important techniques for species automatic identification. Because of not needing prior knowledge, clustering based method is a good choice to accomplish this task. Moreover, the high dimensions of leaf feature are always a challenge for traditional clustering algorithm. While the Sparse Subspace Clustering (SSC) can overcome the defect of traditional method in dealing with the high dimensional data. In this paper we propose to use SSC for leaf clustering. The experiments are performed on the database of leaves with noise and no noise respectively, and compared with some conventional algorithm such as k-means, k-medoids, etc. The results show that the clustering effect of SSC is more accurate and robust than others.

Keywords: Plant leaves · Sparse subspace clustering · Sparse representation · High dimension feature

1 Introduction

Nowadays, with the development of the precision agriculture, the species automatic identification becomes more and more important. Leaf recognition is one of the effective ways to recognize plant species. And the pattern recognition method can improve the efficiency of leaf recognition to a large extent. Until now, scholars have obtained some achievements in this field [1, 2]. These methods generally extract biological characteristics of leaves, such as color, shape, or texture, and then use a classifier to recognize them [3, 4].

Classification algorithm can be divided into supervised learning and unsupervised learning [5, 6]. Representative supervised learning methods, such as Bayesian classifier [7], Support Vector Machine (SVM) [8], Logistic Regression [9] and K-NN [10], etc., are commonly used for classification. But these methods rely on the supervised information mostly. However, in some cases, the acquisition of enough prior information is difficult, so the unsupervised clustering algorithm becomes a kind of alternative way to realize pattern recognition. Commonly used unsupervised learning methods are Gaussian Mixture Model [11], K-means [12], Spectral Clustering (SC) [13], etc. However, these clustering methods need to know the number of subspace in advance or are sensitive to the initialization of data, noises, etc. [14]. So the clustering effects of these algorithms are not so satisfied. And when it encounters the high dimensional data, the

© Springer International Publishing Switzerland 2016
D.-S. Huang and K.-H. Jo (Eds.): ICIC 2016, Part II, LNCS 9772, pp. 55–66, 2016.
DOI: 10.1007/978-3-319-42294-7_5

effect will be worse. Recently, Elhamifar et al. proposed the Sparse Subspace Clustering (SSC) algorithm on the basis of the sparse recovery theory and spectral clustering algorithm [15, 16]. Which constructs the data similarity matrix by sparse representation coefficient, then gets the final clustering results using spectral clustering algorithm. SSC is one of the most effective method in subspace clustering at present [17], and becomes the hot issue in the fields of machine learning, computer vision and image processing in recent years. Many encouraging performances in face recognition, motion segmentation and some other areas have been achieved [18, 19]. To our knowledge, its application for plant leaves identification is not reported at present.

In this paper, we propose to use sparse subspace clustering algorithm for plant identification. Firstly, the basic theory of sparse subspace clustering algorithm is introduced. Then two categories experiments are designed to verify the superiority of SSC. One is performed on the leaf images with no noise pollution, and the other is aimed at the leaf images with Gaussian noise.

2   The Basic Theory of Sparse Subspace Clustering Algorithm

In this section, we introduce the classical SSC method. SSC proposed by Elhamifar [13, 14] regards that each data point lies in a union of subspaces corresponding to several classes or clusters to which the dataset belongs. Therefore, each point can be sparsely represented by other points of a union of subspaces. The SSC then uses sparse representations of data points to cluster the points into separate subspaces. It typically consists of three stages: (1) finding sparse representations of each data point through convex optimization; (2) learning a similarity matrix (i.e., a weight matrix); and (3) clustering the similarity matrix using spectral clustering [20, 21].

Let $\{S_l\}_{l=1}^{n}$ be an arrangement of n subspace of IR^D of dimensions $\{d_l\}_{l=1}^{n}$. Consider a given collection of N data points $\{y_i\}_{i=1}^{N}$ that lie in the union of the n subspaces. Denoting the matrix containing all the data points as

$$Y = [y_1, \ldots, y_N] = [Y_1, \ldots, Y_n]\Gamma \tag{1}$$

Where $Y_l \in IR^{D \times N_l}$ is a rank-d_l matrix of the $N_l > d_l$ points that lie in S_l, and Γ is an unknown permutation matrix. For each data point, it can be represented by some of the other data points. $y_i = Yc_i, c_{ii} = 0$, where $C_i = [c_1 c_2 \cdots c_N] \in R^{N \times N}$, the constraint $c_{ii} = 0$ eliminates the trivial solution of writing a point as a linear combination of itself. In order to obtain the most sparse representation of each data point, one can choose to minimize the $l_0 - norm$ on the convex relaxation process. However, it is a NP-hard problem to solve the (1). Researchers have discovered that if the solution is sparse enough, then the solution of l_0 minimization problem is equal to solve the l_1 minimization problem as [14]:

$$\min \| C \|_1 \ s.t. Y = YC, diag(C) = 0 \tag{2}$$

In practice, however, we consider clustering of data points that are contaminated with noise. In addition, the data is often within the affine subspace rather than a linear subspace. In order to address these problems, the sparse optimization model can be written as

$$\min \| C \|_1 + \frac{\lambda_z}{2} \| Z \|_F^2$$
$$s.t. Y = YC + Z, diag(C) = 0 \tag{3}$$

Where $C = [c_1 c_2 \cdots c_N] \in R^{N \times N}$ is a matrix and $diag(C) \in R^N$ is the vector of the diagonal elements of the matrix C, each column corresponding to the sparse representation of each data point. The sparse coefficient matrix C is applied to spectral clustering algorithm, and realize the data clustering. In the spectral clustering algorithm, the adjacency matrix W of the graph G is built according to the sparse coefficient matrix C. In order to pursue the symmetry of adjacency matrix based on undirected graph G, the adjacency matrix is defined as:

$$W = |C| + |C|^T \tag{4}$$

where $|C|$ represent the absolute value of each element of C. We can get Laplacian matrix W according to the adjacency matrix L

$$L = I - D^{-1/2} W D^{-1/2} \tag{5}$$

where $D \in R^{N \times N}$ is a diagonal matrix with $D_{ii} = \sum_j W_{ij}$. Finally we use the K - means algorithm to get the final clustering results.

To sum up, sparse subspace clustering model can be summarized as Table 1.

3 Design of the Experiments and Results

The leaf images used in experiments are downloaded from the Chinese academy of sciences HeFei intelligence of plant leaf image database (http://www.intelengine.cn/data). The database contains 220 kinds of plants, a total of 16846 leaf images. In this paper, we select 13 different kinds of plant leaves to do the experiment. Each kind of plant contains 50 leaf images. In order to verify the effectiveness of the sparse subspace clustering algorithm, we compare it with other clustering algorithms including spectral clustering, K-Means and K-Medoids etc.

3.1 The Evaluation Standard of Clustering Experiments

In order to objectively evaluate the results, we adopted the accuracy (AC) and the normalized mutual information (NMI) [22, 23], to measure the quality of clustering results. The AC is defined as follows:

$$AC = \frac{\sum_{i=1}^{n} \delta(s_i, map(r_i))}{n} \tag{6}$$

Table 1. Sparse subspace clustering algorithm flow chart

Algorithm: sparse subspace clustering

Input : N noise-free datapoints $\{y_i\}_{i=1}^{N}$

Step 1. Datapoints $\{y_i\}_{i=1}^{N}$ use the sparse subspace clustering model (2), and get sparse coefficient matrix.

Step 2. The each column of coefficient matrix normalize the columns of C as

$$c_i \leftarrow \frac{c_i}{\hbar c_i \, \hbar_{\infty}}$$

Step 3. According to the sparse coefficient matrix to establish similar weighted graph, each data point connect with the data point which is sparse representation. The data points coming from different subspace do not have no connection , weighting matrix $W = |C| + |C|^{\mathrm{T}}$.

Step 4. The Laplacian matrix $L = I - D^{-1/2}WD^{-1/2}$ and get the first n small eigenvalue, then get $N \times n$ matrix U formed by the first n corresponding eigenvectors.

Step 5. Put each row of the U as a vector of n dimension space, and perform the K - means clustering on it.

Output: clusters $Y_1, Y_2, \cdots, Y_n$.

which is used to indicate the labels of clustering results compared with known labels. r_i represents the labels of clustering results, s_i represent the known labels offered. n is the total number of samples, $\delta(s_i, map(r_i))$ is a function. If $s_i = map(r_i)$, then $\delta(s_i, map(r_i)) = 1$, otherwise $\delta(s_i, map(r_i)) = 0$. $map(r_i)$ is a matching function, which well match the labels r_i of the clustering results with the known labels s_i, then concluding AC.

NMI is defined as follows:

$$NMI(C, C') = \frac{MI(C, C')}{\max(H(C), H(C'))} \qquad (7)$$

Which represents normalized mutual information, and measures the clustering performance. Here C represents a known set of real data clustering. C' is a collection of clustering from our algorithm rule, $H(C)$ and $H(C')$ represent the entropy of C and C'. Function MI is defined as:

$$MI(C, C') = \sum_{c_i \in C, c_j' \in C'} p(c_i, c_j') \cdot \log \frac{p(c_i, c_j')}{p(c_i)p(c_j')} \qquad (8)$$

Where $p(c_i)$ is the probability of a random sample of the data sets belonging to clustering c_i, $p(c_j')$ represents the probability of a random sample of the data sets belonging to clustering c_j'. $p(c_i)p(c_j')$ represents the joint probability of the random sample belonging to the clustering c_i and the clustering c_j' in a data set.

The two Criterions, i.e., AC and NMI, range from 0 to 1, and a bigger value indicates a better clustering result. In this paper, during the experiment, we calculate the mean value of AC and NMI and their variance. The smaller the variance, the robust the algorithm is.

3.2   Non-noise Leaf Clustering Result

We select four groups of data to test the effective of the SSC. The experiment design method are listed as follows: the two kinds of similar leaves clustering (data group 1 to 3), two kinds of apparently different leaves clustering (data group 4 to 6), three kinds of similar leaves clustering (data group 7 to 9), and three kinds of apparently different

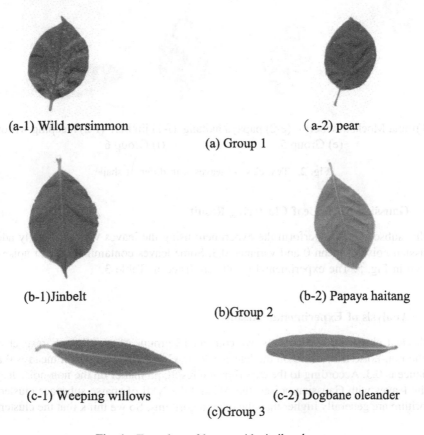

(a-1) Wild persimmon   (a-2) pear

(a) Group 1

(b-1)Jinbelt   (b-2) Papaya haitang

(b)Group 2

(c-1) Weeping willows   (c-2) Dogbane oleander

(c)Group 3

Fig. 1. Two class of leaves with similar shape

leaves clustering (data group 10 to 12). From these four data subsets, we choose 12 group leaves totally to fulfill the experiment (3 groups each subset). During the experiment, we perform the experiment 20 times for each group of data, the average result are listed in Table 2. Some leaves use in the experiment are shown in Figs. 1, 2, 3 and 4.

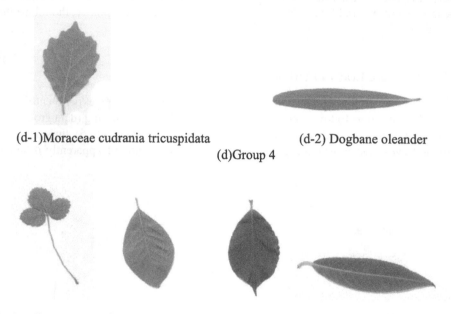

(d-1)Moraceae cudrania tricuspidata   (d-2) Dogbane oleander

(d)Group 4

(e-1)India Mockstrawberry (e-2) papaya haitang (f-1) jin belt (f-2)Weeping willow

(e) Group 5   (f) Group 6

Fig. 2. Two class of leaves with different shape

3.3   Gaussian Noise Leaf Clustering Result

In this subsection, we perform the experiment using the leaves with artificially added Gaussian noise of mean 0 and variance 0.3. Some leaves contaminated with noise are shown in Fig. 5. The experimental results are listed in Table 3.

3.4   Analysis of Experimental Results

In the leaf clustering experiments, we conduct 12 groups of experiments respectively on the non-noise leaves and on the leaves adding Gaussian noise which mean is 0 and variance is 0.3. According to the experimental result, no matter on the non-noise leaves or the leaves with Gaussian noises, the AC and the NMI of sparse subspace clustering algorithm are generally higher than the other algorithms. So we think that the clustering

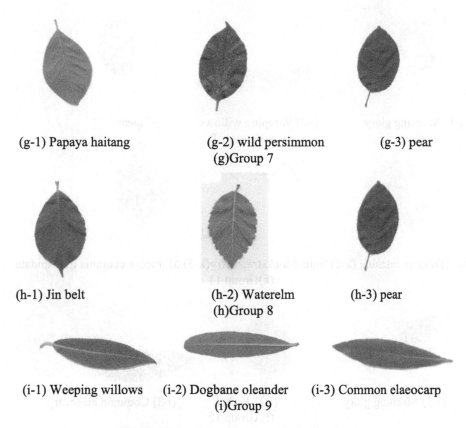

(g-1) Papaya haitang	(g-2) wild persimmon	(g-3) pear
	(g)Group 7	
(h-1) Jin belt	(h-2) Waterelm	(h-3) pear
	(h)Group 8	
(i-1) Weeping willows	(i-2) Dogbane oleander	(i-3) Common elaeocarp
	(i)Group 9	

Fig. 3. Three class of leaves with similar shape

effect of the sparse subspace clustering algorithm is better than the other three algorithms on subject of the leaves clustering.

Although there also exists the situation that AC and NMI of SSC are less than the other algorithms in part of experiments, but the variances are generally little. So we think that sparse subspace clustering algorithm is more stable than the other method for data clustering.

4 Conclusion

In this paper, we propose to apply the classic SSC for leaf recognition. For a variety of different types of leaf, we designed two kinds of experiment, in the first experiment, the leaf images are pure. In the second one, the images are damaged by noise pollution. In order to quantitatively evaluate the clustering effect, the two criterions, AC and NMI, were adopted. The results show that the SSC is more suitable for leaf clustering than other conventional method no matter from precision or robust.

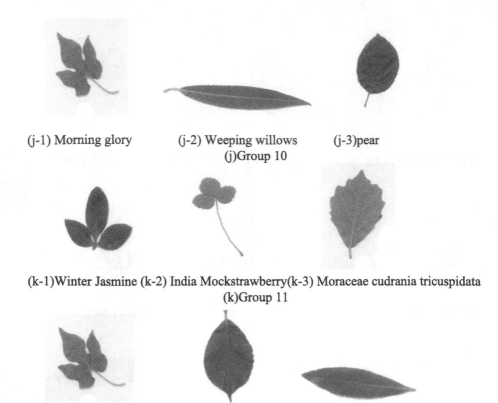

(j-1) Morning glory   (j-2) Weeping willows   (j-3)pear
                                (j)Group 10

(k-1)Winter Jasmine (k-2) India Mockstrawberry(k-3) Moraceae cudrania tricuspidata
                                (k)Group 11

(l-1) Morning glory   (l-2) Jin belt   (l-3) Common elaeocarp
                                 (l) Group 12

Fig. 4. Three class of leaves with different shape

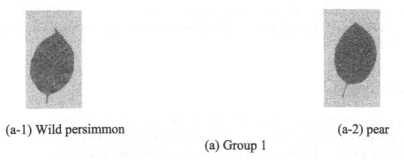

(a-1) Wild persimmon   (a-2) pear
                                (a) Group 1

Fig. 5. Two class of leaves with similar shape

Table 2. The AC(variance) and NMI(variance) values of clustering results

Group number		AC(%)(variance)				NMI(%)(variance)			
		SSC	SC	k-means	k-medoids	SSC	SC	k-means	k-medoids
Two class of leaves with similar shape	1	**71** (.00000)	69.3 (.00011)	67.75 (.00041)	67.75 (.00102)	**13.28** (.00000)	12.32 (.00015)	11.25 (.00020)	11.03 (.00063)
	2	**77** (.00000)	83.9 (.00141)	85.55 (.00045)	82.95 (.00358)	34.38 (.00000)	40.20 (.00536)	45.89 (.00026)	38.79 (.00523)
	3	75 (0)	83.55 (.00011)	78.3 (.00010)	79.85 (.00105)	24.99 (0)	36.07 (.00055)	26.97 (.00014)	30.44 (.00374)
Two class of leaves with different shape	4	97 (0)	96.65 (.00017)	97 (0)	94.46 (.00110)	**83.37** (.00000)	82.12 (.00213)	83.37 (.00000)	75.54 (.01140)
	5	**61** (.00000)	59.6 (.00018)	60.1 (.00025)	60.79 (.00004)	**3.71** (.00000)	2.83 (.00005)	3.27 (.00005)	3.53 (.00002)
	6	83 (.00000)	82.85 (.00031)	86 (.00150)	86 (.00120)	34.80 (.00000)	34.42 (.00148)	51.50 (.00508)	51.50 (.00629)
Three class of leaves with similar shape	7	**50.67** (.00000)	49.57 (.00001)	49.1 (.00002)	48.5 (.00054)	9.56 (.00000)	10.02 (.00002)	9.11 (.00000)	9.52 (.00013)
	8	**61.33** (0)	61.17 (.00359)	60.55 (.00456)	59.93 (.00526)	**30.94** (.00000)	30.48 (.00717)	29.67 (.00412)	28.13 (.00628)
	9	**63.33** (.00000)	55.67 (.00479)	56.3 (.00036)	59.58 (.00167)	**27.63** (0)	20.10 (.00113)	26.74 (.00002)	27.43 (.00064)
Three class of leaves with different shape	10	**52.67** (0)	47.93 (.00008)	51.43 (.00073)	51.9 (.00127)	**20.80** (.00000)	16.71 (.00060)	18.01 (.00016)	15.70 (.00065)
	11	**60.67** (.00000)	54.07 (.00086)	55.74 (.00214)	54.56 (.00138)	**23.33** (.00000)	18.45 (.00011)	22.96 (.00150)	23.30 (.00014)
	12	**72** (.00000)	69.33 (.00880)	61.43 (.00492)	55.24 (.00615)	**39.94** (.00000)	37.64 (.01350)	35.50 (.00273)	37.23 (.00374)

Table 3. The AC(variance) and NMI(variance) values of clustering results

Group			AC(%)(variance)				NMI(%)(variance)			
			SSC	SC	k-means	k-medoids	SSC	SC	k-means	k-medoids
Two class of leaves with similar shape	1		**70.45** (.00021)	69.2 (.00005)	68.7 (.00071)	68.85 (.00132)	**12.76** (.00027)	12.20 (.00011)	12.08 (.00026)	12.75 (.00045)
	2		80.3 (.00043)	85.5 (.00097)	**86.55** (.00031)	81.8 (.00725)	35.90 (.00041)	43.25 (.00363)	43.86 (.00126)	37.45 (.01460)
	3		76.6 (**.00005**)	82.85 (.00018)	77.95 (.00019)	81 (.00043)	25.95 (.00013)	34.27 (.00083)	26.14 (.00041)	31.54 (.00154)
Two class of leaves with different shape	4		97 (0)	96.5 (.00016)	96.75 (.00006)	91.5 (.00075)	**83.37** (.00000)	81.59 (.00193)	82.45 (.00079)	65.51 (.00515)
	5		**61.1** (.00004)	60.6 (.00004)	60.35 (.00030)	58.9 (.00062)	**4.07** (.00002)	3.41 (.00002)	3.41 (.00007)	2.75 (.00015)
	6		74.95 (.00001)	82.4 (.00044)	82.1 (.00000)	83.25 (.00013)	24.48 (.00015)	33.65 (.00190)	34.53 (.00009)	37.63 (.00152)
Three class of leaves with similar shape	7		**49.43** (.00005)	47.87 (.00027)	49.41 (.00003)	48.97 (.00061)	**9.87** (.00000)	8.95 (.00045)	9.22 (.00001)	9.87 (.00009)
	8		59.33 (**.00000**)	59.83 (.00101)	56.86 (.00215)	59.93 (.00624)	**29.37** (.00000)	28.71 (.00141)	29.00 (.00147)	28.87 (.00876)
	9		**64.7** (.00010)	57.67 (.00011)	55.52 (.00141)	58.5 (.00279)	**29.19** (.00018)	27.69 (.00024)	28.98 (.00068)	29.15 (.00027)
Three class of leaves with different shape	10		**51.43** (.00005)	48.33 (.00006)	51.05 (.00056)	50.97 (.00082)	**19.20** (.00031)	15.00 (.00088)	19.18 (.00019)	16.70 (.00038)
	11		**60.17** (.00005)	53.03 (.00147)	56.25 (.00114)	56.87 (.00061)	**23.31** (.00006)	19.10 (.00174)	23.13 (.00128)	22.76 (.00029)
	12		**72** (.00000)	72 (.00000)	68.73 (.00505)	71.63 (.00390)	**39.94** (.00000)	39.94 (.00000)	43.75 (.00011)	44.62 (.00133)

On the other hand, during the experiment, we also find that the clustering results of leaves with multiple branches and irregular shape are not ideal. We think this problem can be solved by extracting multi-features. So future work should be addressed to improve the algorithm from the point of view of the multi-feature, such as shape and texture features.

Acknowledgments. This work was supported by the AnHui University Youth Skeleton Teacher Project(E12333010289), Anhui University Doctoral Scientific Research Start-up Funding (J10113190084), China Postdoctoral Science Foundation (2015M582826), Key Laboratory of Optical Calibration and Characterization/Chinese Academy of Sciences Project and Center of Information Support & Assurance Technology. In addition, this paper is partially supported by Science and Technology Project of Anhui Province (No. 1501b042207).

References

1. Kumar, N., Belhumeur, P.N., Biswas, A., Jacobs, D.W., Kress, W., Lopez, I.C., Soares, J.V.: Leafsnap: a computer vision system for automatic plant species identification. In: Fitzgibbon, A., Lazebnik, S., Perona, P., Sato, Y., Schmid, C. (eds.) ECCV 2012, Part II. LNCS, vol. 7573, pp. 502–516. Springer, Heidelberg (2012)
2. Htike, K.K., Khalifa, O.O.: Comparison of supervised and unsupervised learning classifiers for human posture recognition. In: International Conference on Computer and Communication Engineering, pp. 1–6. IEEE (2010)
3. Isfahani, Z.B.: Comparison of supervised and unsupervised learning classifiers for travel recommendations. J. Global Res. Comput. Sci. **3**(8), 51–55 (2012)
4. Deng, Z.: Supervised learning evidence theory classifier. In: Computer Engineering & Application (2005)
5. Liu, X., Tang, J.: mass classification in mammograms using selected geometry and texture features, and a new SVM-based feature selection method. IEEE Syst. J. **8**(3), 910–920 (2014)
6. Dong, Y., Guo, H., Zhi, W. et al.: Class imbalance oriented logistic regression. In: International Conference on Cyber-Enabled Distributed Computing and Knowledge Discovery, pp. 187–192. IEEE (2014)
7. Kir, B., Oz, C., Gulbag, A.: Leaf recognition using K-NN classification algorithm. In: Signal Processing and Communications Applications Conference (SIU), pp. 1–4. IEEE (2012)
8. Zhong, J., Lichuan, G.U., Tan, J.: Estimating parameters of GMM based on split EM. Comput. Eng. Appl. **48**(34), 28–32 (2012)
9. Wu, W.L.: Study of K-means and K-medoids Algorithms in Clustering Analysis. South China University of Technology, Guangzhou (2011)
10. Wang, S., Chen, F., Fang, J.: Spectral clustering of high-dimensional data via non negative matrix factorization. In: IEEE Conference Publications, pp. 1–8 (2015)
11. Donoho, D.L.: High-dimensional data analysis: the curses and blessings of dimensionality. In: Lecture — Math Challenges of the 21st Century, pp. 1–32 (2000)
12. Na, S., Liu, X., Yong, G.: Research on k-means clustering algorithm: an improved k-means clustering algorithm. In: Third International Symposium on Intelligent Information Technology and Security Informatics, pp. 63-67. IEEE (2010)
13. Elhamifar, E., Vidal, R.: Sparse subspace clustering. CVPR **35**(11), 2790–2797 (2009)

14. Elhamifar, E., Vidal, R.: Sparse subspace clustering: algorithm, theory, and applications. IEEE Trans. Pattern Anal. Mach. Intell. **35**(11), 2765–2781 (2013)
15. Wang, Y., Tang, Y.Y., Li, L.: Minimum error entropy based sparse representation for robust subspace clustering. IEEE Trans. Signal Process. **63**(15), 1 (2015)
16. Gao, M.-M., Chang, T.H., Gao, X.-X.: Research in data stream clustering based on gaussian mixture model genetic algorithm. In: 2nd International Conference on Information Science and Engineering (ICISE), 2010 pp. 3904–3907. IEEE (2010)
17. O'Sullivan, J.A.: Message passing expectation-maximization algorithms. In: IEEE/SP 13th Workshop on Statistical Signal Processing, pp. 841–846. IEEE (2005)
18. Yin, J., Zhang, Y., Gao, L.: Accelerating expectation-maximization algorithms with frequent updates. Diabetes Care **37**(2), 275–283 (2014)
19. Cai, D., He, X., Han, J.: Document clustering using locality preserving indexing. IEEE Trans. Knowl. Data Eng. **17**(12), 1624–1637 (2005)
20. Lu, C.Y.: Sparse representation based face classification and clustering. University of Science and Technology of China, Hefei (2012)
21. Sun, W., Zhang, L., Du, B., et al.: Band selection using improved sparse subspace clustering for hyperspectral imagery classification. IEEE J. Selected Topics Appl. Earth Observations Remote Sens. **8**(6), 2784–2797 (2015)
22. Cai, D., Chen, X.: Large scale spectral clustering via landmark-based sparse representation. IEEE Trans. Cybern. **45**(8), 1669–1680 (2015)
23. Yang, Y., Yang, Y., Shen, H.T., et al.: Discriminative nonnegative spectral clustering with out-of-sample extension. IEEE Trans. Knowl. Data Eng. **25**(8), 1760–1771 (2013)

A Compressed Sensing Based Feature Extraction Method for Identifying Characteristic Genes

Sheng-Jun Li[1], Junliang Shang[1,2(✉)], Jin-Xing Liu[1], and Huiyu Li[1]

[1] School of Information Science and Engineering, Qufu Normal University,
Rizhao 276826, China
{qfnulsj,shangjunliangl10,lihuiyu20}@163.com,
sdcavell@126.com
[2] Institute of Network Computing, Qufu Normal University,
Rizhao 276826, China

Abstract. In current molecular biology, it becomes more and more important to identify characteristic genes closely correlated with a key biological process from gene expression data. In this paper, a novel compressed sensing (CS) based feature extraction method named CSGS is proposed to identify the characteristic genes. Considering the transposed gene expression matrix and class labels as sensing matrix and measurement vector, respectively, CS reconstruction is implemented by basis pursuit algorithm. Top ranking genes with high signal weights are retained as the characteristic genes. Experiments of CSGS are performed on leukemia data set and compared with other sparse methods. Results demonstrate that CSGS is effective in identifying characteristic genes, and is not sensitive to parameters. CSGS could offer a simple way for feature extraction and provide more clues for biologists.

Keywords: Gene expression data · Characteristic genes · Compressed sensing · Feature extraction

1 Introduction

Identifying characteristic genes from gene expression data is necessary as the data usually contain many irrelevant and noisy expressions. And it is effective for early tumor detection and cancer discovery. However, the large amount of gene expression data brings challenges to conventional statistical tools for analyzing and finding inherent correlations between genes and samples. It becomes necessary to reduce the data dimensionality or develop a method that can handle with high-dimensional and small-sample-size problem.

Moreover, many sparse methods have been widely used for gene expression data analysis. Witten et al. proposed a penalized matrix decomposition (PMD) [1], which was used by Liu et al. to analyze plant gene expression data [2] and RNA-seq count data [3]. Luss et al. used SPCA for clustering and feature selection [4]. Though these sparse methods have been successfully implemented on gene expression data, they may

© Springer International Publishing Switzerland 2016
D.-S. Huang and K.-H. Jo (Eds.): ICIC 2016, Part II, LNCS 9772, pp. 67–77, 2016.
DOI: 10.1007/978-3-319-42294-7_6

not give intelligible results on the selection of characteristic genes and the parameter setting is complicated and not easy to use.

Compressed sensing (CS) is a theory that predicts sparse vectors in high dimensions reconstructed from incomplete information [5–7]. It has been successfully used in many fields such as image processing [8], geophysical data analysis [9] and GWAS [10, 11]. Ho et al. [10] confirmed that matrices of human SNP genotypes are good compressed sensors and applied CS to determine nonlinear genetic architecture. Tang et al. [12] proposed a CS-based method to identify the subtyping of gliomas. They demonstrate the advantages of CS methods in compact representation of genomic data, resulting in higher classification accuracies. Li et al. [13] proposed a CS-based two-stage method CSMiner for detecting epistatic interactions. However, CS has not been utilized in identifying characteristic genes.

In this study, we proposed a CS based method, CSGS, for identifying characteristic genes. Considering the transposed gene expression data matrix as a sensing matrix and class label vector as a measurement vector, we select the top ranking coefficients from CS reconstruction. Genes corresponding to the top ranking coefficients are retained as the characteristic genes. To evaluate the validity of our method, experiments applied on leukemia data set are handled by CSGS and other compared methods. The identified genes are appraised using the Gene Ontology tool and correlations between genes and the disease are confirmed by other literatures. By comparing with PMD and SPCA methods, all empirical results show that the novel method outperforms the competitive methods for identifying characteristic genes.

The main contributions of our work are as follows: first, it proposes, for the first time, the idea and method based on CS for identifying characteristic genes; second, our method is insensitive to parameters so it provides a convenient way for biologists.

The rest of the paper is organized as follows. The proposed CS based method CSGS is given. In Sect. 3, we compare our method with other two methods (PMD and SPCA). Finally, the conclusions are given in Sect. 4.

2   Methods

2.1   CS Theory

In CS [5, 7, 14, 15] theory, if a signal $x \in R^N$ is K-sparse, that is only $K \ll N$ coefficients in x are nonzero, it could be recovered stably by its measurements $y \in R^M$ where $y = \Phi x$ and $M \ll N$. Here, we refer to Φ as the $M \times N$ sensing matrix and x as the measurement vector.

The reconstruction procedure searches for $\hat{x}$ with the smallest L_0 norm that is consistent with the observed y, that is

$$\hat{x} = \arg \min \|x\|_0, subject\ to\ y = \Phi x. \tag{1}$$

However, it must require an exhaustive enumeration of all C_N^K possible locations of the nonzero entries in x, and the optimization is a NP-completeness. An alternative to the L_0 norm used in (1) is to use the L_1 norm, defined as $\|x\|_1 = \sum_{n=1}^{N} |x(n)|$. $\hat{X}$ can be reconstructed perfectly by solving basis pursuit (BP) [16]

$$\hat{x} = \arg\min\|x\|_1, subject\ to\ y = \Phi x. \tag{2}$$

Since the L_1 norm is convex, the formula (2) can be seen as a convex relaxation of the formula (1). Random matrices are commonly used for Φ, though non-random matrices can also be used while they satisfy restricted isometry property requirements.

2.2 Compressed Sensing Based Gene Selection Method: CSGS

Based on the theory of CS, we propose a novel method, CSGS, to select characteristic genes from gene expression data.

The gene expression data can be denoted as a matrix D of size $N \times M$, each row of D represents the expression level of a gene in M samples, and each column of D represents the expression level of all the N genes in one sample. It can be described as

$$D = [D_1, D_2, \ldots D_N]^T \tag{3}$$

where $D_i = [d_{i,1}, d_{t,2}, \ldots d_{i,M}]$, $i = 1, 2, \ldots N$.

We transpose gene expression matrix D and consider D^T to be a sensing matrix Φ. The class label vector can be regarded as a measurement vector y. Then we can denote them as

$$\Phi = [\phi_1, \phi_2, \ldots, \phi_N], y = [y_1, y_2, \ldots, y_M]^T, \tag{4}$$

where $\phi_i = D_i$, $D_i = [d_{i,1}, d_{t,2}, \ldots d_{i,M}]$, $y_j \in \{1, 2\}$, $i = 1, 2, \ldots, N$, $j = 1, 2, \ldots, M$.

Characteristic genes are genes associated with diseases or gene-gene interactions. They also can be called informative genes. As characteristic genes are very few relative to the whole gene expression matrix, we can assume that x is sparse. According to $y = \Phi x$, the problem to select characteristic genes associated with diseases can be considered as a problem of CS reconstruction.

As we have stated above, CS reconstruction problem can be implemented by CS formula (2) using reconstruction algorithm of CS theory such as BP algorithm. BP reconstruction algorithm can search as few columns of Φ as possible to approximate y. The measurements y are merely M different randomly weighted linear combinations of the elements of x. So both informative genes and gene-gene interactions can be identified by BP algorithm. In our study, we employ SPGL1 [17] which is an efficient BP reconstruction algorithm to solve the CS reconstruction problem.

By using SPGL1, we obtain $\hat{x}$ from the CS reconstruction and the top ranking coefficients with high signal weights are retained. That is, the characteristic genes are identified based on $\hat{x}$. $\hat{x}$ can be expressed as follows,

$$\hat{x} = [\hat{x}_1, \hat{x}_2, \ldots \hat{x}_N]^T \tag{5}$$

Following the description in [18], the characteristic genes are usually grouped into two classes: up-regulated genes and down-regulated genes, which can be reflected by the positive items and negative items in $\hat{x}$. Here, we only consider the absolute value of the items in $\hat{x}$ to identify the characteristic genes, that is,

$$|\hat{x}| = [|\hat{x}_1|, |\hat{x}_2|, \ldots |\hat{x}_N|]^T \tag{6}$$

The larger the item in $|\hat{x}|$ is, the more informative the gene is. Therefore, we sort the elements in $|\hat{x}|$ in descending order and take the top h genes as the characteristic genes. Here, $h \ll N$ and it can be defined according to the corresponding requirement.

In general, we summarize the CSGS method as follows:

(1) Given the gene expression matrix $D \in R^{N \times M}$, class label vector $y \in R^M$, $h > 0$.
(2) Transpose D and regard it as sensing matrix $\Phi \in R^{M \times N}$.
(3) Obtain $\hat{x}$ via BP reconstruction algorithm solving the $\hat{x} = \arg\min\|x\|_1$, subject to $y = \Phi x$.
(4) Obtain $|\hat{x}|$ by absolute value of $\hat{x}$.
(5) Sort $|\hat{x}|$ in descending order.
(6) Select the genes that have the first h largest entries as characteristic genes.

3 Results and Discussion

To validate the performance of our method, experiments on tumor expression data based on three methods are carried out. In the first subsection, we give the source of gene expression data. How to set the parameters is demonstrated in the next section. Finally, to demonstrate the effectiveness of our method for recognizing the characteristic genes, PMD [1] and SPCA [19] are used for comparison.

3.1 Leukemia Dataset

The leukemia data set has become an important reference in tumor selection study. It consists of 11 cases of acute myelogenous leukemia (AML) and 27 cases of acute lymphoblastic leukemia (ALL) [20]. In this data set, the distinction between AML and ALL, as well as the division of ALL into T and B cell subtypes, is known. The dataset contains 5000 genes in 38 samples.

3.2 Parameters Selection

PMD and SPCA are sparse methods, whose sparse parameters have an enormous influence on the identification accuracy. All the parameters we get from the comparison algorithms are following the results described in their own papers. For fair comparison, 100 genes are identified by three methods.

The control-sparsity parameters of PMD and SPCA are α_1 and γ, respectively. Here, we choose 0.1 as the sparse parameter of PMD and select 0.15 as the sparse parameter for SPCA and they can acquire their best results.

CSGS only need to adjust the iterations to get corresponding number genes. And the adjustment of iterations only influences the sparsity. The selected genes change few with different iterations.

3.3   Gene Ontology (GO) Analysis

The Gene Ontology (GO) enrichment of functional annotation of the identified genes by three methods is detected by ToppFun [21] which is publicly available at http://toppgene.cchmc.org/enrichment.jsp. The analysis of ToppFun provides significant information for the biological interpretation of high-throughput experiments. The p-value cutoff is set to 0.01. As we stated above, 100 genes are identified by the three compared methods.

Table 1 lists the top 14 closely related terms to leukemia corresponding to different methods.

Table 1. The leukemia GO terms corresponding to different methods.

ID	Name	CSGS	PMD	SPCA	Term in Genome
		Input	Input	Input	
		PV	PV	PV	
17092989-SuppTable1	Human Lymphoma Fogel07 33genes	**2.90E−40**	5.28E−35	4.31E−23	33
		20	18	13	
19755675-TableS6	Human Leukemia Li09 419genes	2.74E−33	1.22E−24	**1.22E−37**	410
		33	27	36	
M11197	Housekeeping genes identified as expressed across 19 normal tissues.	3.50E−28	**6.46E−31**	1.09E−18	389
		29	31	22	
12917485-Table8	Human Breast Sotiriou03 706genes	**9.63E−25**	2.52E−23	2.73E−17	513
		29	28	23	
M4872	Genes up-regulated in mature plasma cells compared with plasmablastic B lymphocytes.	**7.34E−20**	None	None	397
		23			
14684422-TableS1	Human Leukemia Chiaretti04 314genes	**5.94E−18**	1.35E−11	8.75E−13	274
		19	14	15	
GO:0006955	Immune response	**7.71E−17**	2.93E−15	**4.14E−17**	1762
		39	38	40	
GO:0001775	Cell activation	1.68E−16	**7.85E−17**	**5.53E−17**	1057
		31	32	32	

(Continued)

Table 1. (*Continued*)

ID	Name	CSGS Input PV	PMD Input PV	SPCA Input PV	Term in Genome
GO:0006952	Defense response	**6.53E−15** 38	6.67E−10 32	None	1901
GO:0098602	Single organism cell adhesion	**2.14E−14** 26	5.26E−13 25	None	836
GO:0034109	Homotypic cell-cell adhesion	**3.15E−14** 22	7.34E−12 20	5.97E−12 20	562
GO:0003823	Antigen binding	**1.71E−11** 11	8.58E−09 9	8.58E−09 9	120
GO:0033218	Amide binding	**2.65E−11** 15	4.07E−09 13	1.50E−04 8	307
GO:0042277	Peptide binding	**8.27E−11** 14	1.17E−09 13	7.36E−05 8	277

'None' denotes that the algorithm cannot give the GO terms; 'PV' denotes the p-values of GO terms associated with leukemia; 'Input' denotes the number of genes associated with the term from input; 'Term in Genome' denotes the number of genes associated with the term in global genome.

In this table, the lowest p-values among the three methods are marked in bold type. It can be found that our method has lower p-values in 11 terms indicating that our method performs better than compared methods. However, in one term (GO: 0001775), PMD and SPCA have lower p-values than CSGS. In term GO: 0006955, CSGS has the same p-value with SPCA and both of them perform better than PMD. In the term 19755675-TableS6, PMD outperforms our method and it is obvious that our method is superior to SPCA in this term. In another term (M11197), SPCA performs better than our method but CSGS has lower p-value than PMD. In the term M4872, only CSGS can identify characteristic genes, and in another two terms (GO: 0006952, GO: 0098602), SPCA cannot identify correlated genes and in addition, CSGS can identify more genes than PMD.

3.4 Correlations Between Genes and Leukemia Data

To further study the performance of CSGS and mine the correlations between the identified genes and leukemia data, the genes recognized by CSGS are verified based on literatures. For simplicity, the top 30 genes identified by CSGS are taken into consideration.

According to [22], there are 210 genes identified related to leukemia. 26 of the top 30 genes selected by CSGS can be found in it. From the literatures, all the top 30 genes are reported to be identified by other papers. For example, MPO is reported in [23] that it is a key regulator of oxidative stress-mediated apoptosis in myeloid leukemic cells; [24] studied the association of GSTP1 gene polymorphism with acute leukaemia; ATP2A3 was reported in [25] suggesting that ATP2A3 gene may not act as a classical tumor suppressor gene but rather haplo-insufficiency of this gene may be enough to change the cell and tissue environment in such a way to predispose to cancer development; [26] analyzed the gene expression profiling study of contribution of GM-CSF and IL8 to the CD44-induced differentiation of acute monoblastic leukemia. The verifications of characteristic genes based on literatures demonstrate that the proposed CS-based method is potentially effective.

The gene functions of the 10 genes are shown in Table 2.

Table 2. Leukemia genes extracted by CSGS.

Gene ID	Gene name	Gene function
4353	MPO	Myeloperoxidase (MPO) is a heme protein synthesized during myeloid differentiation that constitutes the major component of neutrophil azurophilic granules.
2950	GSTP1	This GST family member is a polymorphic gene encoding active, functionally different GSTP1 variant proteins that are thought to function in xenobiotic metabolism and play a role in susceptibility to cancer, and other diseases
6402	SELL	Its defects causes the leukocyte adhesion deficiency.
566	AZU1	Azurophil granules, specialized lysosomes of the neutrophil, contain at least 10 proteins implicated in the killing of microorganisms. This gene encodes a preprotein that is proteolytically processed to generate a mature azurophil granule antibiotic protein, with monocyte chemotactic and antimicrobial activity. It is also an important multifunctional inflammatory mediator. This encoded protein is a member of the serine protease gene family but it is not a serine proteinase, because the active site serine and histidine residues are replaced. The genes encoding this protein, neutrophil elastase 2, and proteinase 3 are in a cluster located at chromosome 19pter. All 3 genes are expressed coordinately and their protein products are packaged together into azurophil granules during neutrophil differentiation.
3576	IL8	Gene expression profiling study of contribution of GM-CSF and IL-8 to the CD44-induced differentiation of acute monoblastic leukemia.
489	ATP2A3	the ATP2A3 gene may not act as a classical tumor suppressor gene, but rather haplo-insufficiency of this gene may be enough to change the cell and tissue environment in such a way to predispose to cancer development.
28639	trbC1	T cell receptor beta constant 1.
6730	SRP68	This gene encodes a subunit of the signal recognition particle (SRP). The SRP is a ribonucleoprotein complex that transports secreted and membrane proteins to the endoplasmic reticulum for processing.
392437	FTLP2	FTL This gene encodes the light subunit of the ferritin protein. Ferritin is the major intracellular iron storage protein in prokaryotes and eukaryotes.
1535	CYBA	Cytochrome b is comprised of a light chain (alpha) and a heavy chain (beta). This gene encodes the light, alpha subunit which has been proposed as a primary component of the microbicidal oxidase system of phagocytes.

For providing more rigorous comparisons of CSGS with other methods, overlaps of identified genes on leukemia data set among CSGS and other three methods, i.e., PMD, SPCA, and PMT-UC [22], are reported by a Venn diagram in Fig. 1. PMT-UC shows 210 genes associated with leukemia and 186 genes are left after we delete the replicated items and items without annotations. It is seen that they have a large overlap and 42 out of 100 genes by CSGS are also identified by PMT-UC. In general, 67 out of 100 genes by CSGS are also identified by other methods. Other 33 genes recognized only by CSGS will also provide new clues for biologists. For example, EGR1 is identified by CSGS and neglected by other three methods and [27] reported that haploinsufficiency of EGR1 leads to the development of myeloid disorders. Further studies with the use of large-scale case-control samples are needed to confirm whether these genes have true associations with leukemia. We hope that, from these results, some clues could be provided for the exploration of causative factors of leukemia.

The results suggest that CSGS is an effective method for identifying characteristic genes on leukemia data set.

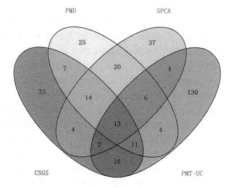

Fig. 1. Venn diagram of CSGS and compared methods on leukemia data. (Color figure online)

3.5   Results on Colorectal Cancer (CRC) Dataset

We also apply CSGS on other TCGA dataset such as Colorectal cancer (CRC) dataset, which is the second largest cause of cancer-related deaths in Western countries. Colorectal carcinogenesis is a multistep process involving apoptosis, differentiation and survival mechanisms. To provide better treatment strategies, there is an urgent need to further understand the precise molecular mechanism of CRC and to identify new prognostic biomarkers and therapeutic targets for colorectal cancer [28].

The characteristic genes we extract from CRC are explored in literatures. For example, [29] shows that ITLN1 had the second highest fold change when comparing early- and late-onset tumors. Compared with the normal colonic mucosa, a higher expression of ITLN1 was in general seen in tumors from the early-onset group, whereas the late-onset group showed a lower ITLN1 expression. KRT20 is an epithelial marker, which has also frequently been used to identify circulating tumor load in CRC patients [30]. Results of [28] shows that POSTN plays an important role in the progression of colorectal cancers.

4 Conclusions

In this paper, based on CS theory, we propose a novel feature extraction method named as CSGS to identify characteristic genes in gene expression data. We transpose gene expression matrix and regard the transposed matrix to be a sensing matrix and select the top ranking genes corresponding to high weight coefficients of reconstruction as the characteristic genes. Results on leukemia data sets demonstrate that CSGS is promising as a feature extraction method compared with PMD and SPCA. Results provide new clues for biologists on the exploration of associated genetic factors with leukemia.

Compared with other methods, CSGS needs fewer parameters, and its parameter influences the results less. Moreover, CSGS is a steady method and it is easy to be used by biologists who are not familiar with algorithms. That is, CSGS offers a simple way for feature extraction. It is believed that more detailed analysis of the CSGS may capture more important features to interpret the mechanism of a complex disease. Moreover, integrative exploratory analysis of multi-omics data can help reveal disease complexities in greater detail. These inspire us to continue working in the future.

Conflict of Interests
The authors declare that there is no conflict of interests regarding the publication of this paper.

Acknowledgments. This work was supported by the National Natural Science Foundation of China (Grant No.61502272, 61572284, 61572283); the Award Foundation Project of Excellent Young Scientists in Shandong Province (BS2014DX004, BS2014DX005); Project of Shandong Province Higher Educational Science and Technology Program (J13LN31); Scientific Research Foundation of Qufu Normal University(XJ201226); the Science and Technology Planning Project of Qufu Normal University (xkj201524); the Elaborate Experiment Project of Qufu Normal University (jp2015005) and the Innovation and Entrepreneurship Training Project for College Students of Qufu Normal University (2015A059).

References

1. Witten, D.M., Tibshirani, R., Hastie, T.: A penalized matrix decomposition, with applications to sparse principal components and canonical correlation analysis. Biostatistics. **10**(3), 515–534 (2009)
2. Liu, J.X., Zheng, C.H., Xu, Y.: Extracting plants core genes responding to abiotic stresses by penalized matrix decomposition. Comput. Biol. Med. **42**(5), 582–589 (2012)
3. Liu, J.-X., Gao, Y.-L., Xu, Y., Zheng, C.-H., You, J.: Differential expression analysis on RNA-Seq count data based on penalized matrix decomposition. IEEE Trans. Nanobiosci. **13** (1), 12–18 (2014)
4. Luss, R., d'Aspremont, A.: Clustering and feature selection using sparse principal component analysis. Opt. Eng. **11**(1), 145–157 (2010)

5. Baraniuk, R.G.: Compressive sensing. IEEE Sign. Process. Mag. **24**(4), 118–120, 124 (2007)
6. Baraniuk, R.G., Cevher, V., Duarte, M.F., Hegde, C.: Model-based compressive sensing. IEEE Trans. Inf. Theor. **56**(4), 1982–2001 (2010)
7. Donoho, D.L.: Compressed sensing. IEEE Trans. Inf. Theor. **52**(4), 1289–1306 (2006)
8. Trocan, M., Tramel, E.W., Fowler, J.E., Pesquet, B.: Compressed-sensing recovery of multiview image and video sequences using signal prediction. Multimedia Tools Appl. **72** (1), 95–121 (2014)
9. Gholami, A., Siahkoohi, H.: Regularization of linear and non-linear geophysical Ill-posed problems with joint sparsity constraints. Geophys. J. Int. **180**(2), 871–882 (2010)
10. Ho, C.M., Hsu, S.D.: Determination of Nonlinear Genetic Architecture Using Compressed Sensing (2014). arXiv preprint arXiv:14086583
11. Vattikuti, S., Lee, J.J., Chang, C.C., Hsu, S.D., Chow, C.C.: Applying compressed sensing to genome-wide association studies. GigaScience **3**(1), 10 (2014)
12. Tang, W., Cao, H., Zhang, J.G., Duan, J., Lin, D., Wang, Y.P.: Subtyping of Gliomaby combining gene expression and CNVs data based on a compressive sensing approach. Adv. Genet. Eng. **1**, 101 (2012)
13. Li, S., Shang, J., Chen, Q., Sun, Y., Liu, J.-X.: A compressed sensing based two-stage method for detecting epistatic interactions. Int. J. Data Min. Bioinf. **14**(4), 354–372 (2016)
14. Duarte, M.F., Eldar, Y.C.: Structured compressed sensing: from theory to applications. IEEE Trans. Sign. Process. **59**(9), 4053–4085 (2011)
15. Huang, H., Misra, S., Tang, W., Barani, H., Al-Azzawi, H.: Applications of Compressed Sensing in Communications Networks (2013). arXiv preprint arXiv:13053002
16. Chen, S.S., Donoho, D.L., Saunders, M.A.: Atomic decomposition by basis pursuit. SIAM J. Sci. Comput. **20**(1), 33–61 (1998)
17. SPGL1: A Solver for Large-Scale Sparse Reconstruction. http://www.cs.ubc.ca/labs/scl/spgl1
18. Kilian, J., Whitehead, D., Horak, J., Wanke, D., Weinl, S., Batistic, O., D'Angelo, C., Bornberg-Bauer, E., Kudla, J., Harter, K.: The Atgenexpress global stress expression data set: protocols, evaluation and model data analysis of UV-B light, drought and cold stress responses. Plant J. **50**(2), 347–363 (2007)
19. Journée, M., Nesterov, Y., Richtárik, P., Sepulchre, R.: Generalized power method for sparse principal component analysis. J. Mach. Learn. Res. **11**, 517–553 (2010)
20. Brunet, J.-P., Tamayo, P., Golub, T.R., Mesirov, J.P.: Metagenes and molecular pattern discovery using matrix factorization. Proc. Natl. Acad. Sci. **101**(12), 4164–4169 (2004)
21. Chen, J., Bardes, E.E., Aronow, B.J., Jegga, A.G.: Toppgene suite for gene list enrichment analysis and candidate gene prioritization. Nucleic Acids Res. **37**(suppl 2), W305–W311 (2009)
22. Wu, M.-Y., Dai, D.-Q., Zhang, X.-F., Zhu, Y.: Cancer subtype discovery and biomarker identification via a new robust network clustering algorithm (2013). http://dx.doi.org/10.1371/journal.pone.0066256
23. Nakazato, T., Sagawa, M., Yamato, K., Xian, M., Yamamoto, T., Suematsu, M., Ikeda, Y., Kizaki, M.: Myeloperoxidase is a key regulator of oxidative stress-mediated apoptosis in myeloid leukemic cells. Clin. Cancer Res. **13**(18), 5436–5445 (2007)
24. Dunna, N.R., Vuree, S., Kagita, S., Surekha, D., Digumarti, R., Rajappa, S., Satti, V.: Association of GSTP1 gene (I105v) polymorphism with acute leukaemia. J. Genet. **91**, 1–4 (2012)
25. Korošec, B., Glavač, D., Volavšek, M., Ravnik-Glavač, M.: Atp2a3 gene is involved in cancer susceptibility. Cancer Genet. Cytogenet. **188**(2), 88–94 (2009)

26. Delaunay, J., Lecomte, N., Bourcier, S., Qi, J., Gadhoum, Z., Durand, L., Chomienne, C., Robert-Lezenes, J., Smadja-Joffe, F.: Contribution of GM-CSF and IL-8 to the CD44-induced differentiation of acute monoblastic leukemia. Leukemia **22**(4), 873–876 (2008)

27. Joslin, J.M., Fernald, A.A., Tennant, T.R., Davis, E.M., Kogan, S.C., Anastasi, J., Crispino, J.D., Le Beau, M.M.: Haploinsufficiency of EGR1, a candidate gene in the del (5q), leads to the development of myeloid disorders. Blood **110**(2), 719–726 (2007)

28. Li, Z., Zhang, X., Yang, Y., Yang, S., Dong, Z., Du, L., Wang, L., Wang, C.: Periostin expression and its prognostic value for colorectal cancer. Int. J. Mol. Sci. **16**(6), 12108–12118 (2015)

29. Ågesen, T., Berg, M., Clancy, T., Thiis-Evensen, E., Cekaite, L., Lind, G., Nesland, J., Bakka, A., Mala, T., Hauss, H.: Clc and Ifnar1 are differentially expressed and a global immunity score is distinct between early-and late-onset colorectal cancer. Genes Immun. **12**(8), 653–662 (2011)

30. Mostert, B., Sieuwerts, A.M., Vries, J.B., Kraan, J., Lalmahomed, Z., van Galen, A., van der Spoel, P., de Weerd, V., Ramírez-Moreno, R., Smid, M.: mRNA expression profiles in circulating tumor cells of metastatic colorectal cancer patients. Mol. Oncol. **9**(4), 920–932 (2015)

Social Computing

Enhancing Link Prediction Using Gradient Boosting Features

Taisong Li[1(✉)], Jing Wang[2(✉)], Manshu Tu[1(✉)], Yan Zhang[1(✉)],
and Yonghong Yan[1(✉)]

[1] The Key Laboratory of Speech Acoustics and Content Understanding Institute
of Acoustics, Chinese Academy of Sciences, Beijing, China
{litaisong,tumanshu,zhangyan,
yanyonghong}@hccl.ioa.ac.cn
[2] National Computer Network Emergency Response Technical
Team/Coordination Center of China (CNCERT/CC), Beijing, China
wangjing@cert.org.cn

Abstract. Link prediction is an important task in social network analysis. The task is to predict missing links in current networks or new links in future networks. The key challenge in link prediction is being lack of features when machine learning methods are applied. Most relevant studies solve the problem by using features derived from network topology. In our work, we propose a novel feature extraction method by employing Gradient Boosting Decision Tree (GBDT), which effectively derive attributes from initial feature set. For GBDT model, input features are transformed by means of boosted decision trees. The output of each individual tree is treated as a categorical input feature to a sparse linear classifier. Extensive experiments demonstrate that the proposed method outperforms a number of mainstream baselines when GBDT features considered. The proposed method is efficient to solve the feature shortage problem in the prediction of links.

Keywords: Link prediction · Social network analysis · Machine learning

1 Introduction

Link prediction is ubiquitous — ranging from computer science to biochemistry, from academic collaboration to online social network via Microblog, Facebook, Twitter, etc. Typically, the problem can be described as: given a snapshot of a network at time t, predict which links will be created at time t + 1, or predict missing links at time t. Most recent works consider predicting future links rather than missing links [1–3]. One of its main applications is collaboration relationship forecast. The link prediction problem can be addressed by using unsupervised methods and supervised methods. Most of the unsupervised methods are based on similarity measure between two nodes, e.g. Common Neighbors (CN) [4], Adamic/Ada (AA) [5], Preferential Attachment (PA) [6], Rooted PageRank (RPR) [1], etc. More recently, researchers adopt machine learning algorithms as supervised methods to assemble features in collaboration network or social network. Many researchers used logistic regression to solve link

© Springer International Publishing Switzerland 2016
D.-S. Huang and K.-H. Jo (Eds.): ICIC 2016, Part II, LNCS 9772, pp. 81–92, 2016.
DOI: 10.1007/978-3-319-42294-7_7

prediction problem [7–9]. Sachan et al. and De Sá et al. applied decision tree model integrate features [10, 11]. Scellato et al. selected random forest and decision tree as state-of-the-art model to exploiting place features [12].

However, current approaches have two important drawbacks. First of all, all the classification and regression methods are based upon feature engineering. This angle of research generally aims at deriving topology features and attributes of nodes from current social network. None of the proposed methods take into account the novel features from other methods. Secondly, to the best of our knowledge, all above examples paid less attention to time slicing in link prediction problem. For instance, to analyze the new links for years 2006 and 2007, Sachan and Ichise used data from six partitions from 2000 to 2005 to create network G [10]. We hypothesis that a better performance can be achieved by predicting one year later rather than two years in the same graph (G). Actually, our experiments empirically support this statement.

Our contributions in this work consist of two major aspects: 1. A new feature extraction method with gradient boosting decision tree (GBDT) [14]: as mentioned before, lacking of features is one of the overheads for the link prediction problem. How to extract more useful information from existing data remains a challenge. In this paper, we propose a novel GBDT-based feature extraction method on the basis of raw features from network topology and node attributes. It compensates information provided by the raw features. Therefore, it enriches the information provided to the prediction process. Besides, for the raw features, not only neighbor based features have we taken into account, but also path based attributes. Our experiments show promising results in terms of improvement in predictive accuracy. 2. A time slice selection approach: stationary time slices, such as predicting two-year links based on a five-year network, are widely employed in this field. On the contrary, we propose a dynamic slice of time to divide training set and testing set, which is more adaptive to different datasets. We have conducted an empirical evaluation of our techniques over two collaboration networks commonly used in link prediction research. Our experiments show that incorporating GBDT features improves the prediction performance of different methods.

The remainder of this paper is organized as follows: Sect. 2 presents the problem definition and some basic notions. Section 3 describes in detail the proposed GBDT-based feature extraction. We introduce our dataset preprocessing, evaluation methodology and experimental setup in Sect. 4. The experimental results and corresponding analysis are presented in Sect. 5. Conclusions are drawn in the last section. Directions of future work are discussed in the same section.

2 Background

A social network can be represented with a graph: $G = <V, E>$, where V refers to a set of nodes, and $E \subseteq V \times V$ represents links between nodes. In co-authorship networks, nodes indicate authors, while a link usually denotes an interaction between authors. Such interactions can refer to both co-authorships and citations. However, in this work, we only take co-authorship into consideration, as we are interested in the evolution of collaborative activities of researchers. In the rest of this paper, we use author and node interchangeably.

In link-prediction problems, the prediction of newly added links are commonly based on features intrinsic to the network itself, e.g. network topology and vertices attributes. In contract, transformed features have been argued that they can improve prediction accuracy to certain extent [14]. In this paper, we are interested in whether adapting feature transforming techniques to enrich inputs for the link-prediction problem can improve the prediction accuracy.

In addition, in link-prediction problems, it is seldom discussed that whether different time slices for training and testing data affect the predictive accuracy. In this paper, we conduct empirical investigations to obtain insights into this issue.

3 Methods

3.1 Features Commonly Used in Link Predictions

In our work, we consider a set of features proposed in previous works that represent first-rate in link prediction [11]. These features are shown in Table 1. They are designed to capture characteristics of network topology. We provide a short description of all the features ranging from neighbor-based similarity to path-based attributes. We also describe our intuitive argument on choosing them as a feature for link prediction problem. Note that, some algorithms are truncated to simple for both the datasets, due to the complexity of computation.

In Table 1, the formula $\varphi(x)$ refers to the collection of neighbors of node x and $|\varphi(x)|$ represents the degree of the node x. Common Neighbors (CN) counts the number of neighbors that two nodes have in common. Jaccards Coefficient (JC) [15] is a normalized metric of common neighbors. It computes the ratio of common neighbors

Table 1. Feature definition

Features	Formula
Common Neighbors (CN)	$S_{xy} = |\varphi(x) \cap \varphi(y)|$
Jaccard's Coefficient (JC)	$S_{xy} = \frac{|\varphi(x) \cap \varphi(y)|}{|\varphi(x) \cup \varphi(y)|}$
Adamic/Adar (AA)	$S_{xy} = \sum_{z \in \varphi(x) \cap \varphi(y)} \left(\frac{1}{\log|\varphi(z)|} \right)$
Preferential Attachment (PA)	$S_{xy} = |\varphi(x)| \times |\varphi(y)|$
Resource Allocation (RA)	$S_{xy} = \sum_{z \in \varphi(x) \cap \varphi(y)} |\varphi(z)|$
Leicht-Holme-Nerman (LHN)	$S_{xy} = \frac{|\varphi(x) \cap \varphi(y)|}{|\varphi(x) \times \varphi(y)|}$
Katz	$Katz = \sum_{l=1}^{\infty} \beta^l A^l$
Rooted PageRank (RPR)	$RPR = (1 - \in)(1 - eD^{-1})^{-1}$
Degree (x)	$|\varphi(x)|$
Degree (y)	$|\varphi(y)|$
Degree (xy)	$|\varphi(x) + \varphi(y)|$

to all neighbors of two nodes. Adamic/Adar (AA) [5] measures similarity between two nodes by assigning more weight to rare shared common neighbors. Preferential Attachment (PA) [6] is based on the idea that new links is more likely to connect to higher-degree nodes. Similar to Adamic/Adar, Resource Allocation (RA) [16] punishes high-degree common neighbors. It accumulates the reciprocal, instead of the logarithm of common neighbors' degree. Therefore, RA performs better for the networks with high average degrees. Leicht-Holme-Nerman (LHN) [17] is similar to Jaccard's Coefficient, but it punishes node pairs that have low percentage of common neighbors. Except for above neighbor based metrics, we have path based attribute Katz and random walk based metric Rooted PageRank. The measure of Katz is defined as follows, where A is an adjacent matrix, l presents all paths with length l and β is an adjustable parameter. The very small β will cause Katz metric much like CN metric because paths of long length contribute very little to final similarities. Rooted PageRank is a modified PageRank metric, where A is an adjacent matrix as well and D is a diagonal matrix with $D_{i,i} = \sum_j A_{i,j}$. The factor $\in$ specifies how likely the algorithm is to visit the nodes neighbors than starting over. To reduce the computational complexity of path based method, we use Truncated Katz and Truncated Rooted PageRank, whose largest hop limited to 2. Finally, we consider the degree of each node of a pair and their sum of degrees as well.

3.2 GBDT Features

The feature transformation has been proposed to enhance prediction accuracy in ads click problems [14]. In this paper, we examine whether such a transformation can improve algorithm performances in link-predictions. From the literature review in the introduction section, we can observe that most features used for link-predictions are extracted from vertices attributes and graph topology. To the best of our knowledge, there has not been such investigation conducted for link-prediction problems.

A decision tree is considered as a classifier in feature transforms [14]. When we take a feature vector as an input to a tree, it is then classified into a leaf node. Then, the value of this leaf node is assigned to 1, others are 0. As a result, the outcome of a decision tree is a binary vector, we then transform this vector into a decimal number. For example, a binary vector [1, 0, 0, 0] is convert into 8 after transformation process. GBDT is a boosting method of decision trees. Unlike bagging methods [18], such as Random Forest, GBDT keeps a long learner tree and grow it sequentially along the gradient descent orientation. There are three steps in feature extraction by using GBDT. Firstly, a feature set, as shown in Table 1, is taken as an input of a decision tree. A number of deviations will occur when a series of feature set entering. Secondly, to fix the deviation, we add another decision tree based on minimizing mean square error of last tree and repeat this step many times. Finally, we get a number of binary vector from all of the trees and transform each vector into a decimal number, which is formed the GBDT features. The new features are improved presentation of initial feature set, because the GBDT model can reduce the loss function continuously and avoid overfitting problem. As it is described above, The GBDT features are dependent on two parameters:

- t: Number of trees. It presents the dimension of GBDT features. To some extent, with the growing of t, the prediction performance increases. However, it also can cause overfitting when the number of tree is too large.
- d: Depth of each tree. It decides the size of each value. Like the parameter of t, overfitting problem will occur when d is too large.

3.3 Time Slices Selection

Link prediction problem can be defined as: given a snapshot of a network for T time periods, predict which links will be created in the following time $T + 1$. Previous works paid less attention to the time slices selection. For instance, papers [8–12] have certain temporal slices of training set and testing set. However, we noted that the length of time T have a significant impact on link prediction at time $T + 1$, and *vice versa*. In this work, we will exploit that given a certain time period T, which time slice of $T + 1$ will get a promising prediction performance. On the other hand, we analyze the influence with dynamic T and fixed $T + 1$.

4 Experiments

4.1 Dataset

To test our link prediction method, we used the Digital Bibliography Library Project (DBLP) database [13] and AMiner dataset [13]. Both of datasets contain bibliographic information on major computer science journals and proceedings. We only derive authors as nodes and co-authorships as edges from dataset. The DBLP network has 77 years evolution history, ranging from 1936 to 2013, but the data increases steady in recent ten years. Hence, we chose only ten years, from 2003 to 2012, for our experiment. The AMiner network has a evolutionary history from 1936 to 2014. We conducted our experiment with 10-year dataset ranging from 2003 to 2012 as well. The statistic properties of both networks are shown in the Table 2.

Table 2. Statistics of networks

Characteristics	DBLP	AMiner
Nodes	991,488	987,930
Edges	3,869,017	3,121,525
Average Degree	7	6
Maximum Degree	1,878	4,851
Triangle Count	8,605,554	5,568,490
Size of LCC	866,329	805,130
Clustering Coefficient	65.36 %	65.28 %
Connected Components	39,956	56,841

4.2   Dataset Preprocessing

Since the dataset is imbalanced and noisy, we conduct undersampling with two steps. We first eliminate isolated nodes and pairs, which do not have any connection with others and have no attribution to link prediction. For instance, some authors never collaborated with others after writing a thesis. We therefore delete such nodes. Secondly, in our project, the positive edges refer to the links in T + 1 but not in period T, and we denoted the set of positive edges as E+. The negative edges denote those pairs that have no links in the time period T and T + 1. We denoted the set of negative edges as E−. The percentage of negative edges is huge, leading to class imbalance and model overfitting. Previous work [19] has proven that the machine learning models, such as logistic regression, would be biased seriously with a tendency of predicting towards negative without undersampling. Therefore, we sample those negative edges randomly from E− to make their numbers the same as the number of the positive edges. These sampling method has commonly employed in previous works for link prediction [7, 11, 19]. Besides, there are other oversampling methods such as Synthetic Minority Over-sampling Technique (SMOTE) [20], we do not consider it, as our network has sufficient positive links.

4.3   Experiment Setup

Candidate Generation. For a network, the amount of potential links equal to $N = |V| \times |V - 1|/2$. When the number of nodes of a network is very high, the processing complexity of each node attributes, network topology features and potential linking pairs is prohibitively high. According to the experiment with Twitter, Yin et al. [21] have found that 90 % of links created are to the users within 2 hops of a given user in the twitter network. Therefore, we conduct predictions only for pairs that are 2-hops away from each other in the co-authorship network. We do not consider the pairs with 1-hop away because we are more interested in those pairs that have never interacted with each other. Table 3 shows the number of positive links and negative links we have labeled for candidate pairs. These candidates derived from graph with 4-year evolution history and labeled with 1-year results. It represents the general situation of two collaboration network with different time slices. From the Table, we can see that positive links vs negative link equals to 1:127 and 1:211, which is extremely imbalance. This imbalance is handled as follows: As mentioned before, we adopt undersampling methods to positive set E^+ and negative set E^-, which generate a number of 40000 examples. From these instances, 20,000 examples are positive links, and another 20,000 examples are negative links.

Table 3. Number of labeled links

Dataset	Positive links	Negative links	Total
DBLP	70,388,002	553,828	70,941,830
AMiner	54,188,452	256,794	54,445,246

Comparison Methods. In our experiment, a variety of learning algorithms are used to study the impact of GBDT features and time slices in the datasets described in the previous section. All algorithms were available in the Scikit-Learn environment [22].

- Gradient Boosting Decision Tree (GBDT).
- Logistic Regression Classifier (LRC).
- Random Forest Classifier (RFC).

Note that, the GBDT model not only used for feature extraction but also employed as a classifier. We train our models with default parameters of Scikit-Learn. The algorithms are evaluated with stratified 10-fold cross-validation. For more reliable results, the cross-validation procedure was executed 10 times for each algorithm and dataset.

Evaluation Metrics. In order to evaluate the link prediction performance with our proposed features, we take three evaluation metrics into consideration:

- Area Under the ROC Curve (AUC): It is frequently employed to classification problem cause it relates to the sensitivity (true positive rate) and the specificity (true negative rate) of a classifier.
- Accuracy (Acc): a distinct evaluation of which pair of node classified into the right class.
- F1 score (F1): a trade-off between Precision and Recall, which is widely used in the Information Retrieval.

All above metrics are in the range of [0, 1], the higher the value, the better the Performance

5 Experiments Results

First of all, we conduct a series of experiments with different temporal slices to investigate the contribution of GBDT features. We derive GBDT features from the initial attributes shown in Table 1. We empirically determine the parameter (t = 19, d = 5) based on the criteria that to make each machine learning model get the best performance. The baseline features for machine learning algorithms are features described in Table 1. We test how well each algorithm can improve when we include GBDT features to the baseline features. We adopt AUC, Accuracy and F1 score to evaluate the supervised learning. Tables 4 and 5 (prefer to Appendix figures) show the results for different supervised methods in terms of AUC, Acc and F1 on two co-authorship networks DBLP and AMiner. The GBDT model performs best regardless of GBDT features taken into account (e.g. GBDT(A)) or not (e.g. GBDT(P)). For DBLP dataset, Fig. 1 shows increasing AUC, Acc and F1 in finding new links when GBDT features were taken into account. What's more, Fig. 1(a) illustrates the improvement with the most appropriate time slice. Figure 1(b) illustrates the one with the worst time slice, and Fig. 1(c) presents the mean improvement of all experiments with different time slices. For AMiner dataset, the results are illustrated in Fig. 3

Table 4. Results for different supervised methods on DBLP dataset

Time slices	Evaluation	GBDT (A)	GBDT (P)	RFC (A)	RFC (P)	LRC (A)	LRC (P)
T = 9, T + 1=1	AUC	**0.9888**	**0.9836**	0.9851	0.9809	0.9876	0.9714
	Accuracy	**0.9728**	**0.9722**	0.972	0.9718	0.9711	0.9138
	F1 score	**0.9726**	**0.9719**	0.9718	0.9718	0.9709	0.9094
T = 5, T + 1=5	AUC	**0.9068**	**0.901**	0.8981	0.8889	0.8979	0.8732
	Accuracy	**0.8436**	**0.8385**	0.8369	0.8286	0.8353	0.7878
	F1 score	**0.8356**	**0.8299**	0.8311	0.8236	0.8273	0.7572
Average of all	AUC	**0.9434**	**0.9389**	0.9368	0.9304	0.9352	0.9168
	Accuracy	**0.8914**	**0.8893**	0.8876	0.8826	0.9352	0.837
	F1 score	**0.8882**	**0.8858**	0.8848	0.8803	0.8841	0.8196

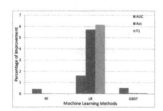 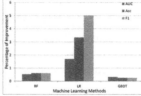 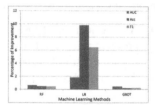

(a) For the best time slice (T=9,T+1=1)   (b) For the worst time slice (T=5,T+1=5)   (c) Average Improvement of all time slices

Fig. 1. Improvement with GBDT features of DBLP dataset (Color figure online)

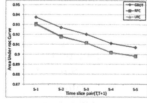 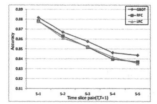 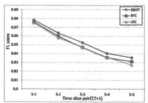

(a) AUC of DBLP Dataset   (b) Accuracy of DBLP Dataset   (c) F1 score of DBLP Dataset

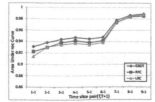 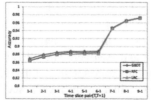 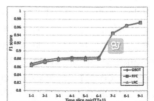

(d) AUC of DBLP Dataset   (e) Accuracy of DBLP Dataset   (f) F1 score of DBLP Dataset

Fig. 2. Variation tendency with different time slices of DBLP dataset (Color figure online)

(prefer to Appendix figures). Both figures indicate that the new features have different positive effect on distinct learning models. Logistic Regression Classifier has a significant improvement with average 5 %, while GBDT get a slight increase less than 1 %, cause the new features is a new presentation of initial features transformed by GBDT model.

Secondly, we will go on our study with advanced analysis of our co-authorship network using different time slices. As described before, GBDT features can be obtained by using our GBDT model. We used that features and initial attributes as input of machine learning models. We predict new links appearing in the period $T + 1$ based on the graph information extracted in the duration T. To analyze the impact of network creation time, we set $T = 1, 2, 3, 4, 5, 6, 7, 8, 9$ and $T + 1 = 1$. In turn, to analyze the impact of prediction time, we set $T = 5$ and $T + 1 = 1, 2, 3, 4, 5$. Figure 2 presents the variation tendency different time slices on DBLP dataset (Aminer dataset shows the same tendency). In the figure, x axis refers to the time slice pair $(T, T + 1)$ and y axis denotes the values of the evaluation metrics. We can observe that the best performances in terms of three machine learning models are experiments with time slice $(T = 9, T + 1 = 1)$. We can also observe that the performance generally improved along with the length of network extraction time when predicting time slice fixed. These results indicate that more details of network structure will contribute to link prediction efficiently. On the other hand, when we fix the information extraction time, the longer time we predict, the worse result we get. These results show that the current network provided limited information to predict future links, and more closer to the current network, more prediction of links can be made.

6 Conclusions and Future Works

In this paper, we proposed a new feature extraction method, named GBDT features, which has never been employed in link prediction problem. We also made an advanced analysis of time slices, which is intrinsically related to the performance of link-prediction problem. The major challenge of the problem is the shortage of contributing attributes, which makes it hard for machine learning algorithms to perform well. Our proposed method solved this problem by deriving new features from original featuring set instead of network topology or node attributes. The new features contribute to all machine learning methods we adopted, especially the logistic regression classifier. On the other hand, we made a further analysis of the impact of time slices. Experiments on two collaboration networks show that the most appropriate time slices are those with a longest T and a shortest $T + 1$, which means more details of network and more closer of predicting time contribute to prediction result efficiently. In the future we would like to do a further research from following three aspects:

- Finding additional features based on network: For machine learning algorithms, one of the important issues is finding effective features. Except for featuring set in Table 1, there might be useful features such as Simrank, Local Path, Hitting Time and so on. We might take those metrics into account in the future work.

- Using weighted network: De Sá and Prudêncio [11] observed that a better performance can be got when we consider the weight of edges in collaboration networks. We can use the information of cooperation or paper citation as additional weights for the edges in the network. Therefore, the complexity of some feature metrics will increase along with the weight. We will employ it with some modification.
- Using deep learning algorithms Deep learning techniques can be used as effective presentation of initial features. Previous works [23, 24] have shown the improvement of link prediction by using DBN and ctRBM. We could make a comparison in the terms of surprised method using the features transformed by deep learning method and GBDT model.

Appendix

Table 5. Results for different supervised methods on Aminer dataset

Time slices	Evaluation	GBDT (A)	GBDT (P)	RFC (A)	RFC (P)	LRC (A)	LRC (P)
T = 9, T + 1=1	AUC	**0.9515**	**0.9503**	0.9463	0.9419	0.941	0.9219
	Accuracy	**0.8888**	**0.8873**	0.8835	0.8792	0.8819	0.8259
	F1 score	**0.8863**	**0.8846**	0.8813	0.8776	0.8792	0.8041
T = 5, T + 1=5	AUC	**0.9289**	**0.9257**	0.9199	0.9146	0.9194	0.9024
	Accuracy	**0.8554**	**0.8529**	0.8467	0.8406	0.8471	0.8137
	F1 score	**0.8514**	**0.8489**	0.8436	0.8375	0.8432	0.793
Average of all	AUC	**0.9429**	**0.9408**	0.936	0.932	0.9327	0.9128
	Accuracy	**0.8772**	**0.8757**	0.8717	0.867	0.9327	0.8186
	F1 score	**0.8751**	**0.8734**	0.8701	0.8656	0.8687	0.8022

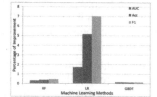 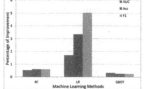 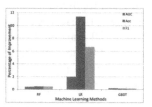

(a) For the best time slice (T=9,T+1=1)   (b) For the worst time slice (T=5,T+1=5)   (c) Average improvement of all time slices

Fig. 3. Improvement with GBDT features of Aminer dataset (Color figure online)

References

1. Liben-Nowell, D., Kleinberg, J.: The link-prediction problem for social networks. J. Am. Soc. Inform. Sci. Technol. **58**(7), 1019–1031 (2007)
2. Lü, L., Zhou, T.: Link prediction in complex networks: a survey. Physica A **390**(6), 1150–1170 (2011)
3. Wang, P., Xu, B., Wu, Y., Zhou, X.: Link prediction in social networks: the state-of-the-art. Sci. China Inf. Sci. **58**(1), 1–38 (2015)
4. Newman, M.E.: Clustering and preferential attachment in growing networks. Phys. Rev. E **64**(2), 025102 (2001)
5. Adamic, L.A., Adar, E.: Friends and neighbors on the web. Soc. Netw. **25**(3), 211–230 (2003)
6. Barabási, A.L., Jeong, H., Néda, Z., Ravasz, E., Schubert, A., Vicsek, T.: Evolution of the social network of scientific collaborations. Physica A **311**(3), 590–614 (2002)
7. Sun, Y., Han, J., Aggarwal, C.C., Chawla, N.V.: When will it happen?: Relationship prediction in heterogeneous information networks. In: Proceedings of the Fifth ACM International Conference on Web Search and Data Mining, pp. 663–672. ACM, February 2012
8. Lu, Z., Savas, B., Tang, W., Dhillon, I.S.: Supervised link prediction using multiple sources. In: 2010 IEEE 10th International Conference on Data Mining (ICDM), pp. 923–928. IEEE, December 2010
9. Rowe, M., Stankovic, M., Alani, H.: Who will follow whom? Exploiting semantics for link prediction in attention-information networks. In: Cudré-Mauroux, P., et al. (eds.) ISWC 2012, Part I. LNCS, vol. 7649, pp. 476–491. Springer, Heidelberg (2012)
10. Sachan, M., Ichise, R.: Using semantic information to improve link prediction results in network datasets. Int. J. Comput. Theory Eng. 71–76 (2011)
11. De Sá, H.R., Prudêncio, R.B.: Supervised link prediction in weighted networks. In: The 2011 International Joint Conference on Neural Networks (IJCNN), pp. 2281–2288. IEEE, July 2011
12. Scellato, S., Noulas, A., Mascolo, C.: Exploiting place features in link prediction on location-based social networks. In: Proceedings of the 17th ACM SIGKDD International Conference on Knowledge Discovery and Data Mining, pp. 1046–1054. ACM, August 2011
13. Tang, J., Zhang, J., Yao, L., Li, J., Zhang, L., Su, Z.: Arnetminer: extraction and mining of academic social networks. In: Proceedings of the 14th ACM SIGKDD International Conference on Knowledge Discovery and Data Mining, pp. 990–998. ACM, August 2008
14. He, X., Pan, J., Jin, O., Xu, T., Liu, B., Xu, T., et al.: Practical lessons from predicting clicks on ads at facebook. In: Proceedings of the Eighth International Workshop on Data Mining for Online Advertising, pp. 1–9. ACM, August 2014
15. Manning, C.D., Raghavan, P., Schütze, H.: Introduction to Information Retrieval, vol. 1, no. 1, p. 496. Cambridge University Press, Cambridge (2008)
16. Zhou, T., Lü, L., Zhang, Y.C.: Predicting missing links via local information. Eur. Phys. J. B **71**(4), 623–630 (2009)
17. Leicht, E.A., Holme, P., Newman, M.E.: Vertex similarity in networks. Phys. Rev. E **73**(2), 026120 (2006)
18. Ho, T.K.: Random decision forests. In: Proceedings of the Third International Conference on Document Analysis and Recognition, 1995, vol. 1. IEEE (1995)
19. Budur, E., Lee, S., Kong, V.S.: Structural Analysis of Criminal Network and Predicting Hidden Links using Machine Learning. arXiv preprint arXiv:1507.05739 (2015)

20. Chawla, N.V., Bowyer, K.W., Hall, L.O., Kegelmeyer, W.P.: SMOTE: synthetic minority over-sampling technique. J. Artif. Intell. Res. **16**, 321–357 (2002)
21. Yin, D., Hong, L., Davison, B.D.: Structural link analysis and prediction in microblogs. In: Proceedings of the 20th ACM International Conference on Information and Knowledge Management, pp. 1163–1168. ACM, October 2011
22. Pedregosa, F., Varoquaux, G., Gramfort, A., Michel, V., Thirion, B., Grisel, O., et al.: Scikit-Learn: machine learning in Python. J. Mach. Learn. Res. **12**, 2825–2830 (2011)
23. Liu, F., Liu, B., Sun, C., Liu, M., Wang, X.: Deep learning approaches for link prediction in social network services. In: Lee, M., Hirose, A., Hou, Z.-G., Kil, R.M. (eds.) ICONIP 2013, Part II. LNCS, vol. 8227, pp. 425–432. Springer, Heidelberg (2013)
24. Li, X., Du, N., Li, H., Li, K., Gao, J., Zhang, A.: A deep learning approach to link prediction in dynamic networks. In: SDM, vol. 14, pp. 289–297 (2014)

Neural Networks

Supervised Learning Algorithm for Spiking Neurons Based on Nonlinear Inner Products of Spike Trains

Xiangwen Wang, Xianghong Lin[✉], Jichang Zhao, and Huifang Ma

School of Computer Science and Engineering,
Northwest Normal University, Lanzhou 730070, China
linxh@nwnu.edu.cn

Abstract. Spiking neural networks are shown to be suitable tools for the processing of spatio-temporal information. However, due to their intricately discontinuous and implicit nonlinear mechanisms, the formulation of efficient supervised learning algorithms for spiking neural networks is difficult, which has become an important problem in the research area. This paper presents a new supervised, multi-spike learning algorithm for spiking neurons, which can implement the complex spatio-temporal pattern learning of spike trains. The proposed algorithm firstly defines nonlinear inner products operators to mathematically describe and manipulate spike trains, and then derive the learning rule from the common Widrow-Hoff rule with the nonlinear inner products of spike trains. The algorithm is successfully applied to learn sequences of spikes. The experimental results show that the proposed algorithm is effective for solving complex spatio-temporal pattern learning problems.

Keywords: Spiking neural networks · Supervised learning · Nonlinear inner products of spike trains · Widrow-Hoff rule

1 Introduction

Recent advances in neurosciences have revealed that neural information in the brain is encoded through precisely timed spike trains, not only through the neural firing rate [1]. Spiking Neural Networks (SNNs) are often referred to as the new generation of neural networks. They have more powerful computing capacity to simulate a variety of neuronal signals and approximate any continuous function, and have been shown to be suitable tools for the processing of spatio-temporal information [2]. Supervised learning in SNNs involves a mechanism of providing the desired outputs with the corresponding inputs. The network then processes the inputs and compares its resulting outputs against the desired outputs. Errors are calculated to control the synaptic weight adjustment. This process occurs over and over until the synaptic weights converge to certain values. When the sample conditions are changed, synaptic weights can be modified to adapt to the new environment after supervised learning. The purpose of supervised learning with temporal encoding for spiking neurons is to make the neurons emit arbitrary spike trains in response to given synaptic inputs.

© Springer International Publishing Switzerland 2016
D.-S. Huang and K.-H. Jo (Eds.): ICIC 2016, Part II, LNCS 9772, pp. 95–104, 2016.
DOI: 10.1007/978-3-319-42294-7_8

At present, researchers have conducted many studies on the supervised learning in SNNs [3, 4], and achieved some results, but many problems remain unsolved. Bohte et al. [5] first proposed a backpropagation training algorithm for feedforward SNNs, called SpikeProp, similar in concept to the BP algorithm developed for traditional Artificial Neural Networks (ANNs). The spike response model (SRM) [6] is used in this algorithm. To overcome the discontinuity of the internal state variable caused by spike firing, all neurons in the network can fire only one single spike. Shrestha et al. [7] analyzed the convergence characteristics of SpikeProp algorithm, and proposed an adaptive learning rate method, the experimental results show that the adaptive learning rate greatly improved the weight convergence and learning speed. Similarly, Ghosh-Dastidar and Adeli [8] put forward a BP learning algorithm named Multi-SpikeProp, with derivations of the learning rule based on the chain rule for a multi-spiking network model. Multi-SpikeProp was applied to the standard XOR problem and the Fisher Iris and EEG classification problems, and the experimental results show that the algorithm has higher classification accuracy than the SpikeProp algorithm. Recently, Xu et al. [9] have extended the Multi-SpikeProp algorithm to allow neurons to fire multiple spikes in all layers. That is, the algorithm can implement the complex spatio-temporal pattern learning of spike trains. The experimental results show that this algorithm has higher learning accuracy for a large number of output spikes. Considering the spike-timing-dependent plasticity (STDP) mechanism of spiking neurons, Ponulak et al. [10, 11] proposed the ReSuMe (Remote Supervised Method) algorithm, which adjusts the synaptic weights according to STDP and anti-STDP processes and is suitable for various types of spiking neuron models. However, the algorithm can only be applied to single layer networks or train readouts for reservoir networks. Sporea and Grüning [12] extended the ReSuMe algorithm to multilayer feedforward SNNs using backpropagation of the network error. The weights are updated according to STDP and anti-STDP rules, and the neurons in every layer can fire multiple spikes. Simulation experiments show that the algorithm can be successfully applied to various complex classification problems and permits precise spike train firing. Based on the ReSuMe algorithm with learning delay, Taherkhani et al. proposed DL-ReSuMe (delay learning remote supervised method) algorithm [13], and put forward a Multi-DL-ReSuMe algorithm [14] for multiple neurons. Mohemmed et al. [15, 16] proposed a SPAN (Spike Pattern Association Neuron) algorithm based on a Hebbian interpretation of the Widrow-Hoff rule and kernel function convolution. Inspired by the SPAN algorithm, Yu et al. [17, 18] proposed a PSD (Precise-Spike-Driven) supervised learning rule that can be used to train neurons to associate an input spatio-temporal spike pattern with a desired spike train. Unlike the SPAN method that requires spike convolution on all the spike trains of the input, the desired output and the actual output, the PSD learning rule only convolves the input spike trains.

For SNNs, input and output information is encoded through precisely timed spike trains, not only through the neural firing rate. In addition, the internal state variables of spiking neurons and error function do not satisfy the continuous differentiability. So, traditional learning algorithms of ANNs, especially the BP algorithm, cannot be used directly, and the formulation of efficient supervised learning algorithms for SNNs is a very challenging problem. In this paper, we present a new supervised learning algorithm for spiking neurons with the nonlinear inner products of spike trains. The rest of this paper is organized as follows. In Sect. 2 we analyze and define the nonlinear inner products of spike trains. In Sect. 3 we derive the learning rule based on the nonlinear inner products of spike trains for spiking neurons. In Sect. 4 the flexibility and power of spiking neurons trained with our algorithm are showcased by a spike sequence learning problem. The conclusion is presented in Sect. 5.

2   Inner Products of Spike Trains

The spike train $s = \{t_i \in \Gamma : i = 1, \ldots, N\}$ represents the ordered sequence of spike times fired by the spiking neuron in the interval $\Gamma = [0, T]$, and can be expressed formally as:

$$s(t) = \sum_{i=1}^{N} \delta(t - t_i) \tag{1}$$

where N is the number of spikes, and $\delta(\cdot)$ represents the Dirac delta function, $\delta(x) = 1$ if $x = 0$ and 0 otherwise.

In order to facilitate the analysis and calculation, we can choose a specific smoothing function h, using the convolution to convert the discrete spike train to a unique continuous function:

$$f_s(t) = s * h = \sum_{i=1}^{N} h(t - t_i) \tag{2}$$

Due to the limited time interval of the corresponding spike train and boundedness of the function $f_s(t)$, we can get:

$$\int_{\Gamma} f_s^2(t)dt < \infty \tag{3}$$

In other words, the function $f_s(t)$ is an element of $L_2(\Gamma)$ space.

For any two given spike trains $s_i, s_j \in s(\Gamma)$, we can define the inner products of the corresponding functions $f_{s_i}(t)$ and $f_{s_j}(t)$ on the $L_2(\Gamma)$ space as follows [19]:

$$F(s_i, s_j) = \langle f_{s_i}(t), f_{s_j}(t) \rangle_{L_2(\Gamma)} = \int_{\Gamma} f_{s_i}(t)f_{s_j}(t)dt \tag{4}$$

The effect of the spike train to the postsynaptic neuron expressed in Eq. 2 is linear. So, considering the biologically plausible nonlinear effect of spiking neuron, we can rewrite the continuous function corresponding to the spike train s as follow:

$$f_s^{\dagger}(t) = g\left(\sum_{i=1}^{N} h(t - t_i)\right) \tag{5}$$

where $g(\cdot)$ is the nonlinear function. So the inner products of the spike trains $s_i, s_j \in S(\Gamma)$ can be extended to the nonlinear form as follow [20]:

$$
\begin{aligned}
F^\dagger(s_i, s_j) &= \left\langle f_{s_i}^\dagger(t), f_{s_j}^\dagger(t) \right\rangle_{L_2(\Gamma)} \\
&= \int_\Gamma f_{s_i}^\dagger(t) f_{s_j}^\dagger(t) dt \\
&= \int_\Gamma g(f_{s_i}(t)) g(f_{s_j}(t)) dt
\end{aligned}
\tag{6}
$$

For the nonlinear function $g(\cdot)$, it can be expressed as:

$$
g(x) = \sigma \left[1 - \exp\left(-\frac{x}{2\sigma^2}\right) \right]
\tag{7}
$$

or

$$
g(x) = \sigma \tanh\left(\frac{x}{\sigma}\right)
\tag{8}
$$

3 Learning Algorithm

The input and output signals of spike neurons are expressed in the form of spike trains; that is, the spike trains encode neural information or external stimuli. The computation performed by single spiking neurons can be defined as a mapping from the input spike trains to the appropriate output spike trains. For a given spiking neuron, we assume that the input spike trains are $s_i \in s(\Gamma)$, $i = 1, \ldots, N$, and the output spike train is $s_o \in s(\Gamma)$. In order to analyze the relationship between the input and output spike trains, we use the linear Poisson neuron model [21]. This neuron model outputs a spike train, which is a realization of a Poisson process with the underlying intensity function estimation. The spiking activity of the postsynaptic neuron is defined by the estimated intensity functions of the presynaptic neurons. The contributions of all input spike trains are summed up linearly:

$$
f_{s_o}(t) = \sum_{i=1}^{N} w_{oi} f_{s_i}(t)
\tag{9}
$$

where the weights w_{oi} represent the strength of the connection between the presynaptic neuron i and the postsynaptic neuron o.

We derive the proposed learning algorithm from the common Widrow-Hoff rule, also known as the Delta rule. For a synapse i, it is defined as:

$$
\Delta w_i = \eta x_i (y_d - y_a)
\tag{10}
$$

where η is the learning rate, x_i is the input transferred through synapse i, and y_d and y_a refer to the desired and the actual neural output, respectively. By substituting x_i, y_d and

y_a with the continuous functions corresponding to the spike trains $f_{s_i}^\dagger(t)$, $f_{s_d}^\dagger(t)$ and $f_{s_o}^\dagger(t)$, a new learning rule for a spiking neuron is obtained:

$$\Delta w_i(t) = \eta f_{s_i}^\dagger(t)\left(f_{s_d}^\dagger(t) - f_{s_o}^\dagger(t)\right) \tag{11}$$

The equation formulates a real-time learning rule and so the synaptic weights change over time. By integrating Eq. 11, we derive the batch version of the learning rule which is under scrutiny in this paper:

$$\begin{aligned}
\Delta w_i &= \eta \int_\Gamma f_{s_i}^\dagger(t)\left(f_{s_d}^\dagger(t) - f_{s_o}^\dagger(t)\right)dt \\
&= \eta\left(F^\dagger(s_i, s_d) - F^\dagger(s_i, s_o)\right)
\end{aligned} \tag{12}$$

where $F^\dagger(\cdot, \cdot)$ is the nonlinear inner products of the spike trains expressed in Eq. 6. In order to facilitate the description, we called the algorithm Nonlinear1 when the nonlinear function $g(\cdot)$ is Eq. 7, and called the algorithm Nonlinear2 when the nonlinear function $g(\cdot)$ is Eq. 8.

4 Simulations

In this section, several experiments are presented to demonstrate the learning capabilities of our learning algorithm. At first, the algorithm is applied to the learning sequences of spikes, by demonstrating its ability to associate a spatio-temporal spike pattern with a desired spike train. Then, we analyze the factors that may influence the learning performance, such as the number of synaptic inputs and the firing rate of spike trains.

To quantitatively evaluate the learning performance, we use the correlation-based measure C [22] to express the distance between the desired and actual output spike trains. The metric is calculated after each learning epoch according to

$$C = \frac{s_d \cdot s_o}{|s_d||s_o|} \tag{13}$$

where s_d and s_o are vectors representing convolution of the desired and actual output spike trains with a Gaussian low-pass filter, $s_d \cdot s_o$ is the inner product, and $|s_d|$, $|s_o|$ are the Euclidean norms of s_d and s_o, respectively. The measure $C = 1$ for the identical spike trains and decreases towards 0 for loosely correlated trains.

In the first experiment, we demonstrate the learning ability of the proposed algorithm to reproduce the desired spatio-temporal spike pattern. Unless stated otherwise, the basic parameter settings are: the number of input neurons is 500 and one output neuron. Initially, the synaptic weights are generated as the uniform distribution in the interval $(0, 0.2)$. Every input spike train and desired output spike train are generated randomly by a homogeneous Poisson process with rate $r = 20$ Hz and $r = 50$ Hz respectively within the time interval of $[0, T]$, and we set $T = 100$ ms here. The results

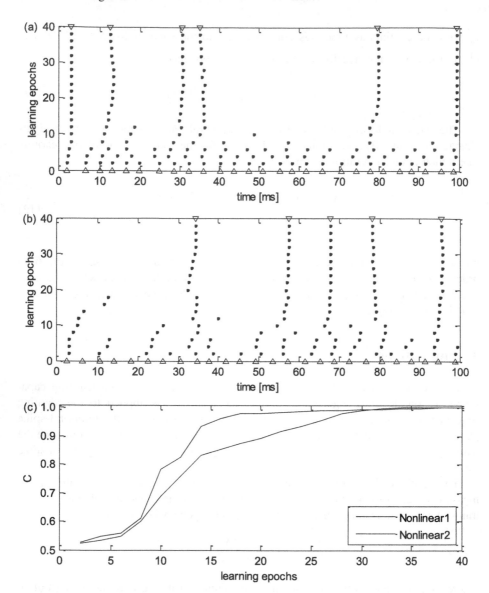

Fig. 1. The learning process of our algorithm. (a) The complete learning process of Nonlinear1. Δ, initial actual output spike train; ∇, desired output spike train; •, actual output spike trains at some learning epochs. (b) The complete learning process of Nonlinear2. (c) The evolution of the learning accuracy of Nonlinear1 and Nonlinear2 with measure C.

are averaged over 50 trials, and on each testing trial the learning algorithm is applied for a maximum of 500 learning epochs or until the measure $C = 1$. The neurons are described by the short term memory SRM [9] so that only the last firing spike contributes to the refractoriness. The parameter values of the SRM neurons used in the experiments are: the time constant of postsynaptic potential $\tau = 2$ ms, the time constant of refractory period $\tau_R = 50$ ms, the neuron threshold $\theta = 1$, and the length of the absolute refractory period $t_R = 1$ ms. The learning rate η of our algorithm is 0.0005.

Figure 1 shows the learning process of our algorithm. Figure 1(a) and (b) show the complete learning process of Nonlinear1 and Nonlinear2 respectively, which include the desired output spike train, the initial output spike train and the actual output spike trains at some learning epochs during the learning process. The evolution of the learning accuracy of Nonlinear1 and Nonlinear2 with measure C in the time interval T is represented in Fig. 1(c). For both Nonlinear1 and Nonlinear2, the measure C increases rapidly at the beginning of the learning, and reached to 1 after 38 and 36 learning epochs, respectively.

Figure 2 shows the learning results with different number of synaptic inputs. The number of synaptic inputs increases from 100 to 1000 with an interval of 100, while the

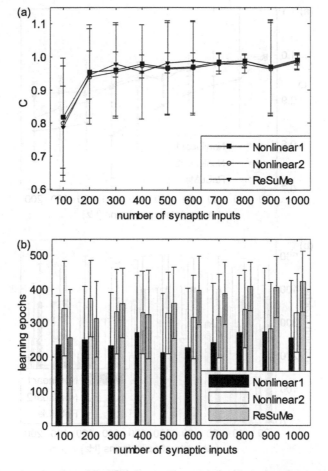

Fig. 2. The learning results with different number of synaptic inputs for our algorithm and ReSuMe algorithm after 500 learning epochs. (a) The learning accuracy with measure C. (b) The learning epochs when the measure C reaches the maximum value.

other settings remain the same. In this experiment, we compare our method with the ReSuMe algorithm [10]. The parameters in the ReSuMe algorithm are the same as those described by Ponulak et al. [10]. The learning rate η of ReSuMe algorithm is 0.005. Figure 2(a) shows the learning accuracy after 500 learning epochs and Fig. 2(b) shows the learning epochs when the measure C reaches the maximum value. Through Fig. 2(a) we can see that both Nonlinear1, Nonlinear2 and ReSuMe can learn with high accuracy. The learning accuracy of our algorithm and ReSuMe increase when the number of synaptic inputs increases gradually. Our algorithm has higher learning accuracy than ReSuMe except the numbers of synaptic inputs are 300, 500 and 600. For example, the measure $C = 0.9792$ for Nonlinear1, $C = 0.9719$ for Nonlinear2 and $C = 0.9552$ for ReSuMe when the number of synaptic inputs is 400. When the number of synaptic inputs is 1000, the measure $C = 0.9916$ for Nonlinear1, $C = 0.9854$ for

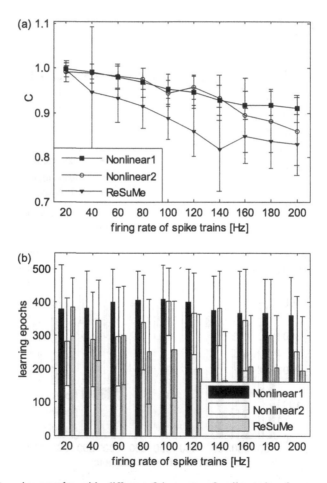

Fig. 3. The learning results with different firing rate of spike trains for our algorithm and ReSuMe algorithm after 500 learning epochs. (a) The learning accuracy with measure C. (b) The learning epochs when the measure C reaches the maximum value.

Nonlinear2 and $C = 0.9871$ for ReSuMe. From Fig. 2(b) we can see that both Nonlinear1 and Nonlinear2 have less learning epochs than ReSuMe. For example, the learning epochs of Nonlinear1 and Nonlinear2 are 213.66 and 328.98 respectively, which are less than 357.70 of ReSuMe when the number of synaptic inputs is 500. When the number of synaptic inputs is 1000, the learning epochs of Nonlinear1 and Nonlinear2 are 256.72 and 331.18 respectively, which are less than 422.02 of ReSuMe.

Figure 3 shows the learning results with different firing rate of spike trains. The firing rate of input and desired output spike trains increases from 20 Hz to 200 Hz with an interval of 20 Hz and the firing rate of input spike trains equals to that of desired output spike trains, while the other settings remain the same. Figure 3(a) shows the learning accuracy after 500 learning epochs and Fig. 3(b) shows the learning epochs when the measure C reaches the maximum value. Through Fig. 3(a) we can see that our algorithm can learn with higher accuracy than ReSuMe algorithm. The learning accuracy of our algorithm and ReSuMe algorithm decrease when the firing rate of input and desired output spike trains increases gradually. For example, the measure $C = 0.9530$ for Nonlinear1, $C = 0.9440$ for Nonlinear2 and $C = 0.8870$ for ReSuMe when the firing rate of input and desired output spike trains is 100 Hz. When the firing rate of input and desired output spike trains is 200 Hz, the measure $C = 0.9093$ for Nonlinear1, $C = 0.8581$ for Nonlinear2 and $C = 0.8289$ for ReSuMe.

5 Conclusions

In this paper, we introduced a new supervised learning algorithm based on nonlinear inner products of spike trains for spiking neurons. Our learning algorithm uses the nonlinear inner products of spike trains to deduce the learning rule. The inner products of spike trains have been used on similarity measures theory for spike trains, but less used for supervised learning algorithms of SNNs at present. The synaptic weight modification rules only depend on the input, actual output and desired output spike trains and do not depend on the specific dynamic of the neuron model. The algorithm is tested on learning sequences of spikes and compared with ReSuMe algorithm. The experimental results indicate that our method is an effective supervised multi-spike learning algorithm for spiking neurons. It can learn spike trains with high accuracy.

Acknowledgement. This work is supported by the National Natural Science Foundation of China (Nos. 61165002 and 61363058), the Natural Science Foundation of Gansu Province of China (No. 1506RJZA127), and Scientific Research Project of Universities of Gansu Province (No. 2015A-013).

References

1. Quiroga, R.Q., Panzeri, S.: Principles of Neural Coding. CRC Press, Boca Raton (2013)
2. Ghosh-Dastidar, S., Adeli, H.: Spiking neural networks. Int. J. Neural Syst. **19**(4), 295–308 (2009)

3. Kasiński, A., Ponulak, F.: Comparison of supervised learning methods for spike time coding in spiking neural networks. Int. J. Appl. Math. Comput. Sci. **16**(1), 101–113 (2006)
4. Lin, X., Wang, X., Zhang, N., et al.: Supervised learning algorithms for spiking neural networks: a review. Acta Electronica Sinica **43**(3), 577–586 (2015)
5. Bohte, S.M., Kok, J.N., La Poutré, J.A.: Error-backpropagation in temporally encoded networks of spiking neurons. Neurocomputing **48**(1), 17–37 (2002)
6. Gerstner, W., Kistler, W.M.: Spiking Neuron Models: Single Neurons, Populations, Plasticity. Cambridge University Press, Cambridge (2002)
7. Shrestha, S.B., Song, Q.: Adaptive learning rate of SpikeProp based on weight convergence analysis. Neural Networks **63**, 185–198 (2015)
8. Ghosh-Dastidar, S., Adeli, H.: A new supervised learning algorithm for multiple spiking neural networks with application in epilepsy and seizure detection. Neural Networks **22**(10), 1419–1431 (2009)
9. Xu, Y., Zeng, X., Han, L., et al.: A supervised multi-spike learning algorithm based on gradient descent for spiking neural networks. Neural Networks **43**, 99–113 (2013)
10. Ponulak, F., Kasinski, A.: Supervised learning in spiking neural networks with ReSuMe: sequence learning, classification, and spike shifting. Neural Comput. **22**(2), 467–510 (2010)
11. Ponulak, F.: Analysis of the ReSuMe learning process for spiking neural networks. Int. J. Appl. Math. Comput. Sci. **18**(2), 117–127 (2008)
12. Sporea, I., Grüning, A.: Supervised learning in multilayer spiking neural networks. Neural Comput. **25**(2), 473–509 (2013)
13. Taherkhani, A., Belatreche, A., Li, Y., et al.: DL-ReSuMe: a delay learning-based remote supervised method for spiking neurons. IEEE Trans. Neural Netw. Learn. Syst. **26**(12), 3137–3149 (2015)
14. Taherkhani, A., Belatreche, A., Li, Y., et al.: Multi-DL-ReSuMe: multiple neurons delay learning remote supervised method. In: 2015 International Joint Conference on Neural Networks (IJCNN), pp. 1–7. IEEE (2015)
15. Mohemmed, A., Schliebs, S., Matsuda, S., et al.: SPAN: spike pattern association neuron for learning spatio-temporal spike patterns. Int. J. Neural Syst. **22**(4), 786–803 (2012)
16. Mohemmed, A., Schliebs, S., Matsuda, S., et al.: Training spiking neural networks to associate spatio-temporal input–output spike patterns. Neurocomputing **107**, 3–10 (2013)
17. Yu, Q., Tang, H., Tan, K.C., et al.: Precise-spike-driven synaptic plasticity: learning hetero-association of spatiotemporal spike patterns. PLoS One **8**(11), e78318 (2013)
18. Yu, Q., Tang, H., Tan, K.C., et al.: A brain-inspired spiking neural network model with temporal encoding and learning. Neurocomputing **138**, 3–13 (2014)
19. Park, I.M., Seth, S., Rao, M., et al.: Strictly positive-definite spike train kernels for point-process divergences. Neural Comput. **24**(8), 2223–2250 (2012)
20. Paiva, A.R.C., Park, I., Principe, J.C.: Inner products for representation and learning in the spike train domain. In: Statistical Signal Processing for Neuroscience and Neurotechnology, vol. 8 (2010)
21. Gütig, R., Aharonov, R., Rotter, S., et al.: Learning input correlations through nonlinear temporally asymmetric Hebbian plasticity. J. Neurosci. **23**(9), 3697–3714 (2003)
22. Schreiber, S., Fellous, J.M., Whitmer, D., et al.: A new correlation-based measure of spike timing reliability. Neurocomputing **52**, 925–931 (2003)

Behavior Prediction for Ochotona curzoniae Based on Wavelet Neural Network

Haiyan Chen[1,2], Aihua Zhang[1(✉)], and Shiya Hu[1]

[1] College of Electronic and Information Engineering,
Lanzhou University of Technology, Lanzhou, China
chenhaiyan@sina.com, {lutzhangah,hushiyaip}@163.com
[2] School of Computer and Communication,
Lanzhou University of Technology, Lanzhou, China

Abstract. Ochotona curzoniae is one of the main biological disasters in the Qinghai-Tibet plateau and adjacent areas in China. Video-based animal behavior analysis is a critical and fascinating problem for both biologists and computer vision scientists. The behavior prediction for Ochotona curzoniae is a basis of Ochotona curzoniae behavior analysis in video recordings. In this paper, a three-layer wavelet neural network is proposed for short-term Ochotona curzoniae behavior prediction. A commonly used Morlet wavelet has been chosen as the activation function for hidden-layer neurons in the feed-forward neural network. In order to demonstrate the effectiveness of the proposed approach, short-term prediction of Ochotona curzoniae behavior in the natural habitat environment is performed, and we analyze the influence on prediction accuracy at various numbers of input neurons. The forecasted results clearly show that wavelet neural network has good prediction properties for Ochotona curzoniae behavior prediction compared with BP neural network. The model can assist biologists and computer vision scientists to create an effective animal behavior analysis method. The principle of our Ochotona curzoniae behavior prediction used wavelet neural network is helpful to other animal behavior prediction and analysis in video recordings.

Keywords: Ochotona curzoniae · Behavior prediction · Wavelet neural network

1 Introduction

Ochotona curzoniae is one of the main creature calamities in Qinghai-Tibet Plateau and its neighborhood area in China. The quantity and population density of Ochotona curzoniae are the reflection of deterioration degree of grassland ecology. It is a critical and interesting problem for biologists to study Ochotona curzoniae behavior for Ochotona curzoniae prevention, grassland protection and ecology protection [1]. Meanwhile, in recent years, the research on Ochotona curzoniae's biology characteristic, morphology, classification, physiology, ecology, behavior, population dynamics and harm for livestock production have caused extensive concern on the zoologists [2]. But the traditional study way to do these is first observe and investigate Ochotona

© Springer International Publishing Switzerland 2016
D.-S. Huang and K.-H. Jo (Eds.): ICIC 2016, Part II, LNCS 9772, pp. 105–116, 2016.
DOI: 10.1007/978-3-319-42294-7_9

curzoniae by biologists in wild inhabiting area for a period of time, then biologists and human observers analyze Ochotona curzoniae's behavior manually. This is a time and labor consuming process, and the observation results of Ochotona curzoniae behavior vary among different observers. Moreover, in biological field, it is a very fundamental problem to understand the habits and characteristics of Ochotona curzoniae. With the development of intelligent surveillance technologies, it has become increasingly popular to study animal behaviors with the assistance of video recordings [3]. This is because video recordings not only can easily gather information about many aspects of the situation in which humans or animals interact with each other or with the environment, but also the video recordings make offline research possible.

Animal behavior study originates from fields including biology, physiology, psychology, neuroscience and pharmacology, toxicology, entomology, animal welfare, and so on [3]. To the best of our knowledge, Ochotona curzoniae's behavior study is still conducted using the traditional way up to now. With the development of intelligent surveillance technologies, we can study Ochotona curzoniae behaviors with the assistance of video recordings.

Over the past years, artificial intelligence has been widely used in predicting of various time variables [4]. Especially, Artificial Neural Network, which is a nonlinear computing approach incented by the learning process of brain has been widely accepted as one of the effective means for modeling a complex time series. Wavelet transform which is an alternative data preprocessing technique has showed excellent performance in time series modeling in the recent years due to its ability to analyze a signal in both time and frequency [5], and overcomes the basic limitations of conventional Fourier transform. Wavelet neural network has the best function approximation ability because the model constricting algorithm is different from common artificial neural network algorithm, it can effectively overcome intrinsic drawbacks of common artificial neural network. Therefore the better prediction performance can be obtained effectively. There are some successful applications of wavelet neural network models which are prepared by the combined use of wavelet transform and neural networks. Lan Gao predicted chaotic time series based on wavelet neural network [6]. A wavelet neural network model is proposed for monthly reservoir inflow prediction by combining the discrete wavelet transform and Levenberg-Marquardt optimization algorithm based feed forward neural networks. Ramana predicted monthly rainfall by combining the wavelet technique with artificial neural network [7]. A novel nonparametric dynamic time delay recurrent wavelet neural network model is presented for forecasting traffic flow in reference [8]. Qulin TAN applied the improved BP neural network model, auxiliary wavelet neural network model and embedded wavelet neural network model to the settlement prediction in one practical engineering monitoring project [9]. An adaptive wavelet neural network is proposed for short-term price forecasting by Pindoriya in the electricity markets [10].

Although wavelet neural network had been used extensively as useful tools for forecasting of various time series variables, as far as we know, wavelet neural network has not been applied to the prediction of animal behavior so far, not been mentioned to Ochotona curzoniae behavior prediction. In this paper, a wavelet neural network model is established based on wavelet artificial neural network theory. We apply this model in a specific behavior prediction system to predict Ochotona curzoniae behavior. The main contributions of our work as follows:

- We first define the basic behavior unit of Ochotona curzoniae including eating running and resting and so on. Then we descript what a basic behavior unit is.
- We subsample the video frames by sample period T and analysis the influence on prediction performance by various sample period T.
- We construct a three-layer wavelet neural network and train the network with Ochotona curzoniae behavior data and predict Ochotona curzoniae behavior by our well trained wavelet neural network.

2 Basic Behavior Units

2.1 The Definition of Basic Behavior Units

To predict the behavior of Ochotona curzoniae, we first need to define the behavior of Ochotona curzoniae and what a basic behavior unit is. The basic behavior units (BBUs) are the behavior primitives of Ochotona curzoniae and higher level analysis will be carried out in terms of these. A BBU can be defined as an activity of Ochotona curzoniae that remains consistent within a period of time, and that can be represented by a set of spatiotemporal image features. Because once the BBUs are recovered from the video sequences of Ochotona curzoniae, we can go further to do higher level analysis, such as analyze complex behaviors of Ochotona curzoniae, which consist of BBUs with spatiotemporal constraints, and influencing factors from the basic behavior model of Ochotona curzoniae.

Our goal is to be able to determine BBUs of Ochotona curzoniae from the features obtained from the previous step at a reasonable accuracy. Domain knowledge of Ochotona curzoniae would always be required as a pre-requisite step before any BBU of Ochotona curzoniae discovery can be done. An appropriate BBU set of Ochotona curzoniae must be found for the particular modeling approach. In our work, a set of BBUs of interest of Ochotona curzoniae are classified by domain experts that might include: running, eating and resting.

Fig. 1. The basic behavior units

We are also interested in such BBUs of Ochotona curzoniae as running, eating, and resting. These behaviors (shown in Fig. 1) can be characterized as follows:

- Running. The body of Ochotona curzoniae stretched. The body and limbs of Ochotona curzoniae as a whole move quickly.
- Eating. The Ochotona curzoniae bends its head down to the ground with slight body motion.
- Resting. The body of Ochotona curzoniae has no obvious movement. The body and limbs of Ochotona curzoniae does not move.

2.2   The Acquisition for Behavior Subsequence

Basic behavior units of Ochotona curzoniae are the primitive actions or activities that happen during a short period of time. So a basic behavior unit of Ochotona curzoniae usually last a period of time. It is not necessary to confirm the behavior for every video frames. We classified the video frames into sequence of BBUs by the sample period T. The selection of T is critical for Ochotona curzoniae behavior analysis. When the sample period T is too small, it is part of a BBU of Ochotona curzoniae, so the sample period T must be large enough to contain one BBU of Ochotona curzoniae. The idea behind this is that the subsequence could encode a time trend which accounts for temporal changes of Ochotona curzoniae behavior [3], as shown in Fig. 2. We used three different sample periods T to analyze the influence on behavior prediction performance, which is selected after trying different lengths and slides one frame at a time in the experiments. We chose T empirically in such a way that it is not too large or too small. If T is too large, it may mix different BBUs of Ochotona curzoniae together, while if it is too small, it would be unable to catch the time trend of Ochotona curzoniae behavior. The choice of T depends on the specific application.

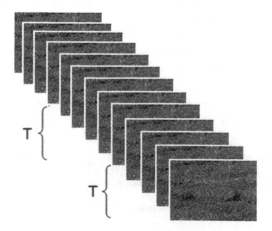

Fig. 2. The demonstration of video frame subsequence

3 The Behaviour Prediction Based on Wavelet Neural Network

3.1 The Construction of Wavelet Neural Networks

Although the concept of artificial neurons was first introduced in 1943 [11], the major applications of artificial neural network have emerged only since the development of the back-propagation method of training in 1986 [12]. The multilayer perceptron trained with BP algorithm is one of the most popular artificial neural network architecture. A multilayer perceptron network can be composed of an input layer, one or more hidden layers of computation nodes and an output layer [13]. Wavelet neural network absorbs the advantages of high resolution of wavelets and the advantages of learning and feed forward of neural network. Wavelet neural network is a kind of neural network based on BP neural network topological structure, which is used the wavelet basic function as the transfer function of the hidden layer nodes. It has two processes that are forward computing (forward propagation) of data stream and backward propagation of error signals. The topological structure of the wavelet neural network we constructed in this paper is shown in Fig. 3.

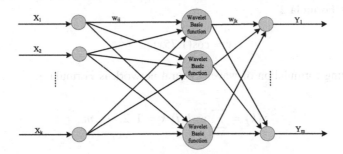

Fig. 3. The topological structure of WNN

We design the wavelet neural network according to the characteristic of Ochotona curzoniae behavior. The designed wavelet neural network including three layers separately is input layer, hidden layer and output layer. The algorithmic flow of our designed wavelet neural network is described as Fig. 4.

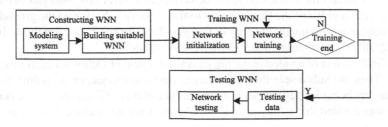

Fig. 4. The algorithmic flow of wavelet neural network

In the wavelet neural network, $X_1, X_2, \cdots, X_k$ are the parameters of input layer, $Y_1, Y_2, \cdots, Y_m$ are the prediction output of output layer in the wavelet neural network, w_{ij} and w_{jk} are the weight of wavelet neural network. $x_i(i = 1, 2, \cdots, k)$ are the input signal sequence. The computing formulation of hidden layer output is described as Formula 1.

$$h(j) = h_j \left[\frac{\sum_{i=1}^{k} w_{ij} - b_j}{a_j} \right] j = 1, 2, \cdots, l \tag{1}$$

In Formula 1, $h(j)$ is the ith node output of hidden layer, w_{ij} is the connection weight between the input layer and the hidden layer, b_j is the translation factor of wavelet basic function h_j, a_j is the scale factor of wavelet basic function h_j, h_j is the wavelet basic function. In this paper, we use Morlet mother wavelet basic function as wavelet basic function in Fig. 3. The choice of the mother wavelet depends on the data to be analyzed [7, 14]. The commonly used mother wavelets are the Daubechies and Morlet wavelet transforms [15]. Morlet wavelets have a more consistent response to similar events, the mathematical formulation of Morlet mother wavelet basic function described as Formula 2.

$$y = \cos(1.75x)e^{-x^2/2} \tag{2}$$

The computing formulation of wavelet neural network is Formula 3.

$$y(k) = \sum_{i=1}^{l} w_{ik}h(i) \ k = 1, 2, \cdots, m \tag{3}$$

In Formula 3, w_{ik} is the weight from hidden layer to output layer, $h(i)$ is the output of ith hidden layer node, l is the number of hidden layer nodes. m is the number of output layer nodes.

3.2   Study Area and Data Collection

The behavior study of Ochotona curzoniae indicates that the behavior of Ochotona curzoniae at some time is relevant to the behaviors at several time intervals before. We collect a section video of the Ochotona curzoniae activity behavior on the ground in the natural habitat environment, locating in Qinghai-Tibet Plateau, east longitude 101°35′ 36″–102°58′15″, northern latitude 33°58′21″–34°48′48″.

We firstly select of a video including an activity cycle of Ochotona curzoniae on the ground. Then we subsample the video frames into a subsequence according the principle that any behavior pattern can last for a period of time. Given the sample period as T, we adopt a video that totally are 5377 frames and we use various T to sample video frames, T separately are 6, 11 and 16. The behavior pattern in each behavior unit is marked by a behavior label. The input data of input layer is the n behavior labels of the n Ochotona curzoniae behavior units before current time. The output data of output layer is the predicted behavior label at the current time. The Ochotona curzoniae behavior prediction can be regarded as using n behavior labels to predict one behavior label, so the output node number of wavelet neural network is 1. The number of hidden layer nodes is six according to the experience value and an optimal value is determined by experiment. The number of input is n, indicating that we use n consecutive behavior labels before predicted behavior label to forecast current behavior label, n separately are 2, 3, 4 and 5.

3.3 The Prediction Results Based on Wavelet Neural Network

In our Ochotona curzoniae behavior prediction experiment, we compare the prediction results to the ground truth to test the prediction accuracy. The ground truth is labeled manually for each frame in behavior subsequence. The number of input and output nodes is determined by the property of the actual input and output variables. However, the number of hidden nodes rests with the complexity of the mathematical property of the problem and is determined usually by trial and error and the modeler. In order to obtain the optimal prediction results, we select the optimal number of hidden layer neurons according to the following principle

$$l < n - 1 \tag{4}$$

$$l < \sqrt{(m+n)} + a \tag{5}$$

$$l = \log_2^n \tag{6}$$

In formulas (4)–(6), l is the number of hidden layer neurons, n is the number of input layer neurons, m is the number of output layer neurons, a is a constant between 0–10. We determined the optimal number of hidden layer neurons according to above principle and experiment.

We select sample period T as 6 firstly, that is to say each behavior unit last for 6 frames. When T = 6, we can totally obtain 1075 behavior label data, we use 715 data to train wavelet neural network and 360 data to test wavelet neural network. The number of input layer nodes separately are 2, 3, 4 and 5, namely 2, 3, 4 and 5 behavior labels before current behavior unit separately are used to predict current behavior label. By the experiments we determine the optimal number of hidden layer neuron separately are 3, 2, 2, 2 when n = 2, 3, 4, 5. Figure 5 is the prediction results of Ochotona curzoniae behavior at T = 6, n = 2, 3, 4, 5.

When sample period T is 11, we determine the optimal number of hidden layer neuron separately are 5, 3, 4, 2 when n = 2, 3, 4, 5 by the experiments, the sample period is increased compared to T = 6, which means the basic behavior unit last for a long time, a behavior unit is likely to contain other behavior pattern. In addition, with

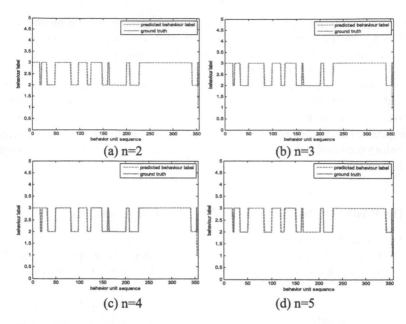

Fig. 5. The behavior prediction results based on WNN (T = 6, n = 2, 3, 4, 5)

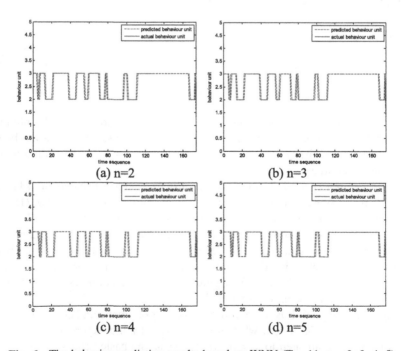

Fig. 6. The behavior prediction results based on WNN (T = 11, n = 2, 3, 4, 5)

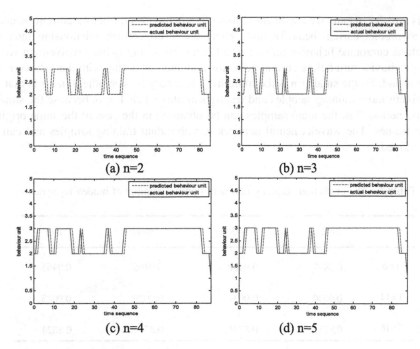

Fig. 7. The behavior prediction results based on WNN (T = 16, n = 2, 3, 4, 5)

the increasing of the sample period T, the training data and test data are reducing. So the prediction accuracy rate is decreasing, but on the whole, the prediction accuracy rate are above 90 % when T = 11 and they are still higher.

When sample period T is 16, we determine the optimal number of hidden layer neuron separately are 3, 2, 3, 3 when n = 2, 3, 4, 5 by the experiments, the sample period is increased compared to T = 11, which means the basic behavior unit last for a more long time, and the current behavior need to be predicted using more behavior units ahead. Although behavior units have stronger time relevance, the current behavior unit is only relevant to a few of succession behavior units before current time, it has nothing with the behavior units more ahead, so the prediction accuracy rates is decreasing. In addition, with the increasing of the sample period T, the training data and test data are reducing. So the prediction accuracy rate is decreasing, the prediction accuracy rate are below 90 % when T = 16. With the increase of n, the prediction accuracy rates also comply with the principle of decline trend basically (Figs. 6 and 7).

We adopt prediction accuracy rate to evaluate prediction performance. The definition of prediction accuracy rate is the ratio that the number of accuracy prediction behavior labels compared to the total number of behavior labels. Table 1 is the prediction accuracy rate with the optimal number of hidden layer neuron. From Table 1, we can get the information that with the number of input nodes increasing, the prediction accuracy rate is decreasing. But for all that, the prediction accuracy rate are above 90 % when T = 6 and T = 11. There are two main reasons for these results. One is n is smaller, means we use fewer consecutive behavior unit before current behavior unit to predict current behavior unit. There are stronger time relationship inter the Ochotona curzoniae behavior units, current behavior unit only has involvement with a few of behavior unit before current behavior unit and has noting with the behavior unit more ahead. So the smaller n is the more high accuracy is. The other reason is that we can obtain more training samples and testing samples when T = 6, because the smaller sample period T is, the more samples can be obtained in the case of the same original video frames. The wavelet neural network has abundant training samples and can be well trained.

Table 1. The prediction accuracy rate with optimal number of hidden layer neuron

	N=2	n=3	n=4	n=5
T=6	0.9497	0.9468	0.9466	0.9465
T=11	0.9040	0.9034	0.9029	0.9023
T=16	0.8750	0.8736	0.8761	0.8824

3.4 The Prediction Results Based on Basic BP Neural Network

In order to illustrate the feasibility effectiveness of Ochotona curzoniae behaviour prediction by wavelet neural network, we compare the prediction results of wavelet neural network with those of fundamental BP neural network. We also adopt three-layer BP neural network that the structure is uniform to the wavelet neural network we constructed in Sect. 3.1. In this paper, we use sigmoid function as activation function. Figure 8 are the prediction results of Ochotona curzoniae behavior at T = 6, n = 2, 3, 4, 5. Table 2 are the prediction accuracy rates used fundamental BP neural network with the optimal number of hidden layer neuron. From Table 2, we can get the information that with the number of input nodes increasing, the prediction accuracy rate is decreasing. But for all that, the prediction accuracy rate are above 90 % when T = 6 and T = 11. Compared with the prediction results used wavelet neural network, some prediction accuracy rates are lower. We can draw the conclusions that the prediction accuracy rates used BP neural network are either equal to the accuracy rates or lower than the accuracy rates used wavelet neural network. So we can obtain more accuracy results in Ochotona curzoniae behaviour prediction by wavelet neural network.

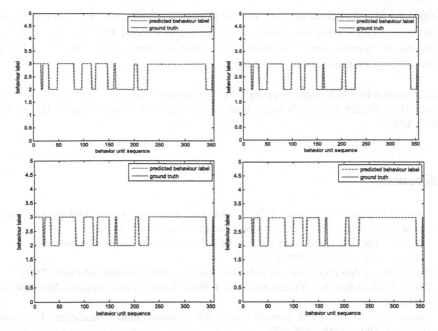

Fig. 8. The behavior prediction results based on BP neural network (T = 6, n = 2, 3, 4, 5)

Table 2. The prediction accuracy rate with BP neural network

	n=2	n=3	n=4	n=5
T=6	0.9469	0.9468	0.9466	0.9465
T=11	0.9040	0.9034	0.9029	0.9023
T=16	0.8750	0.8736	0.8721	0.8235

4 Conclusions

Behavior forecasting of Ochotona curzoniae provides a kind of intelligent method for behavior analysis. The study presented the application of wavelet neural network models for prediction of Ochotona curzoniae behavior in natural habitat environment. The proposed model is a combination of wavelet analysis and artificial neural network. In this paper, our work mainly is for short-term Ochotona curzoniae behavior forecasting because short-term Ochotona curzoniae activity data are easily available to the authors. In order to illustrate the feasibility for Ochotona curzoniae behaviour prediction by wavelet neural network, we compare the prediction results of wavelet neural network with those of fundamental BP neural network. By the analyzing the prediction results used wavelet neural network and the prediction results used BP neural network, we draw a conclusion that it is feasible and effective to forecast the Ochotona curzoniae behaviour by wavelet neural network. We make an attempt for animal intelligent behavior analysis in the video recording.

Acknowledgement. This paper supported by the national natural science fund of China (61362034, 81360229, 61265003) and natural science fund of Gansu province (1310RJY020, 148RJZA020).

References

1. Chen, H.Y., Cao, M.H., Wang, H.Q.: A scheme on the recognition of Ochotona curzoniae based on image process. J. Acta Agrestia Sinica **19**(3), 516–519 (2011)
2. Liu, F.Y., Liu, R.T.: The latest progress of the research situation on Ochotona curzoniae. J. Gard. Agric. **3**, 30–31 (2002)
3. Xue, X.W.: Video-Based Animal Behavior Analysis. The University of Utah (2009)
4. Zhang, Q.H., Albert, B.: Wavelet networks. J. IEEE Trans. Neural Networks **3**(6), 889–898 (1992)
5. Umut, O.: Wavelet neural network model for reservoir inflow prediction. J. Scientia Iranica A. **19**(6), 1445–1455 (2012)
6. Gao, L., Lu, L., Li, Z.: Prediction of chaotic time series based on wavelet neural network. In: MTS/IEEE Conference and Exhibition on Oceans, vol. 4, pp. 2046–2050. IEEE (2001)
7. Ramana, R.V., Krishna, B., Kumar, S.R., et al.: Monthly rainfall prediction using wavelet neural network analysis. J. Water Resour. Manag. **27**(10), 3697–3711 (2013)
8. Jiang, X., Adeli, H.: Dynamic wavelet neural network model for traffic flow forecasting. J. Transp. Eng. **131**(10), 771–779 (2014)
9. Tan, Q.L., Jian, W., Hu, J.P.: Applications of wavelet neural network model to building settlement prediction: a case study. J. Sens. Transducers **169**(4), 1–8 (2014)
10. Pindoriya, N.M., Singh, S.N., Singh, S.K.: An adaptive wavelet neural network-based energy price forecasting in electricity markets. J. IEEE Trans. Power Syst. **23**(3), 1423–1432 (2008)
11. Warren, M.C., Pitts, W.: A logical calculus of ideas immanent in nervous activity. J. Bull. Math. Biophys. **5**(4), 115–133 (1943)
12. Rumelhart, D.E., James, M.C.: Parallel Distributed Processing: Explorations in the Microstructure of Cognition. MIT Press, Cambridge (1986)
13. Pal, S.K., Mitra, S.: Multilayer perceptron, fuzzy sets, and classification. J. IEEE Trans. Neural Networks **3**(5), 683–697 (1992)
14. Krishna, B., Rao, Y., Nayak, P.: Time series modeling of river flow using wavelet neural networks. J. Water Resour. Prot. **3**(1), 50–59 (2011)
15. Benaouda, D., Murtagh, F., Starck, J.L., Renaud, O.: Wavelet-based nonlinear multiscale decomposition model for electricity load forecasting. J. Neurocomputing **70**, 139–154 (2006)

New Filter Design for Static Neural Networks with Mixed Time-Varying Delays

Guoquan Liu$^{(\boxtimes)}$, Shumin Zhou, Xianxi Luo, and Keyi Zhang

Jiangxi Province Engineering Research Center of New Energy Technology
and Equipment, East China University of Technology, Nanchang 330013, China
gqlecit@hotmail.com

Abstract. This paper focuses on designing a H_∞ filter for a class of static neural networks with mixed time-varying delays. Here the mixed time-varying delays contain both discrete and distributed time-varying delays. Based on a Lyapunov-Krasovskii functional combined with a zero equation, a suitable H_∞ filter is obtained for the static neural networks model. The filter can be solved by a linear matrix inequality (LMI). Two numerical examples are presented to validate the proposed method. In addition, the obtained filter can be applied to design the control systems with delays.

Keywords: Filter design · Static neural networks · Lyapunov-Krasovskii functional · Distributed time-varying delay

1 Introduction

In [1], Xu et al. studied a comparison theory for two modeling approaches in neural networks (NNS) system, then the authors further pointed out that NNS system can be classified as local field NNS model or as static neural networks (SNNS) model. During the past decade, the dynamic analysis for SNNS have been focused upon by researchers due to their more real application, such as rainfall forecasting [2], long-term wind speed and power forecasting [3]. Since time delay is unavoidable in implementation of NNS and is often an important source of oscillation and instability, the stability of SNNS with time delays has been become a high priority topic and obtained developed results in reference in the past several years. For example, Cao and Li [4] discussed the global exponential stability problem for static recurrent NNS by constructing a proper Lyapunov functional. Shao [5] solved the delay-dependent stability for SNNS with time-varying delays by linear matrix inequalities techniques. Wang and Wang [6] investigated the Global asymptotic robust stability for SNNS with S-type distributed delays. Based on the Lyapunov-Krasovskii functional approach, the free-weighting matrix method, and the Jensen integral inequality, the delay-dependent globally exponential stability conditions for SNNS were obtained in [7].

As reported in [8–10], an effective way for realizing integrated control systems is to interconnect the spatially distributed components by a computer communication

© Springer International Publishing Switzerland 2016
D.-S. Huang and K.-H. Jo (Eds.): ICIC 2016, Part II, LNCS 9772, pp. 117–129, 2016.
DOI: 10.1007/978-3-319-42294-7_10

network. Such a network introduces randomly varying delays which can degrade the system dynamic performance and be a source of potential instability. The randomly varying delays are also called as distributed delays. In the recent years, the NNS model with distributed delays has received increasing attention. For example, Wang et al. [11] concerned with the problem of globally robust stability analysis for generalized NNS with parameter uncertainties and distributed delays. Li et al. [12] studied the state estimation issue for a class of recurrent NNS with distributed time-varying delays by available output measurements. Lakshmanan et al. [13] obtained a new state estimator for NNS with leakage and distributed delays by a new LKF. Recently, some new results on SNNS with time delays have been published see, e.g. [15–17, 19]. Huang et al. [19] discussed the guaranteed H∞ performance state estimation problem of delayed SNNS via a double-integral inequality and obtained a suitable gain matrix. However, the filter design for SNNS with distributed delays has not been fully studied, there is more space for us and it inspires us to study it. Therefore, it is important to consider the filter issue for SNNS with distributed delays.

Based on the analysis above, this paper studies the filter issue for a class of SNNS with mixed time-varying delays. The mixed time-varying delays contain both discrete and distributed time-varying delays. Via a Lyapunov-Krasovskii functional combined with a zero equation, a suitable H_∞ filter is obtained for the SNNS model. The filter can be solved by an LMI. Two examples show the proposed conditions are effective.

2 Problem Description

Throughout this paper, $\Re^n$ and $\Re^{n \times n}$ denote the n-dimensional Euclidean space and the set of all $n \times n$ real matrices, respectively; for a real symmetric matric Z, $Z > 0$ means that Z is a real symmetric positive definite; "$*$" denotes the term that is induced by symmetry.

Consider the following SNNS with mixed time-varying delays:

$$
\begin{cases}
\dot{x}(t) = -Ax(t) + f(Wx(t - \tau(t)) + J) + B \int_{t-r(t)}^{t} f(x(s))ds + C_1 w(t), \\
y(t) = Cx(t) + Dx(t - \tau(t)) + Ex(t - r(t)) + C_2 w(t), \\
Z(t) = Hx(t), x(t) = \phi(t),\ \forall t \in [-\mu, 0],\ \mu = \max\{\tau, r\},
\end{cases} \tag{1}
$$

where $A = diag\{a_1, a_2, \ldots, a_n\}$ is a constant diagonal matrix with $a_i > 0$, $x(t) = [x_1(t), x_2(t), \ldots, x_n(t)] \in \Re^n$ is the state vector of the neural networks, $\tau(t)$ and $r(t)$ are the discrete delay and the distributed delay, $f(Wx(\cdot)) = [f_1(W_1 x(\cdot)), f_2(W_2 x(\cdot)), \ldots, f_n(W_n x(\cdot))]^T \in \Re^n$ is the activation function, $W = [W_1, W_2, \ldots, W_n] \in \Re^n$ is the delayed connection weight matrix $J = [J_1, J_2, \ldots, J_n]^T \in \Re^n$ is an exogenous input vector, $f(x(t)) = [f_1(x_1(t)), f_2(x_2(t)), \ldots, f_n(x_n(t))]^T \in \Re^n$ denotes the continuous activation function, $w(t)$ is the noise input belonging to $\mathcal{L}_2 \in [0, \infty]$, $y(t) \in \Re^m$ is the network output measurement, B is a distributed connection matrix, C, D, E, C_1, C_2 and H are known real constant matrices with appropriate dimensions. $Z(t) \in \Re^q$, to be estimated, is a linear combination of state, $\phi(t)$ is the initial condition known on $t \in [-\mu, 0]$.

For system (1), the following assumptions are given:

Assumption 2.1. The time-varying delays $\tau(t)$ and $r(t)$ satisfy:

$$0 \le \tau(t) \le \tau, \dot{\tau}(t) \le \tau_d, 0 < r(t) \le r, \dot{r}(t) \le r_d, \tag{2}$$

where τ, τ_d, r, and r_d are constants.

Assumption 2.2. Each neuron activation function $f_i(\cdot)(i = 1, 2, \ldots, n)$ is bound and satisfies the following condition:

$$l_i^- \le \frac{f_i(\alpha_1) - f_i(\alpha_2)}{\alpha_1 - \alpha_2} \le l_i^+, i = 1, 2, \ldots, n, \alpha_1, \alpha_2 \in \Re^n, \alpha_1 \ne \alpha_2, \tag{3}$$

where l_i^-, l_i^+ are some constants (l_i^- and l_i^+ can be positive, negative, and zero).

For the system (1), the following full order filter is constructed for the estimation of $z(t)$

$$
\begin{cases}
\dot{\hat{x}}(t) = -A\hat{x}(t) + f(W\hat{x}(t - \tau(t)) + J) + B \int_{t-r(t)}^{t} f(\hat{x}(s))ds + K(y(t) - \hat{y}(t)), \\
\hat{y}(t) = C\hat{x}(t) + D\hat{x}(t - \tau(t)) + E\hat{x}(t - r(t)), \\
\hat{Z}(t) = H\hat{x}(t), \hat{x}(t) = 0, \forall t \in [-\mu, 0], \ \mu = \max\{\tau, r\}.
\end{cases}
\tag{4}
$$

Define the filter errors $e(t) = x(t) - \hat{x}(t)$, and $z(t) = z(t) - \hat{z}(t)$. Then, the filtering error system can be obtained by combining the filter (4) and the SNNS (1), which is represented as follows:

$$
\begin{cases}
\dot{e}(t) = -(A + KC)e(t) - KDe(t - \tau(t)) - KEe(t - r(t)) + g(We(t - \tau(t))) \\
\quad + B \int_{t-r(t)}^{t} g(e(s))ds + (C_1 - KC_2)w(t), \bar{Z}(t) = He(t),
\end{cases}
\tag{5}
$$

with

$$e(t - \tau(t)) = x(t - \tau(t)) - \hat{x}(t - \tau(t)), e(t - r(t)) = x(t - r(t)) - \hat{x}(t - r(t)),$$
$$g(We(t - \tau(t))) = f((Wx(t - \tau(t)) + J) - f((W\hat{x}(t - \tau(t)) + J),$$
$$g(e(t)) = f(x(t)) - f(\hat{x}(t)).$$

Note that since each function $g_j(\cdot)$ satisfies the Assumption 2.2, hence $g_j(e_j(\cdot))$ and $g_j(W_j e(\cdot))$ satisfy

$$l_i^- \leq \frac{g_i(e_1(\alpha)) - g_i(e_2(\alpha))}{e_1(\alpha) - e_2(\alpha)} \leq l_i^+, i = 1, 2, \ldots, n, e_1(\alpha), e_2(\alpha_2) \in \Re^n, e_1(\alpha_1) \neq e_2(\alpha_2),$$

$$\Gamma_i^- \leq \frac{g_i(W_1 e_1(\alpha)) - g_i(W_2 e_2(\alpha))}{W_1 e_1(\alpha) - W_2 e_2(\alpha)} \leq \Gamma_i^+, i = 1, 2, \ldots, n, W_1 e_1(\alpha), W_2 e(\alpha_1) \in \Re^n,$$

$$W_1 e_1(\alpha_1) \neq W_2 e(\alpha_1).$$

$$(6)$$

Definition 2.3 [14–16]. Given a prescribed level of noise attenuation $\gamma > 0$, find a suitable filter such that the equilibrium point of the filtering error system (5) with $w(t) = 0$ is globally asymptotically stable and $\|z(t)\|_2 < \gamma \|w(t)\|_2$, under zero-initial conditions for all non-zero $w(t) \in \mathcal{L}_2[0, \infty]$. In this case, the filtering error system (5) is said to be globally asymptotically stable with H_∞ performance γ.

Remark 2.4. The Definition 2.1 was first proposed in [14]. Huang, et al. [15], Duan, et al. [16], Huang, et al. [17], Lakshmanan, et al. [18], and Huang, et al. [19] also adopted the same definition to proof their results.

Next, the following lemma is useful in deriving the criterion.

Lemma 2.5. For any constant matrix $\Upsilon > 0$, any scalars a and b with $a < b$, and a vector function $x(t) : [a, b] \to \Re^n$ such that the integrals concerned as well defined, the following holds

$$\left[\int_a^b x(s)ds \right]^T \Upsilon \left[\int_a^b x(s)ds \right] \leq (b - a) \int_a^b x^T(s) \Upsilon x(s)ds. \tag{7}$$

3   Filter Design

In this section, a suitable H_∞ filter is designed for the filtering error system (5).

Theorem 3.1. For given scalars τ, τ_d, r, r_d, and the filter gain matrix K, the filtering error system (5) is globally asymptotically stable with H_∞ performance γ, if there exist matrices $P > 0, R_j = R_j^T > 0, N_j = N_j^T > 0, j = 1, 2, \ldots, 5, M, G$ and diagonal matrices $T_i > 0, i = 1, 2, 3, 4$, such that the following LMI holds:

$$\Psi = [\varphi_{i,j}]_{11 \times 11} < 0, \tag{8}$$

with

$$\varphi_{1,1} = R_1 + R_2 + \tau^2 N_1 + r^2 N_2 - L_1 T_1 - T_1^T L_1^T - \Gamma_1 T_3 - T_3^T \Gamma_1^T - MA - A^T M^T$$
$$- GC - C^T G^T, \varphi_{1,2} = L_2 T_1 - GD - A^T M^T - C^T G^T,$$

$$\varphi_{1,3} = -GE - A^T M^T - C^T G^T, \varphi_{1,4} = M - A^T M^T - C^T G^T, \varphi_{1,5} = MB,$$

$$\varphi_{1,6} = MC_1 - GC_2, \varphi_{1,7} = -A^T M^T - C^T G^T, \varphi_{1,8} = \Gamma_2 T_3 - A^T M^T - C^T G^T,$$

$$\varphi_{1,9} = -A^T M^T - C^T G^T, \varphi_{1,10} = -A^T M^T - C^T G^T, \varphi_{1,11} = P - M - A^T M^T - C^T G^T,$$

$$\varphi_{2,2} = -(1 - \tau_d)R_1 - L_1 T_2 - T_2^T L_1^T - W^T \Gamma_1 T_4 W - W T_4^T \Gamma_1^T W^T - GD - D^T G^T,$$

$$\varphi_{2,3} = -GE - D^T G^T, \varphi_{2,4} = W^T \Gamma_2 T_4 + M - D^T G^T, \varphi_{2,5} = MB - D^T G^T,$$

$$\varphi_{2,6} = MC_1 - GC_2, \varphi_{2,7} = -D^T G^T, \varphi_{2,8} = -D^T G^T, \varphi_{2,9} = L_2 T_2 - D^T G^T,$$

$$\varphi_{2,10} = -D^T G^T, \varphi_{2,11} = -M - D^T G^T, \varphi_{3,3} = -(1 - r_d)R_2 - GE - E^T G^T,$$

$$\varphi_{3,4} = M - E^T G^T, \varphi_{3,5} = MB - E^T G^T, \varphi_{3,6} = MC_1 - GC_2, \varphi_{3,7} = -E^T G^T,$$

$$\varphi_{3,8} = -E^T G^T, \varphi_{3,9} = -E^T G^T, \varphi_{3,10} = -E^T G^T, \varphi_{3,11} = -M - E^T G^T,$$

$$\varphi_{4,4} = -(1 - \tau_d)R_4 - T_4 - T_4^T + M + M^T, \varphi_{4,5} = M + M^T, \varphi_{4,6} = MC_1 - GC_2,$$

$$\varphi_{4,7} = M^T, \varphi_{4,8} = M^T, \varphi_{4,9} = M^T, \varphi_{4,10} = M^T, \varphi_{4,11} = -M, \varphi_{5,5} = -N_3 + MB + B^T M^T,$$

$$\varphi_{5,6} = MC_1 - GC_2, \varphi_{5,7} = B^T M^T, \varphi_{5,8} = B^T M^T, \varphi_{5,9} = B^T M^T, \varphi_{5,10} = B^T M^T,$$

$$\varphi_{5,11} = -M + B^T M^T, \varphi_{6,6} = -\rho^2 \gamma I, \varphi_{6,7} = C_1^T M^T - C_2^T G^T, \varphi_{6,8} = C_1^T M^T - C_2^T G^T,$$

$$\varphi_{6,9} = C_1^T M^T - C_2^T G^T, \varphi_{6,10} = C_1^T M^T - C_2^T G^T, \varphi_{6,11} = C_1^T M^T - C_2^T G^T,$$

$$\varphi_{7,7} = R_3 + \tau^2 T_3 - T_1 - T_1^T, \varphi_{7,11} = -M, \varphi_{8,8} = R_4 - T_3 - T_3^T, \varphi_{8,11} = -M,$$

$$\varphi_{9,9} = -(1 - \tau_d)R_3 - T_2 - T_2^T, \varphi_{9,11} = -M, \varphi_{10,10} = -(1 - \tau_d)R_5, \varphi_{10,11} = -M,$$

$$\varphi_{11,11} = R_5 + r^2 N_4 + \tau^2 N_5 + M + M^T.$$

In this case, the filter gain matrix K of the state estimator is given by $K = M^{-1}G$.

Proof. Consider the following LKF for the filtering error system (5) as

$$V(e(t)) = V_1(e(t)) + V_2(e(t)) + V_3(e(t)), \tag{9}$$

with

$$V_1(e(t)) = 2e^T(t)Pe(t),$$

$$V_2(e(t)) = \int_{t-\tau(t)}^{t} e^T(s)R_1 e(s)ds + \int_{t-r(t)}^{t} e^T(s)R_2 e(s)ds + \int_{t-\tau(t)}^{t} g^T(e(s))R_3 g(e(s))ds$$

$$+ \int_{t-\tau(t)}^{t} g^T(We(s))R_4 g(We(s))ds + \int_{t-\tau(t)}^{t} \dot{e}^T(s)R_5 \dot{e}(s)ds,$$

$$V_3(e(t)) = \tau \int_{-\tau(t)}^{0} \int_{t+\theta}^{t} e^T(s)N_1 e(s)dsd\theta + r \int_{-r(t)}^{0} \int_{t+\theta}^{t} e^T(s)N_2 e(s)dsd\theta$$

$$+ r \int_{-r(t)}^{0} \int_{t+\theta}^{t} g^T(e(s))N_3 g(e(s))dsd\theta + r \int_{-r(t)}^{0} \int_{t+\theta}^{t} \dot{e}^T(s)N_4 \dot{e}(s)dsd\theta\tau$$

$$+ \tau \int_{-\tau(t)}^{0} \int_{t+\theta}^{t} \dot{e}^T(s)N_5 \dot{e}(s)dsd\theta.$$

Under the zero-initial condition, it is obvious that $V(e(t))|_{t=0} = 0$. A performance index is given as follows:

$$J_\infty = \int_0^t \left[\bar{z}^T(s)\bar{z}(s) - \gamma^2 w^T(s)w(s) \right] ds, \qquad t > 0, \tag{10}$$

then, one can get

$$J_\infty \leq \int_0^t \left[\bar{z}^T(s)\bar{z}(s) - \gamma^2 w^T(s)w(s) \right] ds + V(e(t)) - V(e(t))|_{t=0}, \qquad t > 0, \tag{11}$$

then for any $w(t) \in \mathcal{L}_2[0, \infty]$, one can achieve:

$$J_\infty \leq \int_0^t \left[\bar{z}^T(s)\bar{z}(s) - \gamma^2 w^T(s)w(s) + \dot{V}(e(t)) \right] ds. \tag{12}$$

Next, calculating the time-derivative of $V(e(t))$ along the trajectory of the filtering error system (5) is

$$\dot{V}_1(e(t)) = 2e^T(t)P\dot{e}(t), \tag{13}$$

$$\begin{aligned}
\dot{V}_2(e(t)) = &\, e^T(t)R_1 e(t) - (1 - \dot{\tau}(t))e^T(t - \tau(t))R_1 e(t - \tau(t)) \\
&+ e^T(t)R_2 e(t) - (1 - \dot{r}(t))e^T(t - r(t))R_2 e(t - r(t)) \\
&+ g^T(e(t))R_3 g(e(t)) - (1 - \dot{\tau}(t))g^T(e(t - \tau(t)))R_3 g(e(t - \tau(t))) \\
&+ g^T(We(t))R_4 g(We(t)) - (1 - \dot{\tau}(t))g^T(We(t - \tau(t)))R_4 g(We(t - \tau(t))) \\
&+ \dot{e}^T(t)R_4 \dot{e}(t) - (1 - \dot{\tau}(t))\dot{e}^T(t - \tau(t))R_4 \dot{e}(t - \tau(t)), \\
\leq &\, e^T(t)R_1 e(t) - (1 - \tau_d)e^T(t - \tau(t))R_1 e(t - \tau(t)) \\
&+ e^T(t)R_2 e(t) - (1 - r_d)e^T(t - r(t))R_2 e(t - r(t)) \\
&+ g^T(e(t))R_3 g(e(t)) - (1 - \tau_d)g^T(e(t - \tau(t)))R_3 g(e(t - \tau(t))) \\
&+ g^T(We(t))R_4 g(We(t)) - (1 - \tau_d)g^T(We(t - \tau(t)))R_4 g(We(t - \tau(t))) \\
&+ \dot{e}^T(t)R_5 \dot{e}(t) - (1 - \tau_d)\dot{e}^T(t - \tau(t))R_5 \dot{e}(t - \tau(t)),
\end{aligned} \tag{14}$$

$$\dot{V}_3(e(t)) \leq \tau^2 e^T(t)N_1 e(t) - \tau \int_{t-\tau(t)}^t e^T(s)N_1 e(s)ds + r^2 e^T(t)N_2 e(t)$$

$$- r \int_{t-r(t)}^t e^T(s)N_2 e(s)ds + r^2 g^T(e(t))N_3 g(e(t))$$

$$- r \int_{t-r(t)}^t g^T(e(s))N_3 g(e(s))ds + r^2 \dot{e}^T(t)N_4 \dot{e}(t)$$

$$- r \int_{t-r(t)}^t \dot{e}^T(s)N_4 \dot{e}(s)ds + \tau^2 \dot{e}^T(t)N_5 \dot{e}(t) - \tau \int_{t-\tau(t)}^t \dot{e}^T(s)N_5 \dot{e}(s)ds. \tag{15}$$

By using Lemma 2.5, one can get

$$-\tau \int_{t-r(t)}^t e^T(s)N_1 e(s)ds \leq - \left[\int_{t-\tau(t)}^t e(s)ds \right]^T N_1 \left[\int_{t-\tau(t)}^t e(s)ds \right], \tag{16}$$

$$-r \int_{t-r(t)}^t e^T(s)N_2 e(s)ds \leq - \left[\int_{t-r(t)}^t e(s)ds \right]^T N_2 \left[\int_{t-r(t)}^t e(s)ds \right], \tag{17}$$

$$-r \int_{t-r(t)}^t g^T(e(s))N_3 g(e(s))ds \leq - \left[\int_{t-r(t)}^t g(e(s))ds \right]^T N_3 \left[\int_{t-r(t)}^t g(e(s))ds \right], \tag{18}$$

$$-r \int_{t-r(t)}^t \dot{e}^T(s)N_4 \dot{e}(s)ds \leq - \left[\int_{t-r(t)}^t \dot{e}(s)ds \right]^T N_4 \left[\int_{t-r(t)}^t \dot{e}(s)ds \right], \tag{19}$$

$$-\tau \int_{t-\tau(t)}^t \dot{e}^T(s)N_5 \dot{e}(s)ds \leq - \left[\int_{t-\tau(t)}^t \dot{e}(s)ds \right]^T N_5 \left[\int_{t-\tau(t)}^t \dot{e}(s)ds \right]. \tag{20}$$

From (16)–(20), then (15) can be rewritten as follows

$$\dot{V}_3(e(t)) \leq r^2 e^T(t)(N_1 + N_2)e(t) + \tau^2 g^T(e(t))N_3 g(e(t))$$

$$+ r^2 \dot{e}^T(t)(N_4 + N_5)\dot{e}(t) - \left[\int_{t-\tau(t)}^t g(e(s))ds \right]^T N_3 \left[\int_{t-\tau(t)}^t g(e(s))ds \right]. \tag{21}$$

From (6), one can get

$$\begin{aligned} \left[g_i(e_i(t)) - l_i^- e_i(t) \right] \left[g_i(e_i(t)) - l_i^+ e_i(t) \right] &\leq 0, \\ \left[g_i(e_i(t - \tau(t))) - l_i^- e_i(t - \tau(t)) \right] \left[g_i(e_i(t - \tau(t))) - l_i^+ e_i(t - \tau(t)) \right] &\leq 0, \end{aligned} \tag{22}$$

$$[g_i(w_ie_i(t)) - \Gamma_i^- we_i(t)] [g_i(w_ie_i(t)) - \Gamma_i^+ w_ie_i(t)] \le 0,$$
$$[g_i(w_ie_i(t - \tau(t))) - \Gamma_i^- w_ie_i(t - \tau(t))] [g_i(w_ie_i(t - \tau(t))) - \Gamma_i^+ w_ie_i(t - \tau(t))] \le 0. \tag{23}$$

Then, for any $T_j = \mathrm{diag}\{t_{1j}, t_{2j}, \ldots, t_{nj}\} \ge 0, j = 1, 2, 3, 4$, it follows that

$$
\begin{aligned}
0 \le &- 2\sum_{i=1}^{n} t_{i1} [g_i(e_i(t)) - l_i^- e_i(t)] [g_i(e_i(t)) - l_i^+ e_i(t)] \\
&- 2\sum_{i=1}^{n} t_{i2} [g_i(e_i(t - \tau(t))) - l_i^- e_i(t - \tau(t))] [g_i(e_i(t - \tau(t))) - l_i^+ e_i(t - \tau(t))] \\
= &-2g^T(e(t))T_1g(e(t)) + 2e^T(t)L_2T_1g(e(t)) - 2e^T(t)L_1T_1e(t) \\
&- 2g^T(e(t - \tau(t)))T_2g(e(t - \tau(t))) + 2e^T(t - \tau(t))L_2T_2g(e(t - \tau(t))) \\
&- 2e^T(t - \tau(t))L_1T_2e(t - \tau(t)),
\end{aligned}
\tag{24}
$$

$$
\begin{aligned}
0 \le &- 2\sum_{i=1}^{n} t_{i3} [g_i(w_ie_i(t)) - \Gamma_i^- w_ie_i(t)] [g_i(w_ie_i(t)) - \Gamma_i^+ w_ie_i(t)] \\
&- 2\sum_{i=1}^{n} t_{i4} [g_i(w_ie_i(t - \tau(t))) - \Gamma_i^- w_ie_i(t - \tau(t))] \\
&\times [g_i(w_ie_i(t - \tau(t))) - \Gamma_i^+ w_ie_i(t - \tau(t))] \\
= &-2g^T(We(t))T_3g(We(t)) + 2e^T(t)W\Gamma_2T_3g(We(t)) - 2e^T(t)W\Gamma_1T_3We(t) \\
&- 2g^T(We(t - \tau(t)))T_4g(We(t - \tau(t))) + 2e^T(t - \tau(t))W\Gamma_2T_4g(We(t - \tau(t))) \\
&- 2e^T(t - \tau(t))W\Gamma_1T_4We(t - \tau(t)),
\end{aligned}
\tag{25}
$$

where

$$L_1 = \mathrm{diag}\{l_1^- l_1^+, l_2^- l_2^+, \ldots, l_n^- l_n^+\}, L_2 = \mathrm{diag}\{l_1^- + l_1^+, l_2^- + l_2^+, \ldots, l_n^- + l_n^+\},$$
$$\Gamma_1 = \mathrm{diag}\{\Gamma_1^- \Gamma_1^+, \Gamma_2^- \Gamma_2^+, \ldots, \Gamma_n^- \Gamma_n^+\}, L_2 = \mathrm{diag}\{\Gamma_1^- + \Gamma_1^+, \Gamma_2^- + \Gamma_2^+, \ldots, \Gamma_n^- + \Gamma_n^+\}.$$

On the other hand, according to (5), for any appropriately dimensioned matrix M, the following equation holds,

$$
\begin{aligned}
2\zeta^T(t)M[&-(A + KC)e(t) - KDe(t - \tau(t)) - KEe(t - r(t)) \\
&+ g(We(t - \tau(t))) + B\int_{t-r(t)}^{t} g(e(s))ds + (C_1 - KC_2)w(t) - \dot{e}(t)] = 0
\end{aligned}
\tag{26}
$$

where

$$
\begin{aligned}
\zeta^T(t) = &\left[e^T(t), e^T(t - \tau(t)), e^T(t - r(t)), g^T(We(t - \tau(t))), \left(\int_{t-r(t)}^{t} g(e(s))ds\right)^T, \right. \\
&\left. g^T(e(t)), g^T(We(t)), g^T(e(t - \tau(t))), \dot{e}^T(t - \tau(t)), \dot{e}^T(t)\right].
\end{aligned}
$$

Based on the Eq. (26) and inequalities from (13), (14), (21), (24) and (25), one can deduce that

$$\bar{z}^T(t)\bar{z}(t) - \gamma^2 w^T(t)w(t) + \dot{V}(e(t)) \leq \xi^T(t)\Psi\xi(t), \tag{27}$$

where

$$\xi^T(t) = \left[e^T(t), e^T(t-\tau(t)), e^T(t-r(t)), g^T(We(t-\tau(t))), \left(\int_{t-r(t)}^t g(e(s))ds\right)^T, w^T(t),\right.$$
$$\left. g^T(e(t)), g^T(We(t)), g^T(e(t-\tau(t))), \dot{e}^T(t-\tau(t)), \dot{e}^T(t)\right], G = PK.$$

In this case, where $\Psi < 0$, one can ensure the error system (5) with the guaranteed H_∞ performance. As the similar method, when $w(t) = 0$, one can ensure the error system (5) is globally stable.

In the case when $B = 0$ and $E = 0$, the SNNS (1) reduces to

$$\begin{cases} \dot{x}(t) = -Ax(t) + f(Wx(t-\tau(t)) + J) + F_1 w(t), \\ y(t) = -Cx(t) + Dx(t-\tau(t)) + F_2 w(t), \\ Z(t) = Hx(t), x(t) = \phi(t), \forall t \in [-\mu, 0], \ \mu = \max\{\tau, r\}, \end{cases} \tag{28}$$

then, the filtering error system (5) reduces to

$$\begin{cases} \dot{e}(t) = -(A+KC)e(t) - KDe(t-\tau(t)) + g(We(t-\tau(t))) + (C_1 - KC_2)w(t), \\ \bar{Z}(t) = He(t), \end{cases} \tag{29}$$

then, by Theorem 3.1, it is easy to have the following Corollary 3.2.

Corollary 3.2. For given scalars τ, τ_d, r, r_d, and the filter gain matrix K, the filtering error system (29) is globally asymptotically stable with H_∞ performance γ, if there exist matrices $P > 0$, $R_j = R_j^T > 0, j = 1, 3, 4, 5$, $N_j = N_j^T > 0, j = 3, 4, 5$, M, G and diagonal matrices $T_i > 0, i = 1, 2, 3, 4$, such that the following LMI holds:

$$\Psi = [\bar{\varphi}_{i,j}]_{9\times9} < 0, \tag{30}$$

with

$$\bar{\varphi}_{1,1} = R_1 + r^2 N_1 - L_1 T_1 - T_1^T L_1^T - \Gamma_1 T_3 - T_3^T \Gamma_1^T - MA - A^T M^T - GC - C^T G^T,$$

$$\bar{\varphi}_{1,2} = L_2 T_1 - GD - A^T M^T - C^T G^T, \bar{\varphi}_{1,3} = M - A^T M^T - C^T G^T,$$

$$\bar{\varphi}_{1,4} = MC_1 - GC_2, \varphi_{1,5} = -A^T M^T - C^T G^T, \bar{\varphi}_{1,6} = \Gamma_2 T_3 - A^T M^T - C^T G^T,$$

$$\bar{\varphi}_{1,7} = -A^T M^T - C^T G^T, \bar{\varphi}_{1,8} = -A^T M^T - C^T G^T, \bar{\varphi}_{1,9} = P - M - A^T M^T - C^T G^T,$$

$$\bar{\varphi}_{2,2} = -(1 - \tau_d)R_1 - L_1 T_2 - T_2^T L_1^T - W^T \Gamma_1 T_4 W - W T_4^T \Gamma_1^T W^T - GD - D^T G^T,$$

$$\varphi_{2,3} = W^T \Gamma_2 T_4 + M - D^T G^T, \bar{\varphi}_{2,4} = MC_1 - GC_2, \bar{\varphi}_{2,5} = -D^T G^T,$$

$$\bar{\varphi}_{2,6} = -D^T G^T, \bar{\varphi}_{2,7} = L_2 T_2 - D^T G^T, \bar{\varphi}_{2,8} = -D^T G^T, \bar{\varphi}_{2,9} = -M - D^T G^T,$$

$$\bar{\varphi}_{3,3} = -(1 - \tau_d)R_4 - T_4 - T_4^T + M + M^T, \bar{\varphi}_{3,4} = MC_1 - GC_2, \bar{\varphi}_{3,5} = M^T,$$

$$\bar{\varphi}_{3,6} = M^T, \bar{\varphi}_{3,7} = M^T, \bar{\varphi}_{3,8} = M^T, \bar{\varphi}_{3,9} = -M, \bar{\varphi}_{4,4} = -\gamma^2 I,$$

$$\bar{\varphi}_{4,5} = C_1^T M^T - C_2^T G^T, \bar{\varphi}_{4,6} = C_1^T M^T - C_2^T G^T, \bar{\varphi}_{4,7} = C_1^T M^T - C_2^T G^T,$$

$$\bar{\varphi}_{4,8} = C_1^T M^T - C_2^T G^T, \bar{\varphi}_{4,9} = C_1^T M^T - C_2^T G^T, \bar{\varphi}_{5,5} = R_3 + \tau^2 T_3 - T_1 - T_1^T,$$

$$\bar{\varphi}_{5,9} = -M, \bar{\varphi}_{6,6} = R_4 - T_3 - T_3^T, \bar{\varphi}_{6,9} = -M, \bar{\varphi}_{7,7} = -(1 - \tau_d)R_3 - T_2 - T_2^T,$$

$$\bar{\varphi}_{7,9} = -M, \bar{\varphi}_{8,8} = -(1 - \tau_d)R_5, \bar{\varphi}_{8,9} = -M, \bar{\varphi}_{9,9} = R_5 + r^2 N_5 + M + M^T.$$

In this case, the filter gain matrix K of the state estimator is given by $K = M^{-1}G$.

Proof. Construct a Lyapunov-Krasovskii functional for the filtering error system (29)

$$V(e(t)) = 2e^T(t)Pe(t) + \int_{t-\tau(t)}^{t} e^T(s)R_1 e(s)ds + \int_{t-\tau(t)}^{t} g^T(e(s))R_3 g(e(s))ds$$

$$+ \int_{t-\tau(t)}^{t} g^T(We(s))R_4 g(We(s))ds + \int_{t-\tau(t)}^{t} \dot{e}^T(s)R_5 \dot{e}(s)ds$$

$$+ \tau \int_{-\tau(t)}^{0} \int_{t+\theta}^{t} e^T(s)N_1 e(s)dsd\theta + \tau \int_{-\tau(t)}^{0} \int_{t+\theta}^{t} \dot{e}^T(s)N_5 \dot{e}(s)dsd\theta.$$

Then, the rest of proof is similar to the proof of Theorem 3.1 and is omitted here.

4   Numerical Examples

In this section, two examples are given to verify the effectiveness of the result.

Example 4.1. Consider the SNNS (1) where

$$A = \text{diag}\{5 \quad 7\}, B = \begin{bmatrix} 1.3 & 0.21 \\ -0.12 & 0.4 \end{bmatrix}, C_1 = \begin{bmatrix} 0.4 & 0.1 \\ 0.1 & 0.4 \end{bmatrix}, C = \begin{bmatrix} 1.41 & -0.25 \\ 0.5 & 1.1 \end{bmatrix}, D = \begin{bmatrix} 0.8 & 0.14 \\ 0.15 & 1.3 \end{bmatrix},$$

$$E = \begin{bmatrix} 1.61 & -1.26 \\ -0.13 & 2.4 \end{bmatrix}, C_2 = \begin{bmatrix} 0.5 & -0.2 \\ -0.1 & 0.5 \end{bmatrix}, l_1^- = l_2^- = 0.1, l_1^+ = l_2^+ = 0.4,$$

$$\Gamma_1^- = \Gamma_2^- = 0.1, \Gamma_1^+ = \Gamma_2^+ = 0.8, \tau(t) = 0.4 + 0.1\cos(4t), h(t) = 0.3 + 0.7\cos(t).$$

If let $L_1 = 0.04I, L_2 = 0.5I, \Gamma_1 = 0.08I, \Gamma_2 = 0.9I, \tau = 0.5, \tau_d = 0.4, h = 1, h_d = 0.7$, and $W = \text{diag}\{0.4 \quad 0.6\}$, the filter gain matrix K and the performance γ are obtained as follows:

$$K = \begin{bmatrix} 3.1502 & 2.6800 \\ 3.1094 & 4.6385 \end{bmatrix}, \gamma = 43.8259.$$

Therefore, it follows from Theorem 3.1, the filtering error system (5) is globally asymptotically stable.

Example 4.2. Consider the SNNS (1) where

$$A = \text{diag}\{5 \quad 3 \quad 6\}, B = \begin{bmatrix} 1.3 & 0.21 & 0.2 \\ -0.52 & 1.4 & 0.2 \\ 0.2 & 0.1 & 0.8 \end{bmatrix}, C = \begin{bmatrix} 0.73 & -0.11 & 0.31 \\ 0.42 & 0.16 & 0.32 \\ 0.32 & -0.1 & 0.39 \end{bmatrix},$$

$$D = \begin{bmatrix} 0.82 & 0.4 & 0.11 \\ 0.12 & 0.44 & 0.2 \\ -0.2 & 0.1 & 1.65 \end{bmatrix}, E = \begin{bmatrix} 0.53 & 0.65 & -0.1 \\ -0.76 & 1.8 & 0.2 \\ -0.02 & 0.1 & 1.3 \end{bmatrix}, C_1 = 0.3I, C_2 = 0.4I,$$

$l_1^- = l_2^- = l_3^- = 0.1, l_1^+ = l_2^+ = l_3^+ = 0.5, \Gamma_1^- = \Gamma_2^- = \Gamma_3^- = 0.1, \Gamma_1^+ = \Gamma_2^+ = \Gamma_3^+ = 0.7,$
$\tau(t) = 1.3 + 0.1\sin(5t), h(t) = 0.7 + 0.7\sin(t).$

If let $L_1 = 0.05I, L_2 = 0.6I, \Gamma_1 = 0.07I, \Gamma_2 = 0.8I, \tau = 1.6, \tau_d = 0.5, h - 1.4,$ $h_d = 0.7$, and $W = \text{diag}\{1.4 \quad 1.6 \quad 1.7\}$, the filter gain matrix K and the performance γ are obtained as follows:

$$K = \begin{bmatrix} 16.6110 & -7.2478 & 6.7813 \\ -7.2478 & -1.0908 & 2.4715 \\ 6.7813 & 2.4715 & 8.0243 \end{bmatrix}, \gamma = 373.5699.$$

Then, it follows from Theorem 3.1, the filtering error state system (5) is globally asymptotically stable.

5 Conclusion

In this paper, the filter issue is considered for a class of SSNS with mixed time-varying delays. Based on a Lyapunov-Krasovskii functional combined with a zero equation, a suitable H_∞ filter is obtained for the SNNS model. The filter can be solved by an LMI. Two examples are given to verify the effectiveness of the proposed results. Stochastic disturbance and uncertain parameters are inevitable in a variety of neural networks. They may cause oscillation, divergence, and instability to the neural networks. For further research, we plan to analysis the filter design of delayed SNNS with stochastic disturbance and uncertain parameters.

Acknowledgment. This work is supported by the Start-up Foundation for Doctors of East China University of Technology (No. DHBK2012201), the Jiangxi foreign science and technology cooperation plan (No. 20132BDH80007), and the National Natural Science Foundation (Nos. 11565002, 51409047, 61463003, 51567001).

References

1. Xu, Z.B., Qiao, H., Peng, J., Zhang, B.: A comparative study of two modeling approaches in neural networks. Neural Netw. **17**(1), 73–85 (2004)
2. Farajzadeh, J., Fard, A.F., Lotfi, S.: Modeling of monthly rainfall and runoff of Urmia Lake basin using "Feed-forward Neural Network" and "Time Series Analysis" model. Water Resour. Ind. **7**, 38–48 (2014)
3. Barbounis, T.G., Theocharis, J.B., Alexiadis, M.C., Dokopoulos, P.S.: Long-term wind speed and power forecasting using local recurrent neural network models. IEEE Trans. Energy Convers. **21**(1), 273–284 (2006)
4. Li, P., Cao, J.: Stability in static delayed neural networks: a nonlinear measure approach. Neurocomputing **69**(13–15), 1776–1781 (2006)
5. Shao, H.: Delay-dependent stability for recurrent neural networks with time-varying delays. IEEE Trans. Neural Netw. **19**(9), 1647–1651 (2008)
6. Wang, M., Wang, L.: Global asymptotic robust stability of static neural network models with S-type distributed delays. Math. Comput. Model. **44**(1–2), 218–222 (2006)
7. Zheng, C.D., Zhang, H., Wang, Z.: Delay-dependent globally exponential stability criteria for static neural networks: an LMI approach. IEEE Trans. Circ. Syst. II Express Briefs **56**(7), 605–609 (2009)
8. Ray, A.: Performance evaluation of medium access control protocols for distributed digital avionics. J. Dyn. Syst. Meas. Contr. **109**(4), 370–377 (1987)
9. Cooke, K.L., Grossman, Z.: Discrete delay, distributed delay and stability switches. J. Math. Anal. Appl. **86**(2), 592–627 (1982)
10. Luck, R., Ray, A.: An observer-based compensator for distributed delays. Automatica **26**(5), 903–908 (1990)
11. Wang, Z., Shu, H., Liu, Y., Ho, D.W.C., Liu, X.: Robust stability analysis of generalized neural networks with discrete and distributed time delays. Chaos Solitons Fractals **30**(4), 886–896 (2006)
12. Li, T., Fei, S., Zhu, Q.: Design of exponential state estimator for neural networks with distributed delays. Nonlinear Anal. Real World Appl. **10**(2), 1229–1242 (2009)
13. Lakshmanan, S., Park, J.H., Jung, H.Y., Balasubramaniam, P.: Design of state estimator for neural networks with leakage, discrete and distributed delays. Appl. Math. Comput. **218**(22), 11297–11310 (2012)
14. Huang, H., Feng, G.: Delay-dependent H-infinity and generalized H-2 filtering for delayed neural networks. IEEE Trans. Circ. Syst. I Regul. Pap. **56**(4), 846–857 (2009)
15. Huang, H., Feng, G., Cao, J.: Guaranteed performance state estimation of static neural networks with time-varying delay. Neurocomputing **74**(4), 606–616 (2011)
16. Duan, Q., Su, H., Wu, Z.G.: H-infinity state estimation of static neural networks with time-varying delay. Neurocomputing **97**, 16–21 (2012)
17. Huang, H., Huang, T., Chen, X.: Guaranteed H-infinity performance state estimation of delayed static neural networks. IEEE Trans. Circ. Syst. II Express Briefs **60**(6), 371–375 (2013)

18. Lakshmanan, S., Mathiyalagan, K., Park, J.H., Sakthivel, R., Rihan, F.A.: Delay-dependent H-infinity state estimation of neural networks with mixed time-varying delays. Neurocomputing **129**, 392–400 (2014)
19. Huang, H., Huang, T., Chen, X.: Further Result on guaranteed H-infinity performance state estimation of delayed static neural networks. IEEE Trans. Neural Netw. Learn. Syst. **26**(6), 1335–1341 (2015)

Nature Inspired Computing
and Optimization

SMOTE-DGC: An Imbalanced Learning Approach of Data Gravitation Based Classification

Lizhi Peng[1], Haibo Zhang[2], Bo Yang[1(✉)], Yuehui Chen[1], and Xiaoqing Zhou[1]

[1] Shandong Provincial Key Laboratory of Network Based Intelligent Computing, University of Jinan, Jinan 250022, People's Republic of China
{yangbo,yhchen}@ujn.edu.cn
[2] Department of Computer Science, University of Otago, Dunedin, New Zealand

Abstract. Imbalanced learning, an important learning technique to cope with learning cases of one class outnumbering another, has caught many interests in the research community. A newly developed physical-inspired classification method, i.e., the data gravitation-based classification (DGC) model, performs well in many general classification problems. However, like other general classifiers, the performance of DGC suffers for imbalanced tasks. Therefore, we develop a data level imbalanced learning DGC model namely SMOTE-DGC in this paper. An over sampling technique, Synthetic Minority Over-sampling Technique (SMOTE), is integrated with DGC model to improve the imbalanced learning performances. A total of 44 imbalanced classification data sets, several standard and imbalanced learning algorithms are used to evaluate the performance of the proposal. Experimental results suggest that the adapted DGC model is effective for imbalanced problems.

1 Introduction

In many real classification circumstances, the instances show class imbalance distributions: the number of instances in one class (majority class) is outnumbered by the number of instances in another class (minority class). This phenomenon is mainly attributed to the limited instances of the minority class. For example, in Internet traffic classification problems, Web browsing traffic is a dominant type of traffic that occurs in the Internet at each moment [1, 2]. However, capturing enough malicious traffic samples, such as attacks and virus traffic, for training is difficult. Many real-world classification tasks, such as medical diagnosis [3], fraud detection [4], finance risk management [5], network intrusion detection [6], stream classification [7], and bioinformatics [8], have similar diagnosis characteristics. In these imbalanced tasks, the minority class is usually more important than the majority class [9, 17]. Thus, to maximize the recognition rate of the minority class on the premise of considering a good tradeoff for both of the minority and majority classes is the main goal of an imbalanced task.

Most standard classification models are not suitable for imbalanced tasks because such models seek high classification accuracies across the entire dataset. Owing to class

© Springer International Publishing Switzerland 2016
D.-S. Huang and K.-H. Jo (Eds.): ICIC 2016, Part II, LNCS 9772, pp. 133–144, 2016.
DOI: 10.1007/978-3-319-42294-7_11

imbalance, a standard classifier usually classifies a minority class instance incorrectly as the model is over-trained by the majority class instances. Therefore, although a standard classifier can achieve high accuracy in an imbalanced task, the actual performance is poor because its identifying rate of the minority class is low. However, a class imbalance is not the only factor that influences classification performance [10]. Other data distribution characteristics, such as small disjuncts [11], class overlaps [12], noise [13], and borderline samples [14], also hinder classification performance.

A wide range of techniques has been proposed to manage imbalanced classification tasks. These techniques can be classified into three categories: data level approaches, cost-sensitive learning, and ensemble learning. Data level approaches, resampling the imbalanced instance to re-balance the data set, involves two strategies: over-sampling and under-sampling. Over-sampling methods diminish class imbalance by creating new minority class samples [15, 16], while under-sampling methods rebalance class distributions by reducing the number of majority class samples [21, 22]. Cost-sensitive learning considers the varying costs of different misclassification types [17] and the cost matrix during model building. A cost-sensitive learning technique generates a model that has the lowest cost. Many standard classification models, such as C4.5 decision trees [18], artificial neural networks [20], and support vector machines (SVMs) [19], have been modified by this type. Ensemble learning, a dependent technique, is a type of hybrid model that embeds data-level approaches into general classifiers for better classification and generalization. For example, SMOTEBoost [23] combined the SMOTE technique with Boost learning, BalanceCascade [22] combines the under-sampling method with AdaBoost classifiers.

The data gravitation-based classification (DGC) [24] model is a new classification model that is based on Newton's law of universal gravitation. The DGC model refers to a data instance in the data space as a data "particle" and considers the type of "gravitation" between any two data particles in the computation. This gravitation is directly proportional to the product of the "masses" of two data particles and is inversely proportion to the square of the distance between the data particles. By comparing the gravitation from different data classes in the training set, DGC can effectively classify a testing data instance in a simple manner. Simić et al. [25] combined DGC with case-based reasoning to create a hybrid intelligent tool for financial forecasting. Similar classification models inspired by gravitation, such as GBC [27] and CGM [28], have also been presented in recent years. Recently, Reyes et al. [29] present an interesting lazy learning algorithm based on DGC model. They combine the mechanisms of nearest neighbors and gravitation together, and successfully applied the hybrid model for multi label learning.

However, DGC suffers from imbalanced data sets [24], which also affect other general classifiers. In a highly imbalanced training set, the gravitational fields of the minority and majority classes are usually weak and extremely strong, respectively. The strong gravitational field of the majority class attracts most of the minority class samples to the majority class, thus resulting in a low TPR.

Cano et al. have done an excellent work to adapt DGC model for imbalanced tasks [26], and the improved model is called DGC+. They construct a class-independent attribute-class weights matrix instead of the weight vector in the basic DGC model. The attribute-class weights matrix is able to carry the accurate distribution information of different classes. By using the attribute-class weights matrix, DGC+ modify the computation method of gravitations. The modified computation method of gravitation can effectively stress the minority class, and weaken the majority class at the same time. And they use the co-variance matrix adaption evolution strategy (CMA-ES) to search the optimum class-independent attribute-class weights matrix. Empirical studies have shown that DGC+ is effective for standard, noisy, and imbalanced data [26]. However, DGC+ is time-consuming because its model parameters in the attribute-class weights matrix are far more than the parameters of the basic DGC model. And to get higher classification accuracies, DGC+ defines each instance as a single data particle, which also increases the computational complexity.

The objective of this study is to design an imbalanced classification model based on the standard DGC model to improve its behavior for imbalanced problems, especially for high imbalanced tasks.

The rests are organized as follows. Section 2 briefly reviews the standard DGC model including its theoretical basis, classification principles, and feature weighting method. Section 3 introduces SMOTE preprocessing algorithm. Section 4 depicts our experimental setup. Section 5 presents the results and analysis of the imbalanced data sets. And then we summarize and conclude our study in Sect. 6.

2 A Brief Overview on the DGC Model

2.1 Classification Using Data Gravitation

DGC [24] is a classification model that simulates Newton's law of universal gravitation. Basically, the classification idea of DGC is as follows: For a class-unknown data instance, i.e. a testing instance, DGC firstly computes the gravitation from each class in the training set, and then compares the gravitation of different classes. If a certain class exhibits the largest gravitation, the testing instance should be classified into this class.

Concept (Data Particle): Data particle simulates the concept of physical particle in universe. A number of instances with a certain relationship form a data particle, usually the relationship refers to the geometrical neighborhood of these instances. A data particle with only one instance is called atomic data particle. Data particle has two basic properties: data mass and centroid. Data mass is the number of instances in the data particle, and the data centroid is the geometrical center of these instances.

Following the concepts of data mass and centroid, a data particle can be denoted using a pair expression $<m, x>$, where m is its mass, and x is its centroid. When take the class property (feature y) into account, the data particle could be described as a triple expression $<m, x, y>$. It should be noticed that instances from different classes cannot be put into a single particle. Moreover, a testing instance is an atomic particle because its class property is unknown.

Concept (Data Gravitation): Data gravitation is a simulation of physical gravitation in data spaces. Any pair of data particles in a data space have a kind of interacting force, namely data gravitation, and the gravitation is the direct ratio to the product of data mass of the two data particles, and reverse ratio to the square of distance between them, which can be computed by equation:

$$F = \frac{m_1 m_2}{r^2} \tag{1}$$

Where F: Gravitation between two data particles. m1: Data mass of data particle 1. m2: Data mass of data particle 2.r: The Euclidean distance between the two data particle in data space. It can be seen that such a form of data gravitation computation is the same as Newton's law of universal gravitation. However, data gravitation is a kind of scalar without direction, which makes the key difference to physical forces.

For a testing instance (an atomic particle), the gravitation from the particles of a single class can be superposed, and this is the superposition principle:

Lemma 1 (Superposition Principle): Suppose $p_1, p_2, \ldots, p_m$ are m data particles in a data space, and they belong to the same data class. The gravitations they act on another data particle are $F_1, F_2, \ldots, F_m$, and then the composition of gravitation is:

$$F = \sum_{i=1}^{m} F_i \tag{2}$$

On the other hand, the gravitation from different classes can be compared, which forms the simple classification principle of the DGC model.

Lemma 2 (Classification Principle): Suppose c_1, c_2 are two data classes in a training data set. For a given testing instance P (an atomic data particle), the gravitation that data particles in c_1 acts on P is F_1, and F_2 is the gravitation that data particles in c_2 acts on P. If $F_1 > F_2$, then the degree P belongs to c_1 is stronger than that to c_2.

Figure 1 shows the principle of classification with a binary data set.

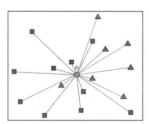

Fig. 1. Classification based on data gravitation.

For a multi-class data set with k classes, if $l_1, l_2, \ldots, l_k$ are the numbers of data particles, respectively, then for a testing instance P, the gravitation that data class i act on it is given by:

$$F_i = \sum_{j=1}^{l_i} \frac{m_{ij}}{|x_{ij} - x|^2} \tag{3}$$

where m_{ij} is the data mass of data particle j in data class i, x_{ij} is its centroid, and x is the centroid of P.

If $F_{i'} = max\{F_1, F_2, ..., F_k\}$, then according to Lemma 2, the test data element belongs to data class i'.

2.2 Weighting Features

DGC uses weight to measure the importance of a feature (input). Suppose there are n features in target problem feature set, and every feature has a weight value, so all feature weights form a n-expression: $<w_1, w_2, ..., w_n>$, We mark it as the feature weight vector w. As the features have been weighted, the distance between two data particles in the data space is not their Euclidean distance in the data space, but the weighted distance:

$$r' = \sqrt{\sum_{i=1}^{n} w_i(x_{1i} - x_{2i})^2} \qquad (4)$$

The calculation of the data gravitation is also changed as:

$$F = \frac{m_1 m_2}{r'^2} \qquad (5)$$

3   Synthetic Minority Over-Sampling Technique (SMOTE)

Due to the class imbalance distributions in imbalanced tasks, standard algorithms usually lost in such tasks. Therefore, as a direct and simple solution, data level algorithms emerged. Strictly, data level approaches belong to preprocessing procedures: they manually increase/decrease instances of the minority/majority classes, aiming at rebalancing the class distributions of a training set.

Synthetic minority over-sampling technique (SMOTE), the most famous data level algorithm, was proposed by Chawla et al. in 2002. SMOTE is an over-sampling method which rebalances an imbalanced data set by increasing the number of the minor class instances. Typically, the way SMOTE algorithm re-balancing a data set can be shown as Fig. 2. In SMOTE algorithm, the minority class is over-sampled by taking each minority class instance and introducing synthetic examples along the line

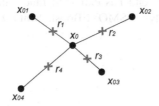

Fig. 2. The way of creating the synthetic data points in the SMOTE algorithm.

segments joining any/all of the k minority class nearest neighbors. As shown in Fig. 2, x0 is the selected instance, x01 to x04 are some selected nearest neighbors and r1 to r4 the synthetic data instances created by the randomized interpolation.

Synthetic instances are generated in the following way: take the difference between the feature vector (sample) under consideration and its nearest neighbor. Multiply this difference by a random number between 0 and 1, and add it to the feature vector under consideration. This causes the selection of a random point along the line segment between two specific features. This approach effectively forces the decision region of the minority class to become more general. An example is detailed in Fig. 2.

In short, its main feature is to form new minority class instances by interpolating between several minority class instances that lie together. Thus, the over fitting problem is avoided and causes the decision boundaries for the minority class to spread further into the majority class space.

4   Empirical Study Settings

4.1   Data Sets

Table 1 shows the characteristics of the binary class data sets where #inst denotes the number of instances of the data set #attrs denotes the number of attributes (features) of the data set and IR is the imbalance ratio. Table 1 is sorted by IR in ascending order. According to the KEEL data set repository 22 of these data sets are low imbalanced with IR < 9.0 and the other 22 data sets are high imbalanced with IR9.0

4.2   Compared Algorithms

As mentioned in the introduction imbalanced classification techniques fall into three categories. Therefore we compare our proposal with carefully selected algorithms covering the three categories:

- **Data level approaches:** As we integrate SMOTE with DGC, we firstly compare SMOTE-DGC with two other SMOTE based approaches: SMOTE-C4.5 and SMOTE-SVM.
- **Cost-sensitive learning:** For this type, C4.5 and SVM are also widely studied base-line algorithms. Therefore, C4.5CS and SVMCS are selected as cost-sensitive learning approaches.
- **Ensemble learning:** Ensemble learning is considered as an effective solution for imbalanced learning tasks, and has extracted much interest. Therefore, we select four approaches for this type: SMOTEBoost, SMOTEBagging, BalanceCascade, and EasyEnsemble.

Table 1. Characteristics of selected binary class data sets

Low imbalanced				High imbalanced			
Data set	#inst	#attr	IR	Data set	#inst	#attr	IR
glass1	214	9	1.82	yeast-2_vs_4	514	8	9.08
ecoli-0_vs_1	220	7	1.86	yeast-0-5-6-7-9_vs_4	528	8	9.35
wisconsinImb	683	9	1.86	vowel0	988	13	9.98
pimaImb	768	8	1.87	glass-0-1-6_vs_2	192	9	10.29
iris0	150	4	2.00	glass2	214	9	11.59
glass0	214	9	2.06	shuttle-c0-vs-c4	1829	9	13.87
yeast1	1484	8	2.46	yeast-1_vs_7	459	7	14.30
habermanImb	306	3	2.78	glass4	214	9	15.46
vehicle2	846	18	2.88	ecoli4	336	7	15.80
vehicle1	846	18	2.90	page-blocks-1-3_vs_4	472	10	15.86
vehicle3	846	18	2.99	abalone9-18	731	8	16.00
glass-0-1-2-3_vs_4-5-6	214	9	3.20	glass-0-1-6_vs_5	184	9	19.44
vehicle0	846	18	3.25	shuttle-c2-vs-c4	129	9	20.50
ecoli1	336	7	3.36	yeast-1-4-5-8_vs_7	693	8	22.10
new-thyroid1	215	5	5.14	glass5	214	9	22.78
new-thyroid2	215	5	5.14	yeast-2_vs_8	482	8	23.10
ecoli2	336	7	5.46	yeast4	1484	8	28.10
segment0	2308	19	6.02	yeast-1-2-8-9_vs_7	847	8	30.57
glass6	214	9	6.38	yeast5	1484	8	32.73
yeast3	1484	8	8.10	ecoli-0-1-3-7_vs_2-6	218	7	39.14
ecoli3	336	7	8.60	yeast6	1484	8	41.40
page-blocks0	5472	10	8.79	abalone19	4174	8	129.44

4.3 Performance Measures

The confusion matrix is the basis in measuring a classification task, wherein rows denote the actual class of the instances and the columns denote the predicted class. Figure 3 shows a typical confusion matrix of a binary classification. True positive (TP) is the number of positive instances that are correctly classified, false positive (FP) is the number of positive instances that are incorrectly classified as negative samples, true negative (TN) is the number of negative instances that are correctly classified, and false negative is the number of negative instances that are incorrectly classified as positive samples. We conduct many types of measures based on the confusion matrix to evaluate classier performance. In this study the following measures are used:

– **Area Under Curve:** The receiver operating characteristic (ROC) curve [30] is a 2D graphical illustration of the trade-off between the TPR and FPR. The TPR is also called sensitivity (Sens), and the FPR is related to another general measure namely specificity (Spec) and they are defined as follows:

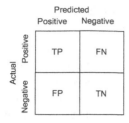

TP: # of positive instances correctly classified

TN: # of negative instances correctly classified

FP: # of negative instances incorrectly classified

FN: # of positive instances incorrectly classified

Fig. 3. Confusion matrix

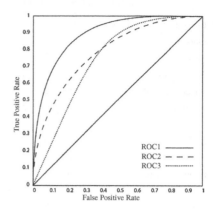

Fig. 4. Example ROC curve

$$Sens = \frac{TP}{TP + FN} \tag{6}$$

$$Spec = \frac{TN}{TN + FP} = 1 - \text{FPR} \tag{7}$$

The ROC curve illustrates the behavior of a classer without considering the class distribution or misclassification cost. An example of an ROC curve is depicted in Fig. 4. The diagonal line represents the trade-off between the sensitivity and 1-specificity for a random model. For a well-performing classifier, the ROC curve needs to be as far to the top left-hand corner as possible, i.e., the area under the ROC curve (AUC) [31] requires to be as large as possible. AUC is computed by the confusion matrix values in relation to the TP rate (TPR) and FP rate (FPR):

$$AUC = \frac{1 + TPR - FRP}{2} \tag{8}$$

Since $Spec = 1 - FPR$ and $Sens = TPR$, we get:

$$AUC = \frac{Sens + Spec}{2} \tag{9}$$

The aforementioned equation shows that *AUC*, which is the mean value of *Sens* and *Spec*, balances the proportion of the correctly predicted positive and negative instances.

- **Geometric Mean:** Geometric mean (GM) is another method for comparing positive and negative classes regardless of their size:

$$GM = \sqrt{Sens \cdot Spec} \tag{10}$$

Similar to *AUC*, *GM* attempts to maximize the accuracy of each class in a balanced manner. *GM* is a performance metric that links both objectives and has increasingly used in imbalanced classifications.

5 Results and Analysis

To show the results in a global view, we firstly give a set of box charts, as can be seen in Figs. 5, 6, 7 and 8. In where, the upper and lower edges of a box are the upper and lower quantiles, respectively; the upper and lower stars are the maximum and minimum values, respectively; the inner horizontal line is the median, and the little square in the box is the mean value. We also give the experimental result tables in Appendix for more detail, including the results of both of the low and high imbalanced data sets.

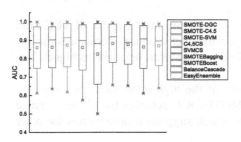

Fig. 5. AUC results (low imbalanced cases) (Color figure online)

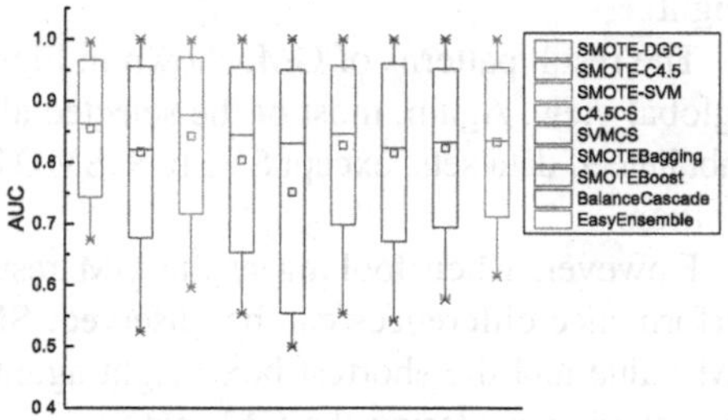

Fig. 6. AUC results (high imbalanced cases) (Color figure online)

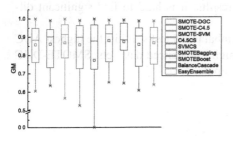
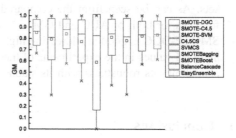

Fig. 7. GM results (low imbalanced cases) (Color figure online)

Fig. 8. GM results (high imbalanced cases) (Color figure online)

The AUC results of low imbalanced data sets shown in Fig. 5 give a monotonous pattern. Most of the compared algorithms, including SMOTE-DGC, do not show significant differences for the low imbalanced data sets. Most boxes, as can be seen from Fig. 5 are approximately located at the same level, and with the similar heights.

The results suggest that most compared algorithms got similar performances for low imbalanced data sets. A deep look at the detailed result table shows the average AUC value of the imbalanced data sets range from 0.82 to 0.89, making small differences. SVMCS, the only abnormally behaved algorithm, output a significantly low average AUC value, and a high result box, which suggests its low performances for the low imbalanced data sets.

The circumstances of the high imbalanced data sets, shown in Fig. 6, are quite different from that of the low imbalanced cases. First of all, the significant differences of the compared algorithms are observed: the upper and low edges of the box along with the little squares in the boxes show more significant differences than that of the low imbalanced data sets. It implies that the selected algorithms performed quite differently for the high imbalanced data sets. SMOTE-DGC, with the highest average AUC value, shows its good performances this time. Another fact can be observed in Fig. 8 is the shortest box height of SMOTE-DGC, which suggests the robustness of this algorithm. Outputting results in sharp contrasts between the SMOTE version and cost-sensitive version, SVM shows surprising behaviors: SMOTE-SVM performed fairly well, while SVMCS got the lowest average AUC value and the highest box height.

The result patterns of GM, shown in Figs. 7 and 8, are similar to that of AUC from a global view. Again, most of the selected algorithms get similar performances for low imbalanced data sets, except SVMCS. SMOTEBoost performs best with its highest box position.

However, when looking at the GM results of the high imbalanced data sets, the performance differences can be observed. SMOTE-DGC achieves the highest average GM value and the shortest box height again, which suggests it outperforms the other algorithms considering the GM metric.

According to the experimental results, the empirical study can be summarized as follows:

- Most imbalanced learning approaches perform similarly for low imbalanced data sets. As can be seen from the compared results, it is hard to find significant differences for both of AUC and GM.
- SMOTE-DGC is good at addressing high imbalanced tasks. It hits the highest AUC and GM values in the experiments of the high imbalanced data sets. SMOTE-DGC also shows its robustness with its stable outputs.

6   Conclusions

We have adapted the DGC model to manage imbalanced classification problems. SMOTE algorithm was applied to rebalance the class distributions for the DGC model. We compared the improved DGC model (SMOTE-DGC) with 8 imbalanced binary classification methods on 44 imbalanced data sets. Experimental results have shown that the two rebalancing mechanisms can effectively overcome the weakness of the basic DGC model for imbalanced tasks. Particularly for highly imbalanced problems our method showed better performances in comparing with the other algorithms.

SMOTE-DGC also showed good generalization ability in the study. We conclude that the SMOTE-DGC model provides an effective means of solving imbalanced problems.

Acknowledgment. This research was partially supported by National Natural Science Foundation of China under grant No. 61472164, No. 61373054, and No. 61203105, the Provincial Natural Science Foundation of Shandong under grant No. ZR2012FM010, and No. ZR2015JL025.

References

1. Zhang, H., Lu, G., Qassrawi, M.T., Zhang, Y., Yu, X.: Feature selection for optimizing traffic classification. Comput. Commun. **35**, 1457–1471 (2012)
2. Peng, L., Zhang, H., Yang, B., Chen, Y., Qassrawi, M.T., Lu, G.: Traffic identification using flexible neural trees. In: Proceeding of the 18th International Workshop of QoS (IWQoS 2012), pp. 1–5 (2012)
3. Mazurowski, M.A., Habas, P.A., Zurada, J.M., et al.: Training neural network classifiers for medical decision making: the effects of imbalanced datasets on classification performance. Neural Netw. Off. J. Int. Neural Netw. Soc. **21**, 427 (2008)
4. Dheepa, V., Dhanapal, R., Manjunath, G.: Fraud detection in imbalanced datasets using cost based learning. Eur. J. Sci. Res. **91**, 486–490 (2012)
5. Brown, I., Mues, C.: An experimental comparison of classification algorithms for imbalanced credit scoring data sets. Expert Syst. Appl. **39**, 3446–3453 (2012)
6. Chairi, I., Alaoui, S., Lyhyaoui, A.: Intrusion detection based sample selection for imbalanced data distribution. In: Proceeding of 2012 Second International Conference on Innovative Computing Technology (INTECH), pp. 259–264 (2012)
7. Ghazikhani, A., Monsefi, R., Yazdi, H.S.: Online neural network model for non-stationary and imbalanced data stream classification. Int. J. Mach. Learn. Cybern. **5**(1), 51–62 (2014)
8. Yu, H., Ni, J., Zhao, J.: ACOSampling: an ant colony optimization-based undersampling method for classifying imbalanced DNA microarray data. Neurocomputing **101**, 309–318 (2013)
9. Zadrozny, B., Elkan, C.: Learning and making decisions when costs and probabilities are both unknown. In: Proceedings of the 7th International Conference on Knowledge Discovery and Data Mining (KDD 2001), pp. 204–213 (2001)
10. Japkowicz, N., Stephen, S.: The class imbalance problem: a systematic study. Intell. Data Anal. J. **6**, 429–450 (2002)
11. Weiss, G.M.: Mining with rarity: a unifying framework. SIGKDD Explor. **6**, 7–19 (2004)
12. García, V., Mollineda, R., Sánchez, J.S.: On the k-NN performance in a challenging scenario of imbalance and overlapping. Pattern Anal. Appl. **11**, 269–280 (2008)
13. Seiffert, C., Khoshgoftaar, T.M., Van Hulse, J., et al.: An empirical study of the classification performance of learners on imbalanced and noisy software quality data. In: Proceedings of IEEE International Conference on Information Reuse and Integration, pp. 651–658 (2007)
14. Napierała, K., Stefanowski, J., Wilk, S.: Learning from imbalanced data in presence of noisy and borderline examples. In: Szczuka, M., Kryszkiewicz, M., Ramanna, S., Jensen, R., Hu, Q. (eds.) RSCTC 2010. LNCS, vol. 6086, pp. 158–167. Springer, Heidelberg (2010)
15. Chawla, N.V., Bowyer, K., Hall, L., Kegelmeyer, W.P.: SMOTE: synthetic minority over-sampling technique. J. Artif. Intell. Res. **16**, 321–357 (2002)

16. Batista, G.E.A.P.A., Prati, R.C., Monard, M.C.: A study of the behavior of several methods for balancing machine learning training data. SIGKDD Explor. **6**, 20–29 (2004)
17. Elkan, C.: The foundations of cost-sensitive learning. In: Proceedings of the 17th IEEE International Joint Conference on Artificial Intelligence (IJCAI 2001), pp. 973–978 (2001)
18. Ting, K.M.: An instance-weighting method to induce cost-sensitive trees. IEEE Trans. Knowl. Data Eng. **14**, 659–665 (2002)
19. Veropoulos, K., Campbell, C., Cristianini, N.: Controlling the sensitivity of support vector machines. In: Proceedings of the International Joint Conference on AI, pp. 55–60 (1999)
20. Zhou, Z.H., Liu, X.Y.: Training cost-sensitive neural networks with methods addressing the class imbalance problem. IEEE Trans. Knowl. Data Eng. **18**, 63–77 (2006)
21. Chan, P.K., Stolfo, S.J.: Toward scalable learning with non-uniform class and cost distributions: a case study in credit card fraud detection. In: Proceedings of the 4th ACM SIGKDD International Conference on Knowledge Discovery and Data Mining, pp. 164–168 (1998)
22. Liu, X.Y., Wu, J., Zhou, Z.H.: Exploratory undersampling for class imbalance learning. IEEE Trans. Syst. Man Cybern. Part B Cybern. **39**, 539–550 (2009)
23. Chawla, N.V., Lazarevic, A., Hall, L.O., et al.: SMOTEBoost: improving prediction of the minority class in boosting. In: Proceedings of the 7th European Conference on Principles of Data Mining and Knowledge Discovery, pp. 107–119 (2003)
24. Peng, L., Yang, B., Chen, Y., Abraham, A.: Data gravitation based classification. Inf. Sci. **179**, 809–819 (2009)
25. Simić, D., Tanackov, I., Gajić, V., Simić, S.: Financial forecasting of invoicing and cash inflow processes for fair exhibitions. In: Corchado, E., Wu, X., Oja, E., Herrero, A., Baruque, B. (eds.) HAIS 2009. LNCS, vol. 5572, pp. 686–693. Springer, Heidelberg (2009)
26. Cano, A., Zafra, S., Ventura, S.: Weighted data gravitation classification for standard and imbalanced data. IEEE Trans. Cybern. **43**(6), 1672–1687 (2013)
27. Parsazad, S., Yazdi, H.S., Effati, S.: Gravitation based classification. Inf. Sci. **220**, 319–330 (2013)
28. Wen, G., Wei, J., Wang, J., et al.: Cognitive gravitation model for classification on small noisy data. Neurocomputing **118**, 245–252 (2013)
29. Reyes, O., Morell, C., Ventura, S.: Effective lazy learning algorithm based on a data gravitation model for multi-label learning. Inf. Sci. **340–341**, 159–174 (2016)
30. Bradley, A.P.: The use of the area under the ROC curve in the evaluation of machine learning algorithms. Pattern Recogn. **30**, 1145–1159 (1997)
31. Huang, J., Ling, C.X.: Using AUC and accuracy in evaluating learning algorithms. IEEE Trans. Knowl. Data Eng. **17**, 299–310 (2005)

Genetic Algorithms

Solving the Static Manycast RWA Problem in Optical Networks Using Evolutionary Programming

Amiyne Zakouni[1], Jiawei Luo[1(✉)], and Fouad Kharroubi[2(✉)]

[1] School of Computer Science and Electronic Engineering, Hunan University,
Changsha, 410082, China
zakouni.amine@gmail.com, luojiawei@hnu.edu.cn
[2] School of Mathematics and Computer Science, Yichun University, Yichun, 336000, China
fouad.kharroubi@gmail.com

Abstract. The Static RWA (Routing and Wavelength Assignment) problem in Optical Networks is a combinatorial optimization problem fit to iterative search methods. In this article we further investigate the static manycast RWA problem in optical networks and solve it using an evolutionary programming (EP) strategy such that the number of the manycast requests established for a given number of wavelengths is maximized. The proposed algorithm solves, approximately, the wavelength assignment problem while a backtracking approach is used to solve the routing issue. We present the details of our proposed algorithm and compare it to another metaheuristic named genetic algorithm (GA). EP shows a 24 % improvement over GA.

Keywords: Backtracking · Evolutionary programming · Genetic algorithm · Manycast · Max-RWA · Metaheuristics · Optical networks

1 Introduction

The future of many technologies such as video conferencing, grid computing, e-Science and peer-to-peer (P2P) will employ a massive amount of data and support for point to multipoint communication. To fortify these accommodations, the next generation Internet will be predicated on optical networks that can provide immense amounts of bandwidth. Manycast [1–3], is a new type of communication that can fortify the point to multipoint nature of future services, in addition to fortifying generally used communication types. The manycast is a generalization of the multicast communication paradigm [4]. Indeed, multicast differs from manycast in that the destinations are determined, though in manycast the destinations must be picked. Manycast is especially useful in grid/cloud computing and e-Science. In the majority of the above cases, we are regularly dealing with a prodigious amount of data.

In these cases, a service provider might have various servers that give an identical service. In particular, consider parallel substance conveyance in an e-Science context. e-Science regularly creates a lot of data [1] that may then be stored in different areas for either backup purposes or so that numerous exploration labs can then process it locally. We can utilize manycast to pick some subset of these locations (e.g., the reduced cost

© Springer International Publishing Switzerland 2016
D.-S. Huang and K.-H. Jo (Eds.): ICIC 2016, Part II, LNCS 9772, pp. 147–157, 2016.
DOI: 10.1007/978-3-319-42294-7_12

storage clusters) to send this data in parallel along a light-tree set up by the network system. This issue can likewise be solved utilizing multicast rather than manycast, yet there are a few disadvantages. Rather than utilizing manycast, where the system would pick the subset of destinations for us, we can utilize multicast by picking k destinations at the source. The primary disadvantage stems from the client point of view. The nodes chosen at the source may be heavily loaded or may provide a slower transfer rate than different nodes in the candidate set. Furthermore, manycast grants the freedom to choose different nodes depending on the state of the network, prompting better execution for client applications and more efficient usage of WDM networks. WDM can provide unprecedented bandwidth, reduce processing costs, and enable efficient failure handling [5, 6]. An end-to-end lightpath has to be established prior to the communication between any two nodes in an optical network. A sequence of lightpath requests arrive over time with each lightpath assigned a random holding time.

These lightpaths should be set up dynamically by deciding a route across the network connecting the source to the destination and allocating a free wavelength along the path. Actual lightpaths cannot be rerouted to accommodate new lightpath requests until they are free, so some of the lightpath requests may be blocked if there are no free wavelengths along the path [7]. Therefore, finding a physical route for each lightpath demand and allocating to each route a wavelength, subject to the constraints, is known as the Routing and Wavelength Assignment (RWA) problem [8–10]. The concept of a lightpath is generalized into that of a light-tree [11], which unlike a lightpath, a light-tree has multiple destination nodes. Thus, a light-tree forms a tree rooted at the source node instead of the typical path created in the physical topology.

The solution to the manycast RWA problem can be either, given a fixed number of wavelengths and a set of manycast requests, to maximize the total number of manycast requests admitted (Max-RWA), or to minimize the number of wavelengths used (Min-RWA), provided that wavelength availability is sufficient to route all of the requests [12]. Given the hard computations of the linear integer program [13], we analyze the problem using metaheuristics. Our objective, given a fixed number of wavelengths, is to maximize the number of manycast requests to be established in a given session or traffic matrix.

The next section reviews the previous work completed on this topic. In Sect. 3, a problem definition and formulation is given. Section 4 proposes our assignment algorithm EP. In Sect. 5, experimental results and a comparison between the proposed approaches is presented. Section 6 discusses the empirical results obtained for the suggested metaheuristics. Finally, conclusions and future work of this paper are drawn in Sect. 7.

2   Previous Work

The RWA problem can be divided into two sub-problems, the path from source to destination - this is the routing part - and the wavelength along the path, which is the wavelength assignment part. Both of these sub-problems are NP complete [14] and tightly linked together. The manycast RWA issue is, therefore, NP complete since it contains the RWA issue as a special case.

Manycast is a special type of multicast communication, in which, from a single source, we must reach a set of destination nodes. These destination nodes are to be selected instead of being given. In fact, there are many previous works that investigate the multicast dilemma. This static multicast RWA is studied in numerous pieces of research [15, 16] targeting the objective of minimizing the blocking probability. Manycast is also a generalization of unicast where the message needs to be delivered to any one of the group. Indeed, there is a wealth of recent work [8, 10, 17–19] that proposes a genetic algorithm and an evolutionary programming to solve RWA problem in the unicast case. While in the manycast case, in numerous previous works, the manycast problem is first presented as quorumcast [2, 20, 21]. In quorumcast, messages are sent to a subset of destinations (quorum pool), which are selected from a set. Charbonneau and Vokkarane [1, 12] propose three heuristics to solve the manycast problem. The objective was to minimize the number of wavelengths required to satisfy all of the manycast requests. In our previous work [3], we propose and compare two metaheuristics, tabu search algorithm and genetic algorithm to solve the static manycast RWA problem by maximizing the number of manycast request established for a given number of wavelengths. In the work [22], an ILP and several heuristics are introduced for solving multi-resource manycast in mesh networks. Few studies, however, tackle the manycast service over optical burst-switched (OBS) networks [23–25].

3   Problem Definition and Formulation

3.1   Problem Definition

Let a network be represented as a graph $G(V, E)$, where V denotes the set of network nodes and E represents the set of unidirectional fibers. Assume that lightpath requests are unidirectional, each carrying W wavelengths. A manycast request is represented as $MR\{s, D_c, k\}$ where s, D_c, k denote the source, the set of candidate destination nodes, and $k \leq |D_c| = m$ is the number of destination nodes needed to reach out of m. If we change the parameters to $k = m = 1$ in the manycast request, we can also perform unicast [1]. Therefore, any algorithm that solves the static manycast RWA problem, in general, should adhere to these following constraints:

1. *Wavelength Continuity Constraint:* The wavelength continuity constraint indicates that a specific request for a source-destination pair should follow a single lightpath [26].
2. *Wavelength Conflict Constraint:* The wavelength conflict constraint affirms that a wavelength might be utilized just once per fiber. In this manner no two signals can cross along the same wavelength in a specific fiber [7].

3.2   Problem Formulation

Let N_{LP_G} be the number of all lightpaths in G. Let $R = (R_i)$ be the vector that contains the request number to which a lightpath belongs. Let N_R be the number of all requests in G. Let *multiplicity*(n) be the number of connection requests desired to be set up for one request. Let β be the sum of all utilized traffic by all requests, as follow:

$\beta = \sum\limits_{n=1}^{N_R} multiplicity(n).$ $\forall k \in \{1, 2 \dots N_R\}.$ Let $D = (d_{ij})$ be the $N_{LP_G} \times N_{LP_G}$ matrix i.e., $d_{ij} = 1$ if lightpaths i and j share a physical link, otherwise $d_{ij} = 0$. Let $T = (T_i)$ be the $1 \times N_R$ vector, i.e., $T_i = \lambda, \forall \lambda \in \{1, 2 \dots W\}$ if the wavelength λ is assigned to the lightpath-tree i, otherwise $T_i = 0$. Let $P = (Pi)$ be the $1 \times N_{LP_G}$ vector, i.e., $P_i = \lambda$, $\forall \lambda \in \{1, 2 \dots W\}$ if the wavelength λ is allocated to the lightpath i, otherwise $P_i = 0$. Let $\phi = 1$, if $T_i \neq 0 \, \forall i \in \{1, 2 \dots N_R\}$, otherwise $\phi = 0$. Our problem can be mathematically formulated as follows:

$$\text{maximize: } F = f(\phi) = \sum_{i=1}^{N_R} \phi_i \quad \text{such as}$$

$$if \ d_{ij} = 1, i \neq j \Rightarrow P_i \neq P_j \ \forall i, j \in \{1, 2, \dots N_R\} \quad (1)$$

$$\sum_{i=1}^{N_R} \phi_i \leq \beta, \forall i \in \{1, 2 \dots N_R\} \quad (2)$$

In constraint (1), wavelengths assigned must be such that no two lightpaths that share a physical link, belonging to different requests, or use the same wavelength on that link. In constraint (2), the sum $\sum\limits_{i=1}^{N_R} \phi_i$ of the elements of P that are different from zero, cannot, under any circumstance, surpass the number β.

4   Our Proposed Metaheuristics

Previous research offers a variety of solutions, from simple to complex metaheuristic algorithms for solving the RWA complication. Here, we extend the same Evolutionary Programming (EP) presented in [8, 10, 19] based on a backtracking approach but this time we utilize it to solve the Static Manycast RWA problem.

4.1   Evolutionary Programming (EP)

The EP is a stochastic optimization procedure that can be applied to a range of difficult combinatorial problems. It's relatively similar to Genetic Algorithm since both them emulate the procedure of natural evolution in order to solve combinatorial optimization problems. For this purpose, EP exploits a myriad of techniques inspired by natural evolution such as selection, mutation, and replacement so that it can create the best near-solutions to optimization issues. Compared with GA, EP has no crossover operator even though it also has the mutation function. The key concepts of the EP explained below:

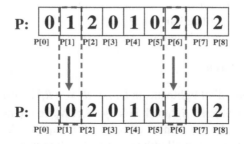

Fig. 1. Mutation phase.

Initial Population: In this phase, each gene in a chromosome solution represents one of the paths generated through a backtracking algorithm so that we can investigate all the possible candidate paths between the source and the destination pairs. These candidate solutions are called chromosomes which take, in our case, the form of bit strings. Each bit position in the chromosome has W possible values. During this step, we initialize the variables that will be used namely: k, D_c, n, P, P_{max} and F_{max}.

Selection: In this step, the chromosomes of the next generation are selected from the current population by evaluating all the chromosomes using a fitness function choosing the best individual.

Mutation: In this stage, the operator randomly inverts some of the bits in a chromosome, which is considered as a random mutation of the new pool. Thus, some randomly chosen elements of the vector P (P contains the best-found solution in terms of the allocated wavelength to the chosen paths for a manycast request) containing the value λ, which represents the wavelength that the lightpath will use, will be randomly changed by a different value of the wavelength λ. In this step, the generated chromosome replaces

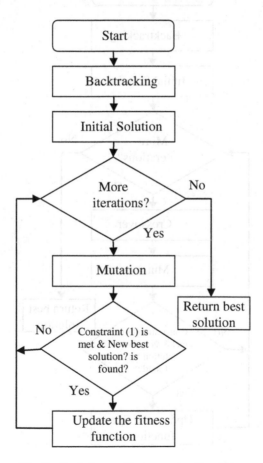

Fig. 2. Evolutionary Programming flow chart.

itself regardless of the fitness function. This concept is shown in Fig. 1. More details about EP can be found in [4, 6, 17–19, 27, 28]. The main working steps of our proposed EP are shown in the general flow (Fig. 2).

4.2 Genetic Algorithm (GA)

The GA is a search technique originally invented by [29] and used in computing to find true or approximate solutions to optimization and search problems. Indeed, this meta-heuristic belongs to the larger class of evolutionary algorithms, which is inspired on process of natural selection and is routinely used to generate useful solutions. Genetic algorithms use biologically-derived techniques such as inheritance, mutation, natural selection, and crossover (or recombination). More details about GA can be found in

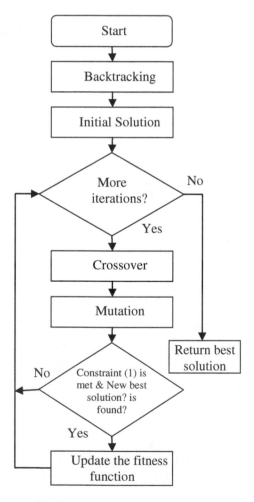

Fig. 3. Genetic Algorithm flow chart.

[3, 4, 6, 17–19, 27–30]. The main working steps of the proposed GA done in our previous work [3] are shown in this following general flow (Fig. 3):

4.3 Backtracking Algorithm

In the work [8, 10, 19] the authors propose a backtracking algorithm for routing unicast demands. This algorithm can extend to the manycast case if a path search is done for every destination one-by-one. Using this method, we have the capacity to investigate all of the possible candidate paths between the source and the destination pairs of the trees. All research in this paper focuses on the extension of the work done on static unicast RWA problems by utilizing the backtracking algorithm. Previous studies focus on k-shortest path [5, 31], which is widely used in the literature to find alternative paths. Hence, by using the backtracking approach, our initial search space will contain not only the k-shortest paths between each source-destination pair, but also all of the possible candidate lightpaths. More details about backtracking can be found in [3, 8, 10, 19].

5   Numerical Results

The hardware we're using for our experiments is an Intel(R) Core(TM) i7- 4790 k CPU 4 GHZ processor with 8 GB RAM, running under Ubuntu 14.04.2. All algorithms were compiled by GCC compiler of Qt Creator 3.4 (based on Qt 5.4 "64Bit"). Our experiment consists of 90 extensive tests. The experiment is executed for 5 manycast groups consisting of 10, 20, 30, 40 and 50 requests, running each test on two different algorithms with the same initial population and parameters. The number of wavelengths, W, chosen for network simulation, is 64, 160, and 320, which are practical values today [32]. Moreover, we use three different destination set sizes (4, 6 and 8). For each request MR, we use 3 different sizes of D_c, and $k = D_c/2$. The maximum number of iterations is fixed at 5000. We run all these simulations on the 14-node NSF network. We assume that splitting capabilities and wavelength converters [33] are not adopted in our case.

As it can be seen from the performance results (Fig. 4), we notice that the number of established sets of manycast sessions increases with the increase of W and decrease when the set of destination is larger:

- When D = 4, the proposed EP gives better results up to 24 % compared to GA technique. However, for other W = 160 and 320, EP and GA algorithms produce the same results.
- When D = 6, we notice that GA outperform EP, specifically when the manycast group sizes is less than 20, this out-performance goes up to 5 %.
- When D = 8, GA shows a decent performance that is close to the EP method when W = 64. Otherwise we observe that GA has a better improvement over EP.
- The time spent by GA or EP to solve the manycast problem increases rapidly depending on the number of manycasts groups as well as the number of the wavelengths. However, the time spent using the EP is three times higher than GA. In fact, since we are dealing with the static case, computations are done offline.
- We should remark that EP shows sufficient performance to solve the manycast RWA problem in many cases. Nevertheless, it takes a while to reach its end. Thus we

consider the time and minimum number of iterations of the best found solution for the future work.

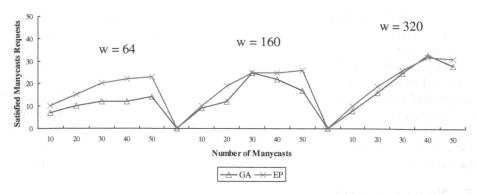

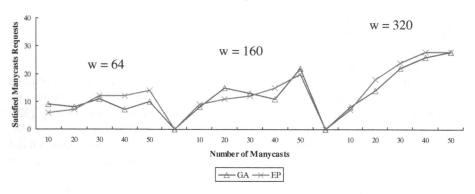

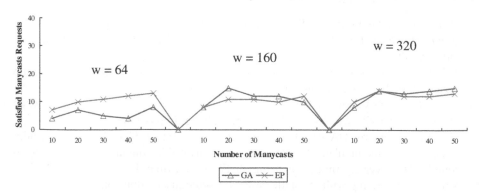

Fig. 4. Satisfied manycast requests for D = 4, 6 and 8.

6 Discussion

In small destination sets EP largely performs better than GA. This out-performance goes up to 24 % improvement over GA, in terms of a solution. However, GA run times are very low compared to the EP approach. GA has shown it can perform better than EP by up to 5 %, especially for large destination sets. This can be related to the crossover site utilized by GA which maintains diversity in the resolution space. Related to the wavelength variation, the results are somewhat predictable, since it is easier for small size manycast requests to be nearly accepted, most notably when the wavelength number is large. In contrast, large size requests often require much of the network resources, in turn many requests could not be satisfied.

Regarding the fairness issue, we observe that GA achieves better fairness among the manycast groups in terms of satisfying all the groups of connections whereas the EP have shown low fairness, specifically when the manycast group increases. For the wavelength reuse issue, we notice that the probability of the re-use of existing wavelength is higher only if the frequency of occurrence of a common physical link is very low, seeing that lately the number of wavelengths increases, so this wavelength reuse problem will be nonexistent.

Our proposed Genetic Algorithm and Evolutionary Programming approaches a satisfactory solution. Our performance evaluation of 90 tests confirms that, although much research attempts to solve the multicast RWA problem, only a few studies try to deal with the manycast RWA issue. It is, therefore, important to develop more, new metaheuristics for solving the manycast RWA problem.

7 Conclusion and Future Work

In this paper we implement and compare two metaheuristic strategies to solve the static manycast RWA problem, with a special focus on maximizing the number of manycast requests established for a given number of wavelengths. The problem was studied for the static case only. We propose two metaheuristics to compute the approximated solutions, in which GA works better when the destination set are larger. This is when we increase the destination set. EP has shown good performances for small destination sets. The routing sub-problem is solved using a backtracking algorithm. The proposed GA and EP in this paper are applicable to a real NSF network. A relevant comparison, counting the performance and the time included, is made between the two metaheuristics, making a total of 90 experiments. The time spent by EP, on average, is three times higher than GA.

This work can be extended by using both splitting capabilities and wavelength converters as we have only considered networks without them. The dynamic manycast problem can be considered for future work for the manycast issue over wavelength-routed networks. The dynamic manycast complication is especially important for services that are related to cloud computing and grid networks, where multiple resources are required. Our work concentrates on NSF networks only. Further research on more metaheuristics in more irregular and more realistic topologies should be conducted. Further study can be conducted with broadcasting communications where many-to-many sessions run simultaneously.

References

1. Charbonneau, N., Vokkarane, V.M.: Tabu search meta-heuristic for static manycast routing and wavelength assignment over wavelength-routed optical WDM networks. In: Proceedings of IEEE International Conference on Communications (ICC 2010), Cape Town, South Africa, 23–27 May 2010
2. Cheung, S.Y., Kumar, A.: Efficient quorumcast routing algorithms. In: Proceedings of IEEE INFOCOM, pp. 840–847 (1994)
3. Zakouni, A., Luo, J., Kharroubi, F.: Genetic algorithm and tabu search algorithm for solving the static manycast RWA problem in optical networks. J. Comb. Optim. (2016). doi:10.1007/s10878-016-0002-3. Springer Science+Business Media, New York
4. Singhal, N.K., Sahasrabuddhe, L.H., Mukherjee, B.: Optimal multicasting of multiple light-trees of different bandwidth granularities in a WDM mesh network with sparse splitting capabilities. IEEE/ACM Trans. Netw. **14**, 1104–1117 (2006)
5. Ramaswami, R.: Optical networking technologies: what worked and what didn't. IEEE Commun. Mag. **44**(9), 132–139 (2006)
6. Jain, R.: Internet 3.0: ten problems with current Internet architecture and solutions for the next generation. In: Proceedings of IEEE MILCOMM, October 2006
7. Skorin-Kapov, N.: Routing and wavelength assignment in optical networks using bin packing based algorithms. Eur. J. Oper. Res. **177**(2), 1167–1179 (2007)
8. Kharroubi, F., He, J., Tang, J., Chen, M., Chen, L.: Evaluation performance of genetic algorithm and tabu search algorithm for solving the Max-RWA problem in all-optical networks. J. Comb. Optim. **30**(4), 1042–1061 (2013). doi:10.1007/s10878-013-9676-y
9. Zang, H., Jue, J.P., Mukherjee, B.: A review of routing and wavelength assignment approaches for wavelength-routed optical WDM networks. Opt. Netw. Mag. **1**(1), 47–60 (2000)
10. Kharroubi, F.: Random search algorithms for solving the routing and wavelength assignment in WDM networks. Ph.D. thesis, Hunan University, June 2014
11. Sahasrabuddhe, L.H., Mukherjee, B.: Light trees: optical multicasting for improved performance in wavelength-routed networks. IEEE Commun. Mag. **37**(2), 67–73 (1999)
12. Charbonneau, N., Vokkarane, V.: Routing and wavelength assignment of static manycast demands over all-optical wavelength-routed WDM networks. J. Opt. Commun. Netw. **2**(7), 427–440 (2010)
13. Krishnaswamy, R.M., Sivarajan, K.N.: Algorithms for routing and wavelength assignment based on solutions of LP-relaxations. IEEE Commun. Lett. **5**(10), 435–437 (2001)
14. Jue, J.P.: Lightpath establishment in wavelength-routed WDM optical networks. In: Ruan, L., Du, D.-Z. (eds.) Optical Networks: Recent Advances, pp. 99–122. Springer, New York (2001)
15. Le, D.D., Zhou, F., Molnar, M.: Minimizing blocking probability for MCRWA problem in WDM networks: exact solutions and heuristic algorithms. IEEE/ OSA J. Opt. Commun. Netw. **7**(1), 36–48 (2015)
16. He, J., Chan, S.-H., Tsang, D.: Routing and wavelength assignment for WDM multicast networks. In: IEEE GLOBECOM, pp. 1536–1540 (2011)
17. Qin, H., Liu, Z., Zhang, S., Wen, A.: Routing and wavelength assignment based on genetic algorithm. IEEE Commun. Lett. **6**(10), 455–457 (2002)

18. Bhanja, U., Mahapatra, S., Roy, R.: An evolutionary programming algorithm for survivable routing and wavelength assignment in transparent optical networks. J. Inf. Sci. **222**, 634–647 (2013)

19. Kharroubi, F., He, J., Chen, L.: Performance analysis of GA, ROA, and TSA for solving the Max-RWA problem in optical networks. In: Optical Fiber Communication Conference, OSA Technical Digest (online) (Optical Society of America, 2014) (2014). paper W2A.48, doi: 10.1364/OFC.2014.W2A.48

20. Low, C.P.: Optimal quorumcast routing. In: Proceedings of IEEE GLOBECOM, vol. 5, pp. 3013–3016, November 1998

21. Wang, B., et al.: An efficient QoS routing algorithm for quorumcast communication. In: Proceedings of IEEE ICNP (2001)

22. She, Q., Kannasoot, N., Jue, J.P., Kim, Y.-C.: On finding minimum cost tree for multi-resource manycast in mesh networks. Elsevier Opt. Switching Netw. **6**(1), 29–36 (2009)

23. Huang, X., She, Q., Vokkarane, V.M., Jue, J.P.: Manycasting over optical burst-switched networks. In: Proceedings of IEEE ICC, pp. 2353–2358 (2007)

24. She, Q., Huang, X., Kannasoot, N., Qiong, Z., Jue, J.P.: Multiresource manycast over optical burst-switched networks. In: Proceedings of IEEE ICCCN, pp. 222–227, August 2007

25. Bathula, B.G., Vokkarane, V.M.: QoS-based manycasting over optical burst-switched (OBS) networks. IEEE/ACM Trans. Netw. **18**(1), 271–283 (2010)

26. Rouskas, G.N., Perros, H.G.: A tutorial on optical networks. In: Gregori, E., Anastasi, G., Basagni, S. (eds.) NETWORKING 2002. LNCS, vol. 2497, pp. 155–193. Springer, Heidelberg (2002)

27. Wei, G.: Study on genetic algorithm and evolutionary programming. In: Proceedings of IEEE International Conference on Parallel, Distributed and Grid Computing, India, pp. 762–766 (2012)

28. Lee, C.Y., Yao, X.: Evolutionary programming using mutations based on the levy probability distribution. IEEE Trans. Evol. Comput. **8**(1), 1–13 (2004)

29. Holland, J.H.: Genetic algorithms. Scientific American, 114–116 (1992)

30. Ouyang, A., Tang, Z., Zhou, X., Xu, Y., Pan, G., Li, K.: Parallel hybrid PSO with CUDA for LD heat conduction equation. Comput. Fluids **110**, 198–210 (2015)

31. Chamberland, S., Khyda, D.O., Samuel, P.: Joint routing and wavelength assignment in wavelength division multiplexing networks for permanent and reliable paths. Comput. Oper. Res. **32**(5), 1073–1087 (2005)

32. Singhal, N.K., Sahasrabuddhe, L.H., Mukherjee, B.: Optimal multicasting of multiple light-trees of different bandwidth granularities in a WDM mesh network with sparse splitting capabilities. IEEE/ACM Trans. Netw. **14**, 1104–1117 (2006)

33. Le, D.D., Zhou, F., Molnar, M.: Minimizing blocking probability for MCRWA problem in WDM networks: exact solutions and heuristic algorithms. IEEE/OSAJ. Opt. Commun. Netw. **7**(1), 36–48 (2015)

Signal Processing

A Control Strategy of Depressing the Voltage Spike During Soft-Switch Based on the Method of PI in Photovoltaic Converter System

Huixiang Xu[⊠] and Nianqiang Li

School of Information Science and Engineering,
University of Jinan, Jinan 250022, China
jd_chuying1990@163.com

Abstract. For the problem of voltage spike which is caused by the MOSFET, and the issue of Variable calculation complex, given in the fly-back photovoltaic converter system (hereinafter abbreviated as PV). A new physical circuit of soft-switch control strategy is proposed which based on PI controller with the algorithm of variable power step to suppress the voltage spike value. According to the theoretical analysis, in this paper, the output voltage of DC part is equal to the input voltage of photovoltaic array and the reflected voltage of secondary side by clamping the voltage of clamped capacitor. The output voltage value from PI controller is used to recovery the leak inductance energy which will help the converter for working on the maximum power point(MPP) steadily. The new design can absorb the leak inductance energy and suppress the voltage spike efficiently. In this paper, for solve the traditional problem, we correct the formula of soft-switch by the method of maximum power point tracking (MPPT) algorithm. The new soft-switch algorithm formula can simplify the demand of physical circuit design through the method of variable power step.

Keywords: Solar energy · Fly-back converter · PI controller · Voltage spike · Soft-switch

1 Introduction

DC/DC conversion is an important part in photovoltaic converter system [1, 2, 6] (PV). It principle is that transform the continuous DC voltage to the intermittent DC voltage by high frequency MOSFEET. The aim is to making converter work under the boundary current mode (BCM) [3]. But the energy loss, which caused by the MOSFET will sharply reduce the inverter efficiency, especially in the high frequency converter system [7]. According to the theory, the soft-switch algorithm can reduce the energy loss to zero [3]. Traditionally, such as mentioned in the Ref. [3], there are multiple variables in the calculation formula of soft-switch complex variable, which will make hardware system more stringent. In Ref. [8], the author has implemented a soft-switch DC/DC converter, which can realize soft switch in the range of load, and diode zero current turn off. But the voltage spike that generate from the leak inductance energy and the MOSFET have not be suppressed, which will bring significant energy loss in the process of switch. In this paper, we design a new regulation circuit to recovery the

© Springer International Publishing Switzerland 2016
D.-S. Huang and K.-H. Jo (Eds.): ICIC 2016, Part II, LNCS 9772, pp. 161–169, 2016.
DOI: 10.1007/978-3-319-42294-7_13

leak inductance energy by the PI controller with the algorithm of variable power step to suppress the voltage spike value. The algorithm combined with the algorithm of MPPT [4, 5] to simplify hardware circuit topology. Compared with others, the algorithms which mentioned in this paper, can improve the efficiency of the conversion better. According to the algorithm of MPPT, we modify the complex calculation variables in the theoretical formula.

2   The Control Strategy of Voltage Spike Depress Based on PI

2.1   The Cause of the Voltage Spike

Compare to the traditional one converter power system for multiple PV modules, the fly-back converter suitable for medium and small power transformation, can be used to one converter and one PV modules connected. It is more easily to work in the Maximum Power Point Tracking (MPPT). The working principle of the fly-back transformer is that the secondary coil do not output power to the load, until the primary coil incentive cut off. Because of the fly-back have an air gap when its designed, so the original boundary energy cannot be 100 % transformation to the secondary side, thus generate the leak inductance. The existence of leak inductance will necessarily produces energy loss, and may cause counter assault if its magnitude is excessive [9–11]. The output power cannot meet the requirements if too much leak inductance energy retention during input, which will be added to the drain-source side of the MOSFET. This problem will lead to the voltage spike if the system not set a reasonable discharged circuit. The voltage curve of the Ugs (the voltage between gate and source of MOS-FET) and the Uds (the voltage between drain and source of MOSFET) has been shown in Fig. 1, approximately, following 30 V voltage spike without the new method.

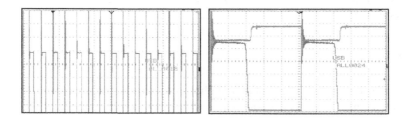

Fig. 1. The voltage of Ugs and Uds

2.2   Circuit Topology and Control Strategy

The physical circuit has been shown in Fig. 2. The circuit includes two parts that are the leak inductance energy recovery and the regulation circuit of PI. Among them, the leakage inductance energy absorption diode (D1) and the clamp capacitor (Ca) are used to absorb and suppress leak inductance energy. The leakage inductance energy feed-back diode (D2), high frequency transformer (T3) and the MOSFET (MOS2) are used to recovery the leak inductance energy, which has been absorbed by the former part.

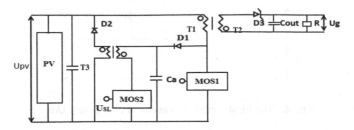

Fig. 2. The physical circuit

When the main power switch (MOS1) conduction, D1 cut-off, the absorb circuit of leak inductance energy do not work. When the MOS1 shut off, D1 turn on, the leak inductance energy will be stored in the clamp capacitance (stored in Ca). When the MOS2 conduction, the rectifier diode (D2) cut-off, the leak inductance energy will be stored in the T3, and when the MOS2 shut off, the energy recovery to the input side by D2.

As the Fig. 3 shown, the control strategy of PI controller is combined with the input voltage of photovoltaic (Upv), grid voltage (Ug), the ratio of transformer (N), and the voltage of clamp capacitance (Uca). The gate voltage (USL) is a waveform of PWM which the frequency fixed and duty-cycle can be changed, in the leak inductance energy feedback circuit. Setting the initial input value of PI controller is equal to the input voltage value of photovoltaic array added the reflected voltage value of the grid (UPV + Ug/N). Measurement the voltage value of clamp capacitor in real-time, and the difference between the set and measured value (e(K)) is the one of input parameters of PI controller. Its output value is the duty-cycle of PWM to control the feedback circuit to absorbing the energy.

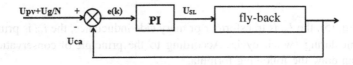

Fig. 3. The control strategy of PI controller

3 Improve the Soft-Switch Formula

Traditionally, the soft switch calculation needs collecting the voltage value and current value at the same time, which will make hardware system more stringent. To solve this problem, in this paper, we have modified the soft-switch algorithm.

The amplitude of the first resonance wave is the largest, with the maximum value Ug/N. So the voltage of drain-source (Uds) reaches the minimum value in the first resonance wave valley, which is Upv − Ug/N. when Upv < Ug/N, the Uds reduce to zero, at this time, the power switch will work under the condition of zero voltage state (ZVS). In ideal condition, the wave of Ugs and Uds has been shown in Fig. 4.

For the PV converter, which worked in the BCM mode, the instantaneous power output (Po) is [3] described as follows:

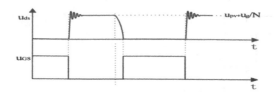

Fig. 4. The ideal condition waveform of Ugs and Uds

$$Po = Pop \sin^2 \phi_G \tag{1}$$

The peak power output (Pop) is described as follows

$$Pop = Ugp \bullet Igp \tag{2}$$

The *Ugp* and *Igp* are peak power grid voltage and grid current respectively.

Set the duty ratio to D. Because of the switching cycle *Ts* is equal to the transformer excitation time added the demagnetization time, so,

$$D \bullet Ts \bullet Upv = \frac{Ug}{N}(1 - D)Ts \tag{3}$$

$$D = \frac{Ug}{N \bullet Upv + Ug} \tag{4}$$

In each switching cycle, the output energy of the transformer is

$$E = L_p \bullet i_{sp}^2 \tag{5}$$

In the Eq. (5), the L_p is transformer primary side inductance, the i_{sp} is primary side peak current during switch cycle. According to the principle of conservation of energy, we can draw the following formula,

$$E = Pop \sin^2 \phi_G Ts \tag{6}$$

Substitute (5) to (7),

$$Ts = Lp \bullet Pop(\frac{\sin \phi_G}{Upv \bullet D})^2 \tag{7}$$

In Eq. (7), the *Pop* is the output power through MPPT algorithm. The equivalent output circuit of the photovoltaic [12] has been shown in Fig. 5. The *Cin* is a large capacity electrolyte capacitor use for balance the instantaneous power of DC side and Ac side. According to the law of conservation of energy, the output energy of photovoltaic (*Epv*) is equal to the input energy of converter (*Ei*) added the energy(*Ec*) which stored in *Cin*:

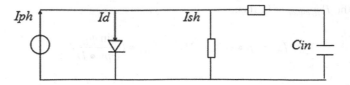

Fig. 5. Equivalent output circuit of the photovoltaic

$$Ec = Epv - Ei \tag{8}$$

Let Tp as a power cycle, Ppv as the average output power of photovoltaic, P_i as the input power of converter, in a power cycle, the above formula can be written as,

$$\frac{1}{2}C_{in}U_{t+Tp}^2 - \frac{1}{2}C_{in}U_t^2 = P_{pv}Tp - \int_t^{t+Tp} p_i dt \tag{9}$$

In the Eq. (9), U_t is the capacitor voltage value instantaneous. U_{t+tp} is the capacitor voltage value after a power cycle. Set time t_0 to t_1, t_1 to t_2, the average output power was P_1, P_2, and the electrolytic capacitor voltage at the time of t_0, t_1 and t_2 are respectively U_0, U_1 and U_2. Take $t_1 = t_0 + T_p$, $t_2 = t_1 + T_p$ substituting (9),

$$\frac{1}{2}C_{in}U_1^2 - \frac{1}{2}C_{in}U_0^2 = P_1 T_p - \int_{t_0}^{t_1} p_i dt \tag{10}$$

$$\frac{1}{2}C_{in}U_2^2 - \frac{1}{2}C_{in}U_1^2 = P_2 T_p - \int_{t_1}^{t_2} p_i dt \tag{11}$$

Because the grid energy is equal in two consecutive power cycles, so

$$\int_{t_0}^{t_1} p_i dt = \int_{t_1}^{t_2} p_i dt \tag{12}$$

subtract (10) from (11):

$$\Delta p = \Delta P_{pv} = P_2 - P_1 = \frac{Cin}{2Tp}(U_2^2 - 2U_1^2 + U_0^2) \tag{13}$$

$\frac{Cin}{2Tp}$ is a constant, Δp is power variation, take $\Delta p = U_2^2 - 2U_1^2 + U_0^2$ and initialize Pop, replace Pop to $Pop_{initial} + \Delta p$,

$$Ts = Lp \bullet (Pop_{initial} + \Delta p) \bullet \left(\frac{\sin\phi_G}{Upv \bullet D}\right)^2 \tag{14}$$

When the $Pop_{initial} = 0$, then,

$$Ts = Lp \bullet (U_2^2 - 2U_1^2 + U_0^2) \bullet (\frac{\sin \phi_G}{Upv \bullet D})^2 \qquad (15)$$

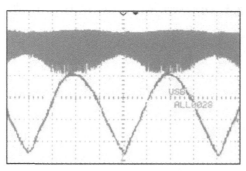

t=2.5ms/lattice

(a) The half wave and driving waveform

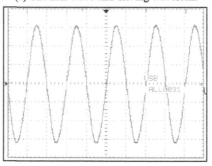

(t=10ms/lattice、peak value=100V/lattice)

(b) The converter AC output voltage

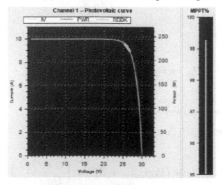

(c) The curve of MPPT and the efficiency

Fig. 6. Experiment waveform

4 The Experimental Result

In this paper, the converter rated capacity is 230 VA, with the input 30 V DC and output 220 V AC. The main power MOSFET is used IRF4321, in which the parasitic capacitance Coss 390PF, the primary induction 5.4 μ H, the leak inductance 0.2 η H, and the ratio of turns is 1:6. The clamp capacitance Ca is 3.3nF. The leak inductance recovery resistance R1 is 100 K. The clamp diode is used ES1D. The switch tube in secondary side is used 17N80, which include the commutation diode C2D05120.

The experiment waveform has been shown in Fig. 6. The half wave and the driving waveform which are measured by the new method, during the commutation of the power MOSFET, has been shown in Fig. 6(a). From the waveform we can found that the MOSFET working in normal condition. As the Fig. 6(b) shown that the grid voltage waveform with the peak value is about 330 V, and very smooth. The efficiency of the maximum power point tracking has been shown in Fig. 6(c), which can reach 99.3 % when full load.

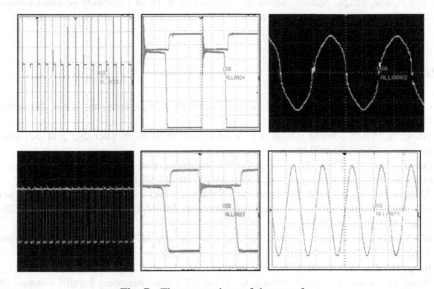

Fig. 7. The comparison of the waveform

5 Conclusions

To solve the problem in tradition design, we correct the formula of soft-switch combined with the method of MPPT, and propose a new method of PI with a variable power step, which will suppress the voltage spike efficiently. The new formula can simplify hardware circuit topology. The new method can improve the conversion efficiency.

The comparison of the waveform before and after modification has been shown in Fig. 7. From the experiment waveform, we can see clearly that the voltage spike has already been suppressed. Meanwhile, the main switch worked under the state of zero voltage switching (ZVS).

In this paper, a new physical circuit of soft-switch control strategy is proposed which based on PI controller with the algorithm of variable power step to suppress the voltage spike value. It's combined with the algorithm of MPPT to simplify hardware circuit topology to improving the efficiency of the conversion.

The new design solves the problem of complication variable calculation, which will largely simplify hardware circuit topology. Theoretical analysis and experimental results show that its can suppress the voltage spike and clamp the drain-source voltage effectively. The design can protect the main power switch by the method of PI controller. We improve the domestic converter efficiency from 94.2 % to 96.2 %, while the world's highest converter efficiency up to 98 % [13].

Acknowledgment. The authors would like to the express thanks for the helpful discussions with Dr Li, and the AMETEK Programmable Power company for providing relevant photovoltaic equipment. The authors also thank Mr. Sun for providing some opinions about the photovoltaic converter system design. Authors would like to thank Han Yu of University of Jinan Computational Intelligence Lab for providing relevant information about this meeting announced.

References

1. Cai, C., Qin, H.: A design of the snubber circuit for flyback converter. Chin. J. Electron Devices **36**(4), 469–472 (2013)
2. Wang, S., Yu, S., Wang, D., et al.: Development of 3KW dispatchable grid connected inverter. ACTA ENERGIAE
3. Gu, J., Wu, H., Chen, G., Xing, Y.: Soft-switching grid-connected PV inverter with interleaved flyback topology. Proc. CSEE **31**(12), 40–45 (2011)
4. Abdelhamid, L., Mourad, H., Sabir, M.: Simulation and experimental design of a new advanced variable step size incremental conductance MPPT algorithm for PV systems. ISA Trans. **62**(08), 6–14 (2015)
5. Ma Y., Guo Q., Zhou X.: A modified variable step size MPPT algorithm. Microcomput. Appl. **34**(17): 78–80, 84 (2015)
6. Jiang, H., Zou, Y.: A soft SPWM resonant inverter. Electric Drive. (06): 14–18 (1996)
7. Wang, Q., Wang, T., Liu, X., Huang, C., Huang, Y.: High frequency parallel resonant DC link soft-switching inverter. Electric Mach. Control **17**(08), 57–62 (2013)
8. Ma, G., Qu, W., Liu, Y.: A novel soft switching bidirectional DC/DC converter and its ZVS condition. Trans. China Electrotechnical Soc. **21**(07), 15–19 (2006)
9. Zhang, H., Zhao, Y.: Design of multifunctional flyback switching power supply. Electric Power Autom. Equip. **31**(01), 113–117 (2011)
10. Lv, L., Xiao, J., Zhong, Z., Shi, Y.: Design of single-ended flyback transformer in high-frequency switching power supply. J. Mag. Mater. Devices **34**(01), 36–38 (2006)
11. Zhu S.: The research and design of grid-connected microinverter based on an interleaved flyback structure. School of Electrical Engineering Shenyang University of Technology (2013)

12. Yang, Y., Zhou, K.: Photovoltaic cell modeling and MPPT control strategies. Trans. China Electrotechnical Soc. (12): 40–45 (2011)
13. Jian, F., Li, Z.: Technology status and future development of PV inverter. High Power Convert. Technol. (03): 5–9 (2014)

A Novel Feature Extraction Method for Epileptic Seizure Detection Based on the Degree Centrality of Complex Network and SVM

Haihong Liu[1,2], Qingfang Meng[1,2(✉)], Qiang Zhang[3], Zaiguo Zhang[4], and Dong Wang[1,2]

[1] School of Information Science and Engineering,
University of Jinan, Jinan 250022, China
ise_mengqf@ujn.edu.cn
[2] Shandong Provincial Key Laboratory of Network
Based Intelligent Computing, University of Jinan, Jinan 250022, China
[3] Institute of Jinan Semiconductor Elements Experimentation,
Jinan 250014, China
[4] CET Shandong Electronics Co., Ltd., Jinan 250101, China

Abstract. Epilepsy is a kind of ancient disease, which is affecting the life of patients. With the increasing of incidence of epilepsy, automatic epileptic seizure detection with high performance is of great clinical significance. In order to improve the efficiency of epilepsy diagnosis, a novel feature extraction method for epileptic EEG signal based on the statistical property of the complex network and an epileptic seizure detection algorithm, which is composed of the extracted feature and support vector machine (SVM) is proposed. The EEG signal is converted to complex network by horizontal visibility graph firstly. Then the degree centrality of complex network as a novel feature is calculated. At last, the extracted feature and SVM construct automatic epileptic seizure detection. A classification experiment of the epileptic EEG dataset is performed to evaluate the performance of the proposed detection algorithm. Experimental results show the novel feature we extracted can distinguish ictal EEG from interictal EEG clearly and the proposed detection algorithm achieves high classification accuracy which can be up to 93.92 %.

Keywords: Epileptic seizure detection · Feature extraction method · Epileptic electroencephalograph (EEG) · Degree centrality · Support vector machine (SVM) · Horizontal visibility graph (HVG)

1 Introduction

Epilepsy is a kind of brain nerve disease which has attracted much more attention of some scholars. Fear, misunderstanding, discrimination and social stigma have surrounded epilepsy patients for many years, which affect the life of patients seriously. Therefore, the prevention and treatment of epilepsy disease are needed urgently. Furthermore, it also has certain social significance to improve the level of people's health.

© Springer International Publishing Switzerland 2016
D.-S. Huang and K.-H. Jo (Eds.): ICIC 2016, Part II, LNCS 9772, pp. 170–180, 2016.
DOI: 10.1007/978-3-319-42294-7_14

At the beginning of the study for epilepsy, it mainly relies on the professional doctor to identify the epileptic EEG signal with the naked eye. This method wastes too much time and the accuracy is not high. Later, a study found that the complex structure of the brain and the irregularity discharge of cerebral cortex neurons make the EEG signal show the nonlinear dynamic characteristics. So, nonlinear time series analysis method is utilized to extract nonlinear information of epileptic EEG signal by relevant scholars. From the information theory perspective, entropy is generalized as the amount of information stored in a more general probability distribution. It can be applied to quantify the complexity of the signal. At beginning, many scholars adopted entropy as the characteristic.

Kannathal found that the approximate entropy of the normal EEG is greater than that of epileptic EEG, which suggests that the EEG during a seizure is more regular and less complex than the normal. Optimized sample entropy [1] was proposed by Y. Song, which removes the calculation redundancy, was applied to seizure detection. N. Kannathal [2] utilized the different entropies to differentiate the different EEG signals. Other nonlinear features [3] were also used for analysis of epileptic EEG signals. As for the traditional nonlinear characteristic commonly used such as approximate entropy and sample entropy of epileptic EEG, the classification perfor- mance is relatively poor and can't satisfy the needs of reality. Recently, complex network theory provides a new way to analyze the time series. In recent years, the statistical property of complex network is appreciated by scholars to study the deep dynamical structure of nonlinear time series. Therefore, the classification of epileptic EEG signal can take advantage of the statistical property of complex network to achieve good results. Jie zhang [4] characterized pseudo-periodic time series through the complex network approach. Zhang and Small [5] proposed the pioneering algo- rithm that converted the pseudo-periodic time series into complex network. They calculated the degree distribution of time series' network and proved that the time series with different dynamics have different degree distribution forms. The literature [6–12] researched the different dynamic time series, applied the directed weighted complex network to study time sequence and further studied the time series of multi-scale characteristic of the complex network. In addition, the complex network theory has wide application in many aspects, such as in finance and economy. In 2008, Lacasa proposed the visibility graph [13, 14] algorithm for the first time, which could convert an arbitrary time series into complex network. They constructed the fractal series' visibility graph and discovered that the power-law degree distribution is related to fractality. The literature [15] studied exchange rate series by visibility graph. The literature [16, 17] used visibility graph to describe the fractional Brownian motions. The literature [18–23] presented that we could utilize the visibility graph to study many kinds of data, such as the financial and economic data. In a word, many studies show that time series can be studied from the perspective of complex network structure by using the visibility graph.

In this paper, we come up with the novel feature extraction method for the epileptic EEG signal based on the statistical property of complex network. The EEG signal is converted to complex network by horizontal visibility graph firstly. Through this step, we can get the time series' complex network. That is to say, we convert the signal from time domain to complex network domain. Then the degree centrality of complex network as a novel feature is calculated. The results show that different state has different characteristic values obviously. At last, the extracted feature and SVM [24, 25] construct automatic epileptic seizure detection. The classification performance of the extracted feature and the detection algorithm are evaluated by a public epileptic EEG dataset.

2 Feature Extraction Methods

This section presents the algorithm of converting the time series to complex network and focuses on introducing the feature extraction method for epileptic EEG. In this paper, we use horizontal visibility graph to construct the EEG signal's complex network firstly. However, most researchers convert the time series to complex network by the proximity network in this part. Then, we construct epileptic EEG feature extraction method based on the degree centrality of complex network combined with SVM.

2.1 Horizontal Visibility Graph

Visibility graph makes it possible to convert any time sequence into complex network domain. Horizontal visibility graph [26] is the improvement of visibility graph. The only difference between them lies in the criteria to determine the connection between nodes. At first, an EEG signal is denoted as $\{x_i\}$ $i \in [1\,M]$, where the x_i is the i_{th} sampling point in the EEG and the length of it is M. In the horizontal visibility graph, we see each sampling point of time series as a node to constitute node set. The connectivity between nodes depends on local convex constraints. For a given time series, as for two sampling point x_i and x_j, there is an edge if $x_k < \min(x_i, x_j)$ for all k with $t_i < t_k < t_j$; if not, there is no edge between the node x_i and x_j. According to this principle, we can get the edge set. As for visibility graph, the criteria is $x_k < x_j + (x_i - x_j)\frac{t_j - t_k}{t_j - t_i}$. That is to say, if the connecting line between this two times samples values can't be separated by any middle time sample point values, this two points are connected. Thus we completed the transformation from time series to complex networks and get the adjacency matrix immediately. For proximity network, firstly the EEG signal is divided up into several non-overlapping cycles and each cycle is treated as a node. Then the distance between two nodes is calculated to measure the similarity between them. Hence, we can get the distance matrix. Secondly, we select an appropriate threshold (th) to determine whether there is an edge between every pair nodes. The rule is defined as $a_{ij} = \begin{cases} 1 & d_{ij} < th \\ 0 & d_{ij} \geq th \end{cases}$; therefore we can get adjacency matrix which is composed of 1 or 0.

In the process of constructing complex network, we adopt the horizontal visibility graph to complete this mapping. This method has some advantages that the traditional methods can't match. It omits the selection of the parameter. As we all know, as long as there is threshold, it will produce errors inevitably. Compared with proximity network, this method omits the choice of the threshold value, thus eliminating the errors produced by choosing threshold. That is to say, the HVG decreases the subjectivity greatly. We can get the complex network corresponded to the time series precisely. This advantage shows the proposed detection algorithm has huge potential for real-time epileptic seizure detection.

2.2 The Degree Centrality of Complex Network

In complex network, the adjacency matrix describes the relationship between the nodes; therefore the connection between nodes reflects the internal relation of the time series. The degree centrality of complex network is defined as

$$c_D(x) = \frac{k(x)}{n-1} \tag{1}$$

In formula (1), $k(x)$ presents the degree of the node (the number of the edge); n presents the number of nodes in complex network; $n-1$ presents the largest possible number of edges of one node. In this paper, we extract the average of the degree centrality as the feature to distinguish the epileptic EEG signal. Later we will perform the experiment to verify the classification result by this feature.

2.3 Support Vector Machine

Support vector machine (SVM) is a kind of classifier. It does well in solving two-class problem. SVM is based on the theories of VC and the structure risk minimum principle. According to the limited sample information, it seeks the best compromise between the complexity of the model and the learning ability in order to get the best generalization ability. The general form of linear discriminant function is $w \cdot x + b = 0$. In order to separate the samples without error and maximize the distance between two groups, an optimal separating hyperplane must satisfy

$$y_i[w \cdot x + b] - 1 \geq 0, \, i = 1, 2, \ldots, n. \tag{2}$$

Meet the above formula and minimize $\|w\|^2$, so this optimization problem can be solved by the saddle point of a Lagrange function with Lagrange multipliers. The optimal discriminant function can be defined as

$$f(x) = \text{sgn}\{(w^* \cdot x) + b^*\} = \text{sgn}\left\{ \sum_{i=1}^{n} \alpha_i^* \cdot y_i(x_i \cdot x) + b^* \right\} \tag{3}$$

Then we replace the inner product by a kernel function K(x, x') in formula (3) to solve the large computational complexity produced by the high dimensions. In this way, the linear separability of projected samples is enhanced and the discriminant function can be rewritten as

$$f(x) = \text{sgn}\left\{\sum_{i=1}^{n} \alpha_i^* \cdot y_i \cdot K(x, x_i) + b^*\right\} \tag{4}$$

There are four kinds of kernel function commonly used. They are linear, radial basis function (rbf), polynomial and sigmoid kernels.

3 Result

3.1 Experiment Details and Data Description

All the simulations were based on a 2.60 GHz quad-core Inter Pentium processor with 4 GB memory. The code was executed in environment of MATLAB 7.0.

In this paper, all of public data is from Department of Epileptology, Bonn University, Germany. The EEG data in dataset D keep track of the interictal periods, while the EEG data in dataset E keep track of the ictal periods. These intracranial EEGs signal were derived from five epileptic patients who experience pre-surgical diagnosis. These single-channel EEG data have 23.6 s time duration and are digitized at 173.6 samples per second at 12-bit resolution. The length of each datum is 4097. The data are described in Fig. 1 in detail. We use these data to certify the efficiency of feature extraction algorithm we proposed.

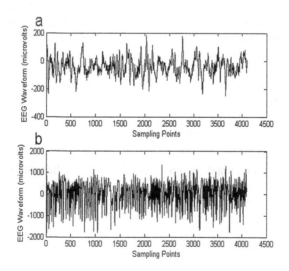

Fig. 1. (a) An interictal EEG sample in dataset D, (b) An ictal EEG sample in dataset E

3.2 Performance Evaluation Parameters

In experiment section, the positive class and the negative class represent the interictal EEG and the ictal EEG respectively. The classification performance is evaluated by using parameters such as sensitivity (SEN), specificity (SPE) and overall accuracy (ACC), which are shown in (5), (6) and (7) respectively.

$$SEN = \frac{TP}{TP + FN},$$ (5)

In formula (5), true positive (TP) stands for the interictal EEG signals recognized by both the detection algorithm and the EEG experts and false negative (FN) is the number of interictal signals labeled epileptic by the detection algorithm. In this paper, SEN means the classification accuracy of the interictal EEG.

$$SPE = \frac{TN}{TN + FP},$$ (6)

In formula (6), true negative (TN) stands for the ictal EEG signals recognized by both the detection algorithm and the EEG experts and false positive (FP) is the number of epileptic signals labeled interictal by the detection algorithm. In this paper, SPE means the classification accuracy of the ictal EEG.

$$ACC = \frac{TP + TN}{TP + FN + TN + FP},$$ (7)

ACC means the number of correctly recognized EEG signals (TP + TN) divided by the total number of EEG signals.

3.3   Classification Experiment Results

We have already mentioned that the statistical property of complex network can reveal the deep dynamical structure of nonlinear time series. The degree centrality is an important property of complex network. It reveals the connection between the nodes and the connection between the nodes reflects the internal relation of the time series. So in this paper, we extract the degree centrality of complex network as the feature to classify the EEG signal.

Figures 2 and 3 depict the mean of degree centrality of complex network in bar graph and line graph respectively. Each data set contains one hundred groups of data. The first one hundred data and the second hundred data belong to the interictal EEG and ictal EEG respectively. Figures 2 and 3 clearly show that the feature values of the two states have obvious difference. We can find the latter value is generally higher than the former. The degree centrality of complex network describes the dynamic characteristics of time series well. By studying the different values of the degree centrality, we can better classify the experiment data.

From the Table 1, we can find that the single feature classification we proposed in this paper is higher than the traditional features, such as approximate entropy and sample entropy. The results indicate that when we use the approximate entropy and sample entropy as the feature, the classification accuracy is 87.25 % and 87.75 % respectively. However, the classification accuracy is up to 92.98 % by the detection

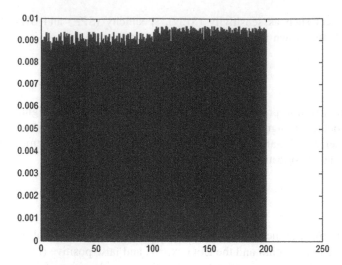

Fig. 2. The bar graph of the mean value of degree centrality

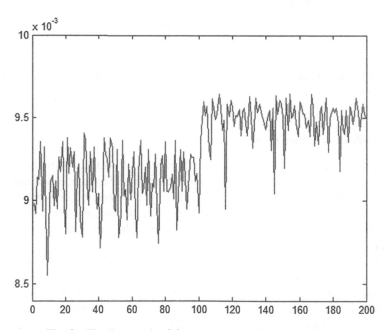

Fig. 3. The line graph of the mean value of degree centrality

algorithm we proposed. From the Table 1, we can find that the classification result of the degree centrality is better than two other widely features, such as approximate entropy and sample entropy.

The Table 2 describes that the classification result of the degree centrality we proposed is better than two other widely features, such as approximate entropy and sample entropy. In order to investigate how the data length affects the performance of

Table 1. The classification results of the proposed feature and two other widely used features for comparison.

Feature	Sample length m	ACC (%)
Approximate Entropy	2048	87.25
Sample Entropy	2048	87.75
Degree of centrality	2048	92.98
	1024	93.00

Table 2. The classification results of the proposed feature combined with SVM and comparison with other features.

Feature	Sample length m	SEN (%)	SPE (%)	ACC (%)
Approximate Entropy + SVM	2048	88.00	90.00	89.00
Sample Entropy + SVM	2048	90.00	92.00	91.00
Degree of Centrality +SVM	2048	89.98	95.98	92.98
	1024	87.92	99.92	93.92

the novel feature, we select 1024 and 2048 sampling points to verify it. The result shows that the data length has little influence on the performance, at least not so far. The classification accuracy is up to 93.92 % when the data length is 1024. In a word, this novel feature we proposed combined with SVM shows the better performance than the traditional features.

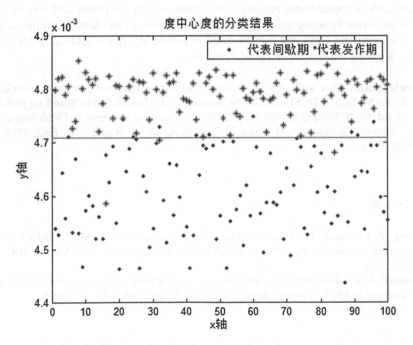

Fig. 4. The best classification result of the degree centrality

Figure 4 shows the best classification result according to the Table 1. Each '*' presents the ictal sample and each '♦' presents the interictal EEG, respectively. From the Fig. 4, we can see the points '*' are higher than the points '♦' except for a few points. Thus most points can be classified correctly and the classification accuracy is up to 93.92 %, which is higher than the traditional methods.

4   Conclusions

There is a strong demand for the development of the automatic epileptic seizure detection algorithms, due to the increased treatment of epilepsy and prevention of the possibility of misreading information by neurologist. The EEG signal records electrical activities of brain and has many complex dynamic properties. In this study, the novel feature is proposed to analyze the nonlinear dynamic characteristics of epileptic EEG signal. This method has the key step of converting which has large influence on classification result. At present, most scholars use the proximity network to convert the time series to complex networks. But the proximity network involves the choice of the threshold value, so it must produce some error inevitably. In this process, we utilize the horizontal visibility graph to complete this transformation. When comparing to proximity network, the horizontal visibility graph decreases the subjectivity greatly. Then the degree centrality of complex network as a novel feature is calculated. The time series with different dynamics have different degree centrality values. At last, the extracted feature and SVM construct automatic epileptic seizure detection. Through the experiment, we could get the better classification results compared to other widely used features such as approximate entropy and sample entropy. Experimental results show that the automatic epileptic seizure detection we proposed can clearly distinguish ictal EEG from interictal EEG and the proposed detection algorithm achieves high classification accuracy which can be up to 93.92 %.

Acknowledgment. This work was supported by the National Natural Science Foundation of China (Grant No. 61201428, 61302128), the Natural Science Foundation of Shandong Province, China (Grant No. ZR2010FQ020, ZR2013FL002), the Shandong Distinguished Middle-aged and Young Scientist Encourage and Reward Foundation, China (Grant No. BS2009SW003, BS2014DX015).

References

1. Song, Y., Crowcroft, J., Zhang, J.: Automatic epileptic seizure detection in EEGs based on optimized sample entropy and extreme learning machine. J. Neurosci. Methods **210**, 132–146 (2012)
2. Kannathal, N., Choo, M.L., Rajendra Acharya, U., Sadasivan, P.K.: Entropies for detection of epilepsy in EEG. Comput. Methods Programs Biomed. **80**, 187–194 (2005)

3. Nurujjaman, M., Narayanan, R., Sekar Iyengar, A.N.: Comparative study of nonlinear properties of EEG signals of normal persons and epileptic patients. Nonlin. Biomed. Phys. **3**, 6 (2009)
4. Zhang, J., Sun, J., Luo, X., Zhang, K., Nakamurad, T., Small, M.: Characterizing pseudoperiodic time series through the complex network approach. Physica D **237**, 2856–2865 (2008)
5. Zhang, J., Small, M.: Complex network from pseudoperiodic time series: topology versus dynamics. Phys. Rev. Lett. **96**, 238701 (2006)
6. Milo, R., Shen-Orr, S., Itzkovitz, S., Kashtan, N., Chklovskii, D., Alon, U.: Network motifs: simple building blocks of complex networks. Science **298**, 824–827 (2002)
7. Small, M., Zhang, J., Xu, X.: Transforming time series into complex networks. In: Zhou, J. (ed.) Complex 2009. LNICST, vol. 5, pp. 2078–2089. Springer, Heidelberg (2009)
8. Xiang, R., Zhang, J., Xu, X.K., Small, M.: Multiscale characterization of recurrence-based phase space networks constructed from time series. Chaos **22**, 013107 (2012)
9. Gao, Z., Jin, N.: Complex network from time series based on phase space reconstruction. Chaos **19**, 033137 (2009)
10. Marwan, N., Donges, J.F., Zou, Y., Donner, R.V., Kurths, J.: Complex network approach for recurrence analysis of time series. Phys. Lett. A **373**, 4246–4254 (2009)
11. Donner, R.V., Zou, Y., Donges, J.F., Marwan, N., Kurths, J.: Ambiguities in recurrence-based complex network representations of time series. Phys. Rev. E **81**, 015101(R) (2010)
12. Gao, Z.K., Jin, N.D.: A directed weighted complex network for characterizing chaotic dynamics from time series. Nonlinear Anal. Real World Appl. **13**, 947–952 (2012)
13. Lacasa, L., Luque, B., Ballesteros, F., Luque, J., Nuno, J.C.: From time series to complex networks: The visibility graph. Proc. Natl. Acad. Sci. USA **105**, 4972–4975 (2008)
14. Lacasa, L., Toral, R.: Description of stochastic and chaotic series using visibility graphs. Phys. Rev. E **82**, 036120 (2010)
15. Yang, Y., Wang, J., Yang, H., Mang, J.: Visibility graph approach to exchange rate series. Phys. A: Stat. Mech. Appl. **388**, 4431–4437 (2009)
16. Ni, X.H., Jiang, Z.Q., Zhou, W.X.: Degree distributions of the visibility graphs mapped from fractional Brownian motions and multifractal random walks. Phys. Lett. A **373**, 3822–3826 (2009)
17. Lacasa, L., Luque, B., Luque, J., Nuño, J.C.: The visibility graph: A new method for estimating the Hurst exponent of fractional Brownian motion. EPL **86**, 30001 (2009)
18. Donges, J.F., Donner, R.V., Kurths, J.: Testing time series irreversibility using complex network methods. EPL **102**, 10004 (2013)
19. Qian, M.C., Jiang, Z.Q., Zhou, W.X.: Universal and nonuniversal allometric scaling behaviors in the visibility graphs of world stock market indices. J. Phys. A Math. Theor. **43**, 33 (2010)
20. Elsner, J.B., Jagger, T.H., Fogarty, E.A.: Visibility network of United States hurricanes. Geophys. Res. Lett. **36**, L16702 (2009)
21. Liu, C., Zhou, W.X., Yuan, W.K.: Statistical properties of visibility graph of energy dissipation rates in three-dimensional fully developed turbulence. Phys. A Stat. Mech. Appl. **389**, 2675–2681 (2010)
22. Tang, Q., Liu, J., Liu, H.: Comparison of different daily streamflow series in US and China, under a viewpoint of complex networks. Mod. Phys. Lett. B **24**, 1541–1547 (2010)
23. Wang, N., Li, D., Wang, Q.: Visibility graph analysis on quarterly macroeconomic series of China based on complex network theory. Phys. A **391**, 6543–6555 (2012)
24. Moguerza, J., Muñoz, A.: Support vector machines with applications. Stat. Sci. **21**, 322–336 (2006)

25. Cristianini, N., Shawe-Taylor, J.: An Introduction to Support Vector Machines and Other Kernel-based Learning Methods. Cambridge University Press, London (2000)
26. Luque, B., Lacasa, L., Ballesteros, F., Liuque, J.: Horizontal visibility graphs: exact results for random time series. Phys. Rev. E **80**, 046103 (2009)

An Improved Rife Algorithm of Frequency Estimation for Frequency-Hopping Signal

Jun Lv, Leying Yun$^{(\boxtimes)}$, and Tong Li

Department of Information Engineering,
Academy of Armored Force Engineering, Beijing, China
1783595517@qq.com

Abstract. In this paper we analysis the performance of Rife algorithm and point out that the performance is poor when the true frequency is much close to quantized frequency of FFT. Aiming at this problem, an improved Rife algorithm is presented. This algorithm can make the signal frequency always located in the center between the two neighboring discrete frequencies by using zoom FFT technique. Then, in order to eliminate the impact of noise for the Rife algorithm, this paper propose to correct the improved Rife algorithm by using the window function for the signal. Simulation result showed that the improved algorithm had high estimation accuracy, strong anti-noise performance and good stability.

Keywords: Rife algorithm · Improved rife algorithm · Zoom FFT · Frequency estimation · Frequency hopping signal

1 Introduction

How to measure and intercept enemy communication signal frequency is a basic task of information warfare because frequency is an important parameter of signal, so there is a very vital significance to estimate frequency fast and high-precision [1]. Many scholars have done research on the problem of sinusoidal signal frequency estimation, which proposed by Abatzoglou [2] through maximum likelihood algorithm is considered to be the optimal estimation method because the error is close to Cramer-Rao Low Bound (CRLB), however the algorithm has the disadvantages of complex computation, and it is difficult to realize real-time processing, so the further application is limited [3].

The DFT has small computations for frequency estimation because it can be achieved by FFT, so it has been widely used in engineering. But the DFT spectrum has energy leakage and picket fence effect, which leads to the large estimated error. And the accuracy of the algorithm is largely dependent on the sample length N. Based on the above problems, a number of scholars have made further research, and proposed a variety of frequency estimation algorithm, such as Rife algorithm [4], Quinn algorithm [5], the energy center of gravity correction method [6], Newton iterative method [7], Frequency correction ratio method [8], Spectrum correction algorithm with triangular window [9] and Three interpolation frequency estimation algorithm and improved algorithm [10], and so on. There are advantages and disadvantages of all these algorithms, it will no longer be described in detail in this paper. Among all these

© Springer International Publishing Switzerland 2016
D.-S. Huang and K.-H. Jo (Eds.): ICIC 2016, Part II, LNCS 9772, pp. 181–191, 2016.
DOI: 10.1007/978-3-319-42294-7_15

algorithms, Rife algorithm is widely used in the field of signal processing because it is simple and easy to realize, it is analyzed in detail in the literature [11], it is pointed out that when the estimated signal frequency is located near the quantization frequency point and the signal noise is comparatively low, there is a big error to estimate the frequency of the signal using Rife algorithm. In this paper, an improved Rife frequency estimation algorithm is presented by using the method of spectrum zooming. Besides, in order to eliminate the impact of noise for the Rife algorithm, this paper propose to correct the improved Rife algorithm by using the window function for the signal. Simulation result showed that the improved algorithm had high estimation accuracy, strong anti-noise performance and good stability.

2 Rife Algorithm

The sinusoidal signal of single frequency represented as:

$$x(t) = a \cos(2\pi f_0 t + \theta_0) \tag{1}$$

where a is amplitude, f_0 is frequency and θ_0 is initial phase.

Sampling $x(t)$ between the tim e $[0, T]$ by equal interval $\Delta t = T/N$, we can get a sequence $x(n)$ of length N. Note $X(k)$ is the DFT of $x(n)$. Because DFT of the real sequence has symmetry, so we can only consider the positive frequency components, the $X(k)$ can be expressed as:

$$X(k) = \frac{a \cdot \sin[\pi(k - f_0 T)]}{2 \sin[\pi(k - f_0 T)/N]} \cdot e^{j[\theta_0 - \frac{N-1}{N}(k - f_0 T)\pi]}, k = 0, 1, 2, \cdots, N/2 - 1 \tag{2}$$

Note the amplitude of the maximum spectral line as $X(k_0)$, so k_0 is the discrete frequency index values of the largest spectral line in the DFT spectrum. The thought of Rife algorithm is to solve the relative bias $\delta(-0.5 \sim 0.5)$ by using the radio of the largest and the second largest spectral line in the DFT spectrum, and modify FFT algorithm, the formula is

$$\hat{f}_0 = (k_0 \pm |\delta|) \cdot \Delta f \tag{3}$$

where $\Delta f = 1/T$, is the frequency resolution for Rife algorithm. The plus or minus is determined according to the relationship between the largest and the second largest spectral line.

3 Further Analysis of Rife Algorithm

3.1 The Deficiency of Rife Algorithm

By analyzing the principle of Rife algorithm we can get that the computational formula through Rife algorithm for frequency estimation is:

$$\hat{f} = \frac{1}{T}[k_0 + r\frac{|X(k_0+r)|}{|X(k_0)| + |X(k_0+r)|}] \tag{4}$$

If $|X(k_0+1)| \leq |X(k_0-1)|, r = -1$; else if $|X(k_0+1)| \geq |X(k_0-1)|, r = 1$.

Under appropriate Signal to Noise Ratio conditions, when the actual frequency f is located in the center area of the largest and the second largest spectral line in the DFT spectrum, the estimation effect of Rife algorithm is better, and the estimation error is much smaller than that obtained by using DFT algorithm directly. On the contrary, when the actual frequency f is close to the largest spectral line in the DFT spectrum, the error is likely to be bigger than that obtained by using the DFT algorithm directly. This feature of Rife algorithm can be summarized as: when the signal frequency f is close to the largest and the second largest spectral line at the midpoint $f(k_0 + r/2)$ of the frequency value, the amplitude of the second largest spectral line $|X(k_0+r)|$ and the largest spectral line $|X(k_0)|$ is very close, Rife algorithm has higher estimation accuracy at this point, as shown in Fig. 1.

When the signal frequency f is close to the frequency value of the largest spectral line, the amplitude of the second largest spectral line $|X(k_0 + r)|$ is small, the estimation accuracy of the Rife algorithm is relatively low at this point, as shown in Fig. 2.

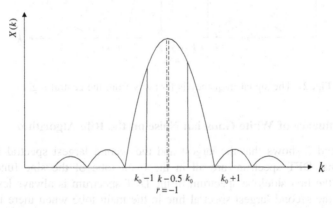

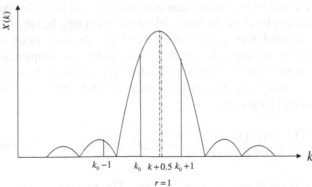

Fig. 1. The signal frequency is located in the central region

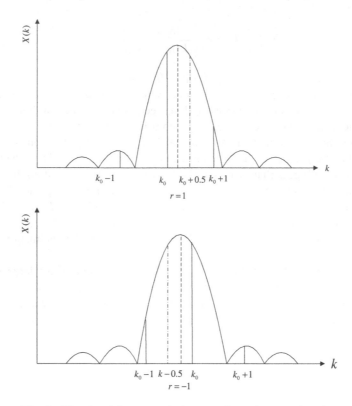

Fig. 2. The signal frequency is far away from the central region

3.2   The Influence of White Gaussian Noise on the Rife Algorithm

The Figs. 1 and 2 shows that the largest and the second largest spectral line of the sampling signal DFT spectrum are all in the main lobe of the sinc function. The amplitude of the first sidelobe spectrum in the DFT spectrum is always less than the amplitude of the second largest spectral line in the main lobe when there is no noise interference, there will be no interpolation direction error occurs at this point. But when there is noise, the amplitude of the first sidelobe spectrum may be larger than that of the second largest spectral line in the main lobe if the absolute value of the relative frequency deviation is relatively small, it will make the frequency interpolation direction error at this time, frequency estimation error is relatively large.

From the literature [11], we can know that the total error of the frequency estimation using Rife algorithm is:

$$\sigma_T^2 = \frac{(1-|\delta|)^2[(1-|\delta|)^2+\delta^2]}{N(SNR_0)\sin c^2(\delta)} + 2\delta^2 erfc\left[\frac{\delta|\sin(\pi\delta)|}{\pi(1-\delta^2)}\sqrt{N(SNR_0)}\right] \tag{5}$$

where $erfc()$ is complementary error function. The root mean square error of the algorithm is:

$$\sigma_f = \sigma_T \cdot (f_s/N) \tag{6}$$

In the interval [0, 0.5], evenly spaced take 20 relative frequency deviation δ, suppose sampling frequency f_s = 200 kHz, sample points N = 1024 and bring them into formula (5) and (6), we can obtain root mean square error curves calculated by Rife algorithm in five different signal to noise ratio conditions, as shown in Fig. 3.

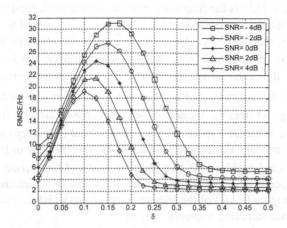

Fig. 3. Frequency root mean square errors of the theoretical calculation with the change of frequency deviation

As can be seen from Fig. 3: (1) Under different signal to noise ratio conditions, there will be a relatively large estimation error when the frequency deviation δ is less than a threshold value δ_F, that's because the amplitude of the first sidelobe spectrum is greater than it of the second largest spectral line in the main lobe lead to interpolation error. The root mean square error of the frequency estimation is maximum when the two amplitudes are equal. At the same time, the higher signal-to-noise ratio, the smaller the interference of noise on the signal, the smaller the threshold δ_F; (2) The higher signal-to-noise ratio, the smaller the root mean square error of the frequency estimated by Rife algorithm. When the frequency relative deviation value δ is located in the interval $\left[\frac{1}{3}, \frac{1}{2}\right]$, five curves tend to level. The same result can be got by the analysis of the other signal-to-noise ratio, it can be considered that in this range, there is a lower root mean square error to estimate frequency by Rife algorithm regardless of the size of the signal-to-noise ratio.

4   An Improved Rife Algorithm

According to analysis in the second quarter, when the signal frequency f is located in the center area of the largest and the second largest spectral line in the DFT spectrum, there is a high performance of the Rife algorithm; But when it is close to the largest spectrum line, the performance of Rife algorithm begin to decline sharply. For this feature of the Rife algorithm, a new method using spectrum zoom to move the position of the second largest spectral line is put forward in this paper, so that the signal frequency is always located in the central area of the largest and the second largest spectral line in the DFT spectrum. And from the theoretical calculation figure, we can get that if $k_0 \Delta f$ is the frequency point matching the largest spectral line, there is a small error of the Rife algorithm when relative frequency deviation δ located in the area of $\left[\frac{1}{3}, \frac{1}{2}\right]$, that is to say, $\left[k_0 + \frac{1}{3}, k_0 + \frac{1}{2}\right]\Delta f$ is the effective area of the Rife algorithm. In a similar way, if $(k_0 + 1)\Delta f$ is the frequency point matching the largest spectral line, the area of relative frequency deviation δ which can lead a small error of the Rife algorithm become $\left[-\frac{1}{2}, -\frac{1}{3}\right]$, that is to say, $\left[k_0 + \frac{1}{2}, k_0 + \frac{2}{3}\right]\Delta f$ is the effective area of the Rife algorithm right now. So the basic idea of this improved algorithm is: define the area $\left[k_0 + \frac{1}{3}, k_0 + \frac{2}{3}\right]$ is the center area of the discrete frequency points corresponding index value k_0 and $k_0 + 1$. Assume $\hat{f}$ is the frequency estimation obtained from the Rife algorithm and judge whether $\hat{f}$ is located in the center area of the largest and the second largest spectral line in the DFT spectrum, take $\hat{f}$ as the final frequency estimation if it is, otherwise have a appropriate tessellate to the DFT spectrum in order to locate the frequency estimation $\hat{f}$ in the center area of the largest and the second largest spectral line in the DFT spectrum, a high estimation accuracy can be maintained in this case.

Assume the amplitude $X(k)$ of the frequency spectrum which is calculated through FFT from the signal sampling sequence $x(n)$ is:

$$X(k) = \sum_{n=0}^{N-1} x(n)e^{-j\frac{2\pi nk}{N}} \quad k = 0, 1, 2, \cdots, N-1 \tag{7}$$

Get the index value k_0 of the largest spectral line and the rough estimated value $k_0\Delta f$ of the signal frequency, calculate the signal frequency estimation $\hat{f}$ according to the formula (3) then. Because there is always exist $|X(k_0)| \geq |X(k_0 + r)|$, so $\hat{f}$ satisfy $|\hat{f} - k_0 f_s/N| \leq \frac{1}{2} \cdot \Delta f$. If $\frac{1}{3} \cdot \Delta f < |\hat{f} - k_0 f_s/N| \leq \frac{2}{3} \cdot \Delta f$, we can think frequency estimation $\hat{f}$ located in the center area of the largest and the second largest spectral line in the DFT spectrum, so $\hat{f}$ is the final frequency estimation. Otherwise, you may need to be revised.

From the spectrum curve after FFT conversion, we can work out $X(k_0 \pm 0.5)$ by using the spectrum zooming method. Suppose the second largest spectral line is located in the left of the largest, we estimate the frequency through $X(k_0 - 0.5)$ and $X(k_0)$. If $X(k_0) \geq X(k_0 - 0.5)$,

$$\hat{f} = \frac{1}{T}\left[k_0 - \frac{|X(k_0 - 0.5)|}{|X(k_0)| + |X(k_0 - 0.5)|}\right] \tag{8}$$

Else

$$\hat{f} = \frac{1}{T}\left[k_0 - \frac{|X(k_0)|}{|X(k_0)| + |X(k_0 - 0.5)|}\right] \tag{9}$$

An Improved Rife Algorithm of Frequency Estimation for Frequency-Hopping Signal   187

In the same way, when the second largest spectral line is located in the right of the largest, we estimate the frequency through $X(k_0)$ and $X(k_0 + 0.5)$. If $X(k_0) \geq X(k_0 + 0.5)$,

$$\hat{f} = \frac{1}{T} \left[k_0 + \frac{|X(k_0 + 0.5)|}{|X(k_0)| + |X(k_0 + 0.5)|} \right] \tag{10}$$

Else

$$\hat{f} = \frac{1}{T} \left[k_0 + \frac{|X(k_0)|}{|X(k_0)| + |X(k_0 + 0.5)|} \right] \tag{11}$$

The frequency estimation $\hat{f}$ which is got by the above methods can be think as the final estimation if $\frac{1}{3} \cdot \Delta f < |\hat{f} - k_0 f_s / N| \leq \frac{2}{3} \cdot \Delta f$; Otherwise need to refine the signal until make the estimation $\hat{f}$ meet the above conditions between $X(k_0 - 0.5)$ and $X(k_0)$, as also as $X(k_0)$ and $X(k_0 + 0.5)$.

In order to eliminate the increasing frequency estimation error caused by the direction of the interpolation error of Rife algorithm, a method that add window handle before to have a DFT transform to the estimated signal is put forward in this paper. The result is an increase in the width of the main lobe in DFT spectrum of the estimative signal, so that there are more spectral lines in the main lobe. At the same time it is more easily to distinguish the second largest spectral line and the third on both side of the largest spectral line, then avoid the error of the direction of the interpolation. In commonly used several kind of window functions, the side lobe attenuation of rectangular window is slow, the energy leakage is serious and the correction precision is low. The side lobe attenuation of hamming window and harming window is quick, the energy is mainly concentrated in the main valve, leakage is less and the correction precision is higher [6]. Therefore, in the process of adding windows, we use non rectangular window in this paper.

The flow chart of the improved Rife algorithm is shown in Fig. 4.

The specific steps are as follows:

(1) process the sampling signal $x(n)$ for adding windows and then do FFT computing, get $X(k)$;
(2) do modular arithmetic to $X(k)$, and calculate the maximum $X(k_0)$;
(3) obtain the estimate value $\hat{f}$ of the signal frequency by Rife algorithm;
(4) calculate the absolute value *dif* of the difference between estimate value $\hat{f}$ and rough estimate value of $k_0 f_s / N$ signal frequency. If $\frac{1}{3} \cdot \Delta f < |\hat{f} - k_0 f_s / N| \leq \frac{2}{3} \cdot \Delta f$, turn to step (6), otherwise continue to step (5);
(5) when the second largest spectral line is located in the left of the largest, do a spectrum refinement and then obtain $X(k_0 - 0.5/i)$, where i is the number of times of spectrum refinement $(i = 1, 2, \cdots \cdots)$. Estimate the signal frequency again by Rife algorithm, obtain the estimate value $\hat{f}$ of the signal frequency. when the second largest spectral line is located in the right of the largest, do a spectrum

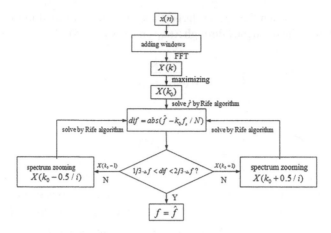

Fig. 4. The flow chart of the improved Rife algorithm

refinement and then obtain $X(k_0 + 0.5/i)$ where i is the number of times of spectrum refinement ($i = 1, 2, \cdots\cdots$). Estimate the signal frequency again by Rife algorithm, obtain the estimate value $\hat{f}$ of the signal frequency, turn to step (4);

(6) The signal frequency estimation $\hat{f}$ obtained by improved Rife algorithm is the final estimate frequency of the signal.

5   Simulation and Result Analysis

Set the simulation signal as

$$x(t) = a\cos(2\pi f_0 t + \theta_0) + w(t) \tag{12}$$

Here: $w(t)$ is white Gaussian noise which mean value is 0 and variance is σ^2. Define signal-to-noise ratio as $\text{SNR} = a^2/\sigma^2$. Sampling frequency $f_s = 200$ kHz, sample points $N = 1024$, sampling interval $\Delta t = 5 \times 10^{-6}$s. Set Carrier frequency $f_0 = f_s/4 = 50$ kHz, order $f_i = f_0 + \delta \cdot \Delta f$. Analysis the effect of frequency estimation using the above signal by improved Rife algorithm, do 500 times Monte Carlo simulation for each frequency point f_i. At the same time, Cramer-Rao low bound of the sinusoidal signal frequency estimation $\hat{f}$ is [12] $\text{var}(\hat{f}) \geq \dfrac{6\sqrt{2}}{(2\pi)^2 \cdot \text{SNR} \cdot (\Delta t)^2 N(N-1)^2}$, the lower limit of variance (CRLB) is $\text{CRLB} = \sqrt{\text{var}(\hat{f})}$.

When the signal-to-noise ratio are 3 dB, 0 dB and −3 dB, estimate the frequency to above simulation signal respectively using Rife algorithm and improved Rife algorithm, we can get root mean square error curves estimated by Rife and improved Rife algorithm under different signal-to-noise ratio are shown in Fig. 5.

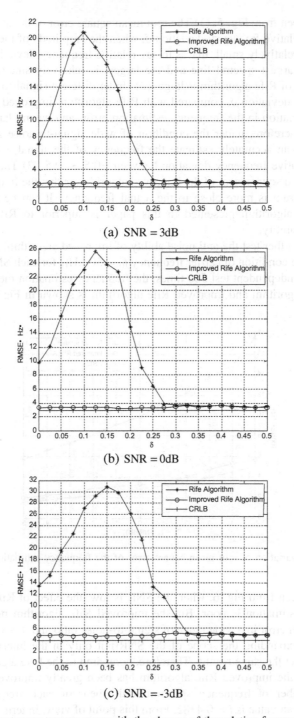

(a) SNR = 3dB

(b) SNR = 0dB

(c) SNR = -3dB

Fig. 5. The root mean square error with the change of the relative frequency deviation

As can be seen from Fig. 5: (1) The root mean square error of the algorithm and the change of the relative frequency deviation δ, when the amplitude of the second largest spectral line is relatively small, that is to say relative frequency deviation δ calculated by the second largest spectral line and the largest is relatively small, the frequency estimation error of Rife algorithm is large. (2) Under different signal-to-noise ratio and same frequency deviation δ, compared with Rife algorithm, improved Rife algorithm has no large variation in the frequency estimation error, and has no large dependence on the SNR. Therefore, under the condition of wide signal-to-noise ratio, improved Rife algorithm can accurately estimate the frequency of the signal, regardless of the value of the relative frequency deviation between $-0.5 \sim 0.5$. (3) The reason for the smaller root mean square error of the improved algorithm is that the frequency after the frequency spectrum is more close to the actual frequency. It can be concluded that improved Rife algorithm presented in this paper is superior to Rife algorithm in accuracy and stability.

In order to further test the anti noise ability of improved algorithm, under the same simulation— the condition of carrier frequency $f_0 = 50$ kHz, for each SNR in the range of $-5 \sim 5$ dB independent test 500 times, the frequency estimation mean square error curve of Rife algorithm and improved Rife algorithm is shown in Fig. 6.

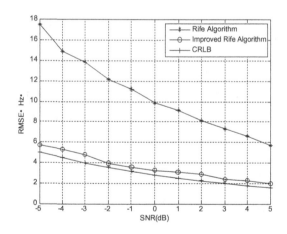

Fig. 6. The variation of root mean square error of the frequency estimation with SNR

As can be seen from Fig. 6, when the SNR is low, the effect of Rife algorithm on the frequency estimation is poor. But the improved Rife algorithm proposed in this paper has a small error in a wide range of signal-to-noise ratio and it is very close to the CRLB. Again expanding the scope of the SNR(not only in the interval from -5 to 5 dB), we can get the same results. Therefore, compared to the Rife algorithm, the anti noise ability of the improved Rife algorithm has been greatly improved. Also in this paper, the number of frequency spectrum refinement of each frequency point is counted, and mean value is $i = 0.4382$. From this point of view, in terms of the amount of computation, improved Rife algorithm is not a lot of increase compared to Rife algorithm.

6 Conclusion

Rife algorithm is a kind of frequency estimation algorithm based on Fast Fourier Transform interpolation. Because of its high computing speed and good real-time performance, Rife algorithm has been widely used in engineering. Based on the theory and simulation analysis of Rife algorithm, an improved Rife algorithm is proposed in this paper. The simulation analysis shows that this improved algorithm is accurate and effective, it can improve the frequency estimation accuracy and stability and the error is close to the CRLB; The improved algorithm is simple and the computation is small, it can effectively solve the contradiction which traditional frequency estimation algorithm can't be able to unify in the estimation accuracy and computation quantity, so it has a great application value in the field of signal processing.

References

1. Wan, L., Yang, X.: A high-accuracy frequency estimation algorithm based on FFT. Electron. Sci. Tech. **23**(10), 79–81 (2010)
2. Wang, H., Zhao, G., Qi, F.: Real-time and accurate single frequency estimation approach. J. Data Acquisition Proc. **24**(2), 208–211 (2009)
3. Qi, G.: Error analysis of frequency and phase estimations based on phase difference of segmented FFTs. J. Data Acquisition Proc. **18**(1), 7–11 (2003)
4. Rife, D., Boorstyn, R.: Single tone parameter estimation from discrete-time Observations. IEEE Trans. Inform. Theory **20**(5), 591–598 (1974)
5. Quinn, B.G., Kootsookos, P.: Threshold behavior of the maximum likelihood estimator of frequency. IEEE Trans. Signal Proc. **42**(11), 3291–3294 (1994)
6. Ding, K., Jiang, L.: Energy centrobaric correction method for discrete spectrum. J. Vibr. Eng. **14**(3), 354–358 (2001)
7. Deng, Z., Liu, Y.: The starting point problem of sinusoid frequency estimation based on newton's method. Acta Electronicasinica **1**(1), 104–107 (2007)
8. Chen, K., Wang, J., Zhang, S.: Spectrum correction based on the complex ratio of discrete spectrum around the main-lobe. J. Vibr. Eng. **21**(3), 314–318 (2008)
9. Zhang, Q., Zhang, P., Zhang, M.: Analysis on conductive vibration of tympanic membrane based on the mesh less method. J. Vibr. Shock **28**(2), 96–98 (2009)
10. Hu, D., Li, D.: Improved algorithm of cubic interpolated frequency estimation. Comput. Eng. Appl. **46**(24), 154–156 (2010)
11. Qi, G., Jia, X.: Accuracy analysis of frequency estimation of sinusoid based on interpolated FFT. Acta Electronicasinica **32**(4), 625–629 (2004)
12. Liu, Y.: Research on High-accurate Frequency Estimation Algorithm. Nanjing University of Science & Technology, Nanjing (2007)

Blind Hyperspectral Unmixing Using Deep-Independent Information

Fasong Wang[1(✉)], Rui Li[2], Jiankang Zhang[1(✉)], and Li Jiang[1]

[1] School of Information Engineering, Zhengzhou University, Zhengzhou, 450001, China
fasongwang@126.com, iejkzhang@zzu.edu.cn
[2] School of Sciences, Henan University of Technology, Zhengzhou, China

Abstract. In linear mixing model (LMM), the endmember fractional abundances should satisfy the sum-to-one constraint, which makes the well-known independent component analysis (ICA) based blind source separation (BSS) algorithms not well suited to blind hyperspectral unmixing (bHU) problem. A novel framework for bHU consulting dependent component analysis (DCA) is presented in this paper. By using the idea of subband decomposition, wavelet packet decomposition based bHU algorithm (termed as SDWP-bHU) is proposed, where the deep independent information of the source signals is exploited to fulfill the endmember signatures extraction and abundances separation tasks. Experiments based on the synthetic data are performed to evaluate the validity of the proposed approach.

Keywords: Blind hyperspectral unmixing (bHU) · Blind source separation (BSS) · Independent component analysis (ICA) · Dependent component analysis (DCA) · Wavelet packet decomposition

1 Introduction

Hyperspectral remote sensors are able to collect spectral with high resolution extending from visible region through the infrared region. With the rapid development of imaging spectroscopy and space technology, remotely sensed hyperspectral images have many potential applications [1, 2]. Generally, due to the limited spatial resolution of state-of-the-art imaging sensors as well as the presence of the combination of different substances into a homogeneous mixture phenomenon, each pixel in the hyperspectral image usually contains more than one pure spectral signature, weighted by their fractional abundances. As an important problem in many hyperspectral image applications, hyperspectral unmixing (HU) attempts to identify the endmember signatures and estimate the corresponding material abundances.

There exist several models to describe the mixing process of hyperspectral signals depending on some crucial factors [1]. As a compromise between model accuracy and tractability, linear mixing model (LMM) has been widely used to solve the HU problem for its implementing efficiently and flexibility in many applications, which means that the spectral radiance upon the sensor location can be considered as a mixture of the endmember radiances weighted by their corresponding fractional abundances. Despite

© Springer International Publishing Switzerland 2016
D.-S. Huang and K.-H. Jo (Eds.): ICIC 2016, Part II, LNCS 9772, pp. 192–201, 2016.
DOI: 10.1007/978-3-319-42294-7_16

the fact that the linear mixing model is not always true, it is always considered as an acceptable model for most real-world scenarios except that special scenarios exhibit strong nonlinearity.

A group of HU algorithms exploiting the LMM have been developed, which are mainly divided into two classes: semi-blind HU approach and blind HU (bHU) method. Most semi-blind HU approaches require some or full knowledge of the existing materials and their spectral, which forms the endmember matrix. If the endmember signature matrix is known, then the estimation of the abundance vector can be considered as a constrained linear least squares problems, which can be solved by classic convex optimization approaches and some advanced efficient modification implementations [3].

For real-world applications, obtaining full knowledge of endmember matrix is not an easy task, which is unlikely in reality to know exactly all the materials in the scene. As a result, in recent years, some efforts were made to develop bHU approaches that may identify the endmember signature matrix and the abundance vector simultaneously without any knowledge about the scene. If we consider the endmember signature matrix as the mixing matrix and the abundance vector as source vector respectively, the bHU problem falls into the scope of the popular blind source separation (BSS) technology, which was comprehensively studied during the last twenty years, especially for the case where the source signals are mutually statistically independent, which is called independent component analysis (ICA).

However, in the framework of bHU, the endmember abundances are not independent for their sum-to-one constraint, as a result, the classic ICA based BSS algorithms are usually not well suited for bHU problem.

Blind separation of dependent source signals is usually called dependent component analysis (DCA) in the BSS society, which has not received much attention in recent years. In this paper, inspired by the existing subband decomposition (SDICA) method [4], we propose a bHU framework called SDWP-bHU, which exploits the fact that the endmember abundances can be divided into some subbands in the transformed domain. We can find out the optimal subbands which make the original source signals as independent as possible, and then based on these subbands, the endmember signature matrix (mixing matrix) can be estimated efficiently. Eventually, the original endmember abundances (source signals) vector is estimated.

The rest of this paper is organized as follows: Sect. 2 introduces the bHU model and briefly reviews some BSS based algorithm for bHU problem. In Sect. 3, we present the proposed SDWP-bHU algorithm. Experimental results are shown in Sect. 4. Finally, in Sect. 5, our main conclusions are outlined.

2 Blind Hyperspectral Unmixing Model

For any fixed pixel l, let $x_m[l]$ denote the hyperspectral camera's measurement at spectral band (sensors) m. Suppose we have M sensors (number of spectral channels), then $\mathbf{x}[l] = [x_1[l], x_2[l], \cdots, x_M[l]]^T \in \mathbb{R}^M$, the LMM can be written as follows,

$$\mathbf{x}[l] = \sum_{i=1}^{N} \mathbf{a}_i s_i[l] + \mathbf{v}[l] = \mathbf{A}\mathbf{s}[l] + \mathbf{v}[l] \qquad (1)$$

for $l = 1, 2, \cdots, L$, where $\mathbf{x}[l]$ contains the random variables assigned to the sensor measurements at different center frequencies, each $\mathbf{a}_i \in \mathbb{R}^M$, $i = 1, 2, \cdots, N$, is called an endmember signature vector, which contains the spectral components of a specific material in the scene indexed by i, N is the number of endmembers or materials in the scene. $\mathbf{A} = [\mathbf{a}_1, \mathbf{a}_2, \cdots, \mathbf{a}_N] \in \mathbb{R}^{M \times N}$ is called the endmember matrix (mixing matrix in BSS). $\mathbf{s}[l] = [s_1[l], s_2[l], \cdots, s_N[l]]^T \in \mathbb{R}^N$ is called the endmember abundance vector (source signals in BSS) at pixel l, $\mathbf{v}[l] = [v_1[l], v_2[l], \cdots, v_M[l]]^T \in \mathbb{R}^M$ denotes an additive permutation at pixel l, which contains the modeling errors, system noise, etc. L denotes the number of pixels.

For a certain pixel l, the endmember abundance $s_i[l]$ represents the fractional area occupied by the ith endmember. Therefore, in the LMM, all the source signals $s_i[l]$ in a special pixel are greater or equal to zero:

$$s_i[l] \geq 0, \ i = 1, 2, \cdots, N, \ \forall l \in \{1, 2, \cdots, L\} \qquad (2)$$

Furthermore, $s_i[l]$ are the percentage contribution of each endmember in the pixel l, they should obey the sum-to-one constraint, that is:

$$\sum_{i=1}^{N} s_i[l] = 1, \ \forall l \in \{1, 2, \cdots, L\} \qquad (3)$$

Obviously, the sum-to-one constraint (3) imposes dependence among random variables $s_i[l]$. That is, every source signal $s_i[l]$ can be expressed as a linear combination of the other $N - 1$ source signals, and this leads to the dependence between different source signals [1].

If the endmember signature $\mathbf{A}$ is known, an estimate of the source signals can be obtained by

$$\hat{\mathbf{s}}[l] = \mathbf{A}^\dagger \mathbf{x}[l] \qquad (4)$$

where $\mathbf{A}^\dagger$ is the Moore-Penrose pseudo inverse of $\mathbf{A}$, i.e., $\mathbf{A}^\dagger = (\mathbf{A}^T \mathbf{A})^{-1} \mathbf{A}^T$. It should be noted that, the estimated source signals contains a noise component, i.e., $\hat{\mathbf{s}}[l] = \mathbf{s}[l] + \tilde{\mathbf{v}}[l]$, where $\tilde{\mathbf{v}}[l] = \mathbf{A}^\dagger \mathbf{v}[l]$ remains Gaussian distribution. In most cases, the endmember signature matrix cannot be accessed easily and HU need to be executed in a blind manner, i.e., without knowing matrix $\mathbf{A}$, which makes the HU problem more complicated and generally called bHU. The problem of bHU falls into the class of BSS problem in signal processing, and the desired results of bHU are to estimate the endmember signature matrix and the corresponding abundances from the observed hyperspectral images, with no or little prior information of the mixing system. Being given little information to solve the problem, bHU is a challenging-but also fundamentally intriguing-problem with many possibilities [1, 5, 6].

3   Proposed SDWP-BHU Framework

In HU problem, the independence property of source signals may not hold as discussed in Sect. 2, and therefore the standard ICA based BSS algorithms cannot give expected results. Among many extensions of the basic ICA model, several researchers have studied the case where the source signals are not statistically independent and we call these models DCA model as a whole.

The first DCA model is the multidimensional independent component analysis (MICA) [7]. Based on this basic extension of the ICA model, there have emerged lots of DCA models and corresponding algorithms, these methods can be divided into two classes, the first one is called statistical approach, such as subband decomposition ICA (SDICA) [4], independent subspace analysis [8], maximum non-Gaussianity method [9] and so on; another class called determined approach which includes non-negative based method [10] and bounded component analysis [11] and so on. In [9], a MaxNG algorithm was introduced to separate dependent sources, moreover, it is implemented to bHU problem, but the method is very time consuming.

Among these algorithms, SDICA assumes that each source signal is represented as the sum of some independent sub-components and dependent sub-components, which have different frequency bands. SDICA model can considerably relax the assumption regarding mutual independence between the original source signals by assuming that the wide-band source signals are generally dependent but some narrow-band sub-components of the source signals are independent [4]. Inspired by this idea, we will give the separation criteria using the subband decomposition based on the wavelet packet decomposition.

In SDICA model, source signals can be represented as:

$$s_i[l] = s_{i,1}[l] + s_{i,2}[l] + \cdots + s_{i,K}[l] \tag{5}$$

where K is the number of subbands, $s_{i,k}[l]$, $i = 1, 2, \cdots, N; k = 1, 2, \cdots, K$ are subband sub-components. And the sub-components are mutually independent for only a certain set of k. The mixed signals are generated from the source signals $s_i[l]$ according to model (1). We assumed that the mixed signals are zero mean by preprocessing procedure. Similar to BSS, the goal of bHU is to find the separation matrix, which can estimate the original source signals

$$y[l] = Wx[l] \tag{6}$$

where $y[l] = \hat{s}[l] \in \mathbb{R}^N$. We shall assume that for certain set of k, subbands in (5) are least dependent or possibly independent [4]. Under presented assumptions, the standard ICA algorithms can be applied to the selected set of k subbands in order to learn separation matrix W as follows:

$$y_k[l] = Wx_k[l] \tag{7}$$

3.1 Preprocessing Transform

Generally, we can use any linear operator $\mathcal{T}$ on $\mathbf{s}[l]$ to extract a set k of proper subbands, that is

$$\mathbf{s}_k[l] = \mathcal{T}_k[\mathbf{s}[l]] \tag{8}$$

where $\mathcal{T}_k$ can represent a linear time invariant filter. Using (8) and subband representation of the source signals (5), applying the operator $\mathcal{T}_k$ on the model (1), one can get

$$\mathbf{x}_k[l] = \mathcal{T}_k[\mathbf{A}\mathbf{s}[l] + \mathbf{v}[l]] = \mathbf{A}\mathbf{s}_k[l] + \mathbf{v}_k[l] \tag{9}$$

Note that the property of transformed signals often yields much better source separation performance than standard BSS, and can work well even in under-determined situations if the sources has sparsities [12].

In this paper, we propose to choose WP [13] as $\mathcal{T}_k[\cdot]$ to obtain optimal subbands representation of the problem (1). Each source and noise signal can be expressed in terms of its WP decomposition coefficients as:

$$s_{ki}^q[l] = \sum_\alpha f_{ki\alpha}^q \varphi_{q\alpha}[l], \quad v_{ki}^q[l] = \sum_\alpha e_{ki\alpha}^q \varphi_{q\alpha}[l] \tag{10}$$

where the indexes q, k, i, α represent the scale level, the subband index, the source index and the shift index, respectively, herein $j = 1, 2, \cdots, 2^q$. $\varphi_{q\alpha}[l]$ is the chosen wavelet and $f_{ki\alpha}^q, e_{ki\alpha}^q$ are the corresponding decomposition coefficients of harmonic signals and noise. If we choose the same representation space as for the mixed signals $\mathbf{x}[l]$, each component of the mixed signals can be written as

$$x_{ki}^q[l] = \sum_\alpha d_{ki\alpha}^q \varphi_{q\alpha}[l] \tag{11}$$

Let $\mathbf{f}_{k\alpha}^q = [f_{k1\alpha}^q, f_{k2\alpha}^q, \cdots, f_{kN\alpha}^q]^T$, $\mathbf{e}_{k\alpha}^q = [e_{k1\alpha}^q, e_{k2\alpha}^q, \cdots, e_{kM\alpha}^q]^T$ and $\mathbf{d}_{k\alpha}^q = [d_{k1\alpha}^q, d_{k2\alpha}^q, \cdots, d_{kM\alpha}^q]^T$ be constructed from the coefficients of the sources, noises and mixtures, respectively. Using the orthogonally property of the wavelets, one obtains

$$\mathbf{d}_k^q = \mathbf{A}\mathbf{f}_k^q + \mathbf{e}_k^q \tag{12}$$

Then, the estimation of the mixing matrix can be performed using the decomposition coefficients $\mathbf{d}_k^q = [d_{k1}^q, d_{k2}^q, \cdots, d_{kM}^q]^T$ of the mixed signals corresponding a special shift index α. Also, when the noise is present, (12) becomes approximately, that is $\mathbf{d}_k^q \doteq \mathbf{A}\mathbf{f}_k^q$, We can conclude that the relationship between decomposition coefficients of the mixtures and the sources is exactly the same as in the original domain of signals, that is $\mathbf{x}_k^q = \mathbf{A}\mathbf{s}_k^q$.

3.2 Separation Criteria

The subbands selection with most independent components $\mathbf{s}_k$ can be done by using the MI measure between the same nodes in the WP trees. Under weak correlation and weak non-Gaussian assumptions, it has been shown in [14] that MI can be approximated via small cumulant approximation of the Kullback-Leibler (KL) divergence (Gram-Charlier expansion of non-Gaussian distributions around normal distribution) as

$$I_k^q(x_{k1}^q, x_{k2}^q, \cdots, x_{kP}^q) \doteq \frac{1}{4} \sum_{\substack{0 \le n,l \le M \\ n \ne l}} c_{nl}^2 + \frac{1}{2} \sum_{r \ge 3} \frac{1}{r!} \sum_{i_1 i_2 \cdots i_r = 1}^{M} c_{i_1 i_2 \cdots i_r}^2 \tag{13}$$

where c_{nl} denotes cross-cumulant or second-order cross-cumulant, $c_{i_1 i_2 \cdots i_r}^2$ denotes square of the related rth order cross-cumulant. Approximation of the joint MI by the sum of pair-wise MI is commonly used in the ICA community in order to simplify computational complexity of the linear instantaneous ICA algorithms [15, 16]. Then, the approximation of MI by the sum of pair-wise MI is described as follows:

$$
\begin{aligned}
I_k^q(x_{k1}^q, x_{k2}^q, L, x_{kM}^q) &\doteq \sum_{\substack{0 \le n,l \le M \\ n \ne l}} I_k^q(x_{kn}^q, x_{kl}^q) \doteq \frac{1}{4} \sum_{\substack{0 \le n,l \le P \\ n \ne l}} c_{nl}^2 \\
&+ \frac{1}{12} \sum_{\substack{0 \le n,l \le M \\ n \ne l}} \sum_{u,v} \left(cum(\underbrace{x_{kn}^q, \cdots, x_{kn}^q}_{u \ times}, \underbrace{x_{kl}^q, \cdots, x_{kl}^q}_{v \ times}) \right)^2 \\
&+ \frac{1}{48} \sum_{\substack{0 \le n,l \le M \\ n \ne l}} \sum_{b,d} \left(cum(\underbrace{x_{kn}^q, \cdots, x_{kn}^q}_{b \ times}, \underbrace{x_{kl}^q, \cdots, x_{kl}^q}_{d \ times}) \right)^2
\end{aligned}
\tag{14}
$$

where $u, v \in \{1, 2\}$, and $u + v = 3$, $b, d \in \{1, 2, 3\}$, and $b + d = 4$, $cum(\cdot)$ denotes cross-cumulants. It has been proven that $I_k^q(x_{k1}^q, x_{k2}^q, \cdots, x_{kM}^q) \ge 0$ and $I_k^q(x_{k1}^q, x_{k2}^q, \cdots, x_{kM}^q) = 0$ when $x_{k1}^q, x_{k2}^q, \cdots, x_{kM}^q$ are mutually statistically independent or pairwise statistically independent. It also follows that the approximation (14) represents a consistent measure of statistical dependence. From the practical reasons, (14) should be rewritten in terms of explicit second-, third- and fourth-order cross-cumulants as follows [15, 16]:

$$I_k^q(x_{k1}^q, x_{k2}^q, \cdots, x_{kM}^q) \doteq \frac{1}{4} \sum_{\substack{0 \le n,l \le P \\ n \ne l}} \text{cum}^2(x_{kn}^q, x_{kl}^q)$$

$$+ \frac{1}{12} \sum_{\substack{0 \le n,l \le P \\ n \ne l}} (\text{cum}^2(x_{kn}^q, x_{kn}^q, x_{kl}^q) + \text{cum}^2(x_{kn}^q, x_{kl}^q, x_{kl}^q)) \tag{15}$$

$$+ \frac{1}{48} \sum_{\substack{0 \le n,l \le P \\ n \ne l}} (\text{cum}^2(x_{kn}^q, x_{kn}^q, x_{kn}^q, x_{kl}^q)$$

$$+ \text{cum}^2(x_{kn}^q, x_{kn}^q, x_{kl}^q, x_{kl}^q) + \text{cum}^2(x_{kn}^q, x_{kl}^q, x_{kl}^q, x_{kl}^q))$$

where $\text{cum}(\cdot)$ in (15) denotes second, third or fourth-order cross-cumulants.

Once the optimal subbands are selected, we can obtain either the estimation of the separation matrix $\mathbf{W}$ or the estimation of the mixing matrix $\mathbf{A}$ by applying standard ICA algorithms on the model $\mathbf{x}_{wp}[l] = \mathbf{As}[l]$. Reconstructed source signals $\hat{\mathbf{s}}$ are estimated from the reconstructed mixed signals $\mathbf{x}_{wp}[l]$, which is obtained by the synthesis operation of the WP transformation.

4   Experimental Results

In this section, in order to confirm the validity and performance of the proposed SDWP-bHU algorithm, simulation experiments using Matlab are presented below.

Four mineral signatures used in the simulations are extracted from USGS spectral library, where the wavelength is 224 ($M = 224$), which forms the signature matrix (mixing matrix) $\mathbf{A} \in \mathbb{R}^{224 \times 4}$. The abundance matrix (source signal) $\mathbf{S} \in \mathbb{R}^{4 \times 1024}$ is generated random using the Matlab software function $\text{rand}(\cdot)$. In order to test the anti-interference ability of the proposed SDWP-bHU algorithm to additive noise, some Gaussian noises are added to the mixtures, where the signal-to-noise ratio (SNR) is defined as

$$\text{SNR} = 10 \log_{10} \frac{E(\mathbf{xx}^T)}{E(\mathbf{vv}^T)} \tag{16}$$

The following performance index is adopted to evaluate the performance of the proposed algorithm called Amari performance index (API) C[17], and $0 \le C \le 1$, the perfect separation implies that $C = 0$. Another performance index is called the signal-to-interface ratio (SIR) between source signals $s_i(t)$ and their estimates $\hat{s}_i(t)$, which is calculated to measure the accuracy of the estimations of the source signals. In conventional BSS algorithms, it can be considered as getting meaningful results when SIR $\ge$ 10dB.

Firstly, we considered the noiseless case. The true endmember signatures are shown in Fig. 1(a), one of the separation results are presented in Fig. 1(b). After 100 simulations, the mean API is 0.0322, the mean SIR of four signatures are: $\text{SIR}_1 = 37.764\text{dB}$, $\text{SIR}_2 = 39.667\text{dB}$, $\text{SIR}_3 = 43.334\text{dB}$, $\text{SIR}_4 = 45.036\text{dB}$.

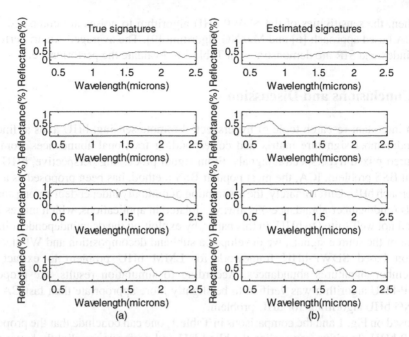

Fig. 1. Comparison of the true endmember signatures and the separated endmember signatures based on SDWP-bHU approach (a) True signatures; (b) Estimated signatures

Table 1. Obtained mean API and mean SIR values for the SDWP-bHU algorithm FastICA and MaxNG algorithms

SNR(dB)	Performance index	Algorithms	Mean values over 100 simulations
30	SIR(dB)	SDWP-bHU	43.26
		fastICA	26.38
		MaxNG	40.29
	API	SDWP-bHU	0.0399
		fastICA	0.0725
		MaxNG	0.0437
15	SIR(dB)	SDWP-bHU	32.46
		fastICA	14.68
		MaxNG	29.74
	API	SDWP-bHU	0.0564
		fastICA	0.2603
		MaxNG	0.0682
5	SIR(dB)	SDWP-bHU	20.38
		fastICA	9.83
		MaxNG	17.49
	API	SDWP-bHU	0.1902
		fastICA	0.478
		MaxNG	0.2114

Then, the sensitivities of the SDWP-bHU algorithm to noises are compared with fastICA based approach [6] and MaxNG algorithm [9]. The averaged Amari performance index and SIR are demonstrated in Table 1 to evaluate the performance.

5   Conclusions and Discussion

As an important research topic of hyperspectral remote sensing, bHU aims at finding the endmember signature matrix and corresponding fractional abundances from the measured mixed hyperspectral signals. From signal processing perspective, bHU is a typical BSS problem. ICA, the most popular BSS method, has been proposed as a tool to approach bHU. Unfortunately, the assumption of mutually independent among source signals (endmember abundances) is always violated in applications, which makes ICA method not well suited to bHU. In this paper, by exploiting the deep independent information of the source signals, we developed a subband decomposition and WP decomposition based SDWP-bHU framework for LMM bHU problem to extract the endmember signatures abundances. According to simulation results, the proposed SDWP-bHU algorithm was verified to be slightly more appropriate than fastICA and MaxNG bHU algorithms for bHU problem.

Based on Fig. 1 and the comparisons in Table 1, one can conclude that the proposed SDWP-bHU algorithm can realize the blind HU and performance slightly better than blind HU fastICA algorithm and MaxNG algorithm.

Acknowledgments. This research is financially supported by the National Natural Science Foundation of China (No.61401401, 61402421, 61571401), the China Postdoctoral Science Foundation (No.2015T80779, 2014M561998) and the open research fund of National Mobile Communications Research Laboratory, Southeast University (No.2016D02).

References

1. Ma, W.-K., Bioucas-Dias, J.M., Chan, T.-H., Gillis, N., Gader, P., Plaza, A., Ambikapathi, A., Chi, C.-Y.: A signal processing perspective on hyperspectral unmixing. IEEE Signal Process. Mag. **31**(1), 67–81 (2014)
2. Bioucas-Dias, J.M., Plaza, A., Dobigeon, N., Parente, M., Du, Q., Gader, P., Chanussot, J.: Hyperspectral unmixing overview: geometrical, statistical, and sparse regression-based approaches. IEEE J. Sel. Top. Appl. Earth Observations Remote Sens. **5**(2), 354–379 (2012)
3. Heylen, R., Burazerovic, D., Scheunders, P.: Fully constrained least squares spectral unmixing by simplex projection. IEEE Trans. Geosci. Remote Sens. **49**(11), 4112–4122 (2011)
4. Zhang, K., Chan, L.W.: An adaptive method for subband decomposition ICA. Neural Comput. **18**(1), 191–223 (2006)
5. Lin, C.-H., Ma, W.-K., Li, W.-C., Chi, C.-Y., Ambikapathi, A.: Identifiability of the simplex volume minimization criterion for blind hyperspectral unmixing: The no pure-pixel case. IEEE Trans. Geosci. Remote Sens. **53**(10), 5530–5546 (2015)
6. Nascimento, J., Dias, J.: Does independent component analysis play a role in unmixing hyperspectral data? IEEE Trans. Geosci. Remote Sens. **43**(1), 175–187 (2005)

7. Cardoso, J.F.: Multidimensional independent component analysis. In: Proceedings of ICASSP 1998, pp. 1941–1944. MIT Press (1998)
8. Szabo, Z., Poczos, B., Lorincz, A.: Separation theorem for independent subspace analysis and its consequences. Pattern Recogn. **45**(4), 1782–1791 (2012)
9. Caiafa, C.F., Salerno, E., Protoa, A.N., Fiumi, L.: Blind spectral unmixing by local maximization of non-Gaussianity. Sig. Process. **88**(1), 50–68 (2008)
10. Ouedraogo, W., Souloumiac, A., Jaidane, M., Jutten, C.: Non-negative blind source separation algorithm based on minimum aperture simplicial cone. IEEE Trans. Signal Process. **62**(2), 376–389 (2014)
11. Erdogan, A.T.: A class of bounded component analysis algorithms for the separation of both independent and dependent sources. IEEE Trans. Signal Process. **61**(22), 5730–5743 (2013)
12. Wang, F.S., Wang, Z.Y., Li, R., Zhang, L.R.: An efficient algorithm for harmonic retrieval by combining blind source separation with wavelet packet decomposition. Digit. Signal Proc. **46**, 133–150 (2015)
13. Mallat, S.: A Wavelet Tour of Signal Processing. Academic Press, New York (1999)
14. Cardoso, J.F.: Dependence, correlation and gaussianity in independent component analysis. J. Mach. Learn. Res. **4**, 1177–1203 (2003)
15. Comon, P., Jutten, C.: Handbook of Blind Source Separation: Independent Component Analysis and Applications. Elsevier, Oxford (2010)
16. Kopriva, I., Sersic, D.: Wavelet packets approach to blind separation of statistically dependent sources. Neurocomputing **71**(7–9), 1642–1655 (2008)
17. Li, Y., Cichocki, A., Amari, S.: Analysis of sparse representation and blind source separation. Neural Comput. **16**(6), 1193–1234 (2004)

Speech Denoising Based on Sparse Representation Algorithm

Yan Zhou[1,2(✉)], Heming Zhao[2], Xueqin Chen[2], Tao Liu[1], Di Wu[2], and Li Shang[1]

[1] School of Electronic and Information Engineering, Suzhou Vocational University,
Suzhou 215104, Jiangsu, China
zhyan@jssvc.edu.cn
[2] School of Electronics and Information Engineering, Suzhou University,
Suzhou 215104, Jiangsu, China

Abstract. A new speech denoising method that aims for processing corrupted speech signal which is based on K-SVD sparse representation algorithm is proposed in this paper. Here, the DCT sparse and redundant representation over dictionary is used for the initial redundant dictionary. In order to analyze the time-frequency characteristics of speech signal clearly, the spectrogram patches are applied as training samples for the sparse decomposition in this approach. However, the training samples need to extend their deployment to arbitrary spectrogram sizes because the K-SVD algorithm is limited in handling small size spectrogram. A global spectrogram was defined prior that forces sparsity over patches in every location in the spectrogram. Afterwards, by using the K-SVD algorithm, the greedy algorithm is used for updating which alternates between dictionary and sparse coefficients. Then a dictionary that describes the speech structure effectively can be obtained. Finally, the corrupted speech signal can be sparsely decomposed under the redundant dictionary. Consequently, the sparse coefficients can be obtained and used to reconstruct the noiseless spectrograms. As a result, the purpose of the separation for the signal and noise is reached. The proposed K-SVD algorithm is a simple and effective algorithm, which is suitable for processing corrupted speech signal. Simulation experiments show that the performance of the proposed K-SVD denoising algorithm is stable, and the white noise can be effectively separated. In addition, the algorithm performance surpasses the redundant DCT dictionary method and Gabor dictionary method. In a word, K-SVD algorithm leads to an alternative and novel denoising method for speech signals.

Keywords: K-SVD algorithm · Speech signal denoising · Speech spectrogram · Sparse redundant dictionary · Sparse representation

1 Introduction

Since the development of speech calculation, much attention has been paid to the feature extraction, but the speech signal usually contains a lot of noise, which influences the speech analysis effect. Therefore, the task of denoising is the premise task. Because the research of speech denoising is important, eliminating the background noise and

© Springer International Publishing Switzerland 2016
D.-S. Huang and K.-H. Jo (Eds.): ICIC 2016, Part II, LNCS 9772, pp. 202–211, 2016.
DOI: 10.1007/978-3-319-42294-7_17

improving the speech quality become a significant research direction at present. The speech signal with noise can be expressed as:

$$y(t) = x(t) + v(t) \tag{1}$$

Here, $y(t)$ is the speech signal with noise which is also named corrupted speech signal, $x(t)$ is the ideal speech signal, $v(t)$ is the additive zero-mean white and homogeneous Gaussian noise. In this paper, it is desire to design an algorithm that can remove the Gaussian White noise $v(t)$ from $y(t)$, getting as close as possible to the original speech signal $x(t)$.

Recently, there are various speech denoising methods, such as the traditional spectral subtraction method [1–3]. However, the probability density function of the speech signal is difficult to estimate. The method of wavelet transform [4–6] is just suitable for the situation of signal mixed with white noise. The linear forecast analysis denoising method [7] can remove the noise composition from speech signal, but at the same time, the phonetic element itself also caused the damage which is resulted from "music noise", which reduces the quality of speech signal. The subspace denoising method [8], and many improvement denoising algorithms. In a word, these methods have a lot of defects and not suitable for conducting the speech signal with low signal to noise ratio.

Sparse decomposition method starts from the point of non-stationary characteristic of speech signal. In recent years, with the development of sparse representation theory, speech signal denoising method based on sparse representation has drawn a lot of research attention [9–14]. It realizes the speech denoising by analyzing the time-frequency characteristic of speech signal reasonably. However, the K-SVD sparse decomposition algorithm [15–17] is simple, flexible and efficient. Moreover, K-SVD redundant dictionary is produced by training speech samples. It can reflect the inherent structure characteristics of speech signal itself. Thus, a speech denoising algorithm that is via sparse and redundant representation over K-SVD dictionary is proposed in this paper.

2 Description of K-SVD Algorithm

2.1 Algorithm Description

Generally speaking, the target of K-SVD learning algorithm is to realize sparse representation adaptively. That is, the signal similar with the training signal all can be sparse represented over this redundant dictionary. The speech redundant dictionary training based on K-SVD learning algorithm includes two steps, one is sparse decomposition, and the other is dictionary atoms updating. When considering this available scheme, the two steps should be alternately performed, thus the redundant dictionary and the sparse matrix can be updated synchronously. The specific steps of speech denoising algorithm based on K-SVD redundant dictionary are described as following.

Step1. Initialization: The training samples and testing samples all should be transformed to corresponding spectrogram. The size of each spectrogram is set as the size of $p \times p$, and doing pretreatment for this spectrograms before dictionary training. The redundant DCT dictionary is employed as the initial dictionary $D = \{d_j\}_{j=1}^{K}$, and the dictionary redundancy is set as γ. Since dictionary learning is limited in handling small spectrogram patches, in this work, a global spectrogram prior that forces sparsity over patches in every location in the spectrogram (with overlaps) is proposed. So the spectrograms should be divided into small overlapping patches randomly.

Step2. Sparse Coding: the initial dictionary must be assumed as fixed firstly in this step. Aligning with this, the corrupted speech signal $Y = \{y_i\}_{i=1}^{N}$ should be decomposed over this initial dictionary by using orthogonal matching algorithm (OMP). Since the iterative termination condition is decided by the decomposition residual. Consequently, the sparse matrix can be got by solving the following formula.

$$\min_{D,X} \sum_i \|x_i\|_0 \quad s.t \|Y - DX\|_F^2 \leq \varepsilon \tag{2}$$

Step3. Dictionary updating: This step is aim to update the redundant dictionary. It is important to fix the sparse matrix X which has been trained at the last step. For the size of K-SVD dictionary is $n \times k$, it is updated column by column and this is an iterative process. Meanwhile, the error E_k is calculated in all samples which have been removed the composition of atom. Notice that error E_k can be calculated as:

$$E_k = Y - \sum_{k=j} d_k x_T^k \tag{3}$$

Here, x_T^k is the k-th line in sparse matrix X of d_k. So that the column error E_R^k of atom d_k can be expressed as:

$$E_R^k = E_k \Omega_k \tag{4}$$

Here, Ω_k is the matrix $N \times |w_k|$, $w_k = \{i | 1 \leq i \leq N, x_T^k \neq 0\}$, w_k is the index group, that is for all the samples been decomposed, which use atom d_k. Then the E_R^k needs to do SVD decomposition, the calculation formula is defined as

$$E_R^k = U\Delta V^T \tag{5}$$

Thus, the first column of U is the updated atom $\tilde{d}_k$. Taken these steps, the updated dictionary can be obtained until it reaches iterative termination condition. When the iterative stops, the updated dictionary is fixed again, it is ready for doing sparse decomposition in the next step. The adaptive redundant K-SVD dictionary can be gotten by repeating steps (3) and (4). After acquiring the redundant dictionary, OMP algorithm is employed to do sparse decomposition for gaining the sparse coefficients.

By this way, the useful spectrogram signal can be reconstructed using these sparse coefficients but except the noise spectrograms, thus, the useful speech part and the noise part can be divided, in short, the whole spectrogram quality can be enhanced.

2.2 Complexity Analysis of K-SVD Algorithm

In evaluating the computational complexity of this algorithm, two stages are considered. The complexity of K-SVD algorithm is mainly measured by calculation times in the progress of generating redundant dictionary. At first sight, in the sparse decomposition stage, the calculation quantity of OMP algorithm implementation is operated. Second, in the stage of dictionary updating, the calculation quantity of the singular value decomposition is considered. Setting J as the iterative times for sparse decomposition and dictionary updating, the dictionary size is set as $n \times k$, and n is the dimensions of the spectrogram, and k is the column of dictionary. At last, setting L as the non-zero number in all sparse vectors x_i, however, the value of L depends heavily on the strength of noise signal. If the noise signal is more strength, instead, the number of sparse representation coefficient is less. Correspondingly, the performance of K-SVD algorithm in restraining noise is much stronger. So that, each spectrogram of size $\sqrt{n} \times \sqrt{n}$, the complexity of all steps requires $O(nkLJ)$.

3 Experimental Results

In order to validate the validity and advantage of the K-SVD algorithm in the denoising progress for corrupted speech signal, the experiments are designed as following. In order to test the behavior of the K-SVD denoising algorithm that uses the adaptive dictionary, here, the speech denoising experiment for the signal with white noise is proposed. The sampling frequency of the test speech samples is 8 kHz and frame length is 128 point. The length of each section is 25 ms and the frame range is set to 25 %. Meanwhile, the time domain speech signal should be mapped into time-frequency spectrogram signal, and the spectrogram size of each signal sample is 256×256.

3.1 The Test Data Samples

One of the clean signals with the length of 25 ms is described in Fig. 1(a). The Gaussian White noise with variance of 25 is added into the clean signal, which is showed in Fig. 1(b). Figure 1(c) and (d) describes the spectrograms for the clean test signal and the corrupted signal respectively.

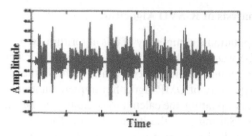

(a) Clean speech signal waveform in time domain

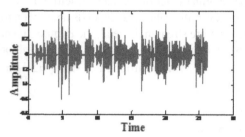

(b) Corrupted speech signal waveform in time domain

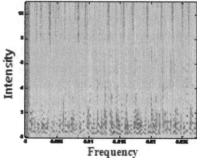

(c) Clean speech signal in frequency domain

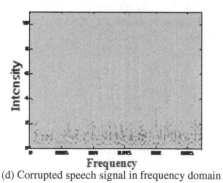

(d) Corrupted speech signal in frequency domain

Fig. 1. The tested speech signal samples ((a) and (b) are the time domain signals, and (c) and (d) are the frequency domain signals)

3.2   The Contrast of Redundant Dictionary

In order to demonstrate the K-SVD redundant dictionary is more suitable for the speech signal processing, this paper designed the following experiment. Furthermore, the redundant DCT dictionary is described for a fairly comparison, this dictionary was also used as the initialization for the K-SVD training algorithm. The K-SVD training patches were taken from an arbitrary set of spectrograms. Figure 2(a) describes the DCT redundant dictionary, Fig. 2(b) shows Gabor redundant dictionary, and Fig. 2(c) shows the

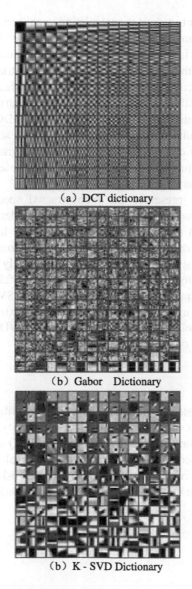

(a) DCT dictionary

(b) Gabor   Dictionary

(b) K - SVD Dictionary

Fig. 2.   Initial DCT dictionary, Gabor dictionary and K-SVD dictionary

new K-SVD redundant dictionary which is trained by the samples. Compared with these three dictionary structures, it can be seen that the difference is obvious. The speech signal in redundant DCT dictionary is approximately sparse. However, the redundant K-SVD dictionary is more suitable for sparse decomposition and it can reflect more speech structure features. The reason is that the redundant K-SVD dictionary is trained by overlapping spectrogram patches.

3.3 Contrast Analysis Under White Noise Environment

In this experiment, in order to reduce the calculation quantity and be suitable for K-SVD algorithm, each spectrogram of 256×256 should be divided into 8×8 small spectrograms randomly (with overlaps). Through the sparse decomposition algorithm, the noise signal is removed when the small spectrogram is reconstructed. Consequently, the purpose of the whole spectrogram denoising is realized. Meanwhile, redundant DCT dictionary denoising algorithm and redundant Gabor dictionary denoising algorithm are applied for contrast. The initial dictionary size of redundant DCT dictionary, redundant Gabor dictionary and redundant K-SVD dictionary are all set for 64×256. K-SVD dictionary is trained on a data-set of 5000 small spectrograms which are randomly chosen. The termination condition is that the iterative executes 200 times. Finally, the redundant K-SVD dictionary can be obtained.

Figure 3 respectively presented the reconstructed spectrogram after different methods. It can be seen that the signal quality after these denoising algorithm can be improved. However, from the figures, it can be obviously found that the method of redundant K-SVD sparse representation has the best performance compared with the others. As a result, this presented denoising algorithm not only gets a good visual effect, but also obtains a fine auditory effect. This is because the useful signal part and the noise signal part are separated qualitatively by the redundant K-SVD sparse representation algorithm. Ultimately, the useful signal can be completely extracted from the corrupted signal. Therefore, the denoising method of redundant K-SVD sparse representation is undisputed superior to the others.

3.4 The PESQ Value

At last, the experiment tested five lie speech samples of different Signal Noise Ratio (SNR), and the PESQ (Perceptual evaluation of speech quality) was introduced to evaluate the speech quality from subjective method. The PESQ score of 4.5 is expressed excellent quality and the score of 0.1 is expressed unacceptable quality. The following Table 1 shows the reconstructed speech signals after denoising by different methods. From the table, we can see that the speech used K-SVD denoising method has got the highest score.

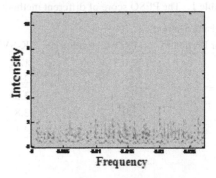

(a) Reconstructed Spectrogram with Gabor dictionary

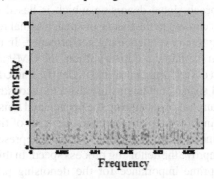

(b) Reconstructed Spectrogram with DCT dictionary

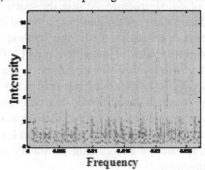

(c) Reconstructed Spectrogram with K-SVD dictionary

Fig. 3. The spectrograms after denoising ((a) is the reconstructed spectrogram with Gabor dictionary, (b) is the reconstructed spectrogram with DCT dictionary, (c) is the reconstructed spectrogram with K-SVD dictionary)

Table 1. The PESQ score of different methods

SNR	Denoising Method		
	Gabor Dictionary	DCT dictionary	K-SVD dictionary
−10	1.1	1.0	1.2
−5	1.3	1.8	2.5
0	2.0	2.1	3.3
5	2.2	2.9	3.8
10	2.7	3.6	4.1

4   Conclusion

A simple method for speech signal denoising which is based on redundant K-SVD dictionary is proposed, according to the theory of speech signal redundant sparse representation. This is leading to a novel denoising performance. In this paper, the K-SVD algorithm is used to train the initial dictionary, then, the OMP algorithm is applied to the sparse decomposition of the corrupted signal under the trained redundant dictionary. Finally, the decomposed sparse coefficients are used to reconstruct the speech signal. After all this steps, the purpose of separating the speech and noise can be achieved. As the spectrogram can ordinary synthesize the characteristics of time domain waveform and spectrum, here, the spectrogram is selected for sparse representation. Significantly, this algorithm can also improve the denoising process speed. In this research, the content of the dictionary is of prime importance for the denoising process. The redundant K-SVD dictionary is obtained by training on the speech signals, so the dictionary can reflects the characteristics of signal structure. At last, the sparse decomposition experimental results have proved the validity of the method proposed in this paper. Moreover, compared to the other speech denoising methods, this method shows a more superior performance whether in objective index or subjective quality. Furthermore, the PESQ is measured, the signal denoised by K-SVD algorithm got a good auditory effect. Although this method has effectively improved the speech signal quality to a certain extent, but there is a large research space that do not be explored in this work. For example, the research of a more complex model that uses several dictionaries switched by signal content, the research of the optimization algorithm for greed algorithm when training the dictionary, the research of a redundant dictionary and more. In a word, all of these are the research direction.

Acknowledgments. This research was supported by: National Natural Science Foundations of China (Grant No. 61372146, 61373098), the Natural Science Foundation of Jiangsu Province under grant (No. BK20131196), the public service platform for the testing and validation of smart LTE terminal devices of suzhou science and technology plan project (No. SZP201310).

References

1. Yoshioka, T, Nakatani, T, Okuno, H.G.: Noisy speech enhancement based on prior knowledge about spectral envelope and harmonic structure. In: 2010 TEEE International Conference on Acoustics Speech and Signal Processing, pp. 4270–4273. TEEE Press, New York (2010)
2. Loizou, P.C.: Speech Denoising: Theory and practice. CRC Press, Boca Raton (2007)
3. Gowreesunker, B.V., Tewfik, A.H.: Learning sparse representation using iterative subspace identification. J. IEEE Trans. Signal Proc. **58**(6), 3055–3065 (2010)
4. Zhang, L., Pan, Q.: On the determination of threshold in threshold based denoising by wavelet transformation. J. Acta Electronica Sinica **29**(3), 400–403 (2001)
5. Yu, X.S.: An improved total variation model with adaptive local constraints and its applications to image denoising, deblurring and inpainting. J. Int. J. Digit. Content Technol. Appl. **5**(12), 170–177 (2011)
6. Loizou, P.C., Gibak, K.: Reasons why current speech-enhancement algorithms do not improve speech intelligibility and suggested solutions. J. IEEE Trans. Audio Speech Language Proc. **19**(1), 47–56 (2010)
7. Yegnanarayana, B., Avendano, C., Hermansky, H., Satyanarayana Murthy, P.: Speech denoising using linear prediction residual. J. Speech Commun. **28**, 25–42 (1999)
8. Tantibundhit, C., Pernkopf, F., Kubin, G.: Joint time-frequency segmentation algorithm for transient speech decomposition and speech denoising. J. IEEE Trans. Audio Speech Denoising **18**(6), 1417–1428 (2010)
9. Donoho, D.L., Johnstone, I.M.: Ideal spatial adaptation by wavelet shrinkage. Biometrika **81**(3), 425–455 (1994)
10. Sigg, C.D., Dikk, T., Buhmann, J.K.: Speech enhancement with sparse coding in learned dictionaries, pp. 4758–4761(2010)
11. Cho, N., Jay Kuo, C.C.: Sparse music representation with source-specific dictionaries and its application to signal separation. J. IEEE Trans. Audio Speech Language Proc. **19**(2), 337–348 (2011)
12. Wen, J., Michael, S.: Scordilis.: Speech denoising by residual domain constrained optimization. J. Speech Commun. **48**, 1349–1364 (2006)
13. Chen, S.S., Donoho, D.L., Saunders, M.A.: Atomic decomposition by basis pursuit. J. SIAM Rev. **43**(1), 129–159 (2001)
14. Lewicki, M.S., Sejnowski, T.J.: Learning redundant representations. J. Neur. Comput. **12**, 337–365 (2000)
15. Zhang, Y.: Hybrid recommendation method IN sparse datasets: combining content analysis and collaborative filtering. J. Int. J. Digital Content Technol. Appl. **6**(10), 52–60 (2012)
16. Aharon, M., Elad, M., Bruckstein, A.M.: K-SVD and its non-negative variant for dictionary design. J. Int. Soc. Optics Photonics, (2005)
17. Elad, M.: Sparse and Redundant Representations: From Theory to Applications in Signal and Image Processing. Springer, New York (2010)

A New Method for Yielding a Database
of Hybrid Location Fingerprints

Yan-Hua Li, Wen-Sheng Tang[(⊠)], Sheng-Chun Wang, and Peng Hui

The Department of Computer Teaching,
Hunan Normal University, Changsha 410081, China
154976552@qq.com

Abstract. Location fingerprinting in wireless LAN positioning used by RSSI (Received Signal Strength Indication) has become one of the most popular indoor positioning technologies. However RSSI is easily affected by environment, it is difficult to improve the accuracy. In this paper, a hybrid fingerprints construction algorithm was presented by RSSI and geomagnetic field which can effectively reduce the inaccuracy compared to the construction location fingerprints that only use RSSI. This article has also presented the C-GPCA (Clustering Grouped Principal Component Analysis) algorithm, which can be used in the further optimization of hybrid fingerprint database, improving the positioning accuracy.

Keywords: Indoor positioning · Cluster analysis · C-GPCA · KNN

1 Introduction

Since the 1970s, IEEE802.11 standard for wireless local area network (WLAN) was released. WIFI positioning technology has been widespread concerned in the industry. Among all indoor positioning technologies, RSS positioning technology has multiple advantages, such as high bandwidth, high speed, high coverage, low cost, and so on. It can achieve many practical personalization features in the application of a short distance, such as patients tracking and monitoring in hospitals, key prisoner monitoring, exhibits security in museum, positioning of shopping carts.

In the complex indoor propagation environment, the indoor positioning algorithm based on geometric features, such as TOA (Time of Arrival) [1], TDOA (Time Difference of Arrival) [2] and AOA (Angle of Arrival) [3], doesn't have a high positioning accuracy for the affection of NLOS propagation, Multi path Propagation and other electronic equipment interference signal [4]. Although the propagation model based access point (AP) position is known, the fast fading propagation channel will cause uncertain RSSI value. As a result, it's not easy to deal with the positioning. The fingerprint positioning method use fingerprint matching technology which can achieve better positioning effect.

The location fingerprinting method generally uses RSSI to achieve target positioning as the cost of localization is relatively low. However the RSSI is easily affected by the indoor complex environment, such as of radio wave propagation distance, interior layout, construction materials, body absorb and climate. It is difficult to

© Springer International Publishing Switzerland 2016
D.-S. Huang and K.-H. Jo (Eds.): ICIC 2016, Part II, LNCS 9772, pp. 212–221, 2016.
DOI: 10.1007/978-3-319-42294-7_18

accurately describe location information with RSSI only. Besides the existing indoor positioning technology is difficult to meet the demand for high-precision positioning [5]. With the development of sensor technology, various sensors, such as a gyro sensor, gravity sensor, an acceleration sensor and geomagnetic sensor, have begun to be widely used in the field of positioning to achieve good positioning effect, particularly those based on the geomagnetic field positioning technology [6]. Literature [7] realized effective positioning system combining RSSI techniques and geomagnetic field positioning technology. The use of the low-cost RSSI as positioning auxiliary information will reduce the matching area of geomagnetic matching and improve matching speed.

Now, many researchers have focused on the offline database construction of positioning fingerprints. There are two main directions. One is the cost of the offline construction of database [8, 9], the other is to overcome RSSI signal interference during the construction [3, 10]. Although the improved fingerprint database has some improvement on positioning accuracy, it still cannot greatly minimize interference to RSSI.

Different from the above methods, this article focuses on the relationship RSSI and geomagnetic fields instead of the improvement of RSSI signal. This paper proposes a hybrid fingerprint algorithm using RSSI and geomagnetic field, which can effectively reduce the inaccuracy caused by just using RSSI. We also refer the design idea from literature [11], using cluster grouped PCA algorithm for the optimization of hybrid fingerprint database, thus further improve the positioning accuracy.

2   Positioning Principle

Location fingerprinting has two phases: offline database construction and online positioning. Offline database construction is to create a fingerprint database while online positioning is to achieve positioning based on the RSSI signal received by targeting algorithms. The procedure of the Location fingerprint database is shown in Fig. 1 [12].

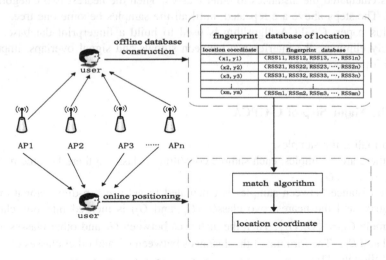

Fig. 1. Location fingerprinting positioning diagram

In online positioning phase, there are multiple matching algorithms. The most fundamental method is K-nearest neighbor algorithm (K Nearest Neighbor, KNN).

KNN algorithm is to select the nearest K (K >= 2) sample points, and then calculate the average coordinates of this K sampling points $(\bar{x}, \bar{y})$, as the target location coordinates.

$$D_i = \sqrt{\sum_{j=1}^{n} \left(rssi_j - RSSI_{mj} \right)^{\alpha}} \quad i = 1, 2, \cdots, k \tag{1}$$

$$(\bar{x}, \bar{y}) = \frac{1}{k} \sum_{i=1}^{k} (x_i, y_i) \tag{2}$$

In Eq. (1), D_i means the distance between sampling point and the mobile terminal in fingerprint matching k is the number of nearest neighbors. When $\alpha = 1$, it's on behalf of the Manhattan distance. When $\alpha = 2$, it's on behalf of Euclidean distance. In Formula (2), (x_i, y_i) represents the location coordinates of sampling points.

3   Construction of the Fingerprint Database

3.1   The Basics of C-GPCA

GPCA is based on the correlation between variables. First, the fingerprint variables are classified by system clustering. Then, the fingerprint variables are grouped according to the classification. At last each group is handled by principal component analysis.

System clustering is classified by the distances between variables or the intimating degree [13]. The main idea is to set some samples as one class, then to predetermine the distances between each class. The nearest two classes are combined into a new class, which is calculated the distances to other classes. Then the nearest two categories are merged. The steps above are repeated, until all the samples become one tree.

In this paper, C-GPCA algorithm is used to build a fingerprint database. It can effectively eliminate the interference between signals and signal overlaps, improving the positioning accuracy.

3.2   The Major Step of C-GPCA

(1) Normalize the samples.
(2) If there are n samples, each sample constitute a class. So there become n classes: $G_1, G_2, \cdots, G_n$
(3) The distance between samples is calculated to get a n * n-dimensional distance matrix, and the nearest two classes (G_i and G_j) is merged into one class G_h, namely $G_h = \{G_i, G_j\}$. Then the distances between G_h and other classes are calculated D_{hk}. The formula of the distances between G_h and other classes $G_k(k \neq h)$ is following [14]:

$$D_{hk} = \min_{x_h \in G_h, x_j \in G_k} d_{hj}$$

$$= \min \left\{ \min_{x_i \in G_p, x_j \in G_k} d_{ij}, \min_{x_i \in G_q, x_j \in G_k} d_{ij} \right\} \tag{3}$$

$$= \min \left\{ D_{pk}, D_{qk}, \right\}$$

Here d_{ij} represents the distance between x_i and x_j.

(4) Calculating the distance and merging them into new classes is repeated, until all the variables become one class.

(5) The classes are grouped according to the classification criteria.

(6) According to each group of variables, the eigenvalues λ_i and the corresponding eigenvectors U_i of the correlation matrix A are calculated [15]

$$[A - \lambda_i] U_i = 0 \tag{4}$$

Here U_i represents the eigenvector of the *No.i* eigenvalue λ_i.

(7) Each group determines the main component score $F_1, F_2, F_3, \cdots, F_m$ based on the cumulative contribution rate which is calculated by $\sum_{i=1}^{k} \lambda_i \Big/ \sum_{i=1}^{m} \lambda_i \geq 80\%$.

(8) Evaluation: Main component analysis is used to extract the first m main components score to get a linear combination. The variance contribution rate of the previous m main components is set as coefficient which is used to construct a comprehensive evaluation function model. In such a case:

$$Y = \alpha_1 F_1 + \alpha_2 F_2 + \cdots + \alpha_m F_m \tag{5}$$

Here F_i represents the score of the *No.i* principal component. Y represents the synthetic evaluation function, which can be sorted according to a comprehensive integrated evaluation function scores for each sample size.

3.3   C-GPCA Algorithm for RSSI Fingerprint

The results of clustering the 10 fingerprint component (AP1, AP2,..., AP10) with RSSI-based fingerprint with group connecting method are shown in Fig. 2.

According to Fig. 2, the entire hybrid fingerprint components can be divided into two categories. The first category is constituted of fingerprint component (AP5, AP9, AP7, AP1, AP4, AP2, AP8). The second category is constituted of fingerprint component (AP6, AP10, AP3). Each category will be called one group. Each group of the fingerprint is operated by principal component analysis. The results are shown in Tables 1 and 2.

As can be seen in Table 1, according to the principal component contribution rate which is more than 80 %, the number of principal components is 4. And the 4 principal component score which is named "F_1', F_2', F_3', F_4'" is calculated separately.

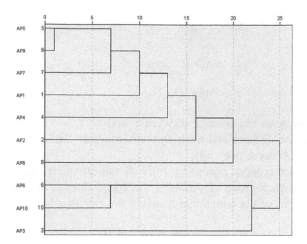

Fig. 2. Clustering dendrogram based on RSSI fingerprint database

Table 1. The contribution rate of the first set of eigenvalues and the accumulated contribution rate

Ingredient	The contribution rate of the correlation matrix eigenvalues and each principal component and the cumulative contribution rate		
	eigenvalue	contribution rate variance	cumulative contribution rate
1	2.595	37.075 %	37.075 %
2	1.473	21.037 %	58.113 %
3	0.933	13.330 %	71.443 %
4	0.855	12.220 %	83.663 %
5	0.699	9.987 %	93.650 %
6	0.299	4.267 %	97.917 %
7	0.146	2.083 %	100.000 %

Table 2. The contribution rate of the second set of eigenvalues and the accumulated contribution rate

Ingredient	The contribution rate of the correlation matrix eigenvalues and each principal component and the cumulative contribution rate		
	eigenvalue	contribution rate variance	cumulative contribution rate
1	1.520	50.668 %	50.668 %
2	1.019	33.959 %	84.627 %
3	0.461	15.373 %	100.000 %

As shown in Table 2, according to the principal component contribution rate which is more than 80 %, the number of principal components is 2. Also the 2 principal component scores which are F_1'' and F_2'' are calculated separately.

Finally, a new comprehensive evaluation function is reconstructed to calculate the composite score named Y for each position of the fingerprint and sorted as follows:

$$Y = 0.37075 \times F_1' + 0.21037 \times F_2' + 0.133 \times F_3' + 0.122 \times F_4'$$
$$+ 0.50668 \times F_1'' + 0.33959 \times F_2''$$

3.4   C-GPCA Algorithm for Hybrid Fingerprint

Cluster the 13 fingerprints of component (AP1, AP2,..., AP10, mag_x, mag_y, mag_z) in the hybrid fingerprint database by connecting the two groups. The results are shown in Fig. 3.

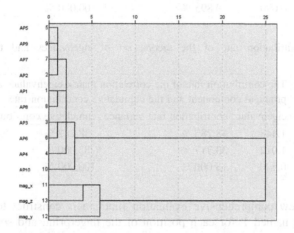

Fig. 3. Clustering dendrogram based on hybrid database

The experimental results show that the entire hybrid fingerprint component will be divided into two categories, namely, (AP1, AP2,..., AP10) and (mag_x, mag_y, mag_z). Then the two fingerprint components are divided into two groups. PCA analysis is executed to the fingerprint component of each group separately. The results are shown in Tables 3 and 4.

As the result in Table 3, according to the principle that the cumulative contribution rate which is more than 80 %, we can determine the number of main components which is 5. Each main component scores named $F_1', F_2', \cdots, F_5'$ is calculated.

As the results in Table 4, according to the principle that the cumulative contribution rate which is more than 80 %, we can determine the number of main components that is 2. Also the 2 principal component score called F_1'' and F_2'' are calculated separately.

Table 3. The contribution rate of the first set of eigenvalues and the accumulated contribution rate

Ingredient	The contribution rate of the correlation matrix eigenvalues and each principal component and the cumulative contribution rate		
	eigenvalue	contribution rate variance	cumulative contribution rate
1	2.755	27.548 %	27.548 %
2	2.407	24.073 %	51.621 %
3	1.213	12.130 %	63.752 %
4	0.919	9.195 %	72.946 %
5	0.808	8.084 %	81.030 %
6	0.637	6.365 %	87.395 %
9	0.196	1.960 %	99.105 %
10	0.090	0.895 %	100.000 %

Table 4. The contribution rate of the second set of eigenvalues and the accumulated contribution rate

Ingredient	The contribution rate of the correlation matrix eigenvalues and each principal component and the cumulative contribution rate		
	eigenvalue	contribution rate variance	cumulative contribution rate
1	1.448	48.282 %	48.282 %
2	1.012	33.717 %	82.000 %
3	0.540	18.000 %	100.000 %

Finally, a new comprehensive evaluation function is construct to calculate the composite score named Y for each position of the fingerprint, and sorted. In such a case:

$$Y = 0.27548 \times F_1' + 0.24073 \times F_2' + 0.1213 \times F_3' + 0.0919 \times F_4'$$
$$+ 0.0808 \times F_5' + 0.48282 \times F_1'' + 0.33717 \times F_2''$$

By calculating the total score Y which is sorted in descending order, selecting the location fingerprint which is larger than average and rebuilding a new fingerprint database, one can remove the fingerprints in corner point or corner jitters, effectively reduce the signal interference, thus will be more conducive to achieve high-precision positioning.

4   Experiment Simulations and Analysis

4.1   Experiment Environment

In this paper, the experimental data is collected in the first floor of Zhishan Building, Hunan Normal University, including the four corridors, one hall. We collected 172 points in the two longitudinal 3 m × 52 m corridors, 248 points in the 4 m × 72 m lateral corridor, 80 points in the 3 m × 47 m lateral corridor and 330 points in the 20 m × 27 m hall. The totally 830 sampling points were evenly distributed in every corner of the hall. The distance between each location is 1 m. For the wireless communication of mobile terminal at any sampling point as well as AP, to ensure the accuracy of the measured signal, we uniformly arranged 10 AP on the first floor. The AP is disposed in both ends of the corridor and hall, as shown in Fig. 4.

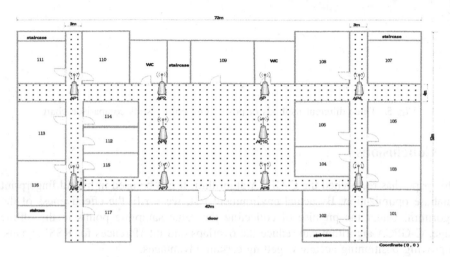

Fig. 4. The first floor plan of Zhishan Building and sampling points

4.2 Simulation Results

In the positioning region, we randomly selected 30 non-reference positions and built positioning fingerprint database with different methods. The average location error of 30 non-reference position is shown in Fig. 5. K is changed from 1 to 5 using KNN. When building a single RSSI-based fingerprint, the location error is between 4.04 m ∼ 5.22 m. When building a single RSSI-based fingerprint, and using C-GPCA algorithm, the location error of the new fingerprint database is between 2.67 m ∼ 3.41 m. When building a new fingerprint database based on RSSI and the geomagnetic field, using C-GPCA algorithm, the location error of the new fingerprint database is between 2.01 m ∼ 2.22 m. From the comparison of the charts, it will achieve better positioning accuracy when there is less signal interferences and overlaps.

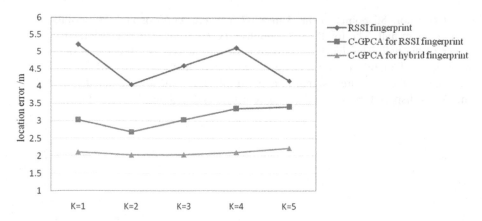

Fig. 5. Three different methods of positioning error results comparison chart

5   Conclusions

This paper has researched the C-GPCA algorithm. It can be used in hybrid fingerprint database optimization. By actual environment test, we verify the effectiveness of the algorithm. Under the premise of collecting the same sampling points in the offline stage, C-GPCA algorithm can reduce the overlaps and interferences for RSSI signals, improving positioning accuracy, getting certain advantages.

Acknowledgment. The authors would like to thank the anonymous reviewers for their helpful and constructive comments. This work is supported by Program for Excellent Talents in Hunan Normal University (Grant no. ET61008) and the Education Reform Item of Hunan Normal University(Grant no [2011]75, [2014]85-39).

References

1. Alavi, B., Pahlavan, K.: Modeling of the TOA-based distance measurement error using UWB indoor radio measurements. IEEE Commun. Lett. **10**(4), 275–277 (2006)
2. Curran, K., Furey, E., Lunney, T., et al.: An evaluation of indoor location determination technologies. J. Location Based Serv. **5**(2), 61–78 (2011)
3. Bahl, P., Padmanabhan, V.N.: RADAR: An in-building RF-based user location and tracking system. In: Proceedings of the Nineteenth Annual Joint Conference of the IEEE Computer and Communications Societies INFOCOM 2000, vol. 2, pp. 775–784. IEEE (2000)
4. Rui, Z., Bang, Z., Zhu, Z.-L., et al.: Overview of indoor locatization techniques and application. Electron. Sci. Technol. **27**(3), 154–157 (2014)
5. Feng, C.: Research and Implementation of a Compressive Sensing-Based Indoor Positioning System Using RSS. Beijing Jiaotong University, Beijing (2011)
6. Liu, W.-Y., Zhang, R.-J., Wang, L., Yang, J., Wang, D.-Y.: Multidimensional fingerprints method for indoor mobile trajectory mapping with geomagnetic information. J. Electron. Inf. Technol. **35**(10), 2397–2402 (2013)

7. Xin, W.: Design and implementation of indoor positioning algorithm based on geomagnetic and RSSI. Hangzhou Dianzi University, Hangzhou (2014)
8. Zhou, W.: An efficient radio map construction method for WIFI positioning, pp. 23–24. Beijing University of Technology, Beijing (2013)
9. Tang, W.-S., Li, S., Kuang, W.-Q.: New algorithm based-on spatial correlation for yielding fingerprints database of RF indoor localization. Comput. Eng. Appl. **44**(23), 226–229 (2008)
10. Feng, C., Au, W.S.A., Valaee, S., et al.: Received-signal-strength-based indoor positioning using compressive sensing. IEEE Trans. Mobile Comput. **11**(12), 1983–1993 (2012)
11. Hou, W.: Discussing to comprehensive evaluation by principal component. Appl. Stat. Manage. **25**(2), 211–214 (2006)
12. Wang, Z.-M., Chen, Z., Pan, C.-H.: Improved fingerprinting algorithm for smart phone indoor positioning. J. Xi'an Univ. Posts Telecommun. **19**(1), 17–20 (2014)
13. He, X.-Q.: Applied Multivariate Statistical Analysis. China Renmin University Press, Beijing (2004)
14. Xiang, D.-J., Li, H.W., Liu, X.Y.: Applied Multivariate Statistical Analysis. University of Geosciences Press, Wuhan (2005)
15. Li, Z.-F., Yang, B.: Feature vector calculation problem in SPSS principal component analysis. Stat. Educ. **3**, 10–11 (2007)

A Note on the Guarantees of Total Variation Minimization

Hao Jiang[1(⊠)], Tao Sun[2], Pei-Bing Du[2], Sheng-Guo Li[1],
Chun-Jiang Li[1], and Li-Zhi Cheng[2,3]

[1] College of Computer, National University of Defense Technology,
Changsha 410072, China
haojiang@nudt.edu.cn
[2] College of Science, National University of Defense Technology,
Changsha 410072, China
[3] The State Key Laboratory of High Performance Computation,
National University of Defense Technology, Changsha 410072, China

Abstract. In this paper, we provide a simplified understanding of the guarantees of 1-dimension total variation minimization. We consider a slightly modified total variation minimization rather than the original one. The modified model can be transformed into an ℓ_1 minimization problem by several provable mathematical tools. With the techniques developed in random sampling theory, some estimates relative to Gaussian mean width are provided for both Gaussian and sub-Gaussian sampling. We also present a sufficient condition for the exact recovery under Gaussian sampling.

1 Introduction

"Reconstructing structured signal from linear sampling is a core research in contemporary signal processing" [1]. Mathematically, the procedure of linear sampling can be presented as follows. Let $x^* \in \mathcal{R}^N$ be the unknown structured signal, $\Phi \in \mathcal{R}^{M \times N}$ be the known random sampling matrix. The observation data $b \in \mathcal{R}^M$ are acquired as

$$b = \Phi x^* + e,$$

where e is the unknown but bounded noise. Under the assumption $\frac{1}{M} \| e \|_1 \leq \sigma$, one can reconstruct the structured signal by employing the following convex program

$$\min_x \| x \|_{\mathcal{K}} \ s.t. \frac{1}{M} \| \Phi x - b \|_1 \leq \sigma, \tag{1}$$

This research is partly supported by the National High Technology Research and Development Program of China (No. 2012AA01A301), National Natural Science Foundation of China (No. 61402495, No. 61170046, No.11401580, No. 61402496, No. 61303189, No.61170049), Science Project of National University of Defense Technology (JC120201) and National Natural Science Foundation of Hunan Province in China (13JJ2001).

D.-S. Huang and K.-H. Jo (Eds.): ICIC 2016, Part II, LNCS 9772, pp. 222–231, 2016.
DOI: 10.1007/978-3-319-42294-7_19

where $\|\cdot\|_\kappa$ is some Minkowski function which can reflect the property of the structure of the signal (in sparse recovery, $\|\cdot\|_\kappa$ is usually set as $\|\cdot\|_1$). An important task in signal processing is estimating the error between the solution of Model (1) and the original signal. Besides this, another one is how many measurements is sufficient for the exact recovery when $\sigma = 0$. Literatures [1–6, 18, 19] show state-of-the-art answers for these questions in terms of Gaussian and sub-Gaussian measurements.

In practical terms, a crowd of signals are not sparse unless being transformed by some operator (may be ill-posed). A typical example of them is signal with a sparse gradient (piecewise constant), which arises frequently from imaging. More precisely, given a signal $x \in \mathcal{R}^N$, let

$$[Dx]_i = x_{i+1} - x_i, \, i = 1, 2 \ldots, N - 1.$$

Then, Dx is sparse if x is piecewise constant. This observation motivates us to reconstruct x^* by solving the following minimization

$$\min_x \| Dx \|_1 \, s.t. \frac{1}{M} \| \Phi x - b \|_1 \leq \sigma. \tag{2}$$

The minimization term $\|Dx\|_1$ is right the Total Variation (TV) of x. TV minimizations [7] have gained increasing attentions in recent years for its important position in signal processing, imaging science and machine learning research [8, 9, 20]. Like the sparse recovery, model solution estimates and sufficient measurements condition are also attractive for the TV minimization recovery. In paper [10] the authors have proved performance guarantees of TV minimization in recovering 1-dimensional signal with sparse gradient support.

In this note, we aim to provide a simplified understanding of performance guarantees of 1-dimensional TV minimization recovery under standard normal Gaussian and sub-Gaussian sampling (proofs in [10] are really very complicated). Actually, paper [10] has pointed out the difficulty in dealing with the TV minimization recovery lies in that D is singular. This property leads to that many perfect reconstruction results in sparse recovery can not be applied (2) directly. A natural idea is that can we slightly modify $\|Dx\|_1$ such that D is nonsingular? Such an idea motivates us to develop a slightly changed TV minimization recovery model. For any $x \in \mathcal{R}^N$, define that $T \in \mathcal{R}^{N \times N}$ satisfy that

$$(Tx)_i := \begin{cases} x_i & \text{if } i = 1, \\ x_i - x_{i-1} & \text{if } 2 \leq i \leq N. \end{cases}$$

Under above definition, the matrix T has an inverse. Obviously, we have $||Tx||\_1 - ||Dx||\_1 = |x\_1|$ is small relative to $||Dx||_1, ||Tx||_1 \approx ||Dx||_1$. A special case is $x_1 = 0$, and then $||Tx||\_1 = ||Dx||\_1$. Given a matrix $\Phi \in \mathcal{R}^{M \times N}$ and $b \in \mathcal{R}^M$, we consider the following modified total variation minimization problem

$$\min_x ||Tx||_1 \text{ s.t.} \frac{1}{M} ||b - \Phi x||_1 \leq \sigma. \tag{3}$$

Let $z := Tx$, Problem (3) can be represented as

$$\min_z ||z||_1 \text{ s.t.} \frac{1}{M} ||b - \Phi T^{-1}z||_1 \leq \sigma. \tag{4}$$

With the help of several sparse reconstruction results, we provide the estimates and sufficient sampling condition for the exact TV minimization recovery.

The rest of the paper is organized as follows. Section 2 introduces notations and some basic needed tools in the convex recovery. Section 3 presents the results. Section 4 concludes the paper.

2 Notations and Several Mathematical Tools

This section collects notations and several tools which play quite important roles in the geometric analysis. More details about these definitions and propositions can be found in [1, 6, 11, 12, 14, 16, 17].

2.1 Notations

We adopt the notations described in this section throughout the paper. Let x and y be two elements of $\mathcal{R}^N$. The ℓ_0, ℓ_1 and ℓ_2 norms of x and y are denoted by $||x||_0 := \#\text{supp}(x)$, $\sum_i |x_i|$ and $||x||_2 := (\sum_i |x_i|^2)^{\frac{1}{2}}$, respectively. The inner product of x and y is defined as $\langle x, y \rangle$.

Let $A \in \mathcal{R}^{N \times N}$ be a matrix, $||A||_2 := \sup_{||x||_2=1} ||Ax||$. Let $\mathcal{K} \subseteq \mathcal{R}^N$ be a set, the convex hull of $\mathcal{K}$ is defined as

$$\text{conv}(\mathcal{K}) = \{\sum_i \lambda_i x_i | x_i \in \mathcal{K}, \sum_i \lambda_i = 1 \text{ and } \lambda_i \geq 0\}$$

The ℓ_1 and ℓ_2 unit ball are defined as $\mathbf{B}_1^N := \{x | ||x||_1 \leq 1\}$ and $\mathbf{B}_2^N := \{x | ||x||_2 \leq 1\}$, respectively. S^{N-1} denotes the spherical surface of $\mathbf{B}_2^N$. For a matrix A, A means its i-th row.

A random variable ξ is sub-Gaussian if $\mathbb{E}exp(\xi^2/\psi^2) \leq e$, for some $\psi > 0$. The smallest ψ is called the sub-Gaussian norm and is denoted by $||\xi||_{\psi^2}$. A random vector $x \in \mathcal{R}^N$ is called sub-Gaussian if all one-dimensional variables $\langle X, u \rangle$, $u \in \mathcal{R}^N$, are sub-Gaussian. The sub-Gaussian norm of x is defined as

$$|| x ||_{\psi^2} := \sup_{u \in \mathcal{R}^N} || \langle x, u \rangle ||_{\psi^2}.$$

A random vector x is called isotropic if $Exx^t = I_N$.

2.2   Mathematical Tools

2.2.1   Gaussian Mean Width

Definition 1 (Gaussian mean width). Let $\mathcal{K} \subseteq \mathcal{R}^N$ be a bounded set, the Gaussian mean width $w(\mathcal{K})$ is defined as

$$w(\mathcal{K}) := E \sup_{u \in \mathcal{K} - \mathcal{K}} \langle g, u \rangle, \tag{5}$$

where $g \sim \mathcal{N}(0, I_d)$ is a standard Gaussian vector in $\mathcal{R}^N$.

The Gaussian mean width is a core tool used in sampling theory. We present some results which will be used.

Proposition 1. For any set $\mathcal{K} \subseteq \mathcal{R}^N$, it holds that

$$w(\mathcal{K}) = w(\text{conv}(\mathcal{K})).$$

Proposition 2. Let $\mathcal{K} \subseteq \mathcal{R}^N$ be a finite subset of $\mathbf{B}_2^N$, it holds that

$$w(\mathcal{K}) \leq C \sqrt{log(|K|)},$$

where C is a positive constant.

2.2.2   Minkowski Function

Definition 2 (Minkowski function). Let $\mathcal{K} \subseteq \mathcal{R}^N$ be a bounded set, the Minkowski function of $\mathcal{K}$ is the non-negative number defined by

$$\|x\|_\mathcal{K} := \inf\{\lambda > 0 : \lambda^{-1} x \in \mathcal{K}\},$$

where x is a point in $\mathcal{R}^N$.

Proposition 3. For any set $\mathcal{K} \subseteq \mathcal{R}^N$, it holds that

$$\mathcal{K} = \{x | \|x\|_\mathcal{K} \leq 1\}.$$

2.2.3   Descend Unite Sphere

Definition 3 (Descend unite sphere). Let $\mathcal{K} \subseteq \mathcal{R}^N$ be a bounded set and $x \in \mathcal{K}$, the descend unite sphere is defined by

$$S(\mathcal{K}, x) := \{\frac{h - x}{\|h - x\|_2} | h \in \mathcal{K}\}.$$

Proposition 4. If x is s-sparse, $\mathcal{K} := ||x||_1 \mathbf{B}_1^N$. Then, we have that

$$w(S(\mathcal{K}, x)) \leq C\sqrt{s\log N}.$$

2.2.4   Estimation Form Linear Gaussian Observations

Proposition 5   (Estimation from Gaussian observation). Assume that x_σ is a solution to

$$\min_x ||x||_{\mathcal{K}} \text{ s.t. } \frac{1}{M}||b - \Phi x||_1 \leq \sigma \tag{6}$$

and $x^* \in \mathcal{K}$. If Φ is generated by the normal standard Gaussian random variable. Then, it holds that

$$\mathbb{E}||x_\sigma - x^*||_2 \leq \sqrt{2\pi}(\frac{w(\mathcal{K})}{\sqrt{M}} + \sigma). \tag{7}$$

Proposition 6   (Estimation from sub-Gaussian observation). Let $\Phi \in \mathcal{R}^{M \times N}$ be sub-Gaussian random matrix and Φ_i are i.i.d., mean zero, isotropic and $||\Phi_i||_{\psi^2} \leq \psi, i = 1, 2, \ldots, M$. Assume that x_σ is a solution to (6) and $x^* \in \mathcal{K}$. Then, it holds that

$$\mathbb{E}||x_\sigma - x^*||_2 \leq C\psi^4(\frac{w(\mathcal{K})}{\sqrt{M}} + \sigma). \tag{8}$$

Propositions 5 and 6 provide expectation errors bound for reconstructing x^* from linear Gaussian or sub-Gaussian random sampling by convex programming (1). The proofs depend on a geometric intuition and are completed with the help of M^* bound [13]. However, these propositions do not imply a sufficient exact recovery condition. To derive the condition, paper [6] interpret the exact recovery as a set escaping through a mesh. In fact, this technique is also used in [2, 15].

Theorem 1 (Escape through a mesh) Let S be a fixed subset of S^{N-1}. Let E be a random subspace of $\mathcal{R}^N$ of a fixed codimension M, drawn from the Grassmanian $G_{n,n-m}$ according to the Haar measure. If $M > w(S)^2$, then

$$S \bigcap E = \emptyset,$$

with the probability $1 - 2.5exp[-(m/\sqrt{m+1} - w(S))^2/18]$.

3 Main Results

Lemma 1. It holds that $||T^{-1}||_2 \sim \mathcal{O}(N)$.

Proof. Easy calculation gives the matrix form of T^{-1} as

$$T^{-1} = \begin{pmatrix} 1 & & & & \\ 1 & 1 & & & \\ 1 & 1 & 1 & & \\ \vdots & \vdots & & \ddots & \ddots \\ 1 & 1 & 1 & \cdots & 1 \end{pmatrix}$$

For any $x = \begin{pmatrix} x_1 \\ x_2 \\ \vdots \\ x_N \end{pmatrix} \in S^{N-1}$, it is trivial to prove that $|x| := \begin{pmatrix} |x_1| \\ |x_2| \\ \vdots \\ |x_N| \end{pmatrix}$ also belongs to

S^{N-1}. Thus, we have that $T^{-1}x = \begin{pmatrix} x_1 \\ x_1 + x_2 \\ \vdots \\ \sum_{i=1}^{N} x_i \end{pmatrix}$ and $T^{-1}|x| = \begin{pmatrix} |x_1| \\ |x_1| + |x_2| \\ \vdots \\ \sum_{i=1}^{N} |x_i| \end{pmatrix}$. From the

Schwartz's inequality,

$$\left(\sum_{i=1}^{j} |x_i|\right)^2 \le j \sum_{i=1}^{j} |x_i|^2 \le j ||x||_2^2 = j.$$

Then, we derive that

$$||T^{-1}x||_2 \le ||T^{-1}|x|||_2 = \sqrt{\sum_{j=1}^{N}\left(\sum_{i=1}^{j}|x_i|\right)^2} \le \sqrt{\sum_{j=1}^{N}j\sum_{i=1}^{j}|x_i|^2} \le \sqrt{\sum_{j=1}^{N}j} \le \sqrt{\frac{N(N+1)}{2}}.$$

On the other hand, consider $\mathbf{1} = \begin{pmatrix} 1 \\ 1 \\ \vdots \\ 1 \end{pmatrix} \in \mathcal{R}^N$. We have $||T^{-1}\mathbf{1}||_2 = \sqrt{\sum_{i=1}^{N} i^2} \sim \mathcal{O}(N^{\frac{3}{2}})$

and $||\mathbf{1}||_2 = \sqrt{N}$. Thus, $||T^{-1}||_2 \ge \frac{||T^{-1}\mathbf{1}||_2}{||\mathbf{1}||_2} \sim \mathcal{O}(N)$.

Lemma 2. Let $\Phi \in \mathcal{R}^{M \times N}$ be the norm standard Gaussian matrix. Then, ΦT^{-1} has the same distribution as $\Phi \Lambda_N$, where $\Lambda_N = diag(\sqrt{N}, \sqrt{N-1}, \ldots, \sqrt{2}, 1)$.

Proof. For any $1 \leq i \leq M$ and $1 \leq j \leq N$, we consider $(\Phi T^{-1})_{i,j}$ and $(\Phi \Lambda_N)_{i,j}$. Then, we obtain that

$$(\Phi T^{-1})_{i,j} = \sum_{k=1}^{N} \Phi_{i,k}(T^{-1})_{k,j} = \sum_{k=i}^{N} \Phi_{i,k} \sim \mathcal{N}(0, N+1-i).$$

On the other hand, we have

$$(\Phi \Lambda_N)_{i,j} = \sqrt{N+1-i}\Phi_{i,j} \sim \mathcal{N}(0, N+1-i).$$

Therefore, they have the same distribution.

From Lemma 1, if $||Tx - Tx^*||_2$ has been estimated as $\mathcal{O}(\varepsilon)$, $||x - x^*||_2$ will be estimated as $\mathcal{O}(N\varepsilon)$. In view of this, we focus on estimating $||Tx - Tx^*||_2$.

Theorem 2. Let $x^* \in \mathcal{R}^n$ be a signal, $\Phi \in \mathcal{R}^{m \times n}$ be a random matrix, and $b \in \mathcal{R}^m$ formed as $b = \Phi x + e$. Under the assumption $\frac{1}{M}||e||_1 \leq \sigma$ and Tx^* is s-sparse, and x_σ is a solution to (3), we have

$$\mathbb{E}||Tx_\sigma - Tx^*||_2 \leq CN\sqrt{\frac{s\log(N)}{M}}||Tx^*||_2 + \sqrt{2N\pi}\sigma, \tag{9}$$

where C is some positive constant.

Proof. From Lemma 2, let $y = \Lambda_N z$, Model (4) can be expressed as

$$\min_y \frac{1}{||Tx^*||_1}||\Lambda_N^{-1}y||_1 \text{ s.t.} \frac{1}{M}||b - \Phi y||_1 \leq \sigma. \tag{10}$$

Let $\mathcal{K} = \{y|\frac{1}{||Tx^*||_1}||\Lambda_N^{-1}y||_1 \leq 1\}$, then the problem above equals to

$$\min_y ||y||_{\mathcal{K}} \text{ s.t.} \frac{1}{M}||b - \Phi y||_1 \leq \sigma. \tag{11}$$

Note that $y^* = \Lambda_N Tx^* \in \mathcal{K}$. From Proposition 5, we have that

$$\mathbb{E}||Tx_\sigma - Tx^*||_2 \leq ||\Lambda_N^{-1}|| \cdot \mathbb{E}||y_\sigma - y^*||_2 \leq \sqrt{\pi N}(\frac{w(\mathcal{K})}{\sqrt{M}} + \sigma). \tag{12}$$

If Tx^* is s-sparse, y^* is also s-sparse. Denote that

$$\widehat{\mathcal{K}} = conv\{\pm e_i\}_{i=1}^{N}$$

Then, we have

$$\{y|||\Lambda_N^{-1}y||_1 \le 1\} \subseteq \sqrt{N}\widehat{\mathcal{K}}$$

That means $\mathcal{K} \subseteq \sqrt{N}||Tx^*||_1\widehat{\mathcal{K}}$. Then, we have

$$w(\mathcal{K}) \le \sqrt{N}||Tx^*||_1 w(\widehat{\mathcal{K}}) \le C||Tx^*||_1 \sqrt{N\log(N)}, \qquad (13)$$

where C is some positive constant. Tx^* is s-sparse, then, we have

$$||Tx^*||_1 \le \sqrt{s}||Tx^*||_2.$$

Therefore,

$$w(\mathcal{K}) \le C||Tx^*||_2 \sqrt{sN\log(N)}.$$

Theorem 3. Let $x^* \in \mathcal{R}^n$ be a signal, $\Phi \in \mathcal{R}^{m \times n}$ be a sub-Gaussian random matrix, and $b \in \mathcal{R}^m$ formed as $b = \Phi x + e$, $\frac{1}{M}||e||_1 \le \sigma$ and Tx^* is s-sparse, and x_σ is a solution to (3). Assume that $\mathbb{E}A_i = 0$, $\mathbb{E}A_iA_i^t = T^2$ and $||\Phi_iT^{-1}||_{\psi^2} \le \psi_N, i = 1,2,\ldots,M$. We have

$$\mathbb{E}||Tx_\sigma - Tx^*||_2 \le C\psi_N^4 \sqrt{\frac{sN\log(N)}{M}}||Tx^*||_2 + C\psi_N^4\sigma, \qquad (14)$$

where C is some positive constant.

Proof. As stated in the proof of Theorem 2, Problem (4) can be reexpressed as (11). Note that

$$\mathbb{E}\Phi T^{-1} = \mathbf{0},$$
$$\mathbb{E}A_iT^{-1}(T^{-1})'A_i^t = \mathbb{E}A_iA_i^tT^{-1}(T^{-1})' = I_N,$$
$$||\Phi_iT^{-1}||_{\psi^2} \le \psi_N, i = 1,2,\ldots,M.$$

Then, from Proposition 6, we have that

$$\mathbb{E}||Tx - Tx^*||_2 \le C\psi_N^4(\frac{w(\mathcal{K})}{\sqrt{M}} + \sigma). \qquad (15)$$

With the estimate of the Gaussian mean width, we can obtain the result.

Theorem 4. If Tx^* is s-sparse and Φ is generated by standard normal Gaussian distribution. Assume that $M \ge Cs\log(N)$. Then, Model (3) reconstructs x^* with high probability, namely $1 - 3e^{-m}$.

Proof. Let $\mathcal{K} := \{z|\|z\|_1 \leq \|z^*\|_1\}$. The noiseless form of Model (4) is

$$\min_z \|z\|_{\mathcal{K}} \text{ s.t.} \Phi T^{-1} z = b.$$

Paper [6] provides a condition of the uniqueness at z^* for the above problem as

$$\mathcal{K} \bigcap \{z|\Phi T^{-1} z = b\} = \emptyset.$$

Easy geometry deduction shows that such condition equals to

$$S(\mathcal{K}, z^*) \bigcap \ker(\Phi T^{-1}) = \emptyset.$$

It is trivial to prove that $\ker(\Phi T^{-1}) = \ker(\Phi)$. Then, the uniqueness requires that

$$S(\mathcal{K}, z^*) \bigcap \ker(\Phi) = \emptyset.$$

From Theorem 1, if $M > w(S(\mathcal{K}, z^*))^2$, $S(\mathcal{K}, z^*) \bigcap \ker(\Phi) = \emptyset$ with high probability. And from Proposition 3, we have

$$M > Cs\log(N).$$

The exact recovery of Model (4) indicates the exact recovery of Model (3).

4 Conclusion

In this paper, we consider both the Gaussian and sub-Gaussian linear sampling for total variation minimization recovery. The analysis is based on a modified TV model. For Gaussian case, we provide a sufficient condition for the exact recovery. The main purpose of this paper is to provide a simplified way to understand the total variation minimization.

References

1. Tropp, J.A.: Convex recovery of a structured signal from independent random linear measurements (2014). arXiv:1405.1102
2. Amelunxen, D., Lotz, M., McCoy, M.B., et al.: Living on the edge: phase transitions in convex programs with random data. Inform. Inf. **3**(3), 224–294 (2014). iau005
3. McCoy, M.B., Tropp, J.A.: Sharp recovery bounds for convex demixing, with applications. Found. Comput. Math. **14**(3), 503–567 (2014)
4. Lerman, G., McCoy, M.B., Tropp, J.A., et al.: Robust computation of linear models by convex relaxation. Found. Comput. Math. **15**(2), 363–410 (2015)
5. Foygel, R., Mackey, L.: Corrupted sensing: novel guarantees for separating structured signals. IEEE Trans. Inform. Theory **60**(2), 1223–1247 (2014)

6. Vershynin, R.: Estimation in high dimensions: a geometric perspective (2014). arXiv:1405.5103
7. Rudin, L.I., Osher, S., Fatemi, E.: Nonlinear total variation based noise removal algorithms. Physica D **60**(1), 259–268 (1992)
8. Cai, J.F., Dong, B., Osher, S., et al.: Image restoration: total variation, wavelet frames, and beyond. J. Am. Math. Soc. **25**, 1033–1089 (2012)
9. Wang, Y., Yang, J., Yin, W., et al.: A new alternating minimization algorithm for total variation image reconstruction. SIAM J. Imag. Sci. **1**(3), 248–272 (2008)
10. Cai, J.F., Xu, W.: Guarantees of total variation minimization for signal recovery. Inform. Inf. **4**(4), 328–353 (2015)
11. Plan, Y., Vershynin, R.: Robust 1-bit compressed sensing and sparse logistic regression: a convex programming approach. IEEE Trans. Inform. Theory **59**(1), 482–494 (2013)
12. Rudelson, M., Vershynin, R.: On sparse reconstruction from Fourier and Gaussian measurements. Commun. Pure Appl. Math. **61**(8), 1025–1045 (2008). iav009
13. Milman, V.: Surprising geometric phenomena in high-dimensional convexity theory. In: European Congress of Mathematics, pp. 73–91. Birkhauser Basel (1998)
14. Gordon, Y.: On Milman's inequality and random subspaces which escape through a mesh in R^n. In: Lindenstrauss, J., Milman, V.D. (eds.) Geometric Aspects of Functional Analys, vol. 1317, pp. 84–106. Springer, Berlin (1988)
15. Donoho, D., Tanner, J.: Counting faces of randomly projected polytopes when the projection radically lowers dimension. J. Am. Math. Soc. **22**(1), 1–53 (2009)
16. Mendelson, S., Pajor, A., Tomczak-Jaegermann, N.: Reconstruction and subgaussian operators in asymptotic geometric analysis. Geometric Func. Anal. **17**(4), 1248–1282 (2007)
17. Vershynin, R.: Introduction to the non-asymptotic analysis of random matrices (2010). arXiv:1011.3027
18. Oymak, S., Thrampoulidis, C., Hassibi, B.: Simple bounds for noisy linear inverse problems with exact side information (2013). arXiv:1312.0641
19. Oymak, S.: Convex Relaxation for Low-Dimensional Representation: Phase Transitions and Limitations. California Institute of Technology, Pasadena (2015)
20. Condat, L.: A direct algorithm for 1D total variation denoising. IEEE Signal Proc. Lett. **20**(11), 1054–1057 (2013)

CDN Strategy Adjustment System Based on AHP

Xi Chen[1], Xie Zhang[1], Zongze Wu[1,2](✉), Youjun Xiang[1],
Shengli Xie[2], and Shuang Li[1]

[1] College of Electronic and Information Engineering,
South China University of Technology, Guangzhou 510641, Guangdong, China
5173204@qq.com, {zhangxie,zzwu,yjxiang}@scut.edu.cn
[2] College of Automation, Guangdong University of Technology,
Guangzhou 510006, Guangdong, China
shlxie@gdut.edu.cn

Abstract. It has been an important research orientation on how to deploy the service strategy intelligently to meet the needs of complicated application scenarios for CDN carriers in order to improve its quality of service (QoS). In order to overcome the deficiency of traditional Static deployment strategy, this paper proposes a CDN Dynamic deployment strategy system based on Analytical Hierarchy Process (AHP). The system will try to structure an open hierarchical architecture and decide the weight of elements influencing the QoS of CDN, and will finally give out the recommendatory solution. It has been proved by the example in the paper that this decision model is efficient in deciding the proposed solution, which is objective and effective.

Keywords: Content delivery network (CDN) · Analytical hierarchy process (AHP) · Deployment strategy · Quality of service

1 Introduction

With the rapid and ongoing development of Internet and multimedia, the Internet traffic has surged significantly. Although the network bandwidth also increases over time, there still remains a bottleneck of QoS due to the lack of guarantee mechanism of Internet user quality of service to deal with heavy traffic load. The Internet "do everything in one's power" service is far from meeting the consumer demands for high quality of services and experiences. In this case, the content distribution network which emerges as the time requires offers endless possibilities to provide distribution service of high quality contents for the users. Development of CDN business and Internet content providers are positively related, and CDN service changes the direction of most Internet content flow to the edge network, thereby reducing the flow of information on the Internet backbone network [2].

CDN refers to the Internet's content distribution network,which is fictitious and an advanced flow distribution network set up on the Internet, and is made up of a nodal server group distributed in different areas. CDN can realize functions such as release of the content of websites to the place closest to the user efficiently and steadily. The basic

© Springer International Publishing Switzerland 2016
D.-S. Huang and K.-H. Jo (Eds.): ICIC 2016, Part II, LNCS 9772, pp. 232–241, 2016.
DOI: 10.1007/978-3-319-42294-7_20

working principle of the CDN is to install specialized nodes at the edge of the Internet and to set up efficient distributive and storage network of the content by utilizing intelligent content routers. The intelligent content routers can lead the users to the nearest, more effective node judged by the users' location and traffic situation of the network. Thus, the burden of the source server will be reduced distinctly, and the website will need less servers, bandwidth, related network facilities, engineers, and so on. In a word, the emergence of CDN extends the capabilities of based network system, which takes advantages of the high-speed network and mature content processing.

With regard to the content delivery network (CDN), the key to improve the quality of CDN services is to reduce the media transmission delay and the utilization of load balancing, which is mainly constrained by the deployment strategy.

Domestic and foreign scholars have done a lot of research on it till date. The existing typical deployment strategy can be divided into two types: static one and dynamic one. Static deployment strategy refers to the model that the deployment strategy is only determined according to the conditions when it (the CDN system) starts working, which is usually based on the fixed access mode. The static deployment strategy only applies to the situation where access mode and the concurrency is known in advance, which is very difficult to adapt to the rapid changes of the system environment and network bandwidth. While dynamic deployment strategy differs from the former one that the strategy based on initial environment of the CDN system will continue to change and adjust according to the system operation status and parameters. The dynamic deployment strategy can reduce the transmission time of the network effectively and improve the performance of the system. However, the study of dynamic data deployment strategy is not a long time.

The exiting content delivery network is mostly based on the static deployment strategy. Many factors influence the network service. The importance of different factors may be different in various situations. When based on the deployment principle of sufficient amount of resources to ensure the absolute availability, static deployment will lead to a spectacular waste of resources, on the other hand it will lead to the lack of scalability to make it difficult to adapt to the environment changes rea-ltime, thereby incapable of guaranteeing the quality of user service. In short, static method cannot calculate the service strategy dynamically and reasonably according to the system operation parameters.

The analytic hierarchy process (AHP) is proposed by T.L.Saaty professor in the early nineteen seventies. It is a simple, flexible method to deal with qualitative problems in quantitative ways. AHP can make the thinking process systematic, scientific and intuitive, which is especially suitable for multi criteria, multi object decisions [3].

Considering that the QoS of CDN is constrained by many factors and most of QoS methods are qualitative, a strategy deployment system based on the AHP is proposed in this paper. The system tries to sort and organize objective and subjective factors into interrelated whole, in which way the empirical data and the objective results can be effectively combined, and a multi-layer analysis model structure is formed. The model ultimately boils down to the relations of different layers.

2 AHP Construction

The AHP model can generally be constructed with the following four steps [4]:

(1) Build the analytic hierarchy structure model;
(2) Establishment of the judgment matrix;
(3) Calculate the weight vector of different layers;
(4) Go through the consistency checking.

The details are as follows:

2.1 Build the Analytic Hierarchy Structure Model

According to the analysis and understanding of the problem, the various factors determining the ultimate goal will be summarized into several hierarchical groups of layers according to whether they share certain characteristics. The factors from the same layer belong to the upper layer, while affect the factors from the lower layer. In this way, you can generally have an analytic hierarchy model that is composed of three layers: the target layer (top layer), criterion layer (middle layer) and program layer (bottom layer).

2.2 Establishment of the Judgment Matrix

In the analytic hierarchy model, the factors on the same level will have a pairwise comparison of to calculate its guidelines on the importance for the upper layer, and then the judgment matrix will be established following the 1–9 quantization scale [5]. The 1–9 quantization scale method is based on a qualitative analysis of people intuition and judgment, which is good at quantifying the critical thinking.

2.3 Calculate the Weight Vector of Different Layers

Try to calculate the relative weight of factors for each judgment matrix. It is divided into two steps: single-level sorting; total sorting. The judgment matrix established in the first step is carried out for the purpose of pairwise judgment results for the same layer. The single-level sorting is to calculate the judgment matrix of all the factors in the same layer relative to the upper layer. Commonly used methods include plot method and the square-root method. While the total sorting takes advantages of the single-level sorting and calculate the judgment matrix of all the factors in the criterion layer with respect to the program layer. The calculation method are the same with that of the single-level sorting.

2.4 Go Through the Consistency Checking

Considering the complexity of objective things and diversity of awareness for people, so in actual use, in order to ensure the rationalization of the AHP conclusion, it's quite necessary to take the consistency checking for the judgment matrix, which helps ensure the logical rationality and scientific reliability of the judgment matrix.

Try to calculate the random consistency ratio, denoted as CR = CI/RI. CR stands for the consistency ratio, and only when CR is less than 0.10 (CR < 0.10) will the matrix be regarded as consistent. Otherwise, the judgment matrix needs to be adjusted so as to satisfy the above requirements. CI stands for the consistency index, which is calculated as follows:

$$CI = (\lambda max - n)/(n - 1)$$

Among which, λmax represents the largest eigenvalue of the judgment matrix; n represents the number of the judgment matrix; while RI is the average consistency

Table 1. The value of random index

n	1	2	3	4	5	6	7	8	9
RI	0	0	0.58	0.9	1.12	1.24	1.32	1.41	1.45

index, and its general value is shown in Table 1 below:

For more details of the principle for the AHP, you can refer to the references [6–7].

3   CDN Strategy Adjustment System Based on AHP

The implementation of the content delivery network consists of two parts: the CDN framework and the CDN management platform. The CDN framework is composed of three layers: the central layer, the core layer and the edge layer, which is responsible for the completion of basic functions to serve the edge users. While the realization of the framework varies, we will not try to elaborate it considering that the realization will not affect the implementation of the strategy adjustment system. The CDN management platform consists of several modules including content management, network management, schedule management, which provides real-time monitoring of CDN operating condition. It also provides fault-bound solutions, which can help CDN operators grasp the real-time condition of the whole network dynamics in order to execute the proactive QoS warning, monitoring and optimizing. By which means, it can apparently shorten troubleshooting time, improve business stability, thereby enhancing the operators' service quality and reducing customers' OPEX (operational costs).

Considering the CDN features, an extra console based on the original management platform is added, which consists of two modules: the monitoring module and the policy adjustment module, as is shown in the left part of Fig. 1. These two modules only take advantages of the information of exiting platforms, so it will not increase the complexity of the whole system. Among them, the monitoring module is responsible for real-time monitoring and collecting information necessary, as well as updating the running status information for each server; while a policy adjustment module deal with adjusting the system's service strategy based on the results of AHP decision tree. And

Advanced CDN Management Platform   **CDN Prototype System**

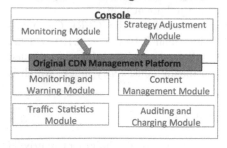
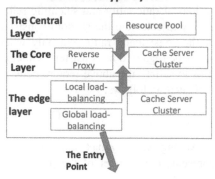

Fig. 1. Architecture of CDN strategy adjustment system based on AHP

the right part of Fig. 1 illustrates the principle and components of universal content delivery network.

A service strategy based on the empirical data or the AHP decision database (the database will be explained later) will be selected once the CDN system starts running. And at the same time, the console also starts working: the monitoring module tries to monitor and collect necessary information, and periodically forwards the information to the AHP decision tree for further disposition; the AHP decision tree module handles the monitoring data, and gives the recommended adjustment programs; the strategy adjustment module will then receive the recommended adjustment programs and dynamically adjust the CDN service strategy; at the same time, the recommended program will be saved to the AHP decision database in accordance with certain rules for subsequent decision-makings and strategy initializations.

4   The Implementation of AHP in CDN Strategy Adjustment System

4.1   Build the AHP-Based CDN Hierarchical Structure Model

In the establishment process of AHP hierarchical structure model, we first need to analyze the index system of the CDN strategy adjustment system, which is an organic whole formed by a number of interrelated and mutually complementary elements.

As there are so many CDN application scenarios, and different scenarios pay different attention to the various factors, so we need to classify the factors into different groups.

Let me take two representative application scenarios as an example: the first scenario is the service aimed at accelerating web pages, big file downloads and so on. Such service are characteristic of small amount of data transferred and frequent TCP/UDP connections per unit time, so the ability to deal with CDN conversation processing is needed; the second scenario go to the streaming service, such as video-on-demand and live video, which usually requires higher bandwidth in order to ensure a stable transmission of video. Another feature of the streaming service is that the TCP/UDP connection requests is less frequent than the former one, but it need a higher throughput capacity.

From the perspective of the empirical data, the factors of CDN network performance can be divided into three parts: the basic network performance, the server information and the service indictors.

The basic network performance refers to some universal network performance parameters, including throughput, response time and rate/bandwidth. CDN throughput usually refers to the successfully transmitted data size in a single unit of time (usually in seconds), which is one of the important performance indicators for CDN servers; Response time refers to the interval when the first packet of the content is received after the user establishes a request via a terminal, as known as end-to-end response time. Response time has a great impact for the user experience. A long response time will gradually lose the user's interest in using the service, so the response time is also a most important CDN service indicator; While the rate/bandwidth is also a basic parameters in the network field, so we will not go further into it.

The server information refers to various running parameters, including but not limited to CPU usage, memory usage, and disk usage for a cluster of CDN servers.

The service indicators refer to some performance parameters to measure the QoS of CDN service, and some parameters can only be obtained by using some specific testing equipment, including MOS value (Mean Opinion Score), MDI value (Media Delivery Index) and the hit rate of files. The MOS value is used to evaluate the quality of CDN service for streaming media, which is an interval of [0,5]. The higher the score is, the better the video quality is; The MDI value is a set of indicators of media services proposed jointly by Cisco and IneoQuest, which is usually used to measure and evaluate IP-based video streaming media transmission quality; CDN hit rate can be simply explained with the formula known as "hit rate = hits/(hits + misses)." The higher the hit rate is, the higher the efficiency of CDN service is. It is trend to cache the contents locally to meet user's access requests, thus reducing the pressure of the source pool, and optimizing network traffic.

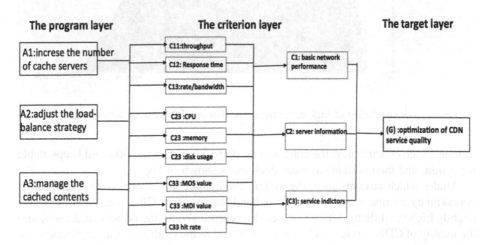

Fig. 2. The hierarchical structure of CDN strategy adjustment system

Then based on the basic steps of AHP, the AHP-based CDN strategy adjustment system model is established, as is shown in Fig. 2.

Figure 2 shows the three available solutions of the program layers: increase the number of the cache server, adjust the load-balance strategy and manage the cached contents. There also exits many other solutions for different scenarios, and the solutions listed above can also be broken down into many specific solutions. For the solution of adjusting the load-balance strategy, there exits different methods, such as round-robin, weighted round-robin and weighted random balanced scheduling algorithm for different application environments. In this paper, for the sake of convenience of calculation, we only list three programs resolutions. This has no effect for elaborating the principle of AHP and the meaning of the combination of AHP and CDN system.

4.2   Establishment of the Judgment Matrix

The adjustment system shall work like this: the monitoring module collects necessary information about the system, and periodically forwards the information to the AHP decision tree, and the decision tree will give the recommendatory solution, then the console will execute the corresponding change.

In order to verify the feasibility of AHP module, we simulate a real scene: the CDN system works just fine, and we use Spirent testing instrumentation to simulate a period

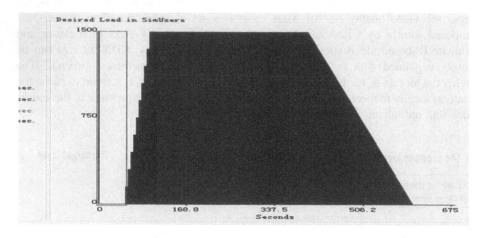

Fig. 3. The simulation of high concurrent situation in CDN strategy adjustment system

of time's high concurrency: the concurrency starts stepped up to 1500, and keeps stable for 5 min, and then sustains a linear decline, as shown in Fig. 3:

Under which circumstance, the system throughput surges due to the sudden high concurrency on one hand, on the other hand the usage of CPU and memory is also slightly higher, while the hit rate of files drops seriously. So the priority of determining the quality of CDN service is $C1 > C3 > C2$; and in the context of this application, we can derive from the empirical data that the priority of solving the C1 is $A2 > A1 > A3$, the priority of solving the C2 is A3 > A1 > A2, the priority of solving the C3 is A3 > A1 = A2.

Based on the above assumptions, we use the 1–9 scale method to go through the pairwise judgment and judgment matrixes can be obtained as follows.

Table 2. The judgment matrix of C1-P

C1	A1	A2	A3
A1	1	1/4	2
A2	4	1	1/8
A3	1/2	1/8	1

Table 3. The judgment matrix of C2-P

C2	C3	G	C1
A1	A1	C1	1
A2	A2	C2	1/5
A3	A3	C3	1/3

With respect to criterion C1, the judgment matrix C1-P is shown in Table 2. With respect to criterion C2, the judgment matrix C2-P is shown in Table 3.

Table 4. The judgment matrix of C3-P

C2	C3	A1	A2
5	3	1	1
1	1/3	1	1
3	1	3	5

Table 5. The judgment matrix of G-C

A3	A1	A2	A3
1/3	1	4	1/3
1/5	1/4	1	1/8
1	3	8	1

With respect to criterion C3, the judgment matrix C3-P is shown in Table 4. With respect to the ultimate target, the judgment matrix G-C is shown in Table 5.

4.3 Calculate the Weight Vector of Every Layer

As illustrated above, the calculation of weight vectors includes two steps: single-level sorting; total sorting. The specific mathematical method will not be elaborated here and I will directly give the calculation result. Before showing the result, the significance of symbols are explained in Table 6.

Table 6. The significance of symbols

Symbol	Meaning
W	the weight vector matrix
Λmax	the largest eigenvalue of the judgment matrix
CI	the consistency index
RI	the average consistency index
CR	the consistency ratio.

With respect to the single-level sorting, the result is as follows:
For the matrix C1-P:

C2-P: W = [0.545,2.1824,0.272]T; λmax = 3; CI = 0; RI = 0.58; CR = 0

Similarly, the result of matrix C2-P, C3-P and G-C is as follows:

C2-P: W = [0.257,0.074,0.669]T; λmax = 3.01; CI = 0.01; CR = 0.015
C3-P: W = [0.127,0.158,0.656]T; λmax = 3.03; CI = 0.01; CR = 0.02

Table 7. The weight of solution layer

	C1	C2	C3	The absolute weight
	0.633	0.106	0.260	
A1	0.182	0.257	0.187	0.191
A2	0.727	0.074	0.158	0.509
A3	0.091	0.669	0.656	0.290

G-C: W = [0.633,0.106,0.260]T; λmax = 3.03; CI = 0.02; CR = 0.03

It's apparent that the result satisfies the consistency condition since CR is less than 0.1(CR < 0.1).

With respect to the total sorting, the result is shown in Table 7:

It turns out from the calculation result that when using the Sprient testing instrumentation to simulate a short period of high concurrency, the AHP decision tree will regard A2 as the recommendatory solution since the absolute weight of A2 is the biggest one, which is to adjust the load-balance strategy. But before that, we also to have a consistency test for the results, in order to ensure the feasibility of the result.

Do the consistency test for the result of the total sorting. The formula for that process is as follows:

$$CR = \frac{\sum_{j=1}^{m} CI(j)a_j}{\sum_{j=1}^{m} RI(j)a_j}$$

The result is that CR = 0.012 < 0.1, so the total sorting is considered to meet the consistency requirement. So now it's time to utilize A2 as the savior for the CDN system to satisfy the needs of high concurrency.

5 Conclusion

Currently, the static deployment model takes a big portion in the CDN strategy adjustment system. The discussion and implementation of the dynamic deployment model is still in its infancy. The proposed method in this paper is based on AHP's strategy to adjust the system and fully takes into account a variety of complex CDN run-time factors, and these factors are mostly qualitative. The method uses AHP to establish the factors model that affect the quality of CDN service, and regularly makes corresponding adjustments based on the decision results, as well as automatically completes the process of policy preservation. This method requires less overhead cost, and is able to fully utilize the existing empirical data to reduce human error and quantify the qualitative factors. So the system is able to provide an intuitive management numerical reference. However, because there are some inherent limitations of the AHP decision tree, the dynamic policy adjustment model should not be confined to this method, we should take advantage of a variety of methods, such as AHM applications, in order to make more accurate and reasonable judgments for the CDN strategic deployment, which is a further research orientation.

Acknowledgement. The work is supported by National Natural Science Foundation of China (61271210) and Major science and technology projects in Guangdong Province (2015B010131014, 2014B010117005).

References

1. Ma, J.: Deployment algorithm of media content delivery system. Control Theory and Control Engineering of University of Science and Technology of China, Anhui (2009)
2. Lie, X.-Y.: The development situation and trend of CDN service. Modern Science and Technology of Telecommunications. Teleinfo Institute China Academy of Telecommunication Research of MIIT **42**(4), 5 (2012)
3. Xie, T.-H.: AHP and its implementation. J. Lanzhou Commercial College **17**(2), 57 (2001)
4. Jing, Q.-Z., Jiang, X.-H., Yang, J.-S., Zhou, Y.-F.: Study on index system of capability of production safety in coal mine based on AHP. China Saf. Sci. J. **16**(9), 74–79 (2006)
5. Xu, S.-B.: Principle of Analytical Hierarchy Process, pp. 15–30. Tianjin University Press, Tian Jin (1998)
6. Jian, Q.-Y.: Mathematic Model, pp. 305–335. Higher Education Press, Beijing (1993)
7. Saaty, T.L.: The Analytic Hierarchy Process. McGraw – Hill, New York (1980)

Pattern Recognition

Detection of Abnormal Event in Complex Situations Using Strong Classifier Based on BP Adaboost

Yuqi Zhang[1], Tian Wang[1(✉)], Meina Qiao[1], Aichun Zhu[2], Ce Li[3], and Hichem Snoussi[2]

[1] School of Automation Science and Electrical Engineering,
Beihang University, Beijing 100191, China
{yuqi4359,meinaqiao}@l26.com, wangtian@buaa.edu.cn
[2] Institute Charles Delaunay-LM2S-UMR STMR 6279 CNRS,
University of Technology of Troyes, 10004 Troyes, France
{aichun.zhu,hichem.snoussi}@utt.fr
[3] College of Electrical and Information Engineering,
Lanzhou University of Technology, No. 287, Langongping Road,
Lanzhou 730050, Gansu, China
xjtulice@gmail.com

Abstract. In order to recognize the abnormal event, such as emergency or panic, happened in public scenes timely, an algorithm based on features extraction and BP Adaboost to detect abnormal frame event from surveillance video of complex situation is proposed. The proposed method detects an abnormal event where people are running, and this panic situation is simulated by the frame in a video. Experiments show that the method can distinguish and detect the abnormal event effectively and efficiently, which has potentiality to be used in the real public monitoring.

Keywords: Abnormal detection · Optical flow · BP Adaboost

1 Introduction

Nowadays monitors are widely used in public so as to ensure public safety. However, technicians in the supervising centers usually cannot caught out incidents in time due to lacking of labor. For example, there is dangerous event such as rob or shooting in a shopping mall, people run towards different directions in a rush. But in the monitoring room of security department, it is really difficult to find the emergency event from the monitoring wall in time. Thus, the public places, such as shopping malls, crowded streets, train stations, and sports fields need intelligent surveillance [1]. In this case, there has been growing interest in the development of abnormal detection. For example, in [2], the researchers built a model to detect pedestrian abnormal behavior in traffic scene.

Commonly, we divide abnormal behaviors [3] into two types: the behavior of small probability or contrary to a priori rule, and the behavior of pattern mismatch with known normal behavior. The current methods mostly focus on the following

© Springer International Publishing Switzerland 2016
D.-S. Huang and K.-H. Jo (Eds.): ICIC 2016, Part II, LNCS 9772, pp. 245–256, 2016.
DOI: 10.1007/978-3-319-42294-7_21

categories: support vector machine (SVM) [4, 5], decision tree [6, 7] and neural network [8, 9]. They usually form a model definition by training the target species, and then compare newly discovered moving objects with the model.

Relevant research work has been conducted in many fields. Take the research [10] on solitary oldies as an example, the authors classified and identified the abnormal action by using Bayesian classifier. The method can also identify the falling action well. Through the detection technology, the intention and the behavior of the oldies can be learned correctly and timely, especially for the elderly abnormal behavior. Vishwakarma [11] used the changes in orientation of Silhouette to detect abnormal behavior, which could be easily used for elderly monitoring application. Also, some relevant research work applied this technology into medical science. In [12], it was used for the recognition of living abnormal cells, which played an important role in studying biological process in the living cells and the interactions between cells and drugs. Rubia [13] focused on abnormal brain activation during cognitive challenge. Adolphs [14] paid attention to psychology by recognizing faces, which might reflect abnormal processing of social information in autism. In [15], researchers studied the recognition of abnormal behaviors like forward fall, backward fall, chest pain, fainting and headache.

In this paper, we focus on the research of abnormal behaviors detection in public. The lawn, indoor and plaza scenes in the UMN dataset [16] are chosen as public places, and a certain amount of frames are selected to construct our experiments. In the experiments, people running towards different directions are considered as abnormal behavior. The optical flow is calculated from successive frames, which represents the movement information. In order to describe the event, the histogram of optical flow orientation descriptor is put forward to fuse the optical flow orientation of the frame. In the rest of the paper, we present that strong classifier based on BP Adaboost [17] is adopted to train the neural network. Then, the experimental results show that the accuracy of the method proposed in this paper is better than other existing methods. At last, we conclude this paper and express some perspectives.

2   Behavior Description

The behavior feature descriptor is presented in this section. Firstly, the motion of the video sequence is extracted by the optical flow method. Then, the optical flow feature of the whole image is fused by the histogram of optical flow orientation descriptor.

2.1   Motion Detection

Optical flow is the instantaneous velocity of the motion of pixels in the image plane. The study of optical flow is to determine the motion of each pixel position using the temporal variation and correlation of the pixel intensity data in the image sequence, the purpose of which is to obtain the motion field that can't be obtained directly from the image sequence, as shown in the Fig. 1.

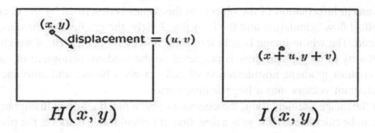

Fig. 1. Pixel offset

Horn and Schunck [18] optical flow method is adopted in this paper. The Horn–Schunck method of estimating optical flow is a global method which introduces a global constraint of smoothness to solve the aperture problem. It is based on the assumption that the intensity is constant, assuming that the variation of the optical flow over the entire image is smooth, which means smooth constraint E_s is minimization:

$$E_s = \int\int [(\frac{\partial u}{\partial x})^2 + (\frac{\partial u}{\partial y})^2 + (\frac{\partial v}{\partial x})^2 + (\frac{\partial v}{\partial y})^2]^2 dxdy \tag{1}$$

Where u and v are the respectively displacements along the x and y coordinate axes. (x, y) is original coordinate and (u, v) is velocity vector.

Minimizing the deviation of optical flow:

$$E_c = \int\int (I_x u + I_y v + I_t)^2 dxdy \tag{2}$$

Where I is the gray value of the pixel point at time t, I_x, I_y and I_t are the derivatives of the gray value at x, y and t.

So the solution to the optical flow field can be transformed into the solution of the following problem:

$$\min \int\int \left\{ (I_x u + I_y v + I_t)^2 + [(\frac{\partial u}{\partial x})^2 + (\frac{\partial \mu}{\partial y})^2 + (\frac{\partial v}{\partial x})^2 + (\frac{\partial v}{\partial y})^2]^2 \right\} dxdy \tag{3}$$

According to the characteristics of the velocity vector of each pixel, the dynamic analysis of the image can be carried out. When there are moving objects in the image, there is relative motion between the object and the background.

2.2   Feature Fusion and Extraction

Although we have got important motion information of optical flow, the information of raw optical flow is noisy and redundant. The bad influence of noises that are mixed in our motion features should be reduced [19]. In our experiments, we adopt histogram of optical flow orientation (HOFO) to provide feature vectors for classification algorithm.

The movement information of the objects in the scenes is described by the distribution of the optical flow orientation and the intensity. Firstly, the gradient of the optical flow is computed. The whole image is split into cells, and then the prescribed weight which is the intensity of the optical flow is projected on the gradient histogram of each cell. Finally, contrast gradient normalization of cells in each block, and combine all the block histogram vectors into a large feature vector.

After the image segmentation, the direction of each patch gradient histogram [20] is proposed to be calculated. $\theta_i(x, y)$ is a direction of optical flow vectors at the pixel (x, y) in frame i,

$$\theta_i(x, y) = \tan^{-1} \frac{dx}{dy} \tag{4}$$

where dx is motion flow vectors in row, dy is motion flow vectors in column.

Divide the orientation of the optical flow into a certain number of bins according to the need, as shown in Fig. 2. Then, according to the gradient direction of each pixel, accumulate them to the histogram by the overlapped blocks.

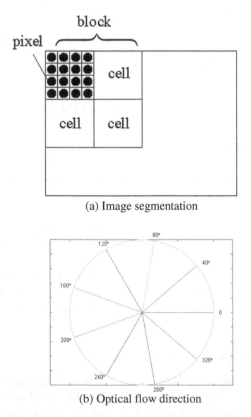

(a) Image segmentation

(b) Optical flow direction

Fig. 2. The histogram of optical flow orientations. (a) Image segmentation (b) Optical flow direction

3 Classification Algorithm

The classification algorithm based on BP Adaboost model is presented in this section. The basic idea of Adaboost algorithm is presented, and then with our feature descriptor the event recognition method is proposed.

3.1 BP Adaboost Model

The thought of Adaboost algorithm is to merge the output of multiple weak classifiers to generate an effective classification. The main steps of the method are as follows: First, the weak learning algorithm and the sample space are given, and the training data of the m group is found out from the sample space. The weight of each training data is $1/m$. Then, with T times iteration computation which use weak learning algorithm, it updates the distribution of the weights of the training data according to the classification results after each run, and will give greater weights for individual data with unsuccessful classification. So it will pay more attention to these individuals in the next iteration in the training process. Weak classifier then gets a classification function sequence $f_1, f_2, \ldots, f_T$, through reduplicative iterations, and assigns a weight to each classification function. If the result of the classification function is better, the corresponding weight is increased. After T times iterations, the final strong classification function F is a combination of the weighted weak classification function. BP Adaboost model, taking BP neural network as weak classifiers, trains the output of the BP neural network prediction samples to get a strong classifier composing of a number of BP neural network weak classifiers, through the Adaboost algorithm.

3.2 Model Building

The algorithm flow chart is shown in Fig. 3. The details is explained bellowing.

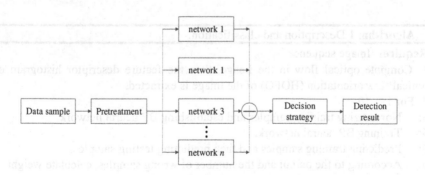

Fig. 3. Algorithm flow chart

Data selection and network initialization: Select m group of training data from the sample space, and initialize the distribution weight $D_t(i) = 1/m$ [21, 22] of the testing data. Then, determine the structure of the neural network according to the dimension of the input and the output of the sample. Next, initialize the weights and thresholds of the BP neural network.

Weak classifier prediction: When training the first t weak classifier, use the training data to train the BP neural network, and forecast the output of the training data. Then, get the sum of forecast errors e_t of forecast sequence $g(t)$. The formula of the sum of errors e_t.

$$e_t = \sum_i D_i(i) \qquad i = 1, 2, \ldots, m \quad (g(t) \neq y) \tag{5}$$

where e_t is the sum of forecast errors, $g(t)$ represents forecast classification results; y represents Expected classification results.

Calculating forecast sequence weight: calculate the weight of the sequences according to the forecast errors e_t of forecast sequence $g(t)$. The formula of weight a_t.

$$a_t = \frac{1}{2} \ln \left\{ \frac{1 - e^t}{e^t} \right\} \tag{6}$$

Testing data and adjusting weights: According to the forecast sequence weight a_t, adjust the weight of training sample in the next run. The formula of adjustment.

$$D_{t+1}(i) = \frac{D_t(i)}{B_t} \times \exp[-a_t y_i g_t(x_i)] \quad i = 1, 2, \ldots, m \tag{7}$$

where B_t is normalization factor, which makes the sum of distribution weights equal to one under the case that weight ratio is invariant.

Strong classification function: After T times training, T groups of weak classification functions $f(g_t, a_t)$ can be obtained. Then strong classification function $h(x)$ is obtained by the combination of T groups of weak classifier functions $f(g_t, a_t)$.

The abnormal event detection method is described in Algorithm 1.

Algorithm 1 Description and classification.

Require: Image sequence
1: Compute optical flow in the image, then the feature descriptor histogram of optical flow orientation (HOFO) of the image is extracted.
2: **For** i=1:10
3:    Normalizing training samples and establishing BP neural network.
4:    Training BP neural network.
5:    Predicting training samples and then predicting testing samples.
6:    According to the output and the number of wrong samples, calculate weight
        a_t (i), then update and normalize D_t.
7: **end**
8: The strong classifier integrates the ten weak classifiers to get the final results.

---

4   Experimental Results

This section presents the results of experiments conducted to illustrate the performance of the BP Adaboost event detection algorithm combining by Adaboost and BP Neural Networks. We set 10 weak classifiers in our work, the receiver operating characteristic (ROC) curve [23] is generated to show the event detection performance. The area under the curve (AUC) is used to estimate the accuracy of the experiments.

The data used in this experiment are respectively recorded in a lawn, indoor and plaza of the UMN dataset. Those figures are selected based on time series from the video. A frame where people are walking in different directions is considered as a normal frame, while a frame where people are running is taken as an abnormal panic frame.

4.1   Abnormal Detection in Lawn Scenes

In this experiment, we use 616 frames for training and 815 frames for testing. In the testing samples, the first 673 frames are normal, the last 142 frames are abnormal. The normal and abnormal examples are show in Fig. 4, the detection results are shown in Fig. 5. The division results shown in the 10 weak classifiers are respectively 679, 713, 707, 699, 644, 705, 697, 666, 698, 700, the strong classifier integrates the results of the 10 weak classifiers into one classification result. The division shown in the result between normal and abnormal behaviors is 672 between the first classification and the eighth classification, which approximates to the real situation (Fig. 6).

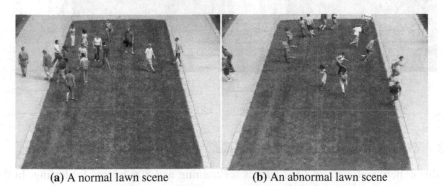

(a) A normal lawn scene   (b) An abnormal lawn scene

Fig. 4. Examples of the normal and abnormal lawn scenes

4.2   Abnormal Detection in Indoor Scenes

In this experiment, we use 1991 frames for training and 2045 frames for testing. In the testing samples, the first 1627 frames are normal, the last 418 frames are abnormal. As shown in Fig. 7, the division results shown in the 10 weak classifiers are respectively 1805, 1645, 1894, 1729, 1544, 1705, 1627, 1572, 1710, 1715. The division shown in the result between normal and abnormal behaviors is 1612 between the fifth classification and the seventh classification.

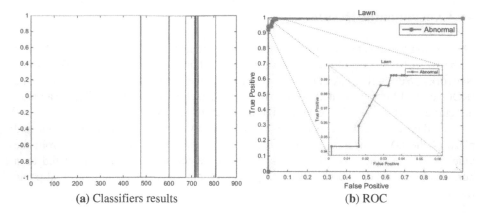

(a) Classifiers results   **(b)** ROC

Fig. 5. The detection results of lawn scene. (a) The blue lines are results of ten weak classifiers. The green line is result of the strong classifier which composes of those ten weak classifiers. (b) Receiver operating characteristic (ROC) curve of lawn scene, the area under the ROC curve (AUC) is 0.9945. (Color figure online)

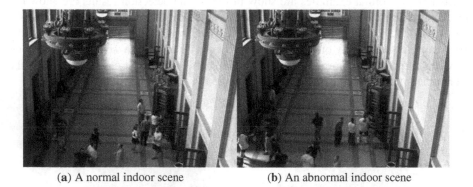

(a) A normal indoor scene   **(b)** An abnormal indoor scene

Fig. 6. Examples of the normal and abnormal indoor scenes.

4.3   Abnormal Detection in Plaza Scenes

In this experiment, we use 1312 frames for training and 806 frames for testing. In the testing samples, the first 719 frames are normal, the last 87 frames are abnormal. In Fig. 9, the division results shown in the 10 weak classifiers are respectively 754, 747, 712, 757, 749, 754, 680, 742, 753, 759. The strong classifier distinguishes the normal and abnormal event at 693 (Fig. 8).

From the experiments in three different scenes, we can see that the proposed event recognition methods can distinguish the states from the optical flow based feature descriptor. Our proposed method can give considerable abnormal detection results (Tables 1 and 2).

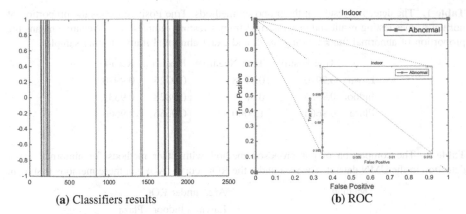

(a) Classifiers results   (b) ROC

Fig. 7. The detection results of indoor scene. (a) Classifiers. (b) Receiver operating characteristic (ROC) curve of indoor scene, the area under the ROC curve (AUC) is 0. 9988. (Color figure online)

(a) A normal plaza scene   (b) An abnormal plaza scene

Fig. 8. Examples of the normal and abnormal plaza scenes.

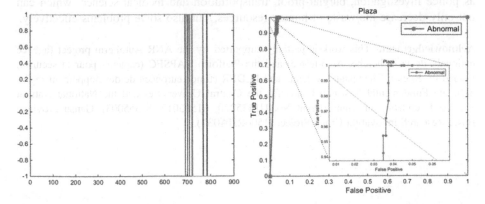

Fig. 9. The detection results of plaza scene. (a) Classifiers. (b) Receiver operating characteristic (ROC) curve of plaza scene, the area under the ROC curve (AUC) is 0. 9820. (Color figure online)

Table 1. The detection results of the proposed methods. True Positive means the proportion of normal frames testing result and exact normal frames testing samples. True Negative means the proportion of abnormal frames testing result and exact abnormal frames testing samples.

	True Positive	True Negative	Precision	Recall
Lawn	0.9970	0.9718	0.9725	0.9970
Indoor	0.9533	0.9833	0.9828	0.9533
Plaza	0.9666	1.0000	0.9666	0.9666

Table 2. The comparison of our proposed method with other methods for abnormal event detection in the UMN dataset. The area under the ROC curve (AUC) in these methods is shown.

Method	Area under ROC		
	Lawn	Indoor	Plaza
Social force [24]	0.96		
Optical flow [24]	0.84		
NN [25]	0.93		
SRC [25]	0.995	0.975	0.964
STCOG [26]	0.9362	0.7759	0.9661
HOFO-BP-Adaboost (Ours)	0.9954	0.9988	0.982

5 Conclusion

In this paper, we proposed a new algorithm to recognize the events, the process of which consists of two sections. First, use optical flow to obtain motion field, and adopt histogram of optical flow orientation (HOFO) to provide feature vectors for classification algorithm. Second, we use strong classifier based on BP Adaboost to train and test behaviors.

In the future work, we plan to include more abnormal behaviors into consideration, such as cluster, fire and collapse. Besides, this technique can apply to many fields, such as police investigation, burglar-proof, transportation and medical science, which can not only decrease the waste of human resources, but also solve problems effectively.

Acknowledgement. This work is partially supported by the ANR AutoFerm project (agence nationale de la recherche, Auto-ferme) and the Platform CAPSEC (capteurs pour la sécurité) funded by Région Champagne-Ardenne and FEDER (fonds européen de développement régional), the Fundamental Research Funds for the Central Universities and the National Natural Science Foundation of China (Grant No. U1435220, 61503017, 61365003), Gansu Province Basic Research Innovation Group Project (1506RJIA031).

References

1. Duque, D., Santos, H., Cortez, P.: Prediction of abnormal behaviors for intelligent video surveillance systems. In: Proceedings of IEEE Symposium on Computational Intelligence and Data Mining (CIDM), pp. 362–367 (2007)
2. Jiang, Q., Li, G., Yu. J., Li, X.: A model based method of pedestrian abnormal behavior detection in traffic scene. In: Proceedings of IEEE First International Smart Cities Conference (ISC2), pp. 1–6 (2015)
3. David, S., Derald, W., Stanley, S., Diane, S.: Understanding Abnormal Behavior. Cengage Learning (2012)
4. Chen, Y., Liang, G., Lee, K., Xu, Y.: Abnormal behavior detection by multi-SVM-based Bayesian network. In: Proceedings of IEEE International Conference on Information Acquisition (ICIA), pp. 298–303 (2007)
5. Palaniappan, A., Bhargavi, R., Vaidehi, V.: Abnormal human activity recognition using SVM based approach. In: Proceedings of IEEE International Conference on Recent Trends in Information Technology (ICRTIT), pp. 97–102 (2012)
6. Sindhu, S., Geetha, S., Kannan, A.: Decision tree based light weight intrusion detection using a wrapper approach. Expert Syst. Appl. **39**(1), 129–141 (2012)
7. Amor, N., Benferhat, S., Elouedi, Z.: Naive Bayes vs decision trees in intrusion detection systems. In: Proceedings of ACM Symposium on Applied Computing, pp. 420–424 (2004)
8. Han, S., Cho, S.: Evolutionary neural networks for anomaly detection based on the behavior of a program. IEEE Trans. Syst. Man Cybern. B Cybern. **36**(3), 559–570 (2005)
9. Li, Y., Liu, L., Wang, E., Zhang, H., Dou, S., Tong, L., Cheng, J., Chen, C., Shi, D.: Abnormal neural network of primary insomnia: evidence from spatial working memory task fMRI. Eur. Neurol. **75**(1–2), 48–57 (2016)
10. Wu, D., Wang, L., Wang, J., Liu, L.: Solitary oldies abnormal action recognition based on MEI. In: Proceedings of IEEE International Conference on Signal Processing, Communications and Computing (ICSPCC), pp. 1–4 (2015)
11. Vishwakarma, D., Kapoor, R., Maheshwari, R., Kapoor, V., Raman, S.: Recognition of abnormal human activity using the changes in orientation of silhouette in key frames. In: Proceedings of 2nd IEEE International Conference on Computing for Sustainable Global Development, pp. 336–341 (2015)
12. Liang, X., Xu, H., Liu, Y.: Recognition of living abnormal cells based on an optical microscope. In: Proceedings of IEEE International Conference on Manipulation, Manufacturing and Measurement on the Nanoscale (3M-NANO), pp. 18–21 (2012)
13. Rubia, K., Smith, A., Brammer, M.J., Toone, B., Taylor, E.: Abnormal brain activation during inhibition and error detection in medication-naive adolescents with ADHD. Am. J. Psychiatry **162**(6), 1067–1075 (2005)
14. Adolphs, R., Sears, L., Piven, J.: Abnormal processing of social information from faces in autism. J. Cogn. Neurosci. **13**(2), 232–240 (2001)
15. Khan, Z., Sohn, W.: Feature extraction and dimensions reduction using R transform and principal component analysis for abnormal human activity recognition. In: Proceedings of IEEE International Conference on Advanced Information Management and Service (IMS), pp. 253–258 (2010)
16. Detection of Events—Detection of Unusual Crowd Activity. http://mha.cs.umn.edu/Movies/Crowd-Activity-All.avi
17. Liu, Y., Chen, Y.-D.: The applied research of motive object tracking based on BP-Adaboost algorithm. J. Hebei Univ. Eng. (Natural Science Edition) **3**, 026 (2012)

18. Horn, B.K., Schunck, B.G.: Determining optical flow. In: Technical Symposium East, pp. 319–331. International Society for Optics and Photonics (1981)
19. Lertniphonphan, K., Aramvith, S., Chalidabhongse, T.: Human action recognition using direction histograms of optical flow. In: Proceedings of IEEE International Symposium on Communications and Information Technologies (ISCIT), pp. 574–579 (2011)
20. Wang, T., Snoussi, H.: Detection of abnormal visual events via global optical flow orientation histogram. IEEE Trans. Inf. Forensics Secur. **9**(6), 988–998 (2014)
21. Liu, X., Wang, J., Yin, M., Edwards, B., Xu, P.: Supervised learning of sparse context reconstruction coefficients for data representation and classification. Neural Comput. Appl. 1–9 (2015)
22. Wang, J., Zhou, Y., Duan, K., Wang, J., Bensmail, H.: Supervised cross-modal factor analysis for multiple modal data classification. In: Proceedings of IEEE International Conference on Systems, Man, and Cybernetics, pp. 1882–1888 (2015)
23. Wang, J., Wang, H., Zhou, Y., McDonald, N.: Multiple kernel multivariate performance learning using cutting plane algorithm. arXiv preprint arXiv:1508.06264 (2015)
24. Mehran, R., Oyama, A., Shah, M.: Abnormal crowd behavior detection using social force model. In: Proceedings of IEEE Conference on Computer Vision and Pattern Recognition (CVPR), pp. 935–942 (2009)
25. Cong, Y., Yuan, J., Liu, J.: Sparse reconstruction cost for abnormal event detection. In: Proceedings of IEEE Conference on Computer Vision and Pattern Recognition (CVPR), pp. 3449–3456 (2011)
26. Shi, Y., Gao, Y., Wang, R.: Real-time abnormal event detection in complicated scenes. In: Proceedings of IEEE International Conference on Pattern Recognition (ICPR), pp. 3653–3656 (2010)

A Comparative Study for the Effects of Noise on Illumination Invariant Face Recognition Algorithms

Guangyi Chen[1(✉)] and Wenfang Xie[2]

[1] Department of Mathematics and Statistics, Concordia University,
1455 de Maisonneuve West, Montreal, QC H3G 1M8, Canada
gchen@alumni.concordia.ca
[2] Department of Mechanical and Industrial Engineering, Concordia University,
1455 de Maisonneuve West, Montreal, QC H3G 1M8, Canada
wfxie@encs.concordia.ca

Abstract. In the literature, the effects of noise on existing face recognition algorithms are neglected, to the best of our knowledge. In this paper, for the first time, we perform an experimental study for the effects of noise on existing illumination-invariant face recognition algorithms. In total, twenty-one algorithms have been included in this study in this paper. We find that, when noise is present in face images, Tan and Triggs' method achieves the highest correct recognition rates for both the extended Yale B face database and the CMU-PIE face database. If face images do not contain noise, isotropic smoothing is preferred because this method obtains the highest average recognition rate (96 %) for the extended Yale B database and 16 out of 21 methods achieve 100 % correct recognition rates for the CMU-PIE face database.

Keywords: Face recognition · Nearest neighbour classifier · Illumination invariant · Gaussian white noise

1 Introduction

Illumination variation presents big challenge to today's face recognition systems, and noise in face images makes even more trouble for face recognition. In the literature, there are two recent survey papers [1, 2] on illumination invariant face recognition. However, both surveys did not consider noise in face images, which could be a problem in face recognition. In general, existing illumination invariant face recognition algorithms can be divided into five major classes: preprocessing, invariant feature extraction, face image modeling, illumination variation learning, and postprocessing. In this work, we mainly study those methods that can be classified as the preprocessing category. According to [2], illumination-preprocessing algorithms can be further classified into three categories: gray-level transformation, gradient and edge extraction, and face reflection field estimation. Even though there are many algorithms for illumination preprocessing in literature, but which one is the best for face recognition when face images are corrupted by noise? So far, there does not exist such works that address this issue.

© Springer International Publishing Switzerland 2016
D.-S. Huang and K.-H. Jo (Eds.): ICIC 2016, Part II, LNCS 9772, pp. 257–267, 2016.
DOI: 10.1007/978-3-319-42294-7_22

In this paper, we conduct a number of experiments in order to determine which algorithm is the best for face recognition when noise is present in face images. We find that when noise is added to Extended Yale B face database [3] and CMU-PIE face database [4], the Tan and Triggs' method [5] achieves the highest recognition rates for all four-noise levels (σ_n = 10, 20, 30, and 40). In addition, isotropic smoothing [6] obtains the highest average recognition rate (96.00 %) for noise-free Extended Yale-B database, and several methods yield perfect classification rates (100 %) for noise-free CMU-PIE database.

2 Existing Methods

We briefly review a number of existing methods for illumination invariant face recognition here. Tan and Triggs [5] worked on illumination invariant face recognition by taking advantages of existing robust illumination normalization, local texture-based face representations, distance transform based matching, kernel-based feature extraction, and multiple feature fusion. Their method achieved state-of-the-art recognition rates on three face databases, which were widely used for testing recognition under bad illumination conditions.

Heusch et al. [6] investigated the face recognition problem under varying illumination conditions in order to improve the performance of face recognition methods. Their experiments achieved big improvement in classification rates and their results were comparable to other state-of-the-art face recognition methods.

Struc and Pavesic [7] developed two new methods for illumination invariant face recognition to overcome the weaknesses of the commonly adopted illumination normalization techniques. They demonstrated that their methods on the Extended Yale-B database were very encouraging.

Park et al. [8] proposed the Retinex method for illumination invariant face recognition. By the well-known Retinex theory, illumination is normally estimated and normalized by smoothing the input image first and then dividing the estimate by the original input image. Their method is tested on Yale face database B, CMU PIE database and AR face database, and it presented consistent and promising results even when ill-illuminated face images were used as a training set.

Gross and Brajovic [9] proposed a new image preprocessing algorithm to compensate for illumination variations in face images. Their algorithm did not involve any training steps, knowledge of 3D face models or reflective surface models. They demonstrated large performance improvements over several standard face recognition algorithms across several public face databases.

Chen et al. [10] proposed a logarithm discrete cosine transform (DCT) for illumination invariant face recognition. They kill a small number of low frequency DCT coefficients and then take the inverse DCT in order to generate illumination invariant faces. However, important features may be removed due to the operation of setting low frequency DCT coefficients to zero.

Oppenheim et al. [11] worked on nonlinear filtering of signals that were products or convolutions of components. They applied their method to image enhancement and audio dynamic range compression and expansion with very encouraging results.

Jabson et al. [12] extended a single-scale center/surround Retinex to a multiscale version, which achieved simultaneously dynamic range compression or color consistency. They also developed a new color restoration method that overcame existing deficiency at the cost of a modest dilution in color consistency.

Wang et al. [13] introduced self-quotient images (SQI) for illumination invariant face recognition:

$$SQI(x,y) = \frac{I(x,y)}{LF * I(x,y)} \tag{1}$$

where * is the convolution operator and LF is a low-pass filter such as Gaussian filter. Illumination invariant property of the SQI algorithm is analyzed according to the Lambertian model.

Freeman and Adelson [14] studied to design filters with arbitrary orientations, allowing one to adaptively steer a filter to any orientation. They presented a number of applications, such as the analysis of orientation and phase, angularly adaptive filtering, edge detection, and shape from shading.

Zhang et al. [15] proposed a wavelet-based face recognition method to detect and eliminate illumination variation in face images. In addition, using wavelet-based denoising algorithm effectively reduced the effect of illumination, and hence the multiscale face structure was constructed.

Du and Ward [16] developed an adaptive region-based image enhancement method for robust face recognition under varying lighting conditions by using discrete wavelet transform (DWT). However, this method introduces artefacts between different regions in face images.

Davidson and Abramowitz [17] investigated the difference of Gaussians as a feature enhancement technique, which performs the subtraction of one blurred version of an original image from another less blurred version. Convolving the original images with Gaussian kernels with different standard deviations can generate the blurred images. Blurring an image using a Gaussian kernel suppresses only high-frequency spatial information. The difference of Gaussians is a band-pass filter, which discard all but a small number of spatial frequencies in the original face image.

Gradient face [18] of a face image I(x, y) is defined as

$$G(x,y) = \arctan(\frac{\partial I(x,y)}{\partial y} / \frac{\partial I(x,y)}{\partial x}), \tag{2}$$

where ∂ means to take the first order derivative. This method is fast to compute and it has very good classification performance in noise-free environment. Nevertheless, it may be sensitive to noise in the face images.

Xie et al. [19] studied face illumination normalization based on both large and small scale features. It is well known that the variations of illumination mainly affect the large-scale features (low-frequency components), but not so much the small-scale features. They claimed that their method produced significantly better recognition rates when compared to state-of-the-art methods.

Weber face [20] calculates the difference of the current pixel intensity value against its neighbours, and then computes the ratio between the summation of the difference and the intensity value of the current pixel:

$$W(x,y) = \arctan(\alpha \sum_{i,j} \frac{I(x_i, y_j) - I(x,y)}{I(x,y)}), \tag{3}$$

where α is a scalar constant, I(x, y) is the center pixel, and $I(x_i, y_j)$ are the neighbour pixels.

More recently, Lai et al. [21] developed a multiscale logarithm difference edgemap method for face recognition under varying lighting conditions. They took the logarithm transform to convert the multiplication of surface albedo and light intensity into an addition, and then they subtracted two neighbour pixels in order to eliminate light intensity. They generated several multiscale edgemaps, which were multiplied by a weight and then all weighted edgemaps were combined to form a robust face map.

Hu [22] proposed a new illumination invariant face recognition method by means of dual-tree complex wavelet transform (DT-CWT). This method first uses the DT-CWT to extract edges from face images, then denoise the DT-CWT coefficients to obtain the multi-scale illumination invariant structures in the logarithm domain, and finally combines the extracted edges and logarithm features.

Nikan and Ahmadi [23] investigated local gradient-based illumination invariant face recognition method using local phase quantisation and multi-resolution local binary pattern fusion. Their experimental results demonstrated the performance improvement of their new method under very bad illumination condition.

Faraji and Qi [24] studied face recognition under varying illumination as well. Their method first performed adaptive homomorphic filtering to reduce illumination effects, then utilized an interpolative enhancement scheme to stretch the filtered image, and finally it generated eight directional edge images and forms an illumination-insensitive face map from the eight edge images.

3   A Comparative Study

We tested different methods for illumination invariant face recognition with Extended Yale Face Database B [3] and CMU Pose, Illumination and Expression (PIE) (CMU-PIE) illumination face database [4]. The first database has faces of 38 subjects with 64 different lighting conditions. There are 2414 available images, which are already cropped and aligned well. These faces are captured under various laboratory - controlled lighting conditions with size 192 × 168. We select one well-lighted face as the single reference and use all the rest 2414 − 38 = 2376 face images as test samples. These faces are classified into 5 subsets according to angles between the light source direction and the camera axis. We normalize the output face maps to 128 × 128 pixels for this face database. Figure 1 shows a sample of face images for one subject in the extended Yale-B database.

The second database contains 41368 images for 68 subjects, where the face images are captured with 13 different poses, 43 different illumination conditions, and

Fig. 1. A sample of face images for one subject in the extended Yale-B database.

4 different expressions. We choose a subset of faces with illumination variations on light intensity and direction in frontal view. There exist 2924 such images in this subset. We normalize the face images to 128×128 pixels for this face database. Figure 2 shows a sample of face images for one subject in the CMU-PIE database.

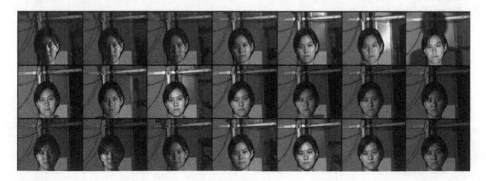

Fig. 2. A sample of face images for one subject in the CMU-PIE database.

Tables 1 and 2 show the correct recognition rates of 21 algorithms for the two face databases without and with Gaussian white noise, respectively. The noisy face image J is generated by adding Gaussian white noise to the original noise-free face image I as follows:

$$J = I + \sigma_n N \tag{4}$$

where σ_n is the noise standard deviation, and N obeys normal distribution with zero mean and unit variance. Our experimental results demonstrate that, by using nearest neighbour (NN) classifier, isotropic smoothing [6] achieves the highest average recognition rate (96.00 %) for the noise-free Extended Yale-B face database. For subset 1

Table 1. The correct classification rates (%) of different methods for illumination invariant face recognition. The best results are highlighted in bold font

Methods	CMU-PIE	Extended Yale-B					
		Subset 1	Subset 2	Subset 3	Subset 4	Subset 5	Average
Adaptive NL Means [7]	**100**	99.11	**100**	83.74	69.20	68.77	84.16
Adaptive Single Scale Retinex [8]	**100**	90.67	**100**	89.67	95.44	93.42	93.84
Anisotropic Smoothing [9]	**100**	89.78	99.12	84.62	76.81	56.30	81.32
LOG-DCT [10]	**100**	97.78	**100**	91.43	83.84	82.35	91.08
Homomorphic [11]	**100**	97.78	**100**	73.85	47.53	45.94	73.02
Isotropic Smoothing [6]	**100**	98.22	**100**	**94.29**	93.54	93.98	**96.00**
Multiscale Retenex [12]	99.78	**100**	**100**	89.67	71.67	73.95	87.06
Multiscale SQI [13]	**100**	98.22	**100**	92.97	92.78	93.70	95.53
NL Means [7]	**100**	99.56	**100**	92.97	87.83	89.64	94.00
Single Scale Retinex [12]	99.63	**100**	**100**	89.45	69.39	71.01	85.97
Single Scale SQI [13]	**100**	93.78	**100**	90.33	95.82	93.70	94.72
Steerable Gaussian [14]	**100**	86.67	**100**	87.91	92.97	93.14	92.14
Wavelet Denoising [15]	**100**	89.78	**100**	87.03	94.87	89.22	92.18
Wavelet Normalization [16]	95.95	98.22	87.28	38.68	38.68	11.22	50.30
Anisotropic Smoothing [9]	**100**	91.11	99.56	86.81	76.62	61.90	83.20
Difference of Gaussian [17]	**100**	95.11	**100**	91.21	96.01	93.00	95.06
Gradient Faces [18]	99.93	97.78	**100**	88.13	85.74	87.54	91.84
Large and Small Scale [19]	97.94	97.78	**100**	**94.29**	92.21	92.72	95.40
Multiscale Weber Faces [20]	**100**	92.00	**100**	90.11	97.15	**94.12**	94.68
Tan and Triggs [5]	**100**	93.33	**100**	91.65	96.77	93.56	95.06
Weber Faces [20]	**100**	92.00	**100**	89.45	**97.34**	93.70	94.50

Table 2. The average correct classification rates (%) of different methods for illumination invariant face recognition with face images corrupted by Gaussian white noise. The best results are highlighted in bold font.

Methods	Extended Yale-B				CMU-PIE			
	Noise standard deviation (σ_n)							
	10	20	30	40	10	20	30	40
Adaptive NL Means [7]	72.97	69.05	64.80	62.54	99.85	99.41	98.97	98.38
Adaptive Single Scale Retinex [8]	73.15	66.72	63.95	61.21	99.48	97.86	95.80	91.60
Anisotropic Smoothing [9]	57.51	44.33	31.63	20.67	98.60	96.39	92.70	88.21
LOG-DCT [10]	77.07	73.68	70.60	67.63	**100**	99.93	99.63	99.26
Homomorphic [11]	57.99	46.54	34.94	23.47	**100**	**100**	99.04	97.49
Isotropic Smoothing [6]	78.87	72.81	68.61	65.20	**100**	99.71	99.19	98.67
Multiscale Retenex [12]	70.29	67.09	65.28	63.41	99.04	98.67	98.16	97.79
Multiscale SQI [13]	79.30	71.24	60.45	47.60	**100**	99.63	99.48	98.53
NL Means [7]	81.06	77.68	74.97	72.58	99.93	99.48	99.19	98.75
Single Scale Retinex [12]	69.50	66.59	65.35	63.61	98.60	98.08	97.72	97.49
Single Scale SQI [13]	75.31	63.00	45.12	30.84	**100**	99.78	99.12	97.86
Steerable Gaussian [14]	78.46	73.15	67.38	62.42	**100**	**100**	99.78	98.82
Wavelet Denoising [15]	72.39	65.21	58.76	52.50	**100**	99.71	98.38	96.68
Wavelet Normalization [16]	49.79	49.93	49.71	49.83	96.09	96.46	96.09	95.80
Anisotropic Smoothing [9]	67.81	58.86	49.01	40.69	99.12	97.27	94.92	91.60
Difference of Gaussian (DoG) [17]	77.73	72.26	66.93	62.32	**100**	**100**	99.85	**99.85**
Gradient Faces [18]	76.79	73.66	71.59	69.75	99.41	99.04	98.60	97.79
Large and Small Scale Features [19]	82.57	77.71	73.44	69.68	99.48	99.56	99.48	98.89
Multiscale Weber Faces [20]	79.80	73.78	68.64	62.66	**100**	**100**	99.71	99.48
Tan and Triggs [5]	**83.39**	**79.24**	**77.02**	**74.96**	**100**	**100**	**100**	**99.85**
Weber Faces [20]	79.43	73.92	68.82	63.20	100	99.93	99.71	99.34

in the Extended Yale-B face database, only single scale Retinex and multiscale Retenex [12] obtain 100 % recognition rates even though it is not difficult to recognition faces in this subset. For subset 2 in the Extended Yale-B face database, 19 out of 21 methods achieve 100 % recognition rates, which is reasonably good. In addition, many methods obtain perfect classification rates (100 %) for the noise-free CMU-PIE face database. When noise is added to the two face databases, the Tan and Triggs' method [5] achieves the highest recognition rates among all methods compared in this paper for all four-noise levels (σ_n = 10, 20, 30, and 40) and both face databases.

Figure 3 shows a comparison among Gradient faces, LOG-DCT, Tan and Triggs, isotropic smoothing, wavelet normalization, and Weber faces for illumination invariant face recognition when Gaussian white noise is added to face images with noise standard deviation σ_n = 0/10/20/30 for the first/second/third/fourth rows. It can be seen that face images with good illumination are not damaged too much by noise and all these six methods generate good illumination invariant face maps. Again, Fig. 4 depicts a comparison among Gradient faces, LOG-DCT, Tan and Triggs, isotropic smoothing, wavelet normalization, and Weber faces for illumination invariant face recognition when Gaussian white noise is added to face images with noise standard deviation $\sigma_n = 0/10/20/30$ for the first/second/third/fourth rows. It can be seen that dark regions in face images are damaged so much by noise that all six methods fail to generate good illumination invariant face maps.

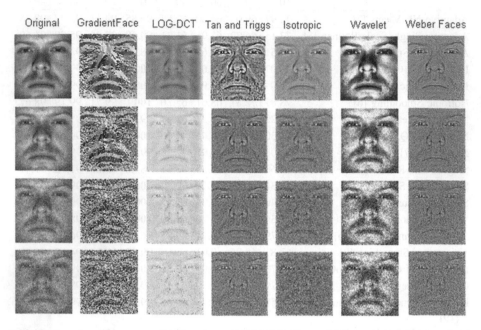

Original   GradientFace   LOG-DCT   Tan and Triggs   Isotropic   Wavelet   Weber Faces

Fig. 3. A comparison among Gradient faces, LOG-DCT, Tan and Triggs, isotropic smoothing, wavelet normalization, and Weber faces for illumination invariant face recognition when Gaussian white noise is added to face images with noise standard deviation $\sigma_n = 0/10/20/30$ for the first/second/third/fourth rows. It can be seen that face images with good illumination are not damaged too much by noise and all these six methods generate good illumination invariant face maps.

It should be pointed out that illumination-preprocessing algorithms normally generate image maps instead of feature vectors. All these face image maps are results of different illumination elimination processes and hence produce different visual appearance. This leads us to illustrate different face image maps for different illumination preprocessing algorithms and different face databases as shown in Figs. 3 and 4.

It should also be pointed out that we did not include the experimental results of four recently published methods [21–24] in this paper because we did not have their Matlab source code. We plan to implement some of these methods in the near future or receive the source code from the authors of these papers.

Original   GradientFace   LOG-DCT   Tan and Triggs   Isotropic   Wavelet   Weber Faces

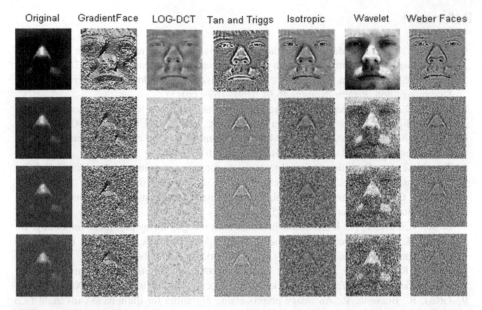

Fig. 4. A comparison among Gradient faces, LOG-DCT, Tan and Triggs, isotropic smoothing, wavelet normalization, and Weber faces for illumination invariant face recognition when Gaussian white noise is added to face images with noise standard deviation $\sigma_n = 0/10/20/30$ for the first/second/third/fourth rows. It can be seen that dark regions in face images are damaged so much by noise that all six methods fail to generate good illumination invariant face maps.

4 Conclusions

In this paper, we have conducted a number of experiments on illumination invariant face recognition. We tested 21 different methods under different illumination conditions using the Extended Yale-B face database and the CMU-PIE face database. Our experiments show that isotropic smoothing [6] produces the highest average recognition rate for the noise-free Extended Yale-B face database, and several methods obtain 100 % classification rate for the noise-free CMU-PIE face database. For noisy face images, the Tan and Triggs' method [5] is the best for all tested noise levels for both the extended Yale B face database and the CMU-PIE face database.

In our future research, we would like to extract features from face images using our previously developed algorithms [25, 26] in order to improve correct recognition rates for both face databases. We would also test the performance of these twenty-one methods under outdoor uncontrolled lighting conditions.

References

1. Zou, X., Kittler J., Messer, K.: Illumination invariant face recognition: a survey. In: Proceedings of the Biometrics: Theory, Applications, and Systems, pp. 1–8 (2007)
2. Han, H., Shan, S., Chen, X., Gao, W.: A comparative study on illumination preprocessing in face recognition. Pattern Recogn. **46**(6), 1691–1699 (2013)
3. Lee, K.C., Ho, J., Kriegman, D.: Acquiring linear subspaces for face recognition under variable lighting. IEEE Trans. Pattern Anal. Mach. Intell. **27**(5), 684–698 (2005)
4. Sim, T., Baker, S., Bsat, M.: The CMU pose, illumination, and expression database. IEEE Trans. Pattern Anal. Mach. Intell. **25**(12), 1615–1618 (2003)
5. Tan, X., Triggs, B.: Enhanced local texture sets for face recognition under difficult lighting conditions. IEEE Trans. Image Process. **19**(6), 1635–1650 (2010)
6. Heusch, G., Cardinaux, F., Marcel, S.: Lighting Normalization Algorithms for Face Verification, IDIAP-com 05-03 (2005)
7. Štruc, V., Pavešić, N.: Illumination invariant face recognition by non-local smoothing. In: Fierrez, J., Ortega-Garcia, J., Esposito, A., Drygajlo, A., Faundez-Zanuy, M. (eds.) BioID MultiComm 2009. LNCS, vol. 5707, pp. 1–8. Springer, Heidelberg (2009)
8. Park, Y.K., Park, S.L., Kim, J.K.: Retinex method based on adaptive smoothing for illumination invariant face recognition. Signal Process. **88**(8), 1929–1945 (2008)
9. Gross, R., Brajovic, V.: An image preprocessing algorithm for illumination invariant face recognition. In: Proceedings of the 4th International Conference on Audio- and Video-Based Biometric Personal Authentication, AVPBA 2003, pp. 10–18 (2003)
10. Chen, W., Er, M.J., Wu, S.: Illumination Compensation and normalization for robust face recognition using discrete cosine transform in logarithmic domain. IEEE Trans. Syst. Man Cybern. Part B **36**(2), 458–466 (2006)
11. Oppenheim, A.V., Schafer, R.W., Stockham, T.G.: Nonlinear filtering of multiplied and convolved signals. Proc. IEEE **56**(8), 1264–1291 (1968)
12. Jabson, D.J., Rahmann, Z., Woodell, G.A.: A multiscale Retinex for bridging the gap between color images and the human observations of scenes. IEEE Trans. Image Process. **6**(7), 897–1056 (1997)
13. Wang, H., Li, S.Z., Wang, Y., Zhang, J.: Self quotient image for face recognition. In: Proceedings of the International Conference on Pattern Recognition, pp. 1397–1400 (2004)
14. Freeman, W.T., Adelson, E.H.: The design and use of steerable filters. IEEE Trans. Pattern Anal. Mach. Intell. **13**(9), 891–906 (1991)
15. Zhang, T., Fang, B., Yuan, Y., Tang, Y.Y., Shang, Z., Li, D., Lang, F.: Multiscale facial structure representation for face recognition under varying illumination. Pattern Recognit. **42**(2), 252–258 (2009)
16. Du, S., Ward, R.K.: Adaptive region-based image enhancement method for robust face recognition under variable illumination conditions. IEEE Trans. Circuits Syst. Video Technol. **20**(9), 1165–1175 (2010)
17. Davidson, M.W., Abramowitz, M.: Molecular expressions microscopy primer: digital image processing – difference of Gaussians Edge Enhancement Algorithm. Olympus America Inc., Florida State University
18. Zhang, T., Tang, Y.Y., Fang, B., Shang, Z., Liu, X.: Face recognition under varying illumination using gradientfaces. IEEE Trans. Image Process. **18**(11), 2599–2606 (2009)
19. Xie, X., Zheng, W.S., Lai, J., Yuen, P.C., Suen, C.Y.: Normalization of face illumination based on large-and small-scale features. IEEE Trans. Image Process. **20**(7), 1807–1821 (2011)

20. Wang, B., Li, W., Liao, Q.: Illumination normalization based on Weber's law with application to face recognition. IEEE Signal Process. Lett. **18**(8), 462–465 (2011)
21. Lai, Z.R., Dai, D.Q., Ren, C.X., Huang, K.K.: Multiscale logarithm difference edgemaps for face recognition against varying lighting conditions. IEEE Trans. Image Process. **24**(6), 1735–1747 (2015)
22. Hu, H.: Illumination invariant face recognition based on dual-tree complex wavelet transform. IET Comput. Vision **9**(2), 163–173 (2015)
23. Nikan, S., Ahmadi, M.: Local gradient-based illumination invariant face recognition using local phase quantisation and multi-resolution local binary pattern fusion. IET Image Proc. **9** (1), 12–21 (2015)
24. Faraji, M.R., Qi, J.: Face recognition under varying illumination based on adaptive homomorphic eight local directional patterns. IET Comput. Vision **9**(3), 390–399 (2015)
25. Chen, G.Y., Bui, T.D., Krzyzak, A.: Invariant pattern recognition using radon, dual-tree complex wavelet and Fourier transforms. Pattern Recogn. **42**(9), 2013–2019 (2009)
26. Chen, G.Y., Xie, W.F.: Contour-based feature extraction using dual-tree complex wavelets. Int. J. Pattern Recognit. Artif. Intell. **21**(7), 1233–1245 (2007)

Algorithms of the Cluster and Morphological Analysis for Mineral Rocks Recognition in the Mining Industry

Olga E. Baklanova[1(✉)] and Mikhail A. Baklanov[2(✉)]

[1] D. Serikbayev East-Kazakhstan State Technical University,
Ust-Kamenogorsk, Kazakhstan
OEBaklanova@mail.ru
[2] Tomsk State University, Tomsk, Russia
baklanov.ma@gmail.com

Abstract. This paper describes an algorithm for automatic segmentation of color images of various ore types, using the methods of morphological and cluster analysis. There are some examples illustrating the usage of the algorithm to solve mineral recognition problems. The effectiveness of the proposed method lies in the area of automatic objects of interest identification inside the image, tuning the parameters of the amount allocated to the segments. This paper contains short description of morphological and cluster analysis algorithms for the mineral recognition in the mining industry.

Keywords: Pattern recognition · Segmentation of color images · Cluster analysis · Morphological analysis · Mineral rocks · Mining industry

1 Introduction

The knowledge of the modal composition of rock or ore is very important to solve the mineralogical and technological issues. Macroscopic examination of the rock is carried out visually with a magnifying glass or microscope followed by description of ores and rock cores [1].

In this case, the following features are determined (ISO 25706-83) [2]:

- The main rock-forming minerals;
- The presence of mineral inclusions, either adversely affecting the durability or decorative;
- Availability of secondary minerals, weathering unstable and loose rocks and minerals, rocks crumble during processing;
- The presence of inclusions of minerals etc. impeding treatment, the nature of their distribution among other rock-forming minerals and quantity;
- The texture and structure formation;
- Fracture;
- Translucency;
- Color.

© Springer International Publishing Switzerland 2016
D.-S. Huang and K.-H. Jo (Eds.): ICIC 2016, Part II, LNCS 9772, pp. 268–278, 2016.
DOI: 10.1007/978-3-319-42294-7_23

Rock micro cracks are determined on the plates with dimensions no less than 200 × 200 mm (thickness 10 mm) with a polished texture of the front surface. Preparation of rock samples is carried out according to ISO 30629 (ISO 30629 - 2011).

For receiving micrographs of rocks the preferred way is to use trinocular microscopes with a camera that does not require an optical adapter with a digital interface USB, controlled by a computer and with the TWAIN support [3].

In offered study we develop a computer vision system for mineral rocks, considering the problems of development of a technique and image recognition technology to assess the qualitative composition of mineral rocks in particular.

2 Materials and Methods

2.1 Methods of Identification of Mineral Rock Images

Consider a sample of slag copper anode as an example (Fig. 1). Micrographs of this sample were kindly provided by the Eastern Research Institute of Mining and Metallurgy of Non-ferrous Metals (Kazakhstan, Ust-Kamenogorsk).

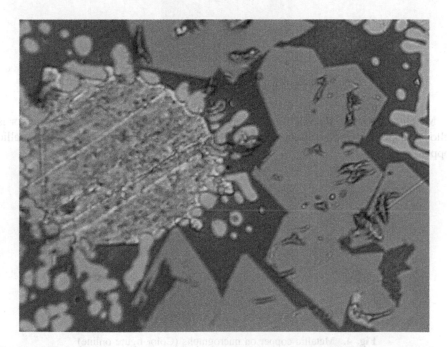

Fig. 1. Micrograph of a sample of slag copper anode, zoomed in 500 times. (Color figure online)

The minerals on micrographs may also be detected by color and shape [4]. Shape of cuprite Cu_2O is round, color is light green. Figure 2 shows the graphical representation of cuprite.

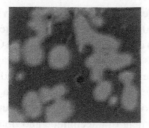

Fig. 2. Cuprite on micrographs (Color figure online)

Color of magnetite Fe_3O_4 on micrographs is dark green. Shape is angular, as expressed by technologists, "octahedral". Figure 3 shows magnetite apart from other minerals picture.

Fig. 3. Magnetite on micrographs (Color figure online)

Metallic copper on the micrographs can be found on the following criteria: color is yellow; shape is round, without flat faces. Figure 4 represents a micrograph metallic copper.

Fig. 4. Metallic copper on micrographs (Color figure online)

Silicate glass - is a dark green mass fills the rest of the space that is left of the other minerals.

This paper is described an algorithm for automatic segmentation of color images of various ore types, using the methods of morphological and cluster analysis. These data indicate that for real micrographs slag samples (and some other minerals) it is possible to use automated qualitative assessment of the mineral composition.

2.2   Methods of Cluster Analysis for Mineral Rocks Images

It is considered the problem of cluster analysis to segment micro-images in mineralogy. In this case, the cluster is uniform in color-luminance characteristics region (segment) if it is a digital image. And according to the specifics of digital images, multiple segments of mineral rocks might be in the same cluster at the same time, and research method determines homogeneity of individual clusters [5].

It is reasonable to use cluster analysis for the problem of segmentation due to two factors: there is only one tuning parameter k denoting the number of clusters that you want to highlight, and the sets of color-brightness characteristics associated with different types of segments on an analyzed image are compact [6].

The classic version of cluster analysis focused on a random selection of centroids is unacceptable for an adequate solution to the problem due to variations in the resulting picture segmentation, which, in turn, depends strongly on the order of submission of observations to the input of the algorithm. As follows from the results of the test image processing, segmentation of each picture is different for the obtained segments.

It is proposed that we develop methods for obtaining initial values of the centroids to solve the problems of inadequate segmentation, and the choice of a set of parameters that form the vector of observations, the most satisfying description of the characteristics of the shared segments.

In the course of the project we have developed methods for obtaining initial values of the centroids of the clusters.

The input parameters of a clustering algorithm are:

-   The number of clusters;
-   The initial values of the centroids of the clusters.

2.3   Methods of the Morphological Analysis of Mineral Shapes

Morphological analysis of objects has got a considerable value. It is necessary in the presence of objects with similar color-brightness characteristics. One of the main stages of the analysis of micrographs of mineral rocks is the analysis of the select objects' form, which aims to establish the differences between the selected objects.

Identification of the classification parameters is one of the primary tasks in pattern recognition [7].

It is offered the following description of the basic model of the object on the basis of morphological features (1):

$$M = <K, Y, C>,$$
(1)

where:

K - cortege of the rock metrics in general;
Y - cortege of the mineral metrics;
C - cortege of the mass metrics without minerals.

Consider in more detail the cortege of rock metrics in general (2):

$$K = <S, P, (Z, Ms, D_s) > , \tag{2}$$

where:

S - area of rock;
P - relation of the area of rock minerals to the area of rock without minerals;
Z - relation of total area of mineral grain to the area of rock;
Ms - average on the area for mineral grain;
D_S - dispersion on the area for mineral grain.

Consider cortege of the mineral metrics (3):

$$Y = <DOS, DOR, DOE > , \tag{3}$$

where:

DOS - the degree of sphericity of the mineral or of its segments;
DOR - the degree of roundness of the mineral or of its segments;
DOE - the degree of elongated thin shape of the mineral or of its segments.

Consider cortege of the mass metrics without minerals, the formula (4):

$$C = <H, S_c, V > , \tag{4}$$

where: H - tone, Sc - saturation; V - value.

3 Algorithms of the Morphological Analysis of Mineral Shapes

3.1 Morphological Algorithm for Objects with Undivided Circles

The basis of the algorithm is the consistent application of morphological operation "erosion" [7] drawn up by the mask of image segments. As a result, a sequence of "erosion" of the image is decomposed into its constituent objects - circumference. It is defined at each stage in the process of "erosion" by the number of objects that are able to separate from each other. When there is nothing left on the binary image, the process ends. The number of circles is defined as the greatest number of distinct objects that were recorded on successive stages of "erosion".

Counting the number of circles is done by the following algorithm:

1. We are provided a binary mask object.

2. Apply morphological operation "erosion" for the binary mask.
3. Count the number of separate circles. The resulting value is recorded in the computer memory.
4. If the number of segments in the image is greater than zero, then go to step 2.
5. Sequentially retrieve from computer memory and determine the greatest among them.

The coefficients of roundness Vadella is determined by the formula (5):

$$DOR = \frac{1}{N} \sum_{i=1}^{N} \frac{r_i}{R},$$

(5)

where:

N - the number of undivided circles;
r_i - radius of undivided circles;
R - radius of the circle entered in a contour.

Figure 5 shows an image of cuprite. You can see in the figure that the object is characterized by adjoining circles.

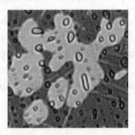

Fig. 5. The object of cuprite on a model micrograph

3.2 Morphological Algorithm for Objects of Round Shape

The algorithm is based on determining the degree of sphericity. Estimating the form of micro-objects numerically is possible using the metric of sphericity coefficient DOS.

The degree of sphericity of the mineral or of its segments is calculated by formula (6):

$$DOS = \frac{2\pi R}{P},$$

(6)

where:

R - radius of round shape;
P - perimeter of the mineral or of its segments.

Perfectly spherical object is characterized by the fact that the circle has the minimum perimeter, therefore, DOS = 1. In the case of winding form factor will decrease and tends to 0 as the object degenerates into a line.

Figure 6 shows an image of metallic copper, where objects have a rounded shape, without flat edges.

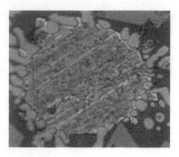

Fig. 6. Object metallic copper on the model micrograph

3.3   Morphological Algorithm for Objects with Elongated Thin Shape

The algorithm is based on the definition of the area object to select objects of small area. Selecting objects of small area is required not only for labelling but also to implement the screening, due to the fact that besides providing micrographs objects represent technological interest, and differentiate a variety of small area artifacts.

The degree of elongated thin shape of the mineral is calculated by formula (7):

$$DOE = \frac{b}{a}, b > a \tag{7}$$

Where:

b - length of a big axis for mineral grain;
a - length of a small axis for mineral grain.

Figure 7 shows an image of delafossite, whose objects have a characteristic needle-like shape.

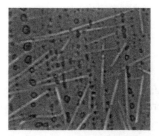

Fig. 7. Object Delafossite on the model micrograph

4   Results and Discussion

Nowadays developed automated image recognition system for assessing the qualitative composition of mineral rocks consists of 7 main subsystems [8]:

(1)   Research and getting micrograph rock.
(2)   Input and identification micrograph rock.
(3)   Pre-processing: improving the quality.
(4)   Definition of image reduction threshold [9].
(5)   Select the feature vector for cluster analysis.
(6)   Cluster analysis of color image to determine the mineralogical composition of rocks [10].
(7)   Morphological analysis of mineral shape to determine the mineralogical composition of rocks.

4.1   Algorithms of Cluster Analysis for Mineral Rocks Images

Segmentation method "k-means" is implemented through a two-step algorithm that minimizes the sum of distances "point-to-centroid" obtained by summing over all K clusters. Another words, the purpose of the algorithm is to minimize variability within clusters and maximize variability between clusters [10].

Algorithm starts with a randomly selected cluster centroid position, and then changes the ownership of points (objects) to clusters, i.e. point moves from one cluster to another in order to get the most significant result.

During the first phase on each iteration all points are rearranged so that they are positioned as close as possible to their centroids, and then converted coordinates centroids of each cluster. This part of the algorithm allows to find quickly, but only an approximately a solution to the problem of segmentation, which is the starting point for the second phase.

During the second stage of the algorithm points are individually subjected to rearrangement in case it reduces the sum of the distances, and the coordinates of the centroids clusters after rearrangement recalculated for each point. Each iteration during the second stage consists of only a single pass through all the points.

For automatic comparison it is used these similarity measures: a measure of the Hausdor measure [11], the function of Liu and Yang [12], the standard deviation and expert evaluation. Initially picture from the scanner enters in the RGB format. Then it might be possible to convert it into CMYK, YIQ, YUV, HSV and XYZ color spaces. To compare the original and segmented images need to submit the last in the form of a pseudo color image. Each region segmented image is painted the average color of the corresponding of the original image. The algorithm for which the value of the function less is best.

Results of testing for test image are presented in Table 1. Similar results were traced and for other test images.

It can be concluded that a feature vector that best satisfies the specifications describing the shared segment must be represented in a HSV color space of the

Table 1. Method of k-means of color spaces for the test image

Color spaces	The Hausdor measure	Function Liu and Yang	Standart deviation	Evaluation expert
CMYK	12.31	8.11	14.55	60 %
YIQ	15.31	10.28	18.06	55 %
YUV	18.39	16.36	20.10	50 %
HSV	1.17	6.12	8.09	85 %
RGB	2.36	3.14	13.25	70 %
XhsYhsZi	2.25	7.16	12.79	75 %

cylindrical passage in the Cartesian coordinate system. Accordingly, the image segment can be represented by a set of points in three-dimensional Cartesian coordinate system. To enhance the stability of the algorithm k, it is necessary to set the initial values of centroid clustering [9].

Once the area is selected it must be recorded in the cluster table. To do this, you must specify the following parameters:

- The name of the cluster;
- Fill color;
- Mineral.

Each cluster includes a certain number of points. Given the ratio of the number of points allocated in each cluster to a number of common points, it is possible to show relative rates of minerals in rock samples.

4.2 Algorithms of Combined Segmentation for Mineral Rocks Images

The search of an object is executed based on combined segmentation while accounting for object classification. For methodologies oriented at the stage of a preliminary search for an object on an image, which precedes classification, segmentation of the entire image using a several variable vector is not effective from the point of computational costs. We can raise the efficiency if we use methodologies based on preliminary search for an object. If we have an object on an image, the segmentation only carried out in the window of a preemptively fixed size that includes the found object; this process considerably improves the speed of automated analysis, by the means of exclusion of low-information parts of an image.

The method of search is founded on the assumption that every object of digitized image can be classified based on previously established finite brightness, shape (size, area). Because of that, to search for an object it's enough to consecutively segment image for every variable separately, using the algorithm of one-dimensional clustering (that doesn't require a lot of time to run) and analysing received resulting segment masks. Correspondingly, the probability of getting an object is higher the more features show up at once in analogous mask coordinates. The amount of clusters for every component and also the choice of specific mask from the aggregate obtained during segmentation are described individually, according to the current task, based on an experiment's results.

To detect joint appearance of several features it makes sense to use the logical operation of conjunction for elements of analysed masks. Result of such an operation would be the mask comprising several local zones corresponding to the sought object, the lack of these will mean the lack of objects of interest on an image.

Because to get the masks we use an efficient algorithm, the processing time is severely reduced. It is also of note that the algorithm is also efficient for the task of solving the task of calculating the formal elements of an image without their classification.

The result of the segmentation is shown in Fig. 8. Various minerals marked in different colors. In this case, the metallic copper is red, magnetite is blue cuprite is orange.

Fig. 8. The result of segmentation (Color figure online)

Considered sample has the following content of useful elements:

- Magnetite - 28.45 %;
- Metallic copper - 18.45 %;
- Cuprite - 7.92 %.

5 Conclusion

In this work we developed the methods and algorithms of computer vision for the purposes of image recognition of mineral spices. It is possible to classify the object sample obtained in the step of scanning the image of mineral rock sample by the algorithm that uses selective classification. Selection process here is the preliminary separation of objects into two groups of samples. This approach is explained by the fact that the presence of objects with similar color-brightness characteristics, but different shapes is possible, and there are objects which have similar color-brightness characteristics. Preliminary determination of group membership allows reducing the computational complexity of classification. The color of the object is determined at the stage of segmentation while sorting by group. The effectiveness of the proposed method lies in the automatic identification of objects of interest on the entirety of the image and in the possibility of tuning the parameters of the algorithm simply by a single value that indicates the amount allocated to the segments. The example of a bundled software, written in C# programming language, was developed for checking the results of the research.

References

1. Chris, P.: Smithsonian Handbooks: Rocks and Minerals. Dorling Kindersley, New York (2002)
2. ISO 25706-83: Interstate standard. Magnifying glasses, types, key parameters. General technical requirements. Date of Introduction 1984-01-01, (1984). http://docs.cntd.ru/document/gost-25706-83
3. Clarke, A.R., Eberhardt, C.N.: Microscopy Techniques for Materials. Science Woodhead Publishing, CRC Press, Cambridge, Boca Raton (2002)
4. Farndon, J.: The Practical Encyclopedia of Rocks and Minerals: How to Find, Identify, Collect and Maintain the World's Best Specimens, with over 1000 Photographs and Artworks. Lorenz Books, London (2006)
5. Mandel, J.: Cluster Analysis, p. 176. Finance and statistics, Moscow (1988)
6. Odell, P.L., Duran, B.S.: Cluster Analysis - A Survey. Springer, Heidelberg (1974)
7. Gonsales, R.C., Woods, R.E.: Digital Image Processing, 3rd edn, p. 976. Pearson Education, Upper Saddle River (2011)
8. Baklanova, O.E., Shvets, O.Y., Uzdenbaev, Z.: Automation system development for micrograph recognition for mineral ore composition evaluation in mining industry. In: Iliadis, L. (ed.) AIAI 2014. IFIP AICT, vol. 436, pp. 604–613. Springer, Heidelberg (2014)
9. Baklanova, O.E., Shvets, O.Y.: Development of methods and algorithms of reduction for image recognition to assess the quality of the mineral species in the mining industry. In: Chmielewski, L.J., Kozera, R., Shin, B.-S., Wojciechowski, K. (eds.) ICCVG 2014. LNCS, vol. 8671, pp. 75–83. Springer, Heidelberg (2014)
10. Baklanova, O.E., Shvets, O.Y.: Cluster analysis methods for recognition of mineral rocks in the mining industry. In: 2014 4th International Conference on Image Processing Theory, Tools and Applications (IPTA), 14–17 October 2014, pp. 273–277 (2015). doi:10.1109/IPTA.2014.7001972
11. Hausdorff, F.: Dimension und äusseres Mass. Math. Ann. **79**(1–2), 157–179 (1918). doi:10.1007/BF01457179
12. Liu, J., Yang, Y.-H.: Multiresolution color image segmentation. IEEE Trans. Image Process. **16**, 689–700 (1994)

A Similarity-Based Approach for Shape Classification Using Region Decomposition

Wahyono, Laksono Kurnianggoro, Yu Yang, and Kang-Hyun Jo[(⊠)]

Intelligent Systems Laboratory, Graduate School of Electrical Engineering,
University of Ulsan, Ulsan 680-749, Korea
{wahyono,laksono,yuyang}@islan.ulsan.ac.kr,
acejo@ulsan.ac.kr

Abstract. Measuring the similarity of two shapes is an important task in human vision systems in order to either recognize or classify the objects. For obtaining reliable results, a high discriminative shape descriptor should be extracted by considering both global and local information of the shape. Taking into account, this work introduces a centroid-based tree-structured (CENTREES) shape descriptor invariant to rotation and scale. Extracting the CENTREES descriptor is started by computing the central of mass of a binary shape, assigned as the root node of tree. The entire shape is then decomposed into b sub-shapes by voting each pixel point according to an angle between point and major principal axis relative to a centroid. In the same way, the central of mass of the sub-shapes are calculated and these locations are considered as level-1 nodes. These processes are repeated for a predetermined number of levels. For each node corresponding to sub-shapes, parameters invariant to translation, rotation and scale are extracted. A vector of all parameters is considered as descriptor. A feature-based template matching with X^2 distance function is used to measure shape dissimilarity. The evaluation of our descriptor is conducted using MPEG-7 dataset. The results justify that the CENTREES is one of reliable shape descriptors for shape similarity.

Keywords: Shape descriptor · Region decomposition · Centroid · Shape matching

1 Introduction

Shape classification plays an important part in a visual object recognition. To do such a task, technically, the various shapes should be known previously considered as templates. Given an unknown shape, our goal is to classify this shape into one of the shape template classes by computing the similarity value of their descriptors, namely shape matching. The similarity computation can be done using some similarity measurements [1] such as Manhattan, Euclidean, X2, and other distance metrics.

The key issue of a shape matching problem is to find an effective shape representation(i.e. shape descriptor) [2]. There are mainly two categories of shape descriptors: contour-based [3–5] and region-based methods [6–10]. These two kinds of method can also be divided into full shape descriptor (e.g. global features) [6–8],

© Springer International Publishing Switzerland 2016
D.-S. Huang and K.-H. Jo (Eds.): ICIC 2016, Part II, LNCS 9772, pp. 279–289, 2016.
DOI: 10.1007/978-3-319-42294-7_24

part-based shape representation (e.g. local features) [9, 11], and combining global and local information [12, 13].

Strategies by decomposing a shape into several sub-regions and extracting local properties were suggested in [9, 11]. Zhang [11] decomposed a shape contour under multiple scales based on a visual perception. For obtaining a contour, they applied morphological operations. Each segment of contour is described by a shape context descriptor. Author [9], rather than extracting features from an edge shape that unsolved occlusion problem, the authors used all points of the shape. First, the shape was decomposed into four sub-shapes based on major and minor principal axes corresponding to its eigenvectors at the centroid of shape. Each sub-shape is subdecomposed into four sub-sub-shapes in the same way. The parameters then are calculated from the all sub-shapes which will represent the final shape descriptor. These parameters are invariant to translation, rotation and scale. However, a high sensitivity of eigenvector direction may affect the matching result when the pixels of a shape are circularly distributed. Thus, a new shape descriptor which utilized the advantages of parameters in [9] namely centroid-based tree-structured (CENTREES) is introduced for shape matching. The CENTREES expands the number of regions and level of sub-shapes in order to reduce this sensitivity impact. The CENTREES is an effective, simple and invariant to rotation, translation and scale. The CENTREES is constructed based on the centroid of a shape and its sub-shapes. The details of this process will be discussed in Sect. 2. To prove the robustness of our descriptor, some experiments are conducted using well-known public dataset in shape matching such as MPEG-7 [14]. The experimental results are then compared with other descriptors. The summary of our experimental results will be presented in Sect. 3. Section 4 will conclude our works.

2   The Proposed Descriptor

This section describes the detail of proposed shape descriptor. Firstly, tree-structured sub-regions decomposition are constructed based on its centroid. In all sub-regions, several parameters which are scale and rotation invariance are calculated. The final descriptor is obtained by combining parameters from all sub-regions. Lastly, the matching cost as well as a matching process for determining the similarity between the shapes is defined.

2.1   Region Decomposition

Shape decomposition procedure splits a given shape into several sub-shapes on the basis of the centroid value. The decomposition process is performed recursively according to division level L and the number of bins, b. Each sub-shape is then represented as node in order to form tree structured as shown on the Fig. 1. For a given shape, the locations of pixel belonging to the shape are regarded as observed vectors of 2-D random vector S. More formally, let define a given shape as S = {p1, p2,..., pN}, where pi = (xi, yi) represents the location of a pixel i and N be the total number of foreground pixels belonging to the shape. The shape decomposition and tree generation are implemented by the following procedure:

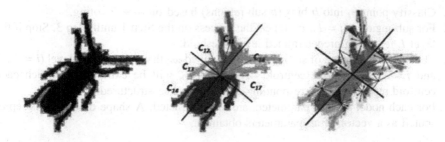

(a) original shape   (b) The first level decomposition (c) The second level

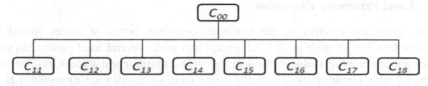

(d) Tree representation of (b)

Fig. 1. Illustration of region decomposition for b equal to 8. The centroid location of each sub-shape is denoted with text and black dot. Bold lines represent the minor and major principal axes. Different color represents different node. A shape is taken from MPEG-7 database. (Color figure online)

1. Calculate the centroid and orientation of S, denoted as C and θ, by:

$$C(\bar{x}, \bar{y}) = \frac{1}{N} \sum_{i=1}^{N} p_i \tag{1}$$

$$\theta = \frac{1}{2} \arctan\left(\frac{2\mu_{11}}{\mu_{20} - \mu_{02}}\right) \tag{2}$$

where

$$\mu_{qr} = \frac{1}{N} \left(\sum_{i=1}^{N} \sum_{j=1}^{N} (x_i - \bar{x})^q (y_i - \bar{y})^r \right) \tag{3}$$

$\bar{x}$ and $\bar{y}$ are centroid in x and y axis respectively, and μ_{qr} is second order central moment. The orientation is needed to obtain rotation invariant.

2. For each location p_i, compute the unsigned angle, α_i, by following formula:

$$\alpha_i = \arctan\left(\frac{x_i - \bar{x}}{y_i - \bar{y}}\right) \tag{4}$$

3. Classify point p_i into b bins (b sub-regions) based on $\alpha_i + \theta$ voting.
4. For sub-region j ($j = 1...b$), repeat the process on the Step 1 until Step 3. Stop if the level L reaches a predetermined level threshold.
5. Thus, the Lb centroid of sub-regions, C_{ij} defined as jth centroid in ith level ($i = 1...L$ and $j = 1...b$) and one centroid of shape, C_{00}, will be extracted in which each centroid represents a corresponding node in the tree-structured.
6. For each node, several parameters are then extracted. A shape descriptor is represented as a vector of all parameters obtained.

2.2   Local Parameter Extraction

Before extracting parameters, the minimum bounding boxes of nodes should be obtained first. For jth node at ith level, except root node, several local parameters such as normal-angle, elongatedness, distance, occupation-ratio, compactness, eccentricity, perimeter-ratio, connectivity, circularity}, and inertia-axis-ratio are extracted. Let us denote P(i, j) be the parent node of node j at level i. These parameters are calculated as following schema:

- **normal-angle(i,j):** the unsigned angle between $C_{ij}C_{P(i,j)}$ and major principal axes corresponding to parent region of node j, where $C_{ij}C_{P(i,j)}$ is vector formed between C_{ij}, the centroid of the node j, and $C_{P(i,j)}$, the centroid of the parent node of the node j. Afterwards, this value should be normalized on the range between 0 and 1.
- **elongatedness(i,j):** the ratio between H_{ij} and W_{ij}, where H_{ij} and W_{ij} are denoted as height and width of minimum bounding boxes of node j at level i, respectively. It is formulated by:

$$E_{ij} = \min\left(\frac{H_{ij}}{W_{ij}}, \frac{W_{ij}}{H_{ij}}\right) \tag{5}$$

- **distance(i,j):** the distance between C_{ij} and its parent's centroid location $C_{P(i,j)}$. In order to make this value be scale invariant, it is divided by M_{i-1}, the maximum length between height and width of minimum bounding box of parent region. Hence, it can be expressed as

$$d_{ij} = \frac{|C_{ij} - C_{P(i,j)}|}{\max\left(H_{P(i,j)}, W_{P(i,j)}\right)} \tag{6}$$

- **occupation-ratio(i,j):** the ratio of R_{ij}, the area of the region corresponding to node j at level i, to B_{ij}, the area of the bounding box corresponding to node j at level i. It can be expressed as following formula

$$O_{ij} = \frac{R_{ij}}{B_{ij}} = \frac{R_{ij}}{H_{ij}W_{ij}} \tag{7}$$

- **compactness(i,j):** the ratio of R_{ij}, the area of the region corresponding to node j at level i, to $R_{P(i,j)}$, the area of the associated region of its parent. Compactness is independent of linear transformations. It can be expressed as follow formula

$$T_{ij} = \frac{R_{ij}}{R_{P(i,j)}} \tag{8}$$

- **perimeter-ratio(i,j):** the ratio between ϕ_{ij}, the perimeter of the bounding box corresponding to node j at level i, and φ_{ij}, the perimeter of the region corresponding to node j at level i. This parameter is useful for handling articulation of outer sub-shape. It can be expressed as in:

$$\kappa_{ij} = \min\left(\frac{\varphi_{ij}}{\phi_{ij}}, \frac{\phi_{ij}}{\varphi_{ij}}\right) \tag{9}$$

- **eccentricity(i,j):** the ratio of major and minor principal axes of the region corresponding to node j at level i. The major and minor principal axes are defined as the major and minor eigenvalues of the region, $\lambda_{b,i}$ and $\lambda_{s,i}$. For obtaining the eigenvalues, first, the covariance matrix of region, Σ_{ij}, is computed as

$$\Sigma_{ij} = \frac{1}{M} \sum_{k=1}^{M} \left(p_k p_k^T - C_{ij} C_{ij}^T\right) \tag{10}$$

where p_k is the location of pixel, and M is the number of pixel on the region. Then, eigenvectors v_k and its corresponding eigenvalues λ_k for $k = 1, 2$ are obtained by solving $\Sigma_{ij} v_k = \lambda_k v_k$. Thus the eccentricity can be express as ratio between small and big eigenvalues, formulated as follow

$$\varepsilon_{ij} = \frac{\lambda_{s,i}}{\lambda_{b,i}} \tag{11}$$

- **circularity-ratio(i,j):** the ratio of A_{ij}, the region area corresponding to node j at level i, to Λ_{ij}, the area of circle where circle has the same perimeter φ_{ij}. It is formulated as

$$r_{ij} = \frac{A_{ij}}{\Lambda_{ij}} \tag{12}$$

Using formula $\Lambda_{ij} = \varphi_{ij}^2/4\pi$, and since 4π is a constant, *circularity-ratio* can be simplified as

$$r_{ij} = \frac{A_{ij}}{\phi_{ij}^2} \tag{13}$$

- **inertia-axis-ratio(i,j):** the ratio between I_{ij}, the axis length of least inertia axis of the region corresponding to node j at level i, and $I_{P(i,j)}$, the axis length of least inertia of the associated region of its parent. Axis of least inertia is defined as unique reference line preserving orientation of shape that passes through centroid.
- **connectivity(i,j):** the ratio of the number of outer pixels that are neighboring with foreground pixel of shape to the region perimeter. The connectivity describes the spatial location of sub-shape related to shapes.

2.3   Invariant Analysis of Parameters

An invariant is a property which remains unchanged when transformations of a certain type are performed to the objects. A robust shape descriptor is required to be transformation invariances in rotation, translation and scale. Thus, this part discusses how these invariants can be achieved from CENTREES.

- The translation invariant can be achieved by using an absolute reference with a fixed location [3]. Our parameters are calculated with respect to a fixed location on the region (i.e. centroid of shape), so it is guaranteed that CENTREES is translation invariant.
- The simplest way to achieve rotation invariant is by calculating all angles of shape with respect to its principal axes. Most CENTREES parameters were computed relative to the shape major principal axis, so our descriptor meets the requirement of rotation invariant. Most extracted parameters are invariant to rotation except distance and connectivity. Connectivity might be sensitive when original shape is slightly rotated and the sub-shape location is not changed.
- CENTREES is scale invariant due to it calculated the normalized parameter distance by dividing with the length of major principal axis. Most extracted parameters are invariant to scale except normal angle and compactness. The bigger portion of sub-shape due to occlusion or irregular structure might obtain different value of normal angle and compactness.

2.4   Similarity Cost Function

In many natural histograms the difference between large bins is less important than the difference between small bins and should be reduced. In shape matching, this property may handle the occlusion problem. Hereafter, in determining the similarity between two shapes, descriptor-based template matching with Chi-Squared (X2) is utilized as matching cost. Chi-Squared (X2) is a histogram distance that takes this into account. The X2 histogram distance comes from the X2 test-statistic where it is used to test the fit between a distribution and observed frequencies.

Defining HS and HSi are the CENTREES descriptor of an unknown shape and ith shapes on the templates. The matching cost between these two shapes, Di is calculated using X2 distance formulated as:

$$D_i = \frac{1}{2} \sum_{k=1}^{M} \left(\frac{[H_s(k) - H_{si}(k)]^2}{H_s(k) + H_{si}(k) + \gamma} \right) \tag{14}$$

where HS(k) and HSi(k) denote the kth element of the CENTREES descriptor of an unknown shape and ith shapes on the templates, respectively, γ is a very small real number to avoid zero division and M represents the vector size. Our goal is to minimize the matching cost for classifying the unknown shape into certain class on the templates. Hence, the unknown shape can be considered to be class i, if the matching cost value Di is minimal among others, formulated as

$$Class(S) = \min_i(D_1, D_2, \ldots, D_N) \tag{15}$$

where N is the number of templates.

3   Experiment

This section describes the experiment detail of our proposed descriptor. The parameters b, the number of region decomposition, and l, the number of decomposition level, were set to be varying (b = {2,4,8,16} and l = {1,2,3}). Note that S(b,l) is denoted as the CENTREES descriptor with b sub-shapes and l level decomposition.

The experiment is conducted using MPEG-7 CE Shape-1 Part-B [14]. It contains of 1400 shape samples, 70 categories, 20 for each category as shown in Fig. 2. The testing set was divided into two cases: (1) including all samples for matching; (2) following [15, 16] which only evaluated their descriptor on a subset of dataset in evaluation experiments. This subset consists of seven distinctive classes: Bone, Hearth, Glass, Fountain, Key, Fork, and Hammer, where each class includes 12 samples. As seen in Fig. 3, the shape classes are very distinct, but the data set shows substantial within-class variations [15]. Our descriptor was compared with several method, such as seven invariant Hu moments (HM), Zernike moments (ZM), and method from [15]. In addition, since some shapes appear flipped and rotated, a modified distance function was applied for computing matching cost. The distance dist(A, B) between a reference shape B on the template database and a query shape A is defined as

$$dist(A, B) = \min\{D(A, B^u), D(A, B^v), D(A, B^h)\} \tag{16}$$

where Bu, Bv, and Bh denote three versions of B: unchanged, vertical flipped, and horizontal flipped. The evaluation protocol is conducted into two protocols:(1) Nearest Neighbor Score. Query shape is assigned as the same class of the nearest similarity from reference shapes; (2) Bull's Eye Score. The retrieval rate is measured by the so-called bull's eye score. Every shape in the database is compared to all other shapes,

Fig. 2. Some of the shapes in the MPEG-7 dataset. One image per class for the first 40 classes (the database has 70 classes).

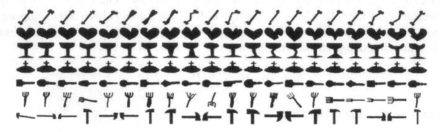

Fig. 3. Subset of MPEG-7 CE Shape-1 Part-B shape which was used in [3, 12].

and the number of shapes from the same class among the 40 most similar shapes is reported. The bull's eye retrieval rate is the ratio of the total number of shapes from the same class to the highest possible number (which is 20 × 1400). Thus, the best possible rate is 100 %.

Figure 4 summaries the comparison result between the proposed descriptor and other methods in term of NN score for 1st until 10th closest matches. As shown in Fig. 4, classification rate of the CENTREES with 4 sub-shapes for the 1st nearest neighbor (NN) score is higher than [9] and [7]. The number of sub-shapes is increased to be 8. It obtains a significant improvement of NN score as much as 6.21 %. The CENTREES descriptor with 8 sub-shapes and 2 level divisions achieves the

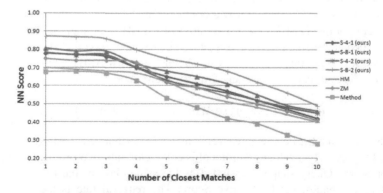

Fig. 4. Result for MPEG-7 CE Shape-1 Part-B Shapes using NN scores and comparison with HM [17], ZM [1], and method from [12].

highest accuracy of 91.86 % comparing with others. Hence, it is observed that increasing the number of sub-shapes and decomposition levels obtains a surprisingly improvement in accuracy. A large number of sub-shapes can extract more local information and spatial configuration of shape.

Table 1 depicts the classification rate of the proposed descriptor using bull's eye score for all 70 classes. If we set the number of sub-shapes equal to 4 and level equal to 2, the CENTREES gains accuracy around 78.75 % that outperforms [3, 20]. In the other hand, if we set the number of sub-shapes equal to 8 with the same level, the CENTREES classification rate is about 87.07 % which outperforms [3, 4, 19, 20], with rate value 85.50 %, 79.92 %, 78.38 %, and 76.51 %, respectively. Yet, the best performance was achieved by Bai et al. [18], with a retrieval rate of 91.61 %.

Table 1. Results for MPEG-7 CE Shape-1 Part-B.

Method	Bulls eye score
Belongie [3]	0.765
Ling [4]	0.855
Bai [18]	0.916
Xie [19]	0.799
Grigorescu [20]	0.783
S-4-2 (Ours)	0.787
Wang [21]	0.903
S-8-2 (Ours)	0.870

Lastly, CENTREES is then evaluated by a subset of MPEG-7 which was used by [15, 16]. Table 2 shows the comparison of classification rate between the CENTREES and other descriptors. Thakoor et al. [15] achieved the highest and the worst accuracy at 97.62 % and 80 %. Bicego et al. [16] reported that the best and the worst accuracy in their experiment are 98.8 % and 92.9 %, which are better than [15]. Meanwhile, in our experiment, CENTREES achieves the best rate at 99.52 % for b = 8 and L = 2, while the lowest accuracy was obtained at 90 % for b = 4 and L = 1.

Table 2. Results in percentage for subset of MPEG-7 CE Shape-1 Part-B Shapes.

Method	CENTREES				HM	ZM	HMM
Class	4–1	8–1	4–2	8–1	[17]	[1]	[12]
Bone	95	95	100	100	85	100	83.3
Hearth	95	100	100	100	90	85	100
Glass	100	100	100	100	100	100	100
Fountain	100	100	100	100	95	100	100
Key	90	85	95	100	85	100	100
Fork	75	80	85	100	65	60	100
Hammer	75	75	80	96.67	100	100	100
Overall	90	90.71	94.29	**99.52**	88.57	94.29	97.6

4 Conclusion

A new descriptor for determining shape similarity, namely centroid-based tree-structured (CENTREES) has been implemented successfully. The CENTREES descriptor combined global and local information of shape by using decomposition approach and extracted high discriminated parameters that are independently to the translation, scale and rotation. The decomposition process was conducted by voting an angle between foreground point and major principal axis respect to centroid position. These processes are repeated for a certain level for forming tree of sub-shapes. The chi-square (X^2) distance function is employed for measuring similarity between the shapes. Based on experimental results, it is observed that if the number of sub-shapes and division levels are getting larger, the classification rate improves significantly. Overall results prove that the CENTREES descriptor could be one of the promising shape descriptors for matching and classifying 2-D shapes.

References

1. Veltkamp, R.: Shape matching: similarity measures and algorithms. In: ICSMA, pp. 188–197 (2001)
2. Bai, X., Rao, C., Wang, X.: Shape vocabulary: a robust and efficient shape representation for shape matching. IEEE Trans. Image Process. **23**(9), 3935–3949 (2014)
3. Belongie, S., Malik, J., Puzicha, J.: Shape matching and object recognition using shape context. IEEE Trans. Pattern Anal. Mach. Intell. **24**(4), 509–522 (2002)
4. Ling, H., Jacobs, D.: Shape classification using the inner-distance. IEEE Trans. Pattern Anal. Mach. Intell. **29**(2), 286–299 (2007)
5. Bai, X., Liu, W., Tu, Z.: Integrating contour and skeleton for shape classification. In: ICCV (2009)
6. Sharvit, D., Chan, J., Hüseyin, T., Kimia, B.: Symmetry-based indexing of image databases. In: Workshop on Content-Based Access of Image and Video Libraries, pp. 56–62 (1998)
7. Hu, M.: Visual pattern recognition by moment invariants. IRE Trans. Inf. Theor. **8**, 179–197 (1962)
8. Kim, Y., Kim, W.: Content-based trademark retrieval system using a visually salient feature. Image Vis. Comput. **16**, 931–939 (1998)
9. Kim, H.K., Kim, J.D.: Region-based shape descriptor invariant to rotation, scale and translation. Signal Process. Image Commun. **16**, 87–93 (2000)
10. Mai, F., Chang, C., Hung, Y.: A subspace approach for matching 2D shapes under affine distortions. Pattern Recogn. **44**(2), 210–221 (2011)
11. Zhang, J., Mu, Z., Ren, Y., Xu, G., Jiang, C.: Shape matching and retrieval via contour multi-scale decomposition. In: ROBIO (2013)
12. Jain, A., Vailaya, A.: Shape-based retrieval: a case study with trademark image databases. Pattern Recogn. **31**(9), 1369–1390 (1998)
13. Wang, X., Feng, B., Bai, X., Liu, W., Latecki, L.: Bag of contour fragments for robust shape classification. Pattern Recogn. **47**(6), 2116–2125 (2014)
14. Jeannin, S., Bober, M.: Description of core experiments for MPEG-7 motions/shape. Technical report, MPEG-7 Seoul (1999)

15. Thakoor, N., Gao, J., Jung, S.: Hidden Markov model-based weighted likelihood discriminant for 2-D shape classification. IEEE Trans. Image Process. **16**(11), 2707–2719 (2007)
16. Bicego, M., Murino, V.: Investigating hidden Markov models capabilities in 2-D shape classification. IEEE Trans. Pattern Anal. Mach. Intell. **26**(2), 281–286 (2004)
17. Toshev, A., Taskar, B., Daniilidis, K.: Shape-based object detection via boundary structure segmentation. Int. J. Comput. Vis. **99**(2), 123–146 (2012)
18. Bai, X., Yang, X., Latecki, L.J., Liu, W., Tu, Z.: Learning context sensitive shape similarity by graph transduction. IEEE Trans. Pattern Anal. Mach. Intell. **32**(5), 861–874 (2009)
19. Xie, J., Heng, P., Shah, M.: Shape matching and modeling using skeletal context. Pattern Recogn. **41**(5), 1756–1767 (2008)
20. Grigorescu, C., Petkov, N.: Distance sets for shape filters and shape recognition. IEEE Trans. Image Process. **12**(7), 729–739 (2003)
21. Wang, J., Bai, X., You, X., Liu, W., Latecki, L.: Shape matching and classification using height functions. Pattern Recogn. Lett. **33**, 134–143 (2012)

Comparison of Non-negative Matrix Factorization Methods for Clustering Genomic Data

Mi-Xiao Hou[1], Ying-Lian Gao[2], Jin-Xing Liu[1,3(✉)],
Jun-Liang Shang[1], and Chun-Hou Zheng[1]

[1] School of Information Science and Engineering,
Qufu Normal University, Rizhao 276826, China
{mixiaohou, shangjunliang110}@163.com,
{sdcavell, zhengch99}@126.com
[2] Library of Qufu Normal University, Qufu Normal University,
Rizhao 276826, China
yinliangao@126.com
[3] Shenzhen Graduate School, Bio-Computing Research Center,
Harbin Institute of Technology, Shenzhen 518055, China

Abstract. Non-negative matrix factorization (NMF) is a useful method of data dimensionality reduction and has been widely used in many fields, such as pattern recognition and data mining. Compared with other traditional methods, it has unique advantages. And more and more improved NMF methods have been provided in recent years and all of these methods have merits and demerits when used in different applications. Clustering based on NMF methods is a common way to reflect the properties of methods. While there are no special comparisons of clustering experiments based on NMF methods on genomic data. In this paper, we analyze the characteristics of basic NMF and its classical variant methods. Moreover, we show the clustering results based on the coefficient matrix decomposed by NMF methods on the genomic datasets. We also compare the clustering accuracies and the cost of time of these methods.

Keywords: Non-negative matrix factorization · Clustering · Genomic data · Dimensionality reduction

1 Introduction

With human's entering the era of big data, massive and high-dimensional data seem to be generated continuously. It is a challenge to reduce the dimensionality of high-dimensional data to achieve the purpose of storing, processing and reconstructing in machine learning and data mining. There are numerous traditional methods to reduce the dimensionality of data, such as Principal Component Analysis (PCA) and Linear Discriminant Analysis (LDA). These methods allow for the existence of negative, which is not applicable in some cases. And they adopted linear dimensionality reduction technology that is not conducive to retaining characteristics of data. As a novel matrix factorization method, NMF overcomes many problems of the traditional

© Springer International Publishing Switzerland 2016
D.-S. Huang and K.-H. Jo (Eds.): ICIC 2016, Part II, LNCS 9772, pp. 290–299, 2016.
DOI: 10.1007/978-3-319-42294-7_25

matrix factorization method and provides a deeper view of the data. NMF can obtain two non-negative matrices to approximate the original data matrix, which reflects the concept of part-based representation in human thought.

NMF method can get the local expression of high-dimensional data by dimensionality reduction. It has been successfully used in bioinformatics, such as genome sequence feature recognition, local feature recognition, biological literature mining. In recent years, many scholars utilized NMF methods to do clustering experiments, such as document clustering, image clustering, tumor clustering. But there are no clearly comparisons of clustering experiments on genomic data, especially on The Cancer Genome Atlas (TCGA) datasets.

The rest of this article is organized as follows: the Sect. 2 introduces a variety of methods of NMF; there are the experiments based on the various NMF methods in Sect. 3; and the last section provides the conclusion of this paper.

2 Methods

In this section, we will introduce the fundamental NMF method and other typical improved NMF methods. And there are three kinds of improved NMF methods: (1) NMF method with manifold learning; (2) NMF method imposed sparse constraints; (3) NMF method with orthogonal constraint.

2.1 Fundamental NMF Method

For basic NMF, it is defined as follows: given a non-negative matrix $\mathbf{X} \in R^{m \times n}$, it is easy to find two non-negative matrices $\mathbf{U} \in R^{m \times k}$ and $\mathbf{V} \in R^{n \times k} (k < \min\{m, n\})$ to approximate $\mathbf{X}$:

$$\mathbf{X} \approx \mathbf{U}\mathbf{V}^{\mathrm{T}}. \tag{1}$$

There are two classical object Eq. [1]. The first one is based on Euclidean distance of two matrices:

$$O_1 = \left\| \mathbf{X} - \mathbf{U}\mathbf{V}^{\mathrm{T}} \right\|^2 = \sum_{i,j} \left(x_{ij} - \sum_{k=1}^{K} u_{ik} v_{jk} \right)^2. \tag{2}$$

The second one is based on Kullback-Leibler (KL) divergence of two matrices:

$$O_2 = D\left(\mathbf{X} \| \mathbf{U}\mathbf{V}^{\mathrm{T}}\right) = \sum_{i,j} \left(x_{ij} \log \frac{x_{ij}}{y_{ij}} - x_{ij} + y_{ij} \right) \tag{3}$$

where $\left[y_{ij} \right] = \mathbf{Y} = \mathbf{U}\mathbf{V}^{\mathrm{T}}$. The purpose of NMF method is minimizing the results of above formulas.

2.2 NMF Method with Manifold Learning

NMF with manifold learning takes the internal spatial structure of the data into account and performs a satisfactory effect compared with many improved NMF methods. Graph Regularized Non-negative Matrix Factorization (GNMF) [2] is a sophisticated method. Assumed that the sample distribution on the low-dimensional manifold embedded high-dimensional Euclidean space based on spectral graph [3] and manifold learning theory [4]. If two data points x_i and x_j are close in the high-dimensional space, then z_i and z_j are also assumed close in low-dimensional manifold according to Locally Linear Embedding (LLE). Considering the neighbor geometry structure of samples, for every point x_i, we can find k nearest neighbors of it and assign the edges to them. Then we should define the weight matrix $\mathbf{W}$. The value of $\mathbf{W}$ usually has three definition methods and different measures can be applied to different situations. Here we choose the most simple and intuitive way: 0–1 weight to represent a measure of the distance between the two points. And it is easy to redefine a new basis represent x_i on low-dimensional space, $\mathbf{Z} = [\mathbf{z_1}, \mathbf{z_2} \cdots \cdots, \mathbf{z_N}]^T$, inside $\mathbf{z_i} = [v_{i1}, v_{i2}, \cdots, v_{ik}]^T$, so European distance is expressed as follows:

$$D(\mathbf{z_i}, \mathbf{z_j}) = \left\| \mathbf{z_i} - \mathbf{z_j} \right\|^2. \tag{4}$$

Combined with the weight matrix, the preliminary target formula can be written:

$$R = \frac{1}{2}\sum\nolimits_{i,j=1}^{N} \left\| \mathbf{z_i} - \mathbf{z_j} \right\|^2 w_{ij} = \sum\nolimits_{i=1}^{N} \mathbf{z_i^T z_i} \mathbf{D}_{ii} - \sum\nolimits_{i,j=1}^{N} \mathbf{z_i^T z_j} w_{ij}$$
$$= Tr(\mathbf{V^T D V}) - Tr(\mathbf{V^T W V}) = Tr(\mathbf{V^T L V}) \tag{5}$$

where $\mathbf{D}_{ii} = \sum_j w_{ij}$ and $\mathbf{L} = \mathbf{D} - \mathbf{W}$, $\mathbf{L}$ is graph Laplacian [3]. The final objective function is defined as follows:

$$O = \left\| \mathbf{X} - \mathbf{UV^T} \right\|^2 + \lambda Tr(\mathbf{V^T L V}) \tag{6}$$

where parameter $\lambda \geq 0$ controls the smoothness of the function. Similar in principle, Manifold Regularized Discriminative Non-negative Matrix Factorization (MD-NMF) [5], the objective function of KL divergence-based algorithm is defined as follows:

$$O = D(\mathbf{X} \| \mathbf{UV^T}) + \frac{\alpha}{2}Tr(\mathbf{UeU^T}) + \frac{\beta}{2}Tr(\mathbf{V^T V})$$
$$+ \frac{\gamma}{2}Tr(\mathbf{V^T}(\mathbf{L_c^{-\frac{1}{2}}})^T \mathbf{L_g} \mathbf{L_c^{-\frac{1}{2}}}\mathbf{V}) \tag{7}$$

where $\alpha \geq 0$, $\beta \geq 0$, $\gamma \geq 0$. $\mathbf{e} = \overline{\mathbf{1}} - \mathbf{I}$, where $\overline{\mathbf{1}}$ is the matrix whose elements are all one and $\mathbf{I}$ is identity matrix with appropriate dimensionality. $\mathbf{L_g}$ indicates the graph Laplacian matrix of samples that have same labels as v_i. $\mathbf{L_c}$ indicates the graph Laplacian matrix of samples that have labels different from v_i. MD-NMF obtains two Laplacian matrices for the sample label and its role has been highlighted.

Graph Regularized Discriminative Non-negative Matrix Factorization (GDNMF) [6] directly joined the label information, which is different from GNMF and MD-NMF. Its objective function can be written as:

$$O = \left\| \mathbf{X} - \mathbf{U}\mathbf{V}^{\mathbf{T}} \right\|^2 + \lambda Tr\left(\mathbf{V}^{\mathbf{T}}\mathbf{L}\mathbf{V}\right) + \gamma \left\| \mathbf{S} - \mathbf{A}\mathbf{V}^{\mathbf{T}} \right\|^2, \tag{8}$$

where λ and γ are nonnegative parameters and $\mathbf{S}$ is a class indicator matrix:

$$S_{i,j} = \begin{cases} 1, & if \ \ d_j{=}i, \\ 0, & othwewise \end{cases} \quad j{=}1,2,...,N \quad i{=}1,2,...,c$$

where d_j represents j-th sample and $d_j \in \{1, 2, \ldots, c\}$, c is the total number of classes. $\mathbf{A}$ is also a non-negative matrix and initialized randomly. After adding the label information, GDNMF method introduced a supervised learning process. The effects of classification or clustering should be better than other methods in theory.

2.3   NMF Method Imposed Sparse Constraints

NMF method with sparse constraints is a common method compared with others. A nature of sparsity is nothing more than to allow the data to be processed to appear more 0 elements. There are many sparse ways to impose constraints on the matrix, such as applying L_0-norm, L_1-norm or $L_{2,1}$-norm constraint to the objective function. Robust Nonnegative Matrix Factorization is one of new methods about NMF, which uses $L_{2,1}$-norm ($L_{2,1}$NMF) [7]. The objection function can be written as:

$$O = \left\| \mathbf{X} - \mathbf{U}\mathbf{V}^{\mathbf{T}} \right\|_{2,1} \quad s.t. \ \ \mathbf{U} \geq 0, \ \mathbf{V} \geq 0. \tag{9}$$

$L_{2,1}$-norm of one matrix $\mathbf{H}$ is defined as follows:

$$\left\| \mathbf{H} \right\|_{2,1} = \sum_{i=1}^{n} \sqrt{\sum_{j=1}^{m} h_{ij}^2} = \sum_{i=1}^{n} \left\| \mathbf{h}^{\mathbf{i}} \right\|_2 \tag{10}$$

where $\left\| \mathbf{h}^{\mathbf{i}} \right\|_2$ is the L_2-norm of i-th row. And $L_{2,1}$-norm shows strong sparsity compared with other sparse methods.

There is another one which called Non-negative Matrix Factorization with Sparse Constraints (NMFSC) [8]. Its objective function is defined as follows:

$$O = \left\| \mathbf{X} - \mathbf{U}\mathbf{V}^{\mathbf{T}} \right\|^2 \tag{11}$$

under sparse constraints:

$$sparseness(u_i) = S_u, \quad \forall i$$

$$sparseness(v_i) = S_v, \quad \forall i$$

where u_i is i-th column of $\mathbf{U}$ and v_i is i-th row of $\mathbf{V}$. Measure of sparseness based on the relationship between L_1-norm and L_2-norm:

$$sparseness(\mathbf{X}) = \frac{\sqrt{n} - (\sum |x_i|)/\sqrt{\sum x_i^2}}{\sqrt{n} - 1} \tag{12}$$

where n is the dimensionality of matrix $\mathbf{X}$. We can selectively apply sparse constraint on $\mathbf{U}$ or $\mathbf{V}$, sparse degree between 0 and 1.

Versatile Sparse Non-negative Matrix Factorization (VSNMF) [9] is a flexible model, which also makes the basis matrix and coefficient matrix with L_1-norm and L_2-norm constraints. But it is different from the form of NMFSC. The target equation is shown below:

$$O = \frac{1}{2}\|\mathbf{X} - \mathbf{U}\mathbf{V}^{\mathbf{T}}\|^2 + \sum_{i=1}^{k}\left(\frac{\alpha_2}{2}\|u_i\|_2^2 + \alpha_1\|u_i\|_1\right) + \sum_{i=1}^{n}\left(\frac{\lambda_2}{2}\|v_i\|_2^2 + \lambda_1\|v_i\|_1\right)$$

$$s.t.\begin{cases} \text{if } t_1 = 1, & \mathbf{U} \geq 0 \\ \text{if } t_2 = 1, & \mathbf{V} \geq 0 \end{cases} \tag{13}$$

where α_1, α_2, λ_1, λ_2 are regulatory factors, $\alpha_1 \geq 0$ controls sparsity of the basis matrix and $\lambda_1 \geq 0$ controls sparsity of the coefficient matrix; $\alpha_2 \geq 0$ controls smoothness of the basis matrix and $\lambda_2 \geq 0$ controls sparsity of the coefficient matrix. And t_1 and t_2 are Boolean variables. The variability of VNMF appears in:

if $\alpha_1 = \alpha_2 = \lambda_1 = \lambda_2 = 0$ and $t_1 = t_2 = 1$, VSNMF becomes standard NMF [10];
if $\alpha_1 = \alpha_2 = \lambda_1 = \lambda_2 = 0$ and $t_1 = 0$, $t_2 = 1$, VSNMF becomes semi-NMF [11];
if $\alpha_1 = \lambda_2 = 0$, α_2, $\lambda_1 \neq 0$ and $t_1 = t_2 = 1$, VSNMF becomes sparse-NMF [12].

The matrix decomposition with sparse constraints on NMF can produce a large number of zero elements. At this time it is very intuitive to do feature selection based on the basis matrix.

2.4   NMF Method with Orthogonal Constraint

Orthogonal Non-negative Factorization (Orth-NMF) [13] method is the most typical NMF method applying orthogonal constraint. As the name suggests, two matrices with orthogonal constraints decomposed by the original matrix. The objective function can be written:

$$O = \|\mathbf{X} - \mathbf{U}\mathbf{V}^{\mathbf{T}}\|^2, \quad s.t. \ \mathbf{U}^{\mathbf{T}}\mathbf{U} = \mathbf{I} \ or \ \mathbf{V}^{\mathbf{T}}\mathbf{V} = \mathbf{I}. \tag{14}$$

While Local Non-negative Matrix Factorization (LNMF) [14] is based on KL divergence, it adopted another method to optimize problem. The key improvement of LNMF is the introduction of the column orthogonality for the $\mathbf{U}$, which can make the $\mathbf{U}$ more sparse but $\mathbf{V}$ become quite not sparse [15, 16]. The objective equation is as follows:

$$O = D(\mathbf{X}||\mathbf{UV^T}) + \alpha \sum_{i,j} b_{ij} - \beta \sum_i h_{ii}, \tag{15}$$

where $\mathbf{B} = [b_{ij}] = \mathbf{U^T U}$, $\mathbf{H} = [h_{ij}] = \mathbf{VV^T}$ and α and $\beta > 0$. The method requires that different bases should be as orthogonal as possible.

3 Experiments

This is the experimental part of clustering. Subsect. 3.1 briefly describes the gene datasets source. In Subsect. 3.2, we provide the data preprocessing methods in detail. The evaluation tactics of clustering are in Subsect. 3.3. There are some tables of clustering accuracies of NMF methods in Subsect. 3.4. And Subsect. 3.5 shows the figures of the time performance of NMF methods.

3.1 Data Corpora

There are experimental data: the cancer genomic datasets processed from TCGA. Cancer gene datasets include Colorectal Cancer (CRC) dataset. CRC dataset contains 65160 features in 281 samples. The samples of them all cover two labels: Normal (negative) and Tumor (positive). CRC contains three sections: Copy Number Variation Data, Gene Expression Data and Methylation Data.

3.2 Data Pretreatment

In our experiments, we utilize two methods to preprocess data: one is normalization method and another is L_2-norm regularization method. Besides, we also carry out experiments on these data without any pretreatments.

3.3 Evaluation Function

The clustering effects are assessed by contrasting labels clustered on the coefficient matrix decomposed by NMF methods with original label. The accuracy (AC) [17] as our evaluation function:

$$AC = \frac{\sum_{i=1}^n \delta(s_i, \mathrm{map}(r_i))}{n} \tag{16}$$

where n is the total number of samples, $\delta(x, y)$ is a delta function that equals to 1 if $x = y$, equals to 0 otherwise. $\mathrm{map}(r_i)$ is mapping function that maps each cluster label r_i to original label s_i.

3.4   Clustering Results

We use k-means method clustering on the original matrix and the coefficient matrix decomposed by NMF methods. And the results are mainly shown in tables. For parameters, the number of iterations is set to 100 for all methods; 0.8 sparse constraint on U and 0 sparse constraint on V in NMFSC method; We set parameters of VSNMF model: $\alpha_2 = 2^{-3}$, $\lambda_1 = 2^{-6}$, $t_1 = t_2 = 1$ and $\alpha_1 = 0$, $\lambda_2 = 0$; U is orthogonal and V is not in Orth-NMF method. Tables 1 and 2 shows the performance comparisons on kinds of genomic datasets: the first row lists the results of data with regularization pretreatment; the second presents the results of data with normalization; the third lists the results of data without any pretreatment.

Table 1. Clustering performance on Gene Expression Data of CRC

Algorithm	Original	NMF	GNMF	MD-NMF	GDNMF	$L_{2,1}$NMF	NMFSC	VSNMF	Orth-NMF	LNMF
Acc_L_2	0.9217	0.8968	1	0.9039	1	0.8541	0.5374	0.8968	0.8861	0.8719
Acc_Norm	0.9217	0.8399	**0.9929**	0.8897	0.8256	0.7687	0.6085	0.8505	0.7794	0.8968
Accuracy	0.9537	0.8185	0.8185	0.8932	0.8470	0.9253	0.8221	0.8221	0.8719	**0.9680**

Table 2. Clustering performance on Methylation Data of CRC

Algorithm	Original	NMF	GNMF	MD-NMF	GDNMF	$L_{2,1}$NMF	NMFSC	VSNMF	Orth-NMF	LNMF
Acc_L_2	0.6975	0.6441	0.9715	0.6797	1	0.6157	0.5658	0.6584	0.6975	0.5943
Acc_Norm	0.5872	0.5409	**0.7687**	0.5338	0.5516	0.6762	0.6833	0.6406	0.5658	0.6477
Accuracy	0.5908	0.6690	**0.9680**	0.5018	0.7438	0.5943	0.6940	0.6334	0.5908	0.5908

In Table 1, most results of these NMF methods are not so good as the clustering effect of original dataset on Gene Expression Data with regularization. But the performances of NMF models with manifold learning are better than some of the other methods, especially the accuracies of the methods based on GNMF and MD-NMF are 1. The clustering effects on normalized dataset and the effects on data with the regularization are almost the same. While on data without pretreatment, except LNMF method has a better effect, the rest have no advantages compared with directly clustering on original dataset. In general, GNMF, GDNMF and LNMF are superior to others on Gene Expression Data of CRC.

On Methylation dataset with regularization, clustering results of the methods based on GNMF and GDNMF are very impressive while others are not as well as expected. We can clearly see that the performances of NMF methods are not prominent except GNMF on dataset of normalized and without pretreatment. GNMF and GDNMF seem to be better in this part.

3.5   Time Performance of NMF Methods

Although some methods have high accuracies, they maybe cost much more time. In order to further assess the characteristics of the NMF methods, we compare their time overhead. The experiments about time cost carried on the same datasets: Gene

Expression Data of CRC. The method of data preprocessing is L_2-norm regularization and other parameters are unchanged. Our experimental environment is the Matlab R2013a with Intel Core i5, 2.60 GHz CPUs, 4 GB RAM and 32-bit OS. We can see that the time cost of NMFSC and $L_{2,1}$ NMF is too much in the figure. And the clustering effects of them are not prominent in clustering experiments. NMF with manifold learning methods such as GNMF and MD-NMF have good clustering effects, but cost slightly higher than other improved methods. GDNMF method is the optimal method, spend less time and have the good clustering results.

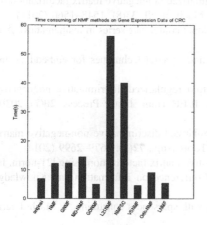

Fig. 1. Time cost of methods on GE of CRC

Through these datasets, we can see that NMF methods with manifold learning perform pretty well especially with sample information in clustering experiments on some datasets in most cases. It can fully illustrate the superiority of supervised learning and importance of data structure. Methods of NMF imposed sparse constraints have no obvious advantages in clustering experiments because of the obtained matrices missing information by sparsity constraints. And the effects of NMF method with orthogonal constraint are between the former two due to they retain some information of data when imposing constraint on matrices to some extent. The type of datasets, data processing methods, setting of parameter will have an impact on the results. What's clearly is that every improvement has applied in different situations. Some may be focused on clustering and some may be focused on feature selection. Here we only give some more reasonable comparisons from some aspects (Fig. 1).

4 Conclusions

In this paper, we introduce some classical NMF methods and provide experiments of clustering on the coefficient matrix decomposed by NMF methods. And we also show rough comparisons from two aspects: clustering accuracies and time performances. In most instances, NMF methods with manifold learning perform pretty well in clustering experiments especially with label information. But the performances of other methods are unstable. Perhaps the type of the datasets and reprocessing methods still influence the effects of the NMF methods. But we really want to give readers some useful references in future works and hope it is helpful for you.

Acknowledgment. This work was supported in part by the NSFC under grant Nos. 61572284, 61502272, 61572283 and 61272339; the Shandong Provincial Natural Science Foundation, under grant Nos. BS2014DX004 and BS2014DX005; Shenzhen Municipal Science and

Technology Innovation Council (No. JCYJ20140417172417174); the Project of Shandong Province Higher Educational Science and Technology Program (No. J13LN31).

References

1. Lee, D.D., Seung, H.S.: Algorithms for non-negative matrix factorization. In: Advances in Neural Information Processing Systems, pp. 556–562 (2001)
2. Cai, D., He, X., Han, J., Huang, T.S.: Graph regularized nonnegative matrix factorization for data representation. IEEE Trans. Pattern Anal. Mach. Intell. **33**(8), 1548–1560 (2011)
3. Chung, F.R.: Spectral graph theory (CBMS regional conference series in mathematics). Am. Math. Soc., vol. 92 (1997)
4. Belkin, M., Niyogi, P.: Laplacian Eigenmaps and spectral techniques for embedding and clustering. NIPS **14**, 585–591 (2001)
5. Guan, N., Tao, D., Luo, Z., Yuan, B.: Manifold regularized discriminative nonnegative matrix factorization with fast gradient descent. IEEE Trans. Image Process. **20**(7), 2030–2048 (2011)
6. Long, X., Lu, H., Peng, Y., Li, W.: Graph regularized discriminative non-negative matrix factorization for face recognition. Multimedia Tools Appl. **72**(3), 2679–2699 (2014)
7. Kong, D., Ding, C., Huang, H.: Robust nonnegative matrix factorization using l21-norm. In: Proceedings of the 20th ACM International Conference on Information and Knowledge Management, pp. 673–682 (2011)
8. Hoyer, P.O.: Non-negative matrix factorization with sparseness constraints. J. Mach. Learn. Res. **5**, 1457–1469 (2004)
9. Li, Y., Ngom, A.: Versatile sparse matrix factorization and its applications in high-dimensional biological data analysis. In: Ngom, A., Formenti, E., Hao, J.-K., Zhao, X.-M., van Laarhoven, T. (eds.) PRIB 2013. LNCS, vol. 7986, pp. 91–101. Springer, Heidelberg (2013)
10. Lee, D.D., Seung, H.S.: Learning the parts of objects by non-negative matrix factorization. Nature **401**(6755), 788–791 (1999)
11. Ding, C., Li, T., Jordan, M.: Convex and semi-nonnegative matrix factorizations. IEEE Trans. Pattern Anal. Mach. Intell. **32**(1), 45–55 (2010)
12. Kim, H., Park, H.: Sparse non-negative matrix factorizations via alternating non-negativity-constrained least squares for microarray data analysis. Bioinformatics **23** (12), 1495–1502 (2007)
13. Ding, C., Li, T., Peng, W., Park, H.: Orthogonal nonnegative matrix t-factorizations for clustering. In: Proceedings of the 12th ACM SIGKDD International Conference on Knowledge Discovery and Data Mining, pp. 126–135 (2006)
14. Li, S.Z., Hou, X.W., Zhang, H., Cheng, Q.: Learning spatially localized, parts-based representation. In: Proceedings of the 2001 IEEE Computer Society Conference on Computer Vision and Pattern Recognition, CVPR 2001, vol. 201, no. 1, pp. I-207–I-212 (2001)
15. Chen, X., Gu, L., Li, S.Z., Zhang, H.-J.: Learning representative local features for face detection. In: Proceedings of the 2001 IEEE Computer Society Conference on Computer Vision and Pattern Recognition, CVPR 2001, vol. 1121, no. 1, pp. I-1126–I-1131 (2001)

16. Pascual-Montano, A., Carazo, J.M., Kochi, K., Lehmann, D., Pascual-Marqui, R.D.: Nonsmooth nonnegative matrix factorization (nsNMF). IEEE Trans. Pattern Anal. Mach. Intell. **28**(3), 403–415 (2006)
17. Xu, W., Liu, X., Gong, Y.: Document clustering based on non-negative matrix factorization. In: Proceedings of the 26th Annual International ACM SIGIR Conference on Research and Development in Informaion Retrieval, pp. 267–273 (2003)

Deep Learning with PCANet for Human Age Estimation

DePeng Zheng, JiXiang Du$^{(\boxtimes)}$, WenTao Fan, Jing Wang,
and ChuanMin Zhai

Department of Computer Science and Technology,
Huaqiao University, Xiamen 361021, China
zhengdepeng8@163.com,
{jxdu,fwt,wroaring,cmzhai}@hqu.edu.cn

Abstract. Human age, as an important personal feature, has attracted great attention. Age estimation has also been considered as complex problem, how to get distinct age trait is important. In this paper, we investigate deep learning techniques for age estimation based on the PCANet, name DLPCANet. A new framework for age feature extraction based on the DLPCANet model. Different from the traditional deep learning network, we use PCA (Principal Component Analysis, PCA) algorithmic to get the filter kernels of convolutional layer instead of SGD (Stochastic Gradient Descent, SGD). Therefore, the model parameters are significantly reduced and training time is shorter. Once final feature has been fetched, we K-SVR (kernel function Support Vector Regression, K-SVR) for age estimation. The experiments are conducted in two public face aging database FG-NET and MORPH, experiments show the comparative performance in age estimation tasks against state-of-the-art approaches. In addition, the proposed method reported 4.66 and 4.72 for MAE (Mean Absolute Error, MAE) for point age estimation using FG-NET and MORPH, respectively.

Keywords: Age estimation · Principal component analysis · Deep learning · DLPCANet model · Kernel function Support Vector Regression (K-SVR)

1 Introduction

Because some factors are uncontrollable in the age process, estimating face age from image has been historically one of the most challenging problems in the computer vision and learning machine field. As illustrated in [1, 2], the process of age estimation is complicated. People with the same age have difference appearances because of variance within the same age, camouflage due to hair, glasses and beards. Besides, it is extremely difficult to collect complete age information for age estimation [3]. However, during the past decades, age estimation has become particularly an interesting topic of computer vision, because of their emerging new applications. In addition, with technology advances in computer science and learning machine, age estimation has been successfully applied in many domains.

© Springer International Publishing Switzerland 2016
D.-S. Huang and K.-H. Jo (Eds.): ICIC 2016, Part II, LNCS 9772, pp. 300–310, 2016.
DOI: 10.1007/978-3-319-42294-7_26

Although age estimation from image is an important computer vision application in academic and industry, but it is mainly by manually designing features for age estimation, the most representative are LBP (Local Binary Patterns, LBP) and SIFT (Scale-invariant feature transform, SIFT) feature and Gabor feature. These feature for age estimation show great success. It mainly depends on the effectiveness of hand-crafted designed feature for specific data and tasks, but designing better feature for new tasks usually requires new domain knowledge. Therefore, most manually features can hardly be adapted to new data and tasks [4].

The goal of this work is to study the age estimation starting from single face images and by means of deep learning, and learning age feature are more discriminative than previous hand-crafted features [5]. Our choice is motivated by the recent advances in fields such as images classification [6] by deep learning.

2 Relative Work

Automatic age estimation caught researchers' attention in 2000 [7]. But research work in this area was from 2006, in which face aging database were available, such as MORPH database [8]. In the earliest, for aging feature representation, Kwon and Lobo [9] used anthropometric information to predict age. Their approach is based on craniofacial development theory and analysis of skin wrinkles [10]. A analogous idea is [11], using a craniofacial growth shape to estimation age, but the approach has a problem which is it can only be used for age group estimation or coarse age estimation, such as, estimation aging belong to young, adult or baby. Geng et al. [12]. proposed an age estimation method named AGES (Aging Pattern Subspace, AGES), which model the long-term aging process of a person and estimates the person's age by minimizing the reconstruction error in corresponded AGES. Manifold learning was also used to extracting age features in [13]. Motivated by the ordinal characteristic of aging process, such as 21 year old's face is much more closely related to the face of 18 year old. In recent years, Deep learning model have demonstrated effective in many fields. Deep learning of most representative is CNN (Convolutional Neural Network, CNN) model, and it was used to learning age feature in [14, 15]. Experiments in the two databases (MORPH and FG-NET), the results achieved success.

The structure of our paper is organized as follows. The proposed approach is illustrated in Sect. 3, where we illustrate a detail description about the method scheme. In Sect. 4, experiment evaluations of the proposed approach and comparison with previous methods are discussed. The last conclusion is given in Sect. 5.

3 Proposed Approach

In this Section, we propose a deep learning model: DLPCANet mode in Fig. 1. Several novelties are shown in this paper. Firstly, we propose a new method for automatic age estimation, which is DLPCANet. As our knowledge, it is the initial that DLPCANet is used to the age estimation. Secondly, we propose DLPCANet model in which the filters kernels are learned from PCA in an unsupervised mode, and avoid parameter tuning.

Thirdly, more importantly, since there is no back propagation for supervised tuning, DLPCANet is much more efficient than existing deep learning network. The DLPCANet model is composed of the convolutional layer, nonlinear processing layer and feature pooling layer. The filter kernel in convolutional layer is learned by PCA, and generated feature maps are aggregated by pooling layer. Top layer output feature maps fed into a K-SVR, this is inspired by the intuition that high level feature combinations and abstract of low features.

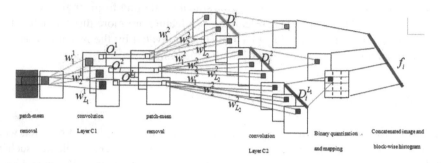

Fig. 1. DLPCANet model

3.1   The Convolutional Filter Layer of DLPCANet

The DLPCANet model contains two convolutional layer, see Fig. 1. The filter kernel in each convolutional layer is learned independently, and the PCA algorithmic is used to learn filters kernel in Convolutional Layer. So we assume that N face images $\{I\}_{i=1}^{N}$ are used as training sample of size $m \times n$, and assuming that the patch size (or two-dimensional filter size) is $p_1 \times p_2$ at all convolutional layer, where p_1 and p_2 are satisfy $1 \leq p_1 \leq m$, $1 \leq p_2 \leq n$. For the *i-th* face image, we take a patch, and all the patches put together to form a matrix of size $p_1p_2 \times m_1n_1$, the matrix denoted as:

$$A_i = \left[a_{i,1}, a_{i,2}, \ldots, a_{i,m_1n_1}\right] \in R^{p_1p_2 \times m_1n_1} \tag{1}$$

In order to introduce competitions between adjacent features within a neighborhood, each column vector in the matrix A_i subtracts the mean value of the corresponding patch to get the matrix $\bar{A}_i = \left[\bar{a}_{i,1}, \bar{a}_{i,2}, \ldots, \bar{a}_{i,m_1n_1}\right] \in R^{p_1p_2 \times m_1n_1}$. By repeating the process for all input images, we can obtain a matrix A

$$A = [\bar{A}_1, \bar{A}_2, \ldots, \bar{A}_N] \tag{2}$$

Then the employ PCA algorithmic is employed to learn the convolution filter kernel from the matrix A, we assume that the number of filter kernel is L_1 in the first convolutional layer. The PCA aims to make the minimizes the reconstruction error of object function, it object function is

$$\min_{V \in R^{p_1 p_2 \times L_1}} \left\| A - V V^T A \right\|_F^2, \ s.t. V^T V = I_{L_1} \tag{3}$$

Where I_{L_1} is a identity matrix, with size is $L_1 \times L_1$. The object function solving process is eigenvalue decomposition method on the matrix AA^T, the convolutional filters are selected as the first L_1 principle eigenvectors of AA^T, donate as:

$$V = [V_1 \ V_2 \ \cdots \ V_{L_1}] \in R^{p_1 p_2 \times L_1} \tag{4}$$

Thus, we can obtain the filter kernel of the convolutional layer of the DLPCANet, and described as:

$$W_l^1 = mat_{p_1 p_2}(V_l) \in R^{p_1 \times p_2}, l = 1, 2, \ldots, L_1 \tag{5}$$

where $mat_{p_1 p_2}(v)$ is a function that transforms $v \in R^{p_1 p_2}$ to a matrix $v \in R^{p_1 \times p_2}$. So the output feature maps of the first convolutional layer of DLPCANet is given by

$$O_{i,l}^1 = \{I_i * W_l^1\} \in R^{m \times n}, i = 1, 2, \ldots, N; l = 1, 2 \ldots, L_1 \tag{6}$$

where $*$ is two-dimensional convolute process, $O_{i,l}^1$ is the first convolution layer output feature map, in order to make $O_{i,l}^1$ having the same size as I_i, so the boundary of I_i is zero-padded before convolution. For each input images $\{I\}_{i=1}^N$ of size $m \times n$, we can get L_1 output feature maps O^1 after the first convolutional layer, donate as:

$$O^1 = [O_{1,1}^1 \ \cdots \ O_{1,L_1}^1 \ \cdots \ O_{N,1}^1 \ \cdots \ O_{N,L_1}^1] \tag{7}$$

In the second convolutional layer, using the feature maps O_1 in the first convolutional layer as input. Assuming that the i-th feature map $O_{i,l}^1$ used as input sample. It is the same as the first convolution layer process, by taking a patch of size $p_1 \times p_2$ to feature maps $O_{i,l}^1$, and obtain a matrix

$$Y_{i,l} = [y_{i,l,1} \ y_{i,l,2} \cdots y_{i,l,m_1 n_1}] \in R^{p_1 p_2 \times m_1 n_1} \tag{8}$$

where $y_{i,l,j}$ denotes the j-th vectorized patch in $O_{i,l}^1$. So for the l-th kernel filter all the input feature maps $\left\{ O_{i,l}^1 \right\}_{i=1}^N$, we can get the following matrix $Y_l = [Y_{1,l} \ Y_{2,l} \cdots Y_{N,l}] \in R^{p_1 p_2 \times N m_1 n_1}$. By repeating this process and concatenate the matrices of the L_1 kernel filter and obtain $Y = [Y_1 \ Y_2 \cdots Y_{L_1}] \in R^{p_1 p_2 \times L_1 N m_1 n_1}$, subtracting each path with mean of patches, for the input Y, we can get

$$\bar{Y} = [\bar{Y}_1 \ \bar{Y}_2 \cdots \bar{Y}_{L_1}] \in R^{p_1 p_2 \times L_1 N m_1 n_1} \tag{9}$$

where $Y_l = [Y_{1,l} \ Y_{2,l} \cdots Y_{N,l}] \in R^{p_1 p_2 \times N m_1 n_1}$, $Y_{i,l} = [y_{i,l,1} \ y_{i,l,2} \cdots y_{i,l,m_1 n_1}] \in R^{p_1 p_2 \times m_1 n_1}$, $i = 1, 2, \ldots, N, l = 1, 2, \ldots, L_1$.

Similar to the first convolutional layer, we assume that the number of filter kernel is L_2 in the second convolutional layer, the PCA algorithmic is used to compute the principle eigenvectores from the matrix $\bar{Y}$. The convolutional filters W_l^2 are selected as the largest L_2 principle eigenvectors. So the second convolution layer output is

$$D^2_{i,l,\tau} = \left\{ O^1_{i,\tau} * W_l^2 \right\}_{l=1}^{L_2}, \tau = 1, 2, \ldots L_1; \; i = 1, 2, \ldots N \tag{10}$$

For each input feature maps $O^1_{i,l}$ of size $m \times n$, we obtain L_2 output feature maps after the second convolutional layer of DLPCANet. Therefore, output feature maps D^2 donate

$$D^2 = [D^2_{1,1,1} \cdots D^2_{1,1,L_2} \cdots D^2_{1,L_1,1} \cdots D^2_{1,L_1,L_2} \cdots D^2_{N,1,1} \cdots D^2_{N,1,L_2} \cdots D^2_{1,L_1,1} \cdots D^2_{1,L_1,L_2}] \tag{11}$$

However, compared with the first convolution layer, the second layer has some difference. We will insert nonlinear process layer when the second convolution layer has been processed. See in Sect. 3.2.

3.2 The Nonlinear Process Layer of DLPCANet

Different with CNN model, the Nonlinear Process Layer of DLPCANet mode uses hashing instead of using Sigmoid and Relu [16]. We binarize the output $D^2_{i,l,\tau}$ of the nonlinear process layer and obtain $Q_{i,l,\tau} = \delta(D^2_{i,l,\tau})$, $l = 1, 2, \ldots L_1$; $i = 1, 2, \ldots N$; $\tau = 1, 2, \ldots L_2$, where $\delta(\cdot)$ is the Heaviside step function, which value is one for positive entries and zero otherwise. Almost repeat the same process, for all the outputs maps D^2, and obtain

$$Q = [Q_{1,1,1} \cdots Q_{1,1,L_2} \cdots Q_{1,L_1,1} \cdots Q_{1,L_1,L_2} \cdots Q_{N,1,1} \cdots Q_{N,1,L_2} \cdots Q_{N,L_1,1} \cdots Q_{N,L_1,L_2}] \tag{12}$$

Around each pixel, we treat the vector of L_2 binary bits as a decimal number, this converts generate a single integer-value feature maps $B_{i,l}$, denote as

$$B_{i,l} = \sum_{\tau=1}^{L_2} 2^{\tau-1} Q_{i,l,\tau}, \; l = 1, 2, \ldots L_1; \; i = 1, 2, \ldots N \tag{13}$$

where the range of $B_{i,l}$ is $[0, 2^{L_2} - 1]$. The final feature maps denote as $B = [B_{1,1} \cdots B_{1,L_1} \cdots B_{N,1} \cdots B_{N,L_1}]$.

3.3 Feature Pooling Layer of DLPCANet

According to our proposed mode structure, using the statistic of block histogram from the nonlinear process layer as output feature. z is the number of blocks when using a block of size $k_1 \times k_2$ to slide each $B_{i,l}, l = 1, 2, \ldots, L_1$, with overlap ratio r. Then we compute the histogram (with 2^{L_2} bins) of value in each block, and concatenate all Z histograms into a vector, which can be defined as $BHist(B_{i,l})$. Through this process, we can get the final feature of the input I_i in the feature pooling layer. It is denoted as $f_i = [BHist(B_{i,1}), BHist(B_{i,2}), \ldots, BHist(B_{i,L_1})] \in R^{(2^{L_2})L_1 Z}$. So by repeating all input the train sample image, we can get the final feature f, we denote as

$$f = [f_1 f_2 \cdots f_N] \in R^{(2^{L_2})L_1 ZN} \tag{14}$$

The final feature f be employed to K-SVR for aging estimation. Our experiment result suggests that overlapping blocks are suitable to age estimation.

4 Experiments

In this section, we would do experiments on public databases for facial aging estimation. We first show two public aging databases. Experiments set are shown in second part. The final is experiment results and analysis.

4.1 Databases

Two databases are used to prove the performance of our propose DLPCANet. The FG-NET aging database is publicly available in the previous works [17]. The database contains 1002 high-resolution color or gray-scale face images composed of 82 multiple-race. Ages range from 0 to 69 years. Most of the images are with large variation of lighting, pose, and expression. Some sample images with FG-NET are displayed in Fig. 2. Another public aging database is MMORPH (Album2) [8]. This databases is much larger than FG-NET. There are 55332 facial images, including six different races (Black, Indian, Asian, White and so on). Ages vary from 16 to 77 years. Samples from MORPH databases are show in Fig. 3.

Fig. 2. The aging faces of sample in the FG-NET databases

Fig. 3. Sample images from the MORPH databases

4.2   Dataset and Experimental Sets

To evaluate the performance of the propose method with aging estimation, the MAE (Mean Absolute Error, MAE) and Cumulative Score (Cs) are used, there are defines as follows: $MAE = \frac{1}{K}\sum_{i=1}^{K} |T_i - P_i|$, where T_i is ground true age, P_i is estimation age, $Cs_a = \frac{K_a}{K} \times 100\%$, K_a represents the numbers of test image whose absolute error of the age estimation and the ground true age is not more than a years, K is the numbers of test image.

Two public age databases, FG-NET databases and MORPH database are used to conduct experiments. For FG-NET databases, we use randomly 80 % data for training, and 20 % data is used for testing. However, for MORPH databases, we randomly split the whole dataset into two independent parts, training sets and testing sets. 5000 images is used for training, 5000 images is used for testing. K-SVR is used to aging estimation. The most important work for K-SVR is choice of kernel function. For complicated and nonlinear aging prediction, suitable kernel function should be employed according to the features of the data sets. In general, kernels including linear, polynomial kernel and Radial Basis Function (RBF) kernel. In our knowledge, the Radial Basis Function (RBF) kernel for SVR is used to aging estimation based on face images.

4.3   Experimental Results

The code is written in Matlab. The DLPCANet are run on Inter (R) Core (TM) i5-4590 CPU @ 3.30 GHz RAM 8.00 GB 64 bit. Training the network on FG-NET images took around 190 s (seconds), Average testing time 0.35 s per test sample. In the following, we show experimental results on public databases for face aging estimation. We will investigate impact of the number of filter kernels and the size of block in our propose method for aging estimation. Finally, we compare with state-of-the-art methods, and it proves that our method can effectively improve performance of aging estimation and robustness.

A. Impact of block overlap ratio of histogram computation on DLPCANet

For the FG-NET database, we set the filter kernels size as $P_1 = P_2 = 5$, and the number of filter is again $L_1 = 13$, $L_2 = 8$, with block size is 5×5. For the MORPH database, the number of filter is $L_1 = 23$, $L_2 = 6$. We just change the block overlap ratio from 0 to 0.5, which to observe the age estimation (MAE) is effected block overlap ratio, it show in Fig. 4. From the MAE curve we can see that the best block overlap ratio is 0.5. For the FG-NET, the MAE is achieve 4.66, and 4.72 in the MORPH.

B. Impact of block size of histogram computation on DLPCANet

In this section, we study the impact of block size on histogram computation for aging estimation. We set the filter kernels size as $P_1 = P_2 = 5$, and the number of filter is $L_1 = 13$, $L_2 = 8$ in the FG-NET. For the MORPH database, the number of filter is $L_1 = 23$, $L_2 = 6$. The block overlap ratio is zero. We increase the block size $k_1 = k_2$

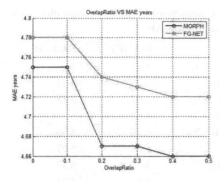

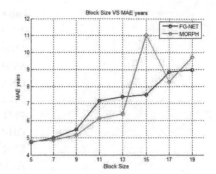

Fig. 4. Impact of block over ratio of histogram computation on DLPCANet (Color figure online)

Fig. 5. The impact of block size of histogram of DLPCANet for aging estimation (Label of absciss-denote [Label Label], such as 5 is [5 5]) (Color figure online)

from 5 to 19 for histogram computation in the top layer, and this block is non-overlapping. The impact of block for age estimation is in Fig. 5. The MAE curve increase with the block size of histogram computation on DLPCANet model. Therefore, we can draw a conclusion that block size has impact on age estimation, and the best size is $k_1 = k_2 = 5$. The best MAE obtained is 4.74 in the FG-NET database. For the MORPH database, the best MAE is 4.78.

C. Impact of the number filter kernel of DLPCANet

The impact of the number of filters kernel of DLPCANet for aging estimation on the testing sets is illustrated in Fig. 6. We set the filter size is $P_1 = P_2 = 5$, block size as $k_1 = k_2 = 5$ and block overlap ratio with 0.5. We can see that MAE is affected by the number of filter kernels. For the FG-NET database, MAE reduce with number of filter kernels, the best of MAE is 4.66, compared to the current best result 4.81 achieved in [18]. For the MORPH database, we can see that MAE reduces more when number of filter kernels increases, the MAE achieves 4.72 less more 4.77 obtained in [19], which is the state-of-the-art method(Convolution Neural Network, CNN). It is importance that our proposed DLPCANet model has fewer parameters than traditional CNN model and is faster. For the MAE curve reduction in Fig. 6, the improvement of age estimation accusation is very evident. This also means the DLPCANet is effective for age estimation of facial images.

Finally, Table 1 shows a summary of the results when DLPCANet is used to aging estimation. In this table, we compare it with recently work. The proposed DLPCANet achieves much lower MAE. Although the MAE from FG-NET database is 4.66, this is more than 4.26 in [15], but our algorithm has fewer parameters and less train time. Particularly, for the MORPH database, the MAE can achieve 4.72, this is less more 4.77 in [15], and in [2, 12, 19, 20]. Part of the age estimation results are display in Fig. 7.

Two public age databases, FG-NET and MORPH databases, we show aging estimation results with random choice ten images. We can see our algorithm performances

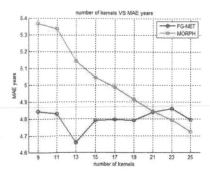

Fig. 6. The impact of the number of filters kernel of DLPCANet for aging estimation (Color figure online)

Table 1. Comparison of different algorithms for age estimation on MAE

Method	FG-NET	MORPH
SVM+FL [18]	4.81	
CNN [15]	4.26	4.77
CA-SVR [19]	4.67	5.88
PLO [21]	4.82	
OHRANK [20]	4.85	5.96
SVR [2]	5.66	5.77
AGES [12]	6.77	8.83
This paper	4.66	4.72

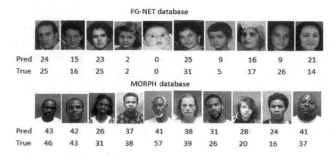

Fig. 7. Part of age estimation result (from FG-NET and MORPH databases)

well in spite of multiple race, different gender and some image with large variation of lighting, pose, and expression.

5   Conclusion and Future Work

In this paper, we have implemented a new age estimation method which is DLPCANet model. A simple deep learning based on PCANet is introduced to solve age estimation problem. The experiments results prove that the proposed model can effectively improve the accuracy in estimation age. For future work, we will study how to improve the estimation age performance and using deep learning techniques for age estimation.

Acknowledgment. This work was supported by the Gran t of the National Science Foundation of China (No. 61175121), the Program for New Century Excellent Talents in University (No. NCET-10-0117), the Grant of the National Science Foundation of Fujian Province (No. 2013J06014), the Program for Excellent Youth Talents in University of Fujian Province (No. JA10006), the Promotion Program for Young and Middle-aged Teacher in Science and Technology Research of Huaqiao University (No. ZQN-YX108), the National Natural Science Foundation of China (61502183), the Scientific Research Funds of Huaqiao University

(600005-Z15Y0016), and Subsidized Project for Cultivating Postgraduates' Innovative Ability in Scientific Research of Huaqiao University (No. 1400214009, 1400214003).

References

1. Geng, X., Zhou, Z.H., Zhang, Y., et al.: Learning from facial aging patterns for automatic age estimation. In: Proceedings of the 14th Annual ACM International Conference on Multimedia, pp. 307–316 (2006)
2. Guo, D.G., Yun, F., Dyer, C.R., et al.: Image-based human age estimation by manifold learning and locally adjusted robust regression. IEEE Trans. Image Process. **17**(7), 1178–1188 (2008). A Publication of the IEEE Signal Processing Society
3. Xin, G., Chao, Y., Zhi, H.Z.: Facial age estimation by learning from label distributions. IEEE Trans. Pattern Anal. Mach. Intell. **35**(10), 2401–2412 (2013)
4. Yoshua, B., Aaron, C., Pascal, V.: Representation Learning: a review and new perspectives. IEEE Trans. Pattern Anal. Mach. Intell. **35**(8), 1798–1828 (2013)
5. Yun, F., Guo, D.G., Huang, T.S.: Age synthesis and estimation via faces: a survey. IEEE Trans. Pattern Anal. Mach. Intell. **32**(11), 1955–1976 (2010)
6. Chan, T.H., Jia, K., Gao, S., et al.: PCANet: a simple deep learning baseline for image classification? IEEE Trans. Image Process. **24**(12), 1 (2014)
7. Lanitis, A., Taylor, C.J., Cootes, T.F.: Toward automatic simulation of aging effects on face images. IEEE Trans. Pattern Anal. Mach. Intell. **24**(4), 442–455 (2002)
8. Ricanek Jr., K., Tesafaye, T.: MORPH: a longitudinal image database of normal adult age-progression. In: IEEE International Conference and Workshops on Automatic Face and Gesture Recognition, pp. 341–345 (2006)
9. Kwon, Y.H., da Vitoria Lobo, N.: Age classification from facial images. Comput. Vis. Image Underst. **74**(1), 1–21 (1999)
10. Alley, T.R.: Social and Applied Aspects of Perceiving Faces. Lawrence Erlbaum Associates, Hillsdale (1988)
11. Ramanathan, N., Chellappa, R.: Modeling age progression in young faces. In: IEEE Computer Society Conference on Computer Vision and Pattern Recognition, pp. 387–394 (2006)
12. Xin, G., Zhi, H.Z., Kate, S.M.: Automatic age estimation based on facial aging patterns. IEEE Trans. Pattern Anal. Mach. Intell. **29**(12), 2234–2240 (2007)
13. Guo, G., Fu, Y., Huang, T.S., et al.: Locally adjusted robust regression for human age estimation. In: IEEE Workshop on Applications of Computer Vision, pp. 1–6. IEEE (2008)
14. Huerta, I., Fernández, C., Segura, C., et al.: A deep analysis on age estimation. Pattern Recogn. Lett. **68**, 239–249 (2015)
15. Wang, X., Guo, R., Kambhamettu, C.: Deeply-learned feature for age estimation. In: 2015 IEEE Winter Conference on Applications of Computer Vision (WACV), pp. 534–541. IEEE (2015)
16. Krizhevsky, A., Sutskever, I., Hinton, G.E.: ImageNet classification with deep convolutional neural networks. In: Advances in Neural Information Processing Systems, pp. 1097–1105 (2012)
17. The FG-NET Aging Database (2010). http://www.fgnet.rsunit.com/, http://www-prima.inrialpes.fr/FGnet/
18. Hadchum, P., Wongthanavasu, S.: Facial age estimation using a hybrid of SVM and Fuzzy Logic. In: 12th International Conference on Electrical Engineering/Electronics, Computer, Telecommunications and Information Technology (ECTI-CON 2015). IEEE (2015)

19. Chen, K., Gong, S., Xiang T., et al.: Cumulative attribute space for age and crowd density estimation. In: IEEE Conference on Computer Vision and Pattern Recognition (CVPR 2013), pp. 2467–2474. IEEE (2013)
20. Chang, K.Y., Chen, C.S., Hung, Y.P.: Ordinal hyperplanes ranker with cost sensitivities for age estimation. In: 2013 IEEE Conference on Computer Vision and Pattern Recognition, pp. 585–592. IEEE (2011)
21. Li, C., Liu, Q., Liu, J., et al.: Learning ordinal discriminative features for age estimation. In: IEEE Conference on Computer Vision and Pattern Recognition, pp. 2570–2577 (2012)

Natural Scene Digit Classification Using Convolutional Neural Networks

Ziqin Wang[1], Peilin Jiang[2,3(✉)], Xuetao Zhang[1,4], and Fei Wang[1]

[1] The Institute of Artificial Intelligence and Robotics, Xi'an Jiaotong University,
No. 28 Xianning West Road, Xi'an 710048, China
wangziqin@stu.xjtu.edu.cn, wfx@mail.xjtu.edu.cn
[2] The School of Software Engineering, Xi'an Jiaotong University,
No. 28 Xianning West Road, Xi'an 710048, China
pljiang@mail.xjtu.edu.cn
[3] National Engineering Laboratory for Visual Information Processing and Application,
Xi'an Jiaotong University, No. 28 Xianning West Road, Xi'an 710048, China
[4] Shaanxi Digital Technology and Intelligent System Key Laboratory, Xi'an Jiaotong University,
No. 28 Xianning West Road, Xi'an 710048, China
xuetaozh@mail.xjtu.edu.cn

Abstract. We used a convolutional neural networks based model to classify scene digits. We proposed the Horizontal and Vertical Feature Block to extract feature from different fields of the input images, which is efficient and has fewer parameters. We introduced a multi-input strategy to add location information to our model, while the traditional methods only use a part of information from the source annotations. More importantly, we released a new dataset for scene digit classification. The new dataset is collected from Baidu street view and mobile photos. The samples in the dataset are from the real world, and they are collected from many kinds of scenes in our daily lives, so that this dataset has huge potential in many applications.

Keywords: Natural scene digit classification · Convolutional neural networks · HV-Block · Multi-input

1 Introduction

Natural scene digit detection is a very hard computer vision task. In the real world, digits appear all over the place. They have different sizes, degrees and fonts. More difficultly, we need find them in many kinds of scenes. As a part of detection tasks, classification is a very important task which is also helpful to detection tasks. We aim to classify the digits in the natural scene in this paper.

Similarly, handwritten character recognition also aim to classify the digits in images. But hand-written character recognition can be considered a solved task for computer vision. However, it's a challengeable problem in the context of complex natural scenes. [11] introduced a digit classification dataset of house numbers extracted from street level images (SVHN). SVHN is a real-world image dataset for developing object recognition

© Springer International Publishing Switzerland 2016
D.-S. Huang and K.-H. Jo (Eds.): ICIC 2016, Part II, LNCS 9772, pp. 311–321, 2016.
DOI: 10.1007/978-3-319-42294-7_27

algorithms. It is similar in flavor to MNIST [17], but comes from a significantly harder, real world problem.

SVHN [11] is widely used, but all of its samples are house numbers in street view images. In order to recognize the digits in most kinds of scenes, we build a new dataset which contains many scenes. The samples are from Baidu Street View images and mobile photos, which contain many scenes in our daily lives. The samples in our dataset are shown in Fig. 1, they are from the images of advertising boards, LED signs, the number of buses, the displays of elevators, the signs in supermarkets, and so on. In this paper, we also propose a new method for scene digit classification. Previous approaches in classifying characters and digits from natural images used multiple hand-crafted features [1] and template-matching [16]. In contrast, convolutional neural networks learn features all the way from the training data, and also learn to classify. [11] demonstrated the superiority of learned features over hand-designed features (Fig. 2).

Fig. 1. Overview of our dataset, which is collected from Baidu street view images and mobile photos

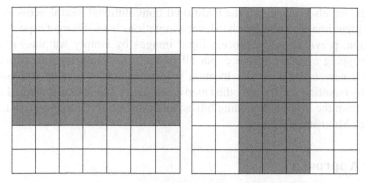

Receptive field of the horizontal branch Receptive field of the vertical branch

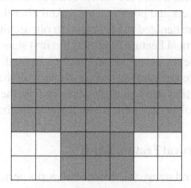

Receptive field of our HV-Block

Fig. 2. The architecture of the horizontal and vertical feature block. Each branch contains two convolutional layers with the kernel size of 5 × 3 and 3 × 1 (3 × 5 and 1 × 3). These two branches are merged by calculating the element-wise sum, and the merged feature maps are fed into a NIN [8] layer

In this paper, we also propose a new method for scene digit classification. Previous approaches in classifying characters and digits from natural images used multiple hand-crafted features [1] and template-matching [16]. In contrast, convolutional neural networks learn features all the way from the training data, and also learn to classify. [11] demonstrated the superiority of learned features over hand-designed features.

Convolutional neural networks [6] are widely used in many fields. Such as image classification [6, 7], object detection [3, 4] and so on [9, 12, 13, 18]. As the convolutional neural networks are trained by Back Propagation algorithm [7], which back propagate the cost layer by layer, it's difficult to train the layers far from the output layer. In order to overcome this problem, we try a method by using horizontal and vertical kernels in convolutional layers which leads to fewer parameters. We introduce the Horizontal and Vertical Feature Block as the first stage of convolutional neural networks to extract features, which improves the performance and need less computation. The HV-Block is described in Sect. 2.1.

Besides, we also notice that the traditional convolutional neural networks need a fixed size input image, while the char-level annotations have different aspect ratio. For classification, previous works process input images by simply warping, padding or directly cropping a square instance, but all of them lose information. However, we introduce a new method to add location information to our convolutional neural networks by inputting a extra smoothed map as a location label. Our smoothed maps are generated from the original bounding boxes, and they are fed into our model. We call this method Multi-input strategy.

2 Our Approach

The traditional Convolutional neural network is composed of stacked stages. Every stage contains convolutional layers and pooling layers. And after these layers, there are fully connected layers, which are denoted as FC layers. In our architecture, we use a two-way block (Horizontal and Vertical Feature Block) as the first stage, which has fewer parameters to optimize and achieves better performance compared with traditional methods. Besides, we also try different methods to make use of the annotation, and finally proposed a method, inputting mask images generated by the annotations to show the exact position of number objects, to reduce the noise in the background.

2.1 Horizontal and Vertical Feature Block

In deep convolutional neural networks, deeper networks perform better if all the layers are trained amply. On the contrary, depth always cause over fitting which make the networks perform badly. In our approach, we proposed Horizontal and Vertical Feature Block which is used in the first stage of a ConvNet. The HV-Block has fewer parameters, so that there is less over fitting in the first two or three layers. And at the same time, it needs computation and has almost the same receptive field as traditional methods, even though the kernel sizes are smaller than the traditional stage.

Our HV-Block is composed of two branches, a horizontal feature branch and a vertical feature branch. Each branch is composed of two convolutional layers with the kernel sizes of 5×2 and 3×1, so that the receptive field of a single branch is 7×2, which is not a square area. However, with padding with zero, the sizes of output feature maps after the two branches are the same. We then merge the two branches by summing of the outputs feature maps. Assuming that there are n output feature maps at the size of $a \times a$ in each branch, we merge them by calculating the element-wise sum as

$$Feature_{sum} = Feature_{Horizontal} + Feature_{Vertical} \tag{1}$$

where $Feature_{sum}$ is the result after merging. Then the merged results are fed into a NIN layer [8], which is equal to a convolutional layer with kernel size of 1×1. As shown in the bottom of Fig. 3, due to the special shape of the receptive fields in horizontal and vertical branch, the final HV-Block also has a unconventional cross-shaped receptive field.

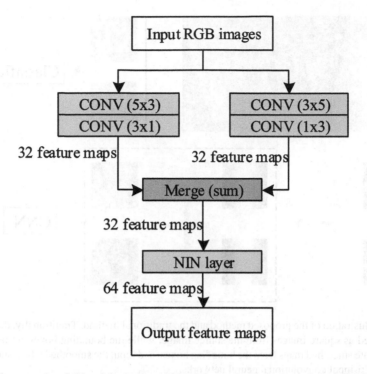

Fig. 3. The receptive fields of the horizontal branch, vertical branch, and the HV-Block. Each branch extracts features from different areas of input images. The HV-Block sees a cross-shaped field instead of a square area

2.2 Char Level Annotation

Both our dataset and SVHN dataset provide char-level annotations by bounding boxes. But many methods need a fixed size input while the bounding boxes are always not square, which can't be wrapped into a fixed size without changing the aspect ratio. A usual method used in [11, 14] is to expand the bounding boxes and generate square digit images. But it adds much irrelevant background information.

In order to reduce interference information, we proposed a novel method to reduce the background information and maintain the information of source images.

As shown in Fig. 4, we firstly crop the digit images and resize to 32×32 according to the expanded bounding box annotation, and generate a binary map at the same size as the digit image (32×32). We initial the binary map using the resized bounding boxes, which can be written as

$$B_{ij} = \begin{cases} 1, & B_{ij} \in Area_{box} \\ 0, & others \end{cases} \tag{2}$$

where B is a binary map and $Area_{box}$ is the area of the resized bounding box.

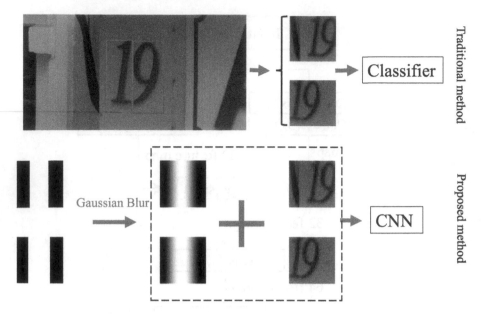

Fig. 4. Illustration of the proposed method and the traditional method. Traditionally, the samples are cropped as square images from the source image, while the bounding boxes are not square. We generate smoothed maps from the bounding boxes, and input the smoothed labels and samples to our multi-input convolutional neural networks.

It means that a pixel in the binary map is set to one if it is in the bounding box, otherwise, it is set to zero.

Then we smoothing the binary map using a Gaussian filter G, and we get the final smoothed map S.

$$S = Binary * G \qquad (3)$$

The smoothed maps are used in our convolutional neural network model, which are described in Sect. 3.2

3   Experiments

3.1   Datasets

Our new dataset. We collected a new dataset for scene digit recognition. SVHN is obtained from house numbers in Google Street View images, while our dataset contains samples from many kinds of scenes. This new dataset contains natural scene digits collected from Baidu street view and mobile photos. All the digits were cropped to the size of 32×32. This dataset is potential and we want that it can help us to build a system which is used to recognize the real-world character. Our dataset contains 15059 digits in total, there are about 1500 digits in each class. We provide char level annotation, and the class annotations are set from 0 to 9 expressing the ten digit. We divided this dataset to three parts, 7000 for training, 2059 for validation and 6000 for testing.

SVHN [11]. We also evaluate our method on the SVHN dataset. SVHN is a real-world image dataset for developing machine learning and object recognition algorithms with minimal requirement on data preprocessing and formatting. The SVHN classification dataset contains 32×32 images with 3 color channels. The dataset is divided into three subsets: training set, extra set and testing set. The extra set is a large set of easy samples and train set is a smaller set of more difficult samples. We construct our validation set the same as [14], 400 per class from training samples and 200 per class from extra samples, yielding a total of 6000 samples. There are more than 70000 training samples and more than 20000 testing sample in SVHN data set, even though there are 531131 additional samples which are less difficult, we didn't use them in training our network.

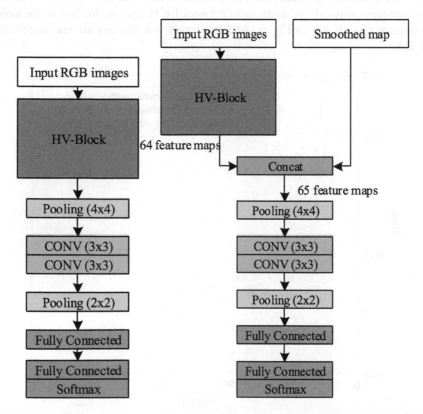

Fig. 5. Illustration of our architecture. Left: Single-input ConvNet with HV-Block. Right: Multi-input ConvNet with HV-Block. We use 96 kernels in the convolutional layers marked in yellow, and use 1024 neurons in the hidden fully connected layers, 10 neurons in the output FC layers (Color figure online)

3.2   Implement Details

Our method is based on the deep convolutional neural networks, there are two main contributions in our approach, the first is the HV-Block, and the second is making use of the generated smoothed label to add the position information into our network.

As shown in the left of Fig. 5, our single-input network uses a HV-Block as the first stage. There are two convolutional layers and two pooling layers following, and we use a fully connected layer and a FC layer with softmax as the output of our network. After all the convolutional layers and the hidden fully connection layer, we use the rectified linear units (ReLU) as the activation function. We also show our multi-input network in the right of Fig. 5. The main architecture is the same as the single-input network, and we concatenate the Smoothed map into the network. In our approach, there are 64 feature maps generated by the HV-Block, and we add the smoothed map label as a feature map, so that there are 65 feature maps as the input of the first pooling layer.

Additionally, we use dropout [15] in all the fully connected layers with the probability of 0.5 to avoid over fitting.

We use stochastic gradient descent algorithm to train our networks. We train both our single-input network and multi-input network for 50 epochs, the loss on the validation dataset was shown in Fig. 6. And the results on the test set are described in Sect. 3.3

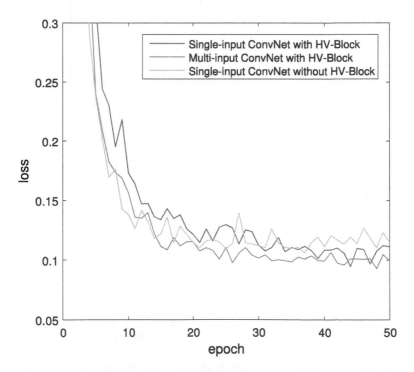

Fig. 6. Loss of three ConvNet on the validation set of our dataset. The HV-Block and the multi-input strategy provide a slight error improvement over traditional architectures (Color figure online)

3.3 Evaluation

In this section, we evaluate our method on our new dataset and the SVHN dataset. As comparative trial on our dataset, we designed a network by replacing the HV-Block with two convolutional layer with kernel sizes of 3×3 and 5×5, we call this network "Single-input ConvNet without HV-Block".

On our new dataset, we used 7000 samples to train our networks for 50 epochs, and chose the parameters which achieve the best performance on the validation set in all epochs. The results on our dataset are shown in Table 1. Results show that our HV-Block improve the results by reducing the error from 2.77 % to 2.55 %, and our Multi-input ConvNet reduce the error from 2.55 % to 2.13 %, which achieved a better result over the single-input network.

Table 1. Evaluation on our dataset. Including our Single-input ConvNet with HV-Block model and Multi-input ConvNet with HV-Block model, the Single-input Con-vNet without HV-Block model is a general convolutional neural networks model

Methods	Error rate on our dataset
Single-input ConvNet without HV-Block	2.77 %
Single-input ConvNet with HV-Block	2.55 %
Multi-input ConvNet with HV-Block	2.13 %

On SVHN dataset, we compare our method with some supervised or unsupervised methods. Such as methods using hand crafted features (histograms-of-Oriented-Gradients features [10], and an off-the-shelf cocktail of binary image features based on [5]) and learned features (stacked sparse auto-encoders and the K-means-based system of [2]). We only use the training set which contains 60000 samples to train our networks, and didn't use the extra set. The results were shown in Table 2. Our method performed better than the other methods.

Table 2. Evaluation on the SVHN dataset. Performance reported by [11, 14] with the additional Supervised ConvNet

Methods	SVHN-Test Accuracy
Single-input ConvNet with HV-Block	94.67 %
Multi-input ConvNet with HV-Block	94.73 %
Binary features (WDCH)	63.3 %
HOG	85.0 %
Stacked sparse auto-encoders	89.7 %
K-means	90.6 %

4   Conclusion

We released a new dataset for scene digit classification. The new dataset is collected from Baidu street view and mobile photos. The samples in the dataset are from the real-world, they are collected from many kinds of scenes, and this dataset has huge potential in many applications.

We also proposed a convolutional neural network based model to classify the natural scene digits. We use the Horizontal and Vertical Feature Block to extract feature in different fields of the input images, which is efficient and has fewer parameters. By comparing our method with the convolutional neural networks model without HV-Block on our dataset, results shows that our HV-Block improve the performance of the CNN model.

We proposed a multi-input strategy to add location information to our model, while the traditional method only uses a part of information of the source an-notations. Similarly, we also evaluate our single-input and multi-input model, we can find that the multi-input strategy improve the performance over the single-input model.

We compared our methods with some other supervised or unsupervised methods, and our methods performed better.

Acknowledgements. This work was supported by the National Natural Science Foundation of China under Grant Nos: 61273366 and 61231018 and the program of introducing talents of discipline to university under grant no: B13043.

References

1. Campos, T.E.D., Babu, B.R., Varma, M.: Character Recognition in Natural Images. Chapman and Hall, London (1992)
2. Coates, A., Ng, A.Y., Lee, H.: An analysis of single-layer networks in unsupervised feature learning. J. Mach. Learn. Res. **15**, 215–223 (2011)
3. Girshick, R., Donahue, J., Darrell, T., Malik, J.: Rich feature hierarchies for accurate object detection and semantic segmentation. In: 2014 IEEE Conference on Computer Vision and Pattern Recognition (CVPR), pp. 580–587 (2014)
4. Khuwuthyakorn, P., Robles-Kelly, A., Zhou, J.: Object of interest detection by saliency learning. In: Daniilidis, K., Maragos, P., Paragios, N. (eds.) ECCV 2010, Part II. LNCS, vol. 6312, pp. 636–649. Springer, Heidelberg (2010)
5. Kimura, F., Wakabayashi, T., Tsuruoka, S., Miyake, Y.: Improvement of hand-written Japanese character recognition using weighted direction code histogram. Pattern Recogn. **30**(8), 1329–1337 (1997)
6. Krizhevsky, A., Sutskever, I., Hinton, G.E.: Imagenet classification with deep convolutional neural networks. In: Pereira, F., Burges, C.J.C., Bottou, L., Weinberger, K.Q. (eds.) Advances in Neural Information Processing Systems, vol. 25, pp. 1097–1105. Curran Associates, Inc., New York (2012). http://papers.nips.cc/paper/4824-imagenet-classification-with-deep-convolutional-neural-networks.pdf
7. LeCun, Y., Boser, B., Denker, J.S., Henderson, D., Howard, R.E., Hubbard, W., Jackel, L.D.: Backpropagation applied to handwritten zip code recognition. Neural Comput. **1**(4), 541–551 (1989)

8. Lin, M., Chen, Q., Yan, S.: Network in network. arXiv preprint arXiv:1312.4400 (2013)
9. Long, J., Shelhamer, E., Darrell, T.: Fully convolutional networks for semantic segmentation. In: 2015 IEEE Conference on Computer Vision and Pattern Recognition (CVPR) (2015)
10. Navneet Dalal, B.T.: Histograms of oriented gradients for human detection, vol. 1, pp. 886–893 (2005)
11. Netzer, Y., Wang, T., Coates, A., Bissacco, A., Wu, B., Ng, A.Y.: Reading digits in natural images with unsupervised feature learning. In: NIPS Workshop on Deep Learning and Unsupervised Feature Learning (2011)
12. Simard, P.Y., Steinkraus, D., Platt, J.C.: Best practices for convolutional neural networks applied to visual document analysis. In: International Conference on Document Analysis and Recognition (2003)
13. Sermanet, P., LeCun, Y.: Traffic sign recognition with multi-scale convolutional networks. In: International Symposium on Neural Networks (2011)
14. Sermanet, P., Chintala, S., LeCun, Y.: Convolutional neural networks applied to house numbers digit classification. In: International Conference on Pattern Recognition. pp. 3288–3291 (2012)
15. Srivastava, N., Hinton, G., Krizhevsky, A., Sutskever, I., Salakhutdinov, R.: Dropout: A simple way to prevent neural networks from overfitting. J. Mach. Learn. Res. **15**(1), 1929–1958 (2014)
16. Yamaguchi, T., Nakano, Y., Maruyama, M., Miyao, H., Hananoi, T.: Digit classification on signboards for telephone number recognition. In: Proceedings of the Seventh International Conference on Document Analysis and Recognition, vol. 1. pp. 359–363 (2003)
17. LeCun, Y., Cortes, C.: The MNIST database of handwritten digits (2005)
18. Zhao, R., Ouyang, W., Li, H., Wang, X.: Saliency detection by multi-context deep learning. In: 2015 IEEE Conference on Computer Vision and Pattern Recognition (CVPR) (2015)

Deep Learning and Shared Representation Space Learning Based Cross-Modal Multimedia Retrieval

Hui Zou, Ji-Xiang Du$^{(\boxtimes)}$, Chuan-Min Zhai, and Jing Wang

Department of Computer Science and Technology, Huaqiao University,
Xiamen 361021, China
{zouhui,jxdu,cmzhai,wroaring}@hqu.edu.cn

Abstract. An increasing number of different multimedia information, including text, voice, video and image, are used to describe the same semantic concept together on the Internet. This paper presents a new method to more efficiently cross-modal multimedia retrieval. Using image and text as an example, we learn the deep learning features of images by convolution neural networks, and learn the text features by a latent Dirichlet allocation model. Then map the two features spaces into a shared presentation space by a probability model in order that they are isomorphic. At last, we adopt centered correlation to measure the distance between them. The experimental results in the Wikipedia dataset show that our approach can achieve the state-of-the-art results.

Keywords: Cross-modal · Cross-media · Retrieval · Deep learning · CNN · Shared presentation space · CC

1 Introduction

With the rapid development of Internet, all kinds of network platforms, including news sites, microblog, social network and video sharing sites, are increasingly changing the method of knowledge acquisition and the style of social cognition. Many scholars have devoted themselves to studying the relationship among different types of multimedia information and got a lot of achievements [1–4].

In order to associate text information with image information, image annotation research has been developed from the manual annotation to automatic annotation since the 70s. Lafferty et al. used conditional random fields (CRF) [5, 6] as a model of image annotation and video annotation.

However the feature space of each two different modal information is often highly heterogeneous that, in recent years, many scholars chose to focus on the correlation between the multimedia information and then obtained the shared representation space between them [7, 8]. Rasiwasia's paper [9] introduced the semantic correlation matching (SCM) algorithm, which used canonical correlation analysis (CCA) to analyse the correlation between text feature space and image feature space and then maximize their coefficient. The Wikipedia Dataset, which was crawled on web pages with large diversity, was used in its experiment. As shown in Fig. 1, it is a random

© Springer International Publishing Switzerland 2016
D.-S. Huang and K.-H. Jo (Eds.): ICIC 2016, Part II, LNCS 9772, pp. 322–331, 2016.
DOI: 10.1007/978-3-319-42294-7_28

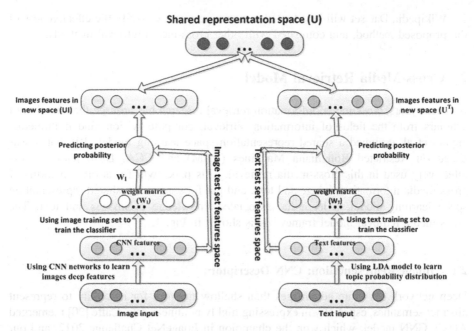

Fig. 1. Cross-media retrieval model

sample selected from Wikipedia Dataset: an article along with the associated image. SIFT and the bag of visual words (BoVW) method was adopted to represent local features of images in cross-modal retrieval in this experiment. The SIFT feature can be effectively used in object retrieval, but couldn't perform well on expressing the rich global content of images.

Feature selection of cross modal information will largely influence the performance of a cross-media retrieval system. The concept of deep learning [10] was first proposed by Hinton et al. in 2006 and the first real deep learning algorithm—convolutional neural networks (CNNs) [11] was put forward by LeCun et al. in 1989. Then CNN has been widely used in image recognition, voice recognition, object detection, behavior detection, etc. [12–14] and made much breakthrough progress. Latent Dirichlet allocation (LDA) [15], a topic model designed by Blein et al., has been extensively used in document classification [16–18] and shown to be very efficient.

Learning good features from images and text and seeking for semantic relationship between two highly heterogeneous feature spaces are two important parts in the study of cross-modal multimedia retrieval. Our paper will combine CNN, LDA and a probability model to build a common representation space as a medium space for images and texts. Both images and text are mapped to this common semantic space, where related images and text will map to the same semantic concept and that it is semantic matching (SM) process. Hence, measuring the similarity between different modals is a good method for us to learn the relationship between them. In view of the two different modal data, we will use an optimized distance metric algorithm to evaluate the similarity.

Wikipedia Dataset will be adopted in our experiment to verify the effectiveness of the proposed method, and compared with other cross-media retrieval methods.

2   Cross-Media Retrieval Model

In recent years, cross-modal information retrieval research has aroused the attention of scholars from the fields of information retrieval, computer vision, and multimedia. Jiquan et al. designed a shared representation space learning model [19] that it was based on Restricted Boltzmann Machines (RBM) and CCA. Deep learning was effectively used in this cross-media retrieval. This paper will focus on two forms of cross-media information, image and text, and put forward a new shared representation space learning model to learn the associated knowledge of images and text. The cross-media retrieval model framework is shown in Fig. 1.

2.1   Image Representation: CNN Descriptors

Deep network has more advantages than shallow network for its ability to represent abstract semantics, especially in expressing highly nonlinear data. Caffe [20] reemerged Alex's CNN model, which won the champion in ImageNet Challenge 2012, and our paper will focus on the experimental data we use to adjust the network parameters of this model, then used it to learn the CNN features of images.

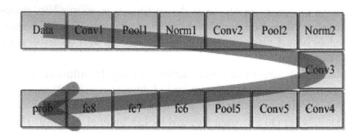

Fig. 2. CNN structure

As shown in Fig. 2, the CNN model has eight layers. Through this network, the image data will be conducted with convolution, pooling and nonlinear transformation, and the last layer is a softmax classification. Convolution layers reduce the parameters of network training by using the method of weight sharing. Pooling layers deal with the results of convolution to provide invariance to translation, rotation, scale and deformation, and also have the effect of data dimension reduction. Rectified linear units (ReLUs) [21] are used to be the activation function in this CNN network:

$$g(x) = \max(0, x)$$

The result of experiment [22] has proved that the trained network has modest sparsity and the two-dimension coordinate graph of ReLUs is closer to the activation model in biology than the graph of sigmoid. Hinton proposed a dropout mechanism in paper [23], and in order to avoid overfitting because of we only have a small quantity of training samples, will need to hidden nodes at a certain probability. And extract the network data from the sixth or seventh layer to express the image and for the next experiment. We will extract data from the sixth and seventh layer in the network separately as images features and use them in the experiment.

2.2 Text Representation: LDA

LDA is a document theme generation model, which includes words, themes, and document, each document in the corpus corresponds to T themes with a multinomial distribution and every theme corresponds to V words with another multinomial distribution:

$$D = \{d_1, d_2, \cdots, d_m\}$$

$$d_i = \{w_1, w_2, \cdots, w_n\}$$

$$w_j \in V = \{v1, v2, \cdots, v_l\}$$

$$T = \{t_1, t_1, \cdots, t_T\}$$

Each document in document collection D can be seen as a sequence composed of multiple words, dictionary V is a collection composed of all the different words, and T is the collection of themes. The probability of each topic of a document represents as p(t|d) and the probability of each word generated by the topic represents as p(w|t). Then we get the probability of each word appears in a document:

$$p(w|d) = p(w|t) * p(t|d)$$

Calculate the probability of every word in each document p(w|d) and then modificate p(w|t) base on p(w|d). If the value of p(w|t) changes, it will in turn affect to the value of p(t|d). We will use p(t|d) to represent the text feature in our experiment. Let $\Re^T$ represents the text features space.

2.3 Shared Representation Space and Similarity Measure Method

Obtain images features and text features and then match these two different modal feature to complete the retrieval between images and text. Given a text (image) query $I_q \in \Re^I \left(T_q \in \Re^T \right)$, the goal of our cross-modal retrieval is to return the closest matched result in the image (text) space through a middle space and a similarity measure method.

2.3.1 Shared Representation Space

Traditional retrieval problems generally would seek for a linear map first:

$$P : \Re^I \rightarrow \Re^T$$

P is an invertible mapping from image space to text space. In the cross-modal retrieval, as different representations were adopted to represent images and text, there is no natural correspondence between $\Re^I$ and $\Re^T$. Simple mapping or calculating the nearest neighbor values cannot be digging inner semantic association between the two heterogeneous space. This paper will use a mechanism to map $\Re^I$ and $\Re^T$ respectively to U^I and U^T:

$$P_I : \Re^I \rightarrow U^I,$$

$$P_T : \Re^T \rightarrow U^T$$

P_I and P_T are both invertible mapping from image space to text space, U^I and U^T are two new spaces and they are isomorphic, such that:

$$U^I = U^T = U$$

Through the two maps above, the two feature spaces map to a shared semantic space U, and U is the Shared representation space.

2.3.2 Probability Model

This article will use a probability model to map two feature spaces to a same shared representation space.

Introduce a vocabulary of semantic concepts: $C = \{c_1, c_2, \cdots, c_k\}$, C represents k types semantic concepts of a document. Use linear classifier to train images training set and text training set respectively, then we get the corresponding weighting matrix W_I and W_T. After that, we use multiclass logistic regression to predict the probability of each test sample belongs to r:

$$P_{C|X}(r|x; w) = \frac{1}{Z(x, w)} \exp(w_j^T x)$$

Where C represents semantic concepts (i.e., the class label), r represents the rth class of k (k is the sum of labels), and X represents $I \in \Re^I$ or $T \in \Re^T$, $Z(x, w) = \sum_r \exp(w_j^T x)$ is a normalized constant. Through this posterior probability formula, we complete the mapping below:

$$P_I : \Re^I \rightarrow U^I, P_T : \Re^T \rightarrow U^T$$

These two maps respectively maps each image feature $I \in \Re^I$ to the a posteriori probability vector $P_{C|I}(r|I)$, maps each text feature $T \in \Re^T$ to the a posteriori probability vector $P_{C|T}(r|T)$, $r \in \{1, 2, 3, \cdots, k\}$. Then we get the images semantic space U^I and text semantic space U^T. The two semantic space is a higher level abstraction than the original feature space, and they are homogeneous spaces and the probability space of semantic concepts. So we can seem the two semantic spaces as the same vector space $(U^I = U^T = U)$, where U is the shared representation space, and that cross-media semantic matching (SM) process will proceed in this space.

2.3.3   Distance Metric

Distance measurement method includes L1 (L1 norm), L2 (L2 norm, Euclidean distance), correlation coefficient (CC), KL-Divergence (KL), k-nearest neighbor (KNN), k-means, etc. These classical methods are effective in measuring distance. For the similarity between two vector similarity in our experiment are different modal feature vector that the similarity between them are also related to the direction of the vectors. So we will try to use an amended center correlation distance metric method in the experiment to complete SM:

$$d_{ij} = -\frac{\sum_{i=1}^{m}\sum_{j=1}^{n}(x_i - \bar{x})(y_j - \bar{y})}{n}, m = n$$

Center correlation main to consider the linear correlation between x_i and y_j, and before computing similarity, x_i and y_j must subtract the average vector, and then calculate the inner product of these two vectors (m and n are the lengths of the two vectors). Negative correlation is used as the distance between two vectors that the larger the correlation is, the smaller the distance will be.

3   The Experimental Results and Analysis

In this section, we will use the proposed retrieval framework for cross-media retrieval experiment to verify the validity of the model and the method of distance measurement.

3.1   Dataset

The Wikipedia Dataset [9] was used in our experiments. It consists of 2866 Wikipedia documents, which are provided by Rasiwasia et al. Each document contains a text and one corresponding related image. All documents are labeled by 10 semantic categories, including Art & architecture, Biology, Geography & places, History, Literature & theatre, Media, Music, Royalty & nobility, Sport & recreation and Warfare. This article adopted the same the training set and testing set as the article [9] that the data set was divided into 2173 training samples, 693 test samples. The goal of our model is to dig the potential relationship between the text and images.

3.2 Caffe Fine Tuning and CNN Feature Extraction

We used a Caffe model in this experiment and then extract a 4096-dimensional feature vector from the input images. To make the CNN features more discriminative for our retrieval model, we fine tuned the parameters of CNN. Since we only have a small quantity of training samples, we adopting the dropout [23] mechanism to reduce overfitting by randomly omitting half of the feature detectors on each training case.

3.3 Analysises of Experimental Results

This experiment used the mean average precision (MAP) as the evaluation index. The value of MAP associated with the rankings of retrieval effect.

a. Validation of the effectiveness of the proposed cross-media retrieval model

We compared our model to three models [9] put forward by Rasiwasia. In these two experiments, we both used normalized correlation (NC), which was performed well in Rasiwasia's experiment, to the measure distance. As shown in Table 1, the comparison results show that the CNN combined with SM used by our model is more effective than the three models proposed by Rasiwasia [9] that the MAP of our I-q-T (image query text) retrieval achieves 0.4019 with the fc6 features, and the MAP of our T-q-I (text query image) retrieval achieves 0.3230 with the fc7 features. Compared to the artificial selected SIFT features, deep learning features can be more effective to express the abstract and deep image semantic. (fc6 is the sixth layer of CNN networks and fc7 is the seventh layer of CNN networks.)

Table 1. The MAP evaluation of multimodal retrieval model

Experiment	Image query text	Text query image	Average
CCA+SIFT [9]	0.249	0.196	0.223
SM+SIFT [9]	0.225	0.223	0.224
SCM+SIFT [9]	0.277	0.226	0.252
SM+CNN(fc6)	**0.4019**	0.3151	0.3585
SM+CNN(fc7)	0.3985	**0.3230**	**0.3608**

b. Validation of the effectiveness of the adopted method of similarity measurement

The distance measurement methods used in our experiment include L1 norm, L2 norm, normal correlation (NC), KL, cosine (CS) and center correlation (CC). Compared with other methods of similarity measurement, the difference is that CC not only considers the direction of the vector of two different modal features, but also eliminates the influence of index dimension after the vectors de-centralised. As shown in Table 2, CC perform the best of all in the experiment with the CNN deep features.

In addition, when using the same distance measurement algorithm, the fc6 features performs better than the fc7 features in I-q-T retrieval mode and the fc7 features performs better than the fc6 features in T-q-I retrieval mode. If consider the effect of

Table 2. Comparison of retrieval results between different measurement methods

Experiment	Distance metric	Image query text	Text query image	Average
SM+CNN(fc6)	L1	0.4006	0.2933	0.3470
	L2	0.3885	0.2793	0.3339
	KL	0.3875	0.2729	0.3302
	NC	0.4019	0.3151	0.3585
	CS	0.3997	0.3256	0.3627
	CC	**0.4169**	0.3517	**0.3843**
SM+CNN(fc7)	L1	0.3969	0.3002	0.3486
	L2	0.3864	0.2873	0.3369
	KL	0.3839	0.2723	0.3281
	NC	0.3985	0.3230	0.3608
	CS	0.3969	0.3295	0.3632
	CC	0.4088	**0.3518**	0.3803

I-q-T mode and T-q-I mode together, the average MAP of fc6 features achieved 0.3843 that they are better to represent images than fc7 features.

c. Comparison with state-of-the-art

In order to further illustrate the advantages of the presented model, Table 3 shows the comparison with state-of-the-art. The first row is a random retrieval experiment. SCM was proposed in paper [9]. MSAE [24] proposed a mapping mechanism based on stacked auto-encoders. CML2R [25] proposed an approach to discover the latent joint representation of pairs of multimodal data. TSRtext and TSRimg were proposed in paper [26]. As shown in Table 3, the proposed model performs the best.

Table 3. Comparison with state-of-the-art

Experiment	Image query text	Text query image	Average
Random	0.118	0.118	0.118
SCM [9]	0.277	0.226	0.252
MSAE [24]	0.187	0.179	0.183
CML^{2}R [25]	0.2330	0.2152	0.2241
TSR$_{text}$ [26]	0.295	0.207	0.251
TSR$_{img}$ [26]	0.322	0.251	0.287
SM+CNN(fc6)	**0.4169**	0.3517	**0.3843**
SM+CNN(fc7)	0.4088	**0.3518**	0.3803

4 Conclusion

Focus on the different cross-media data, we propose a novel deep learning combined with semantic matching model for cross-media retrieval in this paper. We extract deep CNN images features and topic probability distribution of documents, and then map them to a shared representation space via a probability model. The center correlation is used to measure the distance between images and text in the shared space and further to improve the experimental result. Comparison with other experiments demonstrates the effectiveness of our approach. Future work will concentrate on further studying in deep learning, building a larger cross-media dataset and extending the cross-media model so that it can be used to more other cross-media information (e.g. voice, video) retrieval.

Acknowledgement. This work was supported by the Grant of the National Science Foundation of China (No. 61175121, 61502183), the Grant of the National Science Foundation of Fujian Province (No. 2013J06014), the Promotion Program for Young and Middle-aged Teacher in Science and Technology Research of Huaqiao University (No. ZQN-YX108), the Scientific Research Funds of Huaqiao University (No. 600005-Z15Y0016), and Subsidized Project for Cultivating Postgraduates' Innovative Ability in Scientific Research of Huaqiao University (Nos. 1400214009, 1400214003).

References

1. Yang, Y., Xu, D., Nie, F., Luo, J., Zhuang, Y.: Ranking with local regression and global alignment for cross media retrieval. In: International Conference on Multimedia, pp. 175–184 (2009)
2. Srivastava, N., Salakhutdinov, R.R.: Multimodal learning with deep Boltzmann machines. In: Neural Information Processing Systems, pp. 2222–2230 (2012)
3. Lu, X., Wu, F., Tang, S.: A low rank structural large margin method for cross-modal ranking. In: Research and Development in Information Retrieval, pp. 433–442 (2013)
4. Lu, X., Wu, F., Tang, S., Zhang, Z., He, X., Zhuang, Y.: Cross-media semantic representation via bi-directional learning to rank. In: International Conference on Multimedia, pp. 877–886 (2013)
5. Lafferty, J., McCallum, A., Pereira, F.: Conditional random fields: probabilistic models for segmenting and labeling sequence data. In: International Conference on Machine Learning, pp. 282–289 (2001)
6. Xu, X.S., Jiang, Y., Peng, L., Xue, X., Zhou, Z.H.: Ensemble approach based on conditional random field for multi-label image and video annotation. In: International Conference on Multimedia, pp. 1377–1380 (2011)
7. Zhang, Y., Li, G., Chu, L., Wang, S., Zhang, W., Huang, Q.: Cross-media topic detection: a multi-modality fusion framework. In: International Conference on IEEE, pp. 1–6 (2013)
8. Li, L., Jiang, S., Huang, Q.: Learning image vicept description via mixed-norm regularization for large scale semantic image search. In: Computer Vision and Pattern Recognition, pp. 825–832 (2011)
9. Rasiwasia, N., Costa Pereira, J., Coviello, E., Doyle, G., Lanckriet, G.R., Levy, R., Vasconcelos, N.: A new approach to cross-modal multimedia retrieval. In: International Conference on Multimedia, pp. 251–260 (2010)

10. Hinton, G.E., Osindero, S., Teh, Y.W.: A fast learning algorithm for deep belief nets. Neural Comput. **18**(7), 1527–1554 (2006)
11. LeCun, Y., Boser, B., Denker, J.S., Henderson, D., Howard, R.E., Hubbard, W., Jackel, L. D.: Backpropagation applied to handwritten zip code recognition. Neural Comput. **1**(4), 541–551 (1989)
12. Krizhevsky, A., Sutskever, I., Hinton, G.E.: Imagenet classification with deep convolutional neural networks. In: Neural Information Processing Systems, pp. 1097–1105 (2012)
13. Ji, S., Xu, W., Yang, M., Yu, K.: 3D convolutional neural networks for human action recognition. Pattern Anal. Mach. Intell. **35**(1), 221–231 (2013)
14. Razavian, A., Azizpour, H., Sullivan, J., Carlsson, S.: CNN features off-the-shelf: an astounding baseline for recognition. In: Computer Vision and Pattern Recognition Workshops, pp. 512–519 (2014)
15. Blei, D.M., Ng, A.Y., Jordan, M.I.: Latent Dirichlet allocation. J. Mach. Learn. Res. **3**, 993–1022 (2003)
16. Rosen-Zvi, M., Griffiths, T., Steyvers, M., Smyth, P.: The author-topic model for authors and documents. In: Conference on Uncertainty in Artificial Intelligence, pp. 487–494 (2004)
17. Ramage, D., Hall, D., Nallapati, R., Manning, C.D.: Labeled LDA: a supervised topic model for credit attribution in multi-labeled corpora. In: Conference on Empirical Methods in Natural Language Processing, pp. 248–256 (2009)
18. Liu, Y., Niculescu-Mizil, A., Gryc, W.: Topic-link LDA: joint models of topic and author community. In: Annual International Conference on Machine Learning, pp. 665–672 (2009)
19. Ngiam, J., Khosla, A., Kim, M., Nam, J., Lee, H., Ng, A.Y.: Multimodal deep learning. In: International Conference on Machine Learning, pp. 689–696 (2011)
20. Jia, Y., Shelhamer, E., Donahue, J., Karayev, S., Long, J., Girshick, R., Darrell, T.: Caffe: convolutional architecture for fast feature embedding. In: International Conference on Multimedia, pp. 675–678 (2014)
21. Nair, V., Hinton, G.E.: Rectified linear units improve restricted boltzmann machines. In: International Conference on Machine Learning, pp. 807–814 (2010)
22. Li, J., Luo, W., Yang, J., Yuan, X.: Why Does The Unsupervised Pretraining Encourages Moderate-Sparseness. arXiv Preprint arXiv:1312.5813 (2013)
23. Hinton, G.E., Srivastava, N., Krizhevsky, A., Sutskever, I., Salakhutdinov, R.R.: Improving Neural Networks by Preventing Co-adaptation of Feature Detectors. arXiv Preprint arXiv: 1207.0580 (2012)
24. Wang, W., Ooi, B.C., Yang, X., Zhang, D., Zhuang, Y.: Effective multi-modal retrieval based on stacked auto-encoders. Proc. VLDB Endowment **7**(8), 649–660 (2014)
25. Wu, F., Jiang, X., Li, X., Tang, S., Lu, W., Zhang, Z., Zhuang, Y.: Cross-modal learning to rank via latent joint representation. Image Process. **24**(5), 1497–1509 (2015)
26. Ling, L., Zhai, X., Peng, Y.: Tri-space and ranking based heterogeneous similarity measure for cross-media retrieval. In: Pattern Recognition International Conference on IEEE, pp. 230–233 (2012)

Leaf Classification Utilizing a Convolutional Neural Network with a Structure of Single Connected Layer

Xiang He[1(✉)], Gang Wang[1], Xiao-Ping Zhang[1], Li Shang[2], and Zhi-Kai Huang[3]

[1] Institute of Machine Learning and Systems Biology,
College of Electronics and Information Engineering, Tongji University,
4800 Caoan Road, Shanghai 201804, China
Hexiang56@yeah.net
[2] Department of Communication Technology, College of Electronic Information Engineering,
Suzhou Vocational University, Suzhou 215104, Jiangsu, China
[3] Nanchang Institute of Technology, College of Mechanical and Electrical Engineering,
Nanchang 330099, Jiangxi, China

Abstract. Plant plays an important role in human life, so it is necessary to build an automatic system for recognizing plant. Leaf classification has become a research focus for twenty years. In this paper, we propose a single connected layer (SCL) structure adding into the convolutional neural network (CNN). We use this CNN model for plant leaf identification and report the promising results on ICL leaf database. Moreover, we propose some improvement on it to let it perform better. The result shows that our advanced SCL can effectively improve the accuracy of CNN.

Keywords: Convolutional neural network · CNN · Single connected layer · SCL · Dropout · Leaf classification · Gaussian filter

1   Introduction

Plant plays a very important role in human life, we need to set up categorization database and build an automatic system for recognizing plant. Usually, the leaves can be easily gotten and have sufficient planar visible characteristics for differentiating between many species. Leaf classification is meaningful and feasible, and actually has been attracting much attention of many relevant researchers in recent years [1].

As an efficient recognition method, convolutional neural network (CNN) has been getting development and arousing wide attention and has also now become one of the highlights in many fields of science [2].

This paper describes an application for leaf classification that utilizes a convolutional neural network. We propose a one-to-one connection layer that is named as single connected layer (SCL), and we add our SCL into our CNN then make some further improvements. Each kind of CNN has been experimented on ICL leaf database, and the results reflect the rise of the precision.

© Springer International Publishing Switzerland 2016
D.-S. Huang and K.-H. Jo (Eds.): ICIC 2016, Part II, LNCS 9772, pp. 332–340, 2016.
DOI: 10.1007/978-3-319-42294-7_29

2   Convolutional Neural Network (CNN)

The CNN is a feed-forward neural network which can extract the eigen feature from a two-dimensional image. It can directly process grayscale images, and allows image shift and distortion.

In BP neural network, every couple of neurons between neighboring layers are connected, while in CNN, the neighboring layers are not fully connected, but a neuron in Convolutional layer is only connected with some adjacent neurons in neighboring layers according to the local spatial correlation [5]. Another important part of the CNN is the pooling layer to reduce the feature size and the computational complexity. Based on these structure, CNN is a multi-layer neural network on supervised learning. This network model can minimize the loss function via using gradient descent to adjust backwards the weight parameters, and gradually improve the accuracy of the network by iterative training [7].

3   CNN Model Design and Improvement with Dropout and Single Connected Layer

In this paper, we use a CNN with 2 convolution layers and 2 average pooling layers as our original neural network. The size of each convolution kernel is 5*5, and the sampling scales in pooling layers are 2*2. The structure is as the Fig. 1 shows. After the convolution layers and pooling layers is a fully connected single-layer perceptron at last [8].

3.1   Dropout

Dropout is a technology proposed by Hinton et al. [9]. This technology prevents co-adaptation of feature detectors to reduce overfitting in neural network. During neural network learning, dropout let some weights of hidden connection not exert their effects. To explain why dropout work, Srivastava believes that dropout affects like the gender in evolution process, which prevents species from overfitting the current environment and extinction when environment changes [10]. In CNN, dropout can be applied to the last fully connected layers.

3.2   Single Connected Layer (SCL)

In neural network, if we make one-to-one connection between two neighboring layers, then we get a neural network structure that named single connected layer (SCL). The Fig. 2 shows the structure of a single connected layer that has 6 input neurons.

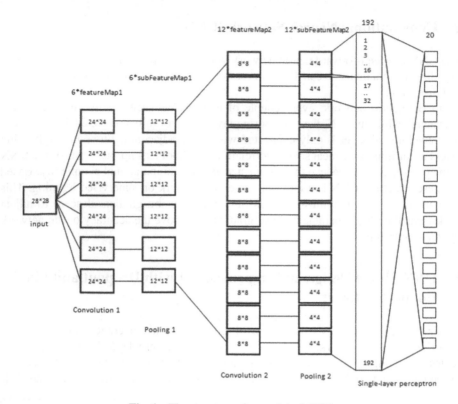

Fig. 1. The structure of our original CNN

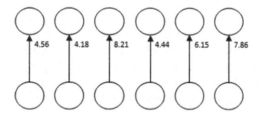

Fig. 2. The structure of a single connected layer

In this SCL, the numbers of connections and output neurons are equal to that of input neurons, and the numbers of trainable parameters is also equal. Let a_{ij}^l denotes the output value of a neuron in the layer l, and w_{ij}^l denotes the weight of a connection between layer l and layer $l - 1$, then the feed forward operation of a SCL can be written as below.

$$a_{ij}^l = w_{ij}^l a_{ij}^{l-1}$$

(1)

During the feed forward process, SCL will have the output values of input neurons multiplied by their output connection's weight. During the error back propagation, because the connection's function in feed forward operation exactly has the derivative equal to the connection's weight, the SCL can have the error of output neurons multiplied by the weight to get the values of input neurons, and the operation can be written as:

$$\delta_{ij}^{l-1} = w_{ij}^l \delta_{ij}^l \tag{2}$$

where δ_{ij}^l denotes an error value in the layer l. After having the error of output neurons multiplied by the values of input neurons and the preset η, the SCL can get the correction values of the connection weights. This operation can be written as:

$$\Delta w_{ij}^l = \eta a_{ij}^{l-1} \delta_{ij}^l \tag{3}$$

where Δw_{ij}^l denotes the correction value. As for matrix calculation, the three operations above are all referred to the product of two corresponding elements from two matrixes with a same size, and that is an operation of dot multiplication in Matlab software. Let n denotes the number of input neurons in SCL, then the computational complexity about SCL is $O(n)$.

In this paper, we add a SCL into the CNN showed by Fig. 1 between the second layer (pooling) and the third layer (convolution), thereby form a CNN as Fig. 3 shows. The initial values of connection weights are all set to be 1 s, so that we need not change other parameters' initial values.

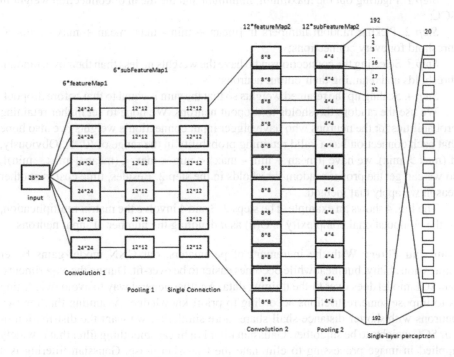

Fig. 3. Structure of our CNN with SCL

Motivation: CNN can reduce the numbers of parameters and the difficulty of training via local sense and parameter sharing. However, in convolution layer, CNN extracts the features of input map by repeating the totally same convolution operation to every area of input map. An image can be directly input into a CNN, and an image of an object always has many pixels outside of the target object. Those pixels compose the background, which shall not interfere the classification result. Therefore, the background pixels should be less important. The above example at least shows that the importance of different pixel in a map can be different. Therefore, distinguishing the neurons is expected to make improvement of accuracy, and we propose the SCL in CNN to realize that idea.

After adding the SCL to our CNN, we do have indeed increased the accuracy on some training data. But SCL does not always work, so we make further advances on the SCL structure.

Dropout: In CNN, dropout is only applied to the last fully connected layers. Dropout works better in hidden neural network layers with large amount of neurons, but in convolution layer an output neuron is only connected with some adjacent input neurons. The dropout directly used in convolution layers can badly disrupt the structure of CNN and decrease the accuracy. But when using dropout in SCL, we find that if we give higher retaining probabilities to the neurons whose input connection's weights are bigger, we get better performance of CNN. In this paper, our dropout method is described as below:

Step 1. Figuring out the maximum, minimum and the mean of connection weight in SCL;

Step 2. Picking random numbers in [mean + min - max, mean + max – min] as threshold for every connections;

Step 3. Selecting the connections who have the weights no less than their own random thresholds and retaining their output neurons;

Step 4. Scaling up the retained weights so that the sum is equal to that before dropout.

We use the random thresholds to dropout neurons. We hope to set higher retaining probabilities for the neurons who have bigger input connection's weights, we also hope that each connection has a valid retaining probability in the range of (0, 1). Obviously, if (max > min), we have [(mean + min – max) < min < max < (mean + max – min)], so we can get the proper random thresholds in the step 2. Besides, time cost is another reason we apply that method.

The Fig. 4 shows an example. The steps 2, 3 and 4 involve the matrix multiplication, so the computational complexity is $O(n)$ as n denoting the number of input neurons.

Gaussian Filter: With the increment of parameters, our CNN model gains better expression ability, but meanwhile becomes easier to be over-fit. During the experiments, our SCL model does over fit the training data. Another general way to avoid over fitting is to impose some restrictions according to priori knowledge. Assuming that the two neurons with shorter distance shall share more similarity, we want the distribution of our SCL weights to be smoother. Gaussian filter is a linear smoothing filter that is widely applied in image processing to eliminate the Gaussian noise. Gaussian filtering is a

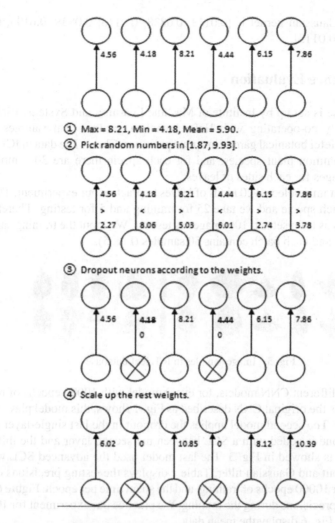

Fig. 4. The process of our dropout method in SCL

process to calculate a weighted means of the input images, so every pixel in output is a weighted means of some adjacent input pixels. The Gaussian kernel can be written as:

$$g(i, j) = c \cdot \exp(-\frac{i^2 + j^2}{2\sigma^2}) \tag{5}$$

where i and j compose the coordinates while the center coordinate is $(0, 0)$. In this paper, we use a Gaussian filter on the SCL weight correction matrix ΔW^l that composed by Δw_{ij}^l in (3), and do smooth the result matrix of SCL weight. Let n denotes the number of input neurons, then $\Delta W^l \in R^n$ and the computational complexity of Gaussian filtering is $O(n)$. In this paper, we use the Gaussian filter with the default kernel of function "fspecial('gaussian')" in Matlab software, whose size is 3 * 3 and sigma is 0.5.

Therefore, our Gaussian kernel is [0.0113, 0.0838, 0.0113; 0.0838, 0.6193, 0.0838; 0.0113, 0.0838, 0.0113].

4   Performance Evaluation

ICL leaf database is set up by Institute of Machine Learning and Systems Biology of Tongji University co-operating with Hefei botanical garden. Leaf samples are all collected in the Hefei botanical garden in Anhui province, China. The data in ICL covers 200 kinds of common plant species, and for each specie there are 30 samples that including 15 images for each sides of leaves.

We have randomly selected 20 kinds of leaves to practice our experiment. There are 30 images for each specie and we take 25 for training and 5 for testing. Therefore we have 500 images as train set and 100 images as test set. We input the training sample in random batches, and each batch contains 50 samples (Fig. 5).

Fig. 5. Image samples in ICL leaf database

There are 4 different CNN models, for every model with 10000 epochs of training. The first model is the original CNN described as Fig. 1 shows, this model plays the role as control group. The second model applies the dropout in the last single-layer perceptron. The third one is added with a SCL between the second layer and the third layer whose structure is showed in Fig. 3. The last model used the advanced SCL with our method of dropout and Gaussian filter. Table 1 displays the testing precisions of those four models after 10000 epochs of training, and the spent time per epoch. Figure 6 shows the curve of test precisions during the training. We practice our experiment for 10 times, and Table 1 and Fig. 6 display the mean data.

Table 1. The test precisions and time spent

	Test precision	T
CNN	82.0 ± 0.27	1.0186 s
CNN + Dropout	89.5 ± 0.90	1.0430 s
CNN + Dropout + SCL	87.7 ± 0.90	1.0458 s
CNN + Dropout + Advanced SCL	**91.9 ± 1.37**	1.0709 s

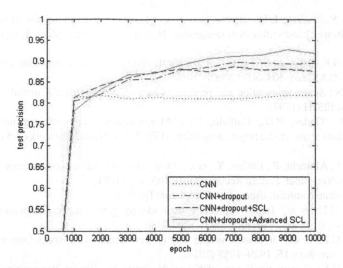

Fig. 6. The test precision curves of the CNNs on the ICL database

As the result reflects, the dropout can greatly improve the accuracy of CNN. Besides, the application of original SCL can also improve the accuracy before 5000 epochs of training. However, it exerts a negative effect later this time. Fortunately, with our method of dropout and Gaussian filter, the advanced SCL helps the precision rise by 2.4 % contrast with the dropout group and gets the best precision.

5   Conclusion

This paper proposes the structure of single connected layer to be applied into CNN. The SCL distinguishes the neurons via its variety of weight value in connections. This paper also proposes two improvement methods on SCL to improve the accuracy of CNN. The experiment on ICL leaf database shows a better performance after the application of the advanced SCL, and proves the positive effectivity.

Acknowledgement. This work was supported by the grants of the National Science Foundation of China, Nos. 61520106006, 61532008, 31571364, 61303111, 61411140249, 61402334, 61472280, 61472173, 61572447, 61373098, and 61572364, China Postdoctoral Science Foundation Grant, Nos. 2014M561513 and 2015M580352.

References

1. Guyer, D.E., Miles, G.E., Gaultney, L.D., et al.: Applocation of machine vision to shape analysis in leaf and plant identification. Trans. ASABE **36**(1), 163–171 (1993)
2. LeCun, Y.: Generalization and network design strategies. Connections Perspect. North-Holland, Amsterdam, 143–155 (1989)

3. LeCun, Y., Jackel, L.D., Bottou, L., et al.: Learning algorithms for classification: a comparison on handwritten digit recognition. Neural networks Stat. Mech. Perspect. **261**, 276 (1995)
4. Hinton, G.E., Salakhutdinov, R.R.: Reducing thr dimensionality of data with neural network. Science **313**(5786), 504–507 (2006)
5. Strom, N.: Sparse connection and pruning in large dynamic artificial neural networks. In: EUROSPEECH (1997)
6. Won, Y., Gader, P.D., Coffield, P.C.: Morphological shared-weight networks with applications to automatic target recognition. IEEE Trans. Neural Networks **8**(5), 1195–1203 (1997)
7. Guyon, I., Albrecht, P., LeCun, Y., et al.: Design of a neural network character recognizer for a touch terminal. Pattern Recogn. **24**(2), 105–119 (1991)
8. https://github.com/rasmusbergpalm/DeepLearnToolbox
9. Hinton, G.E., et al.: Improving neural networks by preventing co-adaptation of feature detectors (2012). arXiv preprint arXiv:1207.0580
10. Hinton, G.E., et al.: Dropout: a simple way to prevent neural networks from overfitting. J. Mach. Learn. Res. **15**, 1929–1958 (2014)
11. Huang, D.S.: Systematic Theory of Neural Networks for Pattern Recognition. Publishing House of Electronic Industry of China, Beijing (1996). (in Chinese)
12. Huang, D.S., Jiang, W.: A general CPL-AdS methodology for fixing dynamic parameters in dual environments. IEEE Trans. Syst. Man Cybern. Part B. **42**(5), 1489–1500 (2012)
13. Wang, X.F., Huang, D.S., Xu, H.: An efficient local Chan-Vese model for image segmentation. Pattern Recogn. **43**(3), 603–618 (2010)
14. Huang, D.S., Du, J.-X.: A constructive hybrid structure optimization methodology for radial basis probabilistic neural networks. IEEE Trans. Neural Netw. **19**(12), 2099–2115 (2008)
15. Huang, D.S.: Radial basis probabilistic neural networks: model and application. Int. J. Pattern Recogn. Artif. Intell. **13**(7), 1083–1101 (1999)
16. Wang, X.-F., Huang, D.S.: A novel density-based clustering framework by using level set method. IEEE Trans. Knowl. Data Eng. **21**(11), 1515–1531 (2009)
17. Huang, D.S., Ip, H.H.S., Chi, Z.-R.: A neural root finder of polynomials based on root moments. Neural Comput. **16**(8), 1721–1762 (2004)
18. Huang, D.S.: A constructive approach for finding arbitrary roots of polynomials by neural networks. IEEE Trans. Neural Netw. **15**(2), 477–491 (2004)
19. Huang, D.S., Ip, H.H.S., Law, K.C.K., Chi, Z.: Zeroing polynomials using modified constrained neural network approach. IEEE Trans. Neural Netw. **16**(3), 721–732 (2005)
20. Zhu, L., Huang, D.S.: Efficient optimally regularized discriminant analysis. Neurocomputing **117**, 12–21 (2013)
21. Zhu, L., Huang, D.S.: A Rayleigh-Ritz style method for large-scale discriminant analysis. Pattern Recogn. **47**, 1698–1708 (2014)

Person Re-identification Based on Color and Texture Feature Fusion

Li Yuan$^{(\boxtimes)}$ and Ziru Tian

University of Science and Technology Beijing, Beijing, China
lyuan@ustb.edu.cn, tianziru122@163.com

Abstract. Due to variations in pose, illumination condition, the appearance of a person differs significantly in two views, which makes person re-identification inherently difficult. In this paper, we propose a feature fusion method for person re-identification, which includes the HSV and color histogram features and the texture feature extracted by the HOG descriptor. The specific process is divided into training phase and recognition phase. In the training phase, we extract the feature descriptors of each image in the reference dataset, and then we learn a subspace in which the correlations of the reference data from different cameras are maximized using canonical correlation analysis (CCA). For re-identification, we extract the feature descriptors of each image in the gallery dataset and the probe dataset. And the feature descriptors are projected onto the CCA subspace and acquire the new feature descriptors. The re-identification is done by measuring the similarity between the gallery image descriptor and the probe image descriptor. Experimental results on standard benchmarking datasets show that the proposed method out performs the state-of-art approaches.

Keywords: Person re-identification · HSV color feature · Color histogram feature · Histogram of oriented gradient · Canonical correlation analysis

1 Introduction

Person re-identification is a task of establishing correspondences between observations of the same person in different cameras [1]. It faces many challenging issues like low resolution frames, the noise of image transmission channel, time-varing light conditions, pose changes and the changes of the angle of cameras [9]. Person re-identification algorithm can be divided into two groups, namely the passive method and the active method, the main difference of the two groups is whether they employ supervised learning techniques in descriptor extraction and matching, the first group does not use this method while the second group does [1]. Recently the active method is widely used and studied, namely appearance-based person re-identification algorithm.

The active methods can be further divided into descriptor based learning algorithm and metric based learning algorithm [1]. Descriptor learning methods attempt to learn

---

This paper is supported by National Natural Science Foundation of China (Grant No. 61300075).

© Springer International Publishing Switzerland 2016
D.-S. Huang and K.-H. Jo (Eds.): ICIC 2016, Part II, LNCS 9772, pp. 341–352, 2016.
DOI: 10.1007/978-3-319-42294-7_30

the most distinctive and stable features to achieve each pedestrian's characteristics. Moreover, color features [2, 6, 11–13], texture features [6], shape features [8, 17] and the fusion of three kinds of features [2, 7] are commonly used. Metric based methods aims to learn the optimal distance measure for the image pairs [3–5], by which the intra-class distances are minimized while the inter-class distances are maximized [1, 9].

The extraction of stable feature descriptors from different views has been extensively researched. In [2], a region-based feature salience algorithm is proposed. More specifically, each person is divided into the upper part and the lower part and the salient color descriptor is proposed. In [6] a reference-based method is proposed for across camera person re-identification. In [11] a method of re-identification is proposed based on unsupervised salience learning and human salience is incorporated in patch matching to find reliable and discriminative matched patches. In [12] propose a method which is based on a symmetry-driven appearance-based descriptor and a matching policy that allows recognizing an individual. The descriptor is dubbed Symmetry-Driven Accumulation of Local Features (SDALFs).In [13], a methodology for re-identification is proposed based on Pictorial Structures (PS). For single image re-identification, this paper adopts PS to localize the parts, extract and match their descriptors. When multiple images of a single individual are available, this paper proposes a new algorithm of Custom Pictorial Structure (CPS). CPS learns the appearance of an individual, improving the localization of its parts. In [17] build a model for human appearance as a function of pose, using training data gathered from a calibrated camera.

Recently, distance learning methods are gaining attentions for re-identification. In [3] formulate person re-identification as relative distance comparison (PRDC) problem in order to learn the optimal similarity measure between a pair of person images. In [5] a transfer local relative distance comparison (t-LRDC) model is formulated to address the open-world person re-identification problem by one-shot group-based verification. The model is designed to mine and transfer useful information from a labelled open-world non-target dataset. In [10], the algorithm, denoted ICT (short for Implicit Camera Transfer), models the binary relation by training a binary classifier with concatenations of pairs of vectors. While the recognition rate is still not ideal.

In order to improve the recognition performance, the difference between the proposed algorithm and the state-of-the-art algorithms is that we propose a fusion feature based person re-identification, which includes HSV color, color histogram features and histogram of oriented gradient texture features. The fusion of HSV color and color histogram can enhance the discrimination of color information. The histogram of oriented gradient texture feature can describe the relationship between the pixels of the image.

This paper is organized as follows. Section 2 introduces the system framework of person re-identification, the extracting process of HSV color feature, color histogram feature and the histogram of oriented gradient texture feature. Additionally, this section also analyzes the correlation of two images from two cameras of the same person and the similarity measure based recognition method. Section 3 introduces several public benchmark datasets (VIPeR, ETHZ(SEQ2), CAVIAR4REID), and the comparative experimental results respectively. Finally, conclusions are drawn and future perspectives are briefed in Sect. 4.

2 Color and Texture Feature Based Person Re-identification

The recognition system in this paper is consisted of two phases: the training phase and the recognition phase. The main purpose of the training phase is to obtain the correlation between the two images which belong to the same pedestrian while coming from two different cameras, namely Cam_A and Cam_B. In this paper, we firstly extract the color and texture features of each image. Then, a reference set containing images from different camera views are used to learn a subspace in which the data from different views are maximally correlated and acquire the transformation matrices w_a and w_b. The Canonical Correlation Analysis (CCA) is used for subspace learning. In the recognition phase, we give the probe and gallery images firstly, then the fusion features are extracted and projected onto the CCA subspace through the learned transformation matrices. Finally, we use the feature descriptor matrices to analysis the similarity among images and get the recognition rate. The proposed system framework is illustrated in Fig. 1.

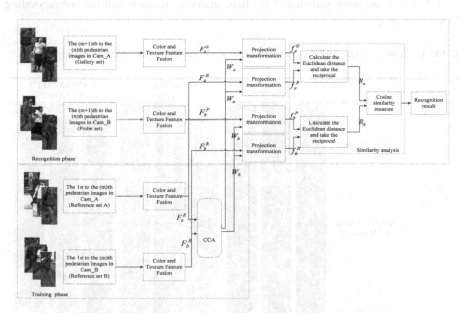

Fig. 1. The proposed system framework of person re-identification

2.1 Color Feature Extraction

Among the proposed color models in the present, RGB color model is one of the most widely used. However the RGB color space is not a visually uniform color space, the distance in the color space cannot represent the human visual color similarity, and therefore the model is not widespread in machine vision application and need to be transformed. In contrast, HSV color space is a uniform color space, and it can better reflect the human perception of color and the identify ability [9]. HSV color space has two characteristics: the luminance component is independent of the color information;

and the hue and saturation components are closely connect with the way of human's perception of color. Therefore, the HSV color space is more suitable for the human visual system to perceive the color characteristics of the image [2, 6].

The conversion formula from RGB color space to HSV color space is:

$$V = \max(R, G, B) \tag{1}$$

$$S = \begin{cases} \frac{V - \min(R,G,B)}{V} & if\ V \neq 0 \\ 0 & otherwise \end{cases} \tag{2}$$

$$H = \begin{cases} 60(G - B)/(V - \min(R, G, B)) & if\ V = R \\ 120 + 60(B - R)/(V - \min(R, G, B)) & if\ V = G \\ 240 + 60(R - G)/(V - \min(R, G, B)) & if\ V = B \end{cases} \tag{3}$$

If H < 0, then H = H + 360. Where $0 \leq V \leq 1, 0 \leq S \leq 1, 0 \leq H \leq 360$.

Table 1 shows some samples of the benchmarking datasets and the corresponding HSV color images.

Table 1. Source images and the corresponding HSV color images on benchmarking datasets

From Table 1 we can see that the HSV color images of different pedestrians have certain difference in HSV color, while the same pedestrian of two images from two cameras has a certain similarity. Therefore, we define the three component values of HSV as the first part of the pedestrian color features, that is f_1.

Table 2 shows the mean quantization of each component of HSV corresponding to the images in Table 1 (in which S and V components are amplified for 40 times). From Table 2 we can see that different pedestrians have different HSV components, while the same pedestrian shares the similarity of HSV components.

Table 2. Mean quantization of HSV components

	Images from Cam_A	H	S	V	Images from Cam_B	H	S	V
VIPeR dataset	1	5	7	15	1	7	4	14
	2	24	6	12	2	24	9	13
	3	5	4	19	3	5	6	18
ETHZ(SEQ2) dataset	1	21	8	29	1	21	8	28
	2	21	9	21	2	21	9	21
	3	21	8	19	3	21	8	19
CAVIAR4REID dataset	1	28	3	19	1	27	3	20
	2	26	6	20	2	28	4	25
	3	24	6	18	3	24	6	20

In order to make full use of the color information of the pedestrians' images, this paper further extracts the HSV color histogram features of pedestrian images. Firstly, we apply the non-equal interval quantization on the HSV space. After the quantilization, the ratio of the H, S, V components is 7:2:2. Secondly the three components are synthesized to one dimensional feature L by $L = 4 \times H + 2 \times S + V + 8$, which has 36 possible values. Then calculate the histogram of L, and return 36 dimensional color histogram. Furthermore, we define color histogram as the second part of the pedestrian color features, namely f_2.

Finally, each image color feature Y is composed of two parts by concatenating f_1 and f_2, namely $Y = [f_1, f_2]^T$.

2.2 Texture Feature Extraction

This paper uses histogram of oriented gradient [2] to extract the texture features of pedestrians. We firstly acquire the gray image, and use the Gamma Correction to adjust the contrast of the image, eliminating the impact of illumination variation. Secondly, the gradient of each pixel of the image is calculated to obtain the silhouette information of the pedestrian, and weaken the illumination interference. Then we divide the image into cells (each cell is of 16 × 16 pixels), and form different blocks by size of 2 × 2 cells. Gradients within each cell were grouped into 9 orientation histogram bins. The HOG descriptor vector for each block was represented by the concatenation of all vectors computed for all of its cells. That is, each block was represented by 36 HOG features (i.e., 4 cells × 9 bins). Finally, the HOG feature of one image is represented as the accumulation of the HOG features of all the blocks. We define the HOG feature vector as the pedestrian texture feature, which is f_3.

Finally, the descriptor of one pedestrian is the fusion of two color features and one texture feature by concatenating of f_1, f_2 and f_3, denote as $F = [f_1, f_2, f_3]$. We denote F_a as the feature descriptor of the image coming from Cam_A, and F_b as the feature descriptor of the image coming from Cam_B.

2.3 Correlation Acquisition

We believe that the two images coming from two cameras while belonging to the same pedestrian should have a certain correlation. In order to get this correlation, we use the method of canonical correlation analysis to calculate two transformation matrices w_a and w_b, which will maximize the correlation coefficient of ρ between the feature descriptor F_a and F_b, and the transformation is $u = w_a^T F_a, v = w_b^T F_b$. The criterion function is given by:

$$\rho = \frac{w_a^T \Sigma_{12} w_b}{\sqrt{w_a^T \Sigma_{11} w_a} \sqrt{w_b^T \Sigma_{22} w_b}} \tag{4}$$

Where Σ_{11} and Σ_{22} is the covariance matrices of F_a and F_b, Σ_{12} is $Cov(F_a, F_b)$.

In the process of calculate the maximum value of ρ, set the constraint condition $w_a^T \Sigma_{11} w_a = 1, w_b^T \Sigma_{22} w_b = 1$ to ensure the uniqueness of the solution. In this condition, we denote the Lagrangian equation as:

$$L = w_a^T \Sigma_{12} w_b - \frac{\lambda}{2} (w_a^T \Sigma_{11} w_a - 1) - \frac{\theta}{2} (w_b^T \Sigma_{22} w_b - 1) \tag{5}$$

Then we should calculate the partial derivation of w_a and w_b respectively and set the partial derivative as 0. We can obtain the equations as:

$$\begin{aligned} \Sigma_{12} w_b - \lambda \Sigma_{11} w_a = 0 \\ \Sigma_{21} w_a - \theta \Sigma_{22} w_b = 0 \end{aligned} \tag{6}$$

According to Eq. (6), $w_a^T \Sigma_{11} w_a = 1, w_b^T \Sigma_{22} w_b = 1$ we can obtain that $\lambda = \theta = w_a^T \Sigma_{12} w_b$. Therefore, we should find the maximum value of λ. According to simplify (6), we can acquire the value of ρ, w_a and w_b at the same time [6, 16].

Here we choose a reference set from VIPeR dataset to illustrate the effect of the transformation matrices w_a and w_b. Figure 2 shows the correlation coefficient of F_a and F_b, Fig. 3 shows the correlation coefficient of f_a and f_b after the projection using w_a and w_b.

From Figs. 2 and 3, we can see that only a few of the same pedestrian's two images' correlation coefficients reach the largest value before canonical correlation analysis. In contrast, the transformation of the feature descriptor from F_a and F_b to

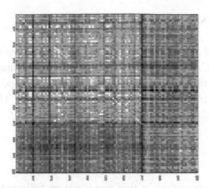   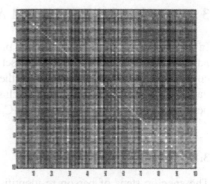

Fig. 2. The correlation coefficient before CCA   **Fig. 3.** The correlation coefficient after CCA

f_a and f_b acquire more maximal values, which means the two images belong to the same pedestrian own the largest correlation coefficients.

2.4 Recognition Based on Similarity Measure

In the recognition phase, we extract the images' features from the Gallery set and the Probe set (i.e. F_a^G and F_b^P), which came from Cam_A and Cam_B. Then the feature descriptors will be projected into the transform matrix space (that is w_a and w_b), and obtain the new feature descriptors (i.e. $f_a^G = w_a^T \cdot F_a^G$ and $f_b^P = w_b^T \cdot F_b^P$). Finally, recognition is done based on the similarity measure.

Similarity measure is mainly divided into two steps. The first step is to calculate the Euclidean distance [8, 9] of f_a^G and f_a^R, coming from the Gallery set and the Reference set respectively (both of them come from Cam_A). Then, calculate the inverse of the Euclidean distance of each image from the Gallery set to every image from Reference set A, and we can acquire the new feature descriptor of the Gallery set (i.e. R_a). Similarly we can obtain the new feature descriptor of the Probe set (i.e. R_b). In the second step, we use the Cosine similarity to calculate the similarity of each image between the Gallery set and the Probe set. We denote:

$$\cos \theta = \frac{\vec{R}_a \cdot \vec{R}_b}{\|R_a\| \cdot \|R_b\|} \tag{7}$$

Compare to other similarity measures, cosine similarity is computationally efficient especially for high dimensional feature descriptors and large datasets.

3 Experimental Results and Analysis

In this section the performance of the proposed algorithm is evaluated on three datasets, namely VIPeR dataset, ETHZ dataset and CAVIAR4REID dataset. These datasets cover several challenging aspects of the person re-identification problem, such as pose variation, illumination variation, occlusions, image blurring and low resolution. The performance is evaluated by the CMC curves.

3.1 Experimental Process

The specific steps of person re-identification algorithm in this paper are as follows:
Training phase:

Step 1: Define a reference set: the images of the $1^{st} \sim m^{th}$ pedestrians. The images from Cam_A form Reference set A. The images from Cam_B form Reference set B.
Step 2: Extract the HSV color feature, color histogram feature and the HOG texture feature of each image from the two Reference set, and form the feature descriptor matrix F_a^R and F_b^R.
Step 3: Calculate the transformation matrix w_a and w_b through CCA, which maximize the correlation between the two feature descriptors.

Recognition phase:

Step 1: Define the $(m+1)^{th} \sim n^{th}$ pedestrian images coming from Cam_A form the Gallery set. The $(m+1)^{th} \sim n^{th}$ pedestrian images coming from Cam_B form the Probe set.
Step 2: Extract the HSV color feature, color histogram feature and the HOG texture feature of each image from the two dataset, and form the feature descriptor matrix F_a^G and F_b^P.
Step 3: Project all the acquired feature descriptor matrices onto the transformation matrix w_a and w_b, and obtain the projected feature descriptor matrices from the Gallery set, the Reference set A, the Probe set and the Reference set B, that is f_a^G, f_a^R, f_b^P and f_b^R.
Step 4: Calculate the Euclidean distance of f_a^G and f_a^R, which come from the Gallery set and the Reference set A. We denote the distances as $\left\{ d_j^G, j = (m+1), (m+2), \ldots, n \right\}$, where $d_j^G = [d(f_j^G, f_1^R), d(f_j^G, f_2^R), \ldots, d(f_j^G, f_m^R)]^T$. Similarly, we calculate the Euclidean distance of f_b^P and f_b^R, which come from the Probe set and the Reference set B. We denote the distances as $\{ d_j^P, j = (m+1), (m+2), \ldots, n \}$, where $d_j^P = [d(f_j^P, f_1^R), d(f_j^P, f_2^R), \ldots, d(f_j^P, f_m^R)]^T$.
Step 5: Calculate the inverse of the Euclidean distance from the Gallery set, that is $R_j^G = [s(d_j^G, j = (m+1), (m+2), \ldots, n)]^T$, then the feature descriptor of the Gallery set can be denoted as $R_a = [R_{m+1}^G, R_{m+2}^G, \ldots, R_n^G]^T$. Similarly, we can acquire the feature descriptor of the Probe set, that is $R_b = [R_{m+1}^P, R_{m+2}^P, \ldots, R_n^P]^T$. Finally, the

identity of a subject in the Probe set is determined by finding a subject k in the Gallery set that is most similar to the probe using cosine similarity between R_a and R_b.

Step 6: According to the similarity measure, count the number of the pedestrian who is recognized correctly. The recognition rate is calculated using the formula of $\frac{\delta}{\zeta} \times 100\%$, where δ is the number of pedestrians recognized correctly, ζ is the number of all the pedestrians.

3.2 Benchmark Datasets Introduction and Performance Comparison of the Proposed Approach with Other Methods

3.2.1 VIPeR Dataset

The VIPeR dataset [6] is designed in a single-shot scenario. It contains image pairs of 632 pedestrians. The images were taken by two cameras with significant view and illumination changes. For each person, a single image is available from each camera view. All the images are normalized to 128 × 48 pixels.

In our experiments, we choose 100 out of 632 pedestrians. The image pairs are randomly divided into two sets with equal number of pedestrians. One is used as the reference set and the other is used for testing. Table 3 shows the comparisons of the proposed method and the state-of-the-art approaches, our approach achieves the highest rank-1 recognition rates. Figure 4(a) shows the CMC curve comparison with other three methods. Compared to the other methods, the proposed method performs better at different ranks.

Table 3. Performance comparison at Rank-1 recognition rate on VIPeR, ETHZ(SEQ2) and CAVIAR4REID

Modality	VIPeR	ETHZ(SEQ2)	CAVIAR4REID
PRDC [3]	15.70	-	-
ICT [10]	15.90	65.00	-
SDALF [12]	19.87	64.40	16.51
CPS [13]	21.84	-	8.50
eSDC_knn [11]	25.63	-	20.45
[17]	21.4	-	-
t-LRDC[5]	23.47	-	15.45
Proposed	**28.85**	**70.59**	**25.00**

3.2.2 ETHZ(SEQ2) Dataset

The ETHZ(SEQ2) dataset [14] contains video sequences of urban scenes captured from moving cameras. It contains 1961 images of 35 pedestrians. The image sizes vary from 80 × 30 to 402 × 181 pixels, and different images of one person have obvious posture changes and serious occlusions. In this paper, the images are resized to 128 × 64 pixels.

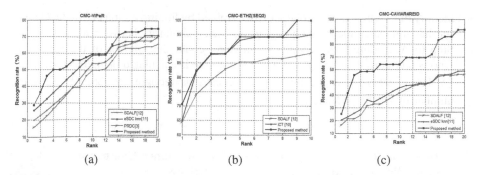

Fig. 4. Comparative performance evaluation on public datasets. (a) Results on VIPeR dataset, (b) Results on ETHZ(SEQ2) dataset, and (c) Results on CAVIAR4REID dataset

We randomly choose 18 persons, and two images per person to build a reference set, and remaining 18 persons, two images per person to build the testing set. The test images from one camera constitute the probe and those from the other camera create the gallery set. Table 3 shows the performance comparison with state-of-the-art methods. Obviously, our approach benefits a better performance. Figure 4(b) shows the comparison of the proposed method with the state-of-the-art methods, our performance is either superior to or same with that of the other methods.

3.2.3   CAVIAR4REID Dataset

The CAVIAR4REID dataset [15] contains 72 unique individuals captured in a shopping center scenario. This dataset was designed to maximize variability with respect to resolution changes, illumination conditions, occlusions and pose changes. The image size ranging from 39×17 to 144×72 pixels, while they are normalized to 128×48 pixels in this paper.

We choose 36 persons to form the reference set, and 36 persons to form the testing set. We compare the rank-1 recognition rate of proposed method and the state-of-the-art on CAVIAR4REID in Table 3. We outperform competing methods at rank-1 on this dataset. Figure 4(c) compares the CMC curves on CAVIAR4REID. Our approach outperforms the other two methods at every rank. The improvement over the state-of-the-art methods at rank-1 is particularly noticeable.

The reason of the proposed method is superior to other methods is that we not only adopt the fusion feature of color and texture, but also extract two kinds of color features which is representative to color images. In addition, we also get two transformation matrices in the training phase, which maximize the same pedestrian's two images' correlation coefficients. In the recognition phase, we use the inverse of the Euclidean distance to process the feature descriptors firstly, and then measure the similarity between the images according to the cosine similarity, which have use the experimental data repeatedly and thoroughly. Thus, the recognition rate of the proposed method has been improved slightly.

4 Conclusion

In this paper we propose a person re-identification method based on feature fusion, which includes color and texture features. In order to improve the distinctiveness of pedestrians' color characteristics, HSV color and color histogram feature are fused to describe the color feature. Moreover, before the similarity measure phase we use the canonical correlation analysis to maximize the similarity of the two images which belong to the same pedestrian. Finally, according to Euclidean distance and cosine similarity we get the similarity between the feature descriptor matrices. From the experimental results, we can see that the method proposed in this paper has achieved a better performance in person re-identification.

However, the result is not quite ideal. The main direction of the future work focus on more stable feature extraction method and precise similarity measure method. In particular, the feature of local area can be considered, and the pedestrians' color feature can be improved by adding spatial information.

References

1. Bedagkar-Gala, A., Shah, S.K.: A survey of approaches and trends in person re-identification. Image Vis. Comput. **32**(4), 270–286 (2014)
2. Geng, Y., Hai-Miao, H.: A person re-identification algorithm by exploiting region-based feature salience. J. Vis. Commun. Image R **29**, 89–102 (2015)
3. Zheng, W.-S., Gong, S., Xiang, T.: Reidentification by relative distance comparison. IEEE Trans. Pattern Anal. Mach. Intell. **35**(3), 635–668 (2013)
4. Ding, S., Lin, L., Wang, G., Chao, H.: deep feature learning with relative distance comparison for person re-identification. Pattern Recogn. **48**, 2993–3003 (2015)
5. Zheng, W.-S., Gong, S., Xiang, T.: Towards open-world person re-identification by one-shot group-based verification. IEEE Trans. Pattern Anal. Mach. Intell. **38**(3), 591–606 (2016)
6. An, L., Kafai, M., Yang, S., Bhanu, B.: Reference-based person re-identification. advanced video and signal based surveillance. In: The 10th IEEE International Conference, pp. 244–249 (2013)
7. Liu, Z., Zhang, Z., Qiang, W., Wang, Y.: Enhancing person re-identification by integrating gait biometric. Neurocomputing **168**, 1144–1156 (2015)
8. Iodice, S., Petrosino, A.: Salient feature based graph matching for person re-identification. Pattern Recogn. **48**(4), 1074–1085 (2015)
9. Saghafi, M.A., Hussain, A., Saad, M.H.M., Tahir, N.M.: Appearance-based methods in re-identification: a brief review. In: IEEE 8th International Colloquium on Signal Processing and Its Applications (CSPA), pp. 404–408(2012)
10. Avraham, T., Gurvich, I., Lindenbaum, M., Markovitch, S.: Learning implicit transfer for person re-identification. In: Fusiello, A., Murino, V., Cucchiara, R. (eds.) ECCV 2012 Ws/Demos, Part I. LNCS, vol. 7583, pp. 381–390. Springer, Heidelberg (2012)
11. Zhao, R., Ouyang, W., Wang, X.: Unsupervised salience learning for person re-identification. In: Proceedings of the IEEE Conference on Computer Vision and Pattern Recognition, pp. 3586–3593 (2013)

12. Bazzani, L., Cristani, M., Murino, V.: Symmetry-driven accumulation of local features for human characterization and re-identification. Comput. Vis. Image Underst. **117**(2), 130–144 (2013)
13. Cheng, D.S., Cristani, M., Stoppa, M., Bazzani, L., Murino, V.: Custom pictorial structures for re-identification. Proc. Br. Mach. Vis. Conf.(BMVC) **1**(2), 68.1–68.11 (2011)
14. Martinel, N., Das, A., Micheloni, C., Roy-Chowdhury, A.K.: Re-identification in the function space of feature warps. IEEE Trans. Pattern Anal. Mach. Intell. **37**(8), 1656–1669 (2015)
15. Lisanti, G., Msi, L., Bagdanov, A.D., Bimbo, A.D.: Person re-identification by iterative re-weighted sparse ranking. IEEE Trans. Pattern Anal. Mach. Intell. **37**(8), 1629–1642 (2015)
16. Hardoon, D.R., Szedmak, S., Shawe-Taylor, J.: Canonical correlation analysis: an overview with application to learning methods. Neural Comput. **16**(12), 2639–2664 (2004)
17. Wu, Z., Li, Y., Radke, R.J.: Viewpoint invariant human re-identification in camera networks using pose priors and subject-discriminative features. IEEE Trans. Pattern Anal. Mach. Intell. **37**(5), 1095–1108 (2015)

Recognition of Mexican Sign Language from Frames in Video Sequences

Jair Cervantes[1(⊠)], Farid García-Lamont[1],
Lisbeth Rodríguez-Mazahua[2], Arturo Yee Rendon[3],
and Asdrúbal López Chau[4]

[1] Posgrado e Investigacion, Autonomous University of Mexico State,
56259 Texcoco, Mexico
jcervantesc@uaemex.mx
[2] Division of Research and Postgraduate Studies, Orizaba Technology Institute,
Orizaba, Veracruz, Mexico
[3] Faculty of Computer Science, Autonomous University of Sinaloa,
Culiacán, Mexico
[4] Centro Universitario UAEM-Zumpango,
Camino Viejo a Jilotzingo continuación calle Rayón, 55600 Zumpango, Mexico

Abstract. The development of vision systems capable to extracting discriminative features that enhance the generalization power of a classifier is still a very challenging problem. In this paper, is presented a methodology to improve the classification performance of Mexican Sign Language (MSL). The proposed method explores some frames in video sequences for each sign. 743 features were extracted from these frames, and a genetic algorithm is employed to select a subset of sensitive features by removing the irrelevant features. The genetic algorithm permits to obtain the most discriminative features. Support Vector Machines (SVM) are used to classify signs based on these features. The experiments show that the proposed method can be successfully used to recognize the MSL with accuracy results individually above 97 % on average. The proposed feature extraction methodology and the GA used to extract the most discriminative features is a promising method to facilitate the communication of deaf people.

Keywords: Mexican Sign Language · Classification · Features

1 Introduction

The development of devices that help to the people with limitations such as diminution of their physical abilities, intellectual or sensory is a current research in many fields. The development of tools to facilitate the communication using the Mexican Sign Language (MSL) between deaf people and hearing people is vital to facilitate interaction, dialogue and information in social and private scopes, in addition to providing access to education and employment. The development of tools to help the overall

© Springer International Publishing Switzerland 2016
D.-S. Huang and K.-H. Jo (Eds.): ICIC 2016, Part II, LNCS 9772, pp. 353–362, 2016.
DOI: 10.1007/978-3-319-42294-7_31

communication between transmitter (deaf people)-receiver [3, 7, 15], and transmitter-receiver (deaf people) [1, 2, 5, 6, 11, 12] is a major challenge of current research.

The actual research is focused in both directions, transmitter-receiver (deaf people) and transmitter (deaf people)-receiver. The first one uses animated characters or avatars in video in which sign language interpreters translate the text. The second one tries to translate a dialogue or identify words in sign language issued by deaf people and translate it by vision systems. To translate a sign language dialogue is not an easy challenge because the signs used in the MSL are signs in movement or static signs at some point in the body or space. Each signal may be composed of five or even seven features; hand configuration, location, direction, orientation, non-manual features, place of articulation and contact point. In addition, each person makes slight and special variations when using the MSL. Some researchers have explored some techniques to solve this problem for the last 15 years. In [4] the authors use shape features extractors to identify signs in the MSL alphabet. However, almost all the signs do not include lateral motion in any directions, because the alphabet signs are the simplest signs in MSL. In [13] the authors used six geometric moments to recognize 70 signs in Japanese Sign Language by segmenting face and hands. In [7] Dreuw develops a system to get text translations from sentences in sign language. Dreuw used signal in video and gets different characteristics from various techniques including PCA, recognition of speed and trajectory of hand. In their experiments the original image is sampled, which could lead to losing important information. In [15] the authors use a combination of geometric characteristics including image intensity, intensity, texture and different types of first and second order derivatives to recognize segmented images. The results reported by Zadehi show errors in the test data of 30 %. Bauer [3] developed a system for recognizing German sign language. In their experiments uses continuous density Hidden Markov Models with one model for each sign. In their approach manual sign language parameters, such as hand shape, hand orientation and location are regarded and used as information for the feature vector.

In pattern recognition, it is necessary that the data-set will be as complete as possible. Some words in MSL cannot be identified by an image. They need to use some frames from the video and extract the information from it. However, to take many frames of a video sequence or use techniques as tracking can makes the classification process computationally prohibitive. This research takes a step in the development of automated translation systems of MSL. An alternative method using frames of video sequences is proposed. The proposed system is used to recognize 249 signs of MSL by using frames. In the proposed methodology are employed geometric features extraction techniques which are invariant to scaling, translation and rotation. A genetic Algorithm is used in order to eliminate all the features that not contribute to the performance.

The rest of the paper is organized as follows: The Sect. 2 describes the methods used for the extraction of features and recognition. The Sect. 2 focuses on describing the proposed algorithm. Section 3 shows the experimental results obtained with the proposed and finally the conclusions are given in Sect. 4.

2 Proposed Method

The aim of research in automatic sign language recognition aims to develop systems with good performance. However, that requires huge amounts of data during training, data under different conditions and with different people. That is, sign language recognition systems require information as representative as possible.

In most cases this is not possible for two reasons, the number of images of signs will always be limited to few people and training time of classifiers. On the other hand, if the images in the training set is incremented then training time is significantly increased.

The greatest difficulty of identification systems sign language is that each sign has a particular way that is defined by; Static hand, Movement of one hand in some direction (ascendant, descendent, lateral), Movement of both hands in different directions, Movement of both hands and use of any part of the body to generate the sign.

All these features should be taken into account because there are signs in the MSL that are similar or they need some movement of both hands and face. Tracking can be a great tool to follow hands and face of a person when try to implement a sign. The principal problem when using tracking is the computational complexity. On the other hand, the movement generated by each sign is the same in all cases, it is like writing any word, always the same letters are used, in the case of hands, always follow the same movement, either ascending, descending or lateral, therefore, to follow some frames of the signal, it is possible to identify the word. The Fig. 1 shows the proposed method.

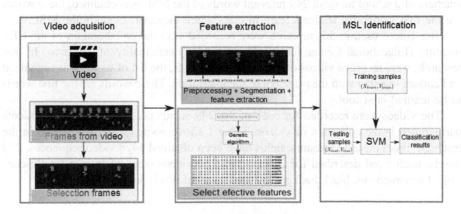

Fig. 1. Proposed methodology

From a video, is possible to obtain many frames by second. The frames define perfectly a sign. The Fig. 2 shown some frames that compose a sign. In the figure is easy to see at the beginning of the sequence the person with his hands at his sides, and then can be seen the movement of his hands generating the sign, to finally return to the starting position. Some of the images are not useful for identifying a signal. In the picture is clear that the first and last frames could be discarded because it is the natural movement of the hands from one rest state to some movement state or from one movement state to a resting state. The interest is in the middle of the sequence of frames, where hands make a sign; therefore focusing on this section of frames is possible to get the best information in order to identify the signs.

However, although a reduction of frames is obtained by eliminating the first and last section of the sequence, the middle section may have too many frames of movement without significant changes. Many frames of this section can be eliminated, is possible to take only a few frames with enough information to get an automatic sign language identifier.

Fig. 2. Frames obtained from the video (Color figure online)

2.1   Dataset

The selected words in this research were chosen by semantic fields, also by how often they are used, or the most common vocabulary words in MSL, including verbs, adjectives, body parts, trades, places, etc. It was done with the help and supervision of teachers of a school for deaf. 249 relevant words of the MSL was obtained, these words are the first words that students learn in class when they start to study the MSL.

The videos used in this research were recorded with the collaboration of the CES institute (Educational Center for the Deaf) in the municipality of Texcoco. In this research, were recorded videos of high school students, the list of words was organized and chosen together with the principal of the institution. These words are the first words to be learned in school.

The videos were recorded for each of the 249 words per student, and 22 students participated in the collection of videos. Table 1 shows some of the words used in the research. From these 15 semantic fields have been obtained 5478 video sequences, 249 words, each word described for 22 persons. The videos were taken in a totally controlled environment, black background and dress of black people.

2.2   Pre-processing

Each video was placed in front of the camera in the same position and orientation at a distance of 2 m. From each video 10 frames were obtained (3200×2400 pixels in JPEG format and RGB Color). The frames where the hands are at rest were eliminated and were obtained only the frames where the hands are not on the sides. From this sequence of frames were obtained 3 frames. The images were re-sized to 0.2 times the size of the original images using cubic interpolation.

Table 1. Semantic fields and words of MSL

	Field	Words
1	Regards	Morning, afternoon, night, thanks, please, see you, bye.
2	Time	Day, hour, week, minutes, seconds
3	Days	Monday, Thursday, Tuesday, Wednesday, Friday,
4	Months	January, February, March, April,
5	School things	Class, notebook, school, pen, exam,
6	Family	wife, husband, father, mother, grandfather,
7	Home things	bedroom, kitchen, lamp, table,
8	Food	pear, apple, orange, dinner, lunch,
9	Clothes	pants, short, dress, shoes,
10	Vehicles	car, motorcycle, bus, taxi,
11	Places	airport, library, downtown, cirque, hospital, hotel,
12	Mexico states	Zacatecas, Yucatan, Veracruz, Quintana roo,
13	Occupations	Teacher, carpenter, mechanic, police, bomber,
14	Verbs	sleep, play, get up, cry, do, listen, smile,
15	Body parts	nose, face, eyes, hands, hair,

Finally, four frames of the middle part of video were taken. It's clear that these frames have to be the most representative of the word. The duration of each video is between 1 and 3 s, depending on the movements of each word and the speed with it is done. To remove small unwanted noise in the images, a Gaussian filter was applied. The Fig. 2 shows the frames obtained and how is possible to eliminate some of them.

2.3 Segmentation

Segmentation is a fundamental task in digital image processing. In this, the pixels within each of the segments are homogeneous; the boundaries between segments correspond to discontinuities in the image and can be used to detect shapes in the image.

There are different segmentation algorithms in the literature, some use spatial information and can produce better results than others. Threshold techniques are an attractive option for local segmentation because of its simplicity. Other image segmentation algorithms attempt to split an image into separate groups of related pixels, called segments. Although arbitrary segmentation of images is dificult, the principles described have been applied successfully to the recognition of sign language. To facilitate the segmentation of images, the videos were taken in an environment fully controlled (black background and people with black shirt and long sleeve).

In the experiments was observed that conventional methods (Thresholding, Otsu, adaptive segmentation) do not work segmenting the images in this research, so a neural network was trained to detect the skin color and segment the frames. In order to obtain a good segmentation even though there are changes in global brightness conditions. The sign region of each image was segmented using three steps: Computation of a high contrast gray value image from an optimal linear combination of the components RGB color. Estimation of a global threshold using the approach Proposed in [8] and finally, fill the holes presented in the segmented binary image.

2.4 Features Extraction

The feature extraction is a critical initial step in any pattern recognition system. The extraction accuracy has a great influence on the final identification results. The proposed system uses only the region of the gesture, determines its limits and calculates properties using the feature extraction. Feature extraction allows us to represent the image using a set of numerical values with high discriminative power, eliminating redundant features and reducing the dimensionality of the image. The characteristics obtained are able to associate very similar ranges to similar images, associate different ranges to different images, besides being invariant to scaling, rotation and translation, allowing the classifier to recognize objects despite having different size, different position and orientation. All these features play an important role in the performance of the algorithm and allow to the classifier discriminate between different classes with high accuracy. Features extracted from each segmented region were divided into two families: geometric and color features.

Geometric Features. Geometric features are one of the most important image features for characterizing an object. Human beings tend to perceive scenes as being composed of individual objects, which can be identified by their shapes. Notice that there exist greater morphological differences in different kinds of signs in the MSL. The selection of the best features is a crucial step for the sign language classification. An efficient sign language classification system must be able to recognize a sign regardless of its location, orientation and size in the field of view, i.e., translation, rotation and scale invariance.

Geometric features provide information on the size and shape of the segmented region. These features can be subdivided in basic features and invariant features. Basic features provide simple information of the segmented region like the area of the hands, perimeter, orientation, eccentricity, minimum bounding rectangle, circularity ratio, ellipse variance, convexity, solidity, projections, centroids, elongation defined by Length and width of the hand and contour curvature among others.

Invariant features provide information invariant to scale, orientation and position. The image moments are widely used as shape features for image processing and classification, which provide a more geometric and intuitive meaning than some simple geometric features. The invariant properties of moments have received considerable attention in recent years, since moments define a simply calculated set of region properties that can be used for classification based on shapes. In this paper, two kinds of moments, Hu geometric moments, Flusser, Gupta, Fourier descriptors and Zernike orthogonal moments, are chosen as the shape features. In the proposed approach, 62 geometric features of the following four groups were extracted from each region.

Color Features. Color features provide information about the color intensity of segmented region. These features can be subdivided in basic intensity features, intensity invariant features, and textural features. Basic intensity features provide information related to the mean, standard deviation of the intensity in the hand area, contrast, local variance, mean first derivative in the boundary, and second derivative in the region. Textural features give a quantitative measure about the spatial arrangement of intensities in the segmented region. This approach is easier to compute and it is widely used. Intensity invariant features. Hu moments and Flusser moments can be used by integrating information of the color variable in the equation to computing moments. These features provide color information in combination with the geometric shape.

Texture features provide information about the spatial arrangement of the colors or intensities in an image. The most important texture features are Haralick features and Local binary patterns. In the proposed approach, 227 features per color channel were extracted, i.e., 227 _ 3 = 681 features for the three color channels: red, green and blue (from RGB color space). In total, 62 geometric and 681 color features, i.e., 743 features, were extracted from each segmented region. Final features obtained were stored in a matrix T of $m \times n$ which contains m images with n features. The features were normalized as:

$$f_{ij} = \frac{T_{ij} - \mu_j}{\sigma_j}$$

where i = 1, and j = 1, μ_j and σ_j are the mean and standard deviation of the jth feature, T_{ij} is the jth feature of ith vector, m is the number of images and n is the number of features. The normalized features have zero mean and a standard deviation equal to one.

2.5   Feature Reduction

The high dimension of images is reduced by extracting discriminative features. However, the extracted features are still too many, and although they may contain very important features, the high dimensional feature vectors include a lot of background (noise) and one would need many more observations to train a model. To reduce the feature dimension and to reduce the number of parameters to be learned in the model, a genetic algorithm is used to choose the most effective features from the entire feature set.

Using the GA, a subset of m features $m < n$ that leads to the smallest classification error is selected. The selected m features were arranged in a new m-vector. The GA selects the best combination of features, maximizing the classification performance.

The initial single chromosome is a chain with all features is $X_{initial} = [x1, x2, \ldots, x773]$ is where $xd = 1$, $d = 1, 2, \ldots, 733$ i.e. all the features are used to train the data-set. Any chromosomal chain with some $x = 0$ is an individual with a different fitness to $X_{initial}$. The aim of the GA is to find an individual to maximize the accuracy of the test set using some of the 733 originally extracted features.

Each iteration of GA, parents are selected for mating using the Roulette Wheel method. Uniform crossover is applied with a swapping probability of 0.5. The crossover probability is set to 0.7. The mutation probability is set to 0.05.

```
GA for Feature reduction
Input: X_initial, size of initial population k, Max genera-
tion gmax. Output: Individual with the best fitness X_best
1. Generate initial population X_N, where X_N∈R^{k×n} and
X_initial ∈X_N.
2. k := 0;
3. for i = 1 to k
4. H_i ← trainSVM(X_Ni|x_d == 1)
5. Obtain fitness from H_i with (X_test; Y_test) :
6. end for
7. Generate new population X_N^g by selection, crossover
and mutation.
8. Add the best individual in the current population
X_best to the newly generated X_N to form the next popula-
tion.
9. k = size of new generation X_N^g
10. Return to 2 if g < gmax or the pre-specified stopping
condition is not satisfied.
11. Obtain positions (1's) of best individual X_best
```

3 Experimental Results

3.1 Classification Metrics

In the experimental results presented in this research the evaluation metrics used for classification and recognition were the following:

1. Classified rate, which is defined by the number of samples correctly classified (CS) and the total number of samples (TS), $Cr = CS = TS$.

2. Sensitivity, defined by the number of positive samples correctly classified and True Positive Samples $Sn = TP = (TP + FN)$.

3. Error rate, defined by the number of samples incorrectly classified (iCS) and samples correctly classified (CS), $Er = iCS = CS$.

3.2 Results

In the experimental results were used 60 % to training (3286) and 40 % to test (2192). The Fig. 3 shows the classified rate results obtained with the 249 signs. In red are described the results obtained using only one frame to identify a sign (OF). In black are shown the results obtained using the proposed method with all the features (CCH) and in blue are shown the results obtained using the proposed method with only the features obtained optimizing with the genetic algorithm (OCH). The mean of classified rate with the proposed method is 0.9704 in comparison with the other methods 0.9191 (CCH) and 0.8565 (OF).

The mean of error rate obtained with the proposed method is 0.0308. The accuracy of the proposed method to identify a sign is very good when is used the proposed method. The separating hyperplane obtained by the SVM in this training phase is the optimal for the data set. However, some authors cite [2, 9, 10, 14], have mentioned that to test the model, on deaf people from which not images were obtained, the accuracy drops considerably, this is because very representative data (signs of left-handed deaf people or idioms) were not learned during the training phase. On the other hand, the sensitivity of the proposed method is 99.03 in comparison with 94.71 and 91.32 obtained when are used all the features and only one frame is used respectively.

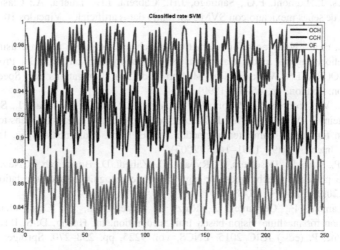

Fig. 3. Classified rate. (Color figure online)

4 Conclusions

This research presents a methodology for recognition of Mexican Sign Language. The proposed system extracts a set of geometric features from several images in video sequences. The combination of the features used and the sequence of images allow obtaining information very representative from images. The proposed system tries to emulate the way the human brain processes movements.

The combination of feature extractors and the method used help to improve the discriminative power of the model. The results show that the signs recognition system helps to improve the classifier's performance by containing different movements of the hands. Moreover, the proposed method extracts the most discriminative features, concentrating on the most important information, leaving out less representative features. Extracting only the necessary features helps to reduce significantly the time to recognize a new sign.

Acknowledgments. The authors would like to acknowledge the support of the Mexican Counsel of Science and Technology and UAEM, Project UAE/3778/CYB.

References

1. Al-Roussan, M., Assaleh, K., Tala'a, A.: Video-based signer-independent Arabic sign language recognition using hidden Markov models. Appl. Soft Comput. **9**(3), 990–999 (2009)
2. Assa, M., Grobel, K.: Video-based sign language recognition using Hidden Markov Models. In: Proceedings of Gesture Workshop, pp. 97–109 (1997)
3. Bauer, B., Hienz, H.: Relevant features for video-based continuous sign language recognition. In: Proceedings of FG 2000. IEEE Computer Society, Washington, D.C., pp. 440–450 (2000)
4. Cervantes, J., Lamont, F.G., Santiago, J.H., Cabrera, J.E., Trueba, A.: Clasificacion del lenguaje de señas mexicano con SVM generando datos artificiales. Vinculos **10**(1), 328–341 (2013)
5. Cole, R., et al.: New tools for interactive speech and language training: using animated conversational agents in the classrooms of profoundly deaf children. In: Proceedings of ESCA/SOCRATES Workshop on Method and Tool Innovations for Speech Science Education, London, pp. 45–52 (1999)
6. Cole, R., Van Vuuren, S., Pellom, B., Hacioglu, K., Ma, J., Movellan, J., Schwartz, S., Wade-Stein, D., Ward, W., Yan, J.: Perceptive animated interfaces: first steps toward a new paradigm for human computer interaction. IEEE Trans. Multimedia Spec. Issue Human Comput. Interact. **91**(9), 1391–1405 (2003)
7. Dreuw, P., Stein, D., Deselaers, T., Rybach, D., Zahedi, D., Bungeroth, J., Ney, H.: Spoken language processing techniques for sign language recognition and translation. Technol. Disabil. **20**(2), 121–133 (2008)
8. García-Lamont, F., Cervantes, J., Ruiz, S., López-Chau, A.: Color characterization comparison for machine vision-based fruit recognition. In: Huang, D.-S, Bevilacqua, V., Premaratne, P. (eds.) ICIC 2015. LNCS, vol. 9225, pp. 258–270. Springer, Heidelberg (2015)
9. Fang, G.L., Gao, W., Zhao, D.B.: Large vocabulary sign language recognition based on fuzzy decision trees. IEEE Trans. Syst. Man Cybernet. **34**(3), 305–314 (2004)
10. Kadous, M.W.: Machine recognition of Auslan signs using PowerGloves: towards large-lexicon recognition of sign language. In: Proceedings of Workshop Integration of Gestures in Language and Speech, pp. 165–174 (1996)
11. San-Segundo, R., Barra, R., Córdoba, R., D'Haro, L.F., Fernández, F., Ferreiros, J., Lucas, J. M., Macías-Guarasa, J., Montero, J.M., Pardo, J.M.: Speech to sign language translation system for Spanish. Speech Commun. **50**, 1009–1020 (2008)
12. San-Segundo, R., Pardo, J.M., Ferreiros, J., Sama, V., Barra-Chicote, R., Lucas, J.M., Sánchez, D., García, A.: Spoken Spanish generation from sign language. Interact. Comput. **22**(2), 123–139 (2010)
13. Tanibata, N., Shimada, N., Shirai, Y.: Extraction of hand features for recognition of sign language words. In: International Conference on Vision Interface, pp. 391–398 (2002)
14. Vamplew, P., Adams, A.: Recognition of sign language gestures using neural networks. Aust. J. Intell. Inf. Process. Syst. **5**(2), 94–102 (1998)
15. Zahedi, M., Dreuw, P., Rybach, D., Deselaers, T., Ney, H.: Using geometric features to improve continuous appearance-based sign language recognition. In: British Machine Vision Conference (BMVC), vol. 3, Edinburgh, UK (2006)

Robust Epileptic Seizure Classification

Farrikh Alzami[1], Daxing Wang[1], Zhiwen Yu[1(✉)], Jane You[2],
Hau-San Wong[3], and Guoqiang Han[1]

[1] School of Computer Science and Engineering,
South China University of Technology, Guangzhou, China
zhwyu@scut.edu.cn
[2] Department of Computing, Hong Kong Polytechnic University,
Hong Kong, China
csyjia@comp.polyu.edu.hk
[3] Department of Computer Science, City University of Hong Kong,
Hong Kong, China

Abstract. A lot of feature vectors and sub-band signals are considered for Epileptic seizure classification. Unfortunately, not all the feature vectors and sub-band signals contribute to the final result. In view of this limitation, we propose a modified Differential Evolution Feature Selection algorithm (MDEFS), which searches the best feature vector subset and the sub-band signals to distinguish three groups of subjects (healthy, ictal and interictal). From the experiment results, it is observed that the bagging method based on the optimal feature subset (the standard deviation attribute in the delta sub-band signal, the time-lag attribute in the delta sub-band signal, fractal dimension in the alpha sub-band signal, the correlation dimension attribute in the alpha sub-band signal and the standard deviation attribute in the beta sub-band signal) selected by MDEFS results in highest classification accuracy of 98.67 %.

Keywords: Epileptic seizure classification · Feature selection · Differential evolution · Bagging · Discrete Wavelet Transform

1 Introduction

Epilepsy is one of the major diseases in the world, especially for children in the developed countries. About 1 % of people worldwide (65 million) have epilepsy, and nearly 80 % of the cases occur in developing countries [1, 2]. Epilepsy is a group of neurological disorders characterized by epileptic seizures. Epileptic seizures are the result of excessive and abnormal cortical nerve cell activity in the brain. Approximately, one out of every three individuals with epilepsy continues to experience frequent epileptic seizure. This seizure poses a serious risk of injury, and also limits the mobility and independence of an individual [3]. Early detection of this disease is crucial for timely treatment to minimize further deterioration. Electroencephalogram (EEG) is usually used for epileptic seizure detection, and epilepsy diagnosis is performed through identification of EEG abnormalities [4]. Since EEG readings are checked by expert neurologist, there are some concerns about accuracy caused by visual fatigue, time and cost for regular examinations. Computer-aided diagnosis

© Springer International Publishing Switzerland 2016
D.-S. Huang and K.-H. Jo (Eds.): ICIC 2016, Part II, LNCS 9772, pp. 363–373, 2016.
DOI: 10.1007/978-3-319-42294-7_32

approaches by EEG time series data analysis offer an efficient solution to epilepsy diagnosis. EEG datasets for epileptic seizure usually contain three groups of subjects, such as (1) healthy subject, (2) epileptic subject during seizure (ictal), and (3) epileptic subject during seizure free interval (interictal).

Due to the non-stationary nature of EEG signals, wavelet analysis is a suitable analysis approach since wavelet transform gives precise frequency information at low frequencies and precise time information at high frequencies [5]. Many research using wavelet transform, especially DWT, to preprocess the EEG signals and decompose them into sub-bands signals. Features which include energy [6, 7], skewness [8], kurtosis [8], max [9, 10], min [9, 10], mean [7–10], median [10], standard-deviation [7, 9, 11], time-lag [12], embedding-dimension [12], approximate-entropy [5, 13, 14], Sample-entropy [14, 15], Fractal-dimension [16, 17], Correlation-dimension [11, 12], Largest-Lyapunov-Exponent [9, 11, 12], etc., are then extracted from those signals. However, not all the feature vectors contribute to the improvement of the performance of the epileptic seizure classification approach. For example, some features, such as Largest Lyapunov Exponent, Embedding dimension and Correlation Dimension, are time consuming to calculate.

In order to address the limitations of traditional epileptic seizure classification approaches, we design a modified Differential Evolution Feature Selection framework (MDEFS) to search for the best subset of feature vectors with respect to the sub-band signals, and identify three groups of epileptic seizure subjects. Our contribution is two-fold: First, a feature selection based epileptic seizure classification approach is proposed to identify different epileptic seizure subjects. Second, the modified differential evolution feature selection algorithm is designed to select a stable feature subset.

The remainder of the paper is organized as follows. Section 2 introduces the process of epileptic seizure classification. Section 3 describes the modified Differential Evolution Feature Selection approach. Section 4 experimentally investigates the performance of the Modified Differential Evolution Feature selection approach and Sect. 5 presents the conclusion.

2   Epileptic Seizure Classification

Figure 1 provides an overview of the feature selection based epileptic seizure classification approach. Specifically, we first adopt a band-pass FIR filter set from 0.5 Hz to 60 Hz to preprocess the EEG raw data, and extract 5 sub-band signals which include delta, theta, alpha, beta and gamma. The sub-band band-limited EEG signal with 0–60 Hz is also considered. As a result, there are six sub-band signals. Then, 15 features are extracted from each sub-band signal, which include energy, skewness, kurtosis, max, min, mean, median, standard-deviation, time-lag, embedding-dimension, approximate-entropy, sample-entropy, fractal-dimension, correlation-dimension and Largest-Lyapunov-Exponent. Next, the modified differential evolution feature selection algorithm (MDEFS) is designed to extract the optimal feature subset. Finally, classifiers which include k nearest neighbor classifier, the Naive Bayes classifier, the decision tree classifier, the support vector machine, the back propagation neural network (BPNN) and the Bagging approach using LMBPNN, are adopted to perform epileptic seizure classification using the features selected by MDEFS.

Fig. 1. An overview of the feature selection based epileptic seizure classification approach

3 Modified Differential Evolution Feature Selection

Traditional Differential Evolution Feature selection (DEFS) approach [18] utilizes the differential evolution optimizer for combinatorial optimization problem, which applies roulette wheel selection. One of the limitation of DEFS for feature selection is that the result is unstable, which means the selected feature vectors are different for each run. This limitation prevents the application of DEFS for epileptic seizure classification. To address the limitation of DEFS, we design the Modified Differential Evolution Feature Selection (MDEFS) to select the stable feature subsets.

Algorithm 1. The modified differential evolution feature selection algorithm

Require:

    Input: a dataset F with n data samples;

          the maximum number of repetitions T;

          the maximum number of generations H;

          the parameters β, γ;

1:   Initialization and parameter setting;

2:   **For** $t = 1$ to T

3:      Generate a population with P individuals $X = \{x_1, x_2, \ldots, x_P\}$ (where the entry value 1 of the indicator vector x_p (where $p \in \{1, \ldots, P\}$) means that the feature is selected, and 0 means that the feature is not selected);

4:      Calculate fitness function $\Phi(x_p)$ for each individual x_p;

5:      **For** $h = 1$ to H

6:         Apply the differential operator to generate a new individual x'_p using the Eqs.(1)(2) based on β;

7:         Apply the cooperation operator using the Eqs.(3) to generate the new individual x''_p based on τ;

8:         **If** there exists redundancy in the new individual x''_p

9:            Compute the relative attribute distribution factor ζ using Eqs(4)-(7);

10:        Put ζ to the roulette wheel to remove the redundancy;

11:      **Endif**

12:      Add the new individual x''_p to the set X;

13:      Maintain the best P individuals;

14:   **End For**

15:   Select the best individual x^* with the largest value of the fitness function from the optimal set X' using Eq. (8)

16: **End For**

17: Obtain T best individuals x^*_t ($t \in \{1, \ldots, T\}$);

18: Calculate the frequency of the individual x^*_t, and sort the individuals x^*_t according to its frequency;

19: Select the individual x^*_t with the largest frequency;

    Output: the selected attribute subset.

Algorithm 1 provides a step-by-step description of the modified differential evolution feature selection algorithm. The input of the MDEFS is a dataset with n sample pairs $F = \{(f_1, y_1), (f_2, y_2), \ldots, (f_n, y_n)\}$, the maximum number of runs T, the maximum number of generations H, and a predefined parameter β. MDEFS will perform T runs. In each run, it first generates a population with P individuals $X = \{x_1, x_2, \ldots, x_P\}$. (where $p \in \{1, \ldots, P\}$ is an indicator vector. The entry value 1 of the indicator vector x_p means that the feature is selected, while 0 of the indicator vector x_p means that the feature is not selected). MDEFS also calculates the fitness function $\Phi(x_p)$ for each individual x_p, which is defined as the accuracy of the classifier on F with respect to the features selected by MDEFS.

Then, MDEFS goes through the generations $h \in \{1, \ldots, H\}$ one by one. In each generation, it applies the differential operator [18] and the cooperation operator. Given three randomly generated indices r_0, r_1 and r_2 (where $r_0, r_1, r_2 \in \{1, \ldots, P\}$), the new individual x_p' is computed as follows:

$$x_p' = x_{r0} + \alpha \times (x_{r1} - x_{r2}) \tag{1}$$

$$\alpha = \frac{\beta \times rand}{\max(x_{r1}, x_{r2})} \tag{2}$$

where $\beta \in [0, 1]$ is a predefined parameter. Given two individuals x_p' and x_p, the cooperation operator is applied as follows:

$$x_{p,l}'' = \begin{cases} x_{p,l}', & \text{if } rand(0, 1) \leq \tau \\ x_{p,l}, & \text{otherwise} \end{cases} \tag{3}$$

where $\tau \in [0, 1]$ is a predefined parameter, and $l \in \{1, \ldots, m\}$, m is the number of attributes.

Next, MDEFS checks if there exists redundancy in the new individual x_p'' using the roulette wheel. If there does not exist redundancy, the new individual x_p'' will be added to the set X directly. Otherwise, the relative attribute distribution factor ζ in the roulette wheel is used to substitute the redundant index. The relative attribute distribution factor ζ supplied to the roulette wheel is computed as follows:

$$\zeta = \left(\Omega^{*h+1}(a_1) - \Omega^{*h}(a_1)\right) \times \Omega^{*h+1}(a_1) + \Omega^{*h}(a_1) \tag{4}$$

$$\Omega^{*h}(a_1) = \frac{\Omega^h(a_1)}{max_l\left(\Omega^h(a_1)\right)} \tag{5}$$

$$\Omega^{*h+1}(a_l) = \frac{\Omega^{h+1}(a_l)}{max_l(\Omega^{h+1}(a_l))} \tag{6}$$

where $\Omega^h(a_l)$ and $\Omega^{h+1}(a_l)$ are the attribute distributions of the attribute a_l in the h-th generation and in the $h + 1$-th generation, respectively. The attribute distribution $\Omega^h(a_l)$ of a_l is calculated as follows:

$$\Omega^h(a_l) = \gamma \times (\frac{\theta_1}{\theta_1 + \theta_2}) + \frac{m - m'}{m} \times (1 - \frac{\theta_1 + \theta_2}{max(\theta_1 + \theta_2)}) \qquad (7)$$

where γ is a positive constant, m' is the number of selected attributes, θ_1 is the frequency of the attribute a_l whose fitness function value is larger than the mean fitness function value of the whole population, and θ_2 is the frequency of the attribute a_l whose fitness function value is less than the mean fitness function value of the whole population. The generated individual using the roulette wheel will be included in the set X. Finally, the individuals in the set X are sorted according to the values of the fitness function, and the best P individuals will be maintained.

After H generations, MDEFS obtains the optimal set X', and selects the best individual x^* with the largest value of the fitness function from X' as follows:

$$x^* = \arg\max_{x_p \in X'} \Phi(x_p) \qquad (8)$$

After T runs, there are T best individuals x_t^* ($t \in \{1, \ldots, T\}$). MDEFS calculates the frequency of the individual x_t^*, and sorts the individuals x_t^* according to its frequency. The individual x_t^* with the largest frequency will be the output of MDEFS, and the corresponding attributes indicated by x_t^* with the largest frequency will be selected by MDEFS as the attribute subset. The classifier is trained by the attribute subset generated by MDEFS, and used to identify different epileptic seizure subjects.

4   Experiment

In our experiments, we adopt the Bonn dataset, which is referred to as SZN, from [19], which includes three classes of epileptic seizure subjects: the healthy subject, the epileptic subject during seizure (ictal) and epileptic subject during seizure free interval (interictal). According to different combinations of sub-band signals, we generate different datasets: (1) the SZN4 dataset includes the sub-band signals delta, theta, alpha and beta. (2) The SZN4BL dataset includes the sub-band signals: delta, theta, alpha, beta and band-limited. (3) The SZN5 dataset includes the sub-band signals delta, theta, alpha, beta and gamma. (4) The SZN6 dataset includes the sub-band signals delta, theta, alpha, beta, gamma and band-limited.

The proposed feature selection based epileptic seizure classification approach is measured by the average accuracy on the dataset. Ten-fold crossover validation is adopted to reduce the effect of randomness. The maximum number of runs T, the maximum number of generations H, the number of individuals in the population P and the parameter τ are set to 150, 100, 50 and 0.5, respectively.

The classifiers in the experiment include the k nearest neighbor classifier (kNN), the Naive Bayes classifier (NB), the decision tree classifier (DT), the pseudo Quadratic

Discriminant Analysis (QDA), the support vector machine (SVM, [20]), the back propagation neural network (BPNN, [21]) and the Bagging approach using LMBPNN (Bagging).

In the following experiments, we first explore the effect of MDEFS. Then, MDEFS is compared with other feature selection techniques for epileptic seizure classification.

4.1 The Effect of MDEFS

In order to explore the effect of MDEFS, we compare the epileptic seizure classification approach based on MDEFS and the approach based on DEFS with respect to different classifiers and different number of selected attributes on the SZN4 dataset, the SZN5 dataset and the SZN6 dataset. Table 1 shows the results obtained by the approaches based on MDEFS and DEFS respectively. It is observed that the approach based on MDEFS outperforms that based on DEFS on most of the different combinations of the conditions. For example, the approach based on MDEFS obtains the better accuracy value 0.96 with respect to the classifier SVM and the number of attributes 5 on the SNZ4 dataset, which is 0.1 larger than that obtained by the approach based on DEFS. The possible reason is that MDEFS adopts the individual with the larger frequency in T runs as the output, which will provide a robust attribute subset and improve the final result.

In addition, the approach based on MDEFS achieves the best performance with an accuracy value of 0.9867 on the SZN4 dataset when the number of selected attributes is set to 5 and the bagging approach is adopted. The corresponding attribute subset includes five attributes: the standard deviation attribute in the delta sub-band signal, the time-lag attribute in the delta sub-band signal, the fractal dimension in the alpha sub-band signal, the correlation dimension attribute in the alpha sub-band signal, and the standard deviation attribute in the beta sub-band signal.

4.2 The Comparison with Other Feature Selection Techniques

In order to explore the performance of different feature selection based epileptic seizure classification approaches, we apply the feature selection techniques on the SZN4, SZN4BL, SZN5 and SZN6 datasets with respect to the number of selected attributes m' from 2 to 8 using all the classifiers. The feature selection techniques include the Relief-F technique (Relief) for the multi-class problem [22], the genetic algorithm (GA), the feature selection approach based on minimum redundancy maximum relevance (mRMR) [23] and our proposed MDEFS approach.

Table 2 shows the best performance of different feature selection based epileptic seizure classification approaches corresponding to the different datasets and the number of selected attributes. For example, Relief-F achieves the best performance on the SZN4 dataset when the number of selected attributes is set to 8. It can be seen that when the best conditions are given, MDEFS obtains the best results with respect to six classifiers, which are the k nearest neighbor classifier (kNN), the naive bayes (NB), the decision tree classifier (DT), the pseudo Quadratic Discriminant Analysis (QDA), the

Table 1. The Comparison of DEFS and MDEFS (where SZN4_2 means that the number of selected attributs is 2 on the SZN dataset)

	KNN		Naïve Bayes		DT		LDA		BPNN		SVM		Bagging	
	DEFS	MDEFS	DEFS	MDEFS	DEFS	MDEFS	DEFS	MDEFS	DEFS	MDEFS	DEFS	MDEFS	DEFS	MDEFS
SZN4_2	**0.9633**	**0.9633**	**0.8600**	**0.8600**	**0.9567**	**0.9567**	**0.9567**	**0.9567**	0.9467	**0.9600**	0.9100	**0.9167**	**0.9700**	**0.9700**
SZN4_3	**0.9707**	0.9700	0.7920	**0.8200**	0.9593	**0.9633**	0.9360	**0.9533**	0.9300	**0.9567**	**0.9100**	0.9067	**0.9600**	**0.9600**
SZN4_4	0.9740	**0.9767**	NaN	**0.9167**	0.9633	**0.9733**	0.9253	**0.9700**	0.9467	**0.9633**	0.8667	**0.9633**	0.9500	**0.9833**
SZN4_5	**0.9767**	**0.9767**	0.9000	**0.9233**	0.9640	**0.9733**	0.9500	**0.9730**	0.9333	**0.9533**	0.8600	**0.9600**	0.9400	**0.9867**
SZN5_2	0.9667	0.9667	**0.9433**	**0.9433**	**0.9600**	**0.9600**	0.9567	**0.9567**	0.9200	**0.9500**	0.9367	**0.9400**	**0.9600**	**0.9600**
SZN5_3	0.9720	**0.9733**	0.8787	**0.9433**	0.9620	0.9600	0.9593	**0.9667**	0.9433	**0.9500**	0.9267	**0.9300**	**0.9467**	**0.9467**
SZN5_4	**0.9680**	0.9667	0.8693	**0.9330**	**0.9573**	0.9567	**0.9513**	0.9500	0.9100	**0.9300**	0.9033	**0.9167**	0.9333	**0.9367**
SZN5_5	0.9660	0.9667	**0.9167**	NaN	**0.9593**	0.9567	0.9273	**0.9567**	**0.9333**	0.9300	0.9100	**0.9233**	**0.9467**	0.9367
SZN6_2	0.9773	**0.9800**	**0.9353**	0.9333	0.9707	**0.9733**	0.9593	**0.9600**	0.9567	**0.9633**	0.9467	**0.9533**	**0.9767**	**0.9767**
SZN6_3	0.9753	**0.9800**	0.9113	**0.9300**	0.9667	**0.9767**	0.9587	**0.9700**	0.9500	**0.9733**	0.9333	**0.9533**	0.9567	**0.9833**
SZN6_4	**0.9773**	0.9667	0.9213	**0.9330**	**0.9673**	0.9567	**0.9560**	0.9500	0.9267	**0.9300**	0.9100	**0.9167**	**0.9367**	**0.9367**
SZN6_5	**0.9753**	0.9667	0.9313	**0.9367**	**0.9640**	0.9567	0.9433	**0.9567**	0.9267	**0.9367**	0.9033	**0.9100**	0.9367	**0.9533**

Table 2. The comparison with other feature selection techniques (where m' is the number of selected attributes for the classification task)

Methods	Datasets	m'	KNN	NB	DT	QDA	BPNN	SVM	Bag
Relief-f	SZN4	8	0.93	0.9167	0.96	0.97	0.96	0.9467	0.9767
GA	SZN5	4	0.9633	0.8533	0.9567	0.96	0.8733	0.8233	0.9533
mRMR	SZN4BL	5	0.9267	0.92	0.9633	**0.9733**	**0.9733**	0.9533	0.9833
MDEFS	SZN4	5	**0.9767**	**0.9233**	**0.9733**	**0.9733**	0.9533	**0.96**	**0.9867**

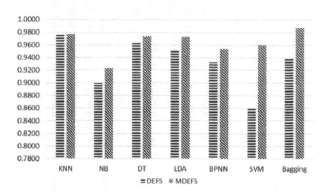

Fig. 2. Accuracy comparison between DEFS and MDEFS using SZN4_5

support vector machine (SVM) and the Bagging approach (Bagging). As a result, MDEFS is the best choice when applied to epileptic seizure classification (Fig. 2).

4.3 The Comparison with Other Epileptic Seizure Classification Approaches

We compare the MDEFS based epileptic seizure classification approaches with other existing epileptic seizure classification approaches on the Bonn dataset, which include the Samanwoy's approach [11], the Guler's approach [9], the Nabeel's approach [24] and the Hsu's approach [12]. Table 3 shows the best results obtained by different epileptic seizure classification approaches. It is observed that our proposed approach achieves the best performance on the Bonn dataset with an accuracy value 0.9867, when compared with its competitors. The possible reason is that our proposed approach not only adopts MDEFS to generate a stable attribute subset, but also uses the bagging technique to generate a set of new datasets and obtain a more accurate and robust final result. The Samanwoy's approach also obtains a satisfactory result.

Table 3. The comparison with other epileptic seizure classification on the Bonn dataset

Methods	m'	Classifier	Accuracy
Samanwoy [11]	9	BPNN	0.967
Guler [9]	24	SVM	0.756
Nabeel [24]	24	QDA	0.842
Hsu [12]	13	SVM	0.876
Our method	5	Bagging + MDEFS	**0.9867**

5 Conclusion

In this paper, we investigate the problem of how to perform robust epileptic seizure classification. The main contribution is a new feature selection based epileptic seizure classification approach and a modified differential evolution feature selection algorithm. The experiment results show that the epileptic seizure classification based on the bagging technique and MDEFS achieves the best performance with the classification accuracy 98.67 %.

The optimal attribute subset selected by our proposed approach includes the standard deviation attribute in the delta sub-band signal, the time-lag attribute in the delta sub-band signal, the fractal dimension in the alpha sub-band signal, the correlation dimension attribute in the alpha sub-band signal and the standard deviation attribute in the beta sub-band signal.

Based on this research, the features [delta-SD, delta-TimeLag, alpha-FD, alpha-CD, beta-SD] that are obtained from Modified Differential Evolution Feature Selection yield the most accurate results among other feature selection methods such as Relief-F, Genetic Algorithm, mRMR and DEFS. These results also indicate that we do not need to use all the sub-band signals for epileptic seizure classification. In addition, we observe that Bagging using LMBPNN is slightly better than other classification approaches for epileptic seizure detection.

Future work would include applying MDEFS with other feature selection to be used in our bagging framework algorithm.

Glossary of Terms/Acronyms

DWT	Discrete Wavelet Transform
Ictal	State or event epileptic seizure
Interictal	State or event between epileptic seizures
SZN	The dataset that seizure is denoted S, healthy Z and interictal is N
LMBPNN	Levenberg-Marquadt Back Propagation Neural Network

References

1. Leppik, I.: Contemporary Diagnosis and Management of the Patient with Epilepsy. Handbooks in Health Care, Newton (2000)
2. Fisher, R., Van Boas, W.E., Blume, W., Elger, C., Genton, P., Lee, P., Engel, J.: Special article epileptic seizures and epilepsy: definitions proposed by the International League Against Epilepsy (ILAE) and the International Bureau for Epilepsy (IBE). Epilepsia 46(4), 470–472 (2005)
3. Shoeb, A.: Application of Machine Learning to Epileptic Seizure Onset Detection and Treatment, Massachusetts (2009)
4. Samanwoy, G.-D., Adeli, H., Dadmehr, N.: Principal component analysis-enhanced cosine radial basis function neural network for robust epilepsy and seizure detection. IEEE Trans. Biomed. Eng. 55(2), 512–518 (2008)
5. Ocak, H.: Optimal classification of epileptic seizures in EEG using wavelet analysis and genetic algorithm. Signal Process. 88, 1858–1867 (2008)
6. Shoeb, A., Edwards, H., Connolly, J., Bourgeois, B., Treves, S., Guttag, J.: Patient-specific seizure onset detection. Epilepsy Behav. 5, 483–498 (2004)
7. Panda, R., Khobragade, P., Jambhule, P., Jengthe, S., Pal, P., Gandhi, T.: Classification of EEG signal using wavelet transform and support vector machine for epileptic seizure diction. In: Proceedings of 2010 International Conference on Systems in Medicine and Biology, pp. 405–408 (2010)
8. Nageswari, M., Banu, U., Kumar, K., Sujith, S.: Feature extraction of ECG using Daubechies wavelet and classification based on fuzzy c-means clustering technique. In: Proceeding of National Conference on Control, Communication and Information Technology, pp. 43–47 (2013)
9. Guler, I., Ubeyli, E.: Multiclass support vector machines for EEG-signals classification. IEEE Trans. Inf. Technol. Biomed. 11(2), 117–126 (2007)
10. Garg, S., Narvey, R.: Denoising & feature extraction of EEG signal using wavelet transform. Int. J. Eng. Sci. Technol. (IJEST) 5(6), 1249–1253 (2013)
11. Samanwoy, G.-D., Adeli, H., Dadmehr, N.: Mixed-band wavelet-chaos-neural network methodology for epilepsy and epileptic seizure detection. IEEE Trans. Biomed. Eng. 54(9), 1545–1551 (2007)
12. Hsu, K.-C., Yu, S.-N.: Detection of seizures in EEG using subband non linear parameters and genetic algorithm. Comput. Biol. Med. 40, 823–830 (2010)
13. Guo, L., Rivero, D., Pazos, A.: Epileptic seizure detection using multiwavelet transform based approximate entropy and artificial neural networks. J. Neurosci. Methods 193, 156–163 (2010)
14. Song, Y., Crowcroft, J., Zhang, J.: Automatic epileptic seizure detection in EEGs based on optimized sample entropy and extreme learning machine. J. Neurosci. Methods 210, 132–146 (2012)
15. Han, L., Wang, H., Liu, C., Li, C.: Epileptic seizure detection using wavelet transform based sample entropy and support vector machine. In: Proceeding of the IEEE International Conference on Information and Automation, pp. 759–762 (2012)
16. Easwaramoorthy, D., Uthayakumar, R.: Analysis of biomedical EEG signals using wavelet transforms and multifractal analysis. In: Communication Control and Computing Technologies (ICCCCT), pp. 544–549 (2010)
17. Khoa, T., Ha, V., Toi, V.: Higuchi fractal properties of onset epilepsy electroencephalogram. Comput. Math. Methods Med. 2012, 1–6 (2012)

18. Khushaba, R., Al-Ani, A., Al-Jumaily, A.: Feature subset selection using differential evolution and a statistical repair mechanism. Expert Syst. Appl. **38**, 11515–11526 (2011)
19. EEG time series download page. http://www.meb.unibonn.de/epileptologie/science/physik/eegdata.html. Accessed 2001
20. Chang, C.-C., Lin, C.-J.: LIBSVM: a library for support vector machines. ACM Trans. Intell. Syst. Technol. (TIST) **2**(3), 27:1–27:27 (2011)
21. Suratgar, A., Tavakoli, M., Hoseinabadi, A.: Modified Levenberg-Marquardt method for neural networks training. World Acad. Sci. Eng. Technol. Int. J. Comput. Electr. Autom. Control Inf. Eng. **1**(6), 1745–1747 (2007)
22. Kononenko, I., Sikonja, M.: Non-Myopic Feature Quality Evaluation with (R)ReliefF. Computational Methods of Feature Selection, pp. 169–191. Chapman & Hall/CRC, Boca Raton (2008)
23. Peng, H., Long, F., Ding, C.: Feature selection based on mutual information: criteria of max-dependency, max-relevance, and min-redundancy. IEEE Trans. Pattern Anal. Mach. Intell. **27**(8), 1226–1238 (2005)
24. Nabeel, A., Thasneem, F., Paul, J.: Detection of epileptic seizure event and onset using EEG. BioMed Res. Int. **2014**, 1–7 (2014)

A Simple Review of Sparse Principal Components Analysis

Chun-Mei Feng[1], Ying-Lian Gao[2], Jin-Xing Liu[1,3(✉)],
Chun-Hou Zheng[1,3], Sheng-Jun Li[1], and Dong Wang[1]

[1] School of Information Science and Engineering,
Qufu Normal University, Rizhao 276826, China
fengchunmei0304@foxmail.com,
{sdcavell,zhengch99,dongwshark}@126.com,
qfnulsj@163.com
[2] Library of Qufu Normal University, Qufu Normal University,
Rizhao 276826, China
yinliangao@126.com
[3] Bio-Computing Research Center, Shenzhen Graduate School,
Harbin Institute of Technology, Shenzhen 518055, China

Abstract. Principal Component Analysis (PCA) is a common tool for dimensionality reduction and feature extraction, which has been applied in many fields, such as biology, medicine, machine learning and bioinformatics. But PCA has two obvious drawbacks: each principal component is line combination and loadings are non-zero which is hard to interpret. Sparse Principal Component Analysis (SPCA) was proposed to overcome these two disadvantages of PCA under the circumstances. This review paper will mainly focus on the research about SPCA, where the basic models of PCA and SPCA, various algorithms and extensions of SPCA are summarized. According to the difference of objective function and the constraint conditions, SPCA can be divided into three groups as it shown in Fig. 1. We also make a comparison among the different kind of sparse penalties. Besides, brief statements and other different classifications are summarized at last.

Keywords: Principal component analysis · Sparse principal component analysis · Rotation · Sparse constraint

1 Introduction

Principal Component Analysis (PCA) as a method for dimension reduction and feature selection has become more and more popular in the fields of biology, medicine, machine learning and bioinformatics [1]. But PCA has two drawbacks: one is that each Principal Component (PC) is a linear combination of all variables; the other is that their loadings are always non-zero, which is difficult to interpret or extract features. In order to solve the shortcomings of PCA, many approaches have been proposed over the last few years. Various literatures about SPCA will be shown in this paper, which can be divided into three categories: different algorithms, the extension of SPCA and theoretical analysis. Besides, there are many literatures about the applications of SPCA.

© Springer International Publishing Switzerland 2016
D.-S. Huang and K.-H. Jo (Eds.): ICIC 2016, Part II, LNCS 9772, pp. 374–383, 2016.
DOI: 10.1007/978-3-319-42294-7_33

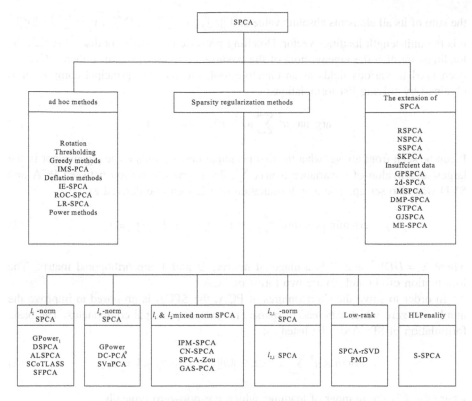

Fig. 1. The classification of SPCA algorithms in this paper.

Richtárik et al. only summarized eight formulations of SPCA [2] and Trendafilov only reviewed several most popular methods in its literature but certainly inadequate. Meanwhile, we have different opinions in many places and more algorithms are added in this paper.

The rest of our paper is organized as follows: Sect. 2 gives some review of PCA and basic formulations of SPCA; Sect. 3 discusses the development of SPCA and gives a figure about the category of various SPCA; we also make some brief statement of many literatures on theoretical analysis and different classifications in Sect. 4.

2 A Notation of PCA and Formulation of Sparse PCA

In this section, we summarize the development of PCA simply. PCA was first mentioned and applied the least squares method to find the first principal component. Besides, PCA can be computed by covariance matrix. Supposing $X = [X_1, \ldots, X_n]^T \in R^{n \times d}$ is a data matrix and where n is the number of samples and d is the number of variables; $\Sigma = \frac{1}{n}X^T X \in R^{d \times d}$ is the data covariance matrix of the original data. $\|x\|_0$ is the number of non-zero entries or l_0 norm, $\|x\|_1$ is the l_1-norm, which is

the sum of its all elements absolute value and $\|x\|_{2,1} = \sum_{i=1}^{n} \sqrt{\sum_{j=1}^{s} x_{ij}^2} = \sum_{i=1}^{n} \|x^i\|_2$.
u is the unit-length loadings vector. Hotelling proposed a significant discovery that the loadings are also the eigenvectors of the covariance matrix [3]. Since then, PCA has been used in various fields as an effective tool and the first principal component is obtained by solving list formulations:

$$\arg\max_u u^T \sum u, \quad s.t. \quad \|u\|_2 = 1. \tag{1}$$

It can be seen from above, what the first principal corresponds is the eigenvector in the largest eigenvalue of covariance matrix $\sum$. Then, the correlation between PCA and SVD was also set up. The new formulation of PCA can be derived as:

$$\arg\min_{d,u,v} \|X - duv^T\|_F^2, \quad s.t. \quad \|u\|_2 = 1, \|v\|_2 = 1, \tag{2}$$

where $X = UDV^T$ and D is a diagonal matrix, U and V are orthogonal matrix. The formulation of (1) and (2) are two forms of PCA.

In order to solve the shortcomings of PCA, the SPCA is proposed to improve the number of zero which can lead to sparse and extract principal components. The basic formulation of SPCA can be listed as:

$$\arg\max_u u^T \sum u \quad s.t. \quad \|u\|_2 = 1, \|u\|_0 \leq k = 1, \tag{3}$$

where the k is the number of loadings which are non-zero typically.

3   Overview of SPCA

3.1   Ad Hoc Methods

In order to obtain SPCA, the first article was proposed based on a technique which names rotation [4]. Jolliffe proposed different rotation technique to ensure receiving sparse loading vectors. What is unfortunate is that the rotation techniques might not promise to produce near zero loadings, especially absolute zero. Other technique like thresholding were proposed. Ma used a new Iterative Thresholding approach for principal components to obtain sparse eigenvectors [5], which are different from Simple Thresholding, Covariance Thresholding and Digonal Thresholding. The other literature proposed several greedy methods to produce a solution for every value of sparsity [6]. But it exits a fatal disadvantage which required a lot of eigenvalues to produce a result that we need. Therefore it makes expensive in computation. d'Aspremont et al. formulated a new semi-definite relaxation to solve the problem and derived a greedy algorithm to compute a series of solution about non-zero coefficient [7]. Another special approach used a branch and bound method to select the number of non-zero loadings [8], which names Information Maximization in Sparse Principal Components Analysis (IM-SPCA). The goal was to give a non-zero weight to variables, as the conditional of little information loss to original PCA. Mackey et al. derived a generalized deflation method to address the problem of iterative SPCA optimization [9]. Wang et al. proposed a new approach to SPCA which used an iterative elimination algorithm (IE-SPCA) to reconsider the problem of sparse [10]. This method was motivated by the Recursive Feature Elimination (RFE) technique which came from simple thresholding method. Kuleshov et al. used Generalized Rayleigh Quotient Iteration (GRQI) based on GPower method to establish a fast algorithms [11]. Another novel method was proposed already, it was the same as the measure [12] based on the power method and introduced a Truncated Power method for sparse eigenvalues problem which names TPower [13]. A new method of SPCA was based on recursive divide-and-conquer approach and we call it Recursive Divide-and-Conquer SPCA (RDC-SPCA) [14]. This method is divided classical SPCA (1) and (2) into a serious simpler sub-problem and added certain constraints can be equal to other methods, like non-negative SPCA [15].

3.2   Sparsity Regularization Methods

By imposing different regularization on the basic formulations of SPCA, the problems, which would arise in the case of pure Lagrange multiplier method or penalty function method. There are three common different ways of penalization: l_0, l_1 and $l_{2,1}$-regularization. In the field of SPCA, it seems l_1 based penalization is performed better than l_0.

There were several methods enforced l_0-regularization. GPower$_0$ method was aimed at extracting sparse PCs by using two single-unit and two block formulations besides enforcing l_0-norm. The formulation was listed as follows [12]:

$$\max_{x \in R^P, u^T u = 1} u^T \Sigma u - \gamma \|u\|_0. \tag{4}$$

Another excellent method applied a tighter approximation to modify the problem as a Difference of Convex functions (DC-PCA) program [27]. SVnPCA was a method of variable selection for SPCA. Ulfarsson et al. applied vector l_0-penalized and penalized Expectation-Maximization (pEM) algorithm to provide an excellent results [16].

Other literature were proposed based on l_1-regularization. A method of Simplified Component Technique Least Absolute Shrinkage and Selection Operator (SCoTLASS) [17] was a l_1-penalized formulation based on Guo et al. and a relaxation of SPCA [18]. d'Aspremont et al. proposed a direct approach (DSPCA) [19] that enhanced the principal components sparse degree by applying semi-definite programming (SDP). The convex relaxation approach can be considered as follows:

$$\max trace(\Sigma X), \quad s.t. \quad trace(X) = 1, 1^T |X|_1 \leq k, X \geq 0. \tag{5}$$

Subsequently, this approach not only computed the first eigenvector of the solution but also recorded its cardinality. Nevertheless, by evaluating a lower bound, they proposed another relaxation algorithm to avoid the necessity of the eigenvalue computations [20]. Aiming at finding sparse loadings and nearly uncorrelated PCs with orthogonal loading vectors, a new formulation was proposed to achieve sparse and a novel Augmented Lagrangian (AL) method to compute a series of non-smooth constrained optimization problems (ALSPCA) [21]. Another approach based on l_1-regularization (GPower$_1$) are equivalent with GPower$_0$ in the previous when replacing l_0 with l_1. Sparse fusion PCA (SFPCA) was proposed by enforcing l_1-norms penalty and a fusion penalty to capture natural "blocking" structures invariables [22]. This approach not only selected different variables for different components, but also encouraged the loadings of highly correlated variables. The proposal aimed to solve the following optimization problem:

$$\min_{A,B}\|X - XBA^T\|_F^2 + \lambda_1 \sum_k^K \|\beta_k\|_1 + \lambda_2 \sum_{k=1}^K \sum_{s<t} |\rho_s, t| |\beta_s, k - \sin g(\rho_s, t)\beta_{t,k}|, \qquad (6)$$
$$s.t. \quad A^T A = I_K,$$

where $\|X - XBA^T\|_F^2 = \sum_{i=1}^n \|X_i - AB^T X_i\|_2^2$, ρ_s, t is the relation between variables X_s. Supposing I_k was a $K \times K$ identity matrix; $B = [\beta_1 \ldots \beta_k]$ and β_k is a d-dimensional column vector; $A = [a_1, \ldots a_k]$ is a $d \times k$ matrix with orthogonal columns and I_k is a $K \times K$ identity matrix.

Four papers were proposed by enforcing l_1 and l_2 mixed regularization [23–25]. Among these, two papers used a new norm which was a convex combination of l_1 and l_2-norms instead of l_2-norm [23, 24]. Qi et al. used a new method of SPCA to replace the former l_0 and l_1-norm in traditional eigenvalue problems, we call this method as SPCA by choice of norm (CN-SPCA) [23]. Besides, an efficient iterative algorithm was proposed to solve the optimization problem. Among these methods we can obtain unrelated principal components or orthogonal loadings. Hein and Bühler used Inverse Power Method (IPM-SPCA) to nonlinear eigenvalue problems on SPCA [24]. Zou et al. (SPCA-Zou) enforced Lasso penalty (l_1) and ridge penalty (l_2) to (2), which solved the following regression-type problem [25]:

$$\min_{A,B}\left\{ \sum_{i=1}^n \|X_i - AB^T X_i\|_2^2 + \lambda \sum_{j=1}^K \|\beta_j\|_2^2 + \sum_{j=1}^k \lambda_{1,j}\|\beta_j\|_1 \right\} \quad s.t. \quad A^T A = I_K. \qquad (7)$$

Leng and Wang proposed a method of Simple Adaptive Sparse Principal Component Analysis (SAS-PCA), which enforced an adaptive Lasso penalty instead of Lasso penalty [26].

Furthermore, only one method for SPCA enforced $l_{2,1}$-penalty [27]. This method was based on SPCA-Zou [25] by modifying the model of regression: enforcing $l_{2,1}$-penalty instead of elastic net. Its advantage can overcome two weaknesses of SPCA-Zou [25]: generally selected feature of SPCA was independent; each PC was diverse. The model listed as follows [27]:

$$\arg\min_{\alpha,\beta} \sum_{i=1}^n \|\Sigma_i \beta\alpha^T - \Sigma_i\|_2^2 + \lambda\|\beta\|_{2,1}, \quad s.t. \quad \alpha^T\alpha = 1. \qquad (8)$$

From the perspective of low rank, two papers also generate sparse results [28, 29]. Shen and Huang proposed a novel approach (SPCA-rSVD) connected with SVD by solving a regularized low rank matrix approximation problem to achieve sparse [29]. Witten et al. proposed a framework which used penalized matrix decomposition (PMD) to compute a rank-k approximation for a matrix [28]. Supposing a matrix $X = \sum_{k=1}^{K} d_k u_k v_k^T$. The author started with a rank-1 approximation to solve list problem:

$$\min_{d,u,v} \frac{1}{2} \left\| X - duv^T \right\|_F^2, \quad s.t. \quad \|u\|_2^2 = 1, \|v\|_2^2 = 1, P_1(u) \leq c_1, P_2(v) \leq c_2, d \geq 0, \quad (9)$$

where u and v are the column of U and V, d is a diagonal element of D, c_1 and c_2 are constants. Papailiopoulos et al. introduced a novel algorithm through low-rank (LR-SPCA) approximations, which has a provable approximation guarantee for SPCA by checking a discrete set of special vector in subspace to compute the k-sparse PC of a semi-definite matrix [30].

The last group is based on HL (h-likelihood) penalty. It was the first time to use HL (h-likelihood) penalty on the SPCA. Lee et al. proposed a new approach (S-SPCA) by shrinking the original data matrix to generate sparse loadings and proved that HL penalty produced the best result in experiments [31]. This method used a new random-effect model for sparse coefficient selection to produce a new function of extremely sparse loading vector. Nevertheless, an important innovation was to obtain shrinking matrix by adding singular-value decomposition.

3.3   The Extension of Sparse PCA

In this section, we introduce a series of other methods for the extension of SPCA. Meng et al. enforced l_1-norm maximization to improve robustness of SPCA (RSPCA), which might replace the traditionally method [32]. Another extension of SPCA is a non-negative SPCA (NSPCA), which means these approaches were based on sparse and added non-negativity [15]. In order to address SPCA with an efficient and simple optimal procedure, Jenatton et al. proposed a structured SPCA (SSPCA), which used a sparse structured signal to face recognition [33]. Tipping et al. proposed sparse kernel PCA (SKPCA), which introduced a nonlinear function and each vector was projected to the high dimensional feature space and principal component analysis in high dimensional space [34]. Under the condition of the original matrix with missing observations or the matrix lack of rank, two papers on insufficient data were proposed [35, 36]. On the basis of SCoTLASS, Lounici presented a method with missing observations, which replaced the original l_1-norm constraint with l_0 [35]. Asteris et al. introduced auxiliary spherical variables and proved that there were many candidate index-sets, as for rank, it contained the optimal index-set [36]. Liu et al. used Gaphics Pocessing Uits (GPUs) to obtain a paralleled method, which can deal with large scale data and this method was called Generalized Power method of SPCA (GP-SPCA) [37]. Xiao developed a two-dimensional SPCA (2d-SPCA) method by computing covariance matrix of two-dimensional image directly rather than convert it into one-dimensional in the past [38]. On the basis of optimization technique, Sharp et al. rewrote Multilinear PCA (MPCA) into multilinear regression forms and relaxed it by l_1 and l_2-norm constraints to be sparse, which names Multilinear Sparse PCA (MSPCA) [39]. Morever, bayesian approach to SPCA is good at obtaining sparse principal component which does not give probability mass to sparse results. From this perspective, Sharp and Rattray gave a novel algorithm (DMP-SPCA) which based on a series of dense message passing algorithms [40]. Sparse Tensor PCA (STPCA) was proposed, which transformed eigen-decomposition to a class of regression problem of SPCA to extract feature [41]. On one hand, Jiang et al. proposed a new approach with graph joint sparse PCA (GJSPCA) to detect and locate abnormal information source for network data streams [42]. On the other hand, for large-sized network, Grbovic et al. divided the large network into overlapped blocks and this method was called Maximum Entropy SPCA (ME-SPCA) [43].

4 Brief Statements and Different Classifications

Johnstone and Lu proposed a methods which was aimed to detect both sparsity and consistency [44]. Besides, we can see a series of conditions of a SPCA method [45]. Cai et al. might depress the multiple regression problems of high dimensional SPCA [46]. Luss et al. proposed a novel dual relaxation using Lagrangian duality which is a smoothing and a first-order gradient scheme for cardinality constraints [47]. Zhang et al. introduced two kinds of relaxations: semi-definite relaxation with l_1 penalization and semi-definite relaxation with l_0 penalization of financial data analysis [48]. In order to obtain the optimal principal subspace estimator in polynomial time, Wang et al. proposed two-stage of SPCA [49]. Gao et al. added some question in PCA to establishment of linkages between frequentist proper ties and Bayesian methodogies [50]. However, there are two types of sparse problems: sparse loadings PCA (slPCA) and sparse variable PCA (svPCA). The first group is zero out loadings and the other is zero out whole variables. Many literatures were based on slPCA to overcome the problem, only four papers described a method for svPCA [16, 51, 52] and Johnstone and Lu based on wavelet transforming the data and preselected threshold to zero [51]. On the contrary, Ulfarsson and Solo used a new penalty to zero out whole variables for SPCA in high dimensions. Moreover, various papers solve the problem with non-convex [6, 17, 19, 25, 29]. The other kind of articles for SVD based on iterative thresholding approaches [5, 25, 29]. In addition, three articles was inspired on alternative methods for SPCA [17, 19, 25].

5 Conclusion

In recent years, various methods for SPCA have been proposed based on different purposes and successfully applied in biology, medicine, machine learning and bioinformatics. In this paper, we give a simple review about the development of SPCA and categorize different algorithms into three groups. The problem of convergence and the consistency of various algorithms or the application of the algorithm to adapt to different scale data in SPCA will be studied in the future.

Acknowledgments. This work was supported in part by the NSFC under grant Nos. 61572284, 61502272, 61572283, 61370163, 61373027 and 61272339; the Shandong Provincial Natural Science Foundation, under grant No. BS2014DX004; Shenzhen Municipal Science and Technology Innovation Council (Nos. JCYJ20140904154645958, JCYJ20140417172417174 and CXZZ20140904154910774).

References

1. Liu, J.-X., Xu, Y., Gao, Y.-L., Zheng, C.-H., Wang, D., Zhu, Q.: A Class-Information-based Sparse Component Analysis Method to Identify Differentially Expressed Genes on RNA-Seq Data. **13**(2), 392–398 (2015)
2. Richtárik, P., Takáč, M., Ahipaşaoğlu, S.D.: Alternating maximization: unifying framework for 8 sparse PCA formulations and efficient parallel codes. arXiv preprint arXiv:1212.4137 (2012)
3. Hotelling, H.: Analysis of a complex of statistical variables into principal components. J. Educ. Psychol. **24**(6), 417 (1933)
4. Jolliffe, I.T.: Rotation of principal components: choice of normalization constraints. J. Appl. Stat. **22**(1), 29–35 (1995)
5. Ma, Z.: Sparse principal component analysis and iterative thresholding. Ann. Stat. **41**(2), 772–801 (2013)
6. Moghaddam, B., Weiss, Y., Avidan, S.: Spectral bounds for sparse PCA: Exact and greedy algorithms. In: Advances in neural information processing systems, pp. 915–922 (2005)
7. d'Aspremont, A., Bach, F.R., Ghaoui, L.E.: Full regularization path for sparse principal component analysis. In: Proceedings of the 24th international conference on Machine learning, pp. 177–184. ACM (2007)
8. Farcomeni, A.: An exact approach to sparse principal component analysis. Comput. Stat. **24** (4), 583–604 (2009)
9. Mackey, L.W.: Deflation methods for sparse PCA. In: Advances in Neural Information Processing Systems, pp. 1017–1024 (2009)
10. Wang, Y., Wu, Q.: Sparse PCA by iterative elimination algorithm. Adv. Comput. Math. **36** (1), 137–151 (2012)
11. Kuleshov, V.: Fast algorithms for sparse principal component analysis based on Rayleigh quotient iteration. In: Proceedings of the 30th International Conference on Machine Learning (ICML-13), pp. 1418–1425 (2013)
12. Journée, M., Nesterov, Y., Richtárik, P., Sepulchre, R.: Generalized power method for sparse principal component analysis. J. Mach. Learn. Res. **11**, 517–553 (2010)
13. Yuan, X.-T., Zhang, T.: Truncated power method for sparse eigenvalue problems. J. Mach. Learn. Res. **14**(1), 899–925 (2013)
14. Zhao, Q., Meng, D., Xu, Z.: A recursive divide-and-conquer approach for sparse principal component analysis. arXiv preprint, arXiv:1211.7219 (2012)
15. Zass, R., Shashua, A.: Nonnegative sparse PCA. In: Advances in Neural Information Processing Systems, pp. 1561–1568 (2006)
16. Ulfarsson, M.O., Solo, V.: Vector sparse variable PCA. IEEE Trans. Sig. Process. **59**(5), 1949–1958 (2011)

17. Jolliffe, I.T., Trendafilov, N.T., Uddin, M.: A modified principal component technique based on the LASSO. J. Comput. Graph. Stat. **12**(3), 531–547 (2003)
18. Tibshirani, R.: Regression shrinkage and selection via the lasso. J. Roy. Stat. Soc. Ser. B (Methodological) **58**, 267–288 (1996)
19. d'Aspremont, A., El Ghaoui, L., Jordan, M.I., Lanckriet, G.R.: A direct formulation for sparse PCA using semidefinite programming. SIAM Rev. **49**(3), 434–448 (2007)
20. d'Aspremont, A., Bach, F., Ghaoui, L.E.: Optimal solutions for sparse principal component analysis. J. Mach. Learn. Res. **9**, 1269–1294 (2008)
21. Lu, Z., Zhang, Y.: An augmented Lagrangian approach for sparse principal component analysis. Math. Program. **135**(1–2), 149–193 (2012)
22. Guo, J., James, G., Levina, E., Michailidis, G., Zhu, J.: Principal component analysis with sparse fused loadings. J. Comput. Graph. Stat. **19**(4), 930–946 (2012)
23. Qi, X., Luo, R., Zhao, H.: Sparse principal component analysis by choice of norm. J. Multivar. Anal. **114**, 127–160 (2013)
24. Hein, M., Bühler, T.: An inverse power method for nonlinear eigenproblems with applications in 1-spectral clustering and sparse PCA. In: Advances in Neural Information Processing Systems, pp. 847–855 (2010)
25. Zou, H., Hastie, T., Tibshirani, R.: Sparse principal component analysis. J. Comput. Graph. Stat. **15**(2), 265–286 (2006)
26. Leng, C., Wang, H.: On general adaptive sparse principal component analysis. J. Comput. Graph. Stat. **18**, 201–215 (2012)
27. Xiaoshuang, S., Zhihui, L., Zhenhua, G., Minghua, W., Cairong, Z., Heng, K.: Sparse Principal Component Analysis via Joint L 2, 1-Norm Penalty. In: Cranefield, S., Nayak, A. (eds.) AI 2013. LNCS, vol. 8272, pp. 148–159. Springer, Heidelberg (2013)
28. Witten, D.M., Tibshirani, R., Hastie, T.: A penalized matrix decomposition, with applications to sparse principal components and canonical correlation analysis. Biostatistics **10**, 515–534 (2009). kxp008
29. Shen, H., Huang, J.Z.: Sparse principal component analysis via regularized low rank matrix approximation. J. Multivar. Anal. **99**(6), 1015–1034 (2008)
30. Papailiopoulos, D.S., Dimakis, A.G., Korokythakis, S.: Sparse PCA through low-rank approximations. arXiv preprint, arXiv:1303.0551 (2013)
31. Lee, D., Lee, W., Lee, Y., Pawitan, Y.: Super-sparse principal component analyses for high-throughput genomic data. BMC Bioinformatics **11**(1), 1 (2010)
32. Meng, D., Zhao, Q., Xu, Z.: Improve robustness of sparse PCA by L 1-norm maximization. Pattern Recogn. **45**(1), 487–497 (2012)
33. Jenatton, R., Obozinski, G., Bach, F.: Structured sparse principal component analysis. arXiv preprint arXiv:0909.1440 (2009)
34. Tipping, M.E., Nh, C.C.: Sparse kernel principal component analysis (2001)
35. Lounici, K.: Sparse principal component analysis with missing observations. In: Houdré, C., Mason, D.M., Rosiński, J., Wellner, J.A. (eds.) High dimensional probability VI, vol. 66, pp. 327–356. Springer, Heidelberg (2013)
36. Asteris, M., Papailiopoulos, D.S., Karystinos, G.N.: Sparse principal component of a rank-deficient matrix. In: IEEE International Symposium on 2011 Information Theory Proceedings (ISIT), pp. 673–677. IEEE (2011)
37. Liu, W., Zhang, H., Tao, D., Wang, Y., Lu, K.: Large-scale paralleled sparse principal component analysis. Multimedia Tools Appl. 1–13 (2014)
38. Xiao, C.: Two-dimensional sparse principal component analysis for face recognition. In: 2010 2nd International Conference on Future Computer and Communication (ICFCC), pp. V2–561-V562-565. IEEE (2010)

39. Lai, Z., Xu, Y., Chen, Q., Yang, J., Zhang, D.: Multilinear sparse principal component analysis. IEEE Trans. Neural Netw. Learn. Syst. **25**(10), 1942–1950 (2014)
40. Sharp, K., Rattray, M.: Dense message passing for sparse principal component analysis. In: International Conference on Artificial Intelligence and Statistics, pp. 725–732 (2010)
41. Wang, S.-J., Sun, M.-F., Chen, Y.-H., Pang, E.-P., Zhou, C.-G.: STPCA: sparse tensor principal component analysis for feature extraction. In: 21st International Conference on 2012 Pattern Recognition (ICPR), pp. 2278–2281. IEEE (2012)
42. Jiang, R., Fei, H., Huan, J.: Anomaly localization for network data streams with graph joint sparse PCA. In: Proceedings of the 17th ACM SIGKDD International Conference on Knowledge Discovery and Data Mining, pp. 886–894. ACM (2011)
43. Grbovic, M., Li, W., Xu, P., Usadi, A.K., Song, L., Vucetic, S.: Decentralized fault detection and diagnosis via sparse PCA based decomposition and Maximum Entropy decision fusion. J. Process Control **22**(4), 738–750 (2012)
44. Johnstone, I.M., Lu, A.Y.: On consistency and sparsity for principal components analysis in high dimensions. J. Am. Stat. Assoc. **104** (2012)
45. Shen, D., Shen, H., Marron, J.S.: Consistency of sparse PCA in high dimension, low sample size contexts. J. Multivar. Anal. **115**, 317–333 (2013)
46. Cai, T.T., Ma, Z., Wu, Y.: Sparse PCA: Optimal rates and adaptive estimation. Ann. Stat. **41** (6), 3074–3110 (2013)
47. Luss, R., Teboulle, M.: Convex approximations to sparse PCA via Lagrangian duality. Oper. Res. Lett. **39**(1), 57–61 (2011)
48. Zhang, Y., d'Aspremont, A., El Ghaoui, L.: Sparse PCA: Convex relaxations, algorithms and applications. In: Anjos, M.F., Lasserre, J.B. (eds.) Handbook on Semidefinite, Conic and Polynomial Optimization, vol. 166, pp. 915–940. Springer, Heidelberg (2012)
49. Wang, Z., Lu, H., Liu, H.: Tighten after relax: Minimax-optimal sparse PCA in polynomial time. Adv. Neural Inf. Process. Syst. **2014**, 3383–3391 (2014)
50. Gao, C., Zhou, H.H.: Rate-optimal posterior contraction for sparse PCA. Ann. Stat. **43**(2), 785–818 (2015)
51. Johnstone, I.M., Lu, A.Y.: Sparse principal components analysis. Unpublished manuscript **7** (2004)
52. Ulfarsson, M.O., Solo, V.: Sparse variable PCA using geodesic steepest descent. IEEE Trans. Sig. Process. **56**(12), 5823–5832 (2008)

Endpoint Detection and De-noising Method Based on Multi-resolution Spectrogram

Jing Zhang[✉]

Cisco School of Informatics, Laboratory of Language Engineering
and Computing, Guangdong University of Foreign Studies,
Guangzhou 510006, China
ha_go@163.com

Abstract. The paper studied endpoint detection algorithm of noisy speech, since the visual differences of spectrogram employed by speech and noise, the paper chose spectrogram endpoint detection methods. Technical difficulties of spectrogram endpoint detection is how to describe the intuitive difference of spectrogram by mathematical amount, according to the descriptive power of autocorrelation coefficients on texture features, the paper described the difference by selecting the autocorrelation function, and proposed column autocorrelation spectrogram detection method. Through the distribution of spectrogram self-correlation function, as the threshold of endpoint detection for the noisy speech, the cut-off point between speech and noise was found out. Since the paper used broadband spectrogram, which employed poor frequency resolution, so there were still residual noise in speech column after autocorrelation spectrum detection, in order to further de-noising in different bands, combined with the multi resolution of empirical mode decomposition (EMD), the paper analyzed the noisy speech by multi-resolution, the target was broken down into different frequency scales and was further analyzed by column autocorrelation spectrogram, experiments shown that the noise reduction effect for noisy speech was ideal.

Keywords: Endpoint detection · Spectrogram · Auto-correlation · Multi-resolution analysis · Empirical mode decomposition

1 Introduction

The so-called speech endpoint detection is to accurately identified the beginning and the end of part of speech in a voice signal, speech data and background noise can be effectively separated, and provide reliable basis for the subsequent processing of the voiceprint recognition system. After many years of painstaking research, the vocal endpoint detection technology has made some achievements in relatively tranquil experimental environment (such as laboratory). With the demand of voiceprint recognition in the society of various areas has become more and more intense;

This work is supported by Guangdong Provincial Science and technology projects#2013B04040 1015.

© Springer International Publishing Switzerland 2016
D.-S. Huang and K.-H. Jo (Eds.): ICIC 2016, Part II, LNCS 9772, pp. 384–393, 2016.
DOI: 10.1007/978-3-319-42294-7_34

Of course, the background noise that the identification system has to face is becoming more and more complex; scholars have begun to study of voiceprint recognition processing algorithms in the noise environment.

We know that in both time domain and frequency domain, we can extract the characteristic parameters which reflect the essential difference between speech and noise. In time domain, the time domain characteristic parameters can be obtained by using short-time auto correlation function [1], higher order statistics [2], fractal [3] and other techniques. While in the frequency domain, the characteristic parameters such as, entropy of information [4, 5], Band variance [6], Characteristic parameters of the spectrum [7] etc. can be extracted.

As several approaches to speech analysis, including time domain analysis and frequency domain analysis, the two have their own advantages and limitations. The advantages of both the time domain analysis and frequency domain analysis are shown as a combination in the spectrogram, and the analysis method is a method which collected the time frequency characteristic of the set of speech signal. And the signal is displayed in the form of image, which can directly reflect the difference of the speech signal and the noise signal. Its appearance has solved the problem of the limitation on restriction of the separation of time domain and the frequency domain. The method of using the endpoint detection of spectrogram has been proposed in references [8, 9]. The technical difficulty of the endpoint detection of the spectrogram is that how come the differences of the speech signal and the noise which is directly reflected by the spectrogram be expressed as a mathematical quantity. Spectrogram is an image representation of speech signal, and some of the techniques of image processing are also suitable for this application. Image correlation algorithm is a kind of structure measurement method based on spatial frequency [10, 11], which can be used to represent the size and the dispersion of the texture elements. By calculating the auto correlation coefficient of the image, then the texture features can be described. By referencing the application of auto-correlation function in image processing proposed in reference [12], this paper took the speech spectrogram auto-correlation function as the measurement of speech and noise, and to calculate the threshold value of the difference between the two signals. In order to make the effect of the endpoint detection effect better in noisy environment, before carrying out the endpoint detection algorithm of the spectrogram, the paper proposed to use of Wiener filtering [13] first to carry out the smoothing pre-treatment and then to carry on the endpoint detection, the detection effect would be better.

There are two kinds of spectrogram, such as Narrow band and broadband [14]. The broad band spectrogram can give the resonance peak frequency of the speech, which has good time resolution, but the frequency resolution is poor. And the narrow band spectrogram has the opposite effect. In consideration the fact that most of the noise in the practical application environment is continuous and broadband, so broadband spectrogram was used in the paper.

Because of the low resolution of the broadband spectrogram, the residual noise in the speech is still remained after the above method. The solution the paper proposed for this problem was to decompose the signal into frequency scale, and then analyzed the spectrogram of the auto-correlation spectrum at each scale. Wavelet analysis has both characteristics of multi-scale and multi-resolution, and can be used in the non-stationary signal. However, the selections of threshold and contraction function are restricted by wavelet basis function, decomposition level [15]. Huang has raised the Empirical Mode Decomposition (EMD) [16] in 1998. The method has been proved to have the characteristics of multi-scale wavelet decomposition, and the effect of noise reduction is equivalent to that of the wavelet method. However, the EMD method is based on the data itself, from the local extreme value of the data, the data is decomposed to overcome the difficulty of selecting the appropriate wavelet basis in wavelet analysis, which makes the method obtain better adaptability.

2   The Endpoint Detection of the Spectrogram Based on Auto Correlation Function

2.1   Spectrogram Analysis

Spectrogram (or sonogram) characterization is the change of voice spectrogram with time, the ordinate is the frequency, and the horizontal axis is time.

Figure 1 shows the basic information of the spectrogram of clean speech and the spectrogram of the speech with noise, and Fig. 2 shows the information of the noise spectrogram.

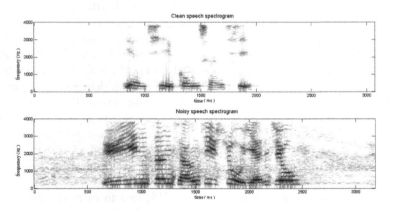

Fig. 1. Spectrogram of Clean Speech and Speech with Noise

Compared with Figs. 1 and 2, it can be seen that there is obvious difference between the spectrogram of the speech and the noise. The speech signal has the shape of stripes, and it has a periodic change in frequency. But no rules could be applied to noise signal; there is no more than a fringe pattern of the law.

2.2   The Auto-Correlation Function in Spectrogram

2.2.1   Column Self-correlation of Speech and Noise

Image auto correlation algorithm is a kind of structure measurement method, whose core idea is to analyze and deal with spatial frequency. The image auto correlation

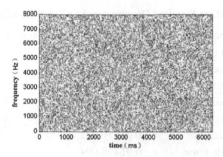

Fig. 2. Spectrogram of the Noise

algorithm can be used to represent the texture element's size and the discrete. For the size of M * N (i = 1, 3, 2... M, j = 1, 2, 3... N () A (I, J), self-correlation is defined [17] as Eq. (1).

$$R(m,n) = \frac{1}{MN} \sum_{i=1}^{M} \sum_{j=1}^{N} A(i,j)A_1(i+m,j+n)_{MN} \tag{1}$$

In Eq. (1), m = 0,1,2,…, n = 0,1,2, M,…, N. A (I, J) is the gray value of I time series, the first j frequency sequence, M is the width of the image, and the N is the height of the image. In this paper, the speech and non-speech spectrogram analysis is based on the self-correlation of the images. Column auto-correlation is obtained from the correlation of the auto-correlation along the column direction, i.e., the result is a frequency sequence, as shown in Eq. (2) [17].

$$R(n) = \frac{1}{MN} \sum_{i=1}^{M} \sum_{j=1}^{N} A(i,j)A(i,j+n)_{MN} \tag{2}$$

2.2.2 Distinguish Between Speech and Non-speech

A suitable algorithm for calculating the adjacent peak position is located in the following experiments.

The speech signals were chosen from the TIMIT speech database, including 630 speakers, each with 10 English phrases. In order to obtain the near peak distance deviation of the speech auto correlation graph, the first 2 sentences of the speech sample were selected from 10 people (6 male and 4 female).

In order to obtain the near peak distance deviation of the noise auto correlation graph, all samples of noise were taken from 20 noise samples of the noisex92. Considering of the voiceprint recognition system was a civil system, so the noise selected 3 samples for white noise, 4 samples for Volvo noise, 3 samples for babble noise, 3 samples for HF channel noise, 3 samples for the pink noise, 2 samples for Factory floor noise1 and 2 samples of Factory floor noise2.

Because the purpose of the research was to obtain the peak distance deviation of the noise and the speech auto correlation graph, this paper chose the statistics as the

Table 1. Mean Standard Deviation and CV of Near Peak Distance

Signal	Mean	Standard deviation	CV
Speech 1	0.018	0.021	1.17
Speech 3	0.013	0.017	1.31
Speech 5	0.015	0.022	1.47
Speech 7	0.016	0.019	1.19
Speech 9	0.012	0.021	1.75
Speech 10	0.013	0.021	1.62
White noise	0.014	0.004	0.29
Volvo noise	0.015	0.007	0.47
Babble noise	0.021	0.01	0.48
HF channel noise	0.015	0.007	0.47
Pink noise	0.015	0.01	0.67
Factory floor noise	0.014	0.008	0.57

average and the standard deviation, as shown in Table 1. In order to make the difference between the speech and noise performance more obvious, the paper added the variation coefficient of Variation Coefficient (CV) in Table 1. Variation coefficient = standard deviation/mean value.

Using the statistical software SPSS22 to analyze the Statistical analysis of data, the T test shows that the variation coefficient is significantly different for the speech and noise auto-correlation diagram of the near peak. So it is necessary to find the critical point of the variation coefficient of the speech and noise. In this paper, the critical point was determined by the distribution function of coefficient of variation.

The threshold is set at the intersection of the relative error curves of the voice and the non-speech, and the coordinates of the points IS 0.91, Therefore, the speech and noise can be judged according to the value of the coefficient of variation, and more than 0.91 for the speech signal. Otherwise it is considered to be noise.

3 Multi-resolution Analysis and EMD Decomposition Algorithm

EMD decomposition process is based on the following assumptions: ① Every signal has at least a maximum and a minimum value; ② time domain interval is decided by the extreme value; ③ lacking extreme value if the data sequence is complete but contains only the inflection point, then it can also be evaluated guide one or more times to represent the extreme points, and the final result can be evaluated by the integration of these components is obtained. The specific method is completed by a "screening" process:

First find all the maximum of the signal $s(t)$ and use cubic spline function to fit envelope-synthesized data sequence, and then find the entire minimum and use cubic spline function to fit to the original data.

Calculate the average of the upper and lower envelope, referred to as $m_1(t)$, the original data series $s(t)$ can be obtained by subtracting the mean value of a new data sequence $h_1(t)$ to remove low frequency, such as Eq. (3).

$$S(t) - m1(t) = h1(t) \tag{3}$$

Because $h_1(t)$ is not generally a component of a sequence of IMF, which requires it to repeat the process. Repeat the above process k times until the definition of $h_1(t)$ in line with IMF requirements, the resulting average value tends to become zero, so the first IMF component $c_1(t)$, appears, which represents the highest frequency component of the signal $s(t)$, such as Eqs. (4), (5).

$$h_{1(k-1)}(t) - m_{1k}(t) = h_{1k}(t) \tag{4}$$

$$c_1(t) = h_{1k}(t) \tag{5}$$

Separate $c_1(t)$ from $s(t)$, the high frequency component $r_1(t)$ is removed to obtain a difference signal, as follows (6).

$$r_1(t) = s(t) - c_1(t) \tag{6}$$

Set $r_1(t)$ as raw data, repeat steps (1), (2) and (3) to give a second IMF component $c_2(t)$, repeat n times to obtain n IMF components. So there would be (7), (8).

$$r_1(t) - c_2(t) = r_2(t) \tag{7}$$

$$r_{n-1}(t) - c_n(t) = r_n(t) \tag{8}$$

When either $r_n(t)$ or $c_n(t)$ satisfies a given condition (usually makes $r_n(t)$ into a monotonic function), the cycle will come to an end, Eq. (9) can be obtained from the above two equations.

$$s(t) = \sum_{j=1}^{n} c_j(t) + r_n(t) \tag{9}$$

Where in (9), $r_n(t)$ as the residual function, represents the average trend of the signals. And each IMF component $c_1(t), c_2(t), \cdots, c_n(t)$ contains a component of signal at different time scales, its scale increases progressively larger. Thus, each component contains a correspondingly different frequency component from high to the bottom segment of the frequency components, which is contained in different frequency band, and change with the signal changes.

The EMD reconstruction formula is (10).

$$x_{lk} = \sum_{j=1}^{k} c_j(t) \tag{10}$$

4 Experimental Results and Analysis

Experimental data were collected by mobile phones in different noisy environment, to record the voice content for the study of speech enhancement technology. The noise environments are: the playground (class), market (music background), and the experimental platform for the MATLAB2010b simulation laboratory.

In the playground, the speech spectrum is shown in Fig. 3. Wiener filtering and column auto correlation spectrum detection results are shown in Fig. 4; Result of the multi-resolution speech spectrum is shown in Fig. 5.

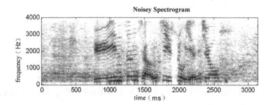

Fig. 3. Spectrum of playground speech

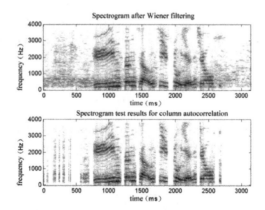

Fig. 4. Test results of Wiener filtering and column autocorrelation in playground

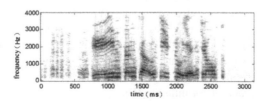

Fig. 5. De-noising results of multi-resolution spectral in playground

The background of the music in market is the speech language spectrum, which is shown in Fig. 6. Wiener filtering and column auto correlation spectrum detection results are shown in Fig. 7. The result of the multi-resolution speech spectrum is shown in Fig. 8.

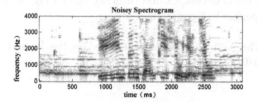

Fig. 6. Speech spectrum of market

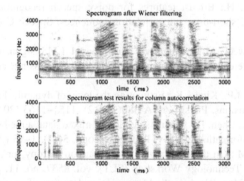

Fig. 7. The detection results of noise spectrum in market

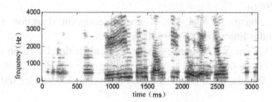

Fig. 8. De-noising results of multi-resolution spectral in market

5 Conclusions

In this chapter, we studied the speech signal pre-processing and endpoint detection method. And since the language spectrum diagram not only reflects the time domain of the speech, but also reflects the frequency domain characteristics of the speech, the paper chose the method of the spectrogram analysis to carry out the endpoint detection. By analyzing the speech signal and the noise signal, the algorithm of the endpoint detection was proposed. In order to make this kind of detection algorithm better and to carry out the preprocessing of the endpoint detection, the experimental results proved that the proposed algorithm had better effect on the endpoint detection.

Due to the broadband spectrogram that the paper had chosen, it had the defect of the frequency resolution. This thesis made use of EMD decomposition algorithm for multi resolution, raising the idea of carrying out detection and analysis of the spectrogram combining EMD decomposition of noisy speech signal, which also enhances the robustness of speaker recognition system.

Acknowledgment. This work is supported by Guangdong Provincial Science and technology projects#2013B040401015.

References

1. You, K.H., Wang, H.: Robust features for noisy speech recognition based on temporal trajectory fitting of short-time autocorrelation sequences. Speech Commun. **28**(99), 13–24 (1999)
2. Haweel, T.I., Haweel, M.T.: Adaptive multichannel LMS signal decoupling. In: IEEE International Conference on Communications, Signal Processing, and their Applications, pp. 1–4 (2015)
3. Sase, T., Ramírez, J.P., Kitajo, K.: Estimating the level of dynamical noise in time series by using fractal dimensions. Phys. Lett. A **380**(11–12), 1151–1163 (2016)
4. Zhang, X., Mei, C., Chen, D.: Feature selection in mixed data: a method using a novel fuzzy rough set-based information entropy. Pattern Recogn. **56**(1), 1–15 (2016)
5. Obin, N., Liu, N.M.: On the generalization of Shannon entropy for speech recognition. In: Spoken Language Technology Workshop (SLT), vol. 8537, no. 11, pp. 97–102 (2012)
6. Qiu-fang, A., Xiao-Jun, W.: A method for endpoint detection of speech using FBV based on harmonious analysis. Comput. Simul. **26**(8), 330–333 (2009)
7. Lareau, J., Lareau, J: Application of Shifted Delta Cepstral Features for GMM Language Identification (2006)
8. Xiang-min, C., Zhang, J., Wei, G.: A speech endpoint detection algorithm based on spectrogram. Audio Eng. **4**(8), 46–49 (2006)
9. Xiao, C., Sun, D., Gao, Y.: A speech enhancement algorithm based on speech spectrogram. Audio Eng. **36**(9), 44–48 (2012)
10. Gonzalez, R.C., Woods, R.E.: Digital Image Processing. Electronics Industry Press, Beijing (2011)
11. Wang, X., Shen, H., Zhang, W.: Image mosaic by using the local autocorrelation algorithm in triangular geometric constraints. Opto-Electron. Eng. **42**(4), 32–37 (2015)
12. Wang, L., Sun, Y.: Gastroscopy image retrieval based on color-texture autocorrelation algorithm. J. Circ. Syst. **16**(2), 46–50 (2011)
13. Soon, I.Y., Koh, S.N.: Speech enhancement using 2-D Fourier transform. IEEE Trans. Speech Audio Process. **11**(6), 717–724 (2003)
14. Zhao, l: Speech Signal Processing. China Machine Press, Beijing (2009)
15. Sun, Yan-kui: Wavelet Transform and Image Processing Techniques. Tsinghua University Press, Beijing (2012)

16. Huang, N.E., Shen, Z., Long, S.R.: The empirical mode decomposition and the Hilbert spectrum for nonlinear and non-stationary time series analysis. R. Soc. London Proc. **454** (1971), 903–993 (1998)
17. Fu, J., Wang, S.W., Cao, X.L.: The research on speech endpoint detection algorithm based on spectrogram row self-correlation. In: 2nd International Conference on Computer Science and Network Technology, pp. 212–216 (2012)

Biometrics Recognition

An Efficient Face Recognition System with Liveness and Threat Detection for Smartphones

Kamlesh Tiwari[1(✉)], Suresh Kumar Choudhary[2], and Phalguni Gupta[3]

[1] Department of CSIS, Birla Institute of Technology and Science Pilani, Rajasthan 333031, India
kamlesh.tiwari@pilani.bits-pilani.ac.in

[2] Department of CSE, Indian Institute of Technology Kanpur, Kanpur 208016, India

[3] National Institute of Technical Teachers' Training and Research, Kolkata 700106,
West Bengal, India
pg@cse.iitk.ac.in

Abstract. This paper proposes a face recognition system with liveness and threat detection on smartphone. Liveness and under-threat situations are decided through eye-blinking and facial expressions. It is designed to handle resource constraints of mobile devices such as low processing power, limited memory, less battery power and low quality of images. It uses Uniform Extended Local Ternary Pattern (UELTP) features for the threat detection. Whereas, Uniform Local Binary Pattern (ULBP) and Binarized Hamming Distance (BHD) are used for liveness detection. The experiments have been conducted on three in-house databases called SmartBioVideo, SmartBioFace and SmartBioThreatFace. Results have found to be promising and time efficient.

Keywords: Biometrics · Face recognition · Liveness detection · Threat detection · Matching · ROC curve

1 Introduction

This paper deals with a problem of designing an efficient face recognition system for smartphones with liveness and under-threat detection from frontal facial images. To the best of our knowledge, there does not exists any general face recognition algorithm, which can differentiate between live face, undet-threat face and photo face. To identify a live face, the vein map of the facial image is the most secured approach, but it requires special and expensive devices. In sequential face images, face movement has no large variation but it has been observed that eye region has large variation due to eye blink. So eye blink can be used to detect liveness of face. Face is the composition of many micro texture patterns. To extract these patterns, local binary pattern and extended local ternary pattern approaches can be used. Incorporation of these light weight processing techniques in smartphones makes face recognition time efficient, which settles the limiting processing power of smartphones.

The incorporation of biometric recognition algorithm in smartphones is challenging as it have limited processing power, storage and energy. There exist approaches like Principal Component Analysis (PCA) [11], Local Binary Patterns (LBP) [2], Local

© Springer International Publishing Switzerland 2016
D.-S. Huang and K.-H. Jo (Eds.): ICIC 2016, Part II, LNCS 9772, pp. 397–406, 2016.
DOI: 10.1007/978-3-319-42294-7_35

ternary Pattern (LTP) [28], Independent Component Analysis (ICA) [5], Hidden Markov Model [23], Neural Network model [16], Speeded Up Robust Features (SURF) [8], Scale-invariant feature transform (SIFT) [10], Linear Discriminant analysis (LDA) [12], Support Vector Machine (SVM) [5], Self Organizing Map (SOM) [13] etc. for face recognition. But, none of them has addressed the problem of under threat detection that could be more likely situation for a smartphone user and is the motivation of this work.

Liveness detection in human face is essential to prevent spoof face. It is done on the basis of mouth movement, 3D properties of face, eyelid movement, eye blink etc. A liveness detection approach based on eye blink using Multiple Gabor Response Waves (MGRW) is proposed in [15]. Conditional random field (CRF) approach is used in [25] that uses 80 video clips to test the system, each video is a duration of 5 s with 30 frames per second. For training, it has considered 1016 labeled close eye images and 1200 open eye images. The closed eye images have been taken from CAS-PEAL-R1 database [3] and Asian PF01 database [6], while open eye images have been taken from FERET database [22]. The single eye blink detection rate, two eye blink detection rate and live face detection rate are 98.2 %, 100 % and 100 % respectively for frontal face without glass images. The impostor detection rate is found to be 98.3 %. Results have been compared with Cascaded Adaboost and HMM approach. A live human face detection application using conditional random field (CRF) and local binary pattern (LBP) is proposed in [21] with eye blink and scene clue respectively. A private database of 96 video clips has been used to analyze the performance. Liveness detection rate is reported to be 99.5 %. In [14], a live human face detection method have been presented using 3D properties of the face, eye blink and mouth movement. The liveness score has been calculated using the rasterflow and eyeflow measurements. The ZJU Eyeblink database [20] has been used to perform the experiment with 80 face videos. The average liveness detection rate is 98.2 % and average false positive rate is found to be 94.7 %. Texture analysis from single images is used in [17]. The NUAA database [26] has been used for testing and the EER is reported o be 2.9 % using LBP approach.

Biometric recognition using smartphones has many challenges because of limited computational power and memory of the smartphones. A scheme in [24], have presented face localization and recognition on smartphones based on selected set of features with size invariant. It has used reference face data as feature vectors for face recognition. The experiment is done on a private small data set. In [19], a face recognition system which has used fisherfaces approach on smartphones with client server approach has been proposed. Training has been done on server side and feature extraction for test image has been done on Droid phone. It has used a small subset of LFW and Essax database as a combination with labeled faces. It has considered 100 images as training set and 110 images as test set. In [7], a fast processing face recognition system on smartphone is proposed which is based on eigenfaces decomposition. The private dataset is used with 126 images as training set and 5 images as test set. The Mahalanobis distance has been used for matching. Fisher face approach is used in [4]. It has used a database contained 45 images for training and 134 images for testing. Technique in [9] proposes a face unlock system for smartphones using OpenCV face detector classifier. Eigenfaces with classifier approach has been considered to recognize the face image. In [27] the performance of Correlation filter approach, PCA approach and Fisherfaces approach has

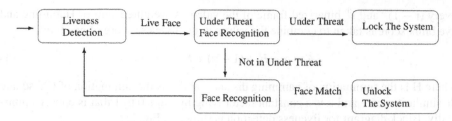

Fig. 1. Block diagram for liveness detection and under threat face recognition application

been analyzed for face verification. In [18], a face verification system on smartphone is implemented which uses optimized correlation filter. A private database has been used to test the system. Superiority of LBP approach for face recognition over PCA, Elastic Bunch Graph Matching and Bayesian classifier has been shown in [1] using FERET database.

The paper is divided into four sections. Current section presents a brief introduction and motivation, whereas next section explains the proposed approach. Experimental results are discussed in Sect. 3 and conclusions are presented in the last.

2 Proposed Approach

The problem of liveness detection and under threat face recognition application has been categorized in three modules (1) Liveness detection, (2) Under Threat Face Recognition, and (3) Frontal Face Recognition. The block diagram of important application modules is shown in Fig. 1.

2.1 Liveness Detection

Liveness detection is based on eye blink score that is computed using histogram features. Hamming distance and Chi square dissimilarity is computed with histogram features of

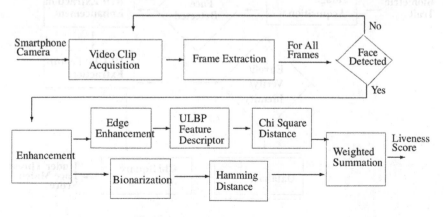

Fig. 2. Proposed liveness detection

every $(i-1)^{th}$ and i^{th} binarized frame of the video. A weighted sum of Hamming and eye blink score is taken to combine the values to get liveness score of eye blink.

$$A = \alpha * H + (1+\alpha) * E \tag{1}$$

where H is the summation of hamming distance, and E is the summation of Chi square dissimilarity score and α is a constant factor lying in range 0 to 1 that is chosen empirically. Block diagram for liveness detection is shown in Fig. 2.

2.2 Threat Recognition

Under threat face recognition involves two aspects. First, we should recognize the identity of user and secondly his state, whether he is in under-threat or not. We propose separate registration for normal and under-threat pose in the database. We propose to use Chi square dissimilarity to match feature vectors extracted by Uniform Extended Local Ternary Pattern (UELTP) of registered and query samples. Chi square dissimilarity score between two global histogram feature vectors H_i and H_j having L_f labels is defined as below:

$$\chi^2(H_i, H_j) = \frac{1}{2} \sum_{k=1}^{L_f} \frac{[H_i(k) - H_i(k)]^2}{H_i(k) + H_i(k)} \tag{2}$$

Block diagram in Fig. 3 shows the proposed under threat face recognition approach. Matching between query and gallery image involves face detection, region of interest (ROI) extraction, enhancement, UELTP feature extraction and distance computation or matching. Enhancement is done in three steps (1) gamma correction to compensate non uniform lightening (2) difference of Gaussian and (3) Contrast equalization. Matching

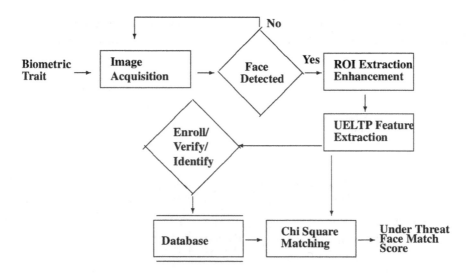

Fig. 3. Block diagram of proposed under threat face recognition

Fig. 4. Regions of two face ROI

is done between the portions of face where most of the expressions are shown during under-threat. Two region of interest (ROI) are cropped from each of the image. Area corresponding to these ROI is show in Fig. 4. Each of ROI are enhanced and histogram feature vectors are calculated for both ROIs. These histogram feature vectors are concatenated into a single global feature vector. Chi square distance metric is used to get the match score using these global feature vectors.

2.3 Frontal Face Recognition

Frontal face recognition is done by using Extended Local Ternary Binary Pattern (ELTBP) feature. Matching score is computed based on Chi-square dissimilarity. Two scores are obtained originating each from the respective two ROIs with respect to their histogram feature vector. To fuse these two match score into a single score, the match score level fusion is used. The match score level fusion is defined as:

$$FusedScore\left(FS_j\right) = \frac{\sum_{i=1}^{T}[T_i(S) * A_{T_i}]}{\sum_{i=1}^{T} A_{T_i}} \quad (3)$$

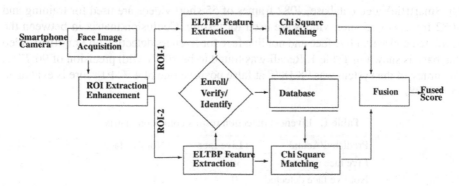

Fig. 5. Proposed face recognition approach

where A_{T_i} is accuracy of trait T_i or approach and $T_i(S)$ is the match score for trait or approach T_i. Block diagram of the face recognition approach is shown in Fig. 5.

The matching process involves the cropping of the face image. ROI enhancement is done for both ROI and histogram feature vectors are calculated for both the ROI. Each ROI has two histogram feature vectors. Two scores are obtained using their respective histogram feature vector match. These scores are melded into a unique global score using their individual accuracy.

3 Experimental Results

Smartphone used in our experiment was XOLO Q1010 having Android version 4.2 (Jelly Bean). Having a 1.3 GHz quad core processor it includes 4 GB of internal memory with rear and from camera of 8 and 2 MP respectively.

To evaluate the effectiveness of the proposed approach, we have used standard performance measures such as correct recognition rate (CRR), equal error rate (EER), cumulative match characteristic curve (CMC), receiver operating characteristic (ROC) curve, precision, recall and F1 are used to evaluate the performance of the system.

3.1 Database

Three in-house databases are created for the purpose using the used for performance analysis and they are SmartBioVideo, SmartBioFace and SmartBioThreatFace. The SmartBioVideo database contains 30 short live face video of students and 25 photo video. Average length of video is 5 s. It contains total 8164 frames. The SmartBioFace database contains frontal face images from 100 students. Each subject has 5 frontal face images. The SmartBioThreatFace database contains face images of 100 students in under-threat as well as normal pose. All images are taken in an unconstrained environment. Each subject has 5 under-threat face images and nearly 5 normal frontal face images.

3.2 Results

On SmartBioVideo database, 4082 frames of 55 short videos are used for training and 4082 frames of 55 short videos are used for testing. Confusion matrix in between the predicted and actual live face and not live face for liveness detection on SmartBioVideo database is shown in Table 1. Recall was found to be 100 % with precision of 96.77 %. Accuracy of the system was 98.18 % at false positive rate of 4 %. F1 score is evaluated to be 0.98.

Table 1. Liveness detection results confusion matrix

Predicted/Acutal	Live face	Not live face
Live face detected	30	1
Not live face detected	0	24

Table 2. Result on SmartBioFace database

Approach	EER (%)	CRR (%)	Fusion strategy
LBP	4.22	99	Feature Level
ULBP	3.97	99	Feature Level
LBP, ULPB	3.96	100	Feature Level
UELTP	3.71	100	Feature Level
ELTP, UELTP	3.22	96.97	Feature Level
LBP, ULBP, UELTP	3.65	100	Feature Level
LBP, ULBP+ELTP, UELTP	3.46	100	Feature an score Level

For SmartBioFace database, Table 2 shows the result of various approaches on this database. LBP, ULBP, ELTP, EULTP approaches are used to get the result on this database with different fusion strategy. Feature level fusion and score level fusion are used. In feature level fusion different feature vectors are concatenated as a single feature vector and in the score level fusion, two score vectors are fused into a single score using weighted sum rule.

Proposed approach has better EER and CRR as compared to other approaches as shown in Table 2 on SmartBioFace database. The proposed approach has EER of 3.46 % and CRR of 100 %. Figure 6(a) shows match score plot of the proposed approach for SmartBioFace database. ROC curve of the proposed approach is shown in Fig. 6(b). Processing time of various steps of proposed approach on SmartBioFace database are shown in Table 3.

For the database of SmartBioThreatFace, 567 images are used as training set and 142 images are used as test set. The uniform ELTP approach is used to get the results on this database. The EER of proposed approach is 18.50 % and CRR is 52.81 % on this database. The CMC curve is shown in Fig. 7(a) while ROC curve of the proposed approach is shown in Fig. 7(b). Processing time on SmartBioThreatFace database for each of the steps on smartphone is shown in Table 3.

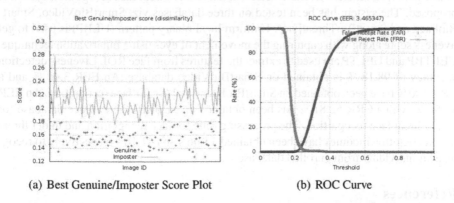

(a) Best Genuine/Imposter Score Plot   (b) ROC Curve

Fig. 6. Best score and ROC for SmartBioFace database (Color figure online)

Table 3. Processing Time for Face Recognition

Database	SmartBioFace	SmartBioThreatFace
Enrollment	156 sec (400 images)	599 sec (567 images)
Face detection	0.02 sec	0.02 sec
ROI extraction and preprocrssing	0.001 sec	0.002 sec
Enhancement	0.002 sec	0.006 sec
Feature Extraction	0.04 sec	0.14 sec
Matching	13.01 sec (400 images)	41 sec (567 images)

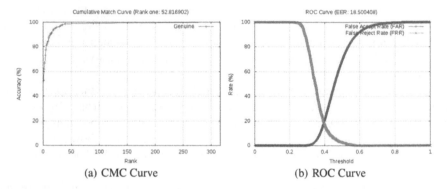

(a) CMC Curve   (b) ROC Curve

Fig. 7. ROC and CMC on SmartBioThreatFace database (Color figure online)

4   Conclusions

In this paper, an efficient liveness detection and under threat face recognition system is proposed. The system has been tested on three databases viz. SmartBioVideo, Smart-BioFace and SmartBioThreatFace. Uniform local binary pattern (ULBP) is used to get liveness score along with capturing the movement of eyes using binarization technique. UELTBP and UELTP are used to extract the features from face ROI. Liveness detection accuracy of 98.18 % is obtained on SmartBioVideo database. An EER 3.46 % and a CRR 100 % have been obtained on SmartBioFace database for face recognition. An EER 18.50 % with a CRR 52.81 % has been obtained on SmartBioThreatFace database for under threat face recognition. The processing time for face recognition and under threat face recognition modules have been obtained 13.06 s and 41.16 s respectively to recognize an individual from enrolled database.

References

1. Ahonen, T., Hadid, A., Pietikäinen, M.: Face recognition with local binary patterns. In: Pajdla, T., Matas, J. (eds.) ECCV 2004. LNCS, vol. 3021, pp. 469–481. Springer, Heidelberg (2004)

2. Ahonen, T., Hadid, A., Pietikainen, M.: Face description with local binary patterns: application to face recognition. IEEE Trans. Pattern Anal. Mach. Intell. **28**(12), 2037–2041 (2006)
3. Zhang, X.H., Shan, S.G., Cao, B., Gao, W., Zhou, D.L., Zhou, D.B.: CAS-PEAL: a large-scale chinese face database and some primary evaluations. J. Comput. Aided Design Comput. Graph. **17**(1), 9–17 (2005)
4. Dave, G., Chao, X., Sriadibhatla, K.: Face recognition in mobile phones. Department of Electrical Engineering Stanford University, USA (2010)
5. D'eniz, O., Castrillon, M., Hern andez, M.: Face recognition using independent component analysis and support vector machines. Pattern Recogn. Lett. **24**(13), 2153–2157 (2003)
6. Dong, H., Gu, N.: Asian face image database pf01. Technical report, Pohang University of Science and Technology (2001)
7. Doukas, C., Maglogiannis, I.: A fast mobile face recognition system for android os based on eigenfaces decomposition. In: Papadopoulos, H., Andreou, A.S., Bramer, M. (eds.) AIAI 2010. IFIP AICT, vol. 339, pp. 295–302. Springer, Heidelberg (2010)
8. Dreuw, P., Steingrube, P., Hanselmann, H., Ney, H., Aachen, G.: SURF-Face: face recognition under viewpoint consistency constraints. In: BMVC, pp. 1–11 (2009)
9. Findling, R.D., Mayrhofer, R.: Towards face unlock: on the difficulty of reliably detecting faces on mobile phones. In: the 10th International Conference on Advances in Mobile Computing and Multimedia, pp. 275–280. ACM (2012)
10. Geng, C., Jiang, X.: Face recognition using sift features. In: 16th IEEE International Conference on Image Processing (ICIP), pp. 3313–3316. IEEE (2009)
11. Gottumukkal, R., Asari, V.K.: An improved face recognition technique based on modular PCA approach. Pattern Recogn. Lett. **25**(4), 429–436 (2004)
12. Kim, H.C., Kim, D., Bang, S.Y.: Face recognition using LDA mixture model. Pattern Recogn. Lett. **24**(15), 2815–2821 (2003)
13. Kohonen, T., Oja, E., Simula, O., Visa, A., Kangas, J.: Engineering applications of the self organizing map. Proc. IEEE **84**(10), 1358–1384 (1996)
14. Kollreider, K., Fronthaler, H., Bigun, J.: Verifying liveness by multiple experts in face biometrics. In: 2008 IEEE Computer Society Conference on Computer Vision and Pattern Recognition Workshops, CVPRW 2008, pp. 1–6. IEEE (2008)
15. Li, J.W.: Eye blink detection based on multiple Gabor response waves. In: International Conference on Machine Learning and Cybernetics, vol. 5, pp. 2852–2856. IEEE (2008)
16. Lin, S.H., Kung, S.Y., Lin, L.J.: Face recognition/detection by probabilistic decision-based neural network. IEEE Trans. Neural Netw. **8**(1), 114–132 (1997)
17. Maatta, J., Hadid, A., Pietikainen, M.: Face spoofing detection from single images using micro-texture analysis. In: 2011 International Joint Conference on Biometrics (IJCB), pp. 1–7. IEEE (2011)
18. Ng, C.K., Savvides, M., Khosla, P.K.: Real-time face verification system on a cell-phone using advanced correlation filters. In: Fourth IEEE Workshop on Automatic Identification Advanced Technologies, pp. 57–62. IEEE (2005)
19. Pabbaraju, A., Puchakayala, S.: Face recognition in mobile devices. Electrical Engineering and Computer Science, University of Michigan (2010)
20. Pan, G., Sun, L., Wu, Z., Lao, S.: Eyeblink-based anti-spoofing in face recognition from a generic webcamera. In: 2007 IEEE 11th International Conference on Computer Vision, ICCV 2007, pp. 1–8. IEEE (2007)
21. Pan, G., Sun, L., Wu, Z., Wang, Y.: Monocular camera-based face liveness detection by combining eyeblink and scene context. Telecommun. Syst. **47**(3–4), 215–225 (2011)

22. Phillips, P.J., Moon, H., Rizvi, S.A., Rauss, P.J.: The FERET evaluation methodology for face-recognition algorithms. IEEE Trans. Pattern Anal. Mach. Intell. **22**(10), 1090–1104 (2000)
23. Samaria, F.S.: Face recognition using hidden Markov models. Ph.D. thesis, University of Cambridge (1994)
24. Schneider, C., Esau, N., Kleinjohann, L., Kleinjohann, B.: Feature based face localization and recognition on mobile devices. In: 9th International Conference on Control, Automation, Robotics and Vision, ICARCV 2006, pp. 1–6. IEEE (2006)
25. Sun, L., Pan, G., Wu, Z., Lao, S.: Blinking-based live face detection using conditional random fields. In: Lee, S.W., Li, S.Z. (eds.) ICB 2007. LNCS, vol. 4642, pp. 252–260. Springer, Heidelberg (2007)
26. Tan, Xiaoyang, Li, Yi, Liu, Jun, Jiang, Lin: Face Liveness Detection from a Single Image with Sparse Low Rank Bilinear Discriminative Model. In: Daniilidis, Kostas, Maragos, Petros, Paragios, Nikos (eds.) ECCV 2010, Part VI. LNCS, vol. 6316, pp. 504–517. Springer, Heidelberg (2010)
27. Venkataramani, K., Qidwai, S., Vijayakumar, B.: Face authentication from cell phone camera images with illumination and temporal variations. IEEE Trans. Syst. Man Cybern. Part C Appl. Rev. **35**(3), 411–418 (2005)
28. Tan, X., Triggs, B.: Enhanced local texture feature sets for face recognition under difficult lighting conditions. IEEE Trans. Image Process. **19**(6), 1635–1650 (2010)

Image Processing

Feature Extraction with Radon Transform
for Block Matching and 3D Filtering

Guang Yi Chen[1](✉) and Wen Fang Xie[2]

[1] Department of Mathematics and Statistics,
Concordia University, Montreal, QC H3G 1M8, Canada
gchen@alumni.concordia.ca
[2] Department of Mechanical and Industrial Engineering,
Concordia University, Montreal, QC H3G 1M8, Canada
wfxie@encs.concordia.ca

Abstract. We propose a novel modification to patch matching in block matching and 3D filtering (BM3D), which is the state-of-the-art in image denoising. The BM3D calculates the distance between two patches by taking the sum of square of the pixel difference. However, when the noise level is very high, this patch matching technique will be less effective. It is well known that Radon transform is very good at suppressing Gaussian white noise and hence in this paper we use it to extract robust features from the two patches for patch matching in BM3D. Experimental results confirm the effectiveness of our proposed modification to BM3D for image denoising in heavily noisy scenarios.

Keywords: Image denoising · Block matching and 3D filtering (BM3D) · Radon transform · Gaussian white noise

1 Introduction

Image denoising is a very important topic in a number of real life applications. Images taken with both digital cameras and conventional film cameras will pick up noise from a number of sources, for example, salt and pepper noise, impulse noise, random noise, etc. Gaussian white noise is the mostly discussed noise in the literature. Images corrupted with this kind of noise can be formulated as follows:

$$B = A + \sigma_n Z \tag{1}$$

where A and B are the noise-free and noisy images, σ_n is the noise standard deviation, and Z is Gaussian white noise with N(0, 1) distribution.

We briefly review several image denoising methods in the literature for image denoising. Sendur and Selesnick [1] proposed a bivariate wavelet denoising method for image denoising. Chen and Kegl [2] incorporated the dual-tree complex wavelet transform (DTCWT) [3] into the ridgelet transform and use it for image denoising. Rajwade et al. [4] developed an image denoising method with higher order singular value decomposition, Dabov et al. [5] invented the block matching and 3D filtering (BM3D) for image denoising and this method is currently the state-of-the-art in the

© Springer International Publishing Switzerland 2016
D.-S. Huang and K.-H. Jo (Eds.): ICIC 2016, Part II, LNCS 9772, pp. 409–417, 2016.
DOI: 10.1007/978-3-319-42294-7_36

literature. Fathi and Naghsh-Nilchi [6] proposed an efficient image denoising method based on a new adaptive wavelet packet thresholding function. Chatterjee and Milanfar [7] developed a patch-based near-optimal technique for image denoising. Chen et al. [11] studied images denoising with feature extraction for patch matching, which is similar to this paper but with worse denoising results than this paper. Kinaus and Zwicker [12] proposed to treat the image denoising task as a simple physical process, which progressively reduce noise by deterministic annealing. Their denoising results are both numerically and visually excellent. Talebi and Milanfa [13] addressed the shortcomings of existing image denoising methods by developing a paradigm for truly global filtering where each pixel is estimated from all pixels in the image. Zuo et al. [14] proposed a texture enhanced image denoising method by enforcing the gradient histogram of the original image. This makes their denoised images look more natural. Fathi and Naghsh-Nihchi [15] developed an efficient image denoising method Based on a new adaptive wavelet packet thresholding function. Cho et al. [16] proposed an image denoising method based on wavelet shrinkage using neighbour and level dependency. Chen et al. [17] studied image denoising by considering neighbor wavelet coefficients in the thresholding processing. Elad and Aharon [18] invented an image denoising technique by using sparse and redundant representations over learned dictionaries.

In this paper, we make small modification to the BM3D algorithm so that it can perform better in the heavily noisy environment. The standard BM3D matches two patches by taking the sum of square of pixel difference. However, when the noise level is high, this approach is less effective. We propose to extract features from the two patches by means of Radon transform and use the extracted features to match the two patches. Our experimental results show that our modification to the BM3D works well for images corrupted by heavy noise.

The organization of this paper is as follows. Section 2 proposes a small modification to the BM3D algorithm by extracting features for patch matching. Section 3 conducts some experiments in order to demonstrate the effectiveness of our modification. Finally, Sect. 4 draws the conclusion of this paper and proposes future research direction.

2 Proposed Modifications

The block matching and 3D filtering (BM3D) [5] is currently the state-of-the-art in image denoising. Given a reference patch, similar patches are found by patch matching. These patches are packed into a 3D array, and then collaborative filtering is performed on each 3D array. Finally, these filtered patches are returned to their original position by weighted average.

The Radon transform is typically discussed in continuum terms, as a mapping from intensity function of the pattern $B(x, y)$ to a function $RT(t, \theta)$ with $t \in R$ and $\theta \in [0, \pi)$, defined as

$$RT(t, \theta) = \int B(x, y)\delta(x\cos\theta + y\sin\theta - t)dxdy \qquad (2)$$

where δ is the Dirac distribution. The discrete Radon transform [8] projects an image onto a number of lines Li passing through the center of the image with different slopes. For every such line, we choose a number of equally spaced points on it and take the sum of those pixel values on a line perpendicular to Li. The slope of the perpendicular lines may become very large when the line Li is parallel to the x-axis. In order to overcome this problem, we add $\pi/4$ to the angle of Li and we only consider angle range of $[\pi/4, 3\pi/4]$ for the lines Li. For this angle range, the slope of the perpendicular lines will not be large. In order to deal with other angles of Li, we rotate the patch by 90° and calculate the Radon coefficients as the unrotated patches. In this way, we can calculate the Radon coefficients reliably. We know that the Gaussian white noise has zero mean, so the Radon transform is robust to this kind of noise.

In this paper, we propose to modify the standard BM3D for patch matching by extracting features from the patches with the Radon transform. Since the Radon coefficients are the sum and hence big, we divide them by the number of those pixels that contributed to this sum. In this way, we can use other steps of the standard BM3D without many modifications. We choose the patches as 33×33 blocks and the extracted features by the Radon transform is 5×5 blocks. Hence, the feature length is 25 for each patch and it should be fast for the patch matching process. However, the computation time of extracting features is extra for our modified BM3D. We can write parallel program to conduct this computation by taking advantages of modern computer clusters or multicore machines. Figures 1 and 2 show the noise-free patch, the noisy patch, and difference between the noisy patch and the noise-free patch, the Radon

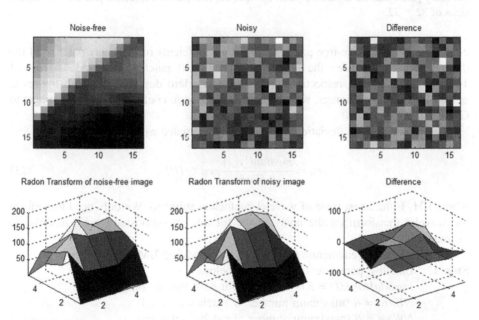

Fig. 1. The noise-free patch, the noisy patch ($\sigma_n = 140$), and difference between the noisy patch and the noise-free patch, the Radon coefficients of the noise-free patch, the Radon coefficients of the noisy patch, and the difference patch between the two Radon coefficient patches for window sizes of 16×16.

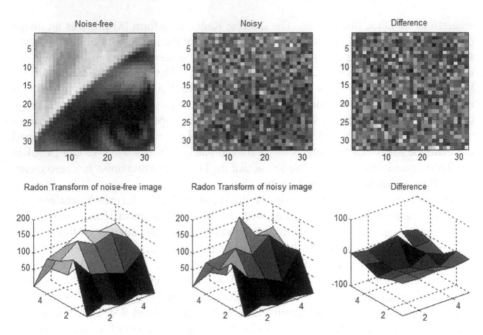

Fig. 2. The noise-free patch, the noisy patch ($\sigma_n = 140$), and difference between the noisy patch and the noise-free patch, the Radon coefficients of the noise-free patch, the Radon coefficients of the noisy patch, and the difference patch between the two Radon coefficient patches for window sizes of 32×32.

coefficients of the noise-free patch, the Radon coefficients of the noisy patch, and the difference patch between the two Radon coefficient patches for window sizes of 16×16 and 32×32, respectively. The noise standard deviation for both figures is $\sigma_n = 140$. From both figures, we can see that the Radon coefficients are very robust to Gaussian white noise.

The noise standard deviation σ_n can be approximated as [10]:

$$\sigma_n = \frac{median(|y_{1i}|)}{0.6745}, y_{1i} \in HH_1.$$ (3)

where HH_1 is the finest scale of wavelet coefficient subband. We only need to perform the wavelet transform on the noisy image for one decomposition scale in order to estimate σ_n.

In summary, we enumerate the steps of our modified BM3D as follows:

Step 1.   Initialize parameters as described below:

lambdaHard3D = 3.0 (threshold for hard thresholding).

NHard = 8 (maximum number of patches retained).

NWien = 8 (maximum number of patches retained).

kHard = 16 (patch sizes).

kWien = 16 (patch sizes).

Step 2.   Given the noisy grayscale image B, perform DTCWT for one scale, and then estimate the noise standard deviation σ_n according to Eq. (3).

Step 3.   If $\sigma_n < 140$, then perform BM3D to B as $B_1 = BM3D(B, \sigma_n)$. Set $\tilde{B} = 255 \times B_1$ because BM3D scales the output image to the range of $[0, 1]$. Stop.

Step 4.   If $\sigma_n \geq 140$, then, for every pixel B_{ij} in the noisy image, take the Radon transform of the patch centered at this pixel and the output Radon coefficients of size 5×5 are saved in a vector of length 25. In this way, we have a Radon features cube of size $M \times N \times 25$ values.

Step 5.   Use these extracted features to match the patches in the noisy image. For every pixel B_{ij} in the noisy image, we use the feature vectors of length 25 to calculate the patch distance. We only keep at most 8 similar patches in this paper. In this way, our modified BM3D should be fast in term of CPU computation time.

Step 6.   Pack the closest patches into a 3D array with regard to pixel B_{ij} in the noisy image.

Step 7.   Threshold the 3D arrays and put the filtered patches to their original locations. And then take the weighted average, just as the standard BM3D. Stop.

The computational complexity of our proposed modification to the standard BM3D is given below. Let us assume the image size to be $M \times N$, the patch size to be $C \times D$, and the search window size $E \times F$. Since we only need to calculate 5×5 Radon coefficients for each patch, the amount of calculation for the whole image is $25 \times \max(C, D) \times M \times N$ flops of operations. Therefore, we save these Radon coefficients as a 3D cube with $M \times N \times 25$ values. Note that we only need to calculate these Radon coefficients once and we save them in computer memory. In order to match the current patch with the patches in the search window, we need $25 \times E \times F \times M \times N$ flops of operations. As a result, the total calculation for our modified BM3D is $COM1 = 25 \times (\max(C, D) + E \times F) \times M \times N$ flops of operation.

It is easy to show that the amount of calculation for the standard BM3D is $COM2 = E \times F \times C \times D \times M \times N$ flops of operations. Hence, our modified BM3D has lower computational complexity than the standard BM3D. In order to show this fact, we give a few examples here. If $C = D = E = F = 32$, then $COM1 = 26400 \times M \times N$ and $COM2 = 1048576 \times M \times N$. If $C = D = E = F = 16$, then $COM1 = 6800 \times M \times N$ and $COM2 = 65536 \times M \times N$. If $C = D = E = F = 8$, then $COM1 = 1800 \times M \times N$ and $COM2 = 4096 \times M \times N$. Under all three examples, our modified BM3D needs less computation than the standard BM3D.

It should be pointed out that our modified BM3D smoothes out the fine details of the denoised images, so it is not good for texture images. On the other hand, the standard BM3D retains more fine features, but it introduces a lot of dirty regions in the denoised images. As a result, there is a trade-off between the two denoising methods with regard to fine features and smooth regions.

The major contribution of our modified BM3D is the following. We have taken advantage of the fact that the Radon transform can suppress the Gaussian white noise, and therefore, better denoising results can be achieved. In addition, the feature size in our modified BM3D is 5×5, which is much smaller than the patch size (33×33). This can be treated as an additional advantage of our modified BM3D. Our experimental results show that our modified BM3D compares favorably to the standard BM3D under heavily noisy environment and they are identical when the noise level is low.

3 Experimental Results

We conducted a number of experiments for reducing the noise in images corrupted with heavy noise. The peak signal to noise ratio (PSNR) is used as a metric to compare different image denoising methods. The PSNR is defined as

$$PSNR = 10 \log_{10}\left(\frac{M \times N \times 255^2}{\sum_{i,j}(B(i,j) - A(i,j))^2}\right) \qquad (4)$$

where $M \times N$ is the number of pixels in the image, and A and B are the noise-free and denoised images.

We utilized our modified BM3D to reduce the noise in several images. Table 1 lists the PSNRs of our modified BM3D (top), the method [11] (middle), and the standard BM3D (bottom) in each cell for images Lena, Boat, Barbara, Cameraman, and Peppers, respectively. It can be seen that our modified BM3D improves the PSNR over the

Table 1. The PSNR of the modified BM3D, the method [11], and the standard BM3D for five test images are given below. The noise standard deviation σ_n goes from 140 to 240. For each cell, the top value corresponds to our modified BM3D, the middle value is for [11], and the bottom value is for the standard BM3D. The highest PSNR value in each cell is highlighted by bold font.

Images	Noise standard deviation (σ_n)					
	140	160	180	200	220	240
Lena	**24.36**	**23.81**	**23.32**	**22.87**	**22.45**	**22.07**
	24.10	23.47	22.91	22.37	21.87	21.41
	23.62	21.95	20.86	19.98	19.15	18.45
Boat	**22.60**	**22.16**	**21.79**	**21.44**	**21.15**	**20.86**
	22.43	21.93	21.49	21.06	20.67	20.30
	22.21	20.92	20.12	19.46	18.81	18.22
Barbara	**21.62**	**21.26**	**20.94**	**20.64**	**20.37**	**20.11**
	21.64	21.20	20.79	20.44	20.09	19.77
	21.31	20.14	19.45	18.78	18.16	17.63
Camera-man	**21.36**	**20.91**	**20.48**	**20.12**	**19.78**	**19.47**
	21.18	20.66	20.18	19.76	19.35	19.00
	21.24	20.04	19.40	18.89	18.34	17.89
Peppers	21.20	**20.68**	**20.21**	**19.76**	**19.39**	**19.07**
	21.20	20.64	20.16	19.72	19.35	18.96
	21.24	19.99	19.32	18.70	18.00	17.35

standard BM3D consistently for heavily noisy images. The only exception is for the Peppers image at $\sigma_n = 140$, where the standard BM3D is better than our modified BM3D. However, the difference between their PSNRs is minor. Figure 3 shows the noise-free images, the noisy images with $\sigma_n = 240$, the denoised images of the standard BM3D, and the denoised images of our modified BM3D for the five test images. It can be seen that our modified BM3D yields better denoising results than the standard BM3D under heavily noisy environment.

Fig. 3. The noise-free images, the noisy images ($\sigma_n = 240$), the denoised image with the standard BM3D, and the denoised images with our modified BM3D for the five test images.

Our modified BM3D is based on the C++ implementation of Lebrun [9]. We use a HP 2000 notebook PC with an AMD 1.6 GHz E-350 processor and 4 GB RAM to conduct the denoising tasks. We notice that our modified BM3D is a bit slower than the standard BM3D due to our inefficient implementation. We could make our program faster by performing some optimization to our code. We leave this task to our future research.

4   Conclusions

In this paper, we have proposed a novel modification to the patch matching process in BM3D by means of feature extraction. We selected the Radon transform for our feature extraction because this transform has the capability of suppressing the Gaussian white noise. We keep all other operations in our modified BM3D the same as the standard BM3D except some parameter changes. Experiments confirmed the improvement of our modified BM3D over the standard BM3D for image denoising under heavily noisy environment.

Future research will be conducted for patch matching by means of affine transform in order to align the similar patches better. After 3D filtering, the aligned patches will perform inverse affine transform and move back to their original location in the images. We will also optimize the C++ code of this paper to make it faster. Based on the analysis in Sect. 2, we believe that our program could be faster than the standard BM3D if we implement it in an efficient way.

Acknowledgments. The authors would like to thank the authors of [1, 5, 9] for posting their denoising software on their websites.

References

1. Sendur, L., Selesnick, I.W.: Bivariate shrinkage with local variance estimation. IEEE Sig. Process. Lett. **9**(12), 438–441 (2002)
2. Chen, G.Y., Kegl, B.: Image denoising with complex ridgelets. Pattern Recogn. **40**(2), 578–585 (2007)
3. Kingsbury, N.G.: Complex wavelets for shift invariant analysis and filtering of signals. J. Appl. Comput. Harmonic Ana. **10**(3), 234–253 (2001)
4. Rajwade, A., Rangarajan, A., Banerjee, A.: Image denoising using the higher order singular value decomposition. IEEE Trans. Pattern Anal. Mach. Intell. **35**(4), 849–862 (2013)
5. Dabov, K., Foi, A., Katkovnik, V., Egiazarian, K.: Image denoising by sparse 3D transform-domain collaborative filtering. IEEE Trans. Image Process. **16**(8), 2080–2095 (2007)
6. Fathi, A., Naghsh-Nilchi, A.R.: Efficient image denoising method based on a new adaptive wavelet packet thresholding function. IEEE Trans. Image Process. **21**(9), 3981–3990 (2012)
7. Chatterjee, P., Milanfar, P.: Patch-based near-optimal image denoising. IEEE Trans. Image Process. **21**(9), 1635–1649 (2012)
8. Kelley, B.T., Madisetti, V.K.: The discrete Radon transform: Part I - theory. IEEE Trans. Image Process. **2**(3), 382–400 (1993)

9. Lebrun, M.: An analysis and implementation of the BM3D image denoising method. Image Process. On Line **2**, 175–213 (2012). http://dx.doi.org/10.5201/ipol.2012.l-bm3d

10. Donoho, D.L., Johnstone, I.M.: Ideal spatial adaptation by wavelet shrinkage. Biometrika **81** (3), 425–455 (1994)

11. Chen, G., Xie, W., Dai, S.-L.: Images denoising with feature extraction for patch matching in block matching and 3D filtering. In: Huang, D.-S., Bevilacqua, V., Premaratne, P. (eds.) ICIC 2014. LNCS, vol. 8588, pp. 398–406. Springer, Heidelberg (2014)

12. Kinaus, K., Zwicker, M.: Progressive image denoising. IEEE Trans. Image Process. **23**(7), 3114–3125 (2014)

13. Talebi, H., Milanfa, P.: Global image denosing. IEEE Trans. Image Process. **23**(2), 755–768 (2014)

14. Zuo, W., Zhang, L., Song, X., Zhang, D.: Gradient histogram estimation and preservation for texture enhanced image denoising. IEEE Trans. Image Process. **23**(6), 2459–2472 (2014)

15. Fathi, A., Naghsh-Nihchi, A.R.: Efficient image denoising method based on a new adaptive wavelet packet thresholding function. IEEE Trans. Image Process. **21**(9), 3981–3990 (2012)

16. Cho, D., Bui, T.D., Chen, G.Y.: Image denoising based on wavelet shrinkage using neighbour and level dependency. Int. J. Wavelets Multiresolut. Inf. Process. **7**(3), 299–311 (2009)

17. Chen, G.Y., Bui, T.D., Krzyzak, A.: Image denoising using neighbouring wavelet coefficients. Integr. Comput.-Aided Eng. **12**(1), 99–107 (2005)

18. Elad, M., Aharon, M.: Image Denoising Via Sparse and Redundant Representations Over Learned Dictionaries. IEEE Trans. Image Process. **15**(12), 3736–3745 (2006)

A Novel Image Steganography Using Wavelet Contrast and Modulus Operation

Weiyi Wei[✉] and Yahong Wen

College of Computer Science and Engineering,
Northwest Normal University, Lanzhou 730070, China
wm9165@qq.com

Abstract. Steganography is the science of hiding data into innocuous objects such that the existence of the hidden data remains imperceptible to an adversary. To obtain larger embedding capacity and imperceptible stegoimages, this paper proposes a new image steganography using wavelet contrast and modulus operation based on the visual characteristics that human are not sensitive to rapid change area and dark area. The method exploits the wavelet contrast value of image block to estimate how many bits will be embedded into the image block, and embeds secrete data by way of modulus operation. Our experimental results show that the proposed approach provides both larger embedding capacity and higher image quality.

Keywords: Image steganography · Wavelet contrast · Modulus operation

1 Introduction

Image data hiding is a kind of camouflage that allow the imperceptible secrete data transmission through covert communication. It is the technology of hiding secret messages within other seemingly innocuous host media or cover media such as images, audio files, texts and so on. Image steganography methods now available mainly include well-know least significant bit [1] and its improved algorithm [2, 3], their all targets are obtaining large capacity steganography and maintaining high fidelity simultaneously.

To improve the embedding capacity, there always be a kind of important scheme which embeds more secret message into rapid change areas and lesser message into smooth area based on human vision system such as BPCS (Bit-plane complexity segmentation) [4], PVD (Pixel-Value Differencing) [5] and SM (Side Match) [6], these improved methods can heighten embedding capacity in some extent, however, there have some respective drawbacks. For example, BPCS method does not take account the weight of each bit plan which influences on the visual quality [7]; SM and PVD methods exploit the difference value between adjoin pixels to estimate how many bits will be embedded into a pixel but it ignores the structure information between pixel and its neighboring pixels.

On the other hand, there also have be some classical methods they focus to improve the visual of the stego-image. Zhang et al. proposed the EMD (Exploit Modification Direction) steganography embedding method which can hide larger payload while

© Springer International Publishing Switzerland 2016
D.-S. Huang and K.-H. Jo (Eds.): ICIC 2016, Part II, LNCS 9772, pp. 418–425, 2016.
DOI: 10.1007/978-3-319-42294-7_37

maintaining a high PSNR value [8]. Liao proposed a secure steganography algorithm based on side match and modular function [9]. In their approaches, the number of bits embedded into a target pixel is still determined by the difference value of its neighboring pixels simply; so as to it ignored the characteristic that human is not sensitive to the noises both in rapid change areas and in darker areas [10]. Fortunately, the wavelet contrast is more suitable for human visual because it takes into account high and low frequency together; so its variety can reflect vision difference of every object in image [11]. Therefore, a novel image steganography using wavelet contrast and modulus operation is proposed in this paper.

The reminder of this paper is organized as follows. Improved wavelet contrast sum is introduced in Sect. 2. Our proposed method is presented in Sect. 3, and the experimental results are show in Sect. 4. The key fundamentals related to our ideas are discussed in Sects. 4.1 and 4.2. Finally, conclusions are given in Sect. 5.

2 Improved Wavelet Contrast Sum

Wavelet transform is a new method of multi-resolution signal analysis. It can express both the summary information and detail for sign. An image is decomposed by 2D wavelet transform with resolution $l-1$, four components can be obtained: the low frequency component A_{l-1}, the horizontal high frequency component D^1_{l-1}, vertical high frequency component D^2_{l-1} and diagonal high frequency component D^3_{l-1}. The low frequency component A_{l-1} is focused on the majority energy of image, the other three are image details.

Pu T defined the wavelet contrast respectively in vertical, horizontal and diagonal direction which take into account the influence of low frequency component and high frequency component for human vision [12]:

$$\text{Vertical Contrast: } C^V_{l-1} = D^1_{l-1}/A_{l-1} \tag{1}$$

$$\text{Horizontal Contrast: } C^H_{l-1} = D^2_{l-1}/A_{l-1} \tag{2}$$

$$\text{Diagonal Contrast: } C^D_{l-1} = D^3_{l-1}/A_{l-1} \tag{3}$$

The variety of wavelet contrast reflects visual difference for every object in image. Later, the sum of wavelet contrast (4), which is proposed according to the characteristic that human vision is relatively insensitive to the region with dramatic changes and darker regions in image [13], it is given by:

$$sum = \sum_{i=1}^{M} \sum_{j=1}^{N} (|D^1_{l-1}(i,j)| + D^2_{l-1}(i,j)| + D^3_{l-1}(i,j)|)/A_{l-1}(i,j) \tag{4}$$

$$k_i = \min(floor(sum_i + 10)/4, 4) \tag{5}$$

Where sum_i and k_i denote wavelet contrast sum and every pixel's embed length in image block i respectively, the constant 10 is obtained by experience. This paper introduces a parameter T to replace constant 10, the improved k_i is given by:

$$k_i = \min(floor(sum_i + T)/4, 4) \tag{6}$$

3 Our Proposed Method

In order to improve the visual fidelity in the stego-image which based on wavelet contrast sum, the paper propose a novel steganography method based on wavelet contrast sum and modular arithmetic. First, it embed secret message by modulus operation, the embed length k_i is obtained by wavelet contrast sum in image $block_i$.

3.1 Hiding the Secret Image

Step 1. Partition the cover image IM with size (M, N) into non overlapping block IM_i (e.g. 8×8);

Step 2. Compute the embed length k_i for every block IM_i, save the k_i with extra 3 bits at most, the cover image IM used in our method are 256 gray scale.

Step 3. Embed k_i bits message into every pixel p_{imn} in IM_i with modulus operation method, p_{imn} is pixel (m, n) in block IM_i. The information function is given:

$$f = (p_{imn} + x) \bmod 2^{k_i} \tag{7}$$

where x is minimum integer satisfied p_{imn} has least variation which is decide by Eqs. 8 and 9.

$$-2^{k_i-1} + 1 \leq x \leq 2^{k_i-1} \tag{8}$$

$$-2^{k_i-1} \leq x \leq 2^{k_i-1} - 1 \tag{9}$$

In order to improve the security, a stochastic sequence s named safe sequence is introduced into expression 7, the improved function is given:

$$f = (s_{imn} + p_{imn} + x) \bmod 2^{k_i} \tag{10}$$

s is a chaotic sequence in this paper, s_{imn} has a range $[0, 2^{k_i} - 1]$.

According to the characteristics of the modular function, when embed k_i bits message into pixel p_{imn}, firstly, convert k_i bits message to decimal d, and find a minimum integer x which satisfied with Eq. 11, then embed x into p_{imn} as expression 12.

$$f = (s_{imn} + p_{imn} + x) \bmod 2^{k_i} = d \tag{11}$$

$$p_{imn}' = p_{imn} + x \tag{12}$$

due to pixel p_{imn} is added with a integer x, the p_{imn}' can beyond the maximum pixel value 255, so the p_{imn} can be adjust with the expression 13 as follow:

$$p_{imn}' = \begin{cases} p_{imn}' + 2^{k_i} & p_{imn}' < 0 \\ p_{imn}' - 2^{k_i} & p_{imn}' > 255 \end{cases} \tag{13}$$

3.2 Extracting the Secret Message

According to characteristics of embedding strategy and modulus operation, the embedded message d can be compute with the expression 12 simply, where d is the message embedded in pixel p_{imn}.

$$d = (a_{imn} + p_{imn}') \bmod 2^{k_i} \tag{14}$$

4   Experimental Results

An excellent steganography has the nature of high payload, better visual fidelity and security. In general, the peak signal-to-noise ratio (PSNR) is employed to evaluate the stego-image quality. If the PSNR is higher than the standard measurement of 30 dB, the secrete message which is stored behind the host image is imperceptible to the human visual system.

In our experiment, secrete message and safe sequence is obtained by Lorenz chaotic sequence, cover images are standard image with size 512×512.

4.1   Embedding Capacity and Visual Fidelity Analysis

Firstly, our experimental results were compared with that of LSB steganography based on wavelet contrast (named LSBWC). For every image block, it's embed length can be obtained by means of its wavelet contrast sum which can reflect its luminance information to a certain extent, meanwhile, the modulus operation is introduced in steganography, so the stego-image's visual fidelity is developed rapidly in the same payload.

Figure 1 shows the stego-image and its difference image comparison in LSBWC and our method. Figure 2 shows histograms from original image, two stego-image with LSBWC and our method; Fig. 3 shows the PSNR comparison with the same embed capacity in LSBWC and our method; Table 1 shows their embed capacity and PSNR

(a)stego-image in LSBWC   (b) difference image in LSBWC

(c) stego-image in our scheme   (d) difference image in our scheme

Fig. 1. Stego image and difference image comparison with LSBWC (T = 8)

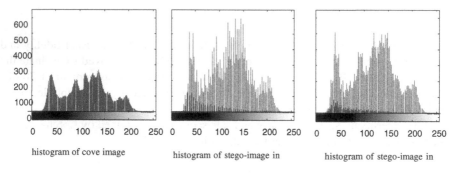

histogram of cove image   histogram of stego-image in   histogram of stego-image in

Fig. 2. The histogram of stego-image comparison with LSBWC

comparison with difference image in LSBWC and our method. It is obvious that stego-image's visual fidelity is improved rapidly.

Secondly, it is contend adaptively that the side and brightness are both combine in our scheme, so it is compared with Liao's method (named PVDM). The embed length of every image block is need to be stored in stego-image, however, it is stored with 3 bits for every image block, and all capacity for the total embed length is 12288bit for image with size 512 × 512. In experimental results comparison, the embed capacity is only the capacity for secret message. Table 2 show embed capacity and PSNR comparison in PVDM method with difference parameter, Table 3 show embed capacity and PSNR in our scheme with difference parameter T. It can be found that the proposed method has high embedding capacity.

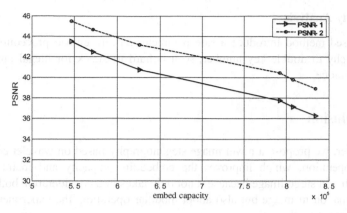

Fig. 3. PSNR comparison with LSBWC and proposed method with T = 8 in Lena

Table 1. PSNR comparison with different T

Image	T = 8			T = 10			T = 12		
	Capacity	PSNR1	PSNR2	Capacity	PSNR1	PSNR2	Capacity	PSNR1	PSNR2
Lena	571904	42.4514	44.65	792448	37.7681	40.4455	834048	36.283	38.8893
Airplane	557888	42.9397	45.0539	788864	37.9528	40.5878	820032	36.7474	39.3188
Peppers	532544	44.1088	46.0363	786816	37.6749	40.7625	794688	37.4031	40.3917
Baboon	705280	38.4382	40.972	873664	35.111	37.6422	967424	32.4248	35.0615
Sailboat	670336	39.809	42.3314	811264	37.0575	39.6592	932480	33.9694	36.4491

Table 2. Embed capacity and PSNR comparison in PVDM method with difference parameter

Image	Scheme (1, 2, 3)		Scheme (2, 3, 4)		Scheme (3, 4, 5)	
	Capacity	PSNR	Capacity	PSNR	Capacity	PSNR
Lena	430966	46.0271	695936	40.2363	974182	33.9198
Airplane	323114	48.2965	585334	42.8307	852512	36.7413
Peppers	280627	50.2084	542457	45.1514	806512	39.2179
Baboon	502576	44.3141	766698	38.4524	1039604	32.3199
Sailboat	368730	37.122	632332	41.4708	904903	35.2787

Table 3. Embed capacity and PSNR in proposed method with difference parameter T

Parameter T	T = 8		T = 10		T = 12		T = 14	
	Capacity	PSNR	Capacity	PSNR	Capacity	PSNR	Capacity	PSNR
Lena	571904	44.65	792448	40.4455	834048	38.8893	1194496	27.6186
Airplane	557888	45.0539	788864	40.5878	820032	39.3188	1052096	34.6298
Peppers	532544	46.0363	786816	40.7625	794688	40.3917	1049152	34.7153
Baboon	705280	40.972	873664	37.6422	967424	35.0615	1156672	30.9895
Sailboat	670336	42.3314	811264	39.6592	932480	36.4491	1086912	33.2723

4.2   Safety Analysis

The proposed method introduce a safe parameter a_{ij} into embed procedure in image steganography to double key number, so it is hard to attack for illegal intruder with listing technique.

5   Conclusion

In this paper, we propose a novel image steganography based on wavelet contrast and modulus operation, which improves the embedding capacity and maintains better visual fidelity in stego image because it not only takes into consideration both edge and light information in image but also adopt modular operation, the experimental results also approved it well.

Acknowledgments. This work is supported by the National Natural Science Foundation of China (71471148), the National Natural Science Foundation of Gansu Province (1506RJZA130).

References

1. Chan, C.K., Cheng, L.: Hiding data in images by simple LSB substitution. Pattern Recogn. **37**(3), 469–474 (2004)
2. Devi, M., Sharma, N.: Improved detection of least significant bit steganography algorithms in color and gray scale images. In: 2014 Recent Advances in Engineering and Computational Sciences (RAECS), pp. 1–5. IEEE (2014)
3. Chutani, S., Goyal, H.: LSB embedding in spatial domain - a review of improved techniques. Int. J. Comput. Appl. **3**(1), 153–157 (2012)
4. Kawaguchi, E, Eason, R.O.: Principles and applications of BPCS steganography. In: Photonics East (ISAM, VVDC, IEMB). International Society for Optics and Photonics, pp. 464–473 (1999)
5. Zhang, X., Wang, S.: Steganography using multiple2base notational system and human vision sensitivity. IEEE Signal Process. Lett. **12**(1), 67–70 (2005)
6. Ramani, G.M.K., Prasad, E.V., Varadarajan, S.: Steganography using BPCS to the integer wavelet transformed image. IJCSNS **7**(7), 293–302 (2007)
7. Zhang, X., Wang, S.: Efficient steganographic embedding by exploiting modification direction. IEEE Commun. Lett. **10**(11), 781–783 (2006)
8. Qinan, L.: Secure steganography algorithm based on side match and modular function. Acta Electronica Sinca **40**(10), 2002–2008 (2012)
9. Toet, A., van Ruyven, J.J., Velaton, J.: Merging thermal and visual images by a contrast pyramid. Opt. Eng. **28**(7), 789–792 (1989)
10. Yang, Z., Mao, S., Chen, W.: New image fusion algorithm based on wavelet contrast. Syst. Eng. Electron. **27**(2), 209–211 (2005)
11. Pu, T., Ni, G.: Contrast-based image fusion using the discrete wavelet transform. Opt. Eng. **39**(8), 2075–2082 (2000)

12. Liu, J., Kang, Z.W., He, Y.: A steganographic method based on wavelet contrast and LSB. Acta Electronica Sinica **35**(7), 1391–1393 (2007)
13. Wei, W.Y.: A new steganography based on wavelet contrast and LSB. Comput. Eng. Appl. **46**(2), 154–155 (2010)

Efficient Specular Reflection Separation Based on Dark Channel Prior on Road Surface

Yao Wang, Fangfa Fu, Jinjin Shi, Weizhe Xu, and Jinxiang Wang[(✉)]

Microelectronics Center, Harbin Institute of Technology, Harbin, 150000, China
jxwang@hit.edu.cn

Abstract. On the busy section of road, vehicle collision could be easily caused by road surface reflection, because road surface reflection always leads to problems in stereo matching, recognition and segmentation. Few attentions have been paid to deal this problem till now. Existing methods rely on a specular free image to detect and estimate specular reflection. Their methods are suitable for image indoor rather than road surface image outdoor. Therefore they are not applicable to road surface reflection. In this paper, a novel method, called dark channel with threshold filter (DCTF) is presented to separate specular reflection from road surface. The method utilizes dark channel prior to roughly get an estimation of the road surface reflection. Then a threshold is proposed which can recover the specular reflection despite of the visual artifacts robustly. Experimental results show that our method significantly outperforms the previous methods in separating specular reflection on road surface.

Keywords: Specular reflection · Road surface · Dark channel prior · Threshold

1 Introduction

The demand for computer vision applications in the fields of video-surveillance, military, and automatic driving, etc. is increasing rapidly. Road segmentation and recognition is crucial to automatic driving. The road reflection as an important factor can influence road segmentation. According to the dichromatic reflection model [1], the reflected light can be divided into two parts: specular and diffuse reflections. The behavior of the specular reflection often leads to problems in stereo matching, segmentation, and recognition, because it captures light source characteristics, creating a discontinuity in the diffuse part. More specifically, specular reflection on road surface, as shown in Fig. 1(a), the red arrow shown in Fig. 1(a) indicates the strong specular reflection. We can see the highlight makes the captured road brighter than normal and creates a discontinuity, which causes difficulties in road segmentation.

Most of the applications simply consider the observed image indoor as a diffuse reflection model, regardless of outdoor road surface specular reflection. One notable work in separating specular reflection is studied by Yang and Tang [10], which shows impressive result but is not applicable to road surface reflection due to ignoring high value of diffuse reflection. Another notable work adopts the dark channel prior [11] to separate specular reflection and employs soft matting [6] to refine the performance.

© Springer International Publishing Switzerland 2016
D.-S. Huang and K.-H. Jo (Eds.): ICIC 2016, Part II, LNCS 9772, pp. 426–435, 2016.
DOI: 10.1007/978-3-319-42294-7_38

(a)

(b)

(c)

(d)

Fig. 1. Specular separation on road surface. (a) Input image. (b) Result of Yang [10]. (c) Result of Kim [6]. (d) Our result. Our result correctly distinguishes between diffuse and specular reflections on road surface, while the previous method inadequately recognizes the background region as specular reflection due to the saturation bias. (Color figure online)

However, the dark channel prior was developed to get specular reflection in the object indoor. The characteristics of specular reflection on the road surface are considerably different from those images. Specifically, compared with indoor object, road surface reflection has very high value of diffuse reflection in the saturation component. Previous approaches are suitable to the object which has very low value of diffuse reflection in the saturation component. Therefore, if previous methods are applied to specular reflection separation on the road surface, blurring artifacts will be caused around whole image as shown in Fig. 1(b) and a restored image will darker than the original image as shown in Fig. 1(c). Those results lack large parts of specular reflection duo to the limitation of low value of diffuse reflection and saturation bias. To overcome these drawbacks, this paper proposes a method to separate specular reflection on the road surface by threshold filter based on dark channel prior as shown in Fig. 1(d). First, we get a rough estimation of specular reflection by dark channel prior and then apply threshold to get real specular reflection.

The rest of this paper is organized as follows. We review some related work in Sect. 2. Rough estimation of specular reflection, threshold filter and dichromatic reflection model are described in Sect. 3. Experimental results of specular reflection separation are presented in Sect. 4 with some conclusions follow in Sect. 5.

2   Related Work

One of the excellent works in separating specular reflection from multiple images was studied in the literature [2, 3]. Nayar [2] and Wolff et al. [13] got a couple of images captured from different polarization angles. In order to separate the specular components, Park and Tou [3] and Tan [14] changed the light source direction to produce two photometric images and used linear basis functions to estimate the intensities of the reflection components. These approaches are of restricted use in indoor environment where the light source is fixed. An efficient way is to change the viewpoint rather than changing the illumination direction. Several approaches employed images taken under a moving light source. For example, Lee [4] put forward a method to detect specular region and Lin [5] treated the specular pixels as outliers by color histogram to remove specular reflection, and matched the remaining diffuse parts in other parts. However, the method is of restricted use in limited specular region. If the specular region is larger than normal size, we cannot consider the large number of pixels involved as outliers. These approaches are moderately useful, yet it may not always applicable in the outdoor conditions.

Kim et al. [6] proposed a more stable solution based on the observation that for most natural images the dark channel of an image provides a pseudo specular reflection result. However, their method again is not suitable for removing specular reflection on road surface. Because indoor light reflection tends to have very high value of specular reflection but low diffuse reflection. While in the outdoor situation, sun light direction is changeable and light reflection tends to have high value specular reflection and high diffuse reflection. The dark channel prior is suitable to low value of diffuse reflection [6].

Some efforts have been made in separating the reflections from a single image. Klinker et al. [7, 15] based on color segmentation to identify specular and diffuse reflections. Mallick [8] and Chung et al. [16] used S and UV channels to represent specular and diffuse components respectively. This SUV space was further used for highlight removal by eroding the specular channel using a single image. Tan et al. [9, 17] successfully removed highlight reflections by using textures. Yang and Tang [10] have developed a fast bilateral [18, 19] filtering approach to refine maximum chromaticity in real-time. Their method shows a satisfactory result in indoor object, however it still fails on the road surface.

All the above methods suffer the same problems that they are not applicable to natural images, such as road surface reflection. To address the limit of the previous approaches, we put forward a simple but efficient specular reflection removing method in this paper. That is, we employ the dark channel prior to get the dark channel of the image and set dark channel image as rough estimation of specular reflection. In order to get real specular reflection, we employ threshold filter to extract real specular reflection form dark channel.

3   Proposed Algorithm

Our proposed algorithm is consisted of rough estimation of specular reflection which we use dark channel prior to get rough estimation of specular reflection, threshold filter by which we can remove incorrect estimation of highlight and dichromatic reflection model will aid us to separate specular reflection from whole images, as shown below.

3.1   Rough Estimate Specular Reflection

In order to remove specular reflection from road surface images, firstly we need to estimate the value of $I_s(x)$ - specular reflection. To begin with, we shortly discuss typical algorithms and their limitation in determining the diffuse reflections. Previous works utilize the pseudo specular-free image to detect the diffuse pixels and estimate the diffuse chromaticity by shifting the maximum or minimum chromaticity of each pixel. However, these approaches are problematic especially when exist high value of diffuse reflections on the road surface since they are all detected as specular reflections. So we use dark channel to estimate rough specular reflection. The rough estimation of specular reflection $I_{dark}(x)$ is expressed as:

$$I^{dark}(\mathrm{x}) = \min_{c \in \{r,g,b\}} I_c(x) \tag{1}$$

3.2   Threshold Filter

Based on the observation a diffuse pixel is likely to have very low intensity in at least one color channel for most of natural images. But only employing dark channel prior is not suitable in the case where the value of diffuse reflection is high under road surface condition.

In order to overcome the limitation of dark channel prior, we employ thresholding method to separate specular reflection. d is the threshold of $I^{dark}(x)$, the value of is ranging from 190 to 205. According to different value of specular reflection, three categories are set: weak, middle strong and strong specular reflection. Only middle strong and strong specular reflection has bad effect on road segmentation. The value of weak specular reflection is from 20 to 190. In order to get middle strong and strong specular reflection, we set value of d is from 190–205. We can distinguish specular reflection from image clearly by using threshold filter. The value of w is from 0.22–0.36 to get good performance. Threshold filter can be described as:

$$\begin{cases} I^{dark}(x) > d \; I^{dark}(x) = I^{dark}(x) * w; \\ I^{dark}(x) <= d \; I^{dark}(x) = 0; \end{cases} \tag{2}$$

3.3   Dichromatic Reflection Model

After threshold filtering, a specular reflection gray image is obtained. According to the dichromatic reflection model [1] which has been widely used in computer graphics, we regard an image as the liner combination of diffuse and specular reflections. We denote the diffuse and specular reflections by $I_d(x)$ and $I_s(x)$, the observed image $I(x)$ is simply expressed as:

$$I(x) = I_d(x) + I_s(x) \tag{3}$$

Therefore a specular-free image $I_d(x)$ is obtained by subtracting $I^{dark}(x)$ from all color channels:

$$I_d(x) = I(x) - I^{dark}(x) \tag{4}$$

This scheme has some benefits against the previous approach [6, 10]. As shown in Fig. 2, our result provides the direct diffuse reflection image. For instance, road surface reflection image as shown in Fig. 2(a), the bottom row in Fig. 2(c) and (d) are very similar to each other and could not get real specular free image result. We could see that their results are not very satisfactory. While the top row in Fig. 2(b) can remove specular reflection efficiently by using proposed method. This is because combining dark channel and threshold together, the specular reflection can be detected exactly and the result becomes close to the real diffuse reflection as the road surface has a very high value in both specular and diffuse reflection.

Fig. 2. Comparisons of high light removal images. (a) Input image. (b) Our result. (c) Result of Kim [6]. (d) Result of Yang [10]. Our result is more likely to (a), in particular, the specular reflection is removed clearly.

Then we use the peak signal-to-noise ratio (PSNR) to evaluate numerical accuracy. For two images I(x) and J(x), the PSNR can be defined as follows:

$$PSNR = 10 \log_{10}((h * w) / \sum_x |I(x) - J(x)|^2) \tag{5}$$

I(x) is the input image in Fig. 3(a), J(x) is our result of separating specular reflection. h and w represent as height and width of image I and J, x is one pixel of these images.

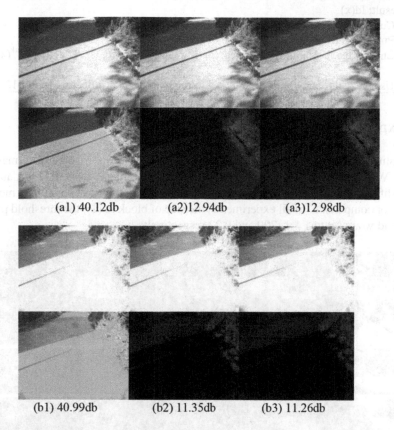

(a1) 40.12db   (a2)12.94db   (a3)12.98db

(b1) 40.99db   (b2) 11.35db   (b3) 11.26db

Fig. 3. Two comparisons of PSNR values of specular reflection removal images. The first row is input image, the second row is PSNR values of proposed method and Kim [6] and Yang [10].

It is widely acknowledged [20] the PSNR values above 40 dB often correspond to almost invisible differences. Image1 and Image2 are in the Fig. 3(a1) and (b1). The PSNR values of images of proposed method are 40.12 and 40.99. Hence we can say our proposed method is more plausible than existing method whose values of PSNR are 12.94, 12.98 and 11.35 and 11.26 separately. The comparison of PSNR value of proposed method and existing method is shown in Table 1.

Table 1. The comparison of PSNR value of proposed method and existing method

Image name	PSNR of proposed method	PSNR of Kim's [6] method	PSNR of Yang's [10] method
Image1	40.12	12.94	12.98
Image2	40.99	11.35	11.26

Algorithm 1:Proposed algorithm for separating specular reflection on road surface
Data: Input image F1
Result: Id(x)
1.Compute the dark channel form input image F1.
2.Put $I^{dark}(x)$ as the rough estimation of specular reflection
3.Employ the threshold filter to get real estimation of specular reflection $I^{dark}(x)*w$ or 0.
4.Use the dichromatic reflection model to remove specular reflection form F1.

4   Experimental Results

We evaluate the performance of the proposed algorithm on six road surface images in Fig. 4. We conducted experiments on several natural images from KITTI database, where diffuse and specular reflections are estimated using previous developed methods [6, 11] for comparison. In all experiments, the size of block in (1), the threshold parameter d and w are set to 1 * 1, 200 and 0.22 respectively.

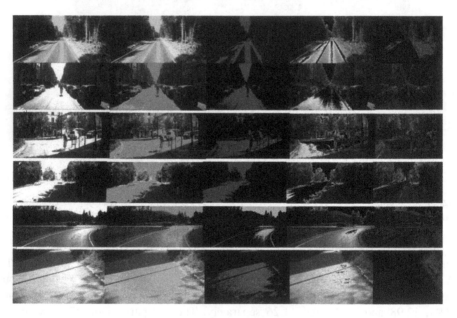

Fig. 4. Road surface images. (a) Input image. (b) and (c) Diffuse and specular reflection of our result. (d) and (e) Specular and diffuse reflection of Yang [10]

Figure 4 visualizes the comparison results for road surface real images, where our algorithm significantly refines the performance of specular reflection separation. The failure of the previous methods shown in (d) and (e) comes from the incorrect detection of diffuse reflection. There methods are desirable in indoor object rather than natural images outdoor.

It is shown dark channel can get rough estimation of specular reflection and threshold filter method can get exactly estimation of specular reflection which help us remove highlight efficiently. Our algorithm employed dark channel prior with threshold in order to achieve more plausible diffuse and specular reflection shown in (b) and (c).

We note that input images (a) of Fig. 4, the value of specular reflection is from 190–250. Our method shows perfect results in these ranges. The method is suitable for strong specular reflection in this paper. For this reason, our algorithm may fail when facing weak specular reflection on the road surface images.

Our method is also suitable for removing highlight on the object, we compare our method with the images provided by [10] in Fig. 5. These images which contain both synthetic and real images have no saturation bias compared to road surface images. In our experiments, the size of block in (1) and the threshold parameter d are set to 1 * 1 and 5 respectively. Owing to low diffuse reflection of object images, w is set to 1.

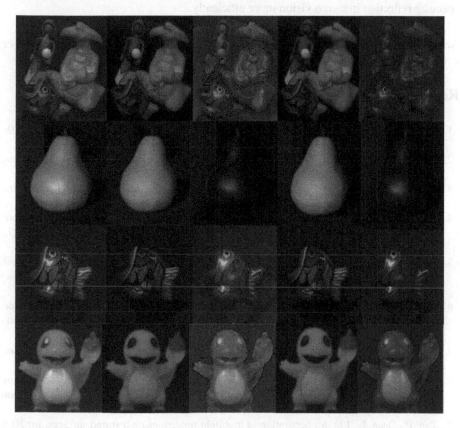

Fig. 5. Specular separation on objects. (a) Input image. (b) and (c) Diffuse and specular reflection by our method. (d) and (e) Diffuse and specular reflection by Yang's [10] method.

Figure 5 visualizes the comparison results for road real images. Our algorithm significantly refines the performance of specular reflection separation. The results show a similar performance to the state of art algorithm.

5   Conclusions

In this paper, we proposed an efficient road surface specular reflection removal algorithm using the dark channel prior with threshold. We first identified the rough estimation of specular reflection by dark channel prior which can capture most parts of highlight on road surface images. Then, we detect real specular reflection by applying threshold filter, which excludes non-specular reflection from dark channel gray image. At last we remove specular reflection from road surface images. Our method is evaluated on real road surface and object images to convince performance. Experimental results showed that the proposed algorithm can remove specular reflection more faithfully in the presence of high value of diffuse and specular reflection, without yielding visual artifacts, than the conventional algorithms [6, 10].

In the future work, we plan to extend the proposed algorithm to remove road surface specular reflection in stereo vision more efficiently.

Acknowledgement. This work was supported by a grant from National Natural Science Foundation of China (NSFC, No. 61504032).

References

1. Shafer, S.A.: Using color to separate reflection components. Color Res. Appl. **10**(4), 210–218 (1985)
2. Nayar, S.K., Fang, X.S., Boult, T.: Separation of reflection components using color and polarization. Int. J. Comput. Vision **21**(3), 163–186 (1997)
3. Park, J.S., Tou, J.T.: Highlight separation and surface orientations for 3-D specular objects. In: Pattern Recognition (1990)
4. Bajcsy, R., Lee, S.W., Leonardis, A.: Detection of diffuse and specular interface reflections and inter-reflections by color image segmentation. Int. J. Comput. Vision **17**(3), 241–272 (1996)
5. Lin, S., Li, Y., Kang, S.B., Tong, X., Shum, H.-Y.: Diffuse-specular separation and depth recovery from image sequences. In: Heyden, A., Sparr, G., Nielsen, M., Johansen, P. (eds.) ECCV 2002, Part III. LNCS, vol. 2352, pp. 210–224. Springer, Heidelberg (2002)
6. He, K., Jin., H., Hadap, S., Kweon, I.: Specular reflection separation using dark channel prior. In: Proceedings of IEEE Conference on Computer Vision and Pattern Recognition, pp. 1460–1467 (2013)
7. Klinker, G.J., Shafer, S.A., Kanade, T.: The measurement of highlights in color images. Int. J. Comput. Vision **2**(1), 7–32 (1988)
8. Mallick, S.P., Zickler, T., Kriegman, D.J., Belhumeur, P.N.: Beyond Lambert: reconstructing specular surfaces using color. In: Conference on Computer Vision and Pattern Recognition (2005)
9. Tan, P., Quan, L., Lin, S.: Separation of highlight reflections on textured surfaces. In: 2006 IEEE Computer Society Conference on Computer Vision and Pattern Recognition, vol. 2. IEEE (2006)
10. Yang, Q., Tang, J., Ahuja, N.: Efficient and robust specular highlight removal (2015)
11. He, K., Sun, J., Tang, X.: Single image haze removal using dark channel prior. IEEE Trans. Pattern Anal. Mach. Intell. **33**(12), 2341–2353 (2011)

12. Levin, A., Lischinski, D., Weiss, Y.: A closed-form solution to natural image matting. PAMI **30**(2), 228–242 (2008)
13. Wolff, L.B., Boult, T.E.: Constraining object features using a polarization reflectance model. IEEE Trans. Pattern Anal. Mach. Intell. **13**(7), 635–657 (1991)
14. Tan, R.T., Ikeuchi, K.: Separating reflection components of textured surfaces using a single image. IEEE Trans. Pattern Anal. Mach. Intell. **27**(2), 178–193 (2005)
15. Klinker, G.J., Shafer, S.A., Kanade, T.: Image segmentation and reflection analysis through color. In: 1988 Orlando Technical Symposium. International Society for Optics and Photonics, pp. 229–244 (1988)
16. Chung, H.-S., Jia, J.: Efficient photometric stereo on glossy surfaces with wide specular lobes. In: IEEE Conference on Computer Vision and Pattern Recognition, CVPR 2008. IEEE (2008)
17. Tan, R.T., Ikeuchi, K.: Reflection components decomposition of textured surfaces using linear basis functions. In: IEEE Computer Society Conference on Computer Vision and Pattern Recognition, CVPR 2005, vol. 1, pp, 125–131. IEEE (2005)
18. Tomasi, C., Roberto, M.: Bilateral filtering for gray and color images. In: Sixth International Conference on Computer Vision. IEEE (1998)
19. Yang, Q., Tan, K.-H., Ahuja, N.: Real-time o(1) bilateral filtering. In: Proceedings of IEEE Conference on Computer Vision and Pattern Recognition, pp. 557–564 (2009)
20. Paris, S., Durand, F.: A fast approximation of the bilateral filter using a signal processing approach. Int. J. Comput. Vis. **81**, 24–52 (2009)

The Scene Classification Method Based on Difference Vector in DCT Domain

Ce Li[1,2(✉)], Ming Li[2], Limei Xiao[2], and Beijie Ren[2]

[1] School of Electronic and Information Engineering,
Xi'an Jiaotong University, Xi'an 710049, China
xjtulice@gmail.com
[2] College of Electrical and Information Engineering,
Lanzhou University of Technology, Lanzhou 730050, China

Abstract. Scene classification is one of the hot research topics in the field of computer vision, it is the basis of the organization and access for a variety of image database, so it has important practical significance. In our previous work, we put forward a novel fast scene classification method via DCT based on the energy concentration and multi-resolution characteristics of DCT coefficients. This paper improved our previous work proposed a scene classification method based on DCT domain using difference vectors. First of all, divided the whole image into the regular grid without repetition, in each grid, do DCT transform with the size of 8 * 8 get the DCT coefficients matrix, extract the AC coefficients in the matrix get the original vectors; Then, selected N images from each category in the database randomly, calculate the average vector of their original vectors, using the original vectors of all images corresponding category subtract the average vector get the difference vectors as the feature vectors; Finally, based on these feature vectors defined above, train classifiers with one-vs.-all support vector machine (SVM). In order to verify the robustness of the proposed algorithm, this paper has built an image database contains eight scene categories according to the OT database, this paper conducted cross validation experiment for the proposed method in the two databases. Experimental results show that the proposed method has higher accuracy and speed in image classification, and has good robustness.

Keywords: Scene classification · DCT · Difference vectors · Multi-resolution · Compressed domain

1 Introduction

The key task of scene classification is to find a relation between the low-level feature description and the high-level knowledge, and then identify the category of the scenes. How the computer automatically identifying the different scenes into the right categories has become a superior interest among the computer vision researches. In the process of feature extraction, whether the mapping relation between local features and scenes existed or not, there are two methods according to that. One is based on low-level features and another is based on high level vocabulary modeling.

D.-S. Huang and K.-H. Jo (Eds.): ICIC 2016, Part II, LNCS 9772, pp. 436–447, 2016.
DOI: 10.1007/978-3-319-42294-7_39

In the method based on low-level features, the texture and color features of scene images are carefully described, and then use the features to validate according to the right classifier. There are global and local feature extractions depending on different sources of low-level features. The global feature extraction uses some eigenvectors obtained by extracting the color and texture features of the whole images to represent the information of the scene. The global method includes Vailaya's [1] hierarchical classification method that combines the low-level features (like the space color pitch as well as proper amount of consistency of edge direction) and binary Bayesian classifier. Oliva and Torralba [2] uses the spatial envelope to get the necessary features for classification. The image is first carried out a Fourier transform and then using Gabor filters in frequency domain at 8 orientations at 4 different scales to extract texture of the image. The square output of each filter is the averaged on a 4×4 grid asthe Gist. Finally use a neural network to classify scene images according to those features. Local feature extraction is based on local block. Then extracting each block to get the features for classification and last according to maximum voting system to decide the categories of the images. Szummer [3] first put forward dividing the image into 16 sub-blocks and then extracting the color and texture as well as frequency information features. Finally use K-NN classifier to separately classify each sub-blocks and using maximum voting strategy to classify the whole image.

The vocabulary modeling based on middle level can resolve the semantic gap between low-level and high-level. This method also can be divided into two kinds. One kind is based on visual vocabulary package and another is local vocabulary concept method. The visual vocabulary package method relays on the initial segmentation in order to obtain regions with information like sky, human, car, grassland, building etc. After that we use local classifier to mark those regions known and with the aid of local information to classify the global scene. Fan [4] use the concept of sensitive significant objection to classify the images. Fredembach [5] extract the feature regions after the segmentation to classify and according to the local classification results to finish the global one. Carson [6] put forward a Blobworld method, which take color and texture together to divide the image into series of regions, by investing the categories and relationships of the regions it is then represent the semantic information of the image. The local vocabulary concept method evaluates the middle-level attributes obtained by the local descriptor around the key points to represent the semantic category of the image. By introducing the middle-level semantic representation it can solve the semantic gap between low and high level. According to the local vocabulary concept method, it does not relay on the initial image segmentation and it uses the local descriptor to represent the scene information. Bag of Words (BoW) [7–9] is one of the mostly used among other methods that it classify the scene by the matching of visual vocabulary histogram. What is to be noticed is the modeling upon the probability models based on BoW, in this way use low dimension latent semantic to represent the high dimension BoW. There are probabilistic latent semantic analysis (pLSA) and Latent Dirichlet allocation (LDA) models in this modeling procession.

All the methods mentioned above improve at the speed and accuracy while in the uncompressed domain. In a more common situation many internet images are all stored and transmitted in the compressed form. A fast scene classification method based on DCT domain can be apply in this compressed images such as JPEG form and this method can greatly improve the speed of classification but with little accuracy decline compared with some classic methods like BoW [7] and spatial envelope method [2]. In order to increase the accuracy, in this paper we proposed a scene classification method based on difference vectors in DCT domain. This method is inspired by [14] with the feature vectors, each every of the category has a much more easy recognized distinctions. With the experiments our method has a more accuracy and faster speed compared with other methods especially in the compressed domain.

The rest of this paper is organized as follows. In Sect. 2, the details of the proposed scene classification based on difference vectors in DCT domain is fully described. Section 3 shows the experiment results with the comparison with some of the classics. Section 4 is the conclusion.

2   A Scene Classification Method Based on Difference Vectors in DCT Domain

The method we proposed consists of the following four steps: (1) Dividing the whole image into 44 sub-image blocks, do the DCT transform with the size of 8 * 8 in each every of the sub-image block. (2) Scanning AC coefficients in each DCT block by three ways, extracting the first 20 AC coefficients, computing the difference vectors and then calculating the average of all the difference vectors in DCT block. Concatenating the difference vectors obtained by the three scanning methods separately as the initial vector of every net. At last concatenating all the initial vectors in the whole image to gain the initial vector of the image. (3) Randomly select N (N = 1, 2, 3 ...) images in every categories in the database and calculate the average of the initial vector accordingly to obtain the mean vector. Then the other images in the database subtract the mean vector above to obtain the difference vector as the feature vector. (4) Use SVM to classify with the feature vector in step three. The frame of the proposed scene classification method is shown in Fig. 1.

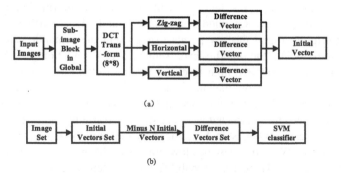

(a)

(b)

Fig. 1. The frame of the proposed scene classification method (a) The process of obtaining the initial vector (b) The process of scene classification.

2.1   Extracting the Initial Vector Based on DCT Coefficients

The two-dimensional DCT transform is of common use in image process, especially in the compressing of the image due to the fact that the DCT has a multi-resolution characteristic. Take a 8 × 8 DCT block as a example, the distribution of DCT coefficients is shown in Fig. 2 with the information that different colors represent DC, LF, MF, and HF with respect from top left corner to bottom right corner which has one DC coefficient and 63 AC coefficients. The DC coefficient is the average energy of the pixels (shown as the average lightness of one image), while the AC coefficients are some detail information like edge, contour and texture in every block. The LF blocks from the top left corner always contain most information while the bottom right blocks have less energy down to zero and thus they are neglected during the quantization of DCT coefficients.

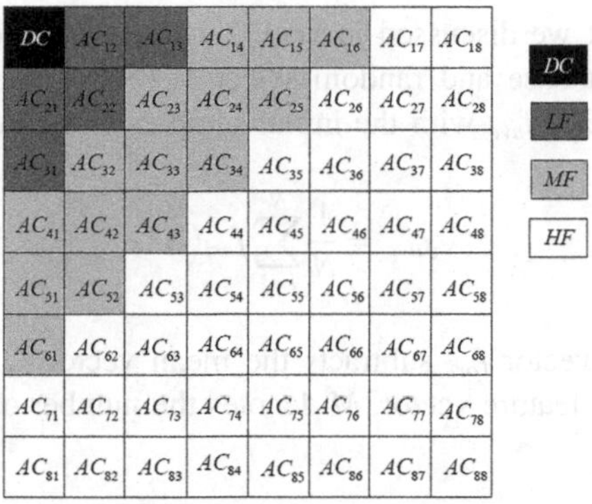

Fig. 2. The distribution of coefficients of 8 * 8 DCT block in DC, LF, MF, HF.

Huang *et al.* have shown the multi-direction and multi-distinguishability of DCT, according to the [15] with the recombination of DCT coefficients. In this paper, we extract the AC coefficients to obtain the initial vectors. Calculating the difference vector is to obtain the relation between DCT coefficients. The steps of feature extraction are as blow:

Step 1: Divide the image into 4 * 4 sub-image blocks.
Step 2: Do DCT transform with the size of 8 * 8 block in each of the sub-image block to obtain the DCT coefficients.
Step3: Scan the DCT coefficients with Zig-zag, horizontal and vertical ways to obtain the textural features in different directions. Extract the first 20 AC coefficients to form three one dimensional vector V_z, V_h and V_v.
Step4: Calculate one-step difference vector $D_k, k = z, h, v$ according to formulation (1) with the three one dimensional vectors acquired in step3.

$$D_k(i) = |[V_k(i)]| - |[V_k(i+1)]| \tag{1}$$

$i = 0, 1, \cdots, n - 1$, n donates the total amount of LF AC coefficients, $[\cdot]$ is the integral operator, $|\cdot|$ is the absolute value operator.

Step 5: Calculate the mean initial difference vectors of all DCT block in one sub-image block $\overline{D_k}, k = z, h, v$ corresponding three scanning ways. Put the mean initial difference vectors of three scanning ways in one sub-image block in cascade, get initial difference vector b of the sub-image block for $b = cat(\overline{D_k}), k = z, h, v$.

Step 6: Put the initial difference vectors of all sub-image blocks in cascade $f_{ori} = cat(b_n), n = 1, 2, \cdots, 16$.

2.2   Extracting the Difference Feature Vectors

The According to what we discussed in Sect. 2.1, calculate the initial vectors f_{ori} for every categories in database and random select N (N = 1, 2, 3 ...) as the samples. Calculate the mean vector f_{ave} with the initial samples in formulation (2).

$$f_{ave} = \frac{1}{N}\sum_{i=1}^{N} f_{ori}(i) \tag{2}$$

The initial feature vector f_{ori} subtracts the mean vector f_{ave}, so we can have the difference vector as the feature vector, M denotes the number of the rest of the images in every category.

$$f_{res}(i) = |f_{ori}(i) - f_{ave}(i)| i = 1, 2 \cdots M \tag{3}$$

In order to see the advantages of difference vector f_{res} over the initial vector f_{ori} more easily, we compare the frequency spectrums of the two in MIT's OT database proposed by Oliva and Torralba [2]. The results are shown in Fig. 3. In order to be of more convictive and generalized, the frequency spectrums are in all average forms. From Fig. 3, the distribution of every category of initial vector is more uniform, and the difference is smaller (see Fig. 3(a)), whereas the distribution of difference vector is much different (different in the coefficient of high energy regions), obvious difference in feature (see Fig. 3(b)). To quantify the different scene category between initial vector and the difference vectors, in this paper we compare the Euclidean distance calculated respectively between scene in every category and the initial vectors of other scene categories as well as the difference vectors. The results are shown in Table 1. The data in Table 1 shows that the difference vectors calculated by the Euclidean distance between scene in every category and other categories are bigger than that of the initial vectors. In this situation the discrimination is more effective and the features are more obvious, this is the reason why we use the difference vector.

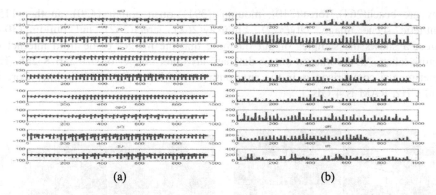

(a)   (b)

Fig. 3. The comparison of frequency spectrums of the initial vector and difference vector.

Table 1. The comparison of Euclidean distance of initial vector and difference vector. In each block, the top number is the Euclidean distance of the initial vector and the bottom number is the Euclidean distance of the difference vector.

	beach	forest	highway	city	mountain	country	street	building
beach	0	369	175	384	181	100	397	329
	0	829	307	1023	764	545	984	936
forest	...	0	267	175	249	307	111	215
		0	851	768	710	596	676	831
highway	...	...	0	269	75	102	304	226
			0	1058	754	553	946	951
city	...	...	...	0	227	311	172	197
				0	803	793	767	920
mountain	...	...	...	...	0	123	288	212
					0	549	873	922
country	...	...	...	...	...	0	341	276
						0	800	852
street	...	...	...	...	...	...	0	245
							0	791
building	...	...	...	...	...	...	...	0
								0

3   Experiments and Analysis

We built a new database modeled on the OT database [2] with same structure in order to verify the effectiveness of our algorithm. Our experiments are carried out on the two database. The OT image database contains 2688 real scene images shown in Fig. 4 with 1472 natural scenes respectively, they are beach(360), forest(328), mountain(374), open country(410), another 1216 synthetic scene image, city(308), highway(260), street(292), tall building(356). Our new database also includes eight same categories as the OT. Our new database contains 2551 images with 1395 natural scenes respectively, they are beach(264), forest(310), mountain(391), open country(430), another 1156 synthetic scene image, city(322), highway(268), street(262), tall building(304). Figure 5 shows the samples of our database that the images come from internet or picture taken. All the images in two database are color images of size 256 × 256. The experimental comparison is in the environment of Matlab 7.0, CPU is Intel Core i6 5.30 GHz, RAM is 4 GB.

Fig. 4. Samples in OT database with 8 categories.

Fig. 5. Samples in New database with 8 categories.

We first test our new database with the DCT method [14] and spatial envelope method [2] and compare the results in OT database for reliability. The results are shown in Table 2, the accuracy decreases about 7 % in our database compared with the OT database. In this way we can find our new database has more complex in scene classification. Figure 6 shows some images that may be classified into the wrong category in our new database.

Table 2. The comparison of accuracy with DCT method and spatial envelope method in OT and New database.

	OT database	New database
DCT (%)	81.43	74.87
Spatial envelope (%)	83.05	76.58

Fig. 6. Some examples of images that may be classified into the wrong category in our new database. From left to right top to down is beach, forest, highway, city, mountain, open country, street and tall building.

3.1   Scene Classification Experiments Settings

In the process of image classification, deciding the category is the last step but is also the most important one. With all the extracted feature vectors we use the optimal classifier to classify the vectors to the right category. The feature vectors in scene images are nonlinear. Support Vector Machine (SVM) classifier is carried out in building a linear decision function in high dimension to represent the nonlinear classification in the original space. SVM has a good generalization ability, the finial classification is decided by few support vectors and has no relationship with the complicity of the algorithm and sample dimension. In a word, SVM is easily carried out and also has robustness.

We carried out four groups of cross validation experiments to testify the effectiveness of difference vector and the robustness of the proposed algorithm.

(1) OT M OT T: Random select N (N = 1, 2, 3 …) images for its initial vectors in OT database and then calculate the average of the initial vectors to obtain a mean vector. The initial vector of the rest of the OT database subtracts the mean vector with the same category respectively to acquire the difference vectors and then do classification.

(2) OT M NEW T: Random select N (N = 1, 2, 3 …) images for its initial vectors in OT database and then calculate the average of the initial vectors to obtain a mean vector. The initial vector of our new database subtracts the mean vector with the same category respectively to acquire the difference vectors and then do classification.

(3) NEW M OT T: Random select N (N = 1, 2, 3 …) images for its initial vectors in our new database and then calculate the average of the initial vectors to obtain a mean vector. The initial vector of the OT database subtracts the mean vector with the same category respectively to acquire the difference vectors and then do classification.

(4) NEW M NEW T: Random select N (N = 1, 2, 3 …) images for its initial vectors in our new database and then calculate the average of the initial vectors to obtain a mean vector. The initial vector of the rest of our new database subtracts the mean vector with the same category respectively to acquire the difference vectors and then do classification.

We train the samples with one-vs.-all SVM, select 100 images of each category as the training samples and leave the rest as the testing samples. The kernel function of SVM is adjusted according to the cross validation and finally choose the histogram intersection kernel function, the results are shown in Fig. 7 with average accuracy of 10 random training and testing.

We also apply our database with the improved spatial envelope feature vector with 520 dimension inspired by Oliva and Torralba [2]. The results are shown in Fig. 8. When N = 1, the accuracy is over 95 % compared to that of the original spatial envelope is 83.05 %. This shows the difference vector is also effective in the spatial envelope method.

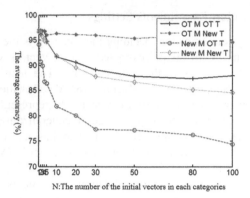

Fig. 7. The accuracy of the cross validation experiments with our proposed method with different number of initial vector. (Color figure online)

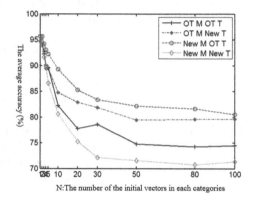

Fig. 8. The accuracy of the cross validation experiments with the improved spatial envelope method with different number of initial vector.

Seen from Figs. 7 and 8, DCT method is of great robustness especially in the OT M NEW T (green line in Fig. 7), the accuracy of NEW M OT T is little lower than the improved spatial envelope method.

It can also be seen from Figs. 7 and 8, the accuracy decreases as the increment of the samples of initial vectors. The accuracy is the highest when N = 1 see the results in Table 3.

Table 3. N = 1, the accuracy of the crossover experiments of the improved spatial envelope method and the proposed method.

Methods	OT M OT T	OT M NEW T	NEW M OT T	NEW M NEW T
Improved spatial envelope (%)	95.87	94.61	95.62	94.75
Ours (%)	96.53	96.92	94.14	96.64

Our proposed algorithm is the highest accuracy (96.92 %) when applied in OT M NEW T. We evaluate our proposed algorithm by using confusion across matrix, the x axis of the table represents the predicted scene class, while the y axis represents true scene categories. The results of DCT based fast scene classification are shown in Fig. 9 (a) with the comparison of our algorithm in Fig. 9(b). The average accuracy of the eight categories is above 93 %. From Fig. 9(a) and (b), our algorithm is outstanding of [14] especially in the tough ones like open country, beach, highway.

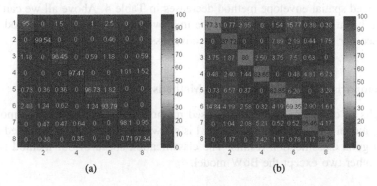

Fig. 9. The pattern of confusion across categories matrix. (a) our proposed method (b) a fast scene classification based on DCT domain [14] from 1 to 8: beach, forest, highway, city, mountain, open country, street and tall building. (Color figure online)

For further testing the robustness of our algorithm, we select some images have nothing in common with the OT and our new database and apply our algorithm with these images. These images contains eight categories with 20 in each category, the size of the images is 256 × 256. Figure 10 shows some samples from that.

Fig. 10. Samples from a new scene database with 8 categories. From left to right top to down: beach, forest, highway, city, mountain, open country, street and tall building.

According to the process mentioned in Sect. 2, calculate the difference vector of all the testing images, N = 1. Training the SVM classifier in OT and our new database, the number of training samples is 100. Use the trained SVM classifier to classify the images from that new scene database images. The result is shown in Table 4. OT denotes the SVM classifier trained in OT and New means the SVM classifier trained in

Table 4. N = 1, the accuracy of the improved spatial envelope method and our proposed method in a new scene database.

Methods	OT	NEW
Improved spatial envelope (%)	95.25	89.59
Our method (%)	96.88	95.85

our new database. The accuracy of our proposed algorithm see no big decline whereas the improved spatial envelope method decreases in Table 4. Above all we can see our algorithm has a better robustness than the improved spatial envelope method, further we can see our algorithm is of better effectiveness.

3.2   Comparison with Other Existed Methods

The results of comparison of the proposed method with spatial envelope model [2], BoW [7] model and Itti's Gist model [18] is shown in Table 5. The proposed method achieved good classification results. The classification speed of our method is higher than the other two except the BoW model.

Table 5. The comparison of accuracy and speed with our proposed method and other classics.

Method	Accuracy (%)	Speed (s/image)
Spatial envelope	83.05	0.670
Bag of words	81.97	0.256
Gist	70.25	0.609
Our method	96.92	0.289

4   Conclusion

This paper proposed a scene classification method based on difference vector in DCT domain. First obtain the initial vector by the multi-resolution and multi-scale characteristic of DCT coefficients. Second acquire the difference vector with more obvious and distinguishing by the initial vector of one category subtracting the mean vector of the respected one. At last do the scene classification by using one-vs.-all SVM classifier. The results show the proposed algorithm achieved better accuracy in classification and faster classification speed, improving the effectiveness in image classification. This method does not need segmentation in scene, weaken the recognition of specific objects, indicates the ability of acquiring the whole information of the whole scene image effectiveness as human. In addition, this method can also be used in compressed domain. The feature vectors of the compressed images based on DCT (JPEG) can be extracted directly in the compressed domain, and in this way no DCT transform means time saving in classification. Compared with the existed method in compressed image classification, the proposed method is easy management for image database especially the image classification based on internet.

Acknowledgments. The paper was supported in part by the China Postdoctoral Science Foundation (2014M550494), the National Natural Science Foundation (NSFC) of China under Grant Nos. (61365003, 61302116), Gansu Province Basic Research Innovation Group Project (1506RJIA031), and Natural Science Foundation of China in Gansu Province Grant No. 1308RJZA274.

References

1. Vailaya, A., Figueiredo, M.: Content-based hierarchical classification of vacation images. In: IEEE International Conference on Multimedia Computing and Systems, pp. 518–523 (1999)
2. Oliva, A., Torralba, A.: Modeling the shape of the scene: a holistic representation of the spatial envelope. Int. J. Comput. Vis. **42**(3), 145–175 (2001)
3. Szummer, M., Picard, R.W.: Indoor-outdoor image classification. In: IEEE International Workshop on Content-Based Access of Image and Video Databases, pp. 42–51 (1998)
4. Fan, J., Gao, Y., Luo, H.: Statistical modeling and conceptualization of natural images. Pattern Recogn. **38**(6), 865–885 (2005)
5. Fredembach, C., Schroder, M., Susstrunk, S.: Eigenregions for image classification. IEEE Trans. PAMI **26**(12), 1645–1649 (2004)
6. Carson, C., Thomas, M., Belongie, S.: Blobworld: a system for region-based image indexing and retrieval. In: Proceedings of International Conference on Visual Information Systems, pp. 509–516 (1999)
7. Quelhas, P., Monay, F., Odobez, J.M.: A thousand words in a scene. IEEE Trans. PAMI **29**(9), 1575–1589 (2007)
8. Lazebnik, S., Schmid, C.: Beyond bags of features: spatial pyramid matching for recognizing natural scene categories. In: IEEE Computer Society Conference on Computer Vision and Pattern Recognition, vol. 2, pp. 2169–2178 (2006)
9. Liu, J.Y., Huang, Y.Z.: Hierarchical feature coding for image classification. Neurocomputing **4**, 22 (2014)
10. Hofmann, T.: Unsupervised learning by probabilistic latent semantic analysis. Mach. Learn. **42**(1/2), 177–196 (2001)
11. Bosch, A., Zisserman, A., Mufioz, X.: Scene classification using a hybrid generative/discriminative approach. IEEE Trans. PAMI **30**(4), 712–727 (2008)
12. Li, F.F., Perona, P.A.: Bayesian hierarchical model for learning natural scene categories. In: IEEE Computer Society Conference on Computer Vision and Pattern Recognition, pp. 524–531 (2005)
13. Blei, D.M., Ng, A.Y., Jordan, M.I.: Latent Dirichlet allocation. J. Mach. Learn. Res. **3**, 993–1022 (2003)
14. Li, C., Li, M.: A novel fast scene classification method via DCT. In: The 2014 7th International Congress on Image and Signal Processing, pp. 752–756 (2014)
15. Huang, X.L., Sun, S.L.: Image retrieval based on DCT compressed domain. Acta Electronica Sinica **30**, 1786–1789 (2002)
16. Huang, X.L., Sun, S.L.: Texture-image classification with rotation invariant in compressed domain. J. Electron. Inf. Technol. 1141–1146 (2002)
17. Sun, L.: Pattern recognition, pp. 155–168. Beijing University of Technology Press (2009)
18. Itti, L., Siagian, C.: Rapid biologically-inspired scene classification using features shared with visual attention. IEEE Trans. PAMI **29**, 300–312 (2007)

Image Compression
Based on Analysis Dictionary

Zongwei Feng[1], Yanwen Chong[1(✉)], Weiling Zheng[1],
Shaoming Pan[1], and Yumei Guo[2]

[1] State Key Laboratory of Information Engineering in Surveying,
Mapping and Remote Sensing, Wuhan University, Wuhan 430079, China
ywchong@whu.edu.cn
[2] School of Electronic Engineering, Xidian University,
Xi'an 710071, Shaanxi, China

Abstract. Along with the extension of the application of the dictionary learned through the synthesis model in the image compression, the time consumption in the sparse representation becomes a key factor restricting the efficiency of the system. Therefore in view of the defect of the synthesis model in the application, combining with the advantages of the analysis model in the sparse representation, we proposed an image block compression model based on analysis dictionary (ALDBCS). In this model, a dictionary which is obtained by using the prior data, is introduced to the process of image compression. The reconstructed simulation experiment proves that the ALDBCS model can not only improve the quality of image reconstruction, but also reduce the consumption of image compression.

Keywords: Blocking compression model · Analysis model · Learning dictionary · Image reconstruction

1 Introduction

With the continuous development of computer vision technology, image as the carrier of visual information attracts the attention of lots of scholars. However, the development of high-resolution and hyperspectral imaging technology bring serious challenges of image transmission and preservation. Recently, the compressed sensing(CS) theory attracted widespread. CS indicates the image can be represented by a small number of observed data, and high precision reconstruction can be achieved via solving the optimization problem, if the image in the transform domain is sparse. In the compressed sensing system, the sparsity of the image in the transform domain is closely related to the image reconstruction quality. Therefore, the research of image sparse representation in the corresponding transform domain, especially the learning dictionary as a transform base, increases gradually. Until now, the sparse representation of image has formed two types: synthesis model and analysis model.

The synthesis model [1–3] defines the sparse representation as: $X = DS$, $s.t. \|s_i\|_0 \leq k$, where D is the over-complete dictionary, is the coefficient matrix of sparse representation under the over-complete dictionary, $s_i \in S$ is the

© Springer International Publishing Switzerland 2016
D.-S. Huang and K.-H. Jo (Eds.): ICIC 2016, Part II, LNCS 9772, pp. 448–457, 2016.
DOI: 10.1007/978-3-319-42294-7_40

coefficient vector corresponding to the $x_i \in X$, x_i is the part of the image X. In the process of solving the optimal sparse representation coefficient, the image k-sparse representation under the over-complete dictionary D is obtained via limiting the non-zero element number of each element s_i less than k. Although the sparse representation under synthesis model has been developed for a long time, it is still in the stage of constant development and improvement. Different from the synthesis sparse model, analysis model which is also called co-sparse analysis model [4–7] attracts a lot of attention of scholars in recent years. Co-sparse analysis model defines the process of sparse representation as: $S = \Omega X$, $s.t. \|s_i\|_0 \le p - l$, where X is the image signal set which is to be sparse representation, Ω is the current analysis dictionary, S is the sparse coefficient matrix of the image signal set X under Ω, s_i is the column vector of sparse coefficient matrix, p is the row number of S, l which is also called joint sparsity is the zero number of s_i, so as to realize the signal sparse representation via limiting the l_0-norm of s_i less than $p - l$.

It is known by $X = DS$, $s.t. \|s_i\|_0 \le k$ and $S = \Omega X$, $s.t. \|s_i\|_0 \le p - l$ that, when both: D and Ω is the square matrix, then both of them can achieve equivalent replacement: $\Omega = D^{-1}$; when both D and Ω is the over-complete dictionary, the $D \in R^{m \times n}(m < n)$ and $\Omega \in R^{p \times q}(p > q)$ cannot achieve equivalent replacement. It is known by $X = DS$, $s.t. \|s_i\|_0 \le k$ and $S = \Omega X$, $s.t. \|s_i\|_0 \le p - l$ that, when obtaining the sparse coefficient matrix of the same dimensions, analysis model contains more number of sub-space, and is more rich and flexible when perform. At the same time, the sparse representation of the analysis model is obtained by the inner product of the dictionary and signal, that is $S = \Omega X$, $s.t. \|s_i\|_0 \le p - l$. This approach reduces the amount of calculation relative to the synthesis model, thereby increasing computing efficiency. Therefore, the image analysis model has more advantages in image sparse representation and image compression. For this reason, analysis model in image processing attracts more and more attention. Rubinstein et al. [8–10] introduced Analysis K-SVD dictionary learning algorithm on the basis of K-SVD dictionary learning algorithm, the algorithm alternating sparse coding and dictionary update to learning analysis dictionary, and it has advantages in image denoising. In the papers [10, 11], an algorithm used to solve the matrix manifold optimization problem was introduced. Where, Simon Hawe et al. proposed the geometric conjugate gradient descent algorithm on the basis of it. References [12–14] introduced the algorithm named simultaneous code word optimization (SimCO) and analyzed the process of learning dictionary use the SimCO. Reference [15] used the sparse observation determinant value as the optimal conditions, then take them into the dictionary learning process. In the paper, the author verified the advantages of the algorithm in the dictionary learning through experiences. Kiechle [16] proposed a bimodal co-sparse analysis model, obtaining the analysis dictionary via the joint gradient descent method based on the flow pattern in the matrix. And the simulation experiment verified the algorithm advantages in image reconstruction. With the development of research, the advantages of analysis dictionary learning in image processing have become increasingly prominent.

In this paper, we take the advantages of analysis model dictionary learning in sparse representation and image processing, and learn from the advantages of image block processing in reference [13], propose an image block compressed sensing algorithm based on analysis dictionary learning (ALDBCS), applying the co-sparse analysis model to the image compression to improve the efficiency of the image processing and the accuracy of image reconstruction.

2 Image Block Compression Based on Analysis Dictionary

Take the time-consuming of learning dictionary in compression system into consideration, this ALDBCS model is divided into two parts which are offline analysis dictionary learning and use learning dictionary to compress image based on the image block compression model. Analysis dictionary learning algorithm is the process of obtaining the analysis dictionary through training data, while compression model based on analysis dictionary expounds how to apply the analysis dictionary learning to the compression process.

2.1 The Analysis Dictionary Learning Algorithm

Via the expression of co-sparse analysis model $S = \Omega X$, we can transform the problem of obtaining the analysis dictionary $\Omega \in R^{p \times m}$ through the given training image data X into solving the following expression:

$$Arg \min_{S,\Omega} ||S - \Omega X||_2, \ s.t. ||S_{:,j}||_0 = p - l \tag{1}$$

where $X \in R^{m \times n}$ is the given training data set, $\Omega \in R^{p \times m}$ is the learning dictionary, p > m represents the dictionary is the over complete dictionary, so that in the case of the same dimension m, The bigger the value of p is, the more the atoms subspace which is chosen in the process of image sparse representation is, which increases the accuracy of sparse representation. $S = \{s_1, s_2, \ldots, s_N\} \in R^{p \times n}$, l is the sparse coefficient matrix, $|| \bullet ||_0$ and $|| \bullet ||_2$ represent l_0-norm and l_1-norm respectively. $||S_{:,j}||_0 = p - l$ represents the number of non-zero elements of the j-th column in S, is the joint sparsity of the current column. $||S - \Omega X||_2$ is the error factor which describes the sparse representation under the learning dictionary. When $||S - \Omega X||_2$ meet the optimized conditions, the optimized dictionary atoms can be obtained.

However, it's inevitably to get such a solution not meet the requirements in the process of solving the expression (1). Therefore, in order to improve the accuracy of solution, expression (1) introduced the limited conditions of the dictionary Ω, which makes each row corresponding to the atoms in the Ω meet $||\Omega_{i,:}||_2 = 1$. In order to verify how well the current dictionary adapt to the original data, the paper introduces $||\hat{X} - X||_2$ to examine the ability of the dictionary adapted to the original data, where $\hat{X} = \Omega^l \Omega X$, Ω is the pseudo-Inverse matrix of the current dictionary Ω, $\hat{X}$ is the reconstruction data acquired under the current dictionaries and sparse coefficient matrix. Therefore, we add $||\hat{X} - X||_2$ to the cost function, because $||\hat{X} - X||_2$ can be transformed into the following form: $Arg \min_{\hat{X}} ||\hat{X} - X||_2 = Arg \min_{\Omega} ||\Omega^l \Omega X - X||_2 \rightarrow Arg \min_{\Omega} ||\Omega^l \Omega - I||_2$ then the expression (1) can be transformed into:

$$Arg \min_{S,\Omega} ||S - \Omega X||_2 + ||\Omega^i \Omega - I||_2, \ s.t. ||S_{:,j}||_0 = p - l$$

$$||\Omega_{i,:}||_2 = 1$$

(2)

It's know by the expression (1) that, solving dictionary atoms and sparse coefficient is the process of achieve two variables S, Ω through the training data set X. Compared with the dictionary learning algorithm based on the synthesis model [14], the above solving process can be divided into two steps: fixing S and updating Ω, fixing Ω and updating S. During the calculation process, the update of the dictionary and the solution of the sparse coefficient can be analogous to the geometrical analysis operator learning algorithm (GOAL) in [10]. To reduce the amount of calculation and dimensions of dictionary, the training data is divided into small blocks with size of $B \times B$ each and finally get the block data X^Γ. The detailed analysis dictionary learning algorithm as shown in Algorithm 1.

Algorithm 1. the analysis dictionary learning algorithm

Input: block data X^Γ base on the training data set X, initial dictionary Ω_0, the
      maximum iterations K, the minimum error ε

Output: the learning dictionary Ω

Algorithm:

    1: initialize parameter i=0, $\Omega^1 = \Omega_0$

    1): calculate the sparse coefficient under the current dictionary:

    $S^i = \Omega^i X^\Gamma$

    2): update the dictionary use the GOAL algorithm: $\Omega^{i+1} \leftarrow \Omega^i$

    3): calculate the value of error factor: $r = ||S - \Omega X||_2 + ||\Omega^i \Omega - I||_2$

    4): if $r \geq \varepsilon$ and the iterations $i \leq K$, then $i = i+1$ and repeat the step 1-3;
        else break the loop;

    2: obtain the final analysis dictionary: $\Omega = \Omega^i$

2.2 The Analysis Dictionary Learning Based Compression Model

Similar to the conventional image compression system, ALDBCS model also includes both encoding side and decoding side two parts. The encoding side is the compression process of the original image data, which can be divided into three parts: the image sparse representation, choosing the optimistic sparse representation coefficients and the atoms subspace and quantization and entropy-code. The decoding side corresponding to the image reconstruction, which mainly includes two parts: entropy decoding and inverse quantization, recover the data. Corresponding to the two parts of the system model, this paper represents ALDBCS model using the following scheme. (the quantization and entropy coding, inverse quantization and entropy decoding modules are not within this paper's research) (Fig. 1).

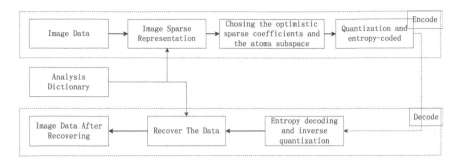

Fig. 1. The analysis dictionary learning based image compression model

The image sparse representation is sparse under the analysis dictionary. It's known by the co-sparse analysis model $S = \Omega X$ that, S is the sparse coefficient matrix of the current data X under the analysis dictionary. Assuming that $X \in R^{m \times n}$, m and n respectively represent the image height and width. It's known by the co-sparse analysis model $S = \Omega X$ that, the size of the dictionary Ω is limited by the X, the dimension of the dictionary Ω meet the condition: $\Omega \in R^{p \times m}(p > m)$. Because of the large dimension of m, the overall dimensions of the dictionary is too large, so that it's not suitable to the storage and computation. In order to reduce the size of the dictionary Ω, the paper utilizes the method of data block processing to obtain block data $X^\Gamma \in R^{d \times Q}, (d = B * B)$, which breaks the compressed data into $d \times Q$ small blocks of data while keeping the data unchanged. What's more, it decreases the dictionary dimensions and reduces large amount of storage space, which is useful to the realization of the system. At the same time, in order to reduce the extra consumption cause by the dictionary learning in image compression, the dictionary learning modules all use offline learning method.

The part of choosing the optimistic sparse representation coefficients and the atoms subspace is to get the optimal sparse coefficient and the corresponding subspaces. It's known by the co-sparse analysis model $S = \Omega X$ that, S^Γ can obtain via the inner product of Ω and X^Γ, which is represented by the expression $S^\Gamma = \Omega X^\Gamma$, therefore, when choosing the optimized sparse coefficient, the paper extract the atom space corresponding to the optimized sparse coefficient as the dictionary of the current reconstruction side while choosing the optimized sparse coefficient according to the elements $s_i^\Gamma \in S^\Gamma$, so as to improve the reconstruction accuracy. When choosing the optimized sparse coefficient, the paper use the Euclidean norm of the original sparse coefficient and the sparse coefficient after optimization as the final loop termination condition. During choosing the optimal coefficients, we chose the coefficients from the s_i^Γ as the descending order. After selecting each coefficient, $||s_i^\Gamma - \hat{s}_i^\Gamma||_2 \leq \sigma$ is calculated, if $||s_i^\Gamma - \hat{s}_i^\Gamma||_2$ is not less than σ, then continue to choose the next sorting coefficient. If $||s_i^\Gamma - \hat{s}_i^\Gamma||_2$ is less than σ, break the loop and finish the process of coefficients optimization selection. Then set the other elements $s_i^\Gamma \in S^\Gamma$ to zero, so as to obtain the final sparse representation coefficient matrix $\hat{S}^\Gamma$. Because most of the coefficient in matrix is 0, compression can be realized during transmission process.

The decoder corresponds to the signal reconstruction part in the system. Because the sparse representation coefficient matrix is obtained by $S^\Gamma = \Omega X^\Gamma$, then $\hat{S}$ can be obtained by $\hat{S} = \Omega_A X^\Gamma$. Therefore, the final reconstructed signal $\hat{X}^\Gamma$ can be obtained by $\hat{X}^\Gamma = \Omega'_A \hat{S}$, where Ω'_A is the pseudo-inverse value of the Ω_A. Recover the reconstructed signal in the form of block, so as to obtain the final reconstructed image $\hat{X}$. Due to the $\hat{X}^\Gamma$ is the data which gets after the process of the original image data through the blocking algorithm, so at the end of the model we should use the inverse-blocking operation to reconstruct the reconstructed data. According to the ALDBCS model, the compression process can be described by the following algorithm.

Algorithm 2. the ALDBCS compression algorithm

Input: initial image data X, analysis dictionary Ω, error coefficient ε
Output: final reconstructed image $\hat{X}$, compression ratio(ratio)

> **Encoder:**
>   1: utilizing the method of block processing to obtain block data $X^\Gamma \in R^{d \times Q}$, the
>   block size is B*B, Q is the block number.
>   2: to obtain the sparse coefficient matrix $S^\Gamma = \Omega X^\Gamma$,
>   3: to optimize the sparse coefficient: if $\| s^\Gamma_i - \hat{s}^\Gamma_i \|_2 \le \sigma$, then break and get the
>   optimized sparse coefficient matrix $\hat{S}^\Gamma$.
>   4: calculate the non-zero elements number (Num) in $\hat{S}^\Gamma$, then ratio=Num/im
>   age size
>
> **Decoder:**
>   1: getting the reconstructed block data according to $\hat{X}^\Gamma = \Omega'_A \hat{S}$ and the
>   available $\hat{S}$
>   2: recovering the reconstructed data in the form of block to get the final
>   reconstructed image $\hat{X}$

3   Experiments

In order to verify the performance of the proposed model, the paper used the standard test image data as test samples to verify the feasibility of the compression model, and made a comparison of the reconstructed results between the proposed algorithm model and the algorithms based on discrete cosine transform (DCT), discrete wavelet transform (DWT) and binary discrete wavelet transform (DDWT) proposed in the reference [13]. To ensure the consistency of the results, we assumed that the block image size is 8 * 8, and simultaneously assumed that the original image number is N, the sampled image number is M, so the sampling rate was defined as M/N. Therefore, in order to keep the consistency of the results, we regarded the sampling rate as compression ratio.

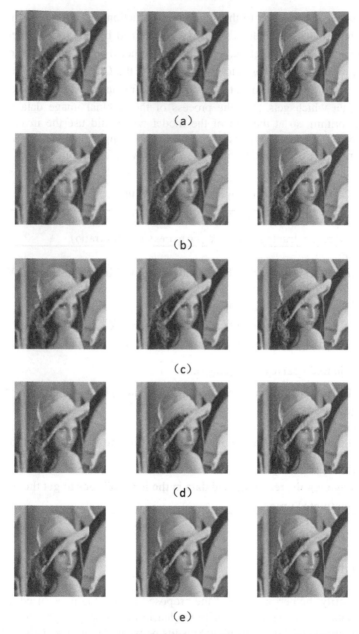

Fig. 2. The reconstructed images under different models at different compression ratio (0.05, 0.10, 0.30): (a) is based on the CT transform and block compressed sensing model; (b) is based on the DCT transform and block compressed sensing model; (c) is based on the DWT transform and block compressed sensing model; (d) is based on the DDWT transform and block compressed sensing model; (e) is based on the proposed model in the paper

And in order to reflect the quality of reconstructed image, in this paper, we took the peak signal to noise ratio (PSNR) and the structure similarity (SSIM) as the research parameters. The performance of current model could be reflected through analyzing the reconstructed results at the same sampling rate.

The following part analyzed the properties of the proposed model from the reconstruction quality and the time-consuming. Fig. 2 show the reconstructed results based on various block compressed sensing models and ALDBCS models. For a more objective analysis on the compression and reconstruction performance based on various block compressed sensing model, Table 1 gives the reconstructed results based on PSNR and SSIM. It's shown in Fig. 2 that, at different sampling rates (M/N = 0.05, 0.10, 0.30), various algorithms can realize compression and reconstruction. The reconstructed images which were based on the transform and block compressed sensing model are smooth but less of details, but the reconstructed images which were based on the ALDBCS model are clear and have rich details, despite existing some burr phenomenon. Table 1 analyzes the reconstructed images of various models using the data, and analyze the effect of compression and reconstruction at different compression ratios under different models. The Table 1 indicates that under the same compression model, the reconstruction quality is improved with the sampling ratio increasing. The reconstruction quality increases faster at a lower compression ratio, while reconstruction quality increases slowly at a higher compression ratio. The result shows that the ALDBCS model is better than the other models at the same sampling ratio. Therefore, both Fig. 2 based on the intuitive visual analysis, and Table 1 based on the data, shows that the ALDBCS model has more advantages in image compression and reconstruction.

Table 1. The reconstruction quality comparison between various model

Models	Sampling ratio(M/N)					
	PSNR/dB			SSIM		
	0.05	0.10	0.30	0.05	0.10	0.30
CT transform	25.24	28.08	32.94	0.79	0.87	0.96
DCT transform	25.17	27.67	32.48	0.78	0.86	0.95
DWT transform	25.14	27.69	32.99	0.80	0.87	0.96
DDWT transform	25.37	28.08	33.46	0.80	0.88	0.96
ALDBCS model	**32.28**	**35.16**	**39.04**	**0.81**	**0.90**	**0.97**

In order to verify the performance of the ALDBCS model, the following experiments compared the reconstructed quality and the efficiency of the compression models, which applies the learning dictionary obtained from analytical model and synthesis model respectively. In the synthesis model, we used the literature [17] K-SVD algorithm as a references, and introduced the learning dictionary into the compression model. To facilitate the recording, our paper designate the above model as *Image Blocking Compressed Sensing based on Learning Dictionary* (BCSLD) The Table 2 compares the quality and time consumption of the two models. The table indicates that under the same blocking method and the same sampling ratio, the ALDBCS model is better than the BCSLD model no mater in the reconstruction quality and the time consumption. In the same sampling ratio, the ALDBCS model has 2-3 dB advantages than the BCSLD model in the reconstructed quality. And in the cost of time, the ALDBCS model has an obvious advantages the BCSLD model. Overall, in the process of image compression, the ALDBCS model has a greater advantage than BCSLD model.

Table 2. The time consumption comparison between various models

Model	Sampling ratio(M/N)					
	PSNR/dB			Time consumption (S)		
	0.05	0.10	0.30	0.05	0.10	0.30
BCSLD	27.36	32.86	37.60	9.64	16.75	**7.33**
ALDBCS	**32.28**	**35.16**	**39.04**	**7.44**	**7.69**	7.80

4 Conclusions

The paper is based on the block compressed sensing, according to the time-consuming of sparse representation, introducing analysis dictionary learning into the block compressed sensing. The sparse coefficient under the analysis model is the inner product of the signal and the dictionary, so that its computation consuming is far less than the other models, which can improve the efficiency of compression and reconstruction. Because the analysis dictionary is obtained by learning, its sparse representation is better than other model's sparse representation. Compared with other compression and reconstruction system based on the transformation, the proposed ALDBCS model is better than other models both in efficiency and reconstruction quality. But because of the time limitation, the paper failed to compare the proposed algorithm with the other novel algorithms. So the proposed algorithm might be the best algorithm. In future, we would like to analysis the proposed algorithm with others and introduce data feature into compression model based on analysis dictionary and optimize the fitness of dictionary atoms subspace.

Acknowledgement. This work has been partially supported by the National Natural Science Foundation of China (Grant No. 61572372 and 41271398), LIESMARS Special Research Funding, and also partially supported by the Fund of SAST (Project No. SAST201425). The funders had no role in study design, data collection and analysis, decision to publish, or preparation of the manuscript.

References

1. Donoho, D.L.: Compressed sensing. IEEE Trans. Inf. Theor. **52**(4), 1289–1306 (2006)
2. Candes, E.J., Romberg, J., Tao, T.: Robust uncertainty principles: exact signal reconstruction from highly incomplete frequency information. IEEE Trans. Inf. Theor. **52** (2), 489–509 (2006)
3. Jiao, L., Yang, S., Liu, F., Hou, B.: Review and prospect of compressed sensing. Chin. J. Electron. **39**(7), 165–1662 (2011)
4. Lian, Q., Shi, B., Chen, S.: Progress in research on dictionary learning models, algorithms and applications. Acta Automatica Sin. **41**(2), 240–260 (2015)
5. Michal, E., Milanfar, P., Rubinstein, R.: Analysis versus synthesis in signal priors. Inverse Probl. **23**(3), 947–968 (2007)
6. Rubinstein, R., Bruckstein, A.M., Michal, E.: Dictionaries for sparse representation modeling. Proc. IEEE **98**(6), 1045–1057 (2010)
7. Nam, S., Davies, M.E., Michal, E., Gribonval, R.: The cosparse analysis model and algorithms. Appl. Comput. Harmonic Anal. **34**(1), 3–56 (2013)
8. Rubinstein, R., Michal, E.: K-SVD dictionary-learning for analysis sparse models. IEEE Int. Conf. Acoust. Speech Signal Process. **22**(10), 540–5408 (2012)
9. Rubinstein, R., Peleg, T., Michal, E.: Analysis K-SVD: a dictionary-learning algorithm for the analysis sparse model. IEEE Trans. Signal Process. **61**(3), 661–677 (2013)
10. Rubinstein, R., Michal, E.: Dictionary learning for analysis-synthesis thresholding. IEEE Trans. Signal Process. **62**(22), 5962–5972 (2014)
11. Ring, W., Wirth, B.: Optimization methods on riemannian manifolds and their application to shape space. Soc. Indian Autom. Manuf. J. Optimi. **22**(2), 596–627 (2012)
12. Simon, H., Martin, K., Klaus, D.: Analysis operator learning and its application to image reconstruction. IEEE Trans. Image Process. **22**(6), 2138–2150 (2013)
13. Dong, J., Wang, W., Dai, W.: Analysis SIMCO: a new algorithm for analysis dictionary learning. In: IEEE International Conference on Acoustic, Speech and Signal Processing (ICASSP), pp. 7193–7197 (2014)
14. Dong, J., Wang, W., Dai, W., Plumbley, M.D., Han, Z., Chambers, J.: Analysis SimCO algorithms for sparse analysis model based dictionary learning. IEEE Trans. Signal Process. **64**(2), 417–431 (2016)
15. Li, Y., Ding, S., Li, Z.: A dictionary-learning algorithm for the analysis sparse model with a determinant-type of sparsity measure. In: Proceeding of the International Conference on Digital Signal Processing, pp. 20–23 (2014)
16. Kiechle, K., Habigt, T., Simon, H.: Martin kleinsteuber.a bimodal cosparse analysis model for image processing. Int. J. Comput. Vis. **114**(2), 33–247 (2015)
17. Michal, A., Michal, E., Alfred, M.B.: K-SVD: an algorithm for designing overcomplete dictionaries for sparse representation. IEEE Trans. Signal Process. **54**(11), 4311–4322 (2006)

An Improved Algorithm Based on SURF for MR Infant Brain Image Registration

Ke Du[1($\boxtimes$)], Stéphane Domas[2], Michel Lenczner[1],
and Guangjin Zhang[3]

[1] FEMTO-ST Institute,
Université de Technologie de Belfort-Montbéliard, Belfort, France
{ke.du,michel.lenczner}@utbm.fr
[2] FEMTO-ST Institute, IUT Belfort-Montbéliard, Belfort, France
stephane.domas@univ-fcomte.fr
[3] Hisense R & D Center, Hisense Electric Co. Ltd., Qingdao, China
zhangguangjin@hisense.com

Abstract. The correct diagnosis of brain diseases is crucial for children with brain disorders. But the complex characteristics of infant brain make the image analysis very complicated. Thus, an accurate image registration is a prerequisite for accurate analysis of MR infant brain images, and it provides valuable information for the diagnosis of doctors. This paper presents our research works on SURF registration algorithm of 2-D MR infant brain images. We firstly describe the original algorithm and analyze its advantages and drawbacks. Then an improved version is proposed, which uses 8-D descriptor vectors with the length of 128. The experiment results show, compared with the original version, our algorithm can achieve more accurate image registration with a little more time consumption. For all the images tested, the increase of correct matching rate varies from a minimum of 5.7 % to a maximum of 14.9 % compared with the classical one.

Keywords: MR image · SURF · Registration · Descriptor vector of features

1 Introduction

In recent years, the incidence of infant brain diseases is rising. Therefore, the correct diagnosis of infant brain diseases in early period has significance for children and it supposes a high quality analysis of the brain structure. For its complex characteristics, it usually requires manual analysis of doctors. Imaging processing technologies are widely used in medical applications. Nevertheless, to diagnose diseases from MR infant brain images is more difficult, it is a promising work.

In order to help doctors in this process, automatic comparisons could be done by a computer, so that it can bring original and valuable information for the diagnosis. The most important thing is to get low error rate. One solution is to use image registration, which is a very important technology in image processing. Its main purpose is to find similarity between two images and get the matching relationship of pixels. How to establish a reasonable correspondence between images is the key point. Until now,

© Springer International Publishing Switzerland 2016
D.-S. Huang and K.-H. Jo (Eds.): ICIC 2016, Part II, LNCS 9772, pp. 458–470, 2016.
DOI: 10.1007/978-3-319-42294-7_41

each image registration algorithm is restricted to one or several classes of images. None of them can be efficiently applied to all images with satisfying performance. For MR infant brain image, the complexity makes it even more challenging.

In this paper, we describe our research works address the problem of images registration applied on 2-D MR infant brain images. The rest of this paper is organized as follows. Section 2 discusses related works including the background of image registration and the categories of existed algorithms. In Sect. 3, we introduce the process of image registration based on descriptor vectors of interest points, especially SURF algorithm. An analysis of its advantages and drawbacks is provided. Section 4 elaborates on the improved SURF algorithm we proposed using 8-D descriptor vectors with the length of 128, which improves the accuracy of registration. Section 5 shows the experiment results. Finally, Sect. 6 gives the conclusions and outlooks.

2 Related Works

Image registration uses a number of similarity measure criteria to establish the relationship between two images, sample image and template image. Then, the parameters of the transformation models must be computed so that the corresponding relationship between pixels in two images can be found. At last, the registration result can be obtained [1, 2]. An example of images in registration is given in Fig. 1.

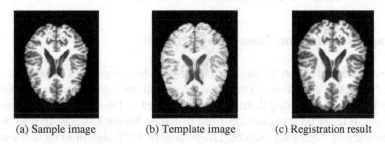

(a) Sample image   (b) Template image   (c) Registration result

Fig. 1. Schematic diagram of image registration

2-D image can be described as a two-dimensional matrix. $I_1(x,y)$ and $I_2(x,y)$ represent the grayscale of pixel $p(x,y)$ in sample image I_1 and template image I_2. The relationship between I_1 and I_2 can be defined as,

$$I_2(x,y) = g(f(I_1(x,y))) \tag{1}$$

Where, f and g are geometric and grayscale transformation function respectively. Registration between images can be achieved by geometric transformation and grayscale transformation. Usually, the grayscale transform is not necessary in practice. The basic framework of image registration is shown in Fig. 2, which is talked in [2].

There are a variety of image registration algorithms. Different ones are applied to different conditions. They can be divided into three categories, including

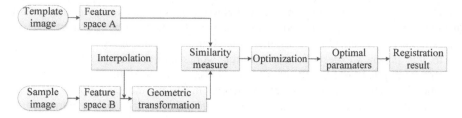

Fig. 2. Flow chat of image registration framework

grayscale-based algorithms, model-based algorithms and feature-based algorithms. The consumption of calculation of feature-based algorithms is much smaller than others', which leads to higher efficiency. Meanwhile, the algorithms have strong robustness regardless of the influence of illumination changes.

There are several reasons make it more challenging to analysis MR infant brain images. Firstly, the newly-born brain is about half large of the adult brain. Secondly, each organization of the brain exhibits different intensities in images at different time. Finally, the colors of gray matter and white matter are changing during infant period. These characteristics make it a good choice to apply registration algorithms based on features on MR infant brain images. The research works presented in this paper focus on obtaining even better performances with feature-based algorithms, by improving the accuracy in the context of 2-D MR infant brain images.

3 Image Registration Based on Features

Many researchers have carried out extensive researches proposed a number of widely used feature-based algorithms. They have a certain increase in either efficiency or accuracy of matching. Among them, SIFT was created by David Lowe and improved in 2004 [3, 4]. It relies on a conversion from matching between two images to building similarities among descriptor vectors. The process of SIFT algorithm includes setup of multi-scale space, extracting interest points, getting descriptor vectors of features, and feature matching. It can handle changes in scale, translation and rotation. Since SIFT algorithm is fast and has so large numbers of applications in the field of image registration, currently, many researchers have put forward their improved SIFT version and achieved good results. In 2006, Herbert proposed the SURF algorithm [5], which is faster and more robust. Although 2-D image registration algorithms based on features are a great achievement, due to the complex structure of brain, especially the infant brain, there are still improvements to find.

3.1 Multi-scale Space Setup

The basis of feature-based image registration algorithm is to extract interest points. Interest points represent significant changes in the image. There are many methods for interest point extraction, such as edge detection and corner detection. However, due to the complex structure of infant brain, methods used on natural images do not work well. Therefore, in order to extract interest points, we need to set up multi-scale space. For some image features are only visible in a particular scale, which can be better to represent characteristics of images. Widely used multi-scale space is Gaussian multi-scale space [6]. It is defined by a Gaussian kernel function,

$$G(x, y, \sigma) = \frac{1}{2\pi\sigma^2} e^{-(x^2 + y^2)/2\sigma^2} \tag{2}$$

For an image $I(x, y)$, its Gaussian multi-scale space can be described as,

$$L(x, y, \sigma) = G(x, y, \sigma) * I(x, y) \tag{3}$$

Where, x, y represent the horizontal and vertical coordinates of 2-D image. σ is the scale parameter, $L(x, y, \sigma)$ is the result in multi-scale space coordinate. The smoothness of the image can be set by σ, where large values corresponds to low resolutions, on the contrary, small values corresponds to high resolutions. Thus, reasonable analysis and calculations can be done according to the image in different resolutions. Focusing on the computation time, Lowe proposed to approximate the Laplacian of Gaussians (LoG) by a Difference of Gaussians (DoG) filter [3].

Although DoG approach was a success, it has been further improved by Herbert in [7] with box filters, where detector of interest points based on Hessian matrix and integral image is used, which reduces the computation time drastically [8].

For a pixel $p(x, y)$ in an image $I(x, y)$, the Hessian matrix is defined as follows,

$$H(I(x, y)) = \begin{bmatrix} \frac{\partial^2 I}{\partial x^2} & \frac{\partial^2 I}{\partial x \partial y} \\ \frac{\partial^2 I}{\partial x \partial y} & \frac{\partial^2 I}{\partial y^2} \end{bmatrix} \tag{4}$$

Working with Gaussian detector done by Lindeberg [9], Hessian matrix takes the following form,

$$H(x, y, \sigma) = \begin{bmatrix} L_{xx}(x, y, \sigma) & L_{xy}(x, y, \sigma) \\ L_{xy}(x, y, \sigma) & L_{yy}(x, y, \sigma) \end{bmatrix} \tag{5}$$

Where, L_{xx}, L_{xy} and L_{yy} are convolutions between the image and second-order Gaussian partial derivatives, taking L_{xx} for an example, as follows,

$$L_{xx} = I(x, y) * \frac{\partial^2 G(x, y, \sigma)}{\partial x^2} \tag{6}$$

The approximation can be made with box filters accompany with integral images. Inside an integral image, as shown in Fig. 3, each pixel is the sum of all the pixels that are above it. After it is computed, it takes only three additions and four memory accesses to calculate the sum of intensities inside a given rectangular region with vertices A, B, C and D of any size, which can be computed as,

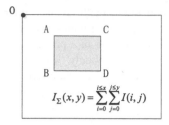

Fig. 3. Schematic diagram of integral image

$$S = A - B - C + D \tag{7}$$

Here, S is used as the value of box filter to make the approximation. Hence, the computational cost is very low for the independence of its size. Similar to Gaussian multi-scale space, box filters with different sizes are used to make convolutions with images. Images at different resolutions constitute the box filter multi-scale space [10].

Interest points, working as matching points, can be either in image space or in scale-space, which include rich image information. Therefore, the quality of the extracted interest points impacts the registration results. In order to select the candidate in multi-scale space, we need to make comparisons between the target interest points and their neighbors. As shown in Fig. 4, the black point in the center is a target interest point, which should be compared with 26 points in its neighborhood.

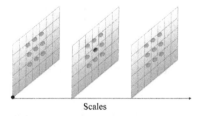

Scales

Fig. 4. Interest points extraction

Interest points are extracted from different scale spaces of the image but one comparison uses only interest points of two scale spaces [11]. Thus, there are a portion of unstable interest points extracted from different scale spaces. In order to remove these unstable points, following derivation is necessary. The scale space function $D(x, y)$ is transformed with Taylor formula in x. Get the first three parts, as follows,

$$D(x, y) = D + \frac{\partial D^T}{\partial x} x + \frac{1}{2} x^T \frac{\partial^2 D}{\partial x^2} x \tag{8}$$

Let $D(x,y) = 0$, then,

$$\hat{x} = -\frac{\partial^2 D^{-1}}{\partial x^2} \cdot \frac{\partial D}{\partial x} \tag{9}$$

With a combination of Eqs. (8) and (9), we can obtain,

$$D(\hat{x}, y) = D + \frac{1}{2}\frac{\partial D^T}{\partial x}\hat{x} \tag{10}$$

$|D(\hat{x},y)|$ is used as the judgement. According to experiment, if $|D(\hat{x},y)| \geq 0.03$, the interest point is regarded as stable, and which can be used to matching.

3.2   Interest Point Description

As the comparison in [12] shows, SURF is invariant to image scaling, blur, and illumination, but partially invariant to rotation and view point changes. Therefore, Haar wavelet is used to detect the orientation. An interest point should be selected in a circular neighborhood and the sum of Haar wavelet responses within a sliding sector window of 60° are calculated, as shown in Fig. 5. Then, the window is rotated by a fixed angle, and the sum is computed once again. After turning for a full circle, the direction with the maximum value is the orientation of the interest point.

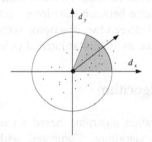

Fig. 5. Determination of orientation in SURF

After getting the orientation, it is possible to extract descriptor vector. In SURF algorithm, a square region in the neighborhood of each interest point is chosen, and it is divided into 4 × 4 small squares, called sub-regions, as shown in Fig. 6(a).

Then, the Haar wavelet responses are computed for each pixel in each sub-region. d_x and d_y represent the horizontal and vertical responses respectively, which are summed up as $\sum d_x$ and $\sum d_y$. In order to take intensity changes into consideration, the sum of the absolute values of the responses is calculated as $\sum |d_x|$ and $\sum |d_y|$. Thus, each sub-region has a 4-D descriptor vector, shown in Fig. 6(b), written as,

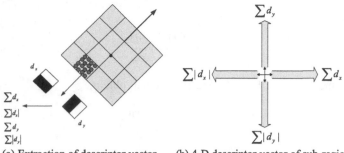

(a) Extraction of descriptor vector   (b) 4-D descriptor vector of sub-region

Fig. 6. Descriptor vector of SURF algorithm

$$v = \left(\sum d_x, \sum d_y, \sum |d_x|, \sum |d_y| \right) \tag{11}$$

In this way, we can obtain a 4-D descriptor vector with the length of 64, which is the classical SURF descriptor. And it can be used in the matching process.

3.3   Interest Point Matching

After getting the descriptor vectors of interest points, a matching between sample and template images can be done. Because of the nature of descriptor vector, the similarity measurement between vectors can be used. To make matching between interest points, we must calculate Euler distance between descriptor vectors of interest points in two images. The shorter distance between two descriptor vectors means a higher degree of similarity, which represents the most similar interest points.

4   Improved SURF Algorithm

SURF algorithm is a registration algorithm based on descriptor vectors of features proposed to improve SIFT algorithm. Compared with SIFT, it focuses on fast matching, thus improves the operating efficiency significantly. But the accuracy can be improved for the following reasons. Firstly, the interest points extracted with box filters are not real corners. Secondly, there are large numbers of interest points on high resolution images including some unstable interest points, which will affect both the efficiency and accuracy. And thirdly, each interest point is described as a 4-D descriptor vector with length of 64 for the 4 × 4 sub-regions. The less amount of computations lead to high efficiency. But the features can be described more accurately, especially for some high required applications. Our work is to improve it by increasing the dimensions of descriptor vector.

d_x and d_y can be regarded as the approximate differential of the image. $d_x > 0$ means the increasing trend in gray gradient of image in positive horizontal direction, while $d_x < 0$ means the decreasing trend. Thus, we express the 4-D descriptor vector in a new way as follows,

$$v = \left(\sum d_{0°}, \sum d_{90°}, \sum d_{180°}, \sum d_{270°} \right) \tag{12}$$

Which includes 4 directions (0°, 90°, 180°, 270°) in the vector, shown in Fig. 7(a).

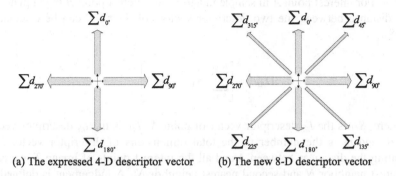

(a) The expressed 4-D descriptor vector   (b) The new 8-D descriptor vector

Fig. 7. The changing in descriptor vector of sub-region

In order to get more details of features in the descriptor vector, four other directions (45°, 135°, 225°, 315°) are added to the vector. Thus, the new descriptor vector becomes an 8-D vector, shown in Fig. 7(b), written as,

$$v = \left(\sum d_{0°}, \sum d_{45°}, \sum d_{90°}, \sum d_{135°}, \sum d_{180°}, \sum d_{225°}, \sum d_{270°}, \sum d_{315°} \right) \tag{13}$$

The value of $d_{45°}$, $d_{135°}$, $d_{225°}$ and $d_{315°}$ can be calculated directly in the same way used to calculate $d_{0°}$, $d_{90°}$, $d_{180°}$ and $d_{270°}$. But it will leads to huge time consumption similar as the case of SURF-128 talked by Herbert in [7]. Here an approximation is made as follows,

$$d_{45°} = \frac{\sqrt{2}}{2} d_{0°} + \frac{\sqrt{2}}{2} d_{90°} \tag{14}$$

$$d_{135°} = \frac{\sqrt{2}}{2} d_{90°} + \frac{\sqrt{2}}{2} d_{180°} \tag{15}$$

$$d_{225°} = \frac{\sqrt{2}}{2} d_{180°} + \frac{\sqrt{2}}{2} d_{270°} \tag{16}$$

$$d_{315°} = \frac{\sqrt{2}}{2} d_{270°} + \frac{\sqrt{2}}{2} d_{0°} \tag{17}$$

Consequently, the vector of a 4 × 4 region contains more detailed information of the image, and the length of which is 128. Actually, the extension and approximation of vector yield better matchings, as shown in experiment section.

Taking accuracy and consumption into account, to achieve interest point matching, we use the ratio of distance between the nearest neighbor and the second nearest neighbor. For interest point A in sample image S and interest point B in template image T, the distance between the two descriptor vectors of A and B can be calculated by Eq. (18).

$$D_{AB} = \left[\sum_{l=0}^{l=m} (S_{Al} - T_{Bl})^2 \right]^{1/2} \tag{18}$$

Where, S_{Al} is the l_{th} descriptor vector of point A, T_{Bl} is the l_{th} descriptor vector of point B, and m is the number of the total dimensions of descriptor vectors. After calculating the distances between A and all the interest points of image T, we can get the nearest neighbor N and second nearest neighbor N'. A judgement is defined as,

$$\theta = D_{AN}/D_{AN'} \tag{19}$$

If θ is less than a certain threshold, A can be matched with B, else there is no matching point in image T. The number of matching points will increase with a larger threshold, and it will also increase the number of mismatching points. Thus, according to experiments, the threshold is usually 0.7. The principle of this method is relatively simple, a little time consuming, highly efficient and perfectly adapted to 2-D MR image matching.

With the 8-D descriptor vector and the rule of matching, the specific registration process can be operated in the following steps.

1. Build the multi-scale space of sample image and template image with box filters of different sizes. Here, the multi-scale space includes three layers, and there are four images in each layer.
2. According to the Hessian matrix approximation, extract the interest points in the middle 10 images, and remove some unstable ones.
3. For the interest points extracted, according to Haar wavelet responses, we can get the orientations of them.
4. Calculate 8-D descriptor vectors of the interest points in sub-regions with the improved SURF algorithm, and get vectors with length of 128 in the neighborhood of 4 × 4 square sub-regions.
5. Achieve the matching of two images with the ratio of distance between the nearest neighbor and the second nearest neighbor.

5 Experimental Results

In order to test the performance of the improved SURF algorithm, the experiment is carried out on a computer with CPU Intel Core i5 2.5 GH, RAM 8.0 GB and Windows7, using Matlab2013b. In the experiment, 5 groups of MR images of different sizes are tested. The sizes of the first 3 groups and last 2 groups are 180 * 260 pixels 200 * 255 pixels respectively. The first group of them is shown in Fig. 8.

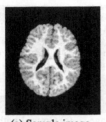 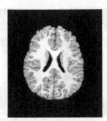

(a) Sample image   (b) Template image

Fig. 8. Images being tested

The interest points are obtained with both the classical SURF algorithm and the improved SURF algorithm. The results of matching between interest points of sample images and template images are shown in Fig. 9. In both image (a) and (b), 60 pairs of interest points are selected and matched. In order to see visually, no matter they match correctly, colored links are used. It is obvious that there are more interest points matched correctly in the improved one. Specific data, including numbers of points matched correctly (NC) and mismatching points (NM), correct matching rate (CR) and time consumption (T) are shown in Table 1.

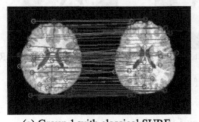 

(a) Group 1 with classical SURF   (b) Group 1 with improved SURF

Fig. 9. The results of interest points matching (Color figure online)

Table 1. Comparison of interest point matching results

	NC	NM	CR/%	T/s
Classical SURF	52	8	86.7	1.112
Improved SURF	57	3	95.0	1.201

With the improved SURF, there are 57 out of 60 interest points matched correctly, 5 more than the classical SURF. The correct matching rate is 95.0 % compared with the classical one with 86.7 %. For the 8-D descriptor vectors with length 128, the time consumption of improved SURF is a little more, from 1.112 s to 1.201 s, increasing by 8.87 %. But compared with the increasing in accuracy, the time consumption is acceptable. The results are shown in Fig. 10, where on the left is the result of classical one, the improved one is on the right and the template image is in the middle.

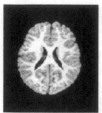 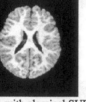 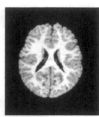

(a) Group 1 with classical SURF   (b) Template image   (c) Group 1 with improved SURF

Fig. 10. The results of image registration

To evaluate the similarity of the images, we make the segmentation according to the results of image registration. The results of segmentation are shown in Fig. 11, where the segmentation result according to classical SURF is on the left, the improved one is on the right and the template image is in the middle.

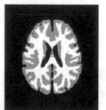 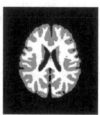 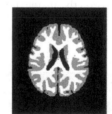 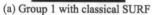

(a) Group 1 with classical SURF   (b) Template image   (c) Group 1 with improved SURF

Fig. 11. The results of image segmentation

Jaccard similarity coefficient is defined as following,

$$J_{sc}(A, B) = \frac{|A \cap B|}{|A \cup B|} \tag{20}$$

Where, A and B represent the set of segmentation results according to template image and image registration result respectively. In this paper, the coefficient is used to evaluate the similarity between the two images. The higher the similarity is, the bigger the value is. In the ideal case, if the two sets are the same, the value should be 1, the maximum value. Different parts of brain in each group of images are chosen as the sets. The Jaccard similarity coefficient of all the groups can be found in Table 2. The result obtained by improved SURF is better than the classical one, where every parts of the brain have higher similarity to the template image for the higher value.

Table 2. Comparison of registration performance

No.	Classical SURF			Improved SURF		
	CR/%	J_{SC}	T/s	CR/%	J_{SC}	T/s
1	86.7	0.834	1.112	95.0	0.931	1.201
2	83.3	0.828	1.156	93.3	0.910	1.217
3	86.7	0.785	1.189	91.6	0.899	1.307
4	78.3	0.789	1.394	90.0	0.851	1.667
5	93.3	0.898	1.446	100.0	0.974	1.613

To make it more convincing, the statistics of accuracy, similarity coefficient and time consumption for all groups of images are shown in Table 2. Generally, as for the higher accuracy, the improved SURF algorithm has better performance on the registration than the classical one. For all the images tested, the increase of correct matching rate varies from a minimum of 5.7 % to a maximum of 14.9 % compared with the classical one. The similarity coefficient is also increased in different degrees, which is consistent with the correct matching rate. The increase of time consumption varies from 5.28 % to 19.6 %, which is still acceptable compared to the gain for registration. Nevertheless, this point constitutes a problem to address in future works.

6 Conclusions and Outlooks

We presented an improved version of SURF algorithm that uses 8-D descriptor vectors with length of 128 to describe the interest points and the ratio of distance between the nearest neighbor and the second nearest neighbor to achieve the matching.

The results have shown that the performance of our improved version is better than the classical one. The important gain in accuracy is due to the use of longer descriptor vectors, which provide more details of images and leads to accurate interest point detection and image registration. The high accuracy is advantageous for cases need high quality of image registration, such as MR infant brain images. As for the commonality of images, the improved SURF algorithm can be used in other different kinds of images, especially color images. Thus, for the future work, in order to widen the application of the improved version, some modifications should be made according to the characteristic of different images, especially to 3D image.

Although the increase of time consumption is acceptable compared with the accuracy, it is still a drawback for many applications, such as on-line computer vision. Therefore, the reduction of time consumption is an important point for future work.

References

1. Rueekert, D., Sonoda, L., Hayes, C., Hill, D.L.G., Leach, M.O., Hawkes, J.: Nonrigid registration using free-form deformations: application to breast MR images. IEEE Trans. Med. Imaging **18**(8), 712–721 (1999)
2. Zitová, B., Flusser, J.: Image registration methods: a survey. Image Vis. Comput. **21**(11), 977–1000 (2003)
3. Lowe, D.G.: Object recognition from local scale-invariant features. In: IEEE International Conference on Computer Vision, Kerkyra, Greece, pp. 1150–1157 (1999)
4. Lowe, D.G.: Distinctive image features from scale-invariant keypoints. Int. J. Comput. Vis. **60**(2), 91–110 (2004)
5. Bay, H., Tuytelaars, T., Van Gool, L.: SURF: speeded up robust features. In: Leonardis, A., Bischof, H., Pinz, A. (eds.) ECCV 2006, Part I. LNCS, vol. 3951, pp. 404–417. Springer, Heidelberg (2006)
6. Kumar, P., Henikoff, S., Ng, P.C.: Predicting the effects of coding non-synonymous variants on protein function using the SIFT algorithm. Nat. Protoc. **4**(7), 1073–1081 (2009)
7. Bay, H., Ess, A., Tuytelaars, T., Van Gool, L.: Speeded-up robust features (SURF). Comput. Vis. Image Underst. **110**(3), 346–359 (2008)
8. Viola, P.A., Jones, M.J.: Rapid object detection using a boosted cascade of simple features. Comput. Vis. Pattern Recogn. **1**, 511–518 (2001)
9. Lindeberg, T.: Scale-space for discrete signals. Pattern Anal. Mach. Intell. **12**(3), 234–254 (1990)
10. Lindeberg, T.: Feature detection with automatic scale selection. Int. J. Comput. Vis. **30**(2), 79–116 (1998)
11. Zhu, Y., Cheng, S., Stanković, V., Stanković, L.: Image registration using BP-SIFT. J. Vis. Commun. Image Represent. **24**(4), 448–457 (2013)
12. Juan, L., Gwun, O.: A comparison of SIFT, PCA-SIFT and SURF. Image Process. **3**(4), 143–152 (2009)

Slippage Estimation Using Sensor Fusion

Thi-Trang Tran and Cheolkeun Ha[✉]

School of Mechanical Engineering, University of Ulsan, Ulsan, South Korea
trantrang286@gmail.com, cheolkeun@gmail.com

Abstract. In this paper, a non-contact slippage estimation approach using sensor fusion is proposed. The sensor consists of a charge-coupled device (CCD) camera and structured light emitter. The slip margin is obtained by estimating very small displacement of the grasped object in consecutive frames sequence captured by CCD camera. In experiments, we apply our approach on a slip-margin feedback control gripper system. The three degree of freedom (DOF) gripper consisting of a CCD camera, structured light and force sensor grasps a target object. The incipient slippage occurs on the contact surface between grip fingers and grasping object when the object is pressed and slid, is estimated by proposed approach. Then, the grip force is immediately controlled by a direct feedback of the estimated slip margin. Consequently, the force is adaptively maintained in order to prevent the object from damage. The proposed approach validity is confirmed by results of experiments.

Keywords: Slippage estimation · Small motion detection · Objects detection

1 Introduction

Tactile receptors distributed on the human skin allow human to sense grasped objects slippage so that human can easily control their hand contact force to prevent the objects from sliding. Imitating skilled human behaviors, many types of tactile sensor have been developed by using electrical resistive, capacitive, electromagnetic or ultrasonic component, piezoelectric, optical component, strain gauges, etc. [1–3]. These sensors structure are complex, they require numerous sensing elements and complicated wiring. Apart from that, vision-based sensors or optical sensors are also developed for tactile sensing [4–6]. In order to dealing with solid objects, the vision-based tactile sensor usually consists of a CCD camera, source light emitter, a transparent acrylic plate, and touchpad [7–9]. These sensors obtain slippage based on the movements of dots printed on the surface of the touchpad captured by the CCD camera. In [10], an approach estimating slippage of grasped flexible object, an elastic object, is proposed. A feature point is drawn on the apex of the elastic sphere and the authors used deformation of the contact area measured by a camera through a transparent plate when an elastic object slides on a rigid plate.

These above contacted sensors can obtain a variety of tactile information such as contact region, slippage, and contact force accurately but many crucial issues remain unresolved. Firstly, these sensors are very expensive and their structures are used to be complex, require many sensing elements and complicated manufacturing. Secondly,

© Springer International Publishing Switzerland 2016
D.-S. Huang and K.-H. Jo (Eds.): ICIC 2016, Part II, LNCS 9772, pp. 471–481, 2016.
DOI: 10.1007/978-3-319-42294-7_42

these sensors still meet difficulty in obtaining various types of information simultaneously for example determining the grasped object position, orientation in
3-dimentional space. Thirdly, the sensor surface of these sensors can easily be damaged
especially the apex contacted with grasped objects, leading to inaccuracy measurement.
In order to solve these above problems, in this paper we propose a non-contact slippage
estimation approach using sensor fusion for slip-margin feedback control gripper
system. The system consists of a structure line light, a CCD camera and gripper holding
target object as described in Fig. 1. When the object is pressed and slid slightly on the
gripper finger, the object slip margin is estimated by calculating the change in object
pose through image frame sequence captured by CCD camera. The grip force is
immediately controlled by a direct feedback of the estimated slippage. The paper is
organized as follows: Sect. 2 presents an overview of proposed slippage estimation in
the slip-margin feedback control gripper system. Proposed non-contact slippage estimation using sensor fusion approach is given in Sect. 3. The experiment results and
conclusion are finally indicated in Sects. 4 and 5, respectively.

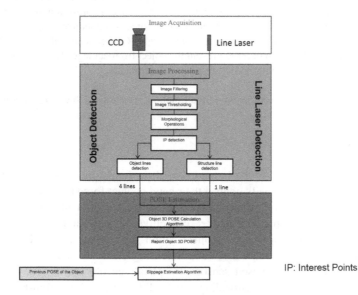

Fig. 1. Slippage estimation using laser and camera approach

2   Overview of Slippage Estimation in the Slip-Margin Feedback Control Gripper System

The slippage estimation using monocular structured light vision system contains a
camera, laser, gripper grasping object. In this paper we use a cylinder object which has
a rectangular yellow shape on top. Because almost all of the object parts are hidden by
grasping gripper except the object top, thus the object top, yellow rectangular, is chosen
as a model for object detection and tracking (Fig. 2a). Suppose that the object size is
unknown and the object color model is Md. Slippage estimation approach is illustrated

in Fig. 1. Firstly, due to the fact that the original image (Fig. 2b) usually contains noise, we smooth the frame sequence with a Gaussian kernel. Then the threshold method is applied to convert RBG images into binary images. During the threshold operation, object model *Md*, a look-up table, is used to convert incoming frame pixels into a corresponding probability. Subsequently, all the small regions are eliminated as they may represent noise by morphological operation.

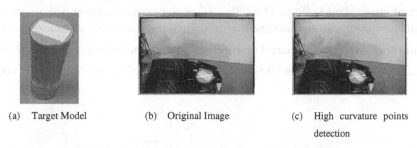

(a)   Target Model   (b)   Original Image   (c)   High curvature points
                                                                            detection

Fig. 2. An example of IP detection. (Color figure online)

In obtained binary image, we discriminates between outer borders and hole borders to detect a list of points, in one way or another, and a curve in an image. IP are then defined by highest curvature points on image curves. According to the target model and system construction, we will find 8 IP (4 IP of the object and 4 IP are intersection of structured light and rectangular) as shown in Fig. 2c. These points will be used for object lines and structured light (laser line) estimation by using line fitting algorithm.

After the image processing process, we obtained 4 object lines l_1, l_2, l_3, l_4 and structured line l_5 are detected in image plane and defined as follows.

$$A_j x + B_j y + C_j = 0$$
$$\text{Subject to } j = \{1, 2, 3, 4, 5\} \tag{1}$$

$[A_j B_j C_j]$ is the unit vector and the index i represents that the line equation corresponding to the line l_j. These lines will be used for object pose calculation.

Finally, slippage is estimated by calculating the difference between object pose in consecutive frames.

3   Proposed Non-contact Slippage Estimation Using Sensor Fusion Approach

The object pose is determined based on the information of detected IP. There are two steps in the pose determination approach: the target reconstruction; the object pose calculation.

There are four coordinate systems in our model as shown in Fig. 3: camera, image, laser and object coordinate systems. CCD camera coordinate system ΣCam is a right hand coordinate system permanently fixed to the camera with its origin ^{Cam}O at the camera origin and the ^{Cam}Z-axis coincides with the camera optical axis, pointing to the light of sight, follows the standard way to decide camera coordinate system. Camera is described by pin-hole model. Let us denote ΣIP image plane coordinate system, whose coordinate is $^{IP}p = [^{IP}x, ^{IP}y, 1]^T$ and the x-axis, y-axis are along the top and left side of image plane. The origin ^{IP}O of ΣIP is at the top left vertex of the image plane and the center of image plane is $^{IP}c = [^{IP}x_c, ^{IP}y_c, 1]^T$. Laser system $\Sigma Laser$ is the right hand coordinate system with the laser line is assumed to lie on the $^{Laser}XZ$ plane and the origin is at the laser origin. ΣTg is target object coordinate system, whose coordinate is $^{Tg}P = [^{Tg}X, ^{Tg}Y, ^{Tg}Z, 1]^T$. The origin ^{Tg}O of ΣTg is at the top left vertex of the target object.

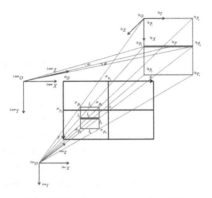

Fig. 3. The model of pose measurement based on sensor fusion.

Suppose that there is a point $^{Cam}S = [^{Cam}X_S, ^{Cam}Y_S, ^{Cam}Z_S]^T$ in camera coordinate system, and its image in image plane is $^{IP}s = [^{IP}x_s, ^{IP}y_s, 1]^T$.

$$w\,^{IP}s = {^{IP}_{Cam}}M\,^{Cam}S \tag{2}$$

$^{IP}_{Cam}M$ can be exactly obtained in the CCD camera calibration process [11, 12].

The physical transformation part between object coordinate (or world coordinate) and camera coordinate is the sum of the effects of some rotation R and some translation t. In homogeneous coordinates, we can combine these within a single matrix as follows

$$^{Cam}_{Tg}W = [^{Cam}_{Tg}R \quad ^{Cam}_{Tg}t] \tag{3}$$

Writing this out, we have:

$$^{Cam}_{Tg}W = \begin{bmatrix} ^{Cam}_{Tg}R_{3\times3} & ^{Cam}_{Tg}t_{3\times1} \\ 0_{1\times3} & 1 \end{bmatrix} \tag{4}$$

Then the relation between image plane and target object coordinate (or world coordinate) can be expressed as follows.

$$^{IP}s = w^{-1}[\,_{Cam}^{IP}M_{3\times3}, 0_{3\times1}]\,_{Tg}^{Cam}W\,^{Tg}S \tag{5}$$

We can choose the object plane as shown in Fig. 3 so that $^{Tg}Z = 0$

$$^{IP}s = w^{-1}\,_{Cam}^{IP}M[r_1 \ r_2 \ r_3 \ t]\begin{bmatrix} ^{Tg}X_s \\ ^{Tg}Y_s \\ 0 \\ 1 \end{bmatrix} = w^{-1}\,_{Cam}^{IP}M[r_1 \ r_2 \ t]\begin{bmatrix} ^{Tg}X_s \\ ^{Tg}Y_s \\ 1 \end{bmatrix} \tag{6}$$

And the problem is needed to be solved is to determine the relative position between camera coordinate and target object coordinate, i.e. we need to find the rotation $_{Tg}^{Cam}R$ and translation $_{Tg}^{Cam}t$.

The non-cooperative target object size is unknown, in order to estimate the object pose in camera coordinate system we need to reconstruct the target object. Suppose that the target width and height are rh, h respectively and the target object lays on ^{Tg}Z-plane. In homogeneous coordinate, the target four vertices are given as following:

$$\begin{cases} ^{Tg}P_1 = [0, 0, 1]^T \\ ^{Tg}P_2 = [0, rh, 1]^T \\ ^{Tg}P_3 = [h, 0, 1]^T \\ ^{Tg}P_4 = [h, rh, 1]^T \end{cases} \tag{7}$$

Their image coordinate and camera coordinate respectively are:

$$\begin{cases} ^{IP}p_i = \left[^{IP}x_{p_i}, {}^{IP}y_{p_i}, 1\right]^T \\ ^{Cam}P_i = \left[^{Cam}X_{P_i}, {}^{Cam}Y_{P_i}, {}^{Cam}Z_{P_i}\right]^T \end{cases} \text{ subject to } i = \{1, 2, 3, 4\} \tag{8}$$

The camera coordinate of the vertices can be calculated as following.

$$\left[^{Cam}P_1, {}^{Cam}P_2, {}^{Cam}P_3, {}^{Cam}P_4\right] = {}_{Cam}^{IP}M^{-1}\left[^{IP}p_1, {}^{IP}p_2, {}^{IP}p_3, {}^{IP}p_4\right]diag(w_1, w_2, w_3, w_4) \tag{9}$$

The target object coordinate of the vertices can be calculated as following.

$$_{Tg}^{IP}W\left[^{Tg}P_1, {}^{Tg}P_2, {}^{Tg}P_3, {}^{Tg}P_4\right] = \left[^{IP}p_1, {}^{IP}p_2, {}^{IP}p_3, {}^{IP}p_4\right]diag(w_1, w_2, w_3, w_4) \text{ with } _{Tg}^{IP}W = {}_{Cam}^{IP}M\left[_{Tg}^{Cam}r_1, {}_{Tg}^{Cam}r_2, {}_{Tg}^{Cam}T\right] \tag{10}$$

Because $^{Tg}P_1, {}^{Tg}P_2, {}^{Tg}P_3$ are not on a same line, the matrix $[^{Tg}P_1, {}^{Tg}P_2, {}^{Tg}P_3]$ is non-singular, so we have:

$$w_4{}^{IP}p_4 = \left[{}^{IP}p_1, {}^{IP}p_2, {}^{IP}p_3\right]diag(w_1,w_2,w_3)\left[{}^{Tg}P_1, {}^{Tg}P_2, {}^{Tg}P_3\right]^{-1}{}^{Tg}P_4 \qquad (11)$$

Substituting (7) into (11) we obtained rectangular constrains:

$$w_4^{-1}[-w_1,w_2,w_3]^T = \left[{}^{IP}p_1, {}^{IP}p_2, {}^{IP}p_3\right]^{-1}{}^{IP}p_4 \qquad (12)$$

Let us denote ${}^{IP}p_j$ subject to $j = \{5,6\}$ be the points of intersections of structure line l_5 and the object lines l_4 and l_2, respectively.

In camera coordinate system and laser coordinate system, the points of intersections of structured-line and object lines are:

$$^{Cam}P_j = w_j \, {}^{IP}_{Cam}M^{-1}{}^{IP}P_j \qquad (13)$$

$$^{Laser}P_j = \left[{}^{Laser}X_{P_j}, 0, {}^{Laser}Z_{P_j}\right] \qquad (14)$$

The physical transformation part between laser projector coordinate and camera coordinate is the sum of the effects of some rotation ${}^{Cam}_{Laser}R$ and some translation ${}^{Cam}_{Laser}T$, which were exactly calculated in CCD camera and laser calibration process [13]. In homogeneous coordinate we can combine these within single matrix as following

$$^{Cam}_{Laser}W = \left[{}^{Cam}_{Laser}R, {}^{Cam}_{Laser}T\right] = \left[{}^{Cam}_{Laser}r_1 \quad {}^{Cam}_{Laser}r_2 \quad {}^{Cam}_{Laser}r_3 \quad {}^{Cam}_{Laser}T\right] \qquad (15)$$

So we have:

$$^{Cam}P_j = {}^{Cam}_{Laser}W^{Laser}P_j = {}^{Laser}X_{P_j}{}^{Cam}_{Laser}r_1 + {}^{Laser}Z_{P_j}{}^{Cam}_{Laser}r_3 + {}^{Cam}_{Laser}T \qquad (16)$$

Combining (13) and (19) we have:

$$w_j \, {}^{IP}_{Cam}M^{-1}{}^{IP}P_j = {}^{Laser}X_{P_j}{}^{Cam}_{Laser}r_1 + {}^{Laser}Z_{P_j}{}^{Cam}_{Laser}r_3 + {}^{Cam}_{Laser}T \qquad (17)$$

Solving (17),

$$\begin{bmatrix} w_j \\ {}^{Laser}X_{P_j} \\ {}^{Laser}Z_{P_j} \end{bmatrix} = \left(\left[{}^{IP}_{Cam}M^{-1}{}^{IP}P_j, -{}^{Cam}_{Laser}r_1, -{}^{Cam}_{Laser}r_3\right]^T \left[{}^{IP}_{Cam}M^{-1}{}^{IP}P_j, -{}^{Cam}_{Laser}r_1, -{}^{Cam}_{Laser}r_3\right]\right)^{-1}$$
$$\left[{}^{IP}_{Cam}M^{-1}{}^{IP}P_j, -{}^{Cam}_{Laser}r_1, -{}^{Cam}_{Laser}r_3\right]^T {}^{Cam}_{Laser}T \qquad (18)$$

Substituting w_j into (13), we obtain

$$^{Cam}P_j = [1,0,0]\left(\left[\,^{IP}_{Cam}M^{-1\,IP}P_j, -\,^{Cam}_{Laser}r_1, -\,^{Cam}_{Laser}r_3\right]^T\left[\,^{IP}_{Cam}M^{-1\,IP}P_j, -\,^{Cam}_{Laser}r_1, -\,^{Cam}_{Laser}r_3\right]\right)^{-1}$$
$$\left[\,^{IP}_{Cam}M^{-1\,IP}P_j, -\,^{Cam}_{Laser}r_1, -\,^{Cam}_{Laser}r_3\right]^T\,^{Cam}_{Laser}T\,^{IP}_{Cam}M^{-1\,IP}P_j \tag{19}$$

In camera coordinate system, the target object plane is determined as following:

$$[m,n,p]\left[^{Cam}X, {}^{Cam}Y, {}^{Cam}Z\right]^T = 1 \tag{20}$$

Hence, the normal vector of the target object plane is:

$$[m,n,p] = [1,1,1]\left[^{Cam}P_1, {}^{Cam}P_2, {}^{Cam}P_3\right]^{-1} \tag{21}$$

We also have:

$$\left[^{Cam}P_1, {}^{Cam}P_2, {}^{Cam}P_3\right]^{-1} = \left(\left[^{IP}P_1, {}^{IP}P_2, {}^{IP}P_3\right]diag(w_1,w_2,w_3)\right)^{-1}\,^{IP}_{Cam}M \tag{22}$$

Combining (20), (21) and (22) we have a constraint

$$[1,1,1]\left(\left[^{IP}P_1, {}^{IP}P_2, {}^{IP}P_3\right]diag(w_1,w_2,w_3)\right)^{-1}\,^{IP}_{Cam}M\left[^{Cam}X, {}^{Cam}Y, {}^{Cam}Z\right]^T = 1 \tag{23}$$

By using fusion of rectangular constraints and constraint in (23) we obtain,

$$w_4 = [1,1,1]\left(\left[^{IP}P_1, {}^{IP}P_2, {}^{IP}P_3\right]diag(b_{11},b_{21},b_{31})\right)^{-1\,IP}[1,0,0]$$
$$\left(\left[\,^{IP}_{Cam}M^{-1\,IP}P_j, -\,^{Cam}_{Laser}r_1, -\,^{Cam}_{Laser}r_3\right]^T\left[\,^{IP}_{Cam}M^{-1\,IP}P_j, -\,^{Cam}_{Laser}r_1, -\,^{Cam}_{Laser}r_3\right]\right)^{-1} \tag{24}$$
$$\left[\,^{IP}_{Cam}M^{-1\,IP}P_j, -\,^{Cam}_{Laser}r_1, -\,^{Cam}_{Laser}r_3\right]^T\,^{Cam}_{Laser}T\,^{IP}P_j$$

With $[-b_{11}, b_{21}, b_{23}]^T = [^{IP}P_1, {}^{IP}P_2, {}^{IP}P_3]^{-1\,IP}P_4$

Substituting (24) into (12), we can calculate w_1, w_2, w_3. The camera coordinate of these vertices can be calculated by using (9).

The real size of the triangular:

$$rh = \left\|^{Cam}P_1\,^{Cam}P_2\right\| \tag{25}$$

$$h = \left\|^{Cam}P_1\,^{Cam}P_3\right\| \tag{26}$$

Because in the triangular plane $^{Tg}Z = 0$, thus we have

$$\left[^{Cam}P_1, {}^{Cam}P_2, {}^{Cam}P_3\right] = \left[^{Cam}_{Tg}r_1, {}^{Cam}_{Tg}r_2, {}^{Cam}_{Tg}t\right]\left[^{Tg}P_1, {}^{Tg}P_2, {}^{Tg}P_3\right] \tag{27}$$

Solving (27), the solution of the relative POSE between the target object coordinate and camera coordinate is

$$\left[{}^{Cam}_{Tg}r_1, \; {}^{Cam}_{Tg}r_2, \; {}^{Cam}_{Tg}t \right] = \left[{}^{Cam}P_1, \; {}^{Cam}P_2, \; {}^{Cam}P_3 \right] \left[{}^{Tg}P_1, \; {}^{Tg}P_2, \; {}^{Tg}P_3 \right]^{-1} \tag{28}$$

$$ {}^{Cam}_{Tg}r_3 = {}^{Cam}_{Tg}r_1 \times {}^{Cam}_{Tg}r_2 \tag{29}$$

We can calculate the center of target object in consecutive frame in camera coordinate system as following

$$ {}^{Cam}P_c = \frac{1}{4} \left[\sum_{i=1}^{4} {}^{Cam}X_{P_i}, \; \sum_{i=1}^{4} {}^{Cam}Y_{P_i}, \; \sum_{i=1}^{4} {}^{Cam}Z_{P_i} \right] \tag{30}$$

Slippage between frame $(k-1)^{th}$ and k^{th}

$$ {}^{Cam}Slip = \left\| {}^{Cam}P_{c,k-1} \; {}^{Cam}P_{c,k} \right\| \tag{31}$$

4 Experimental Results

In this section we verify the accuracy of the proposed system through experiment using a CCD camera and a line laser sensor. The test platform was implemented in C/C++.

4.1 Object Detection and Object Pose Estimation

In this section, we test the proposed approach with numerous known different static object poses in order to estimate the proposed pose estimation error. In Fig. 4, when the object appears in image frame Fig. 4a, the sensor system detects high curvature points as shown in Fig. 4b, and then by using proposed approach, the target object is automatically detected and marked by red overlay as shown in Fig. 4c. Similarly, in Fig. 5, the target object appears in original image Fig. 5a is detected and marked in blue overlay as shown in Fig. 5b. The pose estimation error is shown in Fig. 5c.

4.2 Slippage Estimation

In this section we apply the proposed approach into the slip-margin feedback control gripper system. The three DOF gripper system consists of a CCD camera, structured light and force sensor grasps a target object. The incipient slippage occurs on the contact surface between grip fingers and grasping object when the object is pressed and slid, is estimated by proposed approach. Then, the grip force is immediately controlled by a direct feedback of the estimated slip margin.

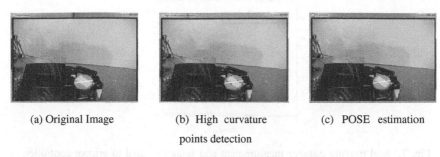

(a) Original Image   (b) High curvature   (c) POSE estimation
                            points detection

Fig. 4. Object detection and POSE estimation through frame sequence.

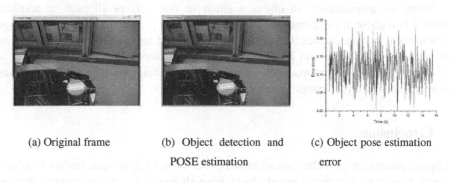

(a) Original frame   (b) Object detection and   (c) Object pose estimation
                            POSE estimation   error

Fig. 5. POSE estimation error

4.2.1 Small Motion Estimation and Error Computation

Suppose that initially target object position value is 0. Figure 6 shows the object position measurement and equivalent computational error. The object motion is estimated by calculating difference in object positions through frame sequence.

4.2.2 Slippage Estimation and Sending Control Signal

Figure 7 shows the object position and slippage information sent to gripper control system through frame sequence.

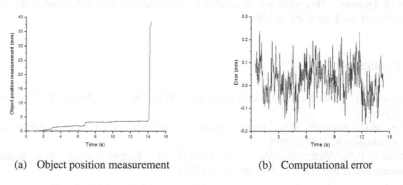

(a)   Object position measurement   (b)   Computational error

Fig. 6. Object position measurement and computational error

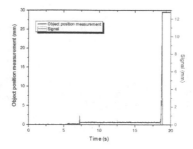

Fig. 7. Real moving distance measurement and sending signal to gripper controller.

When the displacement in object position in one way or slippage in another, between two consecutive frames is greater or equal to a pre-set threshold, the slippage measurement system decides that slippage happens and automatically sends signal to the gripper controller as can be seen in Fig. 7. In practical, the threshold is set at 0.5 mm and the signal information is calculated by rounding object movement distance between 2 consecutive frames to nearest integer number.

5   Conclusion

Slippage estimation provides crucial information for the slip-margin feedback control gripper system to preventing target object from sliding. The current contact slippage measurement sensors generally are very expensive, difficultly in measuring various type of object information and the contact force easily damages to sensor surface. In this paper we proposed a non-contact slippage estimation method by using sensor fusion. The proposed non-contact sensor only requires cheap elements, a camera and laser generator, and simple structure which suitable for compactness of industrial robot. The experimental results shows that the proposed sensor fusion method can measure not only slippage, but also target object pose simultaneously. In industrial environment, we can add rectangular sign on the target object top or simply selecting target object vertices which appear in image plane as indications to measure the target object pose.

Acknowledgement. This work was supported by 2014 Special Research Fund of Mechanical Engineering at the University of Ulsan.

References

1. Shinoda, H.: Contact sensing, a state of the art. J. Robot. Soc. Jpn. **20**(4), 385–388 (2002). in Japanese
2. Lee, M.H., Nicholls, H.R.: Tactile sensing for mechatronics-a state of the art survey. Mechatronics **9**, 1–31 (1999)
3. Dahiya, R.S., Metta, G., Valle, M., Sandini, G.: Tactile sensing—from humans to humanoids. IEEE Trans. Robot. **26**, 1–20 (2010)

4. Ferrier, N.J., Brockett, R.W.: Reconstructing the shape of a deformable membrane from image data. Int. J. Robot. Res. **19**, 795–816 (2000). 14

5. Saga, S., Kajimoto, H., Tachi, S.: High-resolution tactile sensor using the deformation of a reflection image. Sens. Rev. **27**, 35–42 (2007). 15

6. Johnson, M.K., Adelson, E.H.: Retrographic sensing for the measurement of surface texture and shape. In: Proceedings of the IEEE Conference on Computer Vision and Pattern Recognition, pp. 1070–1077, USA (2009)

7. Goro, O., Ashish, D., Norinao, W., Nobuhiko, M.: Vision based tactile sensor using transparent elastic fingertip for dexterous handling. In: Mobile Robots: Perception and Navigation, Chap. 7. InTechOpen (2007)

8. Ito, Y., Kim, Y., Nagai, C., Obinata, G.: Vision-based tactile sensing and shape estimation using a fluid-type touchpad. IEEE Trans. Autom. **9**(4), 734–744 (2012)

9. Ito, Y., Kim, Y., Obinata, G.: Contact region estimation based on a vision-based tactile sensor using a deformable touchpad. Sens. J. **14**, 5805–5822 (2014)

10. Ueda, J., Ikeda, A., Ogasawara, T.: Grip-force control of an elastic object by vision-based slip-margin feedback during the incipient slip. IEEE Trans. Robot. **21**(6), 1139–1147 (2005)

11. Zhengyou, Z.: A flexible new technique for camera calibration. IEEE Trans. Pattern Anal. Mach. Intell. **22**(11), 1330–1334 (2000)

12. Richard, H., Andrew, Z.: Multiple View Geometry in Computer Vision. Cambridge University Press, Cambridge (2004)

13. Tom, B., Steven, M., Richard, G.: Design and calibration of a hybrid computer vision and structured light 3D imaging system. In: Proceeding of the 5th International Conference Automation Robotics and Applications (ICARA), pp. 441–446. IEEE Wellington (2011)

K-SVD Based Image Denoising Method Using Image Residual Information in Different Frequency Bands

Pin-gang Su[1]([✉]), Tao Liu[1], and Zhan-li Sun[2]

[1] Department of Electrical Automation, College of Electronic Information Engineering,
Suzhou Vocational University, Suzhou 215104, Jiangsu, China
{supg,lt}@jssvc.edu.cn
[2] School of Electrical Engineering and Automation, Anhui University,
Hefei 230039, Anhui, China
zhlsun2006@126.com

Abstract. The common image denoising methods only consider how to restore well image information from noise images, but neglect the effects of residual information between restored images and given images. To enhance denoised image's quality, a new image denoising method considering residual information in different frequency bands is discussed in this paper. In this method, an original image is divided into high and low frequency sub-band images by the contourlet transform algorithm. And each sub-band image is first denoised by the K-singular value decomposition (K-SVD) denoising model, thus each residual sub-band image is correspondingly obtained. Further, each residual image is again denoised by K-SVD denoising model. Finally, for each sub-band image denoised and its residual image, the inverse transform of contourlet transform algorithm is used to restore the original image. Compared our method proposed here with common denoising methods of wavelet, contourlet, K-SVD, experimental results show that our method fusing residual information in different frequency bands behaves better denoising effect.

Keywords: Residual image information · Frequency bands · Contourlet transform · K-SVD algorithm · Image reconstruction

1 Introduction

In image processing field, high resolution (HR) images are necessary in application. But, in the processing of obtaining and transmitting images, noise is not avoidable, so, how to denoise is an important and challenge issue to this day [1, 2]. Researchers have being exploring new image denoising techniques. Commonly, denoising methods used are generally divided into three types, such as spatial domain denoising, frequency domain denoising and dictionary learning denoising [2, 3]. Among these methods, dictionary learning based denoising methods are most fashionable at present, which utilize redundant dictionaries to effectively represent image information. These methods mainly pay close attention to using the original noise images to improve the effect of denoising. However, they neglect the influence of residual image information between the original noise image and its denoised version. In 2009, for residual image information,

© Springer International Publishing Switzerland 2016
D.-S. Huang and K.-H. Jo (Eds.): ICIC 2016, Part II, LNCS 9772, pp. 482–492, 2016.
DOI: 10.1007/978-3-319-42294-7_43

Dominique Brunet et al. carried out a series of tests [3], such as independence test, Pearson correlation coefficient test, maximum likelihood ratio significance test, Kolmogorov Smirnov (K-S) test and so on. Test results proved that the residual image not only had distinct correlation with the denoised image, but also had some detail information of the corresponding HR image [4, 5]. Therefore, in order to use adequately the residual image information to enhance the denoising effect. Further, considering the dictionary learning based denoising method's advantages, the K-singular value decomposition (K-SVD) denoising model [6, 7] based image denoising method in different frequency sub-band images is discussed in this paper. K-SVD is a sparse represent method of images. It can learn efficiently over-complete dictionary, especially, for a denoise image, it can synchronously denoise and learn dictionary. In this paper, in our method proposed, first, an original image is divided into high frequency and low frequency sub-band images by using the contourlet transform algorithm. Then, each sub-band image is again denoised by K-SVD model, thus each residual sub-band image is obtained, furthermore, each residual sub-band image is also denoised by K-SVD. Finally, in the inverse transform domain of contourlet, fusing each low and high frequency sub-band image denoised and the corresponding residual sub-band image denoised, the restored result of the original image can be obtained.

2   The Contourlet Transform Method

Contourlet transform was proposed by Do et al. in 2002 [7]. Compared with traditional wavelet method, contourlet transform can offer more flexible multi-resolution and more directional decomposition for images [10] and embody better image structure, because it allows for different directions at each scale. This transform process can be simply generalized as two steps: Firstly, the multi-scale decomposition is realized by utilizing the Laplace pyramid (LP) filter in order to catch all singular points in an image. In each step, LP decomposition gives a sampled low-pass sub-band image of the original and a high-pass one. The process can be iterated on the coarse version. Secondly, for the high-frequency sub-band image, the directional filter bank (DFB) is used to make all singular points in the same direction composed into a contourlet. The DFB is designed to capture the high frequency components of images, and it is efficiently implemented via a l-level tree-structured decomposition that leads to 2^l sub-bands with wedge-shaped frequency partition. The overall result is an image expansion by using basic elements like contour segments. Here, in order to comprehend directly the process of contourlet transform, for a noise version of Lena image used widely in image processing field, assumed that the transform layer number is 2, and the orientation number of each layer is 4, then after contourlet transform, the corresponding low frequency sub-band and high frequency sub-bands of each layer in each orientation can be obtained, shown respectively in Fig. 1. It is clear to see that the low-frequency image contains the majority of energy of the original objection.

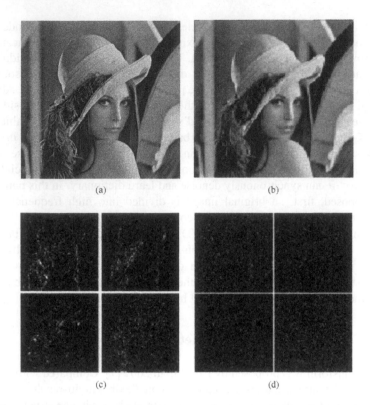

Fig. 1. The noise image of Lena and the corresponding contourlet transform results. (a) Noise image 1 with 5 level of Gaussian noise. (b) Low frequency sub-band. (c) Four high frequency sub-bands of the first Layer. (d) Four high frequency sub-bands of the second Layer.

3   K-SVD Algorithm

3.1   Sparse Representation Idea

K-SVD algorithm is in fact one of sparse representation algorithms [7, 8]. Currently, the theory of sparse representation has been used widely in image research field. Its basic idea is that a natural signal can be represented by compressed methods or by the liner combination of prototype atoms [8]. Usually, these atoms are chosen from a so called over-complete dictionary $D \in \mathfrak{R}^{N \times K}$. Supposed that $x \in \mathfrak{R}^{N}$ is the observed signal, then utilized the sparse representation's idea, x can be approximately represented as $x \approx Ds$, satisfying $\|x - Ds\|_p \leq \varepsilon$, where $s \in \mathfrak{R}^{K}$ is a vector with very few ($<< K$) nonzero entries [8]. In common application, this sparsest representation is the solution of the following formula:

$$(P_0) \min_s \|s\|_0 \ subject\ to\ \|x - Ds\|_2^2 \ \leq \varepsilon. \tag{1}$$

where the symbol $\|\cdot\|_0$ and $\|\cdot\|_2$ are respectively the l^0 and l^2 norm, counting the nonzero entries of a vector. Assumed that x_h and x_l denote respectively a HR and LR image patch vector, according to Eq. (1), x_h can be approximated by the equation of $x_h \approx D_h s$, and x_l can be approximated by $x_l = \Gamma x_h \approx \Gamma D_h s$, here T is the mapping matrix. Thus, the LR dictionary D_l can be calculated by $D_l = TD_h$. Namely, the HR image patch x_h can be restored by using the equation of $x_h = D_h s$. Generally, for an image, to improve its reconstructed version's quality, it is randomly sampled L times with $p \times p$ image patch to obtain the image patch set $X = \{x_1, x_2, \cdots, x_L\} \in \mathfrak{R}^{N \times L}$, thus, the dictionary D can be learned from X, and the optimized problem is described as follows:

$$\{D, S\} = \arg \min_{D,S} \|X - DS\|_F^2 + \lambda \|S\|_1 . \tag{2}$$

subject to $\|d_k\|_2^2 \le 1 \, (k = 1, 2, 3, \cdots, K)$, and d_k is the kth column atoms of the dictionary D, and S denotes the sparse coefficient matrix.

3.2 K-SVD Denoising Model

Let Y denote a clear image patch set, $\tilde{Y}$ denote the noise version of Y. Assumed that the over-complete dictionary D is known and R_{ij} is the extraction mark of lapped image patches, where each image patch $Y_{ij} = R_{ij} Y$ with $p \times p$ pixels in every location has a sparse representation with bounded error. In other words, R is the extraction matrix with the size of $p \times N$ pixels, which extracts the (i, j) block from an image with the size of $N \times N$ pixels. Then, the common K-SVD denoising model is written as

$$J(\hat{D}_{ij}, \hat{Y}) = \arg \min_{s_{ij}, U} \left[\lambda \|Y - \tilde{Y}\|_2^2 + \sum_{i,j} \mu_{ij} \|S_{ij}\|_1 + \sum_{i,j} \|D S_{ij} - R_{ij} Y\|_2^2 \right] \tag{3}$$

To assure the maximum sparseness, the constraint term of $\sum_{i,j} \left(D_{ij}^T D_{ij} \right)$ is considered in Eq. (3). And then, the denoising model is written as follows

$$J(\hat{D}_{ij}, \hat{Y}) = \arg \min_{s_{ij}, U} \left[\lambda \|Y - \tilde{Y}\|_2^2 + \sum_{i,j} \mu_{ij} \|S_{ij}\|_1 + \gamma \sum_{i,j} \left(D_{ij}^T D_{ij} \right) + \sum_{i,j} \|D S_{ij} - R_{ij} Y\|_2^2 \right]. \tag{4}$$

In Eq. (4), the first term controls the degree of the approximation of Y and $\tilde{Y}$ by controlling the relational expression of $\|Y - \tilde{Y}\|_2^2 \le Const \cdot \sigma^2$. And the larger the parameter σ is, the smaller the parameter λ is. The second and the third terms are parts of the image priors that makes sure that, in the constructed image patch set Y, each patch Y_{ij} with the size $p \times p$ pixels in every location has a sparse representation with bounded error, where Y_{ij} is calculated by using the Equation of $Y_{ij} = R_{ij} Y$. Commonly, the optimal relation between σ and λ is thought to be the form of $\lambda_{optimal} = 30/\sigma$. The coefficients μ_{ij} must be locally dependent so as to comply with a set of constraints of the form $\|D S_{ij} - Y_{ij}\|_2^2 \le \varepsilon$. And utilizing the estimated dictionary matrix $\hat{D}$ and sparse coefficient matrix $\hat{S}$, the estimation of noise data $\tilde{Y}$, which is denoted by $\hat{\tilde{Y}}$, can be obtained by the formula of $\hat{\tilde{Y}} = \hat{D}\hat{S}$. Furthermore, the image reconstruction result, denoted by $\hat{Y}$, can be calculated by the following formula:

$$\hat{Y} = \left(\lambda\tilde{Y} + \sum_{i,j} R_{ij}^T \right)^{-1} \left(\lambda\hat{\tilde{Y}} + \sum_{i,j} R_{ij}^T \hat{D}\hat{S}_{ij} \right). \tag{5}$$

4   Our Denoising Method

For a noise image $\tilde{Y}$ ($\tilde{Y} = Y + \eta$, η is the noise), the K-SVD denoising model defined in Eq. (4) is firstly used to denoise $\tilde{Y}$ and the corresponding denoised result can be obtained and denoted by $\hat{\tilde{Y}}$, thus, the residual image $E = \tilde{Y} - \hat{\tilde{Y}}$ can be obtained. The residual image still contains partly useful image information, such as image's texture and edge information and so on. However, many denoising methods, used commonly in image processing field, usually neglect this. In this paper, based on the advantages of K-SVD denoising model described in Eq. (3), and considered the residual sub-band information in low and high frequency domain obtained by contourlet transform method, a new image denoising method is discussed. The steps of our method can be generalized briefly as follows:

Step 1.   Using K-SVD denoising model to denoise a noise image $\tilde{Y}$, and denoting the denoised result as $\hat{\tilde{Y}}$.

Step 2.   Utilizing contourlet transform to decompose respectively the noise image $\tilde{Y}$ and its denoised version $\hat{\tilde{Y}}$ so as to obtain the corresponding low and high frequency sub-band images. Here the number of contourlet transform's layers and that of each layer's orientations is respectively set 2 and 4. Namely, $\tilde{Y}$ and $\hat{\tilde{Y}}$ are respectively transformed into a low frequency sub-band and 8 high frequency sub-bands.

Step 3.   Calculating each residual sub-band image and sampling randomly each sub-band image using a fixed image patch with $p \times p$ pixels to obtain the low and high frequency sub-band image patches set.

Step 4.   Then, for these residual sub-band image patches, using the similarity rule to select those with maximum information and discard those useless image patches. Further, for selected residual image patches, again using K-SVD model to denoise them to obtain denoised residual sub-band images

Step 5.   Superimposing each denoised sub-band image and corresponding each denoised residual sub-band image to obtain final denoised sub-band images.

Step 6.   For denoised sub-band images containing residual information obtained in Step 5, using inverse contourlet transform to restore the original image.

5   Experimental Results and Analysis

For the noise Lena image shown in Fig. 1, its denoised version by our K-SVD denoising model was shown in Fig. 2. For this denoised result, let the number of transform layer be 2 and that of orientation in each layer be 4, then the corresponding low frequency image and high frequency sub-bands obtained by contourlet transform method were also given in Fig. 2. Compared the low frequency image and each high frequency sub-band image shown respectively in Figs. 2 and 1, it is clear to see that, in Fig. 2, the low frequency sub-band denoised has better vision effect, and each high frequency sub-band image's contour is more distinct than that in Fig. 1, especially that of the second layer's each high frequency sub-band obviously better that shown in Fig. 1. As well as much known noise in each high frequency sub-band's background has been obviously reduced, and some details of Lena image can be seen clearly. Otherwise, for low frequency image shown in Fig. 2(b) and each high frequency sub-images shown in Fig. 2(c) and (d), they are again denoised by using K-SVD denoising model, and the denoised result of Fig. 2(b) is shown in Fig. 3, and the denoised results of each high frequency sub-bands are shown in Fig. 4. Clearly, noise existed in high frequency sub-bands of Fig. 2(c) and (d) is further reduced.

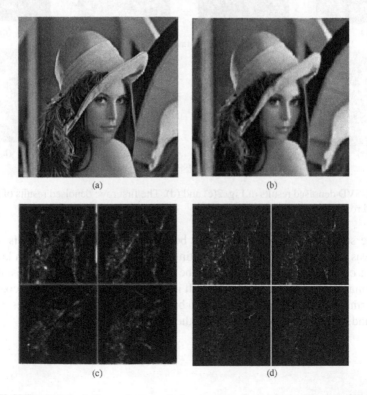

Fig. 2. K-SVD denoised Lena image and corresponding contourlet transform results. (a) Denoised result (5 noise level). (b) Low frequency sub-band of (a). (c) Four high frequency sub-bands of the first Layer. (d) Four high frequency sub-bands of the second Layer.

Fig. 3. K-SVD denoised result of Fig. 2(b).

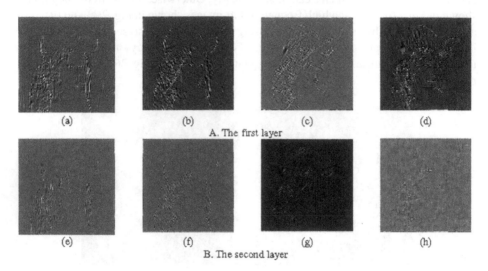

Fig. 4. K-SVD denoised results of Fig. 2(c) and (d). The first row: denoised results of Fig. 2(c), the second row: denoised results of Fig. 2(d).

At the same time, the residual image between the noise image and its denoised version was shown in Fig. 5. At the same time, set the number of transform layer to be 2 and that of orientation in each layer to be 4, the contourlet transform results of the residual image were also shown in Fig. 5. It is clear to see that the image's contour and edge information in residual image still can be distinguished with naked eyes, but in the background, much noise is still existed in the residual image.

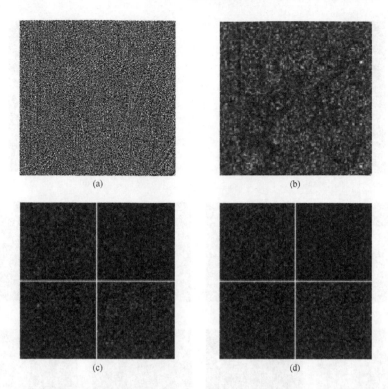

Fig. 5. Residual image and corresponding contourlet transform results. (a) Residual image. (b) Low frequency sub-band of (a). (c) Four high frequency sub-bands of the first Layer. (d) Four high frequency sub-bands of the second Layer.

Otherwise, residual sub-band images and their corresponding denoised results were also given here. But considered the limitation of the paper's length, only the low frequency residual sub-band and the first layer's residual high frequency sub-band images and their denoised results were given, as shown in Fig. 6. Clearly, it is noted that the low frequency residual image still contains some image detail information and the contour features, however, each high frequency sub-band of the residual image still contain much noise so as to distinguish difficultly the image's detail and contour, therefore, high frequency sub-bands are necessary to denoise again. Further, fusing denoised each sub-band image and the corresponding denoised residual sub-band, in inverse contourlet transform domain, the denoised image can be obtained.

To prove that the efficiency of our denoising method, under different noise levels, we also compared our method with other denoising methods, such as contourlet transform, wavelet, common K-SVD denoising model without considering residual sub-band image information. At the same time, to measure the equality of denoised images, the measure criterion, signal noise ratio (SNR) is used here. Due to the length of the paper, we only give several examples of the denoised Lena image corrupted with different additive Gaussian white noise, as shown in Fig. 7. The SNR values for the test images with different noise level σ values (the true noise variance of a noise image is calculated

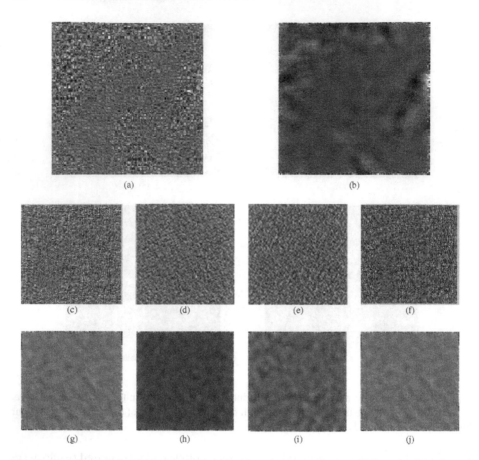

Fig. 6. Residual sub-band images. (a) Residual low frequency image. (b) Denoised version of (a). (c) to (f) are residual high frequency sub-bands of the first Layer. (g) to (j) are denoised version corresponding to (c) to (f).

by the form $\sigma^2 / 255^2$) were listed in Table 1 respectively. From Fig. 5, clearly, when the noise level is small, it is difficult to distinguish denoised results only from visual vision. From Table 1, it is easy to find that when the noise level is small, the SNR value of each denoising method is almost no difference. When the noise level is larger than 15, it is clear to see that, under the same noise level, our method can obtain the largest SNR value, the common K-SVD model is the second, and wavelet is the worse. Especially, when noise level excesses 100, SNR values of wavelet and contourlet are already negative, while the SNR of our method is still positive, and the denoised image's contour still can be clearly distinguished.

Table 1. SNR values of different denoising methods under different noise levels.

Noise level \ SNR	Wavelet	Contourlet	K-SVD	K-SVD considering residual image information	Noise image
5	21.10	20.65	23.98	29.94	19.59
25	12.97	13.61	16.88	19.80	5.71
30	11.95	12.68	15.83	17.89	4.17
50	9.02	9.85	12.63	14.84	0.08
80	5.44	7.12	8.98	10.56	-4.38
100	3.26	5.76	7.73	8.96	-4.38
300	-8.87	0.83	2.52	3.93	-7.71

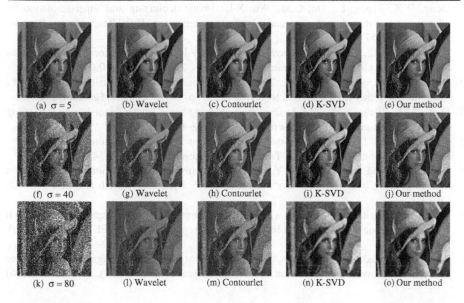

(a) σ = 5   (b) Wavelet   (c) Contourlet   (d) K-SVD   (e) Our method

(f) σ = 40   (g) Wavelet   (h) Contourlet   (i) K-SVD   (j) Our method

(k) σ = 80   (l) Wavelet   (m) Contourlet   (n) K-SVD   (o) Our method

Fig. 7. Denoised results of Lena image with different noise levels using different methods. (a), (f) and (k) are noise images responding to the noise level 5, 40 and 80. (b), (g) and (l) are denoised results of Wavelet method. (c), (h) and (m) are denoised results of Contourlet method. (d), (i) and (n) are denoised results of K-SVD denoising model. (e), (j) and (o) are denoised results of our method proposed here.

6   Conclusions

A new image denoising method considering residual image information using K-SVD denoising model proposed by us in low frequency and high frequency bands is proposed in this paper. For given an image, its corresponding low and high frequency sub-band images are obtained by the contourlet transform method. For a noise image and its denoised version by K-SVD model, the corresponding residual sub-band images in different frequency bands can be calculated. For residual low frequency image and each residual high frequency sub-band, they again denoised by K-SVD model. Finally, using the inverse contourlet transform, the final denoised image can be obtained. Compared with other denoised methods, such as wavelet, contourlet, common K-SVD and so on, experimental results prove that our denoising method outperforms other three methods discussed in this paper, especially for the large noise level, it behaves clear advantage.

Acknowledgement. This work was supported by the grants from National Nature Science Foundation of China (Grant No. 61373098 and 61370109), the grant from Natural Science Foundation of Anhui Province (No. 1308085MF85).

References

1. Dong, W.S., Zhang, L., Shi, G.M., Wu, X.L.: Image deblurring and super-resolution by adaptive sparse domain selection and adaptive regularization. IEEE Trans. Image Process. **20**(7), 1838–1857 (2011)
2. Liu, J., Tai, X.C., Huang, H.: A weighted dictionary learning model for denoising images corrupted by mixed noise. IEEE Trans. Image Process. **22**(3), 1108–1120 (2013)
3. Brunet, D., Vrscay, E.R., Wang, Z.: The use of residuals in image denoising. In: Kamel, M., Campilho, A. (eds.) ICIAR 2009. LNCS, vol. 5627, pp. 1–12. Springer, Heidelberg (2009)
4. Dong, M.K., Jiang, A.M., Sun, J.: Dictionary learning based image denoising method using information fusion of residuals. Microprocessors **1**, 58–62 (2015)
5. Sun, J., Xu, Z.B., Shum, H.Y.: Gradient profile prior and its applications in image super-resolution and enhancement. IEEE Trans. Image Process. **20**(6), 1529–1542 (2011)
6. Lu, J.-Z., Zhang, Q.-H., Xu, Z.-Y., Peng, Z.M.: Image super-resolution reconstruction algorithm using over-complete sparse representation. Syst. Eng. Electron. **34**(2), 403–408 (2012)
7. Do, M.N., Vetterli, M.: Contourlet: a directional multiresolution image representation. In: 9th IEEE International Conference on Image Processing, pp. 357–360. IEEE Press, New York (2002)
8. Aharon, M., Elad, M., Bruckstein, A.: K-SVD: an algorithm for designing overcomplete dictionaries for sparse representation. IEEE Trans. Signal Process. **54**(11), 4311–4322 (2006)

Single Image Super Resolution with Neighbor Embedding and In-place Patch Matching

Zhong-Qiu Zhao[1], Zhen-Wei Hao[1(✉)], Run Su[2], and Xindong Wu[1]

[1] College of Computer and Information, Hefei University of Technology, Hefei, China
1216459032@qq.com
[2] The No. 38 Research Institute of CETC, Hefei, China

Abstract. In this paper, we present a novel image super-resolution framework based on neighbor embedding, which belongs to the family of learning-based super-resolution methods. Instead of relying on extrinsic set of training images, image pairs are generated by learning self-similarities from the low-resolution input image itself. Furthermore, to improve the efficiency of image reconstruction, the in-place matching is introduced to the process of similar patches searching. The gradual magnification scheme is adopted to upscale the low-resolution image, and iterative back projection is used to reduce the reconstruction error at each step. Experimental results show that our method achieves satisfactory performance not only on reconstruction quality but also on time efficiency, as compared with other super-resolution methods.

Keywords: Super-resolution · Neighbor embedding · Self-similarity · In-place matching

1 Introduction

Image super resolution (SR) reconstruction is the process of generating a high-resolution (HR) image by using one or more low-resolution (LR) inputs. As a post-processing technique, it breaks through the limitations of image equipment and environment, and has a wide range of applications in real life such as high-definition television, medical imaging, video surveillance etc. The problem of image SR was first studied by Tsai and Huang in 1980 s [1], and now existing SR methods can be broadly categorized into three categories: interpolated-based, reconstruction-based, and learning-based methods. In this paper, we focus on the learning-based methods which exploit the information from the training data to learn the mapping relationship between the LR and HR image patches.

As a kind of very promising SR technique, learning-based methods were first studied by Freeman et al. [2]. Nowadays, based on different learning models, lots of learning-based methods have been proposed. According to the ways how the mapping relationship is formulated, learning-based SR methods can be divided into: manifold learning [3–6], sparse coding [7–9] and regression-based methods [10–12]. Although learning-based methods can effectively restore the missing details, these methods primarily rely on example data. As the contents of natural images are ever-changing, it is difficult to

© Springer International Publishing Switzerland 2016
D.-S. Huang and K.-H. Jo (Eds.): ICIC 2016, Part II, LNCS 9772, pp. 493–502, 2016.
DOI: 10.1007/978-3-319-42294-7_44

choose a universal training set to fit the reconstruction of all images. When the example data cannot provide accurate information which is similar to the LR input image, lots of artifacts can be easily introduced to the result image. More recently, Glasner et al. [13] proposed a method that exploited patch redundancy in natural image, and realized image SR reconstruction without external database.

In [3], Chang et al. proposed an SR method using neighbor embedding (NE), in which the HR image can be estimated with a set of weighted image patches. Until now, many modified algorithm over the neighbor embedding have been proposed. Bevilacqua et al. [4] proposed to use non-negative least squares to compute the weights of the neighbor patches. Gao et al. [5] proposed a sparse neighbor embedding method which chooses the neighbor patches and computes the corresponding weights simultaneously. Based on the multiscale similarity learning, Zhang et al. [6] accomplished the image SR reconstruction with only the LR input image. Although the image reconstruction methods above can obtain promising reconstruction performance, they need to search a vast example dataset or the whole image for similar patterns, leading to low efficiency in practical applications. Recently, several efforts have been maken to improve time efficiency. With a precomputed projection matrix, Timofte et al. [14] accomplished the image reconstruction and acquired high time efficiency. Based on the in-place examples, Yang et al. [15] proposed to learn a set of regression mapping for fast image SR. Dong et al. [16, 17] used convolution neural network to learn an end-to-end mapping model, and obtained satisfactory reconstruction performance with high time efficiency.

In this paper, we propose a NE-based single image SR scheme which considers both the selection of example data and time efficiency. With the similarity patch redundancy in natural images, our method adopt the multiscale similarity learning to generate the example image pairs using only the input image. And contrary to [6], the proposed method conducted the k-nearest neighbor search with the in-place patch matching. We adopt gradual magnification scheme to upscale the LR image and implement iterative back projection algorithm [18] to correct the reconstruction image at each step.

The remainder of this paper is organized as follows. Section 2 introduces the NE-based approach for SR. Section 3 presents the details of our SR framework. The experimental results and comparisons with several other SR methods are demonstrated in Sect. 4. Finally, several conclusions are drawn in Sect. 5.

2 Neighbor Embedding Approach for Super Resolution

The NE-based SR method was first proposed by Chang et al. [3], which uses a manifold learning called locally linear embedding (LLE) to estimate the HR patches corresponding to the LR ones. Its main idea is that the image patches in different resolutions have similar geometric manifold spaces and each image patch can be represented as the linear combination of its K neighbors.

Let $Y_s = \{y_s^j\}_{j=1}^m$ be the training patches of LR images, $X_s = \{x_s^j\}_{j=1}^m$ be the corresponding HR training patches. We represent the LR test image as a set of overlapping image patches $Y_t = \{y_t^i\}_{i=1}^n$, and $X_t = \{x_t^i\}_{i=1}^n$ are their corresponding HR patches to be estimated. The NE-based SR reconstruction can be summarized as follows:

(1) For each patch $y_t^i \in Y_t$

(a) Find the K nearest neighbors in Y_s in terms of Euclidean distance:

$$N_i = \arg \min_{1 \le j \le m}^{k} \left\| y_t^i - y_s^j \right\|^2 \tag{1}$$

(b) Compute the reconstruction weights of the neighbors that minimize the reconstruction error:

$$\epsilon^i = \left\| y_t^i - \sum_{y_s^j \in N_i} w_{ij} y_s^j \right\|_2^2 \quad s.t. \sum_{y_s^j \in N_i} w_{ij} = 1 \tag{2}$$

(c) Compute the HR image patch using the corresponding HR patches and the reconstruction weights:

$$x_t^i = \sum_{x_s^j \in N_i} w_{ij} x_s^j \tag{3}$$

(2) Construct the HR image by combining the obtained HR image patches.

3   Proposed Algorithm

In this section, we provide the details of our modified algorithm on the NE-based method. The procedure is divided into three steps: (1) Generating example data, (2) NE-based relationship learning with in-place patch matching, (3) SR enhancement using iterative back projection. The proposed SR method is summarized in Algorithm 1. The details in each step of the algorithm are as follows:

3.1   Generating Example Data

It is the most fundamental and important step for learning-based image SR to select the appropriate HR-LR training image pairs. In this paper, utilizing the self-similarity redundancy of the natural image, the HR-LR image pairs can be extracted from the LR input image.

For an observed LR image Y, the problem of image super-resolution can be modeled as:

$$Y = DHX + \sigma \tag{4}$$

where H and D is the operator of blurring and down-sampling, respectively. X is the HR image, and σ is the noise term.

In the process of image reconstruction, we magnify the LR image to the desired size with a gradual scheme. The scale parameter is set to be $k = 0, \dots, \lceil \log(S)/\log(s) \rceil$, where S and s are the overall and gradual up-scaling factors, respectively. To obtain example image pairs, we firstly regard the original LR input image Y as the HR image $X^{(0)}$. Then the corresponding LR image $Y^{(0)}$ can be obtained by down-sampling $X^{(0)}$ and up-scaling the down-sampled image to the size of $X^{(0)}$ using the bicubic interpolation:

$$Y^{(0)} = ((X^{(0)} * B_s) \downarrow_s) \uparrow_s \qquad (5)$$

where B_s is the blurring kernel, $\uparrow_s$ and $\downarrow_s$ are the up-scaling and down-scaling operators with the factor of s. Then $X^{(0)}$ and $Y^{(0)}$ are treated as the initial example image pairs for learning-based algorithm. Iteratively, to obtain the HR image $X^{(1)}$, we produce $Y^{(1)}$ through magnifying $X^{(0)}$ using bicubic interpolation with a factor of s. With the similarity assumption, for each patch in $Y^{(1)}$, we can find several similar patches in $Y^{(0)}$. The HR patch in $X^{(1)}$ can be linearly represented by the image patches in $X^{(0)}$ which correspond to the patches found in $Y^{(0)}$. $X^{(1)}$ and $Y^{(1)}$ are then taken as the HR-LR pairs for succeeding learning. The synthesis processes are repeated until the HR image at scale $\lceil \log(S)/\log(s) \rceil$ is obtained.

3.2   NE-Based Relationship Learning with in-Place Patch Matching

Once the training example pairs are formed, the relationship between the HR and the corresponding LR images can be estimated from the pair mappings. In this paper, we take advantage of the NE-based method for this purpose.

For learning the relationship between HR-LR pairs, K-nearest neighbors (K-NN) search was used to find several of the most similar patches in the training set for each LR image patch. Contrary to previous methods which need to search in a vast external example set or the whole image, we conduct the similar patches search in a restricted local region which is built on the patch matching with small scaling factors is in-place. Thereby, the time efficiency of image reconstruction is improved by narrowing the searching scope. For an image patch at the location (x, y) of $Y^{(k+1)}$, we can find the K most similar patches around its original coordinates (x_r, y_r) in image $Y^{(k)}$, where $x_r = \lfloor x/s + 0.5 \rfloor$ and $y_r = \lfloor y/s + 0.5 \rfloor$. Then we use the LLE algorithm to compute the reconstruction weights between the image patch in $Y^{(k+1)}$ and the matched similar image patches in $Y^{(k)}$ by minimizing the reconstruction error. Then the HR patch can be recovered by using the reconstruction weights and the corresponding HR neighbor patches. Once all the HR patches are generated, the HR image can be obtained through merging HR patches.

3.3   SR Enhancement Using Iterative Back Projection

In each step, due to the unavoidable reconstruction error, the reconstruction result which is fed to next step directly usually leads to the accumulation of errors. Thus, through the iterative back projection (IBP) algorithm we reduce the reconstruction error at each step, and ensure that image reconstruction is conducted along the right direction. The reconstruction error can be formulated as:

$$E(X) = \|(X \otimes H) \downarrow - Y\|^2 \qquad (6)$$

where X is the estimated HR image, Y is the LR input image, and H represents blurring operator.

This problem can be efficiently solved by the gradient descent method, and the updating equation for this iterative method is:

$$X^{t+1} = X^t + ((Y - DHX^t) \uparrow) * p \qquad (7)$$

where t is iteration index, p is back-projection filter, and X^t is the HR image at the tth iteration.

Algorithm 1. Proposed Image Super-Resolution Algorithm

Input: Low-resolution input image Y.
  1. **Initialize:**
  1)   Set the scale parameter $n = \lceil \log(S)/\log(s) \rceil$;
  2)   Set HR image $X^{(0)} = Y$, and the corresponding LR image $Y^{(0)} = (X^{(0)}) \downarrow \uparrow$.
  2. **NE-based reconstruction:**
  for $k = 1$ to n do
  1)   Calculate the LR image $Y^{(k)}$ with the bicubic interpolation of $X^{(k-1)}$;
  2)   For each patch in $Y^{(k)}$, find its K-NN around its original coordinates in $Y^{(k-1)}$, then use LLE algorithm to compute the reconstruction weights;
  3)   Compute HR image patches using the reconstruction weights and the corresponding HR patches in $X^{(k-1)}$;
  4)   Produce the HR image $X^{(k)}$ by merging the HR patches;
  5)   Use the IBP algorithm to reduce the reconstruction error, and obtain the new $X^{(k)}$.
  end for
  3. **Post-Proposing:**
  1)   Use the bicubic interpolation to resize the HR image $X^{(n)}$ to the S times of the LR image size. The non-local means is finally used to suppress the artifacts in the HR image.

Output: High-resolution output image X.

---

4   Experiment Results and Analysis

In this section, the experiments on several benchmark images are shown to validate our proposed method. Several state-of-the-art learning-based SR methods are used as comparison baselines, including NE-based [3], SC-based [7], Glasner's method [13] and Zhang's method [6]. Peak signal-to-noise ratio (PSNR) and structural similarity (SSIM) are adopted to evaluate the objective quality of the SR results, and reconstruction time is used as the indicator of reconstruction efficiency.

4.1 Experimental Settings

In our experiments, the LR images are generated from the original HR images through down-sampling by bicubic interpolation with a factor of 3. Since human vision is sensitive to brightness changes, the SR reconstruction is only conducted in the gray-channel for color images, and the images in the remaining color channels are directly magnified to the desired size using bicubic interpolation. In the reconstruction phase, the LR patch size is set as 5×5 pixels, and 3 pixels are overlapped between adjacent patches. The neighborhood size K is set to be 3, and when conducting the similar patches search across different scales, the search scope is set as 11×11 pixels around the corresponding coordinates. For all the SR experiments, the up-scaling factor S is set to be 3, and we magnify the LR image step-by-step by the factor s which is set to be 1.25. At each step, we use the IBP algorithm to enhance the quality of the reconstruction image, and the number of iterations is set to be 20.

4.2 Experimental Results

Table 1 lists the PSNR and SSIM values of various SR approaches. From Table 1, we can see that our proposed method can obtain the best performance of PSNR and SSIM on most of benchmark images. This fact indicates that with the LR image itself, we can

Table 1. The PSNR and SSIM performance on eight test images of super resolution of 3× magnification

Images \ Methods		NE-based [3]	SC-based [7]	Glasner [13]	Zhang [6]	Proposed
bike	PSNR	23.42	23.87	23.96	24.45	**24.56**
	SSIM	0.741	0.770	0.783	0.792	**0.803**
butterfly	PSNR	24.92	25.51	26.74	**28.18**	27.18
	SSIM	0.830	0.861	0.890	**0.919**	0.909
girl	PSNR	32.79	33.35	33.17	33.52	**33.53**
	SSIM	0.813	0.823	0.825	0.826	**0.829**
plants	PSNR	31.58	32.39	32.85	33.59	**33.72**
	SSIM	0.879	0.902	0.910	0.917	**0.922**
raccoon	PSNR	28.67	28.89	28.79	29.06	**29.15**
	SSIM	0.754	0.759	0.762	0.761	**0.768**
lenna	PSNR	32.23	32.85	32.59	33.51	**33.58**
	SSIM	0.865	0.877	0.879	0.884	**0.887**
baby	PSNR	33.77	34.51	34.29	**35.20**	35.09
	SSIM	0.901	0.917	0.922	0.923	**0.924**
pepper	PSNR	30.67	31.45	32.82	33.36	**34.05**
	SSIM	0.863	0.879	0.885	0.889	**0.893**

recover the HR image and obtain promising results. The reason is that similar patches across different scales are beneficial to recover faithful details.

We also assess the visual quality of the SR methods. Figures 1 and 2 show the SR images obtained by different SR methods on the test images of butterfly and lenna, respectively.

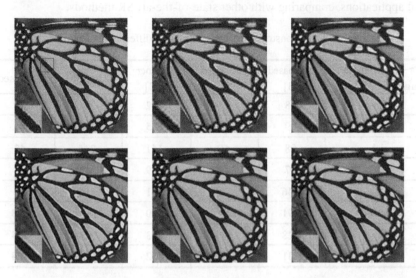

Fig. 1. The SR images(3×) of the butterfly image by various methods. Top row: the ground truth HR image, NE-based [3], SC-based [7]. Bottom row: Glasner [13], Zhang [6], and ours.

Fig. 2. The SR images(3×) of the lenna image by various methods. Top row: the ground truth HR image, NE-based [3], SC-based [7]. Bottom row: Glasner [13], Zhang [6], and ours.

In addition, we use computation time to evaluate the reconstruction efficiency. Table 2 lists the reconstruction time of different SR methods. From the table, we can see that our method takes the least computation time, and possesses the highest reconstruction efficiency. This improvement is due to the adoption of in-place patch matching, which reduces the searching scope for similar patches. Thus, our method is more feasible in real applications, comparing with other state-of-the-art SR methods.

Table 2. The reconstruction time(seconds) of different SR methods.

Methods / Images	NE-based [3]	SC-based [7]	Glasner [13]	Zhang[6]	Proposed
bike	98	63	142	341	**44**
butterfly	99	55	135	331	**44**
girl	98	69	141	345	**45**
plants	97	62	149	336	**44**
raccoon	146	111	241	580	**66**
lenna	391	266	650	2815	**178**
baby	328	285	703	2850	**179**
pepper	331	272	682	2865	**178**

4.3 Experimental Analysis

As one type of learning-based SR methods, the NE-based algorithm is employed in our method to build the relationship between the HR and LR image patches. So, the neighborhood size K is a variable parameter in our method. To study the effect of neighborhood size K on reconstruction performance, we take different K values to test our method on the benchmark images. The PSNR and SSIM results are presented in Fig. 3, from

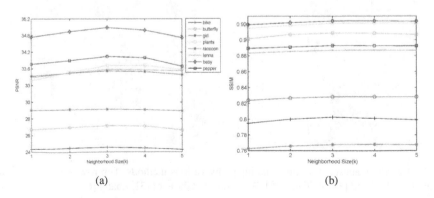

(a)   (b)

Fig. 3. Performance comparison of testing images with different neighborhood sizes. (a) PSNR value with different neighborhood sizes (b) SSIM value with different neighborhood sizes.

which we can see that our method yields better results when the neighborhood size K is set to be three.

5 Conclusion

This paper presents a novel single image SR framework by incorporating both similarity redundancy and in-place patch matching. The similarity redundancy across different scales is used to generate example data for the learning-based reconstruction, and in-place patch matching is utilized to fulfill the similar patches search in order to improve the reconstruction efficiency. To exploit the similarity redundancy across different scales, we magnify the LR image to the desired size with a gradual scheme, and use the IBP algorithm to reduce the reconstruction error at each step. The experimental results indicate that our method can obtain highly satisfactory performance in objective evaluation, while maintaining relatively high reconstruction efficiency.

References

1. Huang, T.S., Tsai, R.Y.: Multi-frame image restoration and registration. Adv. Comput. Vis. Image Process. **1**(2), 317–339 (1984)
2. Freeman, W.T., Jones, T.R., Pasztor, E.C.: Example-based super resolution. IEEE Comput. Graph. Appl. **22**(2), 56–65 (2002)
3. Chang, H., Yeung, D.Y., Xiong, Y.: Super-resolution through neighbor embedding. CVPR **1**, 275–282 (2004)
4. Bevilacqua, M., Roumy, A., Guillemot, C., Morel, M.L.A.: Low-complexity single-image super-resolution based on nonnegative neighbor embedding. In: BMVC, pp. 1–10 (2012)
5. Gao, X., Zhang, K., Tao, D., Li, X.: Image super-resolution with sparse neighbor embedding. IEEE Trans. Image Process. **21**(7), 3194–3205 (2012)
6. Zhang, K., Gao, X., Tao, D., Li, X.: Single image super-resolution with multiscale similarity learning. IEEE Trans. Neural Networks Learn. Syst. **24**(10), 1648–1659 (2013)
7. Yang, J., Wright, J., Huang, T., Ma, Y.: Image super-resolution as sparse representation of raw image patches. In: CVPR, pp. 1–8 (2008)
8. Zeyde, R., Elad, M., Protter, M.: On single image scale-up using sparse-representations. In: Boissonnat, J.-D., Chenin, P., Cohen, A., Gout, C., Lyche, T., Mazure, M.-L., Schumaker, L. (eds.) Curves and Surfaces 2011. LNCS, vol. 6920, pp. 711–730. Springer, Heidelberg (2012)
9. Dong, W., Zhang, L., Shi, G., Wu, X.: Image deblurring and super-resolution by adaptive sparse domain selection and adaptive regularization. IEEE Trans. Image Process. **20**(7), 1838–1857 (2011)
10. Li, D., Simske, S., Mersereau, R.M.: Single image super-resolution based on support vector regression. In: IJCNN, pp. 2898–2901 (2007)
11. Yang, M.C., Chu, C.T., Wang, Y.C.F.: Learning sparse image representation with support vector regression for single-image super-resolution. In: ICIP, pp. 1973–1976 (2010)
12. Yang, M.C., Wang, Y.C.F.: A self-learning approach to single image super-resolution. IEEE Trans. Multimedia **15**(3), 498–508 (2013)
13. Glasner, D., Bagon, S., Irani, M.: Super-resolution from a single image. In: ICCV, pp. 349–356 (2009)

14. Timofte, R., Smet, V.D., Gool, L.V.: Anchored neighborhood regression for fast exampled-based super-resolution. In: ICCV, pp. 1920–1927 (2013)
15. Yang, J., Lin, Z., Cohen, S.: Fast image super-resolution based on in-place example regression. In: CVPR, pp. 1059–1066 (2013)
16. Dong, C., Loy, C.C., He, K., Tang, X.: Learning a deep convolutional network for image super-resolution. In: Fleet, D., Pajdla, T., Schiele, B., Tuytelaars, T. (eds.) ECCV 2014, Part IV. LNCS, vol. 8692, pp. 184–199. Springer, Heidelberg (2014)
17. Dong, C., Loy, C.C., He, K., Tang, X.: Image super-resolution using deep convolutional networks. IEEE Trans. Pattern Anal. Mach. Intell. 38(2), 295–307 (2015)
18. Irani, M., Peleg, S.: Improving resolution by image registration. CVGIP Graphical Models Image Process. 53(3), 231–239 (1991)

A Modified Non-rigid ICP Algorithm for Registration of Chromosome Images

Qian Kou[1(✉)], Yang Yang[1], Shaoyi Du[1], Shuang Luo[1], and Dongge Cai[2]

[1] School of Electronic and Information Engineering, Xi'an Jiaotong University, Xi'an, China
kouqian@stu.xjtu.edu.cn, yyang@mail.xjtu.edu.cn
[2] The Second Affiliated Hospital of Xi'an Jiaotong University, Xi'an 710049, China

Abstract. As an extension of the classic rigid registration algorithm-Iterative Closest Point (ICP) algorithm, this paper proposes a new non-rigid ICP algorithm to match two point sets. Each point in the data set is supposed to match to the model set via an affine transformation. The proposed registration model is built up with a regularization term based on their average affine transformation. For each iteration of our algorithm, firstly correspondences between two point sets are built by the nearest-point search. Then the non-rigid transform parameters between two correspondence point sets are estimated by the proposed method in the closed form. Finally the average affine transformation is updated. A set of challenging data including single and overlapping chromosome images are tested which have significant local non-rigid transformations. Experimental results demonstrate our algorithm has higher accuracy and faster rate of convergence than other algorithms.

Keywords: ICP · Non-rigid registration · Average affine transformation · Chromosome

1 Introduction

Image registration is an important technology for aligning and fusing images of the same scene. It is greatly involved in many fields such as computer vision and medical image processing. Point-matching is a method which is employed in the image registration by registering feature points extracted from two images. With rapid development of image registration, non-rigid point-matching attracts widespread attentions.

The Iterative Closest Point (ICP) algorithm [1–3] is the most popular algorithm for point registration. Because of its accuracy and fast rate of convergence, the ICP algorithm has been used in many fields such as fingerprint and face recognition. But the traditional ICP algorithm only performs well when two matched point sets with rotation and translation transformation. When there are non-rigid transformations, the ICP algorithm can't achieve registration. It also has a low accuracy when there exits noises and outliers.

There are many different algorithms for non-rigid point sets registration. Due to more degrees of freedom and parameters of non-rigid transformation, how to register local non-rigid transformation is still difficult. The algorithms can be classified into two

© Springer International Publishing Switzerland 2016
D.-S. Huang and K.-H. Jo (Eds.): ICIC 2016, Part II, LNCS 9772, pp. 503–513, 2016.
DOI: 10.1007/978-3-319-42294-7_45

groups: not based on the ICP framework and based on the ICP framework. In the first group, Chui and Rangarajan [4] introduced a Robust Point Matching (RPM) based on Thin Plate Spline (TPS). The algorithm uses a global-to-local search and soft assignment for correspondence. This algorithm uses TPS which can be considered as an extension of affine mapping. But it is not suitable when point sets with outliers. And it has high computational complexity. Zheng and Doermann [5] supposed the point matching as an optimization problem to preserve local neighborhood structures. This method believes that local relationship among neighboring points is very important for non-rigid registration. But this constraint needs smooth edges in images. Meanwhile, Jian and Vemuri [6] proposed a non-rigid algorithm GMMREG based on the probabilistic Gaussian mixture model (GMM). In this method, each point set is represented by a mixture of Gaussians and the affine registration is treated as a problem of aligning the mixtures. Myronenko and Song [7] introduced a coherent point drift (CPD) algorithm, in which the affine registration between two point sets is considered as a probability density estimation problem and it fits the GMM centroids of the first point set to the other by maximizing the likelihood. But both of GMMREG and CPD are computationally expensive as one point set must interact with each point in the opposing set. Hasanbelliu et al. [8] proposed a non-rigid registration algorithm by using a generalized correlation measure named correntropy. This algorithm achieves an optimal registration via rigid and non-rigid parameters. The method is robust to impulsive noise. But the correspondence between the two point sets should be known as a priori and two point sets must has same dimension. The computational complexity become high when the point sets are too big. Du et al. [9] introduced a non-rigid algorithm based on graph matching. This algorithm also has high computational complexity.

Some non-rigid algorithms are based on the ICP framework. Amberg et al. [10] supposed there exists an affine transformation for every point of set. When local affine transformation is computed, this algorithm optimizes the registration problem by adding a rigid regularization term and a signing term. But this method needs extra information including the shape of point sets and the sequent of point between two point sets in advance. And its computational complexity is high.

Our paper proposed a modified non-rigid registration algorithm based on the ICP framework. We use an average affine transformation to constrain the affine transformation of each point. The method uses nearest-point search algorithm based on Delaunay triangulation to establish correspondence between two point sets. A set of challenging data including single and overlapping chromosome images are tested which have significant local non-rigid transformations. Comparing experimental results with other algorithm demonstrate our algorithm is a fast and robust technique for non-rigid point sets registration.

2   The ICP Algorithm

Given two point sets, one denotes a model set $\mathbf{M} \triangleq \left\{ \vec{m}_i \right\}_{i=1}^{N_m} \left(N_m \in N \right)$ and the other is a data set $\mathbf{P} \triangleq \left\{ \vec{p}_i \right\}_{i=1}^{N_p} \left(N_p \in N \right)$. The Iterative Closest Point (ICP) algorithm can find a transformation T including translation $\vec{t}$ and rotation R transformations to register two point sets. The formulation can be defined as follow:

$$\min_{R,\vec{t},j\in\{1,2,\cdots,N_m\}} \left(\sum_{i=1}^{N_p} \left\| \left(R\vec{p}_i + \vec{t} \right) - \vec{m}_j \right\|_2^2 \right) \tag{1}$$

The ICP algorithm [1] is accurate and fast for point sets registration. It has two main steps. Firstly, we find the correspondences between two point sets by similarity measure:

$$c(i) = \arg\min_{j\in\{1,2,\cdots,N_m\}} \left(\left\| \left(R_k\vec{p}_i + \vec{t}_k \right) - \vec{m}_j \right\|_2^2 \right) \tag{2}$$

Secondly, we compute the new transformation parameters based on two corresponded point sets:

$$\left(R^*, \vec{t}^* \right) = \arg\min_{R,\vec{t}} \left(\left\| R\left(R_k\vec{p}_i + \vec{t}_k \right) + \vec{t} - \vec{m}_{c(i)} \right\|_2^2 \right) \tag{3}$$

Two main steps are repeated until the stopping criterion is satisfied. Since the ICP algorithm can achieve registration of two point sets with rotation and translation deformations accurately and fast, our algorithm is based on the traditional ICP framework. The details are introduced later.

3   The Non-rigid ICP Model

Given two point sets, one denotes a model set $M \triangleq \left\{ \vec{m}_i \right\}_{i=1}^{N_m} (N_m \in N)$ and the other is a data set $P \triangleq \left\{ \vec{p}_i \right\}_{i=1}^{N_p} (N_p \in N)$. The purpose of non-rigid point sets registration algorithm is to find an optimal transformation parameter X, which makes two point sets M and P can register accurately and fast. There are many ways to denote non-rigid transformation, now we suppose each point in data set P can be registered to a point in model set M by an affine transformation A_i, and $A = (A_1, A_2, \cdots A_{N_p})$. So the optimization problem can be described as follows:

$$\min_{j\in\{1,2,\cdots,N_m\}} \left(\sum_{i=1}^{N_p} \left\| \left(A_i\vec{p}_i + \vec{t}_i \right) - \vec{m}_j \right\|_2^2 \right) \tag{4}$$

To simplify the Eq. (4), we use homogeneous coordinates to indicate optimization problem, so the non-rigid registration can be presented as:

$$\min_{j \in \{1,2,\cdots,N_m\}} \left(\sum_{i=1}^{N_p} \left\| X_i \vec{v}_i - \vec{m}_j \right\|_2^2 \right) \tag{5}$$

where $X_i = (A_i, \vec{t}_i)$ is a m × (m + 1) dimension matrix for affine transformation. Hence all affine transformations between two point sets P and M can be simultaneous as a $(N_p m) \times (m + 1)$ dimension matrix $X = [X_1^T, X_2^T, \cdots, X_{N_p}^T]^T$, and $\vec{v}_i = (\vec{p}_i, 1)^T$.

Obviously, the above-mentioned problem is a high-dimension ill-posed problem. To constrain the solution with a stabilizer is an important part for shape matching because it enforces smoothness on the transformation function. We use a regularization term as the constraint for smoothness of Eq. (5). Amberg et al. [10] proposed a rigid regularization term to constrain the objective function. The rigid regularization term can control affine transformation between two adjacent points closely. But this method needs extra information about two point sets. Hence, we propose an affine constraint to avoid ill-posed problem. We suppose the affine transformation sets are $\{X_i\}_{i=1}^{N_p}$, then the average of affine transformations is $\bar{X} = (1/N_p) \sum_{i=1}^{N_p} X_i$. For avoiding instability of optimization problem, we should control all affine transformations between two point sets not far from $\bar{X}$, so we get an affine variance constraint as follow:

$$\sum_{i=1}^{N_p} \left\| (X_i - \bar{X}) G \right\|_F^2 \tag{6}$$

where G is dimensional matrix and $G = \text{diag}(1, 1, \cdots, 1, \gamma)$. Equation (6) represents the extent of elastic deformation when $G = \text{diag}(1, 1, \cdots, 1, \gamma)$. γ controls the translation in all affine transformation, we choose γ based on the dimension of image points. In our paper, $\gamma = 1$. Hence, the optimization equation can be expressed as follows:

$$\min_{X_i, j \in \{1,2,\cdots,N_m\}} \left(\sum_{i=1}^{N_p} \left\| X_i \vec{v}_i - \vec{m}_j \right\|_2^2 + \alpha \sum_{i=1}^{N_p} \left\| (X_i - \bar{X}) G \right\|_F^2 \right) \tag{7}$$

where α is the coefficient of regularization term, we set parameter α to control the local non-rigid transformation, and in all experiments $\alpha > 0$. When $\alpha \to +\infty$, all affine transformations are the same value. When $\alpha \to 0$, all affine transformations are different. So α is important for local affine transformation, that is to say α plays an important role on correspondences of two points sets.

4   The Modified Non-rigid ICP Algorithm

Traditional ICP algorithm is accurate and fast for registration between two point sets only with translation and rotation. Our paper introduces a modified non-rigid ICP algorithm for registration between two point sets with non-rigid transformation. Because parameter α has a directly influence on affine transformation of every point, so we suppose a given α. The modified non-rigid ICP algorithm also achieves point sets registration by iteration. Every iteration has two main steps.

Firstly, the correspondences are built between two points $\left\{ \vec{v}_i, \vec{m}_{c_k(i)} \right\}$:

$$c_k(i) = \arg\min_{j \in \{1,2,\cdots,N_m\}} \left(\left\| X_{k-1,i}\vec{v}_i - \vec{m}_j \right\|_2^2 \right), \quad i = 1, 2, \cdots, N_p \tag{8}$$

where $\left\{ X_{k-1,j} \right\}_{j=1}^{N_p}$ represent affine transformation in (k-1)th iteration.

Secondly, for control affine transformation of every point not oversize, we get new affine transformation parameters based on $\bar{X}_{k-1}$ in kth step:

$$\bar{X}_{k-1} = \frac{1}{N_p} \sum_{i=1}^{N_p} X_{k-1,i} \tag{9}$$

$$\left\{ X_{k,1}, X_{k,2}, \cdots, X_{k,N_p} \right\} = \arg\min_{X_i \in R^m} \left(\sum_{i=1}^{N_p} \left\| X_i \vec{v}_i - \vec{m}_{c_k(i)} \right\|_2^2 + \alpha \sum_{i=1}^{N_p} \left\| (X_i - \bar{X}_{k-1})G \right\|_F^2 \right) \tag{10}$$

where $\left\{ X_{k,j} \right\}_{j=1}^{N_p}$ represent affine transformation parameters between two point sets $\left\{ \vec{p}_i \right\}_{i=1}^{N_p}$ and $\left\{ \vec{m}_{c_k(i)} \right\}_{i=1}^{N_p}$ in kth step. And $\bar{X}_{k-1}$ is average affine transformation in (k-1)th step.

For establishing correspondence between two point sets, we use nearest-point search algorithm based on Delaunay triangulation for similarity measure. The key of problem is how to compute the affine transformation parameters $\left\{ X_{k,j} \right\}_{j=1}^{N_p}$. Obviously, this problem is a quadratic problem. Firstly, we can get the differential equations of error term and regularization term for the estimation of registration parameters, and to minimize the Eq. (10), according to extremum condition, we get Eq. (11) as follows:

$$0_{m \times (m+1)} = \frac{\partial \left(\sum_{i=1}^{N_p} \left\| X_i \vec{v}_i - \vec{m}_{c_k(i)} \right\|_2^2 + \alpha \sum_{i=1}^{N_p} \left\| (X_i - \bar{X})G \right\|_F^2 \right)}{\partial X_j} = 2X_j \vec{v}_j \vec{v}_j^T - 2\vec{m}_{c_k(i)} \vec{v}_j^T + 2\alpha X_j GG^T - 2\alpha \bar{X}_{k-1} G^T \tag{11}$$

Therefore, we compute the affine transformation $\left\{ X_{k,j} \right\}_{j=1}^{N_p}$ in step k iteration:

$$X_{k,j} = (\vec{m}_{c_k(i)} \vec{v}_j^T + \frac{\alpha}{N_p} \sum_{i=1}^{N_p} X_{k-1,i} G^T)(\vec{v}_j \vec{v}_j^T + \alpha GG^T)^{-1}, \quad j = 1, 2, \cdots, N_p \tag{12}$$

The proposed non-rigid ICP algorithm based on average affine variance constraint is outlined in Table 1 Algorithm procedure.

Table 1. Algorithm procedure

Algorithm procedure: The modified non-rigid ICP algorithm

Input: The two point sets P and M.

Begin: Initialize $\{X_{0,j}\}_{j=1}^{N_p}$ and parameters $\{\alpha_0, \alpha_1, \cdots, \alpha_n\}$, $(\alpha_i > \alpha_{i+1})$.

for $i = 1$ to i_{max} or $\alpha < \alpha_{max}$ do

   $\alpha = \alpha_i$, $X_{i,0,j} = X_{i-1,j}$

   for $k = 1$ to k_{max} or $\frac{1}{N_p}\sum_{i=1}^{N_p}\|X_{i,k,j} - X_{i,k-1,j}\|_F^2 \le \varepsilon^2$ is not satisfied do

   Compute correspondence of two points $\{\vec{p}_l, \vec{m}_{c_k(l)}\}$ using Eq.(8).

   Compute $\{X_{i,k,j}\}_{j=1}^{N_p}$ using Eq.(12).

   $X_{i,j} = X_{i,k,j}$.

End

Output: a transformed $\tilde{P}$ which is best aligns with M.

End

5   Experimental Results

Nowadays, medical images including brain images, chromosome images and so on attract many scholars' concern, a larger number of studies have been proposed and applied for many relevant areas. Our proposed algorithm can be used for register chromosomes which may be used for karyotype analysis. The data of the chromosomes for our experiment is including both single and overlapping, overlapping chromosome images are from Pki descriptive in Ref that provided by the Center of Medical Genetics Research Rajanukul Institutes of Health [11], single chromosome images are download at http://bioimlab.dei.unipad.it [12]. For getting two point set, we do some image preprocessing for chromosomes images including contour extraction algorithms and image filtering.

Firstly, we use Laplacian and Canny contour extraction algorithms to extract the contour points of chromosomes as model set, the results are showed in Fig. 1.

In Fig. 1, there are four different templates of chromosomes, they are overlapping and single chromosomes with irregular deformation contours. We do some preprocessing with chromosomes images including filtering and contour extraction, then get the model shape point set.

Secondly, we get data shape point set by establishing an appropriate template based on some endpoints of the chromosome. We set up a regular elongated shape point set for single chromosome, and set up a regular cruciform shape point set for overlapping chromosome.

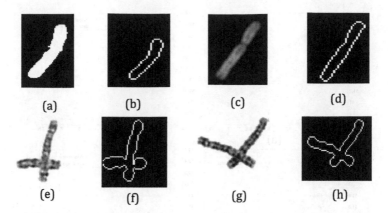

Fig. 1. The chromosomes. (a) Template 1 of chromosome. (b) The contour of template 1. (c) Template 2 of chromosome. (d) The contour of template 2. (e) Template 3 of chromosome. (f) The contour of template 3. (g) Template 4 of chromosome. (h) The contour of template 4.

To verify the fast convergence and robustness of our proposed method, four templates are tested on our algorithm as well as other non-rigid algorithms including affine ICP algorithm and a non-rigid algorithm introduced by Amberg et al. [10] separately. All programs are written in Matlab 7.10.0 and are run on PC with Intel i5-2400 3.10 GHz CPU, 4.00 GB RAM. Below, we introduce the comparative experimental results in detail.

5.1   Registration of Single Chromosome

In the first experiment, we use template 1 of single chromosome contour point sets, the model shape with 64 points, and the data shape with 184 point sets. We use affine ICP algorithm, Amberg's non-rigid algorithm [10] to compare with our algorithm. The results are in Fig. 2.

In Fig. 2, we use our proposed algorithm, affine ICP algorithm and the algorithm introduced by Amberg etc. to register two point sets, the results indicate that affine ICP algorithm can't achieve best registration when two point sets exist significant non-rigid deformation. Amberg's algorithm is also not sensitive for chromosome contour and its rate of convergence is lower than ours. The contrast of convergence rate for these tree algorithms is shown in Table 2.

Table 2.   The RMS and iteration times of different algorithms for single chromosome

Templates	RMS($\times10^{-8}$)			Iteration times		
	Affine ICP algorithm	Amberg's algorithm	Our algorithm	Affine ICP algorithm	Amberg's algorithm	Our algorithm
Template 1	650000	1.5353	0.20873	50	20	11
Template 2	240000	0.8483	0.07296	50	22	11

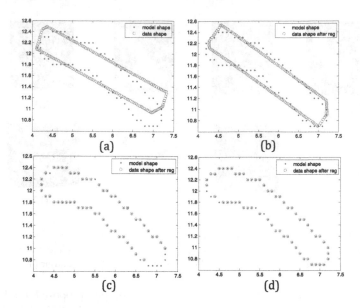

Fig. 2. The registration results of template 1. (a) The two point sets. (b) The result of affine ICP algorithm. (c) The result of Amberg's algorithm. (d) The result of our algorithm.

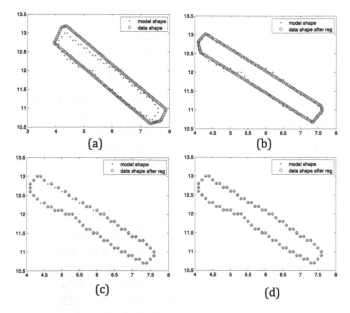

Fig. 3. The registration results of template 2. (a) The two point sets. (b) The result of affine ICP algorithm. (c) The result of Amberg's algorithm. (d) The result of our algorithm.

In the second experiment, we also use another single chromosome to test our algorithm, we use two point sets from template 2, the model shape with 73 points, and the data shape with 184 point sets. The results are shown in Fig. 3.

In Fig. 3, we can also see our algorithm has a better performance than other algorithms for local non-rigid deformation of single chromosome.

Table 2 compares the performance of different registration algorithms quantitatively. We compute the RMS (root-mean-square) error of each iteration. Affine algorithm always have 50 times iterations because we set maximum iterations as 50. The experimental results show that the registration of single chromosome with distinct irregularity edge, our algorithm can register local non-rigid transformation effectively.

5.2   Registration of Overlapping Chromosomes

In the next experiments, we test our algorithm with overlapping chromosomes. In the third experiment, we use two point sets, the model shape with 349 points, the data shape with 457 points. The results are shown in Fig. 4.

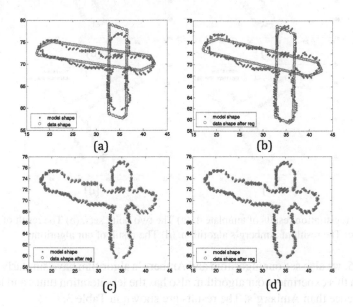

Fig. 4. The registration results of template 3. (a) The two point sets. (b) The result of affine ICP algorithm. (c) The result of Amberg's algorithm. (d) The result of our algorithm

In Fig. 4, the overlapping chromosomes have a complex contour with significant non-rigid deformation which is a challenge for registration algorithm. We can see both the Amberg's algorithm and our algorithm that have good performance, but our algorithm have the less iteration times and lower rate of convergence. The results are shown in Table 3.

Table 3. The RMS and iteration times of different algorithms for overlapping chromosomes

Templates	RMS($\times 10^{-8}$)			Iteration times		
	Affine ICP algorithm	Amberg's algorithm	Our algorithm	Affine ICP algorithm	Amberg's algorithm	Our algorithm
Template 3	4158000	0.03153	0.004589	50	15	11
Template 4	41090000	0.03848	0.004623	50	15	11

In the fourth experiment, we also use overlapping chromosome contour point sets different from template 3, the model shape with 386 points, and the data shape with 435 points. The results are shown in Fig. 5.

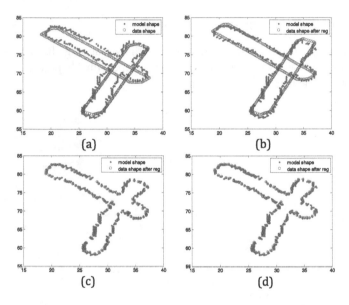

Fig. 5. The registration results of template 4. (a) The two point sets. (b) The result of affine ICP algorithm. (c) The result of Amberg's algorithm. (d) The result of our algorithm.

In Fig. 5, we also see our algorithm and Amberg's algorithm have similarly performance, but in this experiment our algorithm also has the less iteration times and lower rate of convergence than Amberg's. The results are shown in Table 3.

For comparing different algorithms for registration of overlapping chromosomes, we compute the RMS (root-mean-square) error of each iteration, and the results are shown in Table 3.

Table 3 compares the performance of different registration algorithms quantitatively. Affine algorithm always have 50 times iterations because we set maximum iterations as 50. The experimental results show that the registration of overlapping chromosomes with distinct irregularity edge, and when the point sets is lager, our algorithm can register local non-rigid transformation effectively.

6 Conclusion

This paper proposes a non-rigid ICP algorithm using average affine transformation constraint. This algorithm has a better performance on local non-rigid transformation and has a faster rate of convergence than other algorithms. Our algorithm has good application for registration of medical images. In future work, we may continue modifying our algorithm for matching more complex images.

Acknowledgement. This work was in part supported by the National Natural Science Foundation of China (61573274), Postdoctoral Science Foundation of China (2015M582661), Jiangsu Science and Technology Program (BY2014073), and Natural Science Basic Research Plan in Shaanxi Province of China (2015JQ6232).

References

1. Besl, P.J., McKay, N.D.: Method for registration of 3-D shapes. In: Robotics-DL tentative. International Society for Optics and Photonics, pp. 586–606 (1992)
2. Zhang, Z.: Iterative point matching for registration of free-form curves and surfaces. Int. J. Comput. Vis. **13**(2), 119–152 (1994)
3. Chen, Y., Medioni, G.: Object modelling by registration of multiple range images. Image Vis. Comput. **10**(3), 145–155 (1992)
4. Chui, H., Rangarajan, A.: A new point matching algorithm for non-rigid registration. Comput. Vis. Image Underst. **89**(2), 114–141 (2003)
5. Zheng, Y., Doermann, D.: Robust point matching for nonrigid shapes by preserving local neighborhood structures. IEEE Trans. Pattern Anal. Mach. Intell. **28**(4), 643–649 (2006)
6. Jian, B., Vemuri, B.C.: A robust algorithm for point set registration using mixture of gaussians. In: Tenth IEEE International Conference on Computer Vision, ICCV 2005. vol. 2, pp. 1246–1251 (2005)
7. Myronenko, A., Song, X.: Point set registration: coherent point drift. IEEE Trans. Pattern Anal. Mach. Intell. **32**(12), 2262–2275 (2010)
8. Hasanbelliu, E., Sanchez Giraldo, L., Principe, J.C.: Information theoretic shape matching. IEEE Trans. Pattern Anal. Mach. Intell. **36**(12), 2436–2451 (2014)
9. Du, S., Guo, Y., Sanroma, G., Ni, D., Wu, G., Shen, D.: Building dynamic population graph for accurate correspondence detection. Med. Image Anal. **26**(1), 256–267 (2015)
10. Amberg, B., Romdhani, S., Vetter, T.: Optimal step nonrigid ICP algorithms for surface registration. In: IEEE Conference on Computer Vision and Pattern Recognition (CVPR), pp. 1–8 (2007)
11. Saiyod, S., Wayalun, P.: A hybrid technique for overlapped chromosome segmentation of G-band mataspread images automatic. In: 2014 Fourth International Conference Digital Information and Communication Technology and it's Applications (DICTAP), pp. 400–404 (2014)
12. Poletti, E., Grisan, E., Ruggeri, A.: A modular framework for the automatic classification of chromosomes in Q-band images. Comput. Methods Programs Biomed. **105**(2), 120–130 (2012)

Locally Biased Discriminative Clustering Method for Interactive Image Segmentation

Xianpeng Liang[1(✉)], Xiao-Ping Zhang[1], Li Shang[2], and Zhi-Kai Huang[3]

[1] Institute of Machine Learning and Systems Biology,
School of Electronics and Information Engineering,
Tongji University, Caoan Road 4800, Shanghai 201804, China
lxp_liang@tongji.edu.cn
[2] Department of Communication Technology, College of Electronic
Information, Engineering, Suzhou Vocational University, Suzhou, China
[3] College of Mechanical and Electrical Engineering,
Nanchang Institute of Technology, Nanchang 330099, Jiangxi, China

Abstract. Interactive image segmentation is a form of semi-supervised segmentation method by using the user interactive information. It performed well than fully unsupervised segmentation methods. In this paper, we propose a novel interactive image segmentation method, in which a seed vector is used to represent the user scribbles. Then a soft similarity constraint is added to the discriminative clustering model. The soft constraint allows the user to tune the degree which the constraint is satisfied. With respect to the discriminative clustering model, the clustering result is not affected by the assumption to the distribution of the data, and it's easy to add constraint to the clustering variable. The final optimization problem is convex, so it can reach global optimal solution. The proposed method is evaluated on benchmark dataset BSD dataset, and it performs well than state of art methods both in quantitative and qualitative results.

Keywords: Interactive segmentation · Discriminative clustering · Seed vector · Soft constraint · Low rank optimization

1 Introduction

Due to the complexity and diversity of image in real world, fully unsupervised bottom up image segmentation remains a challenge task in computer vision. However, when available supervised prior information is added to the segmentation, it usually can improve the segmentation performance. Interactive segmentation is a semi-supervised segmentation method, in which the user interaction inputs is the supervised prior information. The goal of Interactive segmentation is to segment the foreground objects from the background under the guide of user interaction information.

Recently there are lots of works focus on interactive segmentation [1–11]. According to the type of user input, these works can be divided into two categories: scribble based methods and bounding box based methods. In the scribble based

© Springer International Publishing Switzerland 2016
D.-S. Huang and K.-H. Jo (Eds.): ICIC 2016, Part II, LNCS 9772, pp. 514–522, 2016.
DOI: 10.1007/978-3-319-42294-7_46

methods, user is requested to label part of the pixels both in foreground and background, then image will be segmented using the scribble information. Scribble based method include Graph Cut [12], random walker [13], and Ncuts based method [1, 3, 4, 14], etc. In the bounding box based methods, the user is requested to give a bounding box that is large enough to cover the whole foreground object, and then the segmentation task is done in the region covered by the bounding box. The popular bounding box based methods include: Grab Cut [15], MIL Cut [16], etc.

However, in this paper we only focus on the scribble based methods. Although these scribble based methods are relatively efficient and can achieve relative high precision, there are some obvious limits. For methods based on the variation of the Graph Cut, they heavily rely on the design of different complex energy function [2], which usually is difficult to optimize and the global optimal solution can't be guaranteed. Besides, the user interaction of most scribble based methods is added to the segmentation model in a way of hard constraint which forces the scribble vertexes grouped in the class or separated into different class.

In this paper, we propose a novel scribble based interactive segmentation method called: Locally biased discriminative clustering method (LBDC). LBDC is based on the discriminative clustering segmentation method, and incorporate a soft scribble based constraint. The soft constraint [1] allows user to tune the degree to which the constraint is satisfied, and is robust to situations when user input is not completely accurate. In the frame of discriminative clustering method, a classification principle is used to cluster the pixels, and here a kernel based linear regression model with square loss is used as the clustering model. Besides, a spot of spatial consistency constraint is used to tune the clustering result [17]. There are several merits of the discriminative clustering, first the clustering result isn't affected by the assumption to the distribution of the pixels. Second, it's easy to add constraint to the class variables for discriminative clustering. Here we use a seed vector to represents the user scribble and use the similarity between seed vector and class variable vector as the constraint. Finally, we do experiments on benchmark dataset BSD. Our segmentation accuracy is comparable or better than state of art methods.

The reminder of this paper is organized as follows: Sect. 2 introduces several related works. Our locally biased discriminative model is proposed in Sect. 3. In Sect. 4, experiments are done and compared with other methods. Last we conclude our work in Sect. 5.

2 Related Work

Scribble based interactive segmentation: Boykov and Jolly proposed the Graph Cut [12] which seen the image as a graph and use a energy function to model the min cut between seeded region. The user scribble is described as part of the energy function. The energy function is solved using maflox/min-cut algorithm. In [4], Maji et al. proposed the biased normalized cuts which is based on the Ncuts algorithm. A seed vector is used to represent the user scribbles, and an inequality constraint between seed vector and class variable is added to the object function of Ncuts. This constraint forces the segmentation result toward to the seed vector. [1, 3] are also based on the Ncuts algorithm. They model the user scribble as part of the object function instead of in the form of inequality constraint. Besides, there are other forms of interactive segmentation methods [2, 5, 18].

Discriminative clustering: In [19], Xu et al. first introduced the discriminative clustering method, in which a SVM classification principle is used as the clustering model. The optimal class variable is get when the corresponding classification margin is maximum. Then in [17], Joulin et al. used the discriminative clustering model to do image cosegmentation. The object function is converted to a convex form by convex relaxation, and is solved by a low rank semidefinite program. Besides, there are other type of discriminative clustering models [20].

3   Locally Biased Discriminative Clustering Model

In this section, LBDC is introduced in three parts. First we introduce the scribble based soft constraint. Then the linear regression based discriminative clustering method is described. Last, the mathematical formula of LBDC is presented.

The symbols used in the following are presented. Assume the number of pixels in the image is n, and we use d dimension vector x_i, $i = 1, \ldots, n$ to represent the feature vector of these pixels. The class variable is represented by an n dimension vector y. An n dimension seed vector S is used to record the user input. $\langle \cdot, \cdot \rangle$ is the inner product operator. $|\cdot|$ is the L_1 norm.

3.1   Scribble Based Soft Constraint

In this paper, the type of user interaction is based on the user scribbles. A seed vector (S) is used to record the user input. Specifically, we use S_F to represent the foreground user input and S_B to represent the background user input. If the i-pixel belongs to the foreground user input set, $S_F(i) = 1$, and if it belongs to the background user input set, $S_B(i) = -1$, otherwise, $S_F(i) = 0$, $S_B(i) = 0$.

Different with most of the interactive segmentation methods, we use an inequality based soft constraint to incorporate the user input into the segmentation model.

$$\langle y, S_F \rangle \geq \tau |S_F| \; and \; \langle y, S_B \rangle \leq -\tau |S_B|, \tag{1}$$

where $\langle y, S \rangle$ means the similarity between seed vector and class variable, and $|S|$ means the number of pixels belonging to the user input. Parameter τ is a threshold which can be tuned to decide how similarity of these two variable.

The inequality constraint plays an import role in our LBDC model. It can force the segmentation result biased to prior seed vector, thus it can get more precise result. Besides, by tuning the value of τ, the model is robust to user input in case of inexact user input.

In order to incorporate with the discriminative clustering model introduced in 3.2, formula (1) is converted to an equivalent form:

$$\frac{1}{2}\left(yy^T + 1_n 1_n^T\right)S_F \geq \tau_1 |S_F|S_F, \; and \; \frac{1}{2}\left(Y + 1_n 1_n^T\right)S_B \geq -\tau_2 |S_B|S_B, \qquad (2)$$

where 1_n is an n dimension column vector whose elements are all one.

3.2 Discriminative Clustering

In this part, we introduce the unsupervised discriminative clustering method [17]. Given an image, it will be divided into two class, foreground and background. And we denote the class variable y by: if i-pixel belongs to the foreground, $y_i = 1$, otherwise, $y_i = -1$.

For the features of pixels, we use SIFT descriptors [21], Gabor filters and color histograms [22] to represent the features of pixels. And we use kernel function Φ to translate feature vector x into a high dimension Hilbert space.

Now, we introduce the kernel function based linear regression model:

$$\ell(y,f) = \frac{1}{n}\sum_{j=1}^{n}\left(y_j - f^T\Phi(x_j) - b\right)^2 + \lambda_k \|f\|^2, \qquad (3)$$

where f means the parameter vector, b is the bias term, and parameter λ_k can avoid over fitting.

Given the class variable y, the object function in formula (3) is a function with parameter f. We can get the minimum value of $\ell(y,f)$ by derivative of f, and then let the derivative equal to zero. The optimum solution of f and b is respect to:

$$f^* = \left(\Phi\prod_n \Phi^T + n\lambda_k I_d\right)^{-1}\Phi\prod_n y, b^* = \frac{1}{n}1_n^T(y - \Phi^T f^*). \; \prod_n = I_n - \frac{1}{n}1_n 1_n^T. \quad (4)$$

Combining (4) with (3), we get $g(y)$ which is the minimum value of the loss function in formula (3):

$$g(y) = y^T A y,$$
$$s.t \, A = \Pi_n\left[I_n - \Phi^T\left(n\lambda_k I_n + \Phi\Pi_n\Phi^T\right)^{-1}\Phi\right]\Pi_n, \qquad (5)$$
$$\Pi_n = I_n - \frac{1}{n}1_n 1_n^T.$$

Besides, we add a bit of spatial consistency constraint to the formula (5) to tune the result. The color and position information are used to compute the similarity matrix W. Then we can get the laplacian matrix L defined from the joint block-diagonal similarity matrix.

Thus, the spatial consistency can be added into (5) by the form:

$$\ell(y) = y^T L y \qquad (6)$$

In order to avoid the trivial solution that cluster all the pixels into the foreground class or background class, the constraint to the number of each class is used:

$$\lambda_0 n 1_n \leq \frac{1}{2}\left(yy^T + 1_n 1_n^T\right)1_n \leq \lambda_1 n 1_n \tag{7}$$

Different choice of λ_0 and λ_1 can results in different results, and here we fix $\lambda_0 = 0.05$, $\lambda_1 = 0.95$.

Finally, combining the above formulas (2), (5), (6), (7), we get the object function of the locally biased discriminative clustering model:

$$\min_{y \in \{-1,1\}^n} y^T(A + \mu L)y \tag{8}$$

In the following, we use a symmetric matrix Y to represent the term, yy^T. The elements of Y take the value from $\{-1, 1\}$, the diagonal of Y is 1, and the rank of Y is 1. Then formula (8) is equivalent to the following semidefinite program:

$$\min_Y \ tr(Y(A + \mu L))$$

$$s.t. \ \lambda_0 n 1_n \leq \frac{1}{2}\left(Y + 1_n 1_n^T\right)1_n \leq \lambda_1 n 1_n$$

$$\frac{1}{2}\left(Y + 1_n 1_n^T\right)S_F \geq \tau_1 |S_F| S_F$$

$$\frac{1}{2}\left(Y + 1_n 1_n^T\right)S_B \geq - \tau_2 |S_B| S_B \tag{9}$$

Due to the low rank constraint of matrix Y, formula (9) is not convex. However, we can do convex relaxation by let matrix Y take value from the following set, $\varepsilon = \{Y \in R^{n \times n}, Y = Y^T, diag(Y) = 1_n, Y \succeq 0\}$. Then, formula (9) is a convex semidefinite program problem.

3.3   Optimization of LBDC

In this part, we use a efficient low rank method [23] to solve the above convex semidefinite program (9). The method first transforms the inequality constraint into penalties term by an augmented Lagrangian method. Then, a trust region method is used to solve the low rank solution of the transformed convex function.

4   Experiments

In this section, we will demonstrate the effectiveness of the proposed method by showing both quantitative and qualitative experimental results with user scribbles. We use Berkeley segmentation dataset to test our algorithm. Berkeley segmentation dataset contains five hundred test images with human segmentation masks as the ground-truth data and natural images with abundant colors and complicated textures, which makes it challenging to segment them even with the user scribbles.

For the quantitative evaluation, we compute the accuracy of the segmentation result over the manually labeled mask, which includes both foreground and background, $p = \frac{A(true\, foreground) + A(true\, background)}{A(image)}$, where $A(\cdot)$ means the area of the corresponding region. For the qualitative evaluation, we show that our algorithm performs better than state of arts methods with the same user scribble.

In the following, with respect to the user scribbles, the blue line represent the foreground input and the red line represent the background input. To the segmentation result, the region covered by the green color is the foreground, the region covered by the red color is the background and the white curve represents the contour of the foreground object.

From Fig. 1, we can see that the user scribble in picture (a) only located in the region of the black skirt, so the segmentation result in (b) only regard the part with the skirt as the foreground object. However in (c) the user scribble containing the points in head, body and legs, thus our algorithm can accurately segment the whole woman as the foreground object.

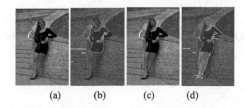

(a)   (b)   (c)   (d)

Fig. 1. The segmentation result by our method. (a) original image with foreground scribble I. (b) the segmentation result with the scribbles in (a). (c) original image with foreground scribble II. (d) the segmentation result with the scribbles in (c).

Figure 2 shows the effect of the using both foreground and background scribbles. For the difficult scene in (a), only using the foreground scribble in (a), the objects can't be fully segmented. The color of upper left corner of (a) is very similar with the color of objects, so it's difficult to distinguish them. However, when the background scribble is used in (c), most part of the upper left corner of (c) is correctly separated into the background. Following, we compare our method with two state of art methods,

(a)   (b)   (c)   (d)

Fig. 2. The segmentation result by our method. (a) original image with foreground scribbles (blue line). (b) the segmentation result with the scribbles in (a). (c) original image with foreground (blue) and background (red) scribbles. (d) the segmentation result with the scribbles. (Color figure online)

(a)   (b)   (c)   (d)

Fig. 3. Foreground extraction results. (a) original image with user scribbles (blue line). (b) random walk. (c) GrabCuts. (d) LBDC. (Color figure online)

Table 1. Quantitative evaluation results on the segmentation results.

Random walk	GrabCut	LBDC
91.43 %	93.78 %	97.75 %

Random walker and GrabCut. Figure 3 shows some sample segmentation results on the BSD dataset (Table 1).

5   Conclusions

In this paper, we propose a novel interactive segmentation method. It combines the discriminative clustering method with a soft user scribble based constraint. We do experiments on BSD dataset, and the experimental results show the effectiveness of our method. However, there remains a challenge when segment the image with complex scene, e.g. the color of the object is very similar with the background scene.

Acknowledgments. This work was supported by the grants of the National Science Foundation of China, Nos. 61520106006, 61532008, 31571364, 61303111, 61411140249, 61402334, 61472280, 61472173, 61572447, 61373098, and 61572364, China Postdoctoral Science Foundation Grant, Nos. 2014M561513 and 2015M580352.

References

1. Chew, S.E., Cahill, N.D.: Semi-supervised normalized cuts for image segmentation. In: Proceedings of the IEEE International Conference on Computer Vision 2015, pp. 1716–1723 (2015)
2. Casaca, W., Nonato, L.G., Taubin, G.: Laplacian coordinates for seeded image segmentation. In: 2014 IEEE Conference on Computer Vision and Pattern Recognition (CVPR), pp. 384–391. IEEE (2014)
3. Shen, J., Du, Y., Li, X.: Interactive segmentation using constrained Laplacian optimization. IEEE Trans. Circuits Syst. Video Technol. 24(7), 1088–1100 (2014)
4. Maji, S., Vishnoi, N.K., Malik, J.: Biased normalized cuts. In: 2011 IEEE Conference on Computer Vision and Pattern Recognition (CVPR), pp. 2057–2064. IEEE (2011)
5. Jian, M., Jung, C.: Interactive image segmentation using adaptive constraint propagation. IEEE Trans. Image Process. Publ. IEEE Signal Process. Soc. 25, 1301–1311 (2016)
6. Wang, X.-F., Huang, D.-S., Xu, H.: An efficient local Chan-Vese model for image segmentation. Pattern Recogn. 43(3), 603–618 (2010)
7. Zhu, L., Huang, D.-S.: Efficient optimally regularized discriminant analysis. Neurocomputing 117, 12–21 (2013)
8. Wang, X.-F., Huang, D.-S.: A novel density-based clustering framework by using level set method. IEEE Trans. Knowl. Data Eng. 21(11), 1515–1531 (2009)
9. Zhao, Y., Huang, D.-S., Jia, W.: Completed local binary count for rotation invariant texture classification. IEEE Trans. Image Process. 21(10), 4492–4497 (2012)
10. Huang, D.-S., Jiang, W.: A general CPL-AdS methodology for fixing dynamic parameters in dual environments. IEEE Trans. Syst. Man Cybern. B Cybern. 42(5), 1489–1500 (2012)
11. Huang, D.-S.: Radial basis probabilistic neural networks: model and application. Int. J. Pattern Recognit Artif Intell. 13(07), 1083–1101 (1999)
12. Boykov, Y.Y., Jolly, M.-P.: Interactive graph cuts for optimal boundary & region segmentation of objects in ND images. In: Proceedings of the Eighth IEEE International Conference on Computer Vision, ICCV 2001, pp. 105–112. IEEE (2001)
13. Grady, L.: Random walks for image segmentation. IEEE Trans. Pattern Anal. Mach. Intell. 28(11), 1768–1783 (2006)
14. Huang, D.-S.: Systematic theory of neural networks for pattern recognition. Publishing House Electron. Ind. China, Beijing 28, 323–332 (1996)
15. Rother, C., Kolmogorov, V., Blake, A.: Grabcut. Interactive foreground extraction using iterated graph cuts. In: ACM Transactions on Graphics (TOG) 2004, vol. 3, pp. 309–314. ACM (2004)
16. Wu, J., Zhao, Y., Zhu, J.-Y., Luo, S., Tu, Z.: Milcut: a sweeping line multiple instance learning paradigm for interactive image segmentation. In: Proceedings of the IEEE Conference on Computer Vision and Pattern Recognition 2014, pp. 256–263 (2014)
17. Joulin, A., Bach, F., Ponce, J.: Discriminative clustering for image co-segmentation. In: 2010 IEEE Conference on Computer Vision and Pattern Recognition (CVPR), pp. 1943–1950. IEEE (2010)
18. Huang, D.-S., Du, J.-X.: A constructive hybrid structure optimization methodology for radial basis probabilistic neural networks. IEEE Trans. Neural Netw. 19(12), 2099–2115 (2008)
19. Xu, L., Neufeld, J., Larson, B., Schuurmans, D.: Maximum margin clustering. In: Advances in Neural Information Processing Systems 2004, pp. 1537–1544
20. Zhang, X.-L.: Convex discriminative multitask clustering. IEEE Trans. Pattern Anal. Mach. Intell. 37(1), 28–40 (2015)

21. Lowe, D.G.: Object recognition from local scale-invariant features. In: The Proceedings of the Seventh IEEE International Conference on Computer Vision, 1999, pp. 1150–1157. IEEE (1999)
22. Hochbaum, D.S., Singh, V.: An efficient algorithm for co-segmentation. In: 2009 IEEE 12th International Conference on Computer Vision, pp. 269–276. IEEE (2009)
23. Journée, M., Bach, F., Absil, P.-A., Sepulchre, R.: Low-rank optimization for semidefinite convex problems. arXiv preprint arXiv:0807.4423 (2008)

Accurate Prior Modeling in the Locally Adaptive Window-Based Wavelet Denoising

Yun-Xia Liu[1,2,3](✉), Yang Yang[4], and Ngai-Fong Law[5]

[1] School of Information Science and Engineering,
University of Jinan, Jinan, China
ise_liuyx@ujn.edu.cn
[2] Shandong Provincial Key Laboratory of Network
Based Intelligent Computing, Jinan, China
[3] School of Control Science and Engineering,
Shandong University, Jinan, China
[4] School of Information Science and Engineering,
Shandong University, Jinan, China
[5] Centre for Signal Processing, Department of Electronic and Information
Engineering, The Hong Kong Polytechnic University, Hong Kong, China

Abstract. The locally adaptive window-based (LAW) denoising method has been extensively studied in literature for its simplicity and effectiveness. However, our statistical analysis performed on its prior estimation reveals that the prior is not estimated properly. In this paper, a novel maximum likelihood prior modeling method is proposed for better characterization of the local variance distribution. Goodness of fit results shows that our proposed prior estimation method can improve the model accuracy. A modified LAW denoising algorithm is then proposed based on the new prior. Image denoising experimental results demonstrate that the proposed method can significantly improve the performance in terms of both peak signal-to noise ratio (PSNR) and visual quality, while maintain a low computation.

Keywords: Image denoising · Orthogonal wavelet transform · Adaptive parameter estimation · Maximum likelihood estimation · Visual quality

1 Introduction

Image denoising has drawn huge research interest [1–3]. It is, in fact, an important engineering application as noise is unavoidable during image capture and transmission. Due to effective modeling of both image and noise in wavelet domain, wavelet methods have been widely studied and have proved to provide good performances [4–12].

Accurate modeling of the wavelet coefficients is of vital importance to the success of the denoising tasks. More and more sophisticated coefficients representation models [8, 13] are adopted for better denoising performance. For example, the locally adaptive window-based (LAW) denoising method [4] adopts a doubly stochastic process (DSP) model [8]. Utilizing the MAP estimator and assuming an exponential density for the local variance, LAWMAP achieves good performance at low computational cost.

© Springer International Publishing Switzerland 2016
D.-S. Huang and K.-H. Jo (Eds.): ICIC 2016, Part II, LNCS 9772, pp. 523–533, 2016.
DOI: 10.1007/978-3-319-42294-7_47

BiShrink [6] incorporates the inter-scale relationship by modeling the joint parent-child distribution. ProbShrink [7] assumes a generalized Laplacian prior and models the denoising problem as detection of "signal of interest". The SURE-Let [9] method, making use of Stein's unbiased risk estimator (SURE) and linear expansion of threshold (Let), is considered to be the state-of-the-art orthogonal wavelet domain denoising method.

LAWMAP is widely studied due to its simplicity and efficiency. It is now returning to researcher's horizon due to recent rapid developments in photo-response nonuniformity noise based image forensics, where it is adopted as the denoiser [14]. There have been extensive discussion on accurate modeling within the LAW framework. For example, non-parametric approach is adopted in [11] for marginal distribution modeling. A bilateral scheme [12] is proposed for better estimation of the local variance.

An accurate prior modeling algorithm that better characterized the local variance filed is proposed in this paper. Firstly, our statistical analysis reveals that the prior modeling in LAWMAP is improper in that the choice of a key parameter λ that characterizes the exponential prior is not optimal. Secondly, we propose a maximum likelihood (ML) method for estimating λ. Goodness of fit experiments are carried out on abundant images to investigate the modeling accuracy of the proposed prior estimation scheme. Results show that our proposed prior is more efficient in characterizing the global subband statistics of wavelet coefficients than the original approach. Thirdly, as an application to image denoising, the modified LAWMAP algorithm gains significant performance improvement as compared to its original approach, with no extra computation cost. These improvements lead the modified LAWMAP an appealing competitor among orthogonal wavelet domain denoisers, especially for texture regions in images.

2   LAWMAP Denoising

Consider the general denoising scenario that the clean wavelet coefficients x are corrupted by zero mean additive Gaussian white noise n with standard deviation σ_n. Then,

$$y(k) = x(k) + n(k), \tag{1}$$

where y is observed noisy coefficients and k is the position index in wavelet domain. The aim of denoising is to recover x as accurate as possible.

2.1   Doubly Stochastic Process Model

In the wavelet domain doubly stochastic process model, wavelet coefficient x and its neighboring coefficients are assumed to be independently drawn from a zero-mean Gaussian distribution $N(0, \theta)$ given their variances θ. The local variance θ is modeled as independently, identically distributed, spatially highly correlated random variables following a distribution $\pi(\theta)$.

Among all $\pi(\theta)$ candidates which characterize the high-peak and heavy-tail properties of local variance, exponential density

$$\pi(\theta) = \lambda e^{-\lambda\theta}, \quad for \quad \theta \geq 0 \tag{2}$$

is adopted [4] as it results in close-form MAP variance estimator, where λ is known as the rate parameter.

2.2 Two-Step Minimum Mean Square Error Estimator

The LAW denoising method [4] consists of two steps: (1) estimation of the variance field for each wavelet coefficient; and (2) a pointwise local wiener filtering. See Fig. 1 for the block diagram of the algorithm.

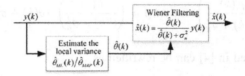

Fig. 1. Block-diagram of the LAW denoising algorithm.

For estimation of the underlying variance field, we can resort to either the ML or the MAP estimation method. Given the current coefficient $y(k)$ and its neighborhoods in a square window $\mathcal{N}(k)$, the ML estimator $\hat{\theta}_{ML}(k)$ is given by

$$\hat{\theta}_{ML}(k) = \max\left(0, \sum_{j \in N(k)} y^2(j) \Big/ M - \sigma_n^2\right) \tag{3}$$

where M denotes the number of coefficients in $\mathcal{N}(k)$. The approximate MAP estimator $\hat{\theta}_{MAP}(k)$ considers also the prior marginal exponential distribution $\pi(\theta)$ as

$$\hat{\theta}_{MAP}(k) = \max\left(0, \frac{M}{4\lambda}\left[-1 + \sqrt{1 + \frac{8\lambda}{M^2}\sum_{j \in N(k)} y^2(j)}\right] - \sigma_n^2\right) \tag{4}$$

has shown to consistently outperform $\hat{\theta}_{ML}(k)$ [4].

In the second step, LAWMAP adopts the minimum mean square error (MMSE) criteria for denoising. Given $\hat{\theta}_{MAP}(k)$, the denoised wavelet coefficient is obtained as,

$$\hat{x}(k) = \frac{\hat{\theta}(k)}{\hat{\theta}(k) + \sigma_n^2} y(k) \tag{5}$$

3 Accurate Prior Modeling of LAWMAP

From the framework of LAWMAP, we see that λ is a key parameter that characterizes the global distribution of the variance field of the subband, thus has a direct impact on the estimation accuracy.

3.1 Analysis of Prior Modeling in LAWMAP

In LAWMAP [4], the parameter λ is set to be the inverse of the standard deviation of wavelet coefficients that was initially denoised using $\hat{\theta}_{ML}(k)$ in Eqs. (3) and (5).

Assume $\hat{\theta}_{ML}(k)$ provides an accurate estimation of the real variance field, i.e., $\hat{\theta}_{ML}(k)$ is very close to $\theta(k)$. From Eq. (1) we get

$$\text{var}[\hat{x}(k)] = \left(\frac{\hat{\theta}_{ML}(k)}{\hat{\theta}_{ML}(k) + \sigma_n^2}\right)^2 \text{var}[y(k)] \approx \left(\frac{\theta(k)}{\theta(k) + \sigma_n^2}\right)^2 (\theta(k) + \sigma_n^2) = \frac{\theta^2(k)}{\theta(k) + \sigma_n^2} \quad (6)$$

Thus the prior λ_1 used in [4] can be rewritten as

$$\lambda_1 = \frac{1}{std[\hat{x}(k)]} = \frac{\sqrt{\theta(k) + \sigma_n^2}}{\theta(k)} \quad (7)$$

An exponential density has a mean value of $1/\lambda$. This indicates that $\pi(\theta)$ is fitted with a variance field whose mean value is

$$\frac{1}{\lambda_1} = \frac{\theta(k)}{\sqrt{\theta(k) + \sigma_n^2}} < \sqrt{\theta(k) + \sigma_n^2} \quad (8)$$

which is related to $\sqrt{\theta(k)}$, i.e., the *local standard deviation*. This **contradicts** with the initial idea of using an exponential prior for modeling the *variance* field. Hence, this leaves room for an improved prior modeling.

3.2 Proposed Method for Estimating λ

As compared with $\hat{x}(k)$, the ML variance estimators $\hat{\theta}_{ML}(k)$ are more informative for estimating λ that characterize the global subband statistics. We thus propose to use maximum likelihood estimator of $\hat{\theta}_{ML}(k)$ within the same subband for estimating λ.

Assume $\{\theta_1, \theta_2, \ldots, \theta_N\}$ are independent exponentially distributed random variables, the maximum likelihood estimator can be proved to be

$$\hat{\lambda}_{ML} = N \bigg/ \sum_{i=1}^{N} \theta_i \quad (9)$$

Notice from Eq. (3) that $\hat{\theta}_{ML}(k)$ can be truncated to zero. This will disturb the ML estimator and thus they are left out. Hence, the proposed prior λ_2 is obtained as,

$$\lambda_2(b) = \frac{N(b)}{\sum_{k=1}^{N(b)} \hat{\theta}_{ML}(k)} \tag{10}$$

where b is the index for bandpass subbands, and $N(b)$ is the number of non-zero $\hat{\theta}_{ML}(k)$ in subband b.

3.3 Goodness of Fit Test for Prior Models λ_1 and λ_2

To investigate the accuracy of the two prior models in representing the global subband statistics, we compare several goodness of fit measures [15] between the empirical histogram H and the two estimated exponential probability density function (pdf)s. As the exponential distribution is continuous, we have to first construct the discrete probability mass functions (pmf) for the estimated $E(\lambda_1)$ and $E(\lambda_2)$. The following four goodness of fit measures are considered:

$$\begin{array}{ll} \text{SS:} & \sum (h_i - e_i)^2 \\ \text{WSS:} & \sqrt{\sum (h_i - e_i)^2 \cdot h_i} \\ \text{KLD:} & \sum h_i \cdot \log(h_i/e_i) \\ \chi^2: & \sum \frac{(h_i - e_i)^2}{e_i} \end{array} \quad i = 1, \ldots, n_b \tag{11}$$

where n_b denotes the total number of bins in the pmf, h_i and e_i are pmf values of the empirical histogram H and the estimated exponential E at bin i respectively. SS is the sum of square errors between empirical histogram and the estimated exponential priors. WSS is a variant of SS, it is enhanced by weighting of the empirical histogram. Kullback-Leibler divergence (KLD) which is a measure originate from information theory depicts extra number of bits needed to code samples from H using codes based on E. χ^2 is the famous Pearson's chi-square statistic. For all of the four measures, smaller values indicate higher modeling quality.

The image set consists of eight 512×512 images: Lena, Barbara, Boat, Peppers, Bicycle, Fingerprint, Flinstones, Al and eight 256×256 images: Baboon, House, Coco, Mit, Bridge, Cameraman, Parrot, Zoneplate (See Fig. 2). These images are diverse in their contents. Some have rich smooth areas such as portrait images, natural scenes, cartoons, some have sharp edges and some are full of textures, e.g. fingerprints, architectures. Square shaped 5×5 window is adopted for spatial adaptivity. Orthogonal wavelet transform with five levels of decomposition and "sym8" filter are used. The two prior modeling schemes are compared for all 15 bandpass subbands.

Table 1 shows four indicators: the minimum, the maximum, the mean and the standard deviation for the image set with $n_b = 256$. From results in Table 1, we see that the proposed λ_2 prior modeling method produces consistent smaller goodness of fit values as compared with the original approach λ_1. This theoretically justifies its fitting

Fig. 2. Image Sets.

Table 1. Goodness of fit measures for prior models λ_1 and λ_2.

Goodness of fit measures		Min	Max	Mean	Std
SS	λ_1	0.0002	0.6455	0.0829	0.1302
	λ_2	**0.0001**	**0.549**	**0.0594**	**0.0963**
WSS	λ_1	0.0022	0.7369	0.1189	0.1649
	λ_2	**0.002**	**0.6472**	**0.0974**	**0.1334**
KLD	λ_1	0.0132	3.9475	0.8365	0.7301
	λ_2	**0.0127**	**3.1989**	**0.6555**	**0.6075**
χ^2	λ_1	0.0211	37.8918	3.604	6.1336
	λ_2	**0.0197**	**30.25**	**2.1792**	**3.8717**

superiority in modeling. Furthermore, similar results have been observed when we tried different parameters for n_b (varing from 256 to 64, 128 and 512) and noise levels (varing from 10 to 100). This indicates that the proposed λ_2 prior modeling method consistently performs better than λ_1, irrespective of the values of n_b and the noise levels.

Figure 3 depicts the histogram of the estimated variance for a typical subband of the "Barbara" image in log scale, together with the two fitted exponential priors. We can clearly see that λ_2 better describes the subband characteristics than λ_1, especially in the modeling of the heavy tail characteristics of the wavelet coefficients. Thus we can expect performance improvement in denoising images due to this accurate prior modeling.

4   Application in Image Denoising

In this part, we apply the proposed prior modeling method in image denosing, and investigate how the accurate modeling of local variance field could benefit LAWMAP image denoising.

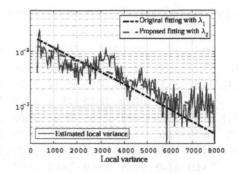

Fig. 3. Comparison of different prior modeling schemes. (Color figure online)

4.1 The Modified LAWMAP Image Denoising Algorithm

In our modified LAWMAP method, the local variances are replaced with those estimated by λ_2 modeling. The algorithm can be summarized as follows:

1. Perform an L-level orthogonal wavelet transform of the noisy image to get the noisy wavelet coefficients y.
2. For each of the wavelet coefficient $y(k)$ within subband b,
   (a) Compute the maximum likelihood estimator $\hat{\theta}_{ML}(k)$ using Eq. (3);
   (b) Estimate exponential prior λ_2 using Eq. (10) and get $\hat{\theta}_{MAP}(k)$ using Eq. (4);
   (c) Estimate the denoised wavelet coefficient $x(k)$ using Eq. (5).
3. Obtain the denoised image via inverse wavelet transform of $\hat{x}$.

4.2 Denoising Performance Comparison with Other Wavelet Algorithms

To compare the proposed denoising method with the state-of-the-art orthogonal wavelet-domain algorithms, we carried out comprehensive experiments on images in Fig. 2. Only results of three images were reported in Table 2 due to the space limitation, where each figure had been averaged over ten noise realizations. Results of other methods were obtained either by programs provided by authors [4, 6, 7, 9, 11] or simulated with parameters suggested in the paper [12]. For the proposed method, orthogonal wavelet transform with five levels of decomposition and "sym8" filter were used. Considering both estimation robustness and spatial adaptivity, window size was set to 5×5.

As is evident in Table 2, the proposed algorithm performs almost the best among the eight methods for different types of images at varying noisy levels. It consistently outperforms LAWML and LAWMAP λ_1 method [4], averaged at 1.66 dB and 0.54 dB, respectively. By accurate modeling, the LAW method with bilateral variance estimation (Law-B) [12] and the proposed method show clear advantage for images with rich textures and details (e.g. "Barbara") over the other competitors. The proposed method is more efficient at noisy circumstances though it reports slight lower PSNR (within 0.02 dB) at low noisy levels. PSNR gain over ProbShrink [7] and the non-parametric

Table 2. PSNR comparison with other orthogonal wavelet domain algorithms (in dB).

Image	Methods	Noise Standard Deviation σ_n								
		10	15	20	25	30	40	50	75	100
Lena	LAWML [4]	34.14	31.99	30.44	29.21	28.20	26.56	25.26	22.77	20.93
	LAWMAP [4]	34.32	32.37	31.01	29.99	29.17	27.89	26.94	25.26	24.12
	BiShrink [6]	34.32	32.47	31.16	30.15	29.34	28.09	27.14	25.44	24.29
	ProbShrink [7]	34.08	32.15	30.81	29.8	28.97	27.68	26.74	25.10	24.04
	SURE-Let [9]	34.56	32.67	31.35	30.32	29.53	28.26	27.33	25.69	**24.61**
	NonPara [11]	34.27	32.09	30.51	29.26	28.23	26.59	25.28	22.77	20.97
	Law-B [12]	34.60	32.48	30.91	29.69	28.68	27.03	25.76	23.35	21.52
	Proposed	**34.63**	**32.75**	**31.43**	**30.41**	**29.59**	**28.32**	**27.37**	**25.70**	24.58
Barbara	LAWML [4]	32.54	30.10	28.44	27.17	26.17	24.62	23.44	21.30	19.75
	LAWMAP [4]	32.58	30.20	28.62	27.43	26.52	25.17	24.2	22.62	21.67
	BiShrink [6]	32.14	29.82	28.23	27.06	26.13	24.81	23.88	22.47	21.65
	ProbShrink [7]	32.34	29.88	28.24	27.02	26.08	24.68	23.69	22.28	21.45
	SURE-Let [9]	32.19	29.67	27.98	26.76	25.84	24.54	23.72	22.52	21.80
	NonPara [11]	32.59	30.24	28.59	27.32	26.3	24.72	23.49	21.30	19.75
	Law-B [12]	**32.95**	**30.61**	28.95	27.70	26.69	25.15	23.95	21.79	20.23
	Proposed	32.94	30.59	**29.01**	**27.84**	**26.91**	**25.52**	**24.52**	**22.84**	**21.91**
Coco	LAWML [4]	35.31	32.92	31.23	29.91	28.79	27.05	25.63	23.05	21.10
	LAWMAP [4]	35.61	33.46	31.96	30.85	29.96	28.57	27.58	25.75	24.49
	BiShrink [6]	36.11	34.09	32.66	31.58	30.71	29.29	28.29	26.47	25.21
	ProbShrink [7]	35.35	33.08	31.47	30.39	29.48	28.20	27.09	25.35	24.20
	SURE-Let [9]	36.22	34.13	32.63	31.55	30.58	29.17	28.07	26.18	24.80
	NonPara [11]	35.33	32.89	31.18	29.75	28.71	26.92	25.44	22.82	20.95
	Law-B [12]	35.87	33.50	31.83	30.54	29.56	27.70	26.17	23.65	21.64
	Proposed	**36.31**	**34.21**	**32.81**	**31.75**	**30.84**	**29.55**	**28.46**	**26.70**	**25.41**

(NonPara) [11] methods are prominent. BiShrink [6] mainly models the parent-child interscale relationship. It performs relatively well for smooth regions (e.g. "Lena", "Coco"), but is inefficient for characterization of high-frequency detailed textures. The proposed algorithm produces significant better PSNR and visual quality in texture regions as compared with the SURE-Let [9] method, which only takes care of the inter-scale relationships. Note that all of these benefit is due to the accurate modeling of local variance field, which maintains the LAW model's simplicity and adds no extra computation cost to the algorithm.

For subjective visual quality comparison, Fig. 4 shows some of the denoised images by different orthogonal wavelet domain denoising methods. As shown in Fig. 4 (i), the texture details in Lena's hat are better preserved and less influenced by artifacts (see parts in the left highlighted ellipses) due to the accurate prior modeling. In Fig. 4 (ii), we can observe that denoised image (h) by the proposed method also demonstrate superiority in denoising smooth regions. Furthermore, from the comparison in Fig. 4 (iii) we can learn that the visual advantage over other methods is more obvious at high

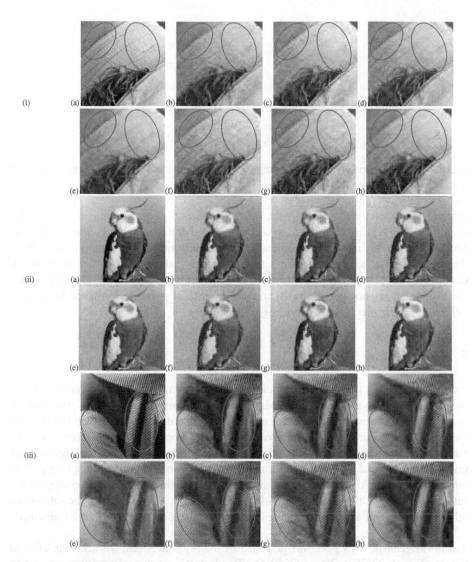

Fig. 4. Subjective quality comparison of various orthogonal wavelet denoising methods for: (i) "Lena" ($\sigma_n = 25$), (ii) "Coco" ($\sigma_n = 25$) and (iii) "Barbara" ($\sigma_n = 50$), where (a) Clean image; (b) LAWMAP [4]; (c) BiShrink [6]; (d) ProbShrink [7]; (e) SURE-Let [9]; (f) NonPara [11]; (g) LawB [12] and (h) the proposed method.

noisy levels. The proposed method recovers the texture details in Barbara's trousers and shawl best as compared with other methods, with least ringing artifacts.

To further test the robustness of the proposed method on different types of images, we have conducted denoising over the image set shown in Fig. 2 at nine different noise levels as in Table 2. Figure 5 depicts the averaged PSNR with respect to the computation time. It is shown that the computational complexities of the proposed λ_2 and the

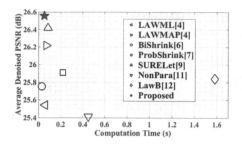

Fig. 5. Denoising performance and computation time comparison for different orthogonal wavelet domain denoising algorithms.

original λ_1 schemes are essentially the same, however, λ_2 improves the average PSNR of about 0.34 dB as compared to λ_1. PSNR advantages over BiShrink [6] and ProbShrink [7] methods are prominent. By accurately modeling of the variance field, the proposed method performs better than the state-of-the-art SURE-Let by 0.14 dB. Computational complexity superiority over NonPara [11] and LawB [12] algorithms are prominent. Hence our proposed method maintains good performance in computational efficiency as well as denoising effectiveness.

5 Conclusions

Accurate modeling of wavelet coefficients is of vital importance in image denoising. Our statistical analysis on the LAWMAP algorithm shows that the prior estimation is not proper in the sense that it cannot model the global distribution of wavelet domain local variance accurately. This motivates us to develop a maximum likelihood estimator for λ. Experimental results show that our proposed method can significantly improve the original algorithm while maintain a low computational complexity. This makes the LAWMAP method a competitive choice for wavelet based denoising. In the future, we will use the parent and child information [9, 10] to capture the inter-scale relationship so that a more accurate estimation of the local variance can be obtained.

Acknowledgement. This work was supported by the National Nature Science Foundation of China (No. 61203269, No. 61305015), Postdoctoral Science Foundation of China (No. 2015M580591). Yun-Xia LIU acknowledges the research scholarships provided by the Centre for Multimedia Signal Processing, Department of Electronic and Information Engineering and the Hong Kong Polytechnic University where the work is partially done.

References

1. Liu, Y., Peng, Y., Qu, H., et al.: Energy-based adaptive orthogonal FRIT and its application in image denoising. Sci. China Inf. Sci. **50**(2), 212–226 (2007)
2. Shen, X., Wang, K., Guo, Q.: Local thresholding with adaptive window shrinkage in the contourlet domain for image denoising. Sci. China Inf. Sci. **56**(9), 61–69 (2013)

3. Liu, Y., Law, N., Siu, W.: Patch based image denoising using the finite ridgelet transform for less artifacts. J. Vis. Commun. Image Represent. **25**(5), 1006–1017 (2014)
4. Mihcak, M., Kozintsev, I., Ramchandran, K., et al.: Low complexity image denoising based on statistical modeling of wavelet coefficients. IEEE Signal Process. Lett. **6**(12), 300–303 (1999)
5. Chang, S., Yu, B., Vetterli, B.: Spatially adaptive wavelet thresholding with context modeling for image denoising. IEEE Trans. Image Process. **9**(9), 1522–1531 (2000)
6. Sendur, L., Selesnick, I.: Bivariate shrinkage functions for wavelet-based denoising exploiting interscale dependency. IEEE Trans. Signal Process. **50**(11), 2744–2756 (2002)
7. Pizurica, A., Philips, W.: Estimating the probability of the presence of a signal of interest in multiresolution single and multiband image denoising. IEEE Trans. Image Process. **15**(3), 645–665 (2006)
8. Tan, S., Jiao, L.: Multivariate statistical models for image denoising in the wavelet domain. Int. J. Comput. Vision **75**(2), 209–230 (2007)
9. Luisier, F., Blu, T., Unser, M.: A new SURE approach to image denoising: interscale orthonormal wavelet thresholding. IEEE Trans. Image Process. **16**(3), 593–606 (2007)
10. Boubchir L., Naitali A., Petit E.: Multivariate statistical modeling of images in sparse multiscale transforms domain. In: 17th IEEE International Conference on Image Processing, Hong Kong, pp. 1877–1880 (2010)
11. Tian, J., Chen, L., Ma, L.: A wavelet-domain non-parametric statistical approach for image denoising. IEICE Electron. Express **7**(18), 1409–1415 (2010)
12. Shi, J., Liu, Z., Tian, J.: Bilateral signal variance estimation for wavelet-domain image denoising. Sci. China Inf. Sci. **56**(6), 83–88 (2013)
13. Liu, J., Moulin, P.: Information-theoretic analysis of interscale and intrascale dependencies between image wavelet coefficients. IEEE Trans. Image Process. **10**(11), 1647–1658 (2001)
14. Chen, M., Fridrich, J., Goljan, M., et al.: Determining Image Origin and Integrity Using Sensor Noise. IEEE Trans. Inf. Forensics Secur. **3**(1), 74–90 (2008)
15. Pi, M., Tong, C., Choy, S., et al.: A fast and effective model for wavelet subband histograms and its application in texture image retrieval. IEEE Trans. Image Process. **15**(10), 3078–3088 (2006)

A Data Fusion-Based Framework for Image Segmentation Evaluation

Macmillan Simfukwe, Bo Peng$^{(\boxtimes)}$, and Tianrui Li

School of Information Science and Technology,
Southwest Jiaotong University, Chengdu 610031, China
Macmillansimfukwe@yahoo.co.uk,
bpeng@home.swjtu.edu.cn, trli@swjtu.edu.cn

Abstract. Image segmentation is an important task in image processing. Nevertheless, there is still no generally accepted quality measure for evaluating the performance of various segmentation algorithms or even different parameterizations of the same algorithm. In this paper, we propose a data fusion-based binary classification framework for image segmentation evaluation. We train and test this framework using a dataset consisting of a variety of image types, their segmentations and respective ground truths, as well as the class labels assigned to each segmentation by human judges. Experimental results show accuracy of up to 80 %.

Keywords: Image segmentation · Segmentation evaluation · Data fusion · Evaluation framework · Classification

1 Introduction

Image segmentation is the domain-dependent partition of an image into homogenous and meaningful constituent parts called segments. It is a prerequisite stage for object detection and other subsequent operations in a computer vision system, thus the quality of the image segmentation results has a direct impact on such a system. Despite tremendous attention being rendered to the development of image segmentation algorithms, less attention has been given to the development of evaluation schemes for their performance. As a result, there is still no satisfactory measure for comparing the performance of various segmentation algorithms, or even different parameterizations of the same algorithm. The ability to assess the quality of segmentation results is essential for developing and improving segmentation algorithms. [1] summarizes the importance of application-independent comparison techniques of image segmentations as follows: (1) need for autonomous selection from among possible segmentations yielded by the same segmentation algorithm; (2) need to place a new or existing segmentation algorithm on a solid experimental and scientific ground; and (3) need to monitor segmentations results on the fly, so that segmentation performance can be guaranteed and consistency maintained.

Image segmentation is a relatively ill-posed problem, and this makes its evaluation difficult. The existing segmentation evaluation methods can be divided into subjective and objective methods. Under subjective methods a human judge uses his intuition to

© Springer International Publishing Switzerland 2016
D.-S. Huang and K.-H. Jo (Eds.): ICIC 2016, Part II, LNCS 9772, pp. 534–545, 2016.
DOI: 10.1007/978-3-319-42294-7_48

assess the performance of the segmentation algorithm. This approach is time consuming and often leads to inconsistent results due to the variations in the visual capabilities of humans. Objective methods can be divided into analytical, empirical goodness and empirical discrepancy methods [2]. Analytical methods do not assess the segmentation results. They directly assess the actual segmentation algorithms from various perspectives such as the algorithm's principle, complexity, efficiency and execution time. Nevertheless, these properties are usually independent of the segmentation results, making it difficult to distinguish between the various segmentation algorithms, in terms of the quality of their results. Empirical goodness and empirical discrepancy methods analyze the segmentation results themselves. The empirical goodness methods evaluate the segmentation results based on some "goodness" parameters which are relevant to the visual properties extracted from the original image and the segmented image. They do so without having any prior knowledge and therefore, they are also referred to as unsupervised evaluation methods. Empirical discrepancy methods, on the other, use a reference image called the ground truth (gold standard) to evaluate the segmentation results. Thus they are also referred to as supervised evaluation methods. During evaluation, the segmentation is compared with the ground truth to assess the level of discrepancy between the two.

In this paper, we propose a data fusion-based framework for image segmentation evaluation. The proposed framework treats the image segmentation evaluation problem as a binary classification problem, by classifying a given segmentation as being either good or bad. We use the evaluation measures as features for the classification task. We also build a new image segmentation evaluation dataset, which contains a variety of images with their respective ground truths and labels assigned to them by human judges (good or bad segmentation). To the best of our knowledge, this is the first work that strives to train a classifier to evaluate a segmentation as being either good or bad.

The remainder of the paper is organized as follows: Sect. 2 presents a background to image segmentation evaluation, Sect. 3 describes our proposed framework and Sect. 4 presents our experimental design and results. In Sect. 5 we conclude and present some possible future work.

2 Related Work

In the past two decades, a number of objective segmentation evaluation methods have been proposed [1, 2]. However, they all have different functional underlying principals and thus make different assumptions about segmentations.

Data fusion has been widely applied in robotics and military fields, with remarkable success. Data fusion is the combination of information from multiple sources to improve application performance. Similarly, we can frame the image segmentation evaluation problem as an evidence fusion problem, by combining the results of different segmentation evaluation measures. Some data fusion-inspired applications in image segmentation evaluation have been reported recently. [3, 4] reported a co-evaluation framework for improving segmentation evaluation by combining various unsupervised evaluation methods. [5] reported an ensemble combination for solving the parameter selection problem in image segmentation. In [6], both supervised and unsupervised evaluation methods are utilized for segmentation evaluation. In all these studies, the evaluation measures are used as features for the classification task, and the same approach is employed in this paper. [7] uses a fusion of unsupervised evaluation measures to address the parameter selection problem in interactive segmentation algorithms.

Our proposed framework evaluates a given segmentation as being either good or bad, while in [3, 4, 6], two given segmentations are compared to each other so as to evaluate which one is better. However, the two given segmentations could both happen to be bad ones, and thus unsuitable for use in the subsequent stages of the machine vision system.

2.1   Evaluation Measures

In this paper, we used the following supervised evaluation measures:

Boundary Displacement Error (BDE) [8]. This is a boundary-based measure which evaluates segmentation quality by calculating the average displacement error of boundary pixels between a given segmentation S and its ground truth G. Given S and G, let B_S be the set of boundary pixels in S and B_G be the set of boundary pixels in G. A distance distribution signature $D_{B_G}^{B_S}$ is the discrepancy between S and G, measured in distance. Define the distance from an arbitrary pixel x in B_G to B_S as the absolute distance from x to all pixels in B_S, $d(x, B_S) = \min\{d_E(x, y)\}, \forall y \in B_S$, where d_E denotes the Euclidean distance between pixels x and y. The discrepancy between B_G and B_S is characterized by the shape of the signature, which is commonly measured by its mean and standard deviation. $D_{B_G}^{B_S}$ having a near-zero mean and a small standard deviation indicates a good quality segmentation.

Probability Rand Index (PRI) [9]. Given a segmentation S and a set of ground truths $G = \{G_1, G_2, \ldots \ldots, G_n\}$, for any pair of pixels (x_i, x_j), we can define labels l_i^S and l_j^S to be the same for the segmentation, i.e. $I(l_i^S = l_j^S)$ if the labels l_i^G and l_j^G for the ground truths are the same, i.e. $I(l_i^G = l_j^G)$. We count the number of times for which this is true and also take note of the number of times for which this is not true, i.e. when $I(l_i^S \neq l_j^S)$ and $I(l_i^G = l_j^G)$ or vice versa. The numbers follow a Bernoulli distribution and give a random variable with the expected value denoted by p_{ij}. The *PRI* is defined by Eq. 1.

$$PRI(S, G) = \frac{1}{\binom{N}{2}} \sum_{i,j} \left[I\left(l_i^S = l_j^S\right) p_{ij} + I(l_i^S \neq l_j^S)(1 - p_{ij}) \right] \tag{1}$$

Where N is the number of pixels and p_{ij} is the ground truth probability that the labels for (x_i, x_j) are the same. PRI takes values in the range [0, 1], where a score value of 1 indicates that the segmentation and the ground truth are identical.

Variation of Information (VOI) [10]. It defines the discrepancy between a segmentation S and its ground truth G in terms of the information difference between them. Equation 2 defines the VOI.

$$VOI(S,G) = H(S) + H(G) - 2I(S,G) \tag{2}$$

Where H represents the respective entropies of S and G, and I represents the mutual information between S and G. If S and G are identical, the value of VOI is equal to zero.

Global Consistency Error (GCE) [11]. Let $M(S,p_i)$ be the set of pixels in segmentation S that contains the pixel p_i and $M(G,p_i)$ be the set of pixels in ground truth G that contains pixel p_i. The local refinement error is defined by Eq. 3.

$$E(S,G,p_i) = \frac{|M(S,p_i) \backslash M(G,p_i)|}{|M(S,p_i)|} \tag{3}$$

The error is equal to zero when S is a refinement of G at pixel p_i. We combine the local refinement errors into GCE as defined by Eq. 4.

$$GCE(S,G) = \frac{1}{n} \min \left\{ \sum_i E(S,G,p_i), \sum_i E(G,S,p_i) \right\} \tag{4}$$

The Hausdorff Distance [12]. Given a segmentation S and its ground truth G, the Hausdorff distance between S and G is given by Eq. 5.

$$
\begin{aligned}
H(S,G) &= \max(h(S,G), h(G,S) \\
&= \max\left(\max_{x \in S} \min_{y \in G} ||x - y||, \max_{x \in G} \min_{y \in S} ||x - y|| \right)
\end{aligned}
\tag{5}
$$

Let G_b denote a set of ground truth objects in image b, S_b denote a set of segmented objects in image b, $S_i \in S_b$ denote the ith segmented object in image b, $G_i \in G_b$ denote a ground truth object that maximally overlaps with S_i in image b. If there is no ground truth object overlapping with S_i, G_i is defined as the ground truth object $G \in G_b$ that has the minimum Hausdorff distance from S_i, $\tilde{G}_i \in G_b$ denotes the ith ground truth object in image b, $\tilde{S}_i \in S_b$ denotes a segmented object that maximally overlaps $\tilde{G}_i$ in image b. If there is no segmented object overlapping with $\tilde{G}_i$, $\tilde{S}_i$ is defined as the segmented object $\tilde{S} \in S_b$ that has the minimum Hausdorff distance from G_i, $G = U_b G_b$ a set of all ground truth objects, $S = U_b S_b$ is a set of all segmented objects, n_S denotes the total number of segmented objects in S, n_G denotes the total number of ground truth objects in G. The measure of shape similarity between all segmented objects in S and all the ground truth objects in G is defined by the object-level Hausdorff Distance, given by Eq. 6.

$$H_{object}(S,G) = \frac{1}{2}\left[\sum_{i=1}^{n_S} w_i H(G_i, S_i) + \sum_{i=1}^{n_S} \tilde{w}_i H(\tilde{G}_i, \tilde{S}_i)\right] \qquad (6)$$

Where $w_i = \frac{|S_i|}{\sum_{j=1}^{N_S}|S_j|}$ and $\tilde{w}_i = \frac{|\tilde{G}_i|}{\sum_{j=1}^{n_G}|\tilde{G}_j|}$. If the segmentation and the ground truth are identical, the object-level Hausdorff distance is equal to zero.

The Dice Index [13]. Given a segmentation S and its ground truth G, the Dice Index is defined by Eq. 7.

$$DI(G,S) = \frac{2|G \cap S|}{|G| + |S|} \qquad (7)$$

Let G_b denote a set of ground truth objects in image b, S_b denote a set of segmented objects in image b, $S_i \in S_b$ denote the ith segmented object in image b, $G_i \in G_b$ denote a ground truth object that maximally overlaps with S_i in image b, $\tilde{G}_i \in G_b$ denote the ith ground truth object in image b, $\tilde{S}_i \in S_b$ denote a segmented object that maximally overlaps with $\tilde{G}_i$ in image b, $G = \cup_b G_b$ is a set of all ground truth objects, $S = \cup_b S_b$ is a set of all segmented objects, n_S denotes the total number of segmented objects in S, n_G denotes the total number of ground truth objects in G. We define the object-level Dice Index as given in Eq. 8.

$$DI_{object}(G,S) = \frac{1}{s}\left[\sum_{i=1}^{n_S} w_i DI(G_i, S_i) + \sum_{i=1}^{n_G} \tilde{w}_i DI(\tilde{G}_i, \tilde{S}_i)\right] \qquad (8)$$

Where $w_i = \frac{|S_i|}{\sum_{j=1}^{N_S}|S_j|}$ and $\tilde{w}_i = \frac{|\tilde{G}_i|}{\sum_{j=1}^{n_G}|\tilde{G}_j|}$. The Dice index values lie in the range $[0, 1]$ and it is equal to 1 when the segmentation is identical to the ground truth.

The F1score [14]. It is defined by Eq. 9.

$$F1score = \frac{2 \times Precision \times Recall}{Precision + Recall} \qquad (9)$$

Where $Precision = TP/(TP + FP)$ and $Recall = TP/(TP + FN)$. The F1score is equal to one if the segmentation and the ground truth are identical.

3 Data Fusion-Based Evaluation Framework

Our data fusion-based binary classification segmentation evaluation framework is presented in Fig. 1. It comprises two phases, namely the training and testing phases.

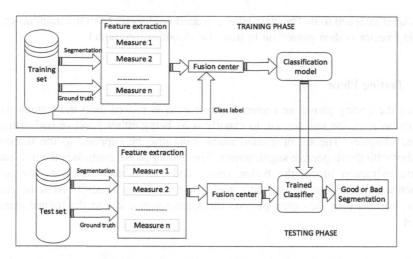

Fig. 1. Data fusion-based segmentation evaluation framework

3.1 Training Phase

The training phase consists of the following blocks: the training data set, the feature extraction block, the fusion center and the resulting classification model. The training dataset contains the segmentations and their respective labels. It also contains the respective ground truths. The training set samples are passed on to the feature extraction block. The feature extraction block consists of the evaluation measures of choice. The measures are used to calculate the scores for the segmentations and these scores are in turn used as feature values for training the classifier.

Feature Extraction Block. The feature extraction block functions as follows: Given a segmentation S, a set of its ground truths $G = \{G_1, G_2, \ldots\ldots, G_n\}$ and a set of evaluation measures $Q = \{f_1, f_2, \ldots\ldots, f_m\}$, we compute the score y for S using Eq. 10.

$$y_i = f_i(S, G_K) \tag{10}$$

Where $f_i \in Q$ and $G_k \in G$. Since there is more than one ground truth for each segmentation, we compute the average score across all the ground truths, using Eq. 11.

$$\bar{y}_i = \frac{1}{n} \sum_{k=1}^{n} f_i(S, G_k) \tag{11}$$

Fusion Center. The scores are forwarded to the fusion center from the feature extraction block. Here the scores are treated as individual feature values. The fusion strategy employed is feature fusion, by simply combining the scores into respective feature sets, $Z_t = [\bar{y}_1, \bar{y}_2, \ldots\ldots, \bar{y}_m]$. Further, We obtain the training vectors, $X_t = \{Z_t, C_t\}$, where t refers to the tth sample (segmentation) and $C_t \in \{C_1, C_2\}$ is the class label assigned to it. The class label is passed on directly from the training set. The training vector is then passed on to train the classification model.

3.2 Testing Phase

During the testing phase, an unseen segmentation is passed onto the framework for evaluation from the test set, i.e. to classify it as being either good or bad, using the trained classifier. The set of ground truths must also be supplied to the framework together with the respective segmentation. The testing phase comprises the test dataset, feature extraction block, the fusion center and the trained classifier. The feature extraction block and the fusion center work in the same way as described in the training phase, but this time they only pass on the feature vector Z_t to the trained classifier model.

3.3 Classification Models

In this paper, we have used five classification algorithms, namely; Decision Tree (DT), Support Vector Machine (SVM), Naïve Bayes (NB), K-Nearest neighbor (KNN) and AdaBoost. Detailed descriptions about these algorithms can be found in [15, 16].

Decision Tree (DT). This is a classification model with the form of a tree structure. Each node is either a leaf node, which indicates the value of the target attribute (class) of examples, or a decision node. A decision node specifies some test to be carried out on a single attribute value, with one branch and a sub-tree for each possible outcome of the test. Their advantage over other supervised learning methods lies in the fact that they represent rules, which are easy to interpret and understand and can also be readily used in a database.

Support Vector Machine (SVM). This is based on structural risk minimization. It finds a separator that maximizes the separation between classes and has demonstrated good performance in the various areas such as handwritten digit recognition, face detection, etc. SVM has excellent classification ability on small datasets and has high robustness.

Naïve Bayesian (NB). It operates on the assumption that the predictors, in our case the evaluation measures, are conditionally independent of each other. For our case two posterior probabilities are computed: the first one is the probability that a given segmentation S is good, given the attribute values, and the second one is the probability that S is bad, given the attribute values. Segmentation S is assigned to the class with the higher posterior probability.

K-Nearest Neighbor (KNN). It assigns the input to the class with most examples among the k neighbors of the input. All the neighbors have equal votes, and the class having the maximum number of voters among the k neighbors is chosen. In the case of ties, weighted voting is taken, and in order to minimize ties k is generally chosen to be an odd number. The neighbors are identified using a distance metric.

AdaBoost. It belongs to a class of supervised learning approaches, collectively known as ensemble learning techniques, which is basically about combining various learners to improve performance. In our study, we have used decision trees as our base learners.

4 Experimental Design and Results

4.1 Dataset Creation

The dataset utilized in our work is derived from the one used in [17]. It contains a variety of images with their corresponding ground truths and in total comprises 1000 different image/segmentation pairs with each pair having between four to ten ground truths. In our study we asked humans to judge a segmentation as being either good or bad, based on how close its segments are to real-world objects, as depicted in the original image. In other words, humans are to judge a given segmentation, depending on how much it resembles the original image. Figure 2 presents samples from the dataset.

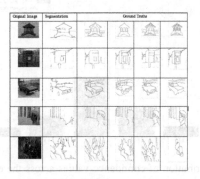

Fig. 2. Samples from the dataset

For each segmentation, eleven human judges were asked to classify it and confidence scores were computed and stored, as follows: given two classes, C_1 and C_2, where C_1 is the class for good segmentations and C_2 is the class for bad segmentations, a human judge can assign a given segmentation S to either C_1 or C_2. Let $P(S \in C_1)$ be the confidence value signifying the likelihood of segmentation S belonging to C_1. The definition of $P(S \in C_1)$ is given by Eq. 12.

$$P(S \in C_1) = \frac{m}{n} \tag{12}$$

Where m is the number of judges' responses indicating that S is a good segmentation and n is the total number of responses, in this case $n = 11$. The confidence value signifying the likelihood of S belonging to C_2 is defined by Eq. 13.

$$P(S \in C_2) = 1 - P(S \in C_1) \tag{13}$$

In order to determine to which class to assign a given segmentation S, the following decision rule is used:

$$S \in \begin{cases} C_1 & if\ P(S \in C_1) > 0.5 \\ C_2 & otherwise \end{cases} \tag{14}$$

Using the procedure just outlined above, we processed the segmentations into a dataset consisting of 433 "good" and 567 "bad" segmentations. Figure 3 summarizes the distribution of confidence values across our dataset. The confidence values show the degree of objectivity regarding the decisions of the human judges. The higher the confidence value, the higher the degree of objectivity. However, we can see that only 50 segmentations have a confidence value equal to one, attesting to the subjective nature of human evaluation.

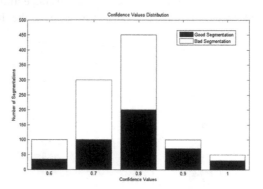

Fig. 3. Distribution of confidence values across the entire dataset

4.2   Results and Analysis

For our experiments, we randomly divided our dataset into 700 training and 300 testing samples. We used the classification models mentioned in Sect. 3.3. We trained the classifiers using individual measures as features, as well as using our proposed framework, i.e. fusion of the measures. The results of testing 300 unseen samples on the trained classifiers are shown in Table 1.

Table 1. Accuracy rates; our framework vs. individual measures

	BDE	PRI	VOI	GCE	F1score	Dice Index	Hausdorff Distance	Proposed Framework
NB	61 %	62 %	38 %	62 %	60 %	55 %	52 %	63 %
SVM	62 %	41 %	38 %	62 %	60 %	55 %	51 %	65 %
KNN	43 %	62 %	49 %	38 %	54 %	57 %	54 %	65 %
DT	–	–	–	–	–	–	–	61 %
AdaBoost	–	–	–	–	–	–	–	80 %

The accuracy rate was calculated by comparing the class label assigned to a given segmentation by the human judges with the class label assigned to such a segmentation by the classifier. The accuracy rate is defined by Eq. 15.

$$Accuracyrate = \frac{D}{W} \times 100\% \tag{15}$$

Where D is the number of times that the class label assigned by the classifier is the same as that assigned by the human judge, for a given segmentation, and W is the total number of test samples, in our case $W = 300$.

We can see from Table 1 that our framework has higher accuracy than all the individual measures, thus confirming the fact that data fusion improves performance. The variations, in terms of the accuracy rate, across the various classifier models, for the same feature (evaluation measure), is due to the fact that different classifier models employ different underlying functional principles and assumptions on the data. Ada-Boost, which is based on the concept of data fusion, has the highest accuracy of 80 %. There are no results from single features for DT and AdaBoost because DT and AdaBoost models need more than one feature as input. Since, this is the first work of this kind, to the best of our knowledge, there are no other results from other studies with which to compare the results of our framework.

The accuracy rate is higher for samples with higher confidence values as shown in Fig. 4. This is an indication that the classification of segmentations by our framework is in agreement with that of the majority of human judges. In other words, the accuracy of our framework is consistent with that of the human judges when their degree of objectivity is high. AdaBoost has the highest accuracy rate in comparison with other classifiers, for all confidence values. However, the difference in terms of time performance between the individual measures and the proposed framework is negligible.

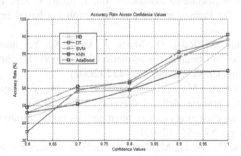

Fig. 4. Accuracy rate with respect to confidence values (Color figure online)

5   Conclusion

In this paper, we have proposed a data fusion-based binary classification framework for image segmentation evaluation, which classifies segmentations as being bad or good. The framework utilizes segmentation evaluation measures as features for the classification task. The accuracy of our framework is consistent with that of humans, provided their degree of objectivity. Experimental results have demonstrated the superiority of our framework over individual evaluation measures, with an accuracy rate of up to 80 %. Different evaluation measures and classification models can be used in our framework directly. In future, we wish to extend the framework to include unsupervised evaluation.

Acknowledgements. This work was supported by the National Science Foundation of China under Grant No. 61202190, and the Science and Technology Planning Project of Sichuan Province under Grant No. 2014SZ0207.

References

1. Zhang, H., Fritts, J.E., Goldman, S.A.: Image segmentation evaluation: a survey of unsupervised methods. Comput. Vis. Image Underst. **110**, 260–280 (2008)
2. Zhang, Y.J.: A survey on evaluation methods for image segmentation. Pattern Recogn. **29**(8), 1335–1346 (1996)
3. Zhang, H., Fritts, J.E., Goldman, S.A.: A co-evaluation framework for improving segmentation evaluation. In: Proceedings of SPIE 5809, Signal Processing, Sensor Fusion, and Target Recognition XIV, pp. 420–430 (2005)
4. Zhang, H., Cholleti, S., Goldman, S.A.: Meta-evaluation of image segmentation using machine learning. In: Proceedings of the 2006 IEEE Computer Society Conference on Computer Vision and Pattern Recognition, pp. 1138–1145 (2006)
5. Wattuya, P., Jiang, X.: Ensemble combination for solving the parameter selection problem in image segmentation. In: da Vitoria Lobo, N., et al. (eds.) SPR&SPR 2008. LNCS, vol. 5342, pp. 392–402. Springer, Heidelberg (2008)
6. Lin, J., Peng, B., Li, T.R., Chen, Q.: A learning-based framework for supervised and unsupervised image segmentation evaluation. Int. J. Image Graph. **14**(3) (2014)
7. Peng, B., Veksler, O.: Parameter selection for graph cut-based image segmentation. In: BMVC (2008)
8. Freixenet, J., Munoz, X., Raba, D., Marti, J., Cufi, X.: Yet another survey on image segmentation: region and boundary information integration. In: Proceedings of European Conference on Computer Vision, pp. 408–422 (2002)
9. Pantofaru, C., Hebert, M.: A comparison of image segmentation algorithms. CMU-RI-TR-05-40, Carnegie Mellon University, 1–32 (2005)
10. Meila, M.: Comparing clusterings – an axiomatic view. In: 22nd International Conference on Machine Learning, pp. 577–584 (2005)
11. Martin, D., Fowlkes, C., Tal, D., Malik, J.: A database of human segmented natural images and its applications to evaluating segmentation algorithms and measuring ecological statistics. In: Proceedings of the Eighth IEEE Conference on Computer Vision, pp. 416–423 (2001)
12. Huttenlocher, D.P., Klanderman, G.A., Rucklidge, W.J.: Comparing images using the Hausdorff distance. IEEE Trans. Pattern Anal. Mach. Intell. **15**(9), 850–863 (1993)
13. Dietenbeck, T., Alessandrini, M., Friboulet, D., Bernard, O.: CREASEG: a free software for the evaluation of image segmentations algorithms based on level-sets. In: Proceedings of the 17th IEEE International Conference on Image Processing, pp. 665–668 (2010)

14. http://www2.warwick.ac.uk/fac/sci/dcs/research/combi/research/bic/glascontest/evaluation/. 15 Oct 2015
15. Alpaydın, E.: Introduction to Machine Learning, 2nd edn. MIT Press, Boston (2010)
16. Theodoridis, S., Koutroumbas, K.: Pattern Recognition, 4th edn. Academic Press, New York (2009)
17. Lin, J., Peng, B., Li, T.R, Chen, Q.: A learning-based framework for image segmentation evaluation. In: 5th International Conference on Intelligent Networking and Collaborative Systems, pp. 691–696 (2013)

Error Based Nyström Spectral Clustering
Image Segmentation

Liu Zhongmin[1], Li Bohao[1(✉)], Li Zhanming[1], and Hu Wenjin[2]

[1] College of Electrical and Information Engineering,
Lanzhou University of Technology, Lanzhou, China
liuzhmx@163.com, libt08@sina.com
[2] School of Math, Northwest University for Nationalities,
Lanzhou 730030, China

Abstract. Spectral clustering algorithm has been a research hotspot in the field of image processing, recent years. Spectral clustering based on the similarity of data while structure of similarity matrix is complex. The calculation of spectral clustering can be very time-consuming, especially in the process of Eigen-decomposition for Laplacian matrix. Nyström extension method could obtain the approximation solution of eigenvectors by using a small amount of sample information, reduce the computational complexity of spectral clustering effectively. Based on the features of image and the error analysis of Nyström a new sampling method is presented. Using Uniform Sampling generates a set of cluster centers at first; then, minimize the error between data and centers by iteration; finally, typical experiment results and analysis are given.

Keywords: Nyström · Spectral clustering · Image segmentation · k-means

1 Introduction

Image segmentation is the technology and process that divide image into characteristic areas and put forward the interested target [1]. It is a classical and crucial problem in the fields of computer vision and image understanding, the key step to realize the research from general image processing into image analysis [2]. Spectral clustering is a novel clustering method which based on the spectral graph theory. Spectral clustering has main advantages of easy implementation and can be used to cluster data with arbitrary shape [3]. Recently spectral clustering algorithm has undergone a rapid development, has wide application in image segmentation and its related fields.

Spectral clustering using the image similarity matrix in image segmentation, divides the image into regions, which makes the pixels in the same regions with high similarity and the different regions with low similarity. Spectral clustering has a satisfactory segmentation effect, however, there are also some obvious disadvantages such as too much time cost and memory usage. The most important steps for spectral clustering are building the similarity matrix and performing Eigen-decomposition on the Laplacian matrix which created based on the similarity matrix [4]. The scale of the similarity matrix in image segmentation is too large, calculation of eigenvalues and eigenvectors make a high computational complexity [5, 6]. For example: to compute an

© Springer International Publishing Switzerland 2016
D.-S. Huang and K.-H. Jo (Eds.): ICIC 2016, Part II, LNCS 9772, pp. 546–556, 2016.
DOI: 10.1007/978-3-319-42294-7_49

image with n pixels and d characteristic dimension, the computational complexity of time about the similarity matrix is $O(d^2 n^2)$, the computational complexity of space is $O(n^2)$, the computational complexity of time about eigenvalues and eigenvectors decomposition as high as $O(n^3)$. The computational complexity of spectral clustering has seriously restricted on its application in image segmentation.

Scholars have made a series of researches to take improvement on spectral clustering, in which the Nyström expansion method is an important application and research method. Nyström extension method could obtain the approximation solution of similarity matrix and eigenvectors by using a small amount of sample information, reduce the computational complexity of spectral clustering effectively. The paper [7] applied the Nyström extension method in spectral clustering at first, and have a good image segmentation effect. In paper [8, 9], the approximation error of Nyström is analyzed in detail, a Nyström algorithm based on k-means sampling is proposed. But k-means sampling has two shortcomings in image segmentation.

(1) The weight of each characteristic value can't be estimated accurately.
(2) k-means clustering need a long time once the number of sampling point is much more.

To solve the problems above, a Nyström algorithm based on the sampling of minimum error is proposed in this paper. We introduce the method of spectral clustering and Nyström in Sect. 2. Start with their mathematical structure to analyze the error of similarity matrix by each data. In Sect. 3, we have error analysis in detail with a new sampling method and image segmentation scheme based on it. In Sect. 4, we show experimental results. After experimental, we found the Nyström algorithm in the segmentation effectiveness compared with the traditional spectral clustering algorithm has significantly improved and have conclude in Sect. 5.

2 Spectral Clustering and Its Improved Algorithm

2.1 Spectral Clustering

Spectral clustering algorithm is based on the spectral graph theory [10]. Graph consists by a number of points and lines that connecting each points. Points represent things, lines known as the weights correspond to relationship between two things. The graph is divided into a number of sub graphs, each sub graph has no intersection, the weight of the truncated lines called the loss function. Spectral clustering achieve the division of the graph by minimizing the loss function.

Set $G(V, E)$ representing a graph, $V = \{v_1, v_2, \ldots, v_n\}$ representing a set of points, E represents a set of edges, and w_{ij} representing the weight between v_i and v_j. Then hypothesis $G(V, E)$ is divided into two sub graph G_1 and G_2, set $q = [q_1, q_2, \ldots, q_n]$ is an n-dimensional vector which used to express division scheme

$$q_i = \begin{cases} c_1 & i \in G_1 \\ c_2 & i \in G_2 \end{cases} \tag{1}$$

The loss function is

$$Cut(G_1, G_2) = \sum_{i \in G_1, j \in G_2} w_{ij} = \frac{\sum_{i=1}^{n} \sum_{j=1}^{n} w_{ij}(q_i - q_j)^2}{2(c_1 - c_2)^2}. \tag{2}$$

Decompose the numerator of formula (2), we have

$$
\begin{aligned}
\sum_{i=1}^{n} \sum_{j=1}^{n} w_{ij}(q_i - q_j)^2 &= \sum_{i=1}^{n} \sum_{j=1}^{n} w_{ij}\left(q_i^2 - 2q_i q_j + q_j^2\right) \\
&= -\sum_{i=1}^{n} \sum_{j=1}^{n} 2w_{ij} q_i q_j + \sum_{i=1}^{n} \sum_{j=1}^{n} w_{ij}\left(q_i^2 + q_j^2\right) \\
&= -\sum_{i=1}^{n} \sum_{j=1}^{n} 2w_{ij} q_i q_j + \sum_{i=1}^{n} 2q_i^2 \sum_{j=1}^{n} w_{ij} \\
&= 2q^T(D - W)q.
\end{aligned} \tag{3}
$$

W is the similarity matrix also called the weight matrix, D is a diagonal matrix

$$D_{ii} = \sum_{j=1}^{n} w_{ij}. \tag{4}$$

Set Laplacian matrix as $L = D - W$,

$$\sum_{i=1}^{n} \sum_{j=1}^{n} w_{ij}(q_i - q_j)^2 = 2q^T L q. \tag{5}$$

Finally, the loss function is

$$Cut(G_1, G_2) = \frac{q^T L q}{(c_1 - c_2)^2}. \tag{6}$$

Since c_1 and c_2 are constant, the problem of the loss function minimization into a problem of $q^T L q$ minimization.

2.2 Nyström Method

Spectral clustering achieved the clustering partition by minimizing $q^T L q$, the realization of the minimum value achieved when q corresponding the second smallest eigenvector of L. Nyström expansion method used to obtain the approximate value of the eigenvectors [11].

$$\int p(y)k(x,y)\Phi_i(y)dy = \lambda_i\Phi_i(x) \tag{7}$$

$p(y)$ is the probability density function, $k(x, y)$ is a positive definite kernel function, λ_i are the eigenvalues and Φ_i are the eigenvectors of the integral equation. Set $\{x_1, x_2, \ldots, x_q\}$ are samplings from $p(y)$, we have

$$\frac{1}{q}\sum_{j=1}^{q} k\left(x, x_j\right)\Phi_i\left(x_j\right) \approx \lambda_i\Phi_i(x). \tag{8}$$

Abstract samples $\{x_1, x_2, \ldots, x_q\}$ from $k(x, x_j)$ again, form the standard eigenvalue decomposition

$$K^{(q)}U^{(q)} = U^{(q)}\Lambda^{(q)}. \tag{9}$$

$K_{ij} = k\left(x_i, x_j\right)$, U is an orthogonal matrix, Λ is angular symmetric matrix,

$$\Phi_i\left(x_j\right) \approx \sqrt{q}U_{ij}^{(q)}, \ \lambda_i \approx \lambda_i^{(q)}/q. \tag{10}$$

Nyström method using different subset sizes q are all approximations to λ_i and Φ_i in the integral Eq. (6). As a result, the Nyström method using a small q can also be deemed as approximating the Nyström method using a large q.

For discontinuous kernel density function, suppose data set $X = \{x_1, x_2, \ldots, x_n\}$ similarity matrix is K, sampling points $Z = \{z_1, z_2, \ldots, z_m\}$, similarity matrix of sampling points is W, the eigenvalues and eigenvectors of K is

$$\Lambda_K \approx \frac{n}{m}\Lambda_z, \Phi_K \approx \sqrt{\frac{m}{n}}E\Phi_Z\Lambda_Z^{-1}. \tag{11}$$

$E_{ij} = k(x_i, z_j)$, Λ_z and Φ_z is the eigenvalues and eigenvectors of W,

$$K = \Phi_K\Lambda_K\Phi_K^{-1} \approx EW^{-1}E \tag{12}$$

Nyström expand method improved the efficiency of the spectral clustering algorithm. By sampling from the data sets, we found that the space complexity of similarity matrix from $O(n^2)$ down to $O(mn)$, and the time complexity of Eigen-decomposition for Laplacian matrix from $O(n^3)$ down to $O(m^2n)$. The significant increase in computing speed makes it possible to split the high resolution image by the spectral clustering algorithm.

3 Error Based Sampling Nyström

Although Nyström expansion method improved the efficiency of the spectral clustering algorithm, the accuracy of image segmentation will be affected if the error between similarity matrix $EW^{-1}E$ and the original similarity matrix K can't be weakened efficiently. In this section, the error between $EW^{-1}E$ and K is analyzed, and an appropriate sampling method is selected to minimize the error, a new Nyström spectral clustering algorithm is presented.

3.1 Error Analyzing

In spectral clustering we chose Gauss function as $k(x, y)$ to communicate the similarity matrix. For Intermediate Value Theorem, Gauss function satisfies the following property:

Exist a constant C_X^k depending on $k(x, y)$ and the sample set X make the inequality holds

$$k(a, b) - k(c, d) \leq C_X^k(\|a - c\| + \|b - d\|). \quad \forall a, b, c, d \tag{13}$$

Based on (13), arrived our conclusion as follows.

At first, we can learn from the formula (12) that the error of similarity matrix is

$$\varepsilon = \left\| K - EW^{-1}E' \right\|_F. \tag{14}$$

$\|\cdot\|_F$ denotes the matrix Frobenious norm.

We tracked at a single data, found that the error of single point is

$$\varepsilon_{I_i, I_j} = \left\| K_{I_i, I_j} - E_{I_i, Z} W^{-1} E_{I_j, Z}' \right\|_F \tag{15}$$

$E_{I_i, Z}$ is the similarity matrix of (X_{I_i}, Z).

If the error of single point could be minimized, we can have a similarity matrix much more close to the original similarity matrix. Compared with ε, minimize ε_{I_i, I_j} show more details of data.

We can find the boundary conditions of ε_{I_i, I_j} as follows.

Suppose p q is the nearest sampling point of I_i and I_j respectively, we define the following matrices

$$A_{I_i, I_j} = K_{I_i, I_j} - W_{pq}; \tag{16}$$

$$B_{I_i, Z} = E_{I_i, Z} - W_{pZ}; \tag{17}$$

$$C_{I_j, Z} = E_{I_j, Z} - W_{qZ}. \tag{18}$$

With inequality (13) and Frobenious norms bounded, we have

$$\left\|A_{I_i,I_j}\right\|_F = \left\|K_{I_i,I_j} - W_{pq}\right\|_F = \left\|k(I_i,I_j) - k(p,q)\right\|$$
$$\leq C_X^k(\|I_i - p\| + \|I_j - q\|) = C_X^k(e_{I_i} + e_{I_j}) \tag{19}$$

$$\left\|B_{I_i,z}\right\|_F = \left\|E_{I_i,z} - W_{pz}\right\|_F = \sum_{j=1}^{m} \left\|k(I_i,q_j) - k(p,q_j)\right\|$$
$$\leq mC_X^k\|I_i - p\| = mC_X^k e_{I_i} \tag{20}$$

$$\left\|C_{I_j,z}\right\|_F = \left\|E_{I_j,z} - W_{qz}\right\|_F = \sum_{i=1}^{m} \left\|k(I_j,q_i) - k(p,q_i)\right\|$$
$$\leq mC_X^k\|I_j - q\| = mC_X^k e_{I_j} \tag{21}$$

Using the bounds on $\left\|A_{I_i,I_j}\right\|_F, \left\|B_{I_i,z}\right\|_F, \left\|C_{I_j,z}\right\|_F$ together with the definition in (15), we have the conclusion (22)

$$\varepsilon_{I_i,I_j} = \left\|K_{I_i,I_j} - E_{I_i,z}W^{-1}E_{I_j,z}'\right\|_F$$
$$= \left\|A_{I_i,I_j} + W_{pq} - B_{I_i,z} + W_{pz}W^{-1}C_{I_j,z} + W_{qz}'\right\|_F$$
$$= \left\|A_{I_i,I_j} + W_{pq} - B_{I_i,z}W^{-1}C_{I_j,z}' - W_{pz}W^{-1}C_{I_j,z}' - B_{I_i,z}W^{-1}W_{qz}' - W_{pz}W^{-1}W_{qz}'\right\|_F$$
$$\leq C_X^k e_{I_i} + e_{I_j} + m^2 C_X^{k2} e_{I_i}e_{I_j}\left\|W^{-1}\right\|_F + mC_X^k e_{I_j}\left\|W_{pz}W^{-1}\right\|_F$$
$$+ mC_X^k e_{I_i}\left\|W^{-1}W_{qz}'\right\|_F + \left\|W_{pz}W^{-1}W_{qz}'\right\|_F + \left\|W_{pq}\right\|_F. \tag{22}$$

From the deduction above, we can see that the final error of the Nyström method is determined by the sampling number, sampling similarity matrix, and the error of the data points from the nearest sampling point.

3.2   Error Based Sampling Algorithm

From the error analysis, we can find that, if the number of samples and the similarity function unchanged, the error between similarity matrix $EW^{-1}E$ and the original similarity matrix K can be optimized by minimize the error between pixel and its nearest sampling point, in image segmentation.

As to gray level images, there are three characteristic dimensions, coordinate x and y, gray value $I(x,y)$. Sampling from gray images, the error of coordinate $\varepsilon_{x,y}$, and the error of gray value $\varepsilon_{I(x,y)}$ must be guaranteed to a minimum. Based on it, a novel sampling method was proposed, which uses uniform sampling for overall planning achieving the minimum $\varepsilon_{x,y}$ and applies the iterative search algorithm for further local optimization.

The sampling steps as follows:

Input: image I, the number of pixels n, the number of samples m, the termination of the iteration parameters σ.

Output: sampling center Q.

Step 1: Make Uniform sampling from I according the m, assigned pixels to nearest sampling points;

Step 2: Calculate the gray gradient in the 3×3 neighborhood of the sampling point, select the point with minimum gradient value as the new sampling point;

Step 3: Assign pixels to each sampling points in the neighborhood $2S \times 2S$ according the distance d

$$S = \sqrt{n/m} \tag{23}$$

$$d = d_{xy} + d_{I(x,y)}; \tag{24}$$

Step 4: Take the average value of pixels that in the same sampling points as new sampling points Q, calculating the error ε between the new sampling points and the old

$$\varepsilon = \varepsilon_{x,y} + \varepsilon_{I(x,y)}. \tag{25}$$

If $\varepsilon \leq \sigma$ take step 5, otherwise step 3;

Step 5: Output Q as sampling points.

It can be seen in the algorithm process, the key point of the algorithm is making full use of the distance between the pixels and the sampling centers in the selection of the data points, so that the error is minimized. According to the k-means clustering algorithm, the Minimal error of coordinate can be obtained by step 1 uniform sampling. Step 2 can prevent the sampling points from the noises and edges. Step 3 and step 4 using iterative algorithm to search for the right error of Gray value, calculated the gray value for each sample center, and change the ascription of pixels from the sampling centers in a certain range, updates the coordinates of sampling centers, minimize the error of gray value.

Improved spectral clustering algorithm steps as follows:

Step 1: Carry out samples according to the algorithm of this paper, constructing sampling similarity matrix W, Similarity matrix between sampling points and pixels E by Gauss's function;

Step 2: Estimate the eigenvalues and eigenvectors by Nyström method, chose eigenvectors corresponding to the first few smallest eigenvalues use $V \in R^{n \times k}$ to represent;

Step 3: Make the matrix V normalized, take Y

$$Y_{ij} = \frac{V_{ij}}{\sqrt{\sum_j V_{ij}}}; \tag{26}$$

Step 4: Each line of the matrix Y is treated as a samples, using k-means algorithms to clustering;

Step 5: Determine the category of the original pixel, the image segmentation complete.

4   Experimental Results and Effectiveness Analysis

In this section, we will make experimental analysis and verify the effectiveness of the algorithm. The proposed algorithm in this paper is applied to gray image segmentation. Artificial image segmentation and Nyström spectral clustering image segmentation by randomly sampling are presented used to test our algorithm. The environment of experimental is Windows10 operating system, i3 Intel processor, 4 GB memory, the experimental platform is MATLAB7.0. In the experiment, All images from the Berkeley Image Segmentation Database, extract m = 100 pixels from each images.

Experimental results as follows.

It is found that on Fig. 1, the strong boundary of the pine separated the image into two regions. Figure 1(b) and (c) shows that the segmentation result using the improved algorithm proposed in this paper is visually indistinguishable from that using the artificial. While, in the result of Random sampling Nyström spectral clustering, a part of the pixels which belong to the background have the same label with the pixels which belong to the foreground.

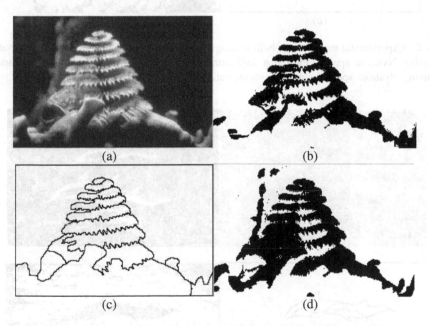

Fig. 1. Experimental results of the pine image. (a) Original image of pine, (b) Error based sampling Nyström spectral clustering segmentation, (c) Artificial segmentation from Berkeley gallery, (d) Random sampling Nyström spectral clustering segmentation.

Figure 2 is the buffalo image segmentation, Fig. 3 is the dolphin image segmentation. By comparison, it is found that on Figs. 2 and 3 due to the complexity of the background, the traditional segmentation method is not effect. Not only a lot of noise exist, false segmentation has taken. While the segmentation result by algorithm in this

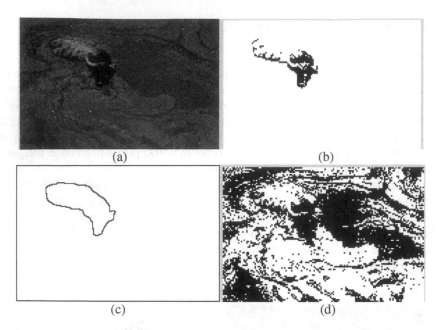

Fig. 2. Experimental results of the buffalo image. (a) Original image of buffalo, (b) Error based sampling Nyström spectral clustering segmentation, (c) Artificial segmentation, (d) Random sampling Nyström spectral clustering segmentation.

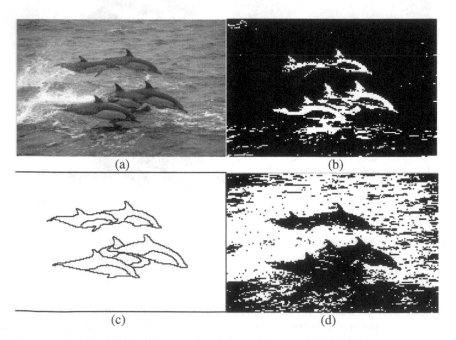

Fig. 3. Experimental results of the dolphin image. (a) Original image of dolphin, (b) Error based sampling Nyström spectral clustering segmentation, (c) Artificial segmentation from Berkeley gallery, (d) Random sampling Nyström spectral clustering segmentation.

paper can be observed more clearly with segmentation contour and image detail. The dolphin obviously distinguish from the background and have less noise. The buffalo is even more clearly, segmentation results are consistent with the results of artificial segmentation basically.

Through the comparison of the experimental results, the validity of the algorithm proposed by this paper is verified. But the algorithm is still insufficient. Our algorithm time consumption more than random sampling significantly as shown in Table 1, and Gauss parameters should be given by artificial as shown in Table 2.

Table 1. Time consumption

Algorithm	Figure 1	Figure 2	Figure 3
Error based	12.175 s	4.300 s	5.820 s
Random	2.717 s	1.186 s	1.555 s

Table 2. Gauss parameters

Feature	Figure 1	Figure 2	Figure 3
Parameters of coordinate	150	750	250
Parameters of gray value	0.3	0.5	0.1

5   Conclusion

In this paper, we developed a spectral clustering image segmentation algorithm based on the error analysis theory. By minimizing the error between Nyström similarity matrix and original similarity matrix, we proposed a new sampling method which minimized the error between the pixel and its nearest sampling point. From the derivation procedures of our algorithm, we have the conclusion that the similarity matrix found by this paper can reflect the image information more realistically. The effectiveness of this algorithm has been verified by standard test images segmentation which from Berkeley. In addition, an analysis of the shortcomings of the algorithms is given, in order to provide a basis for further study.

Acknowledgement. Project supported by the Natural Science Foundation of China (61362034).

References

1. Zhang, Y.: A survey on transition region-based techniques for image segmentation. J. Comput. Aided Des. Comput. Graph. **27**(3), 379–381 (2015)
2. Zhu, Z., Wang, L.: Initialization approach for fuzzy C-means algorithm for color image segmentation. Appl. Res. Comput. **32**(4), 1257–1260 (2015)
3. Shi, J., Maiik, J.: Normalized cuts and image segmentation. IEEE Trans. Pattern Anal. Mach. Intell. **22**(8), 888–905 (2002)

4. Belkin, M.: Laplacian eigen maps and spectral techniques for embedding and clustering. In: Dietterich, T.G., Becker, S., Ghahramani, Z. (eds.) Advances in Neural Information Processing Systems, vol. 14, pp. 585–591. MIT Press, Cambridge (2002)
5. Ng, A.Y., Jordan, M.I., Weiss, Y.: On spectral clustering: analysis and an algorithm. In: Advances in Neural Information Processing Systems. MIT Press, Cambridge (2002)
6. Lu, Z., Carreira-Perpinan, M.A.: Constrained spectral clustering through affinity propagation. In: IEEE Conference on Computer Vision and Pattern Recognition, pp. 1–8 (2008)
7. Fowlkes, C., Belongie, S., Chung, F., Malik, J.: Spectral grouping using Nyström extension. IEEE Trans. Pattern Anal. Mach. Intell. **26**(2), 214–225 (2004)
8. Zhang, K., Tsang, I.W., Kwok, J.T.: Improved Nyström low-rank approximation and error analysis. In: Proceedings of the 25th International Conference on Machine learning, Helsinki, pp. 1232–1239 (2008)
9. Zhang, K., Kwok, J.T.: Clustered Nyström method for large scale manifold learning and dimension reduction. JEEE Trans. Neural Netw. **21**(10), 1576–1587 (2010)
10. Wang, S., Gu, J., Chen, F.: Clustering high-dimensional data via spectral clustering using collaborative representation coefficients. In: Huang, D.-S., Jo, K.-H., Hussain, A. (eds.) ICIC 2015. LNCS, vol. 9226, pp. 248–258. Springer, Heidelberg (2015)
11. Chen, Z., Qiu, Z., Li, J., et al.: Two-derivative Runge-Kutta-Nyström methods for second-order ordinary differential equations. Numer. Algorithms **70**(4), 897–927 (2015)

Supervised Online Dictionary Learning
for Image Separation Using OMP

Yuxin Zhang[✉] and Bo Yuan

Intelligent Computing Lab, Division of Informatics,
Graduate School at Shenzhen, Tsinghua University,
Shenzhen 518055, People's Republic of China
zhang_yuxin@foxmail.com, yuanb@sz.tsinghua.edu.cn

Abstract. In this paper, we propose a new algorithm to perform single image separation based on online dictionary learning and orthogonal matching pursuit (OMP). This method consists of two separate processes: dictionary training for representing morphologically different components and the separation stage. The training process takes advantage of the prior knowledge of the components by adding component recovery error control penalties. The learned dictionaries have lower coherence with each other and better separation ability, which can benefit the separation process in two ways. Firstly, simple sparse coding methods such as OMP can be used to efficiently obtain superior performance. Secondly, well trained dictionaries can lead to satisfactory separation results even when the components are similar. The dictionaries obtained can also serve as good initial inputs for other models using dictionary learning and sparse representation. Experiments on complex images confirm that better results can be achieved efficiently by our method compared to other state-of-the-art algorithms.

Keywords: Online dictionary learning · Image separation · Morphological component analysis · OMP

1 Introduction

Data separation is a kind of fundamental transformation with wide applications in fields such as cosmology, geography and biomedical engineering. As a branch of data separation, the purpose of image separation or decomposition is to identify different components in a single image. A typical objective is to efficiently extract and separate texture and cartoon components mixed in the same image. Single image decomposition has been successfully applied in image processing for tasks such as rain moving from a picture [1] and reflection separation [2].

There are a number of existing methods in the literature that can find a proper separation in different ways. Some of them formulate the problem as matrix factorization and resort to techniques in the field of linear algebra for solutions [3]. Others use algorithms based on numerical methods and variational models [4, 5]. There are also prior-based models such as morphological components analysis (MCA) [6]. Most of

© Springer International Publishing Switzerland 2016
D.-S. Huang and K.-H. Jo (Eds.): ICIC 2016, Part II, LNCS 9772, pp. 557–568, 2016.
DOI: 10.1007/978-3-319-42294-7_50

these algorithms can get reasonable results, but there is still room for improvement with respect to both the speed and performance of the decomposition.

Studies have shown that the sparse representation model can be applied to data separation [7]. The sparsity model has been widely used in the field of signal processing due to its elegant mathematical foundation and ability to depict the essence of natural signals [8]. The core idea of sparse representation is to transform the original signal into a linear combination of a small number of representation atoms. The set of such atoms is called a dictionary. A famous example is the morphological components analysis method [6] for decomposing signals into their building atoms. The success of decomposing mixed signals into desired components relies heavily on the proper choice of dictionaries used in the separation step. Dictionaries must be highly effective in representing their respective components, which means that the signal can be decomposed into a linear combination of a small number of atoms from the dictionary. Traditional dictionaries such as DCT transform, wavelet and shearlet [9] can be very efficient for the separation of certain components from a single signal mixture.

However, the ability of these designed dictionaries to transform signals into sparse representation and extract them from mixed images is limited. In order to separate morphologically distinct components, more flexible dictionaries are in need. To solve this problem, the idea of dictionary learning was introduced. Using dictionary learning to accomplish different signal processing tasks is very popular nowadays. Common dictionary learning methods include MOD [10], K-SVD [11] and online dictionary learning [12]. There are mainly two approaches to incorporating dictionary learning into the process of separation. The first one is to learn dictionaries for each component dataset and then use MCA or sparse coding methods to reconstruct the parts of images containing multiple features [13]. The other one is to combine dictionary learning and the separation process together [14, 15]. These two approaches have their own advantages and disadvantages, which make them suitable for different scenarios. Our method follows the idea of the first strategy and improves upon the performance of the K-SVD based separation method [13] by training better dictionaries.

Note that, in the second method, since the update of the dictionaries and the separation of data are achieved at the same time, the learned dictionaries are of little use for other separation tasks, even if the components are of similar patterns. In real-world applications, this feature can cause unnecessary computational redundancy. By contrast, employing two independent processes can improve the applicability of the dictionaries and reduce the separation time. However, using fixed dictionaries in the separation phase means that dictionaries need to be effective in representing their own components and have minimum correlations with each other. To tackle this challenge, we proposed a supervised dictionary learning method, which aims at learning dictionaries that are suitable for quick and successful separation using sparse coding.

Another issue when applying the second method is the requirement for proper initial dictionaries. Well initialized dictionaries can lead to faster convergence and better final results while poor initialization may cause total failure of the algorithm. In the context of data separation, not only should the initial dictionary have strong representation ability for its aiming component, it is also of great importance that dictionaries for different components are as incoherent as possible with each other. To achieve this objective, some researchers proposed to add incoherence based penalties to the loss function [16] which may be useful in some cases but will compromise the representation ability of the dictionaries. The algorithm proposed in this paper is able to obtain dictionaries that meet all the above requirements to serve as quality initial dictionaries.

2 Algorithm Outline

In this paper, we proposed a novel dictionary learning method for single image separation based on sparse representation. One important assumption of our algorithm is that there is a training dataset for each component. Without loss of generality, we only discuss the situation where the image is composed of two components.

The algorithm consists of two relatively separate processes. Firstly, we use supervised dictionary learning methods to train localized dictionaries for the components. Then the separation of images containing similar components is performed using the learned dictionaries via sparse coding algorithms. In this paper, we used OMP [17] to obtain sparse representations. The problem of decomposing an image into different components can be formulated as:

$$f = u_1 + u_2 \tag{1}$$

where f represents the mixture of component vectors $u_1, u_2 \in \mathbb{R}^m$. In image processing, signals are represented by vectors rearranged from small patches extracted from the original image by either distinct or sliding style. In sparse representation, we assume that each component can be represented sparsely using dictionaries. This leads to the following equation:

$$f = D_1 \alpha_1 + D_2 \alpha_2 = [D_1 \, D_2] \begin{bmatrix} \alpha_1 \\ \alpha_2 \end{bmatrix} \tag{2}$$

where $D_1, D_2 \in \mathbb{R}^{m*n}$ are dictionaries and $\alpha_1, \alpha_2 \in \mathbb{R}^n$ are sparse representations of components u_1, u_2.

In the learning phase, we use the online dictionary learning method to find the dictionaries in Eq. (2). The main contribution of our algorithm is the introduction of the component recovery error control into the original objective function (i.e., use supervision to improve the separation ability of the dictionaries). Suppose β_1 and β_2 are the solutions of the following problems:

$$\{\beta_1, \beta_2\} \triangleq \underset{\beta_1, \beta_2}{\mathrm{argmin}} \, \| f - [D_1 \, D_2] \begin{bmatrix} \beta_1 \\ \beta_2 \end{bmatrix} \|_2^2, \text{ s.t. } \| \beta_1 \|_0 + \| \beta_2 \|_0 < \mu. \tag{3}$$

where $f = u_1 + u_2 + e$, with e representing the noise. Ideally, the reconstructed components should satisfy the following constraints:

$$\| u_1 - D_1 \beta_1 \|_2^2 < \varepsilon_1 \tag{4}$$

$$\| u_2 - D_2 \beta_2 \|_2^2 < \varepsilon_2 \tag{5}$$

where $\varepsilon_1, \varepsilon_2$ are error controlling parameters. To achieve this goal, the objective function of the algorithm has to be modified as follows:

$$\left\{ \widehat{D^1}, \widehat{D^2}, \widehat{\alpha^1}, \widehat{\alpha^2} \right\} = \operatorname*{argmin}_{D^1, D^2, \alpha^1 \alpha^2} \frac{1}{2} \sum_{i=1}^{n} (\| f_i - D^1 \alpha_i^1 - D^2 \alpha_i^2 \|_2^2 + \lambda_1 \| u_i^1 - D^1 \alpha_i^1 \|_2^2$$
$$+ \lambda_2 \| u_i^2 - D^2 \alpha_i^2 \|_2^2), \text{s.t.} \| \alpha^1 \|_0 + \| \alpha^2 \|_0 < \mu. \tag{6}$$

In Eq. (6), λ_1 and λ_2 are the supervision factors, which control the recovery errors of different components. They can be set to different values or kept identical, depending on specific applications. With the ℓ_0 norm constraint, this problem is non-convex. The classical solution is to alternately update between the dictionaries and sparse representations. Here, we solve the above problem by alternately updating the dictionaries and sparse coefficients based on the online dictionary learning method [12] and OMP [17]. The framework of the training process is shown by Algorithm 1.

Algorithm 1: Supervised online dictionary learning for image decomposition

---

Input: data set $X_1 \in \mathcal{R}^{n*m}$, $X_2 \in \mathcal{R}^{n*m}$, parameters $\lambda_1, \lambda_2, \mu_1, \mu_2$
Initialization: randomly select data samples from X_1, X_2 to initialize D^1, D^2
$A_0^1 \leftarrow 0,\ B_0^1 \leftarrow 0, C_0^1 \leftarrow 0, D_0^1 \leftarrow 0, A_0^2 \leftarrow 0,\ B_0^2 \leftarrow 0, C_0^2 \leftarrow 0, E_0^2 \leftarrow 0$
For t=1 to T do:

    Update the coefficients:
    Draw x_t^1 from p(X_1) and x_t^2 from p(X_2)
    Create the mixed signal: $f_t = x_t^1 + x_t^2$
    Sparse coding using OMP:

    $\{\alpha_t^1, \alpha_t^2\} \triangleq \operatorname*{argmin}_{\alpha^1, \alpha^2} \frac{1}{2} \| f_t - D^1 \alpha^1 - D^2 \alpha^2 \|_2^2, s.t. \| \alpha^1 \|_0 + \| \alpha^2 \|_0 < \mu$

    $A_t^1 \leftarrow A_{t-1}^1 + \alpha_t^1 \alpha_t^{1T},\ B_t^1 \leftarrow B_{t-1}^1 + x_t^1 \alpha_t^{1T},$
    $C_t^1 \leftarrow C_{t-1}^1 + \alpha_t^2 \alpha_t^{1T},\ E_t^1 \leftarrow E_{t-1}^1 + f_t \alpha_t^{1T},$
    $A_t^2 \leftarrow A_{t-1}^2 + \alpha_t^2 \alpha_t^{2T},\ B_t^2 \leftarrow B_{t-1}^2 + x_t^2 \alpha_t^{2T},$
    $C_t^2 \leftarrow C_{t-1}^2 + \alpha_t^1 \alpha_t^{2T},\ E_t^2 \leftarrow E_{t-1}^2 + f_t \alpha_t^{2T}$

    Update D_t^1, D_t^2 by Algorithm 2, using D_{t-1}^1, D_{t-1}^2 as warm start:

    $\{D_t^1, D_t^2\} \triangleq \operatorname*{argmin}_{D^1, D^2} \frac{1}{t} \Sigma_{i=1}^t \frac{1}{2} \| f_i - D^1 \alpha_i^1 - D^2 \alpha_i^2 \|_2^2 +$
        $\lambda_1 \| x_i^1 - D^1 \alpha_i^1 \|_2^2 + \lambda_2 \| x_i^2 - D^2 \alpha_i^2 \|_2^2$

End for
Output: dictionaries D^1, D^2.

In the separation stage, the input is the mixed image and learned dictionaries for components of the image. Simple sparse coding algorithms such as OMP can be applied to find the sparse coefficients of the mixture using the combined dictionary. To reconstruct the original components, one just needs to multiply the local dictionaries with their corresponding coefficients and then transform the vectors back to patches. The details of the proposed algorithm are demonstrated in the next section.

3   Implementation Details

(a)   Data Separation Using Sparse Coding Algorithm

In data separation using sparse representation, it is critical that the signal can be represented sparsely. If no limit is put on the number of atoms used by the representation, the process may fail to find parts that belong to different sources. In consideration of this, we need to precisely control the sparsity, leading to the choice of OMP. Also, it has been proved [13] that OMP is better than MCA for separation. The idea of OMP is, in each iteration, to find the atom that is most correlated to the residual and project the residual to the space spanned by the chosen atoms while making the remaining part orthogonal to that space. Details of the algorithm can be found in [17]. In both the training and the separation stages, OMP is applied to find the best representation of the mixture against the combined dictionary:

$$\{\alpha_t^1, \alpha_t^2\} \triangleq \underset{\alpha_t^1, \alpha_t^2}{\mathrm{argmin}} \frac{1}{2} \left\| f_t - [D^1 \ D^2] \begin{bmatrix} \alpha_t^1 \\ \alpha_t^2 \end{bmatrix} \right\|_2^2, \ s.t. \ \| \alpha_t^1 \|_0 + \| \alpha_t^2 \|_0 < \mu \qquad (7)$$

After the representation is retrieved, it is used to update the dictionary atoms in the training stage. In the separation phase, it is for reconstructing the morphologically distinct components $\widehat{u}_1, \widehat{u}_2$.

For different situations, the parameter μ must be chosen accordingly to obtain good performance. Some separation related theories have been proposed to reveal the relation among the incoherence of dictionaries, the sparsity level and the separation error. Generally, the lower the coherence and the higher the sparsity, the better the separation results [18, 19]. Since coherence is more difficult to control, a relatively small μ as the sparsity controller is recommended to get good results.

(b)   Dictionary Training

Proper datasets are essential for training a pair of dictionaries for the two components. In image decomposition, it means that we need some images containing a single pattern from which patches can be drawn to form vectors as data samples. The training requires three datasets, two for the two components and one for mixed signals. The latter one is always generated during the training using the former two datasets.

During each training session, the first step is to find the sparse representation of the mixture using OMP. Then, we use an online dictionary learning method similar to [12] to update the dictionaries. The objective function is as follows:

$$\{\widehat{D^1}, \widehat{D^2}\} \triangleq \underset{D^1, D^2}{\mathrm{argmin}} \frac{1}{t} \sum_{i=1}^{t} \left(\frac{1}{2} \| f_i - D^1 \alpha_i^1 - D^2 \alpha_i^2 \|_2^2 + \lambda_1 \| x_i^1 - D_1 \alpha_i^1 \|_2^2 \right.$$
$$\left. + \lambda_2 \| x_i^2 - D_2 \alpha_i^2 \|_2^2 \right) \qquad (8)$$

Adopting the update process in [12], we use block-coordinate descent with warm starts to update each column of the dictionaries alternately. For example, when updating D^1, with the other dictionary fixed, the objective function becomes:

$$\left\{\widehat{D^1}\right\} \triangleq \underset{D^1}{\text{argmin}} \frac{1}{t} \sum_{i=1}^{t} \left(\frac{1}{2} \parallel r^2 - D^1\alpha_i^1 \parallel_2^2 + \lambda_1 \parallel x_i^1 - D^1\alpha_i^1 \parallel_2^2\right) \tag{9}$$

where $r^2 = f_i - D^2\alpha_i^2$ is the representation residual. Using simple algebraic calculation we have:

$$\begin{aligned}\left\{\widehat{D^1}\right\} \triangleq \underset{D^1}{\text{argmin}} \frac{1}{t} &\left(\left(\frac{1}{2} + \frac{\lambda_1}{2}\right) Tr\left(D^{1^T} D^1 A_t^1\right) - Tr\left(D^{1^T} E_t^1\right) + Tr(D^{1^T} D^2 C_t^1)\right.\\ &\left. - \lambda_1 Tr(D^{1^T} B_t^1)\right)\end{aligned} \tag{10}$$

where $A_t^1 = \sum_{i=1}^{t} \alpha_i^1 \alpha_i^{1^T}, B_t^1 = \sum_{i=1}^{t} x_i^1 \alpha_i^{1^T}, C_t^1 = \sum_{i=1}^{t} \alpha_i^2 \alpha_i^{1^T}, E_t^1 = \sum_{i=1}^{t} f_i \alpha_i^{1^T}$. By calculating the first-order derivatives of Eq. (10), it is clear that the jth atom d_j of dictionary D^1 can be updated by:

$$u_j \leftarrow \frac{1}{\theta}\left(\lambda_1 b_j + e_j - (1 + \lambda_1)D^1 a_j - D^2 c_j\right) + d_j \tag{11}$$

where a_j, b_j, c_j, e_j represent the j_{th} column of auxiliary matrices $A_t^1, B_t^1, C_t^1, E_t^1$, respectively. Thus, the dictionaries can be updated by the algorithm shown below:

Algorithm 2: Dictionary Update

Input: $A_t^1, B_t^1, C_t^1, E_t^1, A_t^2, B_t^2, C_t^2, E_t^2, D_{t-1}^1, D_{t-1}^2, \lambda_1, \lambda_2, \mu_1, \mu_2, \theta$.
Repeat:
   For j=1 to k do:
      Update the j_{th} atom of D^1 using eq. (11):
      $u_j \leftarrow \frac{1}{\theta}\left(\lambda_1 b_j^1 + e_j^1 - (1 + \lambda_1)D_{t-1}^1 a_j^1 - D_{t-1}^2 c_j^1\right) + d_{t-1,j}^1$
      $d_{t,j}^1 \leftarrow \frac{1}{\max(\parallel u_j \parallel_2, 1)} u_j$.
   End for
   For
      Update the j_{th} atom of D^2 using eq. (11):
      $u_j \leftarrow \frac{1}{\theta}\left(\lambda_2 b_j^2 + e_j^2 - (1 + \lambda_2)D_{t-1}^2 a_j^2 - D_{t-1}^1 c_j^2\right) + d_{t-1,j}^2$
      $d_{t,j}^2 \leftarrow \frac{1}{\max(\parallel u_j \parallel_2, 1)} u_j$.
   End for
Until convergence
Return D_t^1, D_t^2.

(c) Parameter Selection

The success of the algorithm relies on choosing proper parameter values. One of the classical challenges frequently encountered in dictionary learning is how to choose the size of the dictionary and the sparsity constraint parameter. Unfortunately, there is no existing theoretical guidance for the selection of these parameters. Although for different separation cases, the optimal parameter values may vary a lot, some general rules can still be drawn from our systematic empirical studies for the selection of patch size, dictionary size, sparsity level, and the weight of the error control penalty.

1. Patch size: normally, smaller patches lead to better results in image denoising, restoration and so on. However, in image decomposition, the size of the patches should be adjusted according to the image components. In our experiments, we compared the results using patch sizes of 10×10 and 20×20.
2. Dictionary size: in sparse representation, over-complete dictionaries are used for better sparse representation. If m is the size of the dictionary and n is the dimension of the signal, then m/n is called the redundancy factor, which describes the over-completeness of the dictionary. Usually, the greater the redundancy factor, the higher the sparsity level of the representation, which in image compression tasks can be very appealing. However, in data separation, higher redundancy may cause higher correlation of the dictionaries. In our experiments, we used twice the dimension of the signals as the number of atoms for one dictionary. This setting of the redundancy factor has been shown to be suitable for most separation cases.
3. Sparsity level: as mentioned earlier, in the proposed algorithm, the sparsity of the transformation to the space of the learned dictionary has to be set in advance. It cannot be too high because the dictionary needs to be specialized for each component. In our experiments, the sparsity level was set to 10 and 20 for dictionaries with size of 100×200 and 400×800 respectively.
4. Error control coefficients (λ_1, λ_2): if these parameter values are too high, the learning of the dictionaries will be difficult to converge. This is mainly because when the values of λ are too large, in the update stage, it will cause over-learning. That is, the dictionary atoms may be changed dramatically so that next time the atoms selected by OMP for the same signal will be significantly different and the separation error cannot be reduced as expected. Using line search, we found that good dictionaries were learned with λ_1, λ_2 equal to 0.7 when the patch size was 10×10.

4 Experiments

We implemented the proposed algorithm using MATLAB 2014 with its core programs coded in C++ and tested it on Windows Server 2012 (64-bit version) with Intel Xeon CPU and 128 GB RAM. We used adaptive MCA [14] and separation algorithms via dictionary learning algorithms including K-SVD [13] and online dictionary learning (ODL) [12] for comparison purpose. Three sets of experiments were conducted to show the effectiveness of our proposed algorithm from different perspectives.

(a) **Experiment 1**

In this experiment, we show the general performance of our algorithm on ten pairs of textures. Fully overlapping patches (10×10) were extracted from 300×300 images for training dictionaries with 200 atoms. During the separation process, 10 different mixtures consisting of similar patterns were used as the inputs. Experiment results measured by PSNR and FSIM are shown in Tables 1 and 2. FSIM was a recently proposed powerful perceptual quality metric [20] for visual quality assessment. Each of the value was averaged over ten individual trials.

Table 1. Separation performance measured by PSNR

PSNR (dB)	Test-1	Test-2	Test-3	Test-4	Test-5	Test-6	Test-7	Test-8	Test-9	Test-10	Avg.
K-SVD	17.61	16.83	16.20	13.47	18.23	17.49	16.36	16.09	13.72	13.78	15.98
ODL	18.18	16.67	16.29	14.53	18.44	17.84	16.43	15.91	14.30	14.63	16.32
Adaptive MCA	16.50	14.75	14.73	13.54	15.55	16.51	15.14	14.18	13.71	14.09	14.87
Ours	18.38	17.00	16.69	15.03	18.44	18.73	16.85	16.02	14.23	15.39	16.68

Table 2. Separation performance measured by FSIM

FSIM	Test-1	Test-2	Test-3	Test-4	Test-5	Test-6	Test-7	Test-8	Test-9	Test-10	Avg.
K-SVD	0.7626	0.7767	0.7935	0.7654	0.7978	0.8179	0.7669	0.8183	0.7864	0.8209	0.7906
ODL	0.7542	0.7843	0.8075	0.7810	0.7958	0.8081	0.7641	0.8182	0.7849	0.8269	0.7925
Adaptive MCA	0.7209	0.7498	0.7800	0.7743	0.7355	0.7398	0.7472	0.7744	0.7672	0.8256	0.7615
Ours	0.7892	0.7891	0.8092	0.7949	0.7953	0.8398	0.7982	0.8125	0.7981	0.8179	0.8044

Tables 1 and 2 show that the PSNR and FSIM values produced by our methods were higher than those by other algorithms in most of the cases and our algorithm achieved the highest average values.

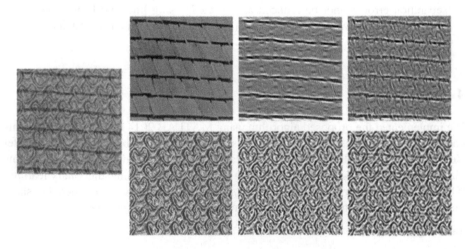

Fig. 1. Separation results on tile and heart-shape textures using our algorithm and K-SVD separation. From left to right: the mixture, the original components, the separated components using the proposed algorithm and K-SVD algorithm, respectively.

From Fig. 1, we can see that the separated components using our algorithm contained less residual from the opposite components and provided better visual effect. Also, the dictionaries learned by the proposed algorithm had lower coherence levels compared to those learned without supervision.

Since the separation methods used in our algorithm, K-SVD and online dictionary learning based separation are identical, their separation times had no significant difference. In Tables 3 and 4 we compared the time consumption by our algorithm and adaptive MCA with different patch sizes. It is clear that our algorithm took much less time compared to adaptive MCA. Although bigger patch means lower separation speed, the quality of the separated images is dramatically improved with the increase of patch size, which can be observed in Fig. 2.

Table 3. Separation time using different methods with patch size 10 × 10

Time (s)	Test-1	Test-2	Test-3	Test-4	Test-5	Test-6	Test-7	Test-8	Test-9	Test-10	Avg.
Adaptive MCA	148.92	147.08	148.45	147.99	172.13	174.15	176.30	174.10	172.69	172.90	163.47
Ours	38.35	38.18	37.66	41.44	40.70	41.71	40.55	39.89	38.60	40.98	39.81

Table 4. Separation time using different methods with patch size 20 × 20

Time (s)	Test-1	Test-2	Test-3	Test-4	Test-5	Test-6	Test-7	Test-8	Test-9	Test-10	Avg.
Adaptive MCA	363.80	356.99	357.50	357.76	471.41	478.63	486.36	484.44	484.89	484.72	432.65
Ours	182.24	184.31	180.57	177.17	177.97	176.46	178.09	178.39	177.74	178.55	179.15

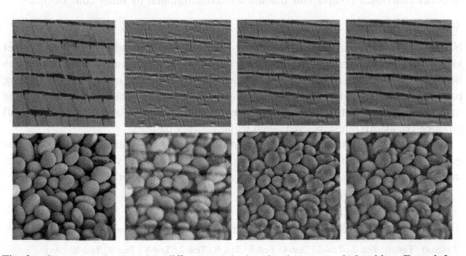

Fig. 2. Separation results using different patch sizes by the proposed algorithm. From left to right: the original components and the separated components using 10 × 10, 20 × 20, 30 × 30 patches, respectively.

(b) **Experiment 2**

In this experiment, it is shown that using our learned dictionaries for algorithm initialization can lead to better separation performance of adaptive MCA. Experimental results are shown in Fig. 3.

Fig. 3. Separation results using different initial dictionaries. From left to right: the mixture, the original components, and adaptive MCA separation results using learned dictionaries by our algorithm and by the K-SVD algorithm.

In the above results, the separation PSNR values were 15.56 (our algorithm) and 14.78 (K-SVD). It is clear that the initial dictionaries learned by our algorithm can lead to better individual components that are less contaminated by other components.

(c) **Experiment 3**

Normally, higher overlapping rate can result in higher separation PSNR and better visual quality. However, high overlapping rate requires more computation, which can be very inconvenient in practice. For instance, the separation using full overlapping patches of a 256 × 256 picture with 8 × 8 patches will take about 60 times more time than using distinct patches. In our experiments, we show that our algorithm can still produce competitive results even when using distinct patches.

Although in Fig. 4 the quality with distinct patches was noticeably inferior to the quality with fully overlapping patches, the computational time shown in Table 5 was tremendously reduced, which can be very appealing in real world applications.

Table 5. Separation time using different patch extraction strategies

Time (s)	Test-1	Test-2	Test-3	Test-4	Test-5	Test-6	Test-7	Test-8	Test-9	Test-10	Avg.
Distinct	0.72	0.62	0.65	0.60	0.53	0.60	0.64	0.75	0.68	0.66	0.64
Sliding	38.35	38.18	37.66	41.44	40.70	41.71	40.55	39.89	38.60	40.98	39.81

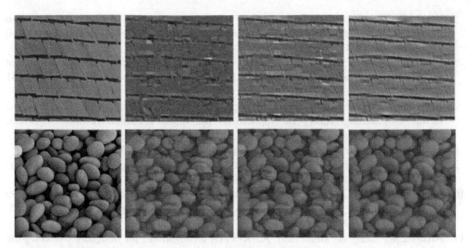

Fig. 4. Comparison of different extraction methods. From left to right: the original components, separation using K-SVD, our algorithm with distinct patches and separation using our algorithm with overlapping patches.

5 Conclusions

We proposed a new method to learn quality dictionaries for single image separation where training datasets are used to perform supervised online dictionary learning. Better trained dictionaries allow the use of simple sparse coding algorithms in the separation phase, which can greatly accelerate the separation process without compromising its performance. In addition, dictionaries learned by our method can also serve as the initial inputs for other methods using sparse representation to achieve better results. Furthermore, our algorithm can produce reasonable results using distinct patches, reducing the separation time greatly. In the future, we will conduct in-depth theoretical analysis of our algorithm and further accelerate the training process.

References

1. Luo, Y., Xu, Y., Ji, H.: Removing rain from a single image via discriminative sparse coding. In: IEEE International Conference on Computer Vision, pp. 3397–3405 (2015)
2. Kong, N., Tai, Y., Shin, J.: A physically-based approach to reflection separation: from physical modeling to constrained optimization. IEEE Trans. Pattern Anal. Mach. Intell. **36** (2), 209–221 (2014)
3. King, B., Atlas, L.: Single-channel source separation using complex matrix factorization. IEEE Trans. Audio Speech Lang. Process. **19**(8), 2591–2597 (2011)
4. Hao, Y., Xu, J., Bai, J., Han, Y.: Image decomposition combining a total variational filter and a Tikhonov quadratic filter. Multidimension. Syst. Signal Process. **26**(3), 739–751 (2015)

5. Wang, G., Pan, Z., Zhao, Z., Sun, X.: The split Bregman method of image decomposition model for ultrasound image denoising. In: 3rd International Conference on Image and Signal Processing, pp. 2870–2875 (2010)
6. Bobin, J., Starck, J., Fadili, J.M., Moudden, Y., Donoho, D.L.: Morphological component analysis: an adaptive thresholding strategy. IEEE Trans. Image Process. 16(11), 2675–2681 (2007)
7. Kutyniok, G.: Data separation by sparse representations. In: Eldar, Y., Kutyniok, G. (eds.) Compressed Sensing: Theory and Applications, pp. 485–517. Cambridge University Press, Cambridge (2012)
8. Elad, M.: Sparse and Redundant Representations. Springer, New York (2010)
9. Liu, S., Hu, S., Xiao, Y.: Image separation using wavelet-complex Shearlet dictionary. J. Syst. Eng. Electron. 25(2), 314–321 (2014)
10. Engan, K., Aase, S.O., Husoy, J.H.: Method of optimal directions for frame design. In: 1999 IEEE International Conference on Acoustics, Speech, and Signal Processing, vol. 5, pp. 2443–2446 (1999)
11. Aharon, M., Elad, M., Bruckstein, A.: K-SVD: an algorithm for designing overcomplete dictionaries for sparse representation. IEEE Trans. Signal Process. 54(11), 4311–4322 (2006)
12. Mairal, J., Bach, F., Ponce, J., Sapio, G.: Online dictionary learning for sparse coding. In: 26th Annual International Conference on Machine Learning, pp. 689–696 (2009)
13. Shoham, N., Elad, M.: Algorithms for signal separation exploiting sparse representations, with applications to texture image separation. In: 2008 IEEE Convention of Electrical and Electronics Engineers, pp. 538–542 (2008)
14. Peyr, G., Fadili, J.M., Starck, J.: Learning the morphological diversity. Soc. Ind. Appl. Math. Imaging Sci. 3(3), 646–669 (2010)
15. Li, Y., Feng, X.: Image decomposition via learning the morphological diversity. Pattern Recogn. Lett. 33(2), 111–120 (2012)
16. Liu, Q., Liu, J., Liang, D.: Adaptive image decomposition via dictionary learning with structural incoherence. In: 2013 IEEE International Conference on Image Processing, pp. 280 284 (2013)
17. Rezaiifar, R.: Orthogonal matching pursuit: recursive function approximation with applications to wavelet decomposition. In: 27th Asilomar Conference on Signals, Systems and Computers, pp. 40–44 (1993)
18. Studer, C., Baraniuk, R.G.: Stable restoration and separation of approximately sparse signals. Appl. Comput. Harm. Anal. 37(1), 12–35 (2014)
19. Studer, C., Kuppinger, P., Pope, G., Bölcskei, H.: Recovery of sparsely corrupted signals. IEEE Trans. Inf. Theory 58(5), 3115–3130 (2012)
20. Zhang, L., Zhang, L., Mou, X., Zhang, D.: FSIM: a fast feature similarity index for image quality assessment. IEEE Trans. Image Process. 20(8), 2378–2386 (2011)

Online Background-Subtraction with Motion Compensation for Freely Moving Camera

Laksono Kurnianggoro, Wahyono, Yang Yu,
Danilo Caceres Hernandez, and Kang-Hyun Jo[✉]

Graduate School of Electrical Engineering, University of Ulsan, Ulsan, Korea
{laksono,wahyono,yuyang,danilo}@islab.ulsan.ac.kr,
acejo@ulsan.ac.kr

Abstract. This paper proposes a background subtraction method for moving camera. The method relies on motion compensation to transfers the background model from the previous frame to the current frame. This motion compensation is carried out using homography transformation where the homography matrix is estimated from the set of point correspondences between previous and current frame. In order to achieve a fast processing speed, optical-flows from grid-based key-points are calculated to define the point correspondences. The background segmentation itself consists of 3 components: background model, candidate background model, and candidate age. Those 3 parameters are used to define the stable pixels which are considered as the background pixels. The proposed method was tested on a public benchmark system and achieved promising result as shown in the experimental report. Moreover, the method is able to work on real time with 56 fps of processing speed.

Keywords: Motion segmentation · Background subtraction · Motion compensation · Homography · Camera motion · Optical flow

1 Introduction

Background subtraction is one of the basic tasks in image processing. It is very useful to narrowing down the testing space for more sophisticated task such as object detection, object recognition, object tracking, action recognition and many more.

There are numerous methods of background-subtraction available nowadays. Some of them only work on static camera such as [1–4] while other can work for both moving camera and static camera such as [5–8].

Some of background subtraction methods requires the background model of specific scene to be presented in advance (e.g. offline learned) which may not suitable for robotic task or any task that require to analyze a new scene. The other bottleneck is that many background-subtraction methods cannot work in real time despite of their great performance in accuracy.

This paper proposes several contributions in order to alleviate several existing problems in the background subtraction research field. Those contributions are listed as follows:

© Springer International Publishing Switzerland 2016
D.-S. Huang and K.-H. Jo (Eds.): ICIC 2016, Part II, LNCS 9772, pp. 569–578, 2016.
DOI: 10.1007/978-3-319-42294-7_51

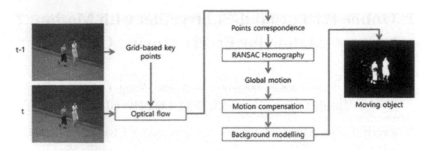

Fig. 1. Scheme of the proposed method. Two consecutive frames are needed to define the global motion which is obtained by examining the homography between them. The homography matrix itself is estimated using RANSAC method by defining it as the best possible transformation of the key-points from the previous frames and their correspondences in the current frame. Having the homography matrix, background model from the previous frame can be transferred to the current frame. After the background model is updated using the image from the current frame, the moving object can be extracted from the scene.

- Proposes a background-subtraction method that work well for both static and moving camera.
- Provides an online method which does not require any offline training. Hence the method is able to works for any new scene.
- Proposes a method that works in real time.

This paper is organized as follows. The introduction of background subtraction method is presented in Sect. 1. Section 2 explains the proposed method, including the motion compensation, background modeling, and segmentation. The Experiments as well as the results are presented in Sect. 3. Finally, a brief of conclusion is available in Sect. 4.

2   The Proposed Method

The scheme of the proposed method is shown in Fig. 1. It needs two images at each frame to define the background model. Firstly, several points are sampled from the image at $t - 1$. In this case, grid-based key points are used instead of automatically generated key points such as Harris corner [9], FAST [10], or SUSAN [11]. The reason behind this decision is that the key-points detector requires some amount of time to provide the result. In contrast, grid-based method provides the key-points instantly. Optical flow is computed for each key-point to obtains the corresponding points in the current frame. Using those pair of key points from previous frame and current frame, the random sample consensus (RANSAC [12]) homography method is employed in order to define the global motion between two consecutive frames. As the homography between two consecutive frames is known, the background model from the previous frame is transferred to the current frame and then update process is performed using the image from the current frame. Finally, the moving objects can be extracted using the background model information.

2.1 Motion Compensation

The motion compensation method requires two consecutive frames in order to estimate the homography matrix (H) between them. Firstly, key-points of the previous image (I_{t-1}) are extracted. A trivial way to obtain the key-points is by manual selection which is easy to program and able to be performed instantly. As shown in Fig. 2(a), key-points are selected with uniform distribution in row and column, 8 points in each rows and 8 points in each column. However, in practice a sufficiently large amount of key-points are used to ensure the accuracy of the homography matrix estimation. The number of key-points also should be decided based on the trade-off between accuracy and computation speed.

| (a) | (b) | (c) |

Fig. 2. Optical flows between two consecutive frames are shown in as green lines (a). Using the estimated homography matrix H, each pixel from the previous image is transformed into the current image location (b). Green box shows the current image, fully opaqued pixels shows the overlapped part of the current image and the compensated image from the previous frame while the semi-transparent part outside green box shows the non-overlapped part of the previous image. As expected, the frame difference between two frames shows the partial parts of the moving object on the scene (c). (Color figure online)

On every key-point in the previous frame, x_{t-1}, optical flow is calculated to define the corresponding points in the current image (x_t). It should be noted that the optical flow defines the flow of x_{t-1} to x_t which can also be defined using feature similarity method such as SIFT [13], SURF [14], and BRIEF [15]. However, in this case the Lucas-Kanade optical flow [16] is sufficient to dealt with this problem and moreover it requires a lower computation load compared to the aforementioned feature-based methods. An example of the optical flow is shown in Fig. 2(a), the line orientation and length represent the direction and magnitude of the optical flow, respectively.

$$x_t = Hx_{t-1} \tag{1}$$

The homography matrix (H) is defined as (1) which is the transformation matrix to transform x_{t-1} to x_t. This problem is solvable using minimum of 4 pairs of point correspondences. Since the point can be wrongly selected (i.e. foreground points instead of background points), the calculated H may not represent the global motion. In this case RANSAC is used to avoid the wrong estimation of H. The idea is to calculate a candidate of homography matrix $\hat{H}$ using a few samples of key-points and then test it to transform the corresponding points as defined in (1) to obtain $\hat{x}_t$. If the error is sufficiently small

(2) then set $\hat{H}$ as the best guess. Otherwise, repeat the process using another set of sample points.

$$\in= \sum_{x \in x_i; \hat{x} \in \hat{x}} \|x - \hat{x}\|_2 \tag{2}$$

Using the homography matrix H, I_{t-1} can be aligned with I_t based on (1) as shown in (3). The aligned image is represented as I_a while the image in previous frame is represented as I_{t-1}. This transformation is useful to transfer the background model from the previous frame to the current frame.

$$I_t(x_t) = I_{t-1}(Hx_{t-1}) \tag{3}$$

2.2 Background Modeling

In this paper, two background models are proposed. The first one is the true background model (M) which is used to perform the background subtraction and the second one is the candidate background model (C). The candidate background model is used to define whether a pixel is stable enough to be a background or not based on its candidate age (α).

Initially, all background values in the background model and the candidate background model are set as uninitialized. A distance measure is used to define the similarity of a pixel in incoming frame ($I(x)$) to $M(x)$ and $C(x)$. This similarity measure is used to define which model needs to be updated. Whenever $I(x)$ is close to $M(x)$ (e.g. $\|I(x) - M(x)\| < \tau_m$; with τ_m is a given threshold) and M is not in uninitialized state, the background model M should be updated as (4) with γ is the learning rate.

$$M_t(x) = (1 - \gamma)M_{t-1}(x) + \gamma I(x) \tag{4}$$

Whenever the first condition is not met, the second condition should be evaluated. If $I(x)$ is close to $C(x)$ or $C(x)$ is uninitialized, $C(x)$ should be updated using the latest pixel value $I(x)$ as well as its age ($\alpha(x)$). The background candidate together with its age acts as an analyzer to define whether the value in a given position is stable enough or fluctuating. If the pixel value is stable enough, then there is a high chance that the pixel is a background pixel. Therefore, whenever the age of a given background candidate is more than a predefined threshold τ_a, the candidate background model ($C(x)$) is copied to the background model ($M(x)$). It should be kept in mind that $\alpha(x)$ should be set to zero whenever $M(x)$ is updated to ensure that the candidate background model is ready to accept a new candidate. If both of the aforementioned condition are not met, $I(x)$ is set as $C(x)$, it means that the pixel is fluctuating and will be considered as background candidate.

In order to deal with the moving camera, the background model from the previous frame should be transferred to the current frame. In this case, (3) is used to transfer M, C, and α from previous frame to the current frame.

2.3   Background Subtraction

The background subtraction method is carried out pixel by pixel on the query image. A pixel is defined as background if its value is close to the background model and foreground otherwise. The foreground mask (F) is defined as (5) with τ_s is the segmentation threshold.

$$F(x) = \begin{cases} 1 & \|I(x) - M(x)\| > \tau_s \\ 0 & otherwise \end{cases} \tag{5}$$

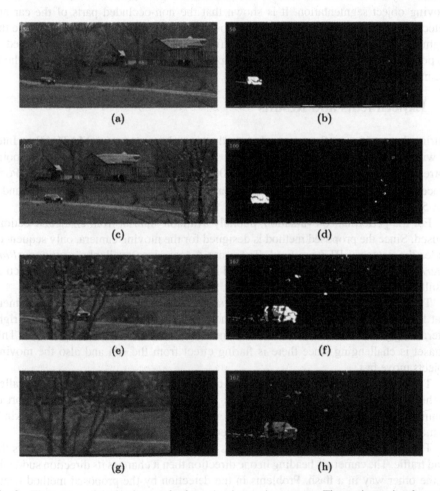

Fig. 3. Illustration of the background subtraction in moving camera. The car is moving from the left scene to the right scene while the camera follows it. The captured scene are shown in several samples (a), (c), and (e) with the corresponding segmentation result shown in (b), (d), and (f) respectively. The background model in the latest frame is shown in (g) and the detected foreground object is shown in (h).

The illustration of the proposed method while working for the moving camera is shown in Fig. 3. Three frames are shown as the example (Figs. 3(a), (c), and (e)) where the car is moving from left of the scene while the camera is following its motion. Their corresponding foreground masks are shown in the right side respectively. In this case, two values of τ_s are used, the lower one is used to define weak foreground such as shadow while the higher value one is used to define the strong foreground.

The background model of the latest frame is shown in Fig. 3(g). It is shown that in the right part of the background model is black colored, it means that the background models in that location are not available yet due to the camera motion. Whenever the background candidate is stable enough, the background model will be available as updated by the value of the background candidate. Figure 3(h) shows the result of the moving object segmentation. It is shown that the non-occluded parts of the car are detected. In practice, pre-processing and post processing are used in order to refine the segmentation result. Here, a 3×3 median filter and a 5×5 Gaussian filter are used to do post processing on the input image. For post-processing, the most common methods are median filter or morphology operations.

3   Experiments and Results

During the experiment, the proposed method was implemented using MATLAB on Intel i7 with 8 GB RAM. The parameters were set as follows, the grid size used for point correspondence is set as 32×32, $\tau_m^2 = 300$, $\tau_s^2 = 3000$, $\tau_\alpha = 15$, and $\gamma = 0.8$. Post-processing was carried out on the input image by applying a 3×3 median filter and a 5×5 Gaussian filter.

For the performance evaluation, publicly available dataset from changeDetection[1] is used. Since the proposed method is designed for the moving camera, only sequences under the category PTZ are used. There are 4 sequences available, *continuousPan*, *intermittentPan*, *twoPositionPTZCam*, and *zoomInZoomOut*. The sample of detection results are shown in Figs. 4 and 5.

The *continuousPan* sequence contains scene from a continuously panning camera that located in the side of a road. The camera was turning to the left and to the right alternatingly at slow speed. In this sequence, cars are the only moving object. This dataset is challenging since there is flaring effect from the sun and also the moving objects move fast.

The *intermittentPan* sequence contains captured scene from a camera that installed in the top of an urban road. The camera is static for a few moment and then start to panning in a slow motion to capture another view of the road. There are several kinds of moving objects in the scene including people, cars, and train.

The *twoPositionPTZCam* sequence was captured using a camera that monitors the road traffic. The camera is heading in one direction then it changes its direction suddenly to the other way in a flash. Problems in the detection by the proposed method occur mainly in the middle of the changes of direction since the homography cannot be computed from the consecutive frame due to the rapid camera movement.

---

[1] http://changedetection.net/.

Fig. 4. Samples of scene from the *zoomInZoomOut* sequence. The camera was zoomed out and then zoomed in, first row shows the scene when the camera was zooming out while the second row shows the scene while the camera was zooming in.

The *zoomInZoomOut* sequence provides a video that was captured using a camera that taking images while zooming out and zooming in. In general, the proposed method works well as shown in Fig. 4. It is able to adapt the situation while the camera is zooming in and zooming out.

According to the experiment result, the proposed method is able to work in real time. As shown in Table 1, the proposed method achieves 56 fps in average for video with QVGA resolution.

Table 1. Processing speed of the proposed algorithm.

Sequence name	Image size	Fps
continuosPan	480×704	16.5
intermittentPan	368×580	25.2
twoPosi-tionPTZCam	340×570	27.3
zoomIn-ZoomOut	240×320	56.9

Performance comparison also provided in Table 2. The performance metrics are listed as follows: recall, specificity, false positive rate, false negative rate, percentage of wrong classification, F-measure, and precision. All of the performance metrics are representing better result for bigger value except for FPR, FNR, and PWC. The data of

Fig. 5. Samples of scene from the PTZ dataset. The first row shows the scene from *continuousPan*, the second row from *intermittentPan*, and the third row from *twoPositionPTZCam*.

Table 2. Performance comparison.

Algorithm	Re	Sp	FPR	FNR	PWC	F-measure	Precission
PAWCS [4]	0.698	0.991	0.009	0.302	1.116	0.462	0.473
SOBS-CF [17]	0.856	0.680	0.320	0.144	31.943	0.037	0.019
EFIC [8]	0.918	0.922	0.078	0.082	7.871	0.584	0.528
MBS V0 [18]	0.577	0.995	0.006	0.423	0.782	0.512	0.499
SharedModel [19]	0.797	0.979	0.021	0.203	2.217	0.386	0.312
SuBSENSE [3]	0.831	0.963	0.037	0.169	3.816	0.348	0.284
CwisarDH [20]	0.336	0.998	0.002	0.664	0.685	0.322	0.482
RMoG [21]	0.641	0.928	0.072	0.359	7.476	0.247	0.221
Proposed	0.713	0.983	0.017	0.287	1.963	0.329	0.402
Proposed (5 × 5 closing)	0.769	0.979	0.021	0.231	2.301	0.318	0.399

the compared methods were taken from the changeDetection website. Currently, PAWCS is the top performer in the changeDetection benchmark. It is shown that the proposed method provides better result in term of Recall while the other measurements are not differs too much compared to the PAWCS results.

4 Conclusion

A background subtraction method has been presented in this paper. The method is able to handle the segmentation in moving camera by using the motion compensation method. Moreover, performance evaluation was conducted in the public benchmark to shows the capability of the proposed method. According to the experiment result, the proposed method achieves a competitive result compared to the top performer in the benchmark and works at 56 fps at QVGA resolution.

References

1. Wahyono, Filonenko, A., Jo, K.H., et al.: Detecting abandoned objects in crowded scenes of surveillance videos using adaptive dual background model. In: 2015 8th International Conference on Human System Interactions (HSI), pp. 224–227. IEEE (2015)
2. Wahyono, Filonenko, A., Jo, K.H., et al.: Illegally parked vehicle detection using adaptive dual background model. In: IECON 2015 - 41st Annual Conference of the Industrial Electronics Society, pp. 002225–002228 (2015)
3. St-Charles, P.L., Bilodeau, G.A., Bergevin, R.: Subsense: a universal change detection method with local adaptive sensitivity. IEEE Trans. Image Process. **24**(1), 359–373 (2015)
4. St-Charles, P.L., Bilodeau, G.A., Bergevin, R.: A self-adjusting approach to change detection based on background word consensus. In: 2015 IEEE Winter Conference on Applications of Computer Vision (WACV), pp. 990–997. IEEE (2015)
5. Kim, S.W., Yun, K., Yi, K.M., Kim, S.J., Choi, J.Y.: Detection of moving objects with a moving camera using non-panoramic background model. Mach. Vis. Appl. **24**(5), 1015–1028 (2013)
6. Yi, K., Yun, K., Kim, S., Chang, H., Choi, J.: Detection of moving objects with non-stationary cameras in 5.8 ms: bringing motion detection to your mobile device. In: Proceedings of the IEEE Conference on Computer Vision and Pattern Recognition Workshops, pp. 27–34 (2013)
7. Yun, K., Choi, J.Y.: Robust and fast moving object detection in a non-stationary camera via foreground probability based sampling. In: 2015 IEEE International Conference on Image Processing (ICIP), pp. 4897–4901. IEEE (2015)
8. Allebosch, G., Deboeverie, F., Veelaert, P., Philips, W.: EFIC: edge based foreground background segmentation and interior classification for dynamic camera viewpoints. In: Battiato, S. (ed.) ACIVS 2015. LNCS, vol. 9386, pp. 130–141. Springer, Heidelberg (2015). doi:10.1007/978-3-319-25903-1_12
9. Harris, C., Stephens, M.: A combined corner and edge detector. In: Alvey Vision Conference, vol. 15, Citeseer, p. 50 (1988)
10. Rosten, E., Drummond, T.W.: Machine learning for high-speed corner detection. In: Leonardis, A., Bischof, H., Pinz, A. (eds.) ECCV 2006, Part I. LNCS, vol. 3951, pp. 430–443. Springer, Heidelberg (2006)
11. Smith, S.M., Brady, J.M.: Susana new approach to low level image processing. Int. J. Comput. Vis. **23**(1), 45–78 (1997)

12. Fischler, M.A., Bolles, R.C.: Random sample consensus: a paradigm for model fitting with applications to image analysis and automated cartography. Commun. ACM **24**(6), 381–395 (1981)
13. Lowe, D.G.: Object recognition from local scale-invariant features. In: Proceedings of the Seventh IEEE International Conference on Computer Vision, vol. 2., pp. 1150–1157. IEEE (1999)
14. Bay, H., Tuytelaars, T., Van Gool, L.: SURF: Speeded Up Robust Features. In: Leonardis, A., Bischof, H., Pinz, A. (eds.) ECCV 2006, Part I. LNCS, vol. 3951, pp. 404–417. Springer, Heidelberg (2006)
15. Calonder, M., Lepetit, V., Strecha, C., Fua, P.: BRIEF: Binary Robust Independent Elementary Features. In: Daniilidis, K., Maragos, P., Paragios, N. (eds.) ECCV 2010, Part IV. LNCS, vol. 6314, pp. 778–792. Springer, Heidelberg (2010)
16. Lucas, B.D., Kanade, T., et al.: An iterative image registration technique with an application to stereo vision. In: IJCAI, vol. 81, pp. 674–679 (1981)
17. Maddalena, L., Petrosino, A.: A fuzzy spatial coherence-based approach to background/ foreground separation for moving object detection. Neural Comput. Appl. **19**(2), 179–186 (2010)
18. Sajid, H., Cheung, S.C.S.: Background subtraction for static & moving camera. In: 2015 IEEE International Conference on Image Processing (ICIP), pp. 4530–4534. IEEE (2015)
19. Chen, Y., Wang, J., Lu, H.: Learning sharable models for robust background subtraction. In: 2015 IEEE International Conference on Multimedia and Expo (ICME), pp. 1–6. IEEE (2015)
20. Gregorio, M., Giordano, M.: Change detection with weightless neural networks. In: Proceedings of the IEEE Conference on Computer Vision and Pattern Recognition Workshops, pp. 403–407 (2014)
21. Varadarajan, S., Miller, P., Zhou, H.: Spatial mixture of gaussians for dynamic background modelling. In: 2013 10th IEEE International Conference on Advanced Video and Signal Based Surveillance (AVSS), pp. 63–68. IEEE (2013)

Computing the Number of Groups for Color Image Segmentation Using Competitive Neural Networks and Fuzzy C-Means

Farid García-Lamont[1]([⊠]), Jair Cervantes[1], Sergio Ruiz[1],
and Asdrúbal López-Chau[1,2]

[1] Centro Universitario UAEM Texcoco, Universidad Autónoma del Estado
de México, Av. Jardín Zumpango s/n, Fraccionamiento El Tejocote,
CP 56259 Texcoco, Estado de México, Mexico
{fgarcial, alchau}@uaemex.mx,
chazarral7@gmail.com, jsergioruizc@gmail.com
[2] Centro Universitario UAEM Zumpango, Universidad Autónoma del Estado
de México, Camino Viejo a Jilotzingo continuación Calle Rayón,
55600 Zumpango, Estado de México, Mexico

Abstract. Fuzzy C-means (FCM) is one of the most often techniques employed for color image segmentation; the drawback with this technique is the number of clusters the data, pixels' colors, is grouped must be defined a priori. In this paper we present an approach to compute the number of clusters automatically. A competitive neural network (CNN) and a self-organizing map (SOM) are trained with chromaticity samples of different colors; the neural networks process each pixel of the image to segment, where the activation occurrences of each neuron are collected in a histogram. The number of clusters is set by computing the number of the most activated neurons. The number of clusters is adjusted by comparing the similitude of colors. We show successful segmentation results obtained using images of the Berkeley segmentation database by training only one time the CNN and SOM, using only chromaticity data.

Keywords: Color characterization · Color spaces · Competitive neural networks

1 Introduction

Image segmentation by color features has been employed for different purposes; for instance, analysis of rocks [1], food [2, 3], medicine [4, 5], among others [6, 7]. The algorithms of related works are based on FCM [8–10]; the FCM require a priori knowledge of the number of clusters the image should be segmented.

In related works the number of clusters is defined by the user [11, 12]; in other ones the number is set by computing the "dominant" colors of the image [10, 13–15]. Basically, the techniques introduced in [13, 14] work as follows: a SOM is trained with the colors of the image, represented in the RGB space, and then the image is processed by the SOM where in a histogram the activation occurrences of each neuron are collected. Finally the number of clusters is obtained by computing the number of peaks of the histogram.

© Springer International Publishing Switzerland 2016
D.-S. Huang and K.-H. Jo (Eds.): ICIC 2016, Part II, LNCS 9772, pp. 579–590, 2016.
DOI: 10.1007/978-3-319-42294-7_52

Using this technique involves the SOM is trained every time a new image is given, which it may be time consuming. On the other hand, most of the works employ the RGB space to represent colors but this space is light sensitive, i.e., two colors with the same chromaticity but with different intensities can be grouped in different clusters [16]. Despite other works employ other color spaces, where the chromaticity is separated from the intensity, the intensity data is also used to process the images, but similar effects are obtained as if the RGB space is employed [10, 11, 14, 17].

Our proposal consist on computing the number of dominant colors by processing the chromaticity of colors, emulating the way the humans perceives colors; that is, humans recognize colors mainly by the chromaticity then by the intensity [16, 18]. Also, humans are capable to recognize the different regions within an image by just identifying the chromaticity features of such sections without using the intensity data of the colors; it is important to mention humans employ the knowledge acquired previously to recognize colors, i.e., they do not need to learn to recognize the colors every time they need to identify a color.

Hence, the contribution of this paper is an approach where an unsupervised neural network (NN) is trained with chromaticity samples of different colors. The image is processed by the NN using the chromaticity data of the colors which is obtained by mapping, previously, the image to the HSV color space. The activation occurrences of each neuron of the NN are collected in a histogram, the number of neurons with the highest activation occurrences, number of clusters, is obtained; the image is segmented using the FCM with the number of clusters obtained.

The paper is organized as follows: in Sect. 2 we introduce our approach for segmentation. The experiments performed and the results obtained are presented in Sect. 3. In Sect. 4 the results are discussed; the paper closes with conclusions and future work in Sect. 5.

2 Proposed Approach

By observing their environment humans can recognize different regions within a scene by the chromaticity features, because humans identify colors mainly by the chromaticity features, later by the intensity [16, 18]. For instance, we can state the color of squares (a) and (b) of Fig. 1 is green because both squares have the same chromaticity although the square (a) is brighter than square (b). On the other hand, we can claim the colors of squares (c) and (d) are different because the chromaticities of both squares are different despite the intensities are the same.

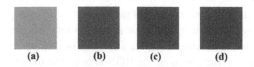

Fig. 1. Color of squares (a) and (b) with the same chromaticity but with different intensities; color of squares (c) and (d) with different chromaticity but with the same intensity (Color figure online)

Humans do not have to learn the colors every time they need to recognize a given color; they just use the knowledge acquired previously. Thus, the steps of our proposal are: (1) train a NN with chromaticity samples of different colors; (2) map the image to the HSV space and extract the chromaticity of each pixel's color of the given image; (3) the chromaticity of each pixel is feeded to the NN trained previously; (4) the activation occurrences of each neuron are collected in a histogram; (5) the number of peaks of the histogram is computed, it defines the number of clusters; (6) the number of clusters is updated by comparing the chromaticity of the neurons with the highest activation occurrences; (7) the image is segmented with the number of clusters obtained in step 6 using FCM.

The number of clusters is updated because there may be neurons with similar chromaticity and occurrence number; that is, the colors these neurons recognize are almost the same, so, they belong to the same section and they must be grouped in the same cluster.

2.1   Chromaticity Characterization

The RGB (Red, Green, Blue) color space is based in a Cartesian coordinate system where colors are points defined by vectors that extend from the origin, where black is located in the origin and white in the opposite corner to the origin, see Fig. 2.

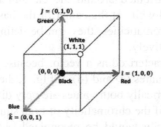

Fig. 2. RGB color space (Color figure online)

The color of a pixel p is written as a linear combination of the basis vectors red, green and blue [16]:

$$\phi_p = r_p\hat{i} + g_p\hat{j} + b_p\hat{k} \qquad (1)$$

Where r_p, g_p and b_p are the red, green and blue components, respectively. The orientation and magnitude of a color vector defines the chromaticity and the intensity of the color, respectively [16]. This color space is sensible to non-uniform illumination; even if two vectors with the same chromaticity but with different intensities, they represent different colors.

Thus, we employ the HSV (Hue, Saturation, Value) color space to represent colors because the chromaticity is decoupled from the intensity [16]; also, in [18] is claimed this space emulates the human perception of color. Figure 3 shows the cone shaped of the HSV space, where the color of a pixel p in the HSV space is written as [16]:

$$\varphi_p = [h_p, s_p, v_p] \tag{2}$$

Where h_p, s_p and v_p are the hue, saturation and value components, respectively. The hue is the chromaticity, saturation is the distance to the glow axis of black-white, and value is the intensity. The real ranges of hue, saturation and value are $[0, 2\pi]$, $[0, 1]$, and $[0, 255]$, respectively.

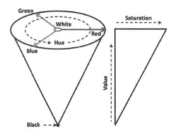

Fig. 3. HSV color space (Color figure online)

The chromaticity is distributed around the circumference of the cone; black is located at the cone's tip and white at the center of the base at the bottom of the cone. Black and white are not chromaticities, they can be defined as a low value color and low saturation color, respectively.

The chromaticity is characterized as a vector because of the case when the hue is almost 0 or 2π. Consider squares (c) and (d) of Fig. 1, their hue values are $\pi/100$ and $19\pi/10$, respectively. Numerically both values are very different but the chromaticities of both squares are similar; if the chromaticity of both squares is classified only by the scalar values, the chromaticity would be recognized as if they were very different. Thus, this problem is overcome as follows; let φ_p be the color of a pixel in the HSV space represented as in Eq. (2), the chromaticity is modeled as:

$$\psi_p = [\cos h_p, \sin h_p] \tag{3}$$

2.2   Neural Network Training

In this study we perform experiments with a CNN and a SOM. These kind of unsupervised NNs are based on finding the index of the winning neuron before external stimuli. The difference between these NNs lies in the training. In a CNN, only the weight vector of the winning neuron is updated; while in a SOM, where the neurons are set in a specific array, the weight vector of the winning neuron and the weight vectors of the neighbor neurons are updated. The weight vectors of the neurons are updated with the Kohonen learning rule [19].

Because of the fuzzy nature of color, it is not possible to recognize all the colors of the spectrum; hence, the color spectrum is "divided in" a finite amount of colors. The number of colors the NN can recognize depends on its size; in this paper we perform tests with a CNN with 16 neurons and a SOM also with 16 neurons set in a 4 × 4 array.

Therefore, both NNs can recognize up to 16 different colors. Both NNs are trained with the elements of the set Ψ built with chromaticity samples as follows:

$$\Psi = \left\{ \psi_k = [\cos \theta_k, \sin \theta_k] \middle| \theta_k = \frac{2\pi}{256} k : k = 0, 1, \ldots, 255 \right\} \tag{4}$$

2.3   Computing the Number of Clusters

Computing the number of clusters or sections of the image involves performing the following operations to each pixel of the image:

1. The pixel $\phi = [r, g, b]$ is mapped to the HSV space obtaining the vector $\varphi = [h, s, v]$
2. Verify if the color of the pixel is black by comparing its intensity with respect to a threshold value. That is, if $v < \delta_v$ then the occurrence for black is collected in the histogram, go to step 6.
3. If the color is not black then it is verified if the color of the pixel is white by comparing its saturation with respect to a threshold value. In other words, if $s < \delta_s$ then the occurrence for white is collected in the histogram, go to step 6.
4. If $v > \delta_v$ and $s > \delta_s$ then the color is a chromaticity and modeled it as Eq. (3).
5. The vector ψ is processed by the NN; the index of the winning neuron is collected in the histogram.
6. End.

Where δ_s and δ_v are the threshold values for saturation and value, respectively. Experimentally we found the best values are $\delta_s = \mu_s - \sigma_s$ and $\delta_v = \mu_v - \sigma_v$; where μ_s and μ_v are the mean of the saturation and intensity values of the image, respectively; σ_s and σ_v are the standard deviation of the saturation and intensity values of the image, respectively.

The number of clusters is defined by the number of bins of the histogram greater than zero. In the histogram there are also collected the activation occurrences of the neurons which are activated a few times; it implies that there are irrelevant small parts within the image that are considered they form important parts of the image.

Hence, there must be selected just the bins whose number is large enough to represent a significant section within the image. Thus, let P be the set of indexes of the normalized histogram bins whose number is greater or equal than threshold δ_H:

$$P = \{k \in \mathbb{N} | H(k) \geq \delta_H\} \tag{5}$$

Where $H(k)$ is the value of the normalized histogram at bin k. Therefore, the total of clusters c is the number of elements of the set P; in other words:

$$c = |P| \tag{6}$$

2.4   Adjusting the Number of Clusters

As stated before, there may be neurons with almost the same activation occurrences and with similar chromaticity. That is, there is a section within the image with a common color that activates two neurons several times; therefore, such section of the image is segmented in two parts.

Due to the colors are almost the same it means they must be grouped in the same cluster. Thus, the colors of the neurons with the highest activation occurrences are compared by computing the orientation difference between the weight vectors of the neurons as follows:

$$\Delta\theta_{i,j} = \cos^{-1}\left(\frac{\mathbf{w}_i \cdot \mathbf{w}_j^T}{\|\mathbf{w}_i\|\|\mathbf{w}_j\|}\right), \forall i,j \in P, i \neq j \tag{7}$$

Where $\Delta\theta_{i,j}$ is the orientation difference between the weight vectors $\mathbf{w}_i$ and $\mathbf{w}_j$ of neurons i and j, respectively. If the orientation difference is smaller than the threshold value δ_θ then the number of clusters must be reduced. That is:

$$\text{if } \Delta\theta_{i,j} < \delta_\theta \text{ then } c_a = c_a + 1 \tag{8}$$

Where c_a is the number of similar colors. The final number of clusters c_t is:

$$c_t = c - c_a \tag{9}$$

Finally, the image is segmented using the FCM with c_t clusters. It is important to mention the image is processed with the colors represented in the RGB space.

3   Experiments and Results

Lately the Berkeley Segmentation Database[1] (BSD) is becoming the benchmark for testing segmentation algorithms of color images. We select randomly a set of 12 images, shown in Fig. 4, from the 300 color images the BSD contains. For experiments we employ the following threshold values: $\delta_H = 0.001$ and $\delta_\theta = \pi/6$, for both NNs.

Figure 5 shows the images obtained using the CNN. Table 1 shows the number of clusters computed and the final number of clusters obtained for segmentation for each image processed by the CNN.

The final number of clusters is computed with Eq. (9); i.e., the total of clusters minus the number of similar colors.

Figure 6 shows the images obtained using the SOM with 4 × 4 neurons. Table 2 shows the number of clusters computed and the final number of clusters obtained for

---

[1] http://www.eecs.berkeley.edu/Research/Projects/CS/vision/bsds/.

Fig. 4. Images employed for experiments; original images taken from the Berkeley image database (Color figure online)

Fig. 5. Images obtained after processing the images of Fig. 4 using the competitive neural network (Color figure online)

Table 1. Number of clusters computed and final number of clusters obtained for segmentation using the competitive neural network

Image	c	c_t	Image	c	c_t
1	7	6	7	8	6
2	8	6	8	8	6
3	7	5	9	7	5
4	6	4	10	9	7
5	8	6	11	7	5
6	6	4	12	7	5

segmentation for each image processed by the SOM. The final number of clusters is computed as Eq. (9).

By observing Tables 1 and 2, the SOM recognizes more colors than the CNN; but also, there are reduced more similar colors using the SOM. It is notable the appearance difference between the images of Figs. 5 and 6, where there are more colors or sections in the images obtained using the SOM. In the following section there are discussed and analyzed the results obtained.

Fig. 6. Images obtained after processing the images of Fig. 4 using the self-organizing map (Color figure online)

Table 2. Number of clusters computed and final number of clusters obtained for segmentation using the self-organizing map

Image	c	c_t	Image	c	c_t
1	8	7	7	11	6
2	10	7	8	12	8
3	8	5	9	9	6
4	7	6	10	13	9
5	11	6	11	10	6
6	7	4	12	10	7

4 Discussion

We claim, from the images shown in Sect. 3, it is possible to estimate the number of colors within an image by processing it using a NN trained with chromaticity samples, without training it every time an image is processed. The segmentation of the images using both NNs is, to some extent, similar because there are computed almost the same number of clusters; but the SOM recognizes more colors than the CNN. We perform a discussion about it in Sect. 4.1

On the other hand, the colors of the images, processed by the FCM, are represented in the RGB space, but in this space the chromaticity is sensitive to the intensity, we discuss in Sect. 4.2 how the segmentation is affected if the vectors are normalized and then processed by the FCM.

4.1 Comparison Between Both Neural Networks

Despite both NNs are trained with the same training set and with the same number of neurons, the number of colors they recognize is different although they process the same images. The segmentation of the images 1, 2, 3, 5, 6, 7, 9, 10, 11 and 12 obtained using both NNs is very similar because the final numbers of clusters are the same or almost the same. For instance, in image 1 it is computed 6 clusters using the CNN and 7 employing the SOM; in image 10 it is computed 7 clusters employing the CNN and 9 using the SOM; in image 12 it is computed 5 clusters using the CNN and 7 employing the SOM. In other images it is computed the same final number of clusters using both NNs; specifically images 3, 5, 6 and 7.

The images where the appearance difference is more notable, depending on the NN employed, are the images 4 and 8. In the image 4 processed by the CNN, the color corresponding to the grass area has the same green hue; but in the same section of the image, processed by the SOM, is segmented with two different kinds of green hue. In image 8, the grass and the house are segmented with the same color using the CNN; while using the SOM the grass and the house are segmented with different colors.

The SOM recognizes more colors, but when the number of clusters is adjusted, there are obtained values similar to the number of clusters obtained using the CNN. The plausible explanation is the architecture of the SOM emulates the geometric shape of the HSV about the chromaticity, that is, the base of the cone, see Fig. 3.

4.2   Normalizing the RGB Color Vectors

When the FCM are applied to the images, the colors of the pixels are represented in the RGB space. Despite several colors with the same chromaticity can be grouped in different clusters if their intensities vary. For instance, the image 8 of Figs. 5 and 6, the area corresponding to the water is segmented in three parts because of the intensity differences, despite the chromaticity is almost the same. Also, in image 12 of Fig. 6 and image 9 obtained with both NNs, the sky area is segmented with two intensity levels.

If only the chromaticity data is processed the segmentation changes significantly; that is, the color vectors are normalized and then they are processed by the FCM. In this way the color vectors have the same magnitude, therefore the same intensity; but the orientation does not change, so, the chromaticity does not change. Figure 7 shows the images obtained employing the CNN and normalizing the color vectors before they are processed by the FCM.

The colors of several sections are defined better, they are more homogeneous and it is easier to appreciate them. For example, in image 7 the section corresponding to the stone wall has only one kind of hue; in image 9 the sky is segmented homogeneously with the same blue hue. The segmentation of image 8 resembles the image segmentation obtained with the SOM; in the same image, using the CNN, the hue of the grass area is alike to the hue of the house, but in the same image of the Fig. 7 the same area is segmented in green hue, similar to the image obtained using the SOM.

In image 12 there are two areas segmented with different kinds of green hue, which it is not appreciated in the same image shown in Figs. 5 and 6, besides the hue of the sections are more homogeneous.

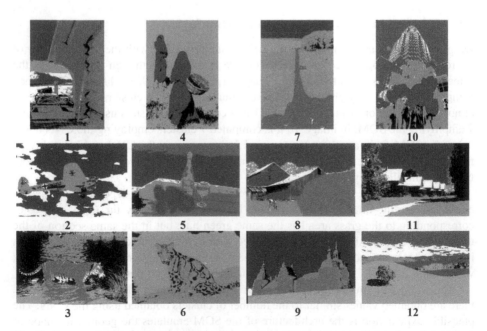

Fig. 7. Images obtained using the competitive neural network by normalizing the color vectors of the pixels before being processed by the FCM (Color figure online)

5 Conclusions and Future Work

We have introduced a proposal to compute the number of clusters a color image must be segmented using fuzzy c-means. We employ a competitive neural network and a self-organizing map, trained with chromaticity samples of several colors. The chromaticity data of each image's pixel is feeded to the neural networks, in a histogram there are collected the activation occurrences of each neuron. The number of the most activated neurons represents the amount of colors within the image; therefore, the number of sections or clusters. This number is adjusted by eliminating similar colors.

The performances of both neural networks are alike, although the self-organizing map recognizes more colors, when the number of clusters is adjusted, there are obtained almost the same values for both neural networks. The number of colors within the image is computed by employing only the chromaticity data; the image segmentation success without training the neural networks every time a new image is given.

As future work it is necessary to perform experiments with more images and with different threshold values δ_H and δ_θ, and sizes of the neural network, so as to find optimal values for these parameters that influence the segmentation; with and without normalizing the color vectors, as shown in images of Fig. 7, the segmentation seems to be more accurate if the color vectors are normalizing before they are processed with the fuzzy c-means. The quantitative evaluation of our segmentation proposal can be computed using the probabilistic rand index and variation of information metrics along with the Berkeley segmentation database, which are becoming the standard metrics and benchmark, respectively, to compute the performance of segmentation algorithms of color images [11].

References

1. Lepistö, L., Kuntuu, I., Visa, A.: Rock image classification using color features in Gabor space. J. Electron. Imaging **14**(4), 1–3 (2005)
2. Pathare, P., Linus, U., Al-Said, F.: Colour measurement and analysis in fresh and processed foods: a review. Food Bioprocess Technol. **6**(1), 36–60 (2013)
3. Santos, J., Rodrigues, F.: Applications of computer vision techniques in the agriculture and food industry: a review. Eur. Food Res. Technol. **235**(6), 989–1000 (2012)
4. Abbas, A.A., Guo, X., Tan, W.H., Jalab, H.A.: Combined spline and B-spline for an improved automatic skin lesion segmentation in dermoscopic images using optimal color channel. J. Med. Syst. **38**, 80 (2014)
5. Goffredo, M., Schmid, M., Conforto, S., Amosori, B., D'Alessio, T., Palma, C.: Quantitative color analysis for capillaroscopy image segmentation. Med. Biol. Eng. Comput. **50**(6), 567–574 (2012)
6. Guan, T., Zhou, D., Xu, C., Liu, Y.: A novel RGB Fourier transform-based color space for optical microscopic image processing. Robot. Biomimetics **1**, 16 (2014)
7. Ozturk, O., Aksac, A., Ozyer, T., Alhajj, R.: Boosting real-time recognition of hand posture and gesture for virtual mouse operations with segmentation. Appl. Intell. **43**(4), 786–801 (2015)

8. Kim, J.Y.: Segmentation of lip region in color images by fuzzy clustering. Int. J. Control Autom. Syst. **12**(3), 652–661 (2014)
9. Guo, Y., Sengur, A.: A novel color image segmentation approach based on neutrosophic and modified fuzzy c-means. Circuits Syst. Sig. Process. **32**(4), 1699–1723 (2014)
10. Balasubramaniam, P., Ananthi, V.P.: Segmentation of nutrient deficiency in incomplete crop images using intuitionistic fuzzy c-means clustering. Nonlinear Dyn. **83**(1), 849–866 (2016)
11. Mujica-Vargas, S., Gallegos-Funes, F.J., Rosales-Silva, A.J.: A fuzzy clustering algorithm with spatial robust estimation constraint for noisy color image segmentation. Pattern Recogn. Lett. **34**(4), 400–413 (2013)
12. Nadernejad, E., Sharifzadeh, S.: A new method for image segmentation based on fuzzy c-means algorithm on pixonal images formed by bilateral filtering. Sig. Image Video Process. **7**(5), 855–863 (2013)
13. Khan, A., Ullah, J., Jaffar, M.A., Choi, T.S.: Color image segmentation: a novel spatial fuzzy genetic algorithm. Sig. Image Video Process. **8**(7), 1233–1243 (2014)
14. Khan, A., Jaffar, M.A., Choi, T.S.: SOM and fuzzy based color image segmentation. Multimedia Tools Appl. **64**(2), 331–344 (2013)
15. Omran, M., Salman, A., Engelbrecht, A.P.: Dynamic clustering using particle swarm optimization with application in image segmentation. Pattern Anal. Appl. **8**(4), 332–344 (2006)
16. Gonzalez, R.C., Woods, R.E.: Digital Image Processing, 2nd edn. Prentice Hall, Upper Saddle River (2002)
17. Liu, Z., Song, Y.Q., Chen, J.M., Xie, C.H., Zhu, F.: Color image segmentation using nonparametric mixture models with multivariate orthogonal polynomials. Neural Comput. Appl. **21**(4), 801–811 (2012)
18. Ito, S., Yoshioka, M., Omatu, S., Kita, K., Kugo, K.: An image segmentation method using histograms and the human characteristics of HSI color space for a scene image. Artif. Life Robot. **10**(1), 6–10 (2006)
19. Kohonen, T.: The self-organizing map. Proc. IEEE **78**(9), 1464–1480 (1990)

Improved Parallel Gaussian Elimination Algorithm in Magnetotelluric Occam's Inversion

Yi Xiao[1,2(✉)], Pengdong Gao[2], and Yongquan Lu[2]

[1] College of Computer Science, Communication University of China,
Beijing 100024, China
louisxcode@yahoo.com
[2] Key Laboratory of Media Audio and Video,
Communication University of China, Beijing 100024, China

Abstract. An improved parallel Gauss algorithm is put forward in MT Occam. Through analysing the process of the triangle, the eliminations of coefficient matrix and column matrix are merged. To avoid repeated calculation, column matrix elimination uses the intermediate result of coefficient matrix calculation directly. By defining two parameters, back substitution can use the result of coefficient matrix immediately. Meanwhile, the elimination triangle is divided to make the algorithm accord with the threads limit of the device. By using OpenCL the improved algorithm is implemented and applied to more different platforms. The experiments use two models under TE mode and TM mode to analysis the speedup. The results reveal that the improved algorithm can achieve higher speedup with solving large coefficient matrix size. Because air layer is added in TE mode, the coefficient matrix band expands, the triangle elements increase, and the speedup rises substantially. But the initialization time will account for a large proportion when solving smaller matrix size.

Keywords: Parallel Gaussian elimination · OpenCL · MT Occam

1 Introduction

Gaussian elimination algorithm is intensive-computation part of magnetotelluric Occam's inversion (MT Occam). In the forward calculation, the algorithm solves the stiffness matrix which is built by finite element method [1]. The stiffness matrix is symmetric and all the non-zero elements are confined within a band. Thus it is compressed using band storage structure.

Implementing the parallel algorithm is mainly distributed structure, multi-core shared memory structure and accelerator structure. In distributed structure, parallel Gaussian elimination algorithm divides the coefficient matrix into blocks and allocates the blocks to different computational nodes. This is suitable for solving coefficient matrix with large degree [2–4]. In multi-core shared memory structure, SIMD vector instructions are used to solve coefficient matrix on single computational node. Great performance can be obtained to deal with smaller matrix scale [5]. By combining characteristics of distributed and shared memory structure, more cluster resources can

© Springer International Publishing Switzerland 2016
D.-S. Huang and K.-H. Jo (Eds.): ICIC 2016, Part II, LNCS 9772, pp. 591–600, 2016.
DOI: 10.1007/978-3-319-42294-7_53

be used to solve large-scale matrix [6]. Recently, GPU accelerator is widely used in the matrix solution [7]. By designing proper parallel algorithm for GPU, you can get higher acceleration than the multi-core structure [8].

In the previous work, we designed a parallel Gaussian elimination algorithm to accelerate the forward response calculation, and implemented the algorithm using CUDA FORTRAN on GPU [9]. The parallel algorithm use pivot row as an offset to determine the location of matrix elements. By dividing the algorithm process into fine-grained pieces, tasks are mapped to GPU threads. Therefore, the number of thread depends on the size of the band. However, when solving large band stiffness matrix, the number of thread exceeds the limit of some computing devices. Thus the calculation process needs to be divided further. Meanwhile, CUDA Fortran only runs on NVIDIA CUDA graphics card. In this paper, we further analysis the original parallel algorithm and perform the improved algorithm. The improved algorithm merges the computation of coefficient matrix elimination and column matrix elimination. The elimination triangle is further divided. The improved parallel algorithm implemented in OpenCL to expand the range of computing device.

2   Relative Works

2.1   OpenCL

OpenCL is a framework for heterogeneous systems and manage by the Khronos Group [10]. At present, OpenCL is gained widespread support. For example, Intel, NVIDIA, AMD, Altera and many other companies offer OpenCL development tools. OpenCL abstract the hardware as various kinds of models. Thus the algorithm can be divided and mapped to different model to solve. In those models, system is abstracted as platform, computer devices are divided into host and compute device, the memory is organized to a hierarchy. Figure 1 shows the main corresponding relation between OpenCL and CUDA in execution model and memory model.

OpenCL provides a set of runtime kernel API and programming model to use the computing devices. OpenCL, different with CUDA, requires program to choose the

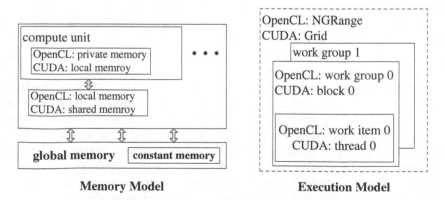

Fig. 1. Relation between OpenCL and CUDA in different model

platform and create a context according to the device you want to use. The program compiles kernel code at run time based on the executing device. Therefore OpenCL consumes more time than CUDA on in the program initialization [11, 12].

2.2 Elimination Triangle

Gaussian elimination algorithm in MT Occam is based on elimination triangle iterative calculation. For iteration, elements are confined within a triangle which is called elimination triangle as shown in Fig. 2 [13]. This structure is controlled by the parameters L and j in Eq. (1). The primary row, the upper boundary of the triangle, moves down from the first row (row 1) of stiffness matrix to the last (row N) with the elimination computing. In Gaussian elimination stiffness matrix is called coefficient matrix. In the triangle, we define the number of primary row is 0, L is the row under the primary row, j is the elements of L-th row. If the band of the coefficient matrix is D, the number of pivot row elements is D and the triangle contains $D(D+1)/2$ elements. When L = 1 the L-th row in the triangle contains D − 1 elements.

$$S_{m+L,j}^{(m)} = S_{m+L,j}^{(m-1)} - \left(S_{m,L+1}^{(m-1)} \Big/ S_{m,1}^{(m-1)}\right) \times S_{m,j+L}^{(m-1)} \ (1 \leq L \leq D-1, \ 1 \leq j \leq D-L) \quad (1)$$

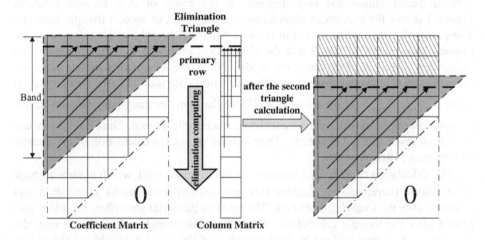

Fig. 2. The calculation process of elimination triangle

3 Improved Parallel Algorithm

3.1 Algorithm Analysis

Gaussian elimination in forward calculation can be divided into three parts: the coefficient matrix elimination, column matrix elimination and back substitution. The parameter is defined as follow: S is coefficient matrix, RM is column matrix, N is the number of rows and D is band. Equation (1) is the coefficient matrix elimination process, Eq. (2) is the column matrix elimination process and Eq. (3) is the back substitution process. In equation, $S_{m+L,j}^{(m)}$ represents j-th the element in (m + L)-th row of coefficient matrix, when calculating the m-th triangle. Unlike in Eqs. (1) and (2), the primary row in Eq. (3) moves from the last row of coefficient matrix to the first. Therefore the first step of back substitution is to calculate $RM_N = RM_N/S_{N,1}$ and defined as 0-th iteration. The parameter NBLIM is defined by the special structure of stiffness matrix the lower right corner of which are 0, as shown in Fig. 2.

$$RM_{m+L}^{(m)} = RM_{m+L}^{(m-1)} - \left(S_{m,L+1}^{(m-1)}\middle/S_{m,1}^{(m-1)}\right) \times RM_m^{(m-1)} \quad (1 \le L \le D - 1) \tag{2}$$

$$\begin{cases} RM_{N-m}^{(m)} = RM_{N-m}^{(m-1)}\middle/S_{N-m,1} - \sum_{j=2}^{NBLIM} \left(S_{N-m,j}/S_{N-m,1}\right) \times RM_{N-m+j-1}^{(m-1)} \\ \qquad\qquad\qquad (2 \le j \le D) \\ RM_N^{(0)} = RM_N^{(0)}\middle/S_{N,1} \end{cases} \tag{3}$$

From the formula of Gaussian, the elimination processes of coefficient and column matrix have same direction and primary row. If the primary row is m-th row, the column matrix elimination only depends on the result of m-th triangle iteration. Figure 2 shows the process of elimination and the result of second triangle iteration. The process of elimination starts from row 1. After the second triangle calculation, the primary row moves to row 3 and the elimination of row 1 to 2 has finished. The movement direction of primary row is defined as the elimination direction. Equations (1) and (2) have the same elimination direction. Equation (2) uses the elimination result $S_{m,L+1}^{(m-1)}\big/S_{m,1}^{(m-1)}$ at the same time. Under the same primary row, column matrix elimination can be calculated after coefficient matrix elimination. Thus Eq. (2) can use the partial result of Eq. (1) directly. Therefore we merge the calculation of the triangle and column matrix together.

The calculation of $RM_m/S_{m,1}$ depends on the triangle result, which is same as back substitution. Therefore, we combine this part with column matrix elimination and calculate after the triangle elimination. The same as the serial algorithm, Eq. (3) is also placed after the triangle calculation. When the m-th triangle calculation is end, the elements of m-th row will not be used directly and the position will be substituted by the result of Eq. (4).

$$S_{m,j} = S_{m,j}/S_{m,1} \quad (2 \le j \le D) \tag{4}$$

Equation (5) shows the improved process of primary row calculation in the triangle. Equation (6) is merged calculation of column matrix. After completing m-th iteration, column matrix elimination uses the result of primary row directly and back substitution completes part of the calculation.

$$S_{m,j}^{(m)} = \begin{cases} 1\middle/S_{m,1}^{(m-1)} & j = 1 \\ S_{m,j}^{(m-1)}\middle/S_{m,1}^{(m-1)} & 2 \le j \le D \end{cases} \tag{5}$$

$$RM_{m+L}^{(m)} = \begin{cases} S_{m,1}^{(m)} \times RM_{m+L}^{(m-1)} & L = 0 \\ RM_{m+L}^{(m-1)} - S_{m,L+1}^{(m)} \times RM_m^{(m-1)} & 1 \leq L \leq D-1 \end{cases} \tag{6}$$

3.2   Implementation and Optimization

The parameter ID is an array which stores offset of each element in the triangle without primary row. The offset defined in ID is distance from corresponding elements to the first element of primary row. The parameter numID represents the number of elements ID, nfr is the number of frequencies, P is an array using shared memory, offset(m) is offset value of m-th row in coefficient matrix. The improved parallel Gaussian elimination algorithm is as follow.

```
for n=0 to nfr-1 par-do
   (1) set S and RM to corresponding offset
   (2) for m=0 to N-1 do
   (2.1) read m-th primary row to P[D]
   (2.2) for i=0 to numID par-do
         idx=ID[i]
         S[offset(m)+idx]-=P[idx/D]/P[0]*P[idx%D]
      endfor
   (2.3) for i=0 to D-1 par-do
         t=(a*P[i]+!a)/P[0]
         S[offset(m)+i]=t
         RM[m+i]=a*RM[m+i]+c*RM[m]*t
      endfor
   endfor
endfor
```

Let the coefficient matrix start from row 0 to correspond with the actual implementation. When step (2) calculates row $N - D + 1$, the elements of the triangle start to reduce. A boundary condition is added to limit the data read and write. Figure 3 shows the triangle of $(N - D + 2)$-th primary row, where elements are represented by a dotted line. In this process, with the primary row moving down, the triangle exceeds the boundary of coefficient matrix. Therefore, a position should be set to control the memory access. Because the rows from 0-th to $N - D$-th have whole elimination triangle structure, we divide step (2) into two parts to reduce the conditional judgment: from row 0 to $N - D$ and row $N - D + 1$ to $N - 1$. The position of exclamation mark in Fig. 3 is the matrix border.

When numID is greater than the device limitation, the triangle can be divided equally. If the maximum threads number of device is E, elements of the triangle can be divided into (numID/E + 1) parts, and no data is related between those parts. By using the parameter a and c, Eqs. (1) and (2) are merged to reduce the branch in step (2.3). The definition is shown in Eq. (7). The algorithm uses a and c to adjust process for each thread - the 0 thread calculates back substitution and the other threads calculate elimination.

$$a = \begin{cases} 0 & threadID = 0 \\ 1 & threadID \neq 0 \end{cases} \quad c = \begin{cases} 1 & threadID = 0 \\ -1 & threadID \neq 0 \end{cases} \tag{7}$$

The main details of algorithm kernel function implement are as follow:

```
threadID=get_local_id(0);
/* calculate the offset of each coefficient matrix and
 * column matrix */
offset=get_group_id(0)*height*width;
offset_rm=get_group_id(0)*height;
/* calculate the border of each coefficient matrix */
border=(get_group_id(0)+1)*height*width-width+2;
/* calculate the parameter a and c */
a=!(!threadID);
c=(!threadID)<<1-1;
/* the elimination of row 0 to N-D controlled by i */
for(i=0;i<=nnode-nband;i++){
  /* read the primary row */
  if(threadID<nband) do step(2.1);
  barrier(CLK_LOCAL_MEM_FENCE);
  for(j=0;j<loop_num;j++){
    idx=id[j*E+threadID]; /*get index of the triangle*/
    do step(2.2);
  }
  if(threadID<nband) do step(2.3);
  barrier(CLK_LOCAL_MEM_FENCE);
}
/* the elimination of row N-D+1~D-2
 * the boundary condition will be added */
for(i=1;i<nband-1;i++){
  ...
  if((idx+offset)<border) do step(2.2);
  ...
}
do back substitution
```

The coefficient matrix of each frequency is mapped to work-groups. The offset of each matrix is calculated by get_group_id function. To ensure that all the data are read into the shared memory, synchronization for work-group is called after step (2.1). The same for step (2.3), next iteration depends on the coefficient matrix and the column matrix calculation. The parameter loop_num is block number of the triangle. The value of loop_num is based on the number of elements and the number of device limitation threads. For unequal situation, loop_num also needs to be limited by numID.

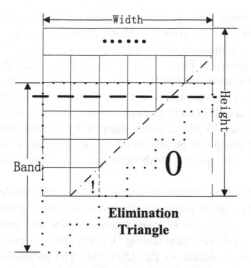

Fig. 3. The triangle on matrix boundary

4 Experimental Results

The experiments are performed on a server with ubuntu 14 operating system, equipped with an Intel E5-2620 CPU, 64G main memory and a Tesla K80 GPU. MT Occam serial algorithm is provided by university of California 3.0 version source. Model parameters are the same as 1549-model and 889-model. Stiffness matrices in TE and TM mode are calculated respectively. We count solution time under different frequency number and calculate the results for the serial algorithm and parallel algorithm. The results compared with previous parallel algorithm are given.

The speedup of parallel algorithm on different model is shown in Fig. 4. Compare with 4 curves of 889-model and 1549-model, a higher speedup can be obtained in solving a larger scale. In TE mode, the air layer is added to make expand stiffness matrix size. Thus the speedup of TE mode is higher than TM mode. But curve 889_TE

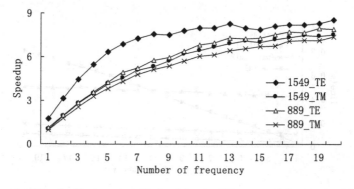

Fig. 4. The speedup of algorithm in different model scales

is above 1549_TM curve slightly, that is because air layers affect the band size and increase the elements of the triangle. More elements can be solved in parallel. Therefore the band is significant for acceleration performance.

Another feature shown in Fig. 4 is speedup declining when the parallel algorithm solves 14 or more frequencies stiffness matrix. The 1549_TE curve decline obviously. The reason for this phenomenon is that the frequencies are mapped to different word-group. In the actual implementation process, word-group is corresponding to SMX. Because K80 has 13 SMX, when mapped 13 or more word-groups to SMX, more than one word-group will be scheduled on the same SMX. Therefore accelerated effect affected in 14 word-groups. With the expansion of computation scale, speedup will picked up.

Figures 5, 6 and 7 show the calculation time of the 1549 model TE and TM mode and 889 model TM mode by using the improved and the original parallel algorithm, where kernel-curve is the time of matrix transmission and improved solution and parallel-curve is kernel-curve's time adding OpenCL environment creation. The data of 889 model TE mode is similar to the 1549-TM. The paralle2-curve has the same meaning as kernel-curve while representing the original parallel algorithm. Serial-curve represents the serial algorithm. As shown in figures, the improved algorithm consumes less time, but it needs to initialize OpenCL environment.

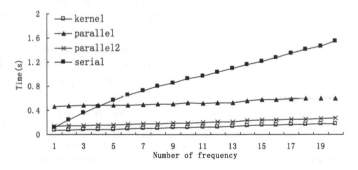

Fig. 5. Comparison with the original parallel algorithm under 1549-TE

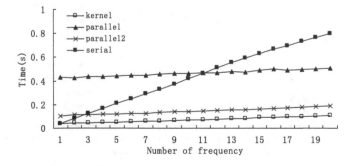

Fig. 6. Comparison with the original parallel algorithm under 1549-TM

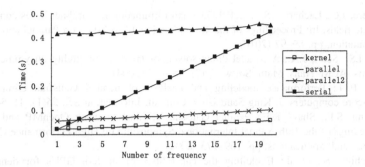

Fig. 7. Comparison with the original parallel algorithm under 889-TM

In Figs. 6 and 7, the initialization time is far greater than the algorithm, especially for the 889_TM curve. The results reflect that 889_TM's acceleration is insufficient to hide the additional time consuming. In the actual implementation, the algorithm initializes only once. The proportion of the initialization time will decrease with the expansion of computing scale.

5   Conclusion

By merging the elimination of coefficient matrix and column matrix, the improved parallel algorithm increases the local variable utilization and reduces memory access. In the basis of the elimination triangle, the computation is divided to accommodate the different device limitations. The improved algorithm obtains a higher speedup. In our experimental results, the improved parallel algorithm can further reduce the computing time and improve the performance of the algorithm. Especially for the large band stiffness matrix, performance will be more effective. This reveals that the matrix band is a significant parameter affecting the algorithm acceleration performance. But the algorithm relies on the synchronization mechanism of the work-group. When the number of frequencies exceeds hardware limitations, the speedup will decline. OpenCL spends more time on environment initialization than CUDA FORTRAN. But it is rather little for the whole time of MT Occam. Therefore, using OpenCL or CUDA FORTRAN is based on the specific implementation environment and the matrix scale and hardware must be considered in choosing speedup algorithm.

References

1. DeGroot-Hedlin, C., Constable, S.: Occam's inversion to generate smooth, two-dimensional models from magnetotelluric data. Geophysics **55**(12), 1613–1624 (1990)
2. Cosnard, M., Marrakchi, M., Robert, Y., Trystram, D.: Parallel Gaussian elimination on an MIMD computer. Parallel Comput. **6**(3), 275–296 (1988)

3. Faugère, J.C., Lachartre, S.: Parallel Gaussian Elimination for Gröbner bases computations in finite fields. In: Proceedings of the 4th International Workshop on Parallel and Symbolic Computation, pp. 89–97 (2010)
4. Duff, I.S., Scott, J.A.: A parallel direct solver for large sparse highly unsymmetric linear systems. ACM Trans. Math. Softw. **30**(2), 95–117 (2004)
5. Sibai, F.N.: Performance modeling and analysis of parallel Gaussian elimination on multi-core computers. J. King Saud Univ. Comput. Information Sci. **26**(1), 41–54 (2014)
6. McGinn, S.F., Shaw, R.E.: Parallel Gaussian elimination using OpenMP and MPI. In: Proceedings of the 16th Annual International Symposium on High Performance Computing Systems and Applications, pp. 169–173 (2002)
7. Barrachina, S., et al.: Exploiting the capabilities of modern GPUs for dense matrix computations. Concurr. Comput. Pract. Exp. **21**(18), 2457–2477 (2009)
8. Che, S., Li, J., Sheaffer, J.W., Skadron, K., Lach, J.: Accelerating compute-intensive applications with GPUs and FPGAs. In: Symposium on Application Specific Processors, pp. 101–107 (2008)
9. Xiao, Y., Liu, Y.: GPU Acceleration for the Gaussian elimination in magnetotelluric occam inversion algorithm. In: Proceedings of the 4th International Conference on Computer Engineering and Networks, pp. 123–131 (2015)
10. Stone, J.E., Gohara, D., Shi, G.: OpenCL: a parallel programming standard for heterogeneous computing systems. Comput. Sci. Eng. **12**(3), 66–73 (2010)
11. Fang, J., Varbanescu, A.L., Sips, H.: A comprehensive performance comparison of CUDA and OpenCL. In: 2011 International Conference on Parallel Processing, pp. 216–225 (2010)
12. Karimi, K., Dickson, N.G., Hamze, F.: A performance comparison of CUDA and OpenCL. arXiv preprint arXiv:1005.2581 (2010)
13. Zhu, J., Taylor, Z.R.L., Zienkiewicz, O.C.: The Finite Element Method: Its Basis and Fundamentals, 6th edn, pp. 9–10. Butterworth-Heinemann, Burlington (2005)

Extraction of Feature Points on 3D Meshes Through Data Gravitation

Chengwei Wang[1,2], Dan Kang[1,2], Xiuyang Zhao[1,2(✉)], Lizhi Peng[1,2],
and Caiming Zhang[3]

[1] School of Information Science and Engineering, University of Jinan,
Jinan 250022, People's Republic of China
[2] Shandong Provincial Key Laboratory of Network Based
Intelligent Computing, Jinan 250022, People's Republic of China
cmqingji@sina.cn, ujn_kangdan@qq.com,
xiuyangzhao@gmail.com, plz@ujn.edu.cn
[3] School of Computer Science and Technology, Shandong University,
Jinan 250101, People's Republic of China
czhang@sdu.edu.cn

Abstract. Feature points are particularly simple elements which demonstrate a model efficiently and availably; nonetheless, the points on 3D models cannot be extracted completely yet. Therefore, we propose a new algorithm based on data gravitation to extract the feature points on 3D meshes. First, we select the point with the maximum Gaussian curvature as the initial feature point set. Then, we use farthest point sampling to calculate the farthest distance from feature point set, and add this point into feature point set. Next we use the farthest distance to calculate data gravitation and select the point with the largest data gravitation until the farthest distance is smaller than a given threshold. Finally we get the feature points set on 3D meshes. In our experiments, we compare our algorithm with other algorithms. Results show that our algorithm can capture feature points effectively; consequently, the set of feature points reflects the features of 3D meshes precisely. Moreover, our algorithm is simple and is therefore easy to implement.

Keywords: Data gravitation · Feature points · 3D mesh · Gaussian curvature

1 Introduction

The feature points of 3D meshes are points that can represent geometry characteristics or texture saliencies for models. In fact, such points constitute a particularly simple method of representing geometry in a model and are not influenced by changes in coordinate systems. The use of feature points can also save much time and memory during calculation; hence, it's easy to study 3D meshes via these points. Given these advantages, researchers have kept investigating feature points to date. Feature points extraction is a significant aspect of computer graphics research, and such points are widely applied to various fields, including mesh model registration [1, 2], recognition [3], segmentation [4], and skeleton extraction [5]. Thus, extracting the feature points on 3D mesh models plays an important role in model post-processing and provides a strong support basis for other complex operations.

© Springer International Publishing Switzerland 2016
D.-S. Huang and K.-H. Jo (Eds.): ICIC 2016, Part II, LNCS 9772, pp. 601–612, 2016.
DOI: 10.1007/978-3-319-42294-7_54

Feature points denote a set of points whose geometric properties change significantly on models. Such properties generally include the normal vector and the curvature of a point. Many researchers have focused on the mathematical methods related to the geometric characteristics of meshes, e.g., curvature, and we can generally extract feature points according to the curvature of each point on meshes [6, 7]. This kind of methods is developed early and occasionally involves surface fitting that requires much calculation; moreover, this method considers only the local property and not the global impact. Therefore, the feature points obtained may not accord with people's cognition; a feature point in a local zone may not be salient in a global zone. Lee et al. [8] proposed the concept of mesh saliency to solve the aforementioned problem; nonetheless, few researchers study saliency on 3D models.

In the current study, we examine original mesh data in the preprocessing stage and then propose an algorithm based on data gravitation [9]. Slight gravitation is observed among points; this occurrence is similar to the mutual gravitation among celestial bodies. We incorporate the gravitation from a central point to its neighbors; this factor is known as Gravitation-Value. When this value is large, the feature of the central point is salient; thus, we maintain large Gravitation-Value points and eliminate small ones when this value is used to extract feature points. Then, we compare our algorithm with other algorithms by three errors. Our contributions are twofold: (1) we propose a novel algorithm that combines data gravitation and the Farthest point sampling (FPS) [10] to extract the feature points on 3D meshes; and (2) we present a feature point extraction evaluation method based on hypothesis testing.

This paper is organized as follows. Related works are introduced in Sect. 2. The concept of data gravitation is presented in Sect. 3. Section 4 describes the process of feature point extraction through our algorithm. In Sect. 5, the results of experiments are discussed and are compared with those of other algorithms. Section 6 presents the study conclusion.

2   Related Work

As mentioned previously, early studies were based on geometric properties such as the normal vector and the curvature of a point. Milroy et al. [11] first estimated the normal vector at each point based on the orthogonal cross section model; second, they used the normal and the neighbors' information to fit a quadratic surface. Third, these researchers calculated the principal curvature and principal direction of the surface. Finally, they set these points as feature points whose principal curvatures were excellent in the principal direction. Yang and Lee [12] developed a method in which the quadratic parametric surface was computed to estimate the local curvature characteristics of the surface. The least squares method was applied to minimize the Euclidean distance among measured points, including their neighbors and quadratic parametric surfaces. Upon deriving the parameter equation for the surface, the curvature of a point was easily calculated; Finally, the researchers set the feature points such that the curvature of each point was greater than a given threshold. Ohtake et al. [13] employed the implicit surface fitting algorithm for models to estimate principal curvatures and their directional derivative. Subsequently, these researchers extracted points with extreme principal curvatures in the corresponding principal direction to serve as feature points. Nonetheless, such methods based on a single geometric property are prone to overlooking the global features on models, as mentioned previously. Therefore, we must develop a feature point extraction method based on both local and global characteristics.

Farthest point sampling (FPS) is an early method that was utilized to extract sample points. FPS can effectively extract feature points, but there are also many redundancies; as a result, the valid feature points are no longer obvious. Ruggeriet et al. [2] applied a characteristic function with the Laplace–Beltrami operator (LBO) and selected all the key points of an LBO feature vector as the sampling points. Kim et al. [14, 15] determined the average geodesic distance function by using the point with the maximum average geodesic distance field as a feature point. Gebal et al. [16] employed the auto spread function (the auto diffusion function or ADF) that calculated the amount of remaining heat on the body after a short time in a certain unit. The local extremes of ADF in particular were observed in the sharp point of the object features. The aforementioned methods can describe the local and global features on a 3D model, and they are not affected by rigid transformation; thus, they are stable [17]. Isometric-deformed objects can be efficiently matched with such techniques [18, 19]. However, these methods are based on isometric-invariant properties, and identifying such properties is time consuming.

Rodolà et al. [20] introduced a sampling method for 3D surfaces that was general enough for use with a wide range of tasks. This approach adopts a new notion labeled as relevance to determine a point; this technique is a sound alternative for many scenarios because of its generic purpose. Kaiser et al. [21] presented point descriptors for 3D point clouds called the covariance matrix pyramids, which facilitated the comparison of unstructured and unequal numbers of points. Liu et al. [22] proposed a novel method to select the key points from a depth map with a 3D freeform shape that is inspired by retinex theory. Depths were transformed via the Hotelling method and normalized to reduce their dependence on a particular viewpoint. Then, adaptive smoothing was applied to find key points with locally maximal depths. These methods are strongly theoretical.

3   Data Gravitation

3.1   Origin of Data Gravitation

Data gravitation is derived from physical universal gravitation [9, 23]. Universal gravitation is a type of force between any two objects in the universe; this force obeys the universal law of gravitation. This law indicates that the gravitation between two objects is proportionate to their masses and is inversely proportional to the distance between two objects. According to the features presented above, the universal gravitation law can be described as follows:

$$F = \frac{G * M * m}{r^2} \tag{1}$$

G is gravitational constant, M and m are the masses of two objects separately, and r is the Euclidean distance between these objects.

3.2   Definition of Data Gravitation

Peng et al. [9] define data gravitation as the similarity among data particles and a type of scalar. This important factor differs from physical force; the gravitation from different data classes can be compared in the same data particle. The gravitation of data in the same class follows the superposition principle.

These researchers suggest that data gravitation can be successfully applied to data classification and feature selection; thus, we are motivated to extract the feature points on 3D meshes via this method.

According to the definition of data gravitation, we can detect the points on 3D meshes as data particles in practical experiments. These data points possess many geometric properties, such as normal, curvature, and color properties. The property of a feature point (e.g., the Gaussian curvature) can be regarded as the mass of the point. Given a reasonable gravitational constant, we calculate the data gravitation between two points.

4   Feature Points Based on Data Gravitation

In the present study, we set the Gaussian curvature as the mass. The Gaussian curvature is computed with the algorithm developed by Taubin [24]. Data gravitation that we use is defined as follows:

$$F = \frac{G * c_1 * c_2}{r^2} \qquad (2)$$

r is the geodesic distance between two vertices v_1 and v_2. c_1 and c_2 are the Gaussian curvatures of v_1 and v_2, respectively, and v_2 is a neighbor of v_1. G is constant; thus, we set its value to 1 in our experiments.

Our entire algorithm is described as follows.

Input: the global point set X, triangle set T, Gaussian curvature C, neighbor radius threshold r_0.

Output: feature point set $X'' = \{x_1'', x_2'', \ldots, x_n''\}$.

Steps:

(1) For every point, the Gaussian curvature is normalized as follows:

$$c_x' = \frac{c_x - c_{min}}{c_{max} - c_{min}} \qquad (3)$$

c_{min} and c_{max} are the minimum and maximum Gaussian curvatures of all points.

(2) We divide all points into two parts: the feature points set and the non-feature points set. We utilize the point with the maximum Gaussian curvature as the initial point.

(3) We use farthest point sampling to get the farthest geodesic distance from the feature points set, then add the point into the feature point set. r denotes the farthest geodesic distance.

(4) The neighbors of x are calculated within a radius of r, i.e., $N(x) = \{x_{neigh1}, x_{neigh2}, \ldots, x_{neighn}\}$.

(5) If the normalized Gaussian curvature c'_x is larger than α, we use Eq. (4) to calculate the corresponding data gravitation.

$$F(x) = \sum_{y \in N(x)} \frac{G*(c'_x + \alpha)^\beta * c'_y}{r^2} \tag{4}$$

Otherwise, we apply Eq. (5) for the same purpose:

$$F(x) = \sum_{y \in N(x)} \frac{G*c'_x * c'_y}{r^2} \tag{5}$$

where c'_x and c'_y are the Gaussian curvatures of v_x and of its neighbor, respectively. α and β are the control parameters that we aim to make the Gaussian curvature more significant.

(6) We believe that a feature point displays significant data gravitation. We add the point with the maximum data gravitation into the feature point set. Then return (3) until r is smaller than r_0.

5 Experiment and Analysis

5.1 Evaluation Methods

An effective benchmark is described in *A Benchmark for 3D Interest Point Detection Algorithms* [25, 26]. This publicly available benchmark provides a complete measurement method for various types of feature point extraction algorithms. We employ the measurement method proposed by Dutagaci et al. [27] to test the results of our algorithm, including false positive error (FPE), false negative error (FNE) and weighted miss error (WME). The measurement method involves computing three errors between the feature points and the ground truth data. The smaller the error is, the more effective the algorithm is.

A denotes the feature points that are detected by an algorithm on model M, and G represents the set of ground truth interest points. Given a point g in G, the geodesic neighborhood of radius r is calculated as $G_r(g) = \{p \in M | d(g, p) \leq r\}$. $d(g, p)$ is the geodesic distance between points g and p, and the parameter r indicates the localization error tolerance. If a detected point $a \in A$ is detected in $G_r(g)$, then C is assigned to reflect these correctly detected points.

If a point g is observed in G but not in C, this point is called false negative. The FNE given localization error tolerance r is determined as follows:

$$FNE(r) = \frac{N_G - N_C}{N_G} \tag{6}$$

where N_G and N_C are the numbers of points in G and C, respectively.

If a point a is detected in A but not in G, this point is known as false positive. The FPE is obtained as follows given localization error tolerance r:

$$FPE(r) = \frac{N_A - N_C}{N_A} \tag{7}$$

where N_A is the number of detected points in A.

WME is a metric that depends on human subjects. If a point detected by an algorithm is also fluently selected by human subjects, then the algorithm has a low WME. Therefore, an effective algorithm should display low FNE, FPE, and WME values.

5.2   Experiment Results

Figure 1 shows a part of experiment models: ant, armadillo, bird_3, camel, hand2, octopus and screwdriver. This figure also depicts the results of the algorithms and of our data gravitation method. The models represent different classes of objects. Figure 2 presents the FNE, FPE, and WME results of different algorithms for a part of test models rendered in Fig. 1, and Fig. 3 depicts the average FNE, FPE, and WME values for our experimental models.

Models: Fig. 1 illustrates 8 columns that correspond to different algorithms' results, and 7 rows that represent 7 models. Each row shows the diverse feature points from different methods on the same model. The first column reflects the ground truth data. Human subjects mark the 3D interest points on a set of 3D models in a web-based subjective experiment. We set the marked set as a standard test set, and all the points detected by the algorithms are compared with the ground truth data. Thus, we can demonstrate that if the error between the standard and the detected data is small, the algorithm is effective. The columns display the results of Heat Kernel Signature (HKS), mesh saliency, 3D-SIFT, 3D-Harris, salient points, SD-corners, and data gravitation from left to right. DG is the abbreviation of data gravitation. In addition, we use different colors to distinguish the feature points of the different methods. The method HKS varies significantly from the others in that most of the methods detected more points than the ground truth data but the number of feature points generated with HKS is considerably less the ground truth. The points selected by HKS usually correspond to the tips of the extremities of the models, such as the feature points on the ears, mouth, and foot of the model camel shown in Fig. 1(d). Therefore, the method overlooks fine components that were selected by the human subjects, including the intersections of the neck and the camel's body. Other methods, including mesh saliency, 3D-Harris, and salient points, also do not detect these parts. Hence, HKS can detect global saliency

•Ground truth ◦HKS •Mesh saliency •3D-SIFT •3D-Harris ◦Salient points ◦SD-corners •DG

(a) Ant

(b) Armadillo

(c) Bird_3

(d) Camel

(e) Hand2

(f) Octopus

(g) Screwdirver

Fig. 1. Ground truth data (first column), and the interest points detected by the algorithms: HKS [17] (second column), mesh saliency [8] (third column), 3D-SIFT [29] (fourth column), 3D-Harris [30] (fifth column), salient points [31] (sixth column), SD-corners [32] (seventh column), and the proposed data gravitation method (eighth column). (Color figure online)

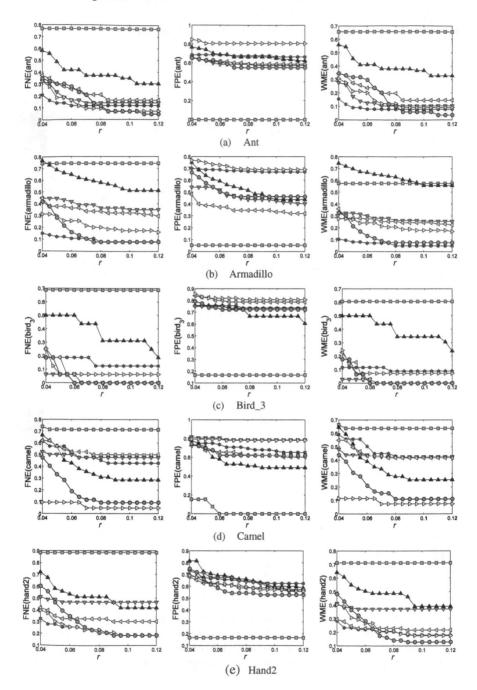

Fig. 2. FNE, FPE, and WME graphs of algorithms for the test models

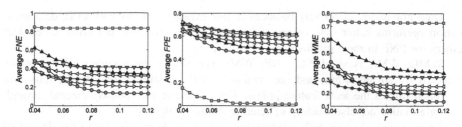

Fig. 3. Average values graphs of FNE, FPE, and WME

effectively, and its results are also selected in the ground truth data; however, this approach may overlook local details on occasion. By contrast, SD-corners, and our method are effective in such cases and can detect both global and local saliencies effectively (e.g., Fig. 1(f) and (g)).

The data gravitation and SD-corners approaches typically capture slightly more points than the ground truth data, unlike the other methods. Therefore, these two methods detect many accurate feature points in the process.

Following the intuitive comparison of all methods with the ground truth on certain models, we can draw the rough conclusion that data gravitation is an effective and advantageous method.

Figure 2 shows FNE, FPE, and WME values of a part of models in Fig. 1.

FNE: The FNE values of certain methods are high, they are close to 1 generally. As mentioned previously, FNE indicates the number of points overlooked by an algorithm. If the FNE is high, then the algorithm has missed many of the points in ground truth data; otherwise, the algorithm detects most of the ground truth points. When r is low, the points detected by the algorithm are close to the ground truth data. If the distance between the two points is greater than the value of r, then the two points are not regarded as a pair of successfully captured points and a high FNE value is generated. Thus, the FNE values of all algorithms are usually high in the beginning; when r eventually increases, the distance becomes lenient and an algorithm can detect additional points. As a result, FNE decreases.

The methods of mesh saliency, SD-corners, and data gravitation report low FNE values in most graphs. This finding indicates that these methods can usually locate feature points effectively. The 3D-SIFT performance is inferior to that of the other methods due to the inaccuracy of the voxel structure. Data gravitation improves significantly when r is greater. When the FNE decreases, the detected points are more precise and the corresponding algorithm improves.

FPE: FPE reveals the number of points that are selected by mistake. If an algorithm selects many points that are not included in the ground truth data, then its FPE is high; that of data gravitation is typically low. Although the methods of mesh saliency and SD-corners can obtain a low FNE, they also report a high FPE. These techniques capture considerably more points than the ground truth data; therefore, the probability of capturing false points is high as well given that the ground truth data are constant. That is, an algorithm may generate a high FPE when its FNE is low such as HKS. This is consistent with model analysis depicted in Fig. 1.

Data gravitation effectively balances in our experiments. As with FPE, data gravitation performs better than others, meanwhile data gravitation performs better than others for FNE in most graphs (e.g., Fig. 2(a), (e)).

WME: WME is similar to FNE; WME can reflect the accuracy of the points detected by an algorithm, and our method usually obtains a low WME; thus, our approach misses the least feature points. That is, the feature points detected through data gravitation are also fluently selected by human subjects.

These values decline with r increasing, which is clearly similar to the degree of accuracy of the detected feature points. Our detected points are close to the ground truth data when the error tolerance r is low.

An effective algorithm must report low FNE, FPE, and WME values. Thus, the algorithm detects the most points in the ground truth data and does not add many unimportant points. Figure 3 presents the average FNE, FPE, and WME values of models in experiments. Figure 3 shows that data gravitation performs better than others with r increasing, so our method can extract feature points that are closest to ground truth.

5.3   Analysis of Experiment Results

In order to test results of algorithms on all models, we observe the slight differences between two methods via the statistical method known as the Wilcoxon Signed Ranks Test [28]. This test is an effective and typical method of inferring the population condition based on sample results. Our sample consists of 10 models; all models included in the website [25] are set as the population.

We generate two hypotheses:

H_0:   We obtain the same result with two methods.

H_1:   We generate distinguished results via both methods.

We use Statistical Product and Service Solutions (SPSS) software to calculate p-value. We set α to 0.05; if p is smaller than 0.05, then we disregard H_0 and the two methods differ significantly. Otherwise, these approaches generate the same effects. In our experiments, when r is low, all methods do not differ considerably; this finding is in line with the scenarios described above. As r increases, our method performs better than others. After hypothesis testing, we can draw the conclusion that the results obtained with the population are similar to those of the 10 models. Thus, our method can exert positive effects on all models.

6   Conclusion

In this paper, we have introduced a new method named data gravitation to extract feature points on 3D models. This approach is derived from universal gravitation; therefore, this technique is easily understood and implemented. We obtain the vertices on 3D meshes to represent planets in the universe; then, we use our method to calculate the data gravitation of each vertex as with the universal gravitation of each planet. We then select points with significant data gravitation as feature points. Our experiments show that unlike other methods, our algorithm can effectively achieve our goal and can be applied to a wide range of localization error tolerance r values. The experiment results effectively reflect the characteristics of a model, and the results of hypothesis testing are comparable to or are even better than those of other methods.

Acknowledgement. This research was supported by National Natural Science Foundation of China under contract (Nos. 61373054, 61203105, 61472164), Natural Science Foundation of Shandong Province under contract (Nos. ZR2012FQ016, ZR2012FM010).

References

1. Pan, X., Zhang, G.D., Zhou, C., Zhou, Q.: Three-point matching strategy for 3D deformable shapes. J. Zhejiang Univ. Technol. **41**(5), 539–544 (2013)
2. Ruggeri, M., Patane, G., Spagnuolo, M., Saupe, D.: Spectral-driven isometry-invariant matching of 3D shapes. Int. J. Comput. Vis. **89**(2–3), 248–265 (2010)
3. Lipman, Y., Funkhouser, T.: Möbius voting for surface correspondence. ACM Trans. Graph. **28**, 72–84 (2009)
4. Yue, D.: Research on 3D mesh segmentation and skeleton algorithms based on feature points. Harbin Institute of Technology (2011)
5. Bai, X., Latecki, L., Liu, W.: Skeleton pruning by contour partitioning with discrete curve evolution. IEEE Trans. Pattern Anal. Mach. Intell. **29**, 449–462 (2007)
6. Liu, S., Zhou, R., Zhang, L.: Feature line extraction from triangular mesh model. J. Comput. Aided Des. Comput. Graph. **15**, 444–453 (2003)
7. Ma, L., Xu, Y., Li, Z.: Extracting feature points for scattered points based on Gauss curvature extreme point. J. Syst. Simul. **20**, 2341–2344 (2008)
8. Lee, C.H., Varshney, A., Jacobs, D.: Mesh saliency. In: ACM SIGGRAPH 2005, pp. 659–666 (2005)
9. Peng, L., Yang, B., Chen, Y., Abraham, A.: Data gravitation based classification. Inf. Sci. **179**(6), 809–819 (2009)
10. Eldar, Y., Lindenbaum, M., Porat, M., Zeevi, Y.: The farthest point strategy for progressive image sampling. IEEE Trans. Image Process. **6**, 1305–1315 (1997)
11. Milroy, M., Bradley, C., Vickers, G.: Segmentation of a wraparound model using an active contour. Comput. Aided Des. **29**(4), 299–320 (1997)
12. Yang, M., Lee, E.: Segmentation of measured point data using a parametric quadric surface approximation. Comput. Aided Des. **31**(7), 449–457 (1999)
13. Ohtake, Y., Belyaev, A., Seidel, H.: Ridge-valley lines on meshes via implicit surface fitting. ACM Trans. Graph. **23**(3), 609–612 (2004)
14. Kim, V., Lipman, Y., Funkhouser, T.: Blended intrinsic maps. ACM Trans. Graph. **30**(4), 79:1–79: 12 (2011)
15. Kim, V., Lipman, T., Chen, X., Funkhouser, T.: Möbius transformations for global intrinsic symmetry analysis. Comput. Graph. Forum **29**(5), 1689–1700 (2010)
16. Gebal, K., Bærentzen, J., Aanæs, H., Larsen, R.: Shape analysis using the auto diffusion function. Comput. Graph. Forum **28**(5), 1405–1413 (2009)
17. Sun, J., Ovsjanikov, M., Guibas, L.: A concise and provably informative multi-scale signature based on heat diffusion. In: Eurographics Symposium on Geometry Processing (SGP), pp. 1383–1392 (2009)

18. Ovsjanikov, M., Merigot, Q., Memoli, F., Guibas, F.: One point isometric matching with the heat kernel. Comput. Graph. Forum 29(5), 1555–1564 (2010)
19. Sun, J., Chen, X., Funkhouser, T.: Fuzzy geodesics and consistent sparse correspondences for deformable shapes. Comput. Graph. Forum 29, 1535–1544 (2010)
20. Rodolà, E., Albarelli, A., Cremers, D., Torsello, A.: A simple and effective relevance-based point sampling for 3D shapes. Pattern Recogn. Lett. 59, 41–47 (2015)
21. Kaiser, M., Xu, X., Kwolek, B., Sural, S., Rigoll, G.: Towards using covariance matrix pyramids as salient point descriptors in 3D point clouds. Neurocomputing 120, 101–112 (2013)
22. Liu, Y., Martin, R., Dominicis, L., Li, B.: Using Retinex for point selection in 3D shape registration. Pattern Recogn. 47(6), 2126–2142 (2014)
23. Peng, L., Zhang, H., Yang, B., Chen, Y.: A new approach for imbalanced data classification based on data gravitation. Inf. Sci. 288(20), 347–373 (2014)
24. Qi, B.: Curvatures estimation and the improvement of Taubins method on triangular mesh. Dalian University of Technology (2008)
25. A Benchmark for 3D Interest Point Detection Algorithms. http://www.itl.nist.gov/iad/vug/sharp/benchmark/3DInterestPoint/
26. Song, R., Liu, Y., Martin, R., Rosin, P.L.: 3D point of interest detection via spectral irregularity diffusion. Comput. Vis. 29, 695–705 (2013)
27. Dutagaci, H., Cheung, C., Godil, A.: Evaluation of 3D interest point detection techniques via human-generated ground truth. Vis. Comput. 28, 901–917 (2012)
28. Agresti, A., Finlay, B.: Statistical Methods for the Social Sciences, 4th edn, pp. 181–203. Publishing House of Electronics Industry, Beijing (2011)
29. Godil, A., Wagan, A.: Salient local 3D features for 3D shape retrieval. In: Proceedings of SPIE (2011)
30. Sipiran, I., Bustos, B.: Harris 3D: a robust extension of the Harris operator for interest point detection on 3D meshes. Vis. Comput. 27, 963–976 (2011)
31. Castellani, U., Cristani, M., Fantoni, S., Murino, V.: Sparse points matching by combining 3D mesh saliency with statistical descriptors. In: Proceedings of Eurographics, pp. 643–652 (2008)
32. Novatnack, J., Nishino, K.: Scale-dependent 3D geometric features. In: Proceedings of ICCV, pp. 1–8 (2007)

An Improved Ultrasound Image Segmentation Algorithm for Cattle Follicle Based on Markov Random Field Model

Jun Liu[1,2] and Bo Guan[1,2(✉)]

[1] College of Computer Science and Technology,
Wuhan University of Science and Technology, Wuhan, China
[2] Hubei Province Key Laboratory of Intelligent Information Processing
and Real-Time Industrial System, Wuhan, China
ljwhcn@qq.com, hbguan@foxmail.com

Abstract. In this paper, we proposed an improved ultrasound image segmentation algorithm for cattle follicle based on Markov random field model. According to the original ultrasound image dataset, we removed the speckle noise in ultrasound images by anisotropic diffusion filtering algorithm on the first step, and used the image enhancement technology to enhance the contrast of target area, then combined with an improved k-means algorithm for initial segmentation to realize basic classification of image pixels. As for the discontinuous over segmentation, we used area rule to remove the discontinuous over-segmentation region. Compared to the traditional MRF algorithm, this new algorithm has more accurate segmentation of the target area, better segmentation effect. The improved k-means algorithm to make initial segmentation for MRF model can also avoid initial clustering center to be selected randomly in comparison with the traditional k-means algorithm.

Keywords: Cattle follicle · Ultrasound image segmentation · Segmentation accuracy · Stability

1 Introduction

In the field of image segmentation, the ultrasound image segmentation has been an important and challenging subject. Based on the analysis of the ultrasound image of cattle follicle, this paper aims to find the exact timing of the ovulation period which will be of great significance to the cattle's breeding.

MRF algorithm is widely used in image segmentation field. Li et al. [1] applied MRF algorithm to extract an infrared target fast and accurately. In the paper of [2], a MRF model was proposed to segment myocardium in Magnetic Resonance Imaging. Yousefi et al. [3] used a hybrid of MRF and social algorithms to segment brain tissue in MR images. Li and Liu [4] developed a multi-resolution MRF model with variable potentials in wavelet domain to segment texture image. Mridula et al. [5] applied combining GLCM (gray level co-occurrence matrix) features and MRF model to segment color textured images.

© Springer International Publishing Switzerland 2016
D.-S. Huang and K.-H. Jo (Eds.): ICIC 2016, Part II, LNCS 9772, pp. 613–623, 2016.
DOI: 10.1007/978-3-319-42294-7_55

In the process of ultrasound image scanning, the effective area of the scanner is often presented as a sector, which increases the difficulty and complexity of the image segmentation. In order to get a better segmentation of the target area, we choose the ROI (region of interest) by hand to do pre-treatment and segmentation, which deduce the degree of complexity. To make sure the accuracy of the experiment, all images in the experiment must be chosen the same ROI randomly, namely, we use the same ROI in the dataset.

2 Background

2.1 Markov Random Field (MRF)

MRF was first applied for image processing field in the mid-1980s (Geman and Geman (1984)) [6].

Markov random process means that the state of a certain point is only related to a former state of this point and independent of other states during a randomized procedure. If we regard the image pixel and its surrounding pixel as a two-dimensional random field, we can define $S = \{(i,j)|1 \leq i \leq M, 1 \leq j \leq N\}$ as a discrete 2-D rectangular lattice site [7], where the width and height of image is M pixel and N pixel.

Let $\wedge = \{1, 2, \ldots L\}$ be the state space in the image, where L is the number of dividing the image into different areas.

If we define $X = \{x_s | s \in S\}$ as a random field of the image to be segmented, then we can learn that $x_s \in \wedge$.

Similarly, we can define the neighborhood system of MRF as following:

Assuming $\delta = \{\delta(s) | s \in S\}$ is defined as a collection of image neighborhood system on the S. If it satisfies the following three conditions, we can take $\delta(s)$ as image neighborhood system of s, where r is one of the adjacent pixels of s.

Condition 1: $\delta(s) \in S$,
Condition 2: $s \notin \delta(s)$,
Condition 3: we can derive $r \in \delta(s)$ according to this formula $\forall s, r \in S, s \in \delta(r)$. At the same time, the former can be derived from the latter.

Based on the definition above, we can draw the definition of the Nth-order symmetric neighbor [8] as below (1) shown.

$$\delta^{(n)}(s) = \{s + r, r \in \eta_s^n\}, \eta_s^1 = \{(0, -1), (0, 1), (-1, 0), (1, 0)\} \tag{1}$$

Where η_s^n refers to the symmetric neighbor set of site s [8], $\delta^{(n)}(s)$ is about the image Nth-order symmetric neighbor [8] of pixel point s. The following Table 1 is for 5th-order symmetric neighbor [8].

The more the order of the defined neighborhood system is, the higher the complexity of the algorithm operation will be. Thus, in this paper, we only need to define the second-order symmetric neighbor of the image. According to the second-order symmetric neighbor, we can define its sub groups as follows.

Table 1. 5th-order symmetric neighbor

5	4	3	4	5
4	2	1	2	4
3	1	s	1	3
4	2	1	2	4
5	4	3	4	5

We define the image by a pixel s or the collection of s and its neighborhood system as a child group c, where we use C to represent a collection of all the sub groups of c.

In conclusion, under the condition of $\delta(s) \in S$, if the random fields $X = \{X_s, s \in S\}$ satisfy two conditions below, then X is a MRF which takes $\delta(s)$ as the neighborhood system.

Condition 1: $P\{X = x\} > 0, \forall x \in \wedge, P\{X = x\}$ represents a probability of a value.

Condition 2:
$$P\{X_s = x_s | X_r = x_r, r \neq s, \forall r \in \delta(s)\}$$
$$= P\{X_s = x_s | X_r = x_r, \forall r \in \delta(s)\}.$$

2.2 Gibbs Distribution and the Optimal Segmentation Criterion

We define $\delta(s), s \in S$ as the neighborhood system on S, then the joint probability distribution of the random field has the following formula (2) [1, 9, 10] form.

$$P(X = x) = \frac{1}{Z}\exp\{-U(x)\} \tag{2}$$

Where $U(x) = \sum_{c \in C} V_c(x)$ is energy function, $Z = \sum_x \exp\{-U(x)\}$ is the normalization constant partition function.

So we can deduce the following formula (3) [1, 9, 10]:

$$P(X = x) = \frac{\exp\left\{-\sum_{c \in C} V_c(x)\right\}}{\sum_x \exp\left\{-\sum_{c \in C} V_c(x)\right\}} \tag{3}$$

Where $V_c(x)$ is the only sub group potential function related with each picture element value in the sub group c.

From the Gibbs distribution above we can obtained that if the random field and its neighborhood system consist of a MRF, we can derive the following formula (4) according to the Gibbs distribution [9, 10].

$$P(x_s|x_r) = \frac{\exp\left\{-\left(\sum\limits_{c\in C} V_c(x_s|x_r)\right)\right\}}{\sum\limits_{x_s=1}^{L} \exp\left\{-\left(\sum\limits_{c\in C} V_c(x_s|x_r)\right)\right\}}, r \in \delta(s) \tag{4}$$

According to Bayes rule [11] in the following formula (5) and the maximum of a posteriori criterion (MAP) [11, 12], we can know the optimal segmentation rules as the following formula (6) [11, 12] shows.

$$P(B|A) = \frac{P(A|B) * P(B)}{P(A)} \tag{5}$$

$$\hat{X} = \arg\max_X P(X|Y) = \arg\max_X \frac{P(Y|X) * P(X)}{P(Y)} \tag{6}$$

Where $Y = \{y_i, i \in S\}$ is a collection of image pixels, y_i is a certain pixel in the image Y, $X = \{x_i, i \in S\}$ is the collection of corresponding category field after segmentation. If the image is segmented into L categories area, we can deduce $x_i = 1, 2, 3 \ldots L$.

For (5), because $P(A)$ is a fixed constant for a given image to be segmented [11, 12], we can derive the optimal segmentation effect as the following formula (7) [11, 12] shown.

$$\arg\max_X P(Y|X) * P(X) \tag{7}$$

Thus, according to the above formulae (2), (3), (4), (5), (6) and (7), we can derive the follow formulae (8) and (9) [1].

$$P(X) = \prod_{i=1}^{M*N} P(x_i) = \frac{\prod\limits_{i=1}^{M*N} \exp\left(-\sum\limits_{c\in C} V_c(x_i)\right)}{\sum\limits_{x_i=1}^{L} \exp\left(-\sum\limits_{c\in C} V_c(x_i)\right)}, V_c(x_i) = \begin{cases} 0 & x_i = x_j \\ \beta & x_i \neq x_j \end{cases} \tag{8}$$

In this formula, β is coupling constant. M and N are the width and the length of the images respectively. x_j is a pixel point in the second-order symmetric neighbor of x_i.

$$P(Y = y|X = x) = \prod_s P(Y = y_s|X = x_s)$$

$$= \prod_s \frac{1}{\sqrt{2\pi}\delta_m} \exp\left\{-\frac{(y_s - \mu_m)^2}{2\delta_m^2}\right\} \tag{9}$$

Where δ_m and μ_m are respectively the variance and the mean in the m-th category area.

3   Proposed Method

We often need to make an initial segmentation before using MRF. In the original initial segmentation by k-means algorithm, the initial clustering centers need to be artificially set by random, which will reduce the accuracy of the initial image segmentation.

Instead of traditional k-means method, we improve the algorithm for the ROI. In the improved algorithm, according to the value of the image's pixel gray value, three relatively stable initial clustering centers are determined, based on which the image's pixel centers are segmented into three types to avoid setting the initial clustering center point. Later the pixel's centers are finely segmented though the application of MRF model. After many experiments, it is found that the segmentation effect will be the best when the value of parameter β of the formula (8) is 1. At the same time, we decide to apply second-order adjacent system of MRF model and adopt the area method to remove the excess and discrete over-segmentation area to reduce the degree of complexity and improve the efficiency of segmentation. The main steps of improving the segmentation algorithm are as follows:

Step1: Get the ROI region of initial image and record coordinates information of the ROI region in the initial image.
Step2: Do image enhancement processing for the ROI region to enhance contrast of the target area.
Step3: Handle the ROI region by the anisotropic diffusion filtering algorithm [13, 14] to remove the speckle noise in ultrasound image.
Step4: Achieve the initial segmentation and divide the initial segmentation image pixels into three categories by the improved k-means clustering [15] algorithm. The improved algorithm is as follows:

  a. Initialization. Find the initial center point v1, v2, v3 by fuzzy c-means clustering algorithm. We can find three relatively stable center points without setting a random initial cluster centers in the k-means algorithm when applying the k-means. The initial point is selected randomly in the traditional k-means algorithm.
  b. Iteration. The basis of the classification is the distance from the center point. The data will be allocated to the nearest cluster center according to the similarity criterion in the distance.
  c. Termination condition. $J(m,k) < = 0.0005$ or reach a record of 100 times of iterations.

$$J(m,k) = \max_{k=1,2,3} \left\{ abs \left(\frac{mean(\sum_{i=1}^{m} x_{i,k}) - u_k}{u_k} \right) \right\} \tag{10}$$

Where $J(m,k)$ refers to the maximum value obtained though the formula (10) among the m data sources around the k-th clustering center. $\sum_{i=1}^{m} x_{i,k}$ refers to the sum of the m data sources around the k-th cluster center. If the step b is first iteration, the default value of $u_k, k \in \{1, 2, 3\}$ is 1.

d. Update clustering center. If the conditions for the termination of the step c are not satisfied, then $v_k = mean\left(\sum_{i=1}^{m} x_{i,k}\right), u_k = v_k, k \in \{1, 2, 3\}$, jumping to step b to continue the calculation.

Step5: Figure out the value of $P(X)$ in the formula (7).

Step6: Figure out the value of $P(Y|X)$ in the formula (7).

Step7: The maximum probability of each pixel point is in the classification result of step4 according to the calculation results of step5 and step6, which is segmentation's result of pixels point in the image.

Step8: The corresponding state value and classification results calculated by Step5, Step6 and Step7 are used as the state value of next iteration classification.

Step9: Repeat step5, step6, step7 and step8 these four steps until the classification state of image tends to be stable or reach the largest number of iterations to jump out of the loop to execute the next step.

Step10: After iterations above, the first classes' pixel value of the segmented area is assigned to 255, and the other two classes' pixel value is assigned to 0. Thus, the segmentation area will be a binary image area.

Step11: For the discontinuous multi block closed ROI area after the segmentation of step9, we will get a continuous closed area by area rule. Namely, we only retain the larger area and remove the small area by the area rule.

Step12: Restoring segmented ROI area in the original image by means of coordinate translation according to record coordinate information in step1.

4 Experimental Results

Figure 1 shows the original ultrasound image of cattle follicle. After the segmentation of ultrasound images of cattle follicle by our improved algorithm, the segmentation results of the ROI region in a slice are shown in Figs. 2 and 3 show the result of ROI final segmentation in the original ultrasound image.

The experimental environment of this algorithm is based on Windows 7 operating system, and using matlab2012a for the program encoding. In order to evaluate the result of our algorithm for image segmentation, we tend to compare the area of algorithm partition with the gold standard split manually by experts. TP represents the overlap part of the algorithm partition and gold standard. FP means the area not included by gold standard but the algorithm. FN stands for the area not included by algorithm but the golden standard. TN refers to the area segmented by algorithm but not included by gold standard. As shown in Fig. 4.

The similarity criteria (SI) [11, 17], overlap fraction (OF) and extra fraction (EF), [11, 18] are used to evaluate the effect of segmentation [11]. Those formulas are shown in the (11) [11], (12) [11], (13) [11] below.

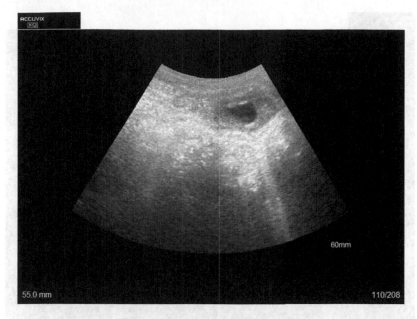

Fig. 1. Original ultrasound image

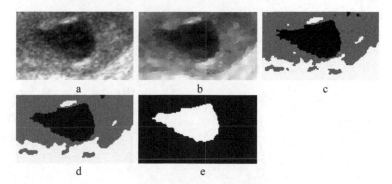

Fig. 2. a, b, c, d, e are results of ROI region after treatment of image enhancement, image denoising, initial segmentation and MRF segmentation and treatment of area rule respectively.

$$SI = \frac{2 \times TP}{2 \times TP + FP + FN} \tag{11}$$

$$OF = \frac{TP}{TP + FN} \tag{12}$$

$$EF = \frac{FP}{TP + FN} \tag{13}$$

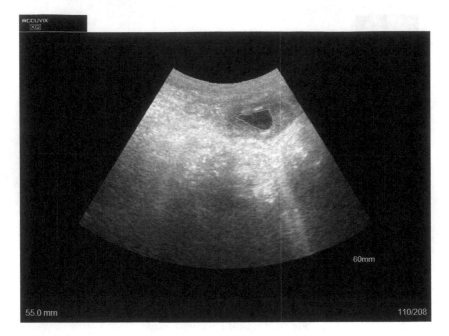

Fig. 3. Final segmentation result

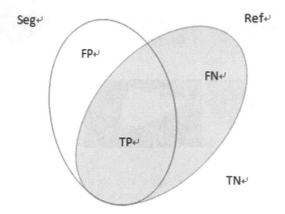

Fig. 4. Comparison of a segmentation (Seg) with the reference (Ref) image. "Seg" stands for the area of current algorithm partition; "Ref" refers to the area of gold standard.

The interval between each tangent plane is 0.5 mm when the ultrasound probe scanned cattle follicle. We make the experimental comparison through two periods of two groups of cattle follicle ultrasonic image set. The results are shown in Table 2 below. ID number from 1 to 8 and ID number from 9 to 19 stand for the two periods of two groups of ultrasound images respectively.

Table 2. Experimental segmentation results

ID	Name	TP (pixel)	FP (pixel)	FN (pixel)	SI	OF	EF
1	COW_77_07142010_0000_R_110	1582.125	177.375	27.125	0.939	0.983	0.110
2	COW_77_07142010_0000_R_111	1673.000	43.375	326.625	0.900	0.837	0.022
3	COW_77_07142010_0000_R_112	1702.750	69.000	163.750	0.936	0.912	0.037
4	COW_77_07142010_0000_R_113	1523.125	95.125	256.875	0.896	0.856	0.053
5	COW_77_07142010_0000_R_114	1492.875	187.500	162.375	0.895	0.902	0.113
6	COW_77_07142010_0000_R_115	1210.875	327.375	192.750	0.823	0.863	0.233
7	COW_77_07142010_0000_R_116	1535.125	279.125	259.125	0.851	0.856	0.156
8	COW_77_07142010_0000_R_117	1019.250	334.000	322.750	0.756	0.760	0.249
9	COW_77_07162010_0001_R_84	1450.250	102.750	114.125	0.930	0.927	0.066
10	COW_77_07162010_0001_R_85	1479.500	222.875	104.125	0.900	0.934	0.141
11	COW_77_07162010_0001_R_86	1638.125	56.750	242.750	0.916	0.871	0.030
12	COW_77_07162010_0001_R_87	1698.125	32.500	242.125	0.925	0.875	0.017
13	COW_77_07162010_0001_R_88	1252.250	125.750	102.500	0.916	0.924	0.093
14	COW_77_07162010_0001_R_89	1163.000	215.500	41.250	0.901	0.966	0.179
15	COW_77_07162010_0001_R_90	1131.000	138.000	53.250	0.922	0.955	0.117
16	COW_77_07162010_0001_R_91	826.000	413.000	9.500	0.796	0.989	0.494
17	COW_77_07162010_0001_R_92	980.750	189.625	18.625	0.904	0.981	0.190
18	COW_77_07162010_0001_R_93	858.625	93.875	37.125	0.929	0.959	0.105
19	COW_77_07162010_0001_R_94	814.500	161.125	22.875	0.899	0.973	0.192

To order to have a good segmentation effect, the value of SI and OF should be close to 1 as soon as possible, and the value of EF should be close to 0 [11, 16].

From the experimental results of Table 2, we can learn that the value of SI and OF is nearly 90 % and the value of EF is close to 0.1 in most images slice. Namely, the algorithm is effective and stable.

5 Conclusions

This paper presents an improved MRF model to segment the cattle follicle ultrasound image. The algorithm is effective and stable. As the grey value of cattle follicle ultrasound image are not uniform, we improved the traditional k-means method to avoid selecting the initial clustering center point in the process of applying the k-means method. After the initial segmentation, we use the MRF model for further segmentation. For the discontinuous over-segmentation region, based on the characteristics of follicle area biggest in the over-segmentation regions, we propose to use the area rule to remove it.

In conclusion, the segmentation of cattle follicle ultrasound image helps to find the best time for cattle reproduction and promotes the development of the animal husbandry.

Acknowledgement. This work was supported by the National Natural Science Foundation of China (Grant No. 31201121, No. 61373109 and No. 61403287), the Natural Science Foundation of Hubei Province (Grant No. 2014CFB288) and Open foundation of Hubei Province Key Laboratory of Intelligent Information Processing and Real-time Industrial System (Grant Nos. ZNSS2013A001 and ZNSS2013A004).

References

1. Li, Y., Mao, X., Feng, D., Zhang, Y.: Fast and accuracy extraction of infrared target based on Markov random field. Sig. Process. **91**(5), 1216–1223 (2011)
2. Cordero-Grande, L., Vegas-Sánchez-Ferrero, G., Casaseca-de-la-Higuera, P., San-Román-Calvar, J.A., Revilla-Orodea, A., Martín-Fernández, M., Alberola-López, C.: Unsupervised 4D myocardium segmentation with a Markov Random Field based deformable model. Med. Image Anal. **15**(3), 283–301 (2011)
3. Yousefi, S., Azmi, R., Zahedi, M.: Brain tissue segmentation in MR images based on a hybrid of MRF and social algorithms. Med. Image Anal. **16**(4), 840–848 (2012)
4. Li, Q., Liu, G.: Multi-resolution Markov random field model with variable potentials in wavelet domain for texture image segmentation. In: 2010 International Conference on Computer Application and System Modeling (ICCASM), vol. 9, p. V9-342 (2010)
5. Mridula, J., Kumar, K., Patra, D.: Combining GLCM features and markov random field model for colour textured image segmentation. In: 2011 International Conference on Devices and Communications (ICDeCom), pp. 1–5 (2011)
6. Cao, Y., Luo, Y., Yang, S.: Image denoising based on hierarchical Markov random field. Pattern Recogn. Lett. **32**(2), 368–374 (2011)
7. Qin, A.K., Clausi, D.A.: Multivariate image segmentation using semantic region growing with adaptive edge penalty. IEEE Trans. Image Process. **19**(8), 2157–2170 (2010)
8. Ye, X.F., Zhang, Z.H., Liu, P.X., Guan, H.L.: Sonar image segmentation based on GMRF and level-set models. Occan Eng. **37**(10), 891–901 (2010)
9. Monaco, J.P., Tomaszewski, J.E., Feldman, M.D., Hagemann, I., Moradi, M., Mousavi, P., Madabhushi, A.: High-throughput detection of prostate cancer in histological sections using probabilistic pairwise Markov models. Med. Image Anal. **14**(4), 617–629 (2010)
10. Roche, A., Ribes, D., Bach-Cuadra, M., Krüger, G.: On the convergence of EM-like algorithms for image segmentation using Markov random fields. Med. Image Anal. **15**(6), 830–839 (2011)
11. Khayati, R., Vafadust, M., Towhidkhah, F., Nabavi, M.: Fully automatic segmentation of multiple sclerosis lesions in brain MR FLAIR images using adaptive mixtures method and Markov random field model. Comput. Biol. Med. **38**(3), 379–390 (2008)
12. Yang, X., Clausi, D.A.: Evaluating SAR sea ice image segmentation using edge-preserving region-based MRFs. IEEE J. Sel. Top. Appl. Earth Obs. Remote Sens. **5**(5), 1383–1393 (2012)
13. Gupta, A., Tripathi, A., Bhateja, V.: Despeckling of SAR images via an improved anisotropic diffusion algorithm. In: Satapathy, S.C., Udgata, S.K., Biswal, B.N. (eds.) Proceedings of Int. Conf. on Front. of Intell. Comput. AISC, vol. 199, pp. 747–754. Springer, Heidelberg (2013)
14. Tauber, C., Batatia, H., Ayache, A.: A robust speckle reducing anisotropic diffusion. In: 2004 International Conference on Image Processing, 2004, ICIP 2004, vol. 1, pp. 247–250 (2004)

15. Khan, S.S., Ahmad, A.: Cluster center initialization algorithm for K-means clustering. Pattern Recogn. Lett. **25**(11), 1293–1302 (2004)
16. Anbeek, P., Vincken, K.L., van Osch, M.J., Bisschops, R.H., van der Grond, J.: Probabilistic segmentation of white matter lesions in MR imaging. NeuroImage **21**(3), 1037–1044 (2004)
17. Zijdenbos, A.P., Dawant, B.M., Margolin, R.A., Palmer, A.C.: Morphometric analysis of white matter lesions in MR images: method and validation. IEEE Trans. Med. Imaging **13**(4), 716–724 (1994)
18. Stokking, R., Vincken, K.L., Viergever, M.A.: Automatic morphology-based brain segmentation (MBRASE) from MRI-T1 data. NeuroImage **12**(6), 726–738 (2000)

Three-Dimensional Cement Microstructure Texture Synthesis Based on CUDA

Kun Tang[1], Bo Yang[1(✉)], Lin Wang[1], Xiuyang Zhao[1], Yueqi Wang[2],
and Haixiao Zhang[3]

[1] Shandong Provincial Key Laboratory of Network Based Intelligent
Computing, University of Jinan, Jinan 250022, China
yangbo@ujn.edu.cn
[2] Viterbi School of Engineering, University of Southern California,
2727 Ellendale Place, Los Angeles, CA 90007, USA
[3] Shandong Inspur Software Co., Ltd., Jinan 250101, China

Abstract. Three-dimensional reconstruction of cement is becoming increasingly important in cement hydration. Although many physical experiments have been conducted on cement hydration, and various algorithms have been developed to simulate the cement hydration for a long term, few algorithms have been developed to synthesize the three-dimensional microstructure of cement. Thus, we improve the Tree-structure Vector Quantization algorithm, which is effective in Unified Device Architecture. Experimental results indicate that the synthesis process in the proposed method is shorter and easier to implement which providing the same outcomes.

Keywords: Three-dimensional cement microstructure · Texture synthesis · Compute unified device architecture · Particle swarm optimization

1 Introduction

Research on cement hydration is crucial in foreground and production manufacturing to improve the performance of concrete and cement. In particular, three-dimensional reconstruction and visualization techniques are effective approaches to examine the performance of cement hydration comprehensively [1, 2].

Establishing the three-dimensional microstructure of cement is one of the most important issues analyzing cement performance. However, only a few algorithms exhibit good estimation and prediction performance for cement. For example, Wei [3, 4] used the parameterless method based on a sample image, however, some of the results produced some blurs. Kopf [5] proposed a method that integrated two-dimensional optimization and histogram matching into the synthesis of solid texture. Although this approach is applicable to many type of texture, its high complexity is believed to restrict the widespread application of this method. Developing an effective algorithm will significantly contribute to further studies on cement performance. Our method is more effective and specific than TSVQ [6] algorithm because of two reasons. First, the isotropic feature of cement differs from other type of texture. Therefore, building the Gaussian pyramid is not necessary in the proposed method, which makes the algorithm

© Springer International Publishing Switzerland 2016
D.-S. Huang and K.-H. Jo (Eds.): ICIC 2016, Part II, LNCS 9772, pp. 624–634, 2016.
DOI: 10.1007/978-3-319-42294-7_56

more effective and simpler to implement. Second, the TSVQ algorithm requires a large amount of data to perform the calculation during the texture synthesis of three-dimensional microstructure. It requires necessitates high performance computation, which can sometimes cause problems. The proposed algorithm achieves the same outcomes as that of the TSVQ algorithm, but it is simpler and more convenient to implement. The three-dimensional microstructure also helps in analyzing cement hydration, which shortens the process period and improves the synthesis of cement.

Although the three-dimensional data can be obtained through microtomography, the details of their evolution cannot be determined accurately. Identifying different ages is extremely difficult using the TSVQ algorithm, particularly in a probabilistic environment and with unprofessional equipment. This problem has been addressed by our algorithm, which is based on Markov Random Field (MRF), MRF is the key aspect of our research. The paper is organized as follows:

- The simplicity and effectiveness of the proposed approach determining the three-dimensional microstructure of cement are described with respect to the TSVQ algorithm.
- The serial program of the texture synthesis for the three-dimensional microstructure of cement is analyzed.
- Reliable three-dimensional figures are obtained, which cost less than using CUDA architecture.

The other sections of this paper are described as follows. Related works are presented in Sect. 2. Section 3 briefly describes the TSVQ algorithm. The experimental results and conclusion are discussed in Sects. 4 and 5 respectively.

2 Related Works

Efros and Leung proposed the model of a synthetic method based on MRF during an Institute of Electrical and Electronics Engineers conference in 1999 [7]. Their algorithm assumes that, based on the MRF model, the probability distribution of an arbitrary pixel is only related to the neighborhood points and not to the rest of the points. In most image synthesis cases, their algorithm produces ideal results. The performance of texture synthesis depends on the size of the neighborhood window. Thus, repeated experiments are necessary. However, this algorithm is time consuming, which is its main drawback.

Wei and Levo [6] developed a similar and also used L type neighborhood. Their algorithm does not utilize the probability function synthesize points. Instead, the neighborhood matching results are used directly as the basis of synthetic points instead. The efficiency of this algorithm is improved by adopting multiple resolutions.

Ashikhmin [8] proposed a synthesis algorithm that specifically focus on natural textures and this algorithm achieved a certain degree of progress. The efficiency of this algorithm was improved because the searching range was reduced, and the algorithm would only search the neighborhood. However, this approach is less effective for smooth images.

Xu and Yu [9] proposed multi-seeds fast texture synthesis algorithm. They adopted the principle of spiral linear search and added multiple seeds into the texture synthesis sample. These steps improved the speed of synthesis.

The image quilting algorithm [10] can synthesize more than one pixel for texture synthesis each time. Thus, it is more efficient than the previous method. However, when this approach is used, texture synthesis becomes complicated and gaps easily result, which reduces the quality of the synthesis.

Liang [11] proposed an algorithm of real-time texture synthesis. The algorithm is based on synthetic methods with the optimized KD tree, quad tree pyramids, and principal component analysis to achieve the objective of real-time texture synthesis.

Cohen [12] presented a new texture synthesis algorithm based on the concept of Wang. This approach synthesizes texture images according to the rules based on the set of Wang.

Kwatra [13] proposed a texture synthesis algorithm based on graph cut. This algorithm adopts texture block replication and handles the overlapping areas of texture blocks. Although this method has achieved considerable process, it is computationally complex and extremely time-consuming.

The difficulties faced by obtaining three-dimensional cement microstructure and the acceleration achieved by high performance tools inspire us to explore synthesizing texture of three-dimensional cement microstructure for cement hydration using CUDA.

3   Methodology

An algorithm which is derived from MRF texture models was proposed by Wei. The algorithm generates texture through a deterministic searching process. It accelerates the synthesis process using tree-structured vector quantization. His method uses a Gaussian pyramid with level L from low to high resolutions. The Gaussian pyramid is built through successive filtering and sampling operations. Each pyramid level, except for the level with the highest resolution, is a blurred and decimated version of the original image. To improve the algorithm, the original solid of cement is scanned directly instead of using the Gaussian pyramid. The scanning process is illustrated in Fig. 1, which simplified the synthesis process.

Fig. 1. The process scanning of 3D cement solid

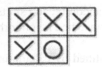

Fig. 2. Causal neighborhood of two-dimensional

Fig. 3. 3D causal neighborhood

Firstly, the two-dimensional neighborhood must be extended to higher dimensions. The two-dimensional L-type neighborhood is shown in Fig. 2, it reveals the neighborhood of "O", the value of "O" is determined by "X". As the isotropic feature of cement, causal neighborhood is used to 3D cement texture synthesis. The causal three-dimensional neighborhood is presented in Fig. 3. Parallel CUDA is adopted to speed up the texture synthesis of three-dimensional cement. The analysis indicates that the serial process can seek for highly repeated parts, which is time consuming.

Using MRF as the texture model, we aim to generate cement concrete solid that is similar to the example three-dimensional solid. Our algorithm is briefly described as follows. Table 1 lists the symbols for the reference.

Table 1. Table of symbols

Symbol	meaning
A_1	Input 2D texture image of cement
A_2	Output 2D texture image of cement
C_1	Input 3D solid of cement
C_2	Output 3D solid of cement
g_i	Value of a pixel in A_1
G_i	Value of a pixel in C_1
O_i	Value of a pixel in A_2
o_i	Value of a pixel in C_2
p_i	An output pixel in A_2 or C_2
N(p)	Neighborhood around the pixel p

First, the model generates C_2 randomly, and starts scanning from the first point p in C_2. The model can search the best matched neighborhood from C_1 until all the pixels in C_2 are traversed.

The best neighborhood is identified by calculating the Euclidean distance. A point in C_1, g_i is a pixel of $N(p)$ in A1. o_i is a pixel of $N(p)$ in A_2. d is the sum of the distance between g_i and o_i. The Eq. 1 is shown as follows,

$$d = \sqrt{\sum_{i=1}^{n} |g_i - o_i|} \tag{1}$$

The Eq. 2 for Euclidean distance of three-dimensional solid of cement is represented as follows.

$$D = \sqrt{\sum_{i=1}^{n} |G_i - O_i|} \tag{2}$$

The best matching neighborhood, which has the minimum value D, $D_{min} < D$, can be established. The gray information of G_i is assigned to the O_i in G_i. The procedure is repeated until finding the best matching neighborhood is identified. Using this method in spite of the good simulation results, while because of the huge amount of data. Therefore CUDA can improve the speed of the algorithm.

3.1 Analysis of the Serial Process

Before parallelizing the algorithm, the analyzes the serial process to identify the time-consuming component. The distribution of the time cost ratio is shown in Fig. 4. The result indicates that found that randomly generating a new three-dimensional solid structure, calculating the Euclidean distance and searching for the best matching neighborhood are major steps in the serial process. The percentage of calculating the Euclidean distance is 87.86 %.

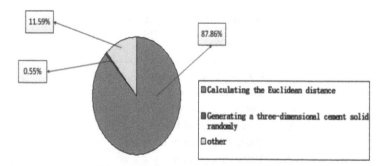

Fig. 4. The time consumption analysis of serial process (Color figure online)

The size of the original solid cement is assumed as $60 \times 60 \times 60$ and that of the neighborhood is $23 \times 23 \times 23$; hence, 2.8×10^{14} calculations are necessary to determine the Euclidean distance to create a cement solid. The calculation process can be parallelized using CUDA architecture.

3.2 Parallelization

The fraction, which requires extensive computation, can be parallelized to decrease the amount of computation significantly. Under this condition, the analysis and design of a parallel algorithm are as follows.

The size of a neighborhood should be set. The neighborhood of a new point can be acquired from the original solid. All the data are copied from the memory to graphics memory. The Euclidean distance, which can be parallelized using the GPU thread, is calculated. The Euclidean distance represents the neighborhood distance between the original and output three-dimensional solid. In the following step, the data in the graphics memory can be transmitted to the memory. The best matching neighborhood is searched. The gray value of the new synthesis point is replaced by the center point. This step is repeated until the three-dimensional solid is synthesized. The process is described in detail in Fig. 5.

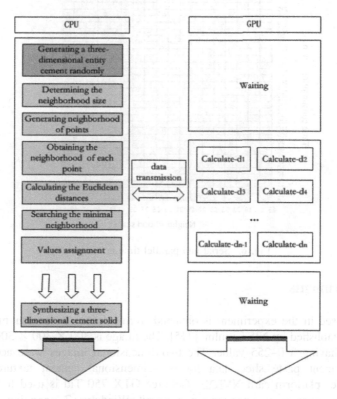

Fig. 5. Parallel temporal diagram

Table 2 presents the speed ratio of the parallel and serial processes. The practice shows that the efficiency of three-dimensional texture synthesis has increased by 1292.4375 %. The efficiency decreases exponentially when CUDA is used. Therefore our algorithm exhibits high efficiency.

The experimental result indicates that the parallel process is over a thousands times higher than the serial process and can be significantly improved by increasing the size of the cement solid, as shown in Fig. 6.

Table 2. Time cost of serial and parallel

Size	Run time (hours)		Speedup
	GPU parallel	CPU serial	
30 × 30 × 30	0.0240	31.0425	1293.4375

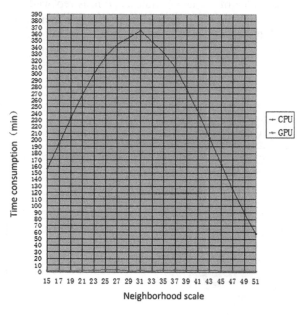

Fig. 6. Serial and parallel time consuming

4 Experiments

The data used in the experiment is obtained from http://visiblecement.nist.gov [14], which is established by NIST [online] [15]. The image is 300 × 300 × 300 bytes with each byte having a 0–255 value. The two-dimensional images were acquired from original cement paste slice data for two-dimensional cement texture synthesis. The graphics platform card NVIDIA GeForce GTX 750 Tid is used for the experiments. The program is operated on a 64-bit Windows 7 operating system and Intel^{R}Core^{TM}2 Quad GPU. Moreover, the process and the method for synthesizing the three-dimensional solid texture of cement are expounded, based on Microsoft Visual Studio which is adopted as the development platform, uses the programming language C. Finally, 3DMed [16] is used to export the result of the experiments.

First, the TSVQ algorithm is implemented to synthesize the two-dimensional image of cement. The neighborhood types are shown in Fig. 7. The size of the neighborhood is denoted as N × N. The output pictures of the experiment results are depicted in Fig. 8.

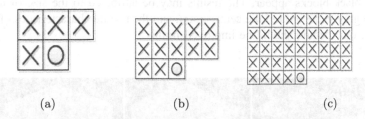

(a)   (b)   (c)

Fig. 7. The neighborhood sizes are (a) 3 × 3, (b) 5 × 5, (c) 9 × 9

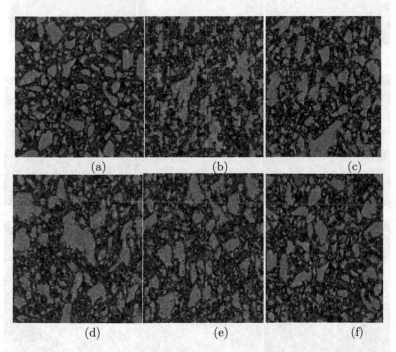

Fig. 8. A comparison of texture synthesis results using different size of two-dimensional neighborhood in TSVQ algorithm. (a) is the sample texture The neighborhood sizes are (b) 3 × 3, (c) 5 × 5, (d) 7 × 7, (e) 9 × 9, (f) 51 × 51. All images shown are of size 100 × 100.

The quality of the output with a 5×5 neighborhood is poorer than that of (a) picture. However, the 7×7 and 51×51 neighborhoods produce better results than the other sizes. The quality of the outcome is undulatory, but does not exhibit a linear relationship. The results of three-dimensional cement microstructure are presented in Fig. 9. Overall, the three-dimensional cement microstructure with a size of $19 \times 19 \times 19$ is better than the others. A smaller neighborhood such as a $13 \times 13 \times 13$ neighborhood, produces inferior results. The outcomes are similar to those of two-dimension texture, which increase unstably. In a $21 \times 21 \times 21$ neighborhood, black blocks appear. The results may be attributed to the size of neighborhood, the gray value around the center point, and the texture characteristics of cement, etc. Thus, our method should be improved further.

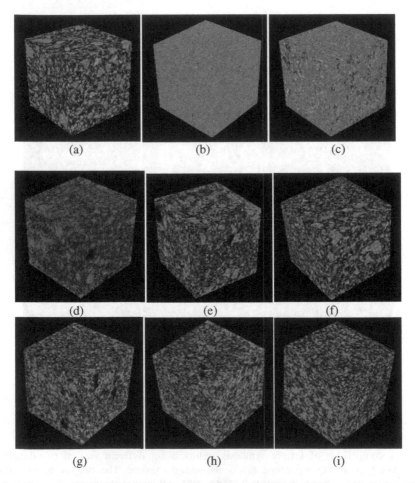

Fig. 9. Synthesis of 3D cement results with different neighborhood sizes. (a) is the original three-dimensional of cement. The neighborhood size of (b) $13 \times 13 \times 13$, (c) $15 \times 15 \times 15$, (d) $17 \times 17 \times 17$, (e) $19 \times 19 \times 19$, (f) $21 \times 21 \times 21$, (g) $23 \times 23 \times 23$, (h) $25 \times 25 \times 25$, (i) $51 \times 51 \times 51$. All images shown are of size $60 \times 60 \times 60$.

5 Conclusion

The three-dimensional synthesis process is an extremely difficult and complicated problem in computer graphics. The main objective of this study is to provide a simple yet efficient method for texture synthesis. Although studies on cement have effective method has not yet been presented and relevant studies remain minimal. Various natural phenomena can be studied for the 3D texture synthesis of cement. Therefore, an algorithm that is sufficiently simple and low cost will be effective.

This study presents that fast synthesis of three-dimensional solid in CUDA and analyzes the influence of neighborhood size on physical composition of cement. It can simulate different ages of cement hydration. The experiment result that, assuming that the original physical dimensions of the three-dimensional cement structure is $N \times N \times N$, the best neighborhood is $N/3$, it begins to slowly increase the size of the neighborhood. The result of the synthesis of the three-dimensional cement structure continues to improve. However, the quality of the result is reduced when $N \geq N/3$.

Our algorithm takes advantage of the TSVQ algorithm for causal neighborhood, and appropriately accelerates the algorithm using CUDA parallel architecture. In contrast with the serial process, determining the optimal neighborhood size significantly increases the efficiency of the three-dimensional simulation of cement. This approach provides a basis for the synthesis of three-dimensional cement solid synthesis using 2D images.

Acknowledgments. This work was supported by National Natural Science Foundation of China under Grant No. 61573166, No. 61572230, No. 61373054, No. 61472164, No. 81301298, No. 61302128, No. 61472163. Shandong Provincial Natural Science Foundation, China, under Grant ZR2015JL025. National Key Technology Research and Development Program of the Ministry of Science and Technology under Grant 2012BAF12B07-3. Science and technology project of Shandong Province under Grant No. 2015GGX101025. Jinan Youth Science & Technology Star Project under Grant No. 2013012.

References

1. 袁润章. 胶凝材料学. 武汉工业大学出版社, 武汉 (1996)
2. Chen, W., Brouwers, H.J.H.: Mitigating the effects of system resolution on computer simulation of Portland cement hydration. Cem. Concr. Comp. **30**, 779–787 (2008)
3. Wei, L.Y.: Texture synthesis from multiple sources. In: ACM SIGGRAPH 2003 Sketches and Applications. ACM, New York (2003)
4. Qin, X.J., Yang, Y.H.: Aura 3D textures. IEEE Trans. Visual Comput. Graphics **13**, 379–389 (2007)
5. Witkin, A., Kass, M.: Reaction-diffusion textures. Comput. Graph. **25**(4), 299–308 (1991)
6. Wei, L.-Y., Levoy, M.: Fast texture synthesis using tree-structured vector quantization. In: Akeley, K. (ed.) SIGGRAPH 2000, Computer Graphics Proceedings, pp. 479–488. ACM Press/ACM SIGGRAPH/Addison Wesley Longman (2000)
7. Wei, L.Y.: Texture synthesis by fixed neighborhood searching. Ph.D. dissertation. Stanford University, Stanford (2002)

8. Ashikhmin, M.: Synthesizing natural textures. In: 2001 ACM Symposium on Interactive 3D Graphics, March, pp. 217–226 (2001)

9. 徐晓刚,于金辉,马利庄.多种子快速纹理合成.中国图形图像学报. **7**(10), 995–999 (2002)

10. Efros, A.A., Freeman, W.T.: Image quilting for texture synthesis and transfer. In: Proceedings of SIGGRAPH 2001, pp. 341–347. ACM Press, New York (2001)

11. Liang, L., Liu, C., Xu, Y., Guo, B., Shum, H.Y.: Real-time texture synthesis by patch-based sampling. Technical Report MSR-TR-2001-40, Microsoft Research, March 2001

12. Cohen, M.F., Shade, J., Hiller, S., et al.: Wang tiles for image and texture generation. In: Annual Conference Series ACM SIGGRAPH 2003, pp. 286–294, San Diego (2003)

13. Kwatra, V., Schodl, A., Essa, I., Turk, G., Bobick, A.: Graphcut textures: image and video synthesis using graph cuts. ACM Trans. Graph. **22**(3), 277–286 (2003)

14. Visible Cement Dataset. http://visiblecement.nist.gov

15. Bentz, D.P., Mizell, S., Satterfield, S., Devaney, J., George, W., Ketcham, P., Graham, J., Porterfield, J., Quenard, D., Vallee, F., Sallee, H., Boller, E., Baruchel, J.: The visible cement data set. NIST J. Res. **107**(2), 137–148 (2002)

16. Key Laboratory of molecular Imaging, Chinese Academy of Sciences. http://www.3dmed.net

Multiphase Image Segmentation
Based on Improved LBF Model

Ji Zhao[(⊠)], Huibin Wang, and Han Liu

School of Software, University of Science and Technology Liaoning,
Anshan, Liaoning, China
zhaoji_1974@126.com

Abstract. In view of the problem of low efficiency of image segmentation with intensity inhomogeneity and the problem of the multi object image can't be segmented, a new multi-phase image segmentation algorithm based on HLBF model is proposed. The application of magnetic resonance imaging in medicine is used to demonstrate the validity of the model. The proposed model replaces the Gauss kernel function in the original LBF model with the new kernel function to improve the time efficiency. Meanwhile, the HLBF model is further integrated into the variational level set of multi-phase image segmentation strategy to achieve the segmentation of multi-phase image with intensity inhomogeneity. Experimental results show the efficiency of the proposed method. The proposed model has advantages over the traditional segmentation method in terms of time efficiency and accuracy.

Keywords: Image segmentation · Intensity inhomogeneity · Level set · Multi-phase image · Kernel function

1 Introduction

In recent years, brain disease has become one of the diseases with high incidence. In order to analyze brain structure better, brain magnetic resonance imaging technology is widely used in medical field. So the accurate segmentation of magnetic resonance imaging for medical diagnosis is very important. In the actual scene, brain imaging will be affected by the device itself and the magnetic field which result the intensity inhomogeneity of the organization in different parts of the magnetic resonance imaging. In addition, the brain map is usually made of dark gray, white, light gray and black, but the use of the previous fitting of the global information of the segmentation model is not achieved.

So far, some segmentation methods are used in the field of image processing. In 1987, Kass et al. [1] proposed the first parameter active contour model, namely Snake model which is also one of the most classical parameter active contour models. Mumford and Shah [2] proposed a region segmentation model, namely Mumford-Shah model. Chan and Vese [3] improved Mumford-Shah model, namely C-V model, which assume that the image is divided into two parts, and the target is different from the

© Springer International Publishing Switzerland 2016
D.-S. Huang and K.-H. Jo (Eds.): ICIC 2016, Part II, LNCS 9772, pp. 635–644, 2016.
DOI: 10.1007/978-3-319-42294-7_57

background. In 2002, Vese and Chan [4] improved the original C-V model by using the multiphase level set framework, namely piecewise constant model (PC model) which uses all gray information of the inner and the exterior, so it is robust to the image with noise and weak edges. But PC model will produce the error segmentation when the gray level is not uniform. To overcome this problem, Tsai et al. [5] proposed piecewise smooth model (PS model) which is applied to the multiphase images of intensity inhomogeneity. But the level set function need to be added to the two partial differential equations, which greatly increases the amount of computation. Li et al. [6] introduced the local information of the image into the model of region segmentation, the proposed model based on the region of scalable energy is called LBF (Local Binary Fitting) model. The new model can get more accurate segmentation results for images with intensity inhomogeneity, but only the two phase images can be segmented. In addition, the model needn't reinitialize the level set which speeds up the evolution to some extent. But LBF model is still difficult to meet the requirements of high real-time applications. Chen et al. [7] proposed a method for brain image segmentation based on FCM and level set. Although this method can be used for the accurate segmentation of brain tissues, it is not suitable for images with intensity homogeneity.

For accurate and effective segmentation of MRI, a multiphase segmentation algorithm is proposed. In order to improve the segmentation effect, the image is preprocessed by the method of denoising and histogram equalization to enhance the gray contrast of the image. For the disadvantages of the LBF model calculation, the reasons for low efficiency of the segmentation is analyzed and found out. A new type of kernel function is proposed, the HLBF model is constructed and the validity of the method is demonstrated by experiments. A multiphase image segmentation algorithm based on HLBF model by combining the N level set function is proposed to segment the 2N phase image.

2 Improved LBF Model

LBF model proposed by Li et al. [6] was integrated with local area information. Its essence is the introduction of weight kernel function in the energy functional by the size of its value to control the fitting of the regional area [8, 9].

The Gauss kernel function is used in the LBF model, and its general expression is as follows:

$$K_\sigma(x) = \frac{1}{(2\pi)^{1/2}\sigma} e^{-|x|^2/2\sigma^2} \tag{1}$$

In the process of image segmentation, the evolution of the curve depends on the kernel function to some extent. LBF model show that the model with Gaussian kernel function not only has general characteristics kernel function, but also has inflection point and "3σ" characteristics.

1. First, the two order derivative of Gauss kernel function:

$$K_\sigma''(x) = \frac{1}{(2\pi)^{1/2}\sigma^5}(x^2 - \sigma^2)e^{-x^2/2\sigma^2} \tag{2}$$

When $x = \sigma$, $K_\sigma''(x) = 0$. There is a turning point in the area where the function value will be changed.

2. When $|x| \geq 3\sigma$, $K_\sigma(x) \cong 0$;

These two features are the characteristics of the Gauss kernel itself, which is also the factor that causes the low efficiency of the curve evolution in the segmentation process. So this paper will propose a new type of kernel function, expression is as follows:

$$K = \frac{1}{4} - \frac{1}{2\pi}\arctan(x^2) = \frac{\pi - 2\arctan(x^2)}{4\pi} \tag{3}$$

The graph features of the new kernel function and the Gauss kernel function are shown in Fig. 1.

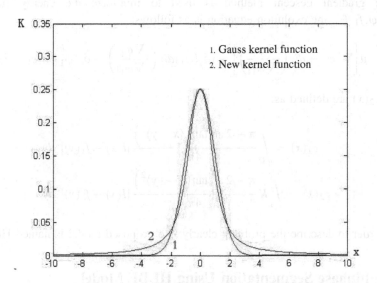

Fig. 1. Comparison chart of New Kernel and Gauss Kernel function

By comparison, the new kernel function is similar to the image features of Gauss kernel function and the change of the Gauss kernel function is slow, and the expression is relatively simple.

Therefore, we replace the original Gauss kernel function with proposed new kernel function in the LBF model, and the energy function is defined as:

$$E^{HLBF}(\varphi_1, f_1, f_2) = \lambda_1 \int \left[\int \frac{\pi - 2 \arctan\left((x - y)^2\right)}{4\pi} |I(y) - f_1(x)|^2 \right.$$

$$\left. * \left(\frac{1}{2} + \frac{1}{\pi} \arctan\left(\frac{\varphi_1(y)}{\varepsilon} \right) \right) dy \right] dx + \lambda_2 \int \left[\int \frac{\pi - 2 \arctan\left((x - y)^2\right)}{4\pi} |I(y) - f_2(x)|^2 \right.$$

$$\left. * \left(\frac{1}{2} + \frac{1}{\pi} \arctan\left(\frac{\varphi_1(y)}{\varepsilon} \right) \right) dy \right] dx + \mu \int \frac{1}{2} (|\nabla \varphi_1(x)| - 1)^2 dx + v \int \delta(\varphi_1(x)) |\nabla \varphi_1(x)| dx$$

(4)

$f_1(x) f_2(x)$ are given as:

$$f_1(x) = \frac{K(x) * (I(x) H(\varphi_1(x)))}{K(x) * H(\varphi_1(x))} \tag{5}$$

$$f_2(x) = \frac{K(x) * (I(x)(1 - H(\varphi_1(x))))}{K(x) * (1 - H(\varphi_1(x)))} \tag{6}$$

The gradient descent method is used to minimize the energy functional $E^{HLBF}(\varphi, f_1, f_2)$, the evolution equation is as follows:

$$\frac{\partial \varphi_1}{\partial t} = \mu \left[\nabla^2 \varphi_1 - div \left(\frac{\nabla \varphi_1}{|\nabla \varphi_1|} \right) \right] + v \delta_\xi(\varphi_1) div \left(\frac{\nabla \varphi_1}{|\nabla \varphi_1|} \right) - \delta_\xi(\varphi_1)(\lambda_1 e_1 - \lambda_2 e_2) \tag{7}$$

$e_1(x)$, $e_2(x)$ are defined as:

$$e_1(x) = \int_\Omega \frac{\pi - 2 \arctan\left((x - y)^2\right)}{4\pi} |I(x) - f_1(y)|^2 dy$$

$$e_2(x) = \int_\Omega K \frac{\pi - 2 \arctan\left((x - y)^2\right)}{4\pi} |I(x) - f_2(y)|^2 dy$$

(8)

In order to describe the problem clearly, the proposed model is named HLBF.

3   Multiphase Segmentation Using HLBF Model

At present, the method of multiphase level set segmentation can be classified into four categories [10]: N level set function partition N phase image, N level set function partition 2N phase image, 1 level set function partition N + 1 phase image, N − 1 level set function partition N phase image.

For multiple target, a new multiphase image segmentation model is proposed based on HLBF model by combining the N level set function to segment the 2N phase image.

For example, the two level set is divided into four phase inhomogeneous image energy functional as follows:

$$
\begin{aligned}
E^{HLBF}(\varphi_1,\varphi_2,f_{11},f_{10},f_{01},f_{00}) = \lambda_{11} \int & \left[\int \frac{\pi - 2\arctan\left((x-y)^2\right)}{4\pi} |I(y) - f_{11}(x)|^2 \right. \\
& \left. * \left(\frac{1}{2} + \frac{1}{\pi}\arctan\left(\frac{\varphi_1(y)}{\varepsilon}\right)\right)\left(\frac{1}{2} + \frac{1}{\pi}\arctan\left(\frac{\varphi_2(y)}{\varepsilon}\right)\right) dy \right] dx \\
+ \lambda_{10} \int & \left[\int \frac{\pi - 2\arctan\left((x-y)^2\right)}{4\pi} * |I(y) - f_{10}(x)|^2 \left(\frac{1}{2} + \frac{1}{\pi}\arctan\left(\frac{\varphi_1(y)}{\varepsilon}\right)\right) \right. \\
& \left. * \left(\frac{1}{2} + \frac{1}{\pi}\arctan\left(\frac{1-\varphi_2(y)}{\varepsilon}\right)\right) dy \right] dx + \lambda_{01} \int \left[\int \frac{\pi - 2\arctan\left((x-y)^2\right)}{4\pi} \right. \\
& * |I(y) - f_{01}(x)|^2 \left(\frac{1}{2} + \frac{1}{\pi}\arctan\left(\frac{1-\varphi_1(y)}{\varepsilon}\right)\right) \\
& \left. * \left(\frac{1}{2} + \frac{1}{\pi}\arctan\left(\frac{\varphi_2(y)}{\varepsilon}\right)\right) dy \right] dx + \lambda_{00} \int \left[\int \frac{\pi - 2\arctan\left((x-y)^2\right)}{4\pi} \right. \\
& \left. * |I(y) - f_{00}(x)|^2 \left(\frac{1}{2} + \frac{1}{\pi}\arctan\left(\frac{1-\varphi_1(y)}{\varepsilon}\right)\right)\left(\frac{1}{2} + \frac{1}{\pi}\arctan\left(\frac{1-\varphi_2(y)}{\varepsilon}\right)\right) dy \right] dx \\
+ \mu \int & \frac{1}{2}(|\nabla \varphi_1(x)| - 1)^2 dx + \mu \int \frac{1}{2}(|\nabla \varphi_2(x)| - 1)^2 dx \\
+ \nu \int & \delta(\varphi_1(x))|\nabla \varphi_1(x)| dx + \nu \int \delta(\varphi_2(x))|\nabla \varphi_2(x)| dx
\end{aligned}
$$

$$(9)$$

For the determination of f, the following convolution model is proposed:

$$
f_{11}(x) = \frac{K(x) * (I(x)(H(\varphi_1(x))H(\varphi_2(x))))}{K(x) * (H(\varphi_1(x))H(\varphi_2(x)))} \tag{10}
$$

$$
f_{10}(x) = \frac{K(x) * (I(x)(H(\varphi_1(x))(1 - H(\varphi_2(x)))))}{K(x) * (H(\varphi_1(x))(1 - H(\varphi_2(x))))} \tag{11}
$$

$$
f_{01}(x) = \frac{K(x) * (I(x)((1 - H(\varphi_1(x)))H(\varphi_2(x))))}{K(x) * ((1 - H(\varphi_1(x)))H(\varphi_2(x)))} \tag{12}
$$

$$
f_{00}(x) = \frac{K(x) * (I(x)((1 - H(\varphi_1(x)))(1 - H(\varphi_2(x)))))}{K(x) * ((1 - H(\varphi_1(x)))(1 - H(\varphi_2(x))))} \tag{13}
$$

In order to clearly describe the implementation process of the algorithm, The algorithm flow diagram is shown in Fig. 2.

4 Experimental Results and Analysis

Algorithm development platform is Visual C ++ 6.0 and Matlab R2013a.

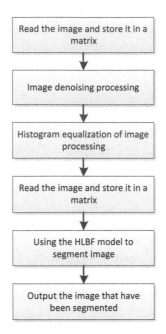

Fig. 2. Flow diagram of the algorithm

4.1 Algorithm Evaluation Criteria

Kappa coefficient [11, 12] is defined as:

$$K = \frac{N \sum_{i=1}^{r} m_{ij} - \sum (m_{i+} \cdot m_{+i})}{N^2 - \sum (m_{i+} \cdot m_{+i})} \tag{14}$$

Parameter description: N is the total number of samples. R is the number of rows of the matrix. m_{i+} is the sum of the i lines of the matrix. m_{+i} is the sum of the i column of the matrix.

When the value of Kappa is 0.81–1, the segmentation accuracy is almost the same. When the value of Kappa is 0.61–0.80, the segmentation accuracy is high. When the value of Kappa is 0.41–0.60, the accuracy of the segmentation is moderate. When the value of Kappa is 0.21–0.40, the segmentation accuracy is general. When the value of Kappa is 0.0–0.20, the segmentation accuracy is very low.

4.2 Experimental Comparison

Contrast experiment 1: MRI of brain 1 is axial weighted image, TR value is low, so the image clarity is relatively low (Fig. 3).

Illustration: (a) is initial contour; (b) is the result of LBF model. It can be seen that the gray matter is not completely separated. (c) is the result of PC multiphase model.

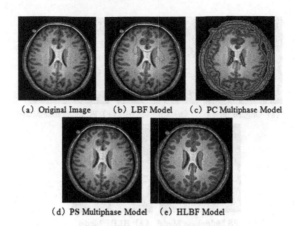

(a) Original Image   (b) LBF Model   (c) PC Multiphase Model

(d) PS Multiphase Model   (e) HLBF Model

Fig. 3. Comparison of segmentation results of brain 1

Obviously the results of the model are not satisfactory. (d) is the result of PS multiphase model. It is better than the first two methods of segmentation results. But the accuracy of the segmentation of individual parts of the brain gray matter and cerebrospinal fluid is general. (e) is the result of HLBF multiphase model. This model can segment the brain white matter, gray matter and cerebrospinal fluid accurately.

Contrast experiment 2: MRI of brain 2 is axial weighted image, TR value is high, so the image clarity is relatively high (Fig. 4).

Illustration: (a) is initial contour; (b) is the result of LBF model. Although the white matter of the brain is roughly divided, some of the regions still are segmented by mistake. The cerebrospinal fluid was almost not segmented. The gray areas are not divided in the middle of the brain. (c) is the result of PC multiphase model. The segmentation effect of the model is not ideal. (d) is the result of PS multiphase model. Compared with the previous two methods, the segmentation result is improved, but the cerebral gray matter is under-segmentation, and the individual regions of the cerebrospinal fluid are not separated. (e) is the result of HLBF multiphase model. The segmentation results are satisfactory.

In summary, The LBF model can't segment the multiphase image of intensity inhomogeneity effectively. Although the PC multiphase model is suitable for the multiphase images, it will produce the wrong segmentation when the image is not uniform. The PS multiphase model further solves the phenomenon of image with intensity inhomogeneity, but the model is affected by the initial contour, even if the same image will get different results. Through a large number of experiments, the proposed algorithm can segment the magnetic resonance images of the brain accurately.

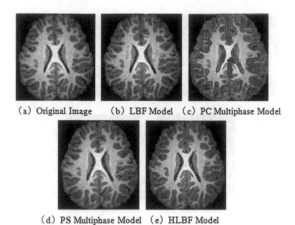

(a) Original Image   (b) LBF Model (c) PC Multiphase Model

(d) PS Multiphase Model (e) HLBF Model

Fig. 4. Comparison of segmentation results of brain 2

4.3   Segmentation Evaluation

4.3.1   Efficiency of Segmentation Time

Figure 5 shows the segmentation time of four MRI images by LBF, PC, PS and HLBF methods respectively.

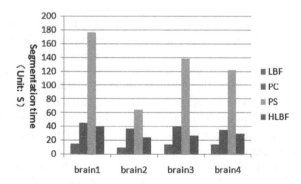

Fig. 5. Comparison of segmentation time (Color figure online)

It is obvious that the new model is shorter than the PC and PS. The reason is that the PS multiphase model needs to be initialized level set function as a symbol distance function in the segmentation process. The proposed model includes the penalty term, which guarantees the evolution of the level set function as SDF, which reduces the time loss of the process. In addition to the above aspects, the PC model is based on the global information of the algorithm, which will lead to more time consumption.

In addition, although the LBF model is shorter than the proposed model, the brain can't be divided into different parts. The reason is that the LBF model contains only one level set function, so the time is shorter.

4.3.2 Accuracy of Segmentation

The segmentation accuracy will be based on the evaluation criteria of Kappa coefficients. The segmentation accuracy of four kinds of brain maps by four different segmentation models are demonstrated respectively (Fig. 6).

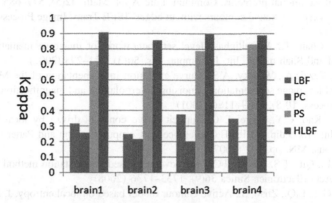

Fig. 6. Comparison of segmentation accuracy (Color figure online)

We obtain the segmentation accuracy and time of the 80 experimental images. The mean time of LBF model is 13.48875s. The PC multiphase model is 41.24713s. The PS multiphase model is 125.49055s. The HLBF multiphase model is 32. 756325s. The mean value of Kappa coefficient in LBF model is 0.3271. The PC multiphase model is 0.2093. The PS multiphase model is 0.7558. The HLBF multiphase model is 0.8957. We find that the proposed HLBF model has obvious advantages in time efficiency or accuracy for the segmentation of brain image.

5   Conclusion

By improving the original LBF model, the new kernel function is replaced by the new Gauss kernel function, namely HLBF model. The experimental results show that the proposed method is more efficient than the original LBF model and the time efficiency is improved by 30 %–50 %. In order to segment the structure of the MRI of brain, the multiphase image segmentation strategy is further integrated into the variational level set. Experimental results show that the proposed algorithm can segment the brain white matter, gray matter and cerebrospinal fluid and is better in time efficiency and accuracy.

Acknowledgement. This research is funded by the Education Department of Liaoning Province Foundation grant Number LJQ2014033 and University of Science and Technology Liaoning Foundation grant Number 2013RC08.

References

1. Kass, M., Witkin, A., Terzopoulos, D.: Snake active contour models. Int. J. Comput. Vis. **1**(4), 321–331 (1988)
2. Mumford, D., Shah, J.: Optimal approximations by piecewise smooth functions and associated variational problems. Commun. Pure Appl. Math. **42**(5), 577–685 (1989)
3. Chan, T., Vese, L.: Active contours without edges. IEEE Trans. Image Process. **10**(2), 266–277 (2001)
4. Vese, L., Chan, T.: A multiphase level set framework for image segmentation using the Mumford and Shah model. Int. J. Comput. Vis. **50**(3), 271–293 (2002)
5. Tsai, A., Yezzi, A., Willsky, A.S.: Curve evolution implementation of the Mumford-Shah functional for image segmentation, denoising, interpolation, and magnification. IEEE Trans. Image Process. **10**(8), 1169–1186 (2001)
6. Li, C.M., Kao, C.Y., Gore, J. C: Implicit active contours driven by local binary fitting energy. In: Proceeding of IEEE Conference on Computer Vision and Pattern Recognition, Minneapolis, MN, pp. 1–7 (2007)
7. Chen, Z.B., Qiu, T.S., Su, R.: FCM and level set based segmentation method for brain MR images. Acta Electronica Sinica **36**(9), 1733–1736 (2008)
8. Pan, G., Gao, L.Q., Zhao, S.: Active contour model based on local entropy. J. Image Graph. **12**(1), 78–85 (2013)
9. Oliver, G., Klaus, D.T., Volkmar, L.: Prior shape level set segmentation on multistep generated probability maps of MR datasets for fully automatic kidney parenchyma volumetry. IEEE Trans. Med. Imaging **33**(2), 312–325 (2012)
10. Lin, Y: Research on Key Techniques of Image Segmentation Based on Level Set Method. Harbin Engineering University (2010)
11. Wang, J: Study on the Application of Kappa Coefficient in the Consistency Evaluation. Sichuan University (2006)
12. Salah, M.B., Mitiche, A., Ayed, I.B.: Efficient level set segmentation with a kernel induced data term. IEEE Trans. Image Process. **19**(2), 220–223 (2010)

Multi-scale Spectrum Visual Saliency Perception via Hypercomplex DCT

Limei Xiao[1], Ce Li[1,2(✉)], Zhijia Hu[1], and Zhengrong Pan[1]

[1] College of Electrical and Information Engineering,
Lanzhou University of Technology, Lanzhou 730050, China
xjtulice@gmail.com
[2] School of Electronic and Information Engineering,
Xi'an Jiaotong University, Xi'an 710049, China

Abstract. Based on the salient object of human visual perception inconsistent scale, this paper proposes a multi-scale spectrum visual saliency perception with hypercomplex discrete cosine transform. In hypercomplex image color space to build parallel computing model, the method use HDCT to extract local spectral feature in an image. Meanwhile, the sparse energy spectrum calculated by hypercomplex discrete cosine transform on local image region was taken as visual stimulation signal. Then a visual saliency measurement was taken on both this region and its neighbor regions. Finally, the multi-sacle normalization was on the visual saliency response. The subjective and objective experimental results on the public saliency perception datasets demonstrated that both the precision and time cost of the proposed approach were better than the other state-of the-art approaches.

Keywords: Hypercomplex DCT · Visual saliency · Multi-scale · Paralleled computation

1 Introduction

Visual saliency perception is an important cognitive mechanism, which is formed in the long-term of natural selection and evolution process for adapting to complex external environment with the limited number of neurons. To percept the saliency visual information is an important content in computer vision, pattern recognition, image processing and understanding, and multimedia processing. It also has broad applications in video surveillance [1], digital copyright protection, media processing, graphic design, communication, entertainment and so on. However, visual perception in sparse representation of the objective visual data at the same time is still a relatively subjective problem. The characteristics of different polymerization process or prior knowledge can lead to different saliency perception. This makes visual perception to be one of the challenging problems in biological visual computing. Visual perception computing has become one of the hotspots in the related research.

Visual attention has two kinds of mechanisms: generation and discriminant. The generation of data driven approach pays attention to the characteristics of contrast. It is a kind of primary visual perception. Hence it is difficult to reflect the "wholeness" and

© Springer International Publishing Switzerland 2016
D.-S. Huang and K.-H. Jo (Eds.): ICIC 2016, Part II, LNCS 9772, pp. 645–655, 2016.
DOI: 10.1007/978-3-319-42294-7_58

"semantic" of the target. Itti's [2] saliency model is the most famous model using data driven. The basic idea is to extract various characteristics from a single input image, such as color, bright, direction etc. Each characteristic's attention image is formed through Gaussian Pyramid and the central-surrounding operator (Center-surround). Then the saliency graph is acquired by normalization. On this basis, through the winner-take-all neural network, the focus point is guided to transfer.

In recent years, a large number of researchers used generation method in the saliency target detection which has made a great progress of saliency testing computing framework. Most visual saliency algorithms built a lot of visual perception calculation models from information theory [3], the graph model [4], Bayesian inference [5], and optimal decision [6], conditional random field [7] and signal processing [8–12], and other theoretical tools.

Task-driven discriminant method is a kind of advanced visual processing mechanism. Due to it is excessive dependent on specific and abstract high-level visual characteristics, the model at this stage remains to be a breakthrough on the implementation mechanism. Navalpakkam and Itti [13] put forward to Itti's model [2] based on the introduction of Top-Down a priori knowledge according to the characteristics of the target detection and recognition, and considered the characteristics of the different weights to find the target objects. The basic idea of this model is to use a priori knowledge of long-term memory to detect the matching degree of related task target in the short-term memory. By studying the underlying data characteristics to compute the biased visual attention, and find the most salience object in a scene. Finally according to task correlation of identifying the target to update the memory, the process of image understanding and parsing finishes. Because of the difference of human knowledge structure, the definition of task is different. There are many discriminative models for calculating the attention. Moran Cerf et al. [14] proposed a method of saliency target detection for the face detection task. Abdollahian and Delp [15] utilized the priori knowledge that the photographer's attention change is the camera focus movement, and built a relationship model between the camera movement and the saliency graph model.

From the perspective of signal processing, the saliency perception signal is taken as a filtering process for a kind of important signal perception in visual signals. In recent years, the signal frequency domain analysis method has been widely introduced into the visual perception work significantly. Hou [8] proposed a visual perception method based on amplitude spectrum of the residual. This method thinks that the difference of the natural image statistics spectrum and the image structure spectrum is saliency target highlight part. Guo [9] noted that the key factors that cause visual stimuli is the phase spectrum but not the amplitude spectrum. Li et al. [10] presented a visual perception model using the optimal scale entropy based on the hypercomplex Fourier transform spectrum. This model supposes that a saliency target in a particular scale space has the most saliency degree. But it ignores the case that the saliency of multi-scale fusion and multiple targets is not necessarily in the same scale space. Fang [11] proposed by hypercomplex Fourier transform spectrum saliency perception locally. But it ignored the position differences between the local information of saliency impact while in computational vision saliency. In addition, Boris [12] used the HDCT pulse spectrum to provide a saliency perception model. Although the algorithm considered the discrete cosine hypercomplex transform in parallel to redundancy, but similar to those in literature [10], is actually of the result in a single scale space.

Based on the above observation, this paper proposes to use the hypercomplex DCT to form the visual stimulation signal for parse energy spectrum of the local image block. We measure the spectral characteristics of the surrounding image block. The final visual saliency perception algorithm is finished by normalization under the multi-scale. Through the experiment tests, the proposed algorithm has higher perception ability.

2   Hypercomplex DCT

The development of the technology of hypercomplex frequency domain analysis makes the researchers naturally begin to pay close attention on the discrete cosine transform method under the Hypercomplex domain. The visual perception [12] and template matching [16] work are applied.

Here, we take a pure imaginary hypercomplex $h(x, y)$ of a M × N three-channel color images I as an example, and introduce the definition and implementation of hypercomplex discrete cosine transform (HDCT). The HDTC have the left and right exchange in two forms. They are defined respectively as follows [16]:

(1) Hypercomplex discrete cosine left transform

$$
F_{HDCT}^L(h(x, y)) = H^L(u, v)
$$
$$
= \lambda^M(u)\lambda^N(v) \sum_{x=0}^{M-1} \sum_{y=0}^{N-1} u_q \cdot h(x, y)\alpha_{u,x}^M \alpha_{v,y}^N
$$
(1)

(2) Hypercomplex discrete cosine right transform

$$
F_{HDCT}^R(h(x, y)) = H^R(u, v)
$$
$$
= \lambda^M(u)\lambda^N(v) \sum_{x=0}^{M-1} \sum_{y=0}^{N-1} h(x, y)\alpha_{u,x}^M \alpha_{v,y}^N \cdot u_q
$$
(2)

where, u_q in Eq. (1) of type (2) is the coordinate axes for HDCT transformation. It is a unit pure imaginary hypercomplex with $u_q^2 = 1$. (u, v) is corresponding to the M × N space coordinates in the corresponding frequency domain. Same as the DCT, the parameters λ and α can be defined:

$$
\lambda_u^M = \begin{cases} \sqrt{1/M}, & for \quad u = 0; \\ \sqrt{2/M}, & for \quad u \neq 0; \end{cases}
$$
(3)

$$
\lambda_v^N = \begin{cases} \sqrt{1/N}, & for \quad v = 0; \\ \sqrt{2/N}, & for \quad v \neq 0; \end{cases}
$$
(4)

$$\alpha_{u,x}^M = \cos\left[\frac{\pi}{M}\left(x + \frac{1}{2}\right)u\right] \tag{5}$$

$$\alpha_{v,y}^N = \cos\left[\frac{\pi}{N}\left(y + \frac{1}{2}\right)v\right] \tag{6}$$

After the HDCT transformation, image hypercomplex $h(x, y)$ is still a hypercomplex. Same as hypercomplex discrete Fourier transform, HDCT also has the corresponding left and right transformation, the definitions are respectively:

$$IF_{HDCT}^L(H^L(u, v)) = h(x, y)$$
$$= \sum_{u=0}^{M-1}\sum_{v=0}^{N-1} \lambda^M(u)\lambda^N(v)u_q \cdot H^L(u, v)\alpha_{u,x}^M\alpha_{v,y}^N \tag{7}$$

$$IF_{HDCT}^R(H^R(u, v)) = h(x, y)$$
$$= \sum_{u=0}^{M-1}\sum_{v=0}^{N-1} \lambda^M(u)\lambda^N(v)H^R(u, v)\alpha_{u,x}^M\alpha_{v,y}^N \cdot u_q \tag{8}$$

here, the parameters λ and α are defined consistent with HDCT, $u_q = -\sqrt{1/3}i_1 - \sqrt{1/3}i_2 - \sqrt{1/3}i_3$.

3   Multi-scale HDCT Visual Saliency Perception Algorithm

Visual scale characteristics are the basic property of human visual. The description of the image detail is different under different scales. With different information entropy, the human eye's visual stimulus energy input is also not the same. To this end, we introduce the visual scale information to the proposed multi-scale HDCT visual perception algorithm. The algorithm framework is shown in Fig. 1.

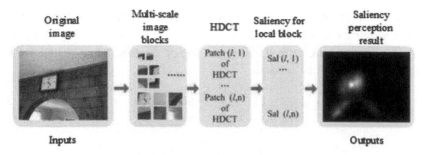

Fig. 1. The framework of the multi-scale spectrum visual saliency perception with hypercomplex DCT.

3.1   Multi-scale Image Block Acquiring

The input image signal is set as $q(x, y)$, and is broken down into multiple scales of different image blocks $q = \{b_1^l, b_2^l, \ldots, b_N^l\}$, which for the scale series $l = \{1, 2, 3\}$ (image block at all levels, respectively 4×4, 8×8, 16×16 pixels). At the same scale, the relations among block are $b_i^l \cap b_{i+1}^l \neq \emptyset$, which means image block is a kind of overlapping method. It is generally considered that the overlapping image block way is advantage to obtain the complex texture structure. For each image block $b \in q$, we can set its surrounding space neighborhood collection as $S(b^l) = \{s_1, s_2, \ldots, s_m\}$, and $S(b^l) \cap b^l \neq \emptyset$. This structure can be regarded as a central-surround structure in the multi-scale space. It is a kind of simulation of human visual receptive field model [8].

Each image block b_i^l can be described as Eq. (9) in the HSV color space by hypercomplex. The real item is zero in the hypercomplex:

$$b^l(x, y) = b_H^l i_1 + b_S^l i_2 + b_V^l i_3 \tag{9}$$

Where, (x, y) is the pixel coordinates of the image block. l is the level of the image block. For the scales of an image block, b_H^l, b_S^l, b_V^l, are the three components in HSV color space for each image block. i_1, i_2, i_3 are the three pure imaginary components in the hypercomplex.

3.2   The Calculation of the Image Block HDCT Transform Spectrum

In Eq. (9), we can further solve the corresponding HDCT coefficient. The frequency coefficient $b^l(x, y)$ of the neighborhood image block $S(b^l)$ can be solved by the HDCT left or right transformation Eq. (1) or (2).

$$\begin{cases} B^l(u, v) = HDCT\left(b^l(x, y)\right) \\ QS^l(b^l) = HDCT\left(S(b^l)\right) \end{cases} \tag{10}$$

Where, (u, v) is the coordinates in the frequency domain after the transformation for image block $b^l(x, y)$. $(u, v) \in (M, N)$ is the image size.

Based on the definition of spectrum energy in DCT, each image's block HDCT spectrum energy E_B^l is

$$E_B^l = \sum_{k=1}^{K} \left[B^l(u, v)\right]^2 \tag{11}$$

Where, K is the number of corresponding HDCT coefficient for all levels of image block. The block size are 4×4, 8×8, and 16×16; respectively. E_B^l is the HDCT spectrum energy for level l image block.

3.3   The Difference of HDCT Spectrum Energy Between Local Blocks

The difference of HDCT energy spectrum between the ith image block and its adjacent P numbers image block is defined as follows:

$$diff(B_i) = \sqrt{\sum_{j=1}^{P} [(\log(E_B^i + 1) - \log(E_B^j + 1))^2] \Big/ dis^2(i,j)} \qquad (12)$$

Where, $dis(i,j)$ is for the Euclidean distance between the ith image block and the jth image block. Hence, ith image block of saliency degree is the $diff(B_i)$ under the corresponding scale.

3.4   Visual Saliency Normalized Calculation

The visual saliency perception graph after the full image normalization is:

$$Sal(x,y) = f_{gaussian} * [\frac{1}{L}\sum_{l=1}^{L} sm^l(x,y)] \qquad (13)$$

Where, L is the total levels for multi-scale image. $f_{gaussian}$ is the image Gaussian smoothing function. The Gaussian kernel has $\sigma = 3$.

The above mentioned multi-scale HDCT spectrum energy visual saliency perception algorithm can be summarized as follows:

Step1: read input image $q(x, y)$;
Step2:
    for $l = 1$ to 3

    (1) Take image $q(x, y)$'s lth scale level image block b_i^l.
    (2) Transfer image block b_i^l to HSV color space.
    (3) Construct hypercomplex $b^l(x, y)$ according to the Eq. (9)
    (4) Parallel solve HDCT left or right transformation $B^l(u, v)$ for $b^l(x, y)$.
    (5) Calculate according to the Eq. (11) and its energy spectrum difference of neighborhood.

    end for

Step3: Normalize result of step 2;
Step4: Use $f_{gaussian}$ Gaussian smoothing for the results of step 3 to obtain the final saliency perception response output;

4   Experimental Results

4.1   Experiment Design

In order to test the performance of the proposed model, in this section we designed two sets of comparison experiments: eye movement attention simulation and the saliency degree perception experiment for static natural images. The test data set are respectively: (1) the eye attention simulation data using ND. Bruce eye test images, and the average of 20 human eyes' movement images [3], 240 images (681 × 511 pixels) in total; (2) static natural images saliency perception data with images from ground-truth image data set of XD. Hou [8]. The number of this data set is not many, but it includes a variety of resolution and size images. The image contents are relatively rich. The saliency targets in the image is divided into the single objective and multiple objectives, etc. It is one of the classical set of visual saliency test image. At the same time, the evaluation data in this image data set is tested by four people who draw the contour area on saliency region manually. The algorithm testing software and hardware are: MATLAB under Windows XP 2007; CPU: Intel Core2 2.2 GHz and 2 G memory.

4.2 Comparison with Eye Movement Simulation Testing

In this section, the proposed algorithm is compared on the ND. Bruce eye test database with the classical Itti's algorithm [2], Hou's the spectral residual SR algorithm [8], Guo's hypercomplex phase spectrum algorithm PQFT [10], the maximum information entropy AIM algorithm [3] and single scale QDCT algorithm [12]. The subjective experimental results are shown in Fig. 2.

From the Fig. 2, we can see that the visual perception response of the proposed algorithm is more conform to the eye movement results than other algorithms. The PQFT algorithm has a poor ability of the saliency perception in complex scene. SR

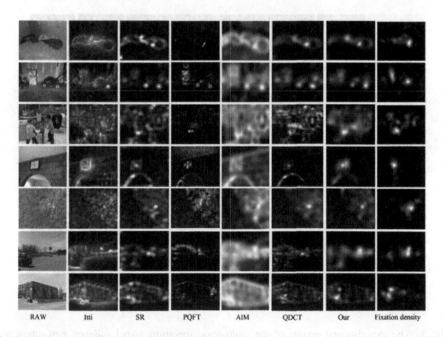

Fig. 2. The eye movement simulation detection results of the multi-scale spectrum visual saliency perception with hypercomplex DCT and other algorithms.

and Itti algorithm performed in generally. AIM, under the effect of long-term visual stimulation basis function, has a relatively wide response output domain. The QDCT pulse spectrum must pay more attentions to details, which inhibits the contrast of the saliency area and the surrounding scene to some extent. The proposed algorithm in this paper, using the spectrum energy difference of the central-surrounding image block under multi-scale by hypercomplex DCT, has a good visual perception effect. But because it is a kind of local contrast algorithm, in scene for the image with complex scene structure, such as the shopping scene at the third line and parking panorama at the sixth line in Fig. 2, the eye movement simulation precision is weaker. Overall, the proposed algorithm is superior to other algorithms.

4.3   Natural Images Marked Degree of Awareness

The proposed algorithm in this paper are compared with the above five kinds of saliency algorithm in Hou image data sets [8]. The subjective experiment results are shown in Fig. 3. From the subjective data shown in Fig. 3, we can see that the proposed algorithm is better consistent with the manually annotated saliency area. But as a local saliency method, there is still the problem that for large saliency object shown in the seventh line in Fig. 3, the output response precision is limited. At the same time, we also can see that introduction of the local block processing method of multi-scale on the one hand improved the precision of perception, but on the other hand also increased the output noise.

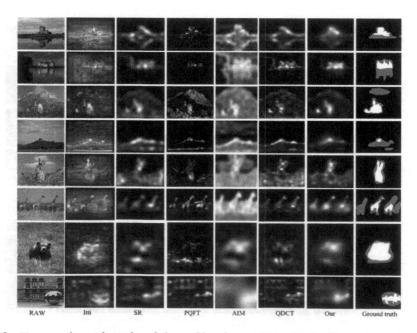

Fig. 3. The experimental results of the multi-scale spectrum visual saliency perception with hypercomplex DCT and other algorithms.

4.4 The Objective Evaluation Data

In order to quantify the experimental results and the average detection error objectively, we combined the above experimental results for objective data analysis. We can see that the proposed algorithm is superior to other five kinds of algorithm from the P-R curve (as shown in Fig. 4).

These subjective and objective experimental results show that the proposed algorithm has good robustness. The perceived saliency areas (or objects) are consistent with the artificial annotated visual perception targets and eye movement data.

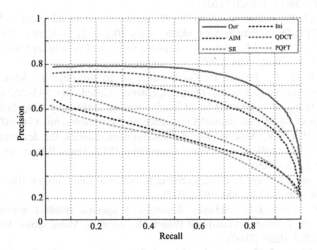

Fig. 4. The comparisons for average recall/percentage among the multi-scale spectrum visual saliency perception with hypercomplex DCT and other algorithms. (Color figure online)

5 Conclusions

Based on visual perception problem in HDCT domain with a parallel computing framework, we design a new visual perception model significantly using HDCT's elimination of redundant features. The proposed algorithm approximate the visual stimuli of different image block as HDCT spectrum energy. The visual saliency is defined according to the spectrum energy stimulus differences between image block and its neighborhood image blocks. The saliency response is output through the visual multi-scale saliency normalized fusion. Finally, through a series of experiments proved that the subjective and objective performance of the proposed algorithm has the advantages of high precision and good stability. In the future work, we will consider introducing texture, movement information and advanced prior features to the proposed model, which will further improve the accuracy of algorithm for the saliency perception.

Acknowledgment. The paper was supported in part by the National Natural Science Foundation (NSFC) of China under Grant Nos. (61365003, 61302116), National High Technology Research and Development Program of China No. 2013AA014601, China Postdoctoral Science Foundation (2014M550494), and Gansu Province Basic Research Innovation Group Project (1506RJIA031).

References

1. Borji, A., Itti, L.: State-of-the-art in visual attention modeling. IEEE Trans. Pattern Anal. Mach. Intell. **35**(1), 185–207 (2013)
2. Itti, L., Koch, C., Niebur, E.: A model of saliency-based visual attention for rapid scene analysis. IEEE Trans. Pattern Anal. Mach. Intell. **20**(11), 1254–1259 (1998)
3. Bruce, N.D., Tsotsos, J.K.: Saliency, attention, and visual search: an information theoretic approach. J. Vis. **9**(3), 1–24 (2009)
4. Harel, J., Koch, C., Perona, P.: Graph-based visual saliency. In: Advances in Neural Information Processing Systems, Montreal, Quebec, Canada, pp. 545–552 (2006)
5. Ma, Y.F., Zhang, H.J.: Contrast-based image attention analysis by using fuzzy growing. In: Proceedings of ACM International Conference on Multimedia, pp. 374–381 (2003)
6. Gao, D.S., Han, S.Y., Vasconcelos, N.: Discriminant saliency, the detection of suspicious coincidences and applications to visual recognition. IEEE Trans. Pattern Anal. Mach. Intell. **31**(6), 989–1005 (2009)
7. Liu, T., Yuan, Z.J., Sun, J., et al.: Learning to detect a salient object. IEEE Trans. Pattern Anal. Mach. Intell. **33**(2), 353–367 (2011)
8. Hou, X.D., Zhang, L.Q.: Saliency detection: a spectral residual approach. In: IEEE Conference on Computer Vision and Pattern Recognition, Minneapolis, Minnesota, USA, pp. 2280–2287. IEEE (2007)
9. Guo, C.L., Zhang, L.M.: A novel multiresolution spatiotemporal saliency detection model and its applications in image and video compression. IEEE Trans. Image Process. **19**(1), 185–198 (2010)
10. Li, J., Levine, M.D., An, X.J., et al.: Visual saliency based on scale-space analysis in the frequency domain. IEEE Trans. Pattern Anal. Mach. Intell. **35**(4), 996–1010 (2013)
11. Fang, Y.M., Lin, W.S., Lee, B.S., Lau, C.T., Chen, Z.Z., Lin, C.W.: Bottom-up saliency detection model based on human visual sensitivity and amplitude spectrum. IEEE Trans. Multimedia **14**(1), 187–198 (2012)
12. Schauerte, B., Stiefelhagen, R.: Predicting human gaze using quaternion DCT image signature saliency and face detection. In: IEEE Workshop on the Applications of Computer Vision, Breckenridge, Colorado, USA, pp. 137–144. IEEE (2012)
13. Navalpakkam, V., Itti, L.: An integrated model of top-down and bottom-up attention for optimizing detection speed. In: IEEE Conference on Computer Vision and Pattern Recognition, pp. 2049–2056 (2006)
14. Cerf, M., Harel, J., Wolf, E., et al.: Predicting human gaze using low-level saliency combined with face detection. In: Advances in Neural Information Processing Systems, pp. 241–248 (2007)

15. Abdollahian, G., Delp, E.J.: Finding regions of interest in home videos based on camera motion. In: IEEE International Conference on Image Processing, pp. 545–548 (2007)
16. Wei, F., Hu, B.: Quaternion discrete cosine transform and its application in color template matching. In: International Congress on Image and Signal Processing, Sanya, Hainan, China, pp. 252–256. IEEE (2008)

Information Security

An Efficient Conjunctive Keyword Searchable Encryption Scheme for Mobile Cloud Computing

Tao Lin[1(✉)], Zexian Sun[2], Hexu Sun[2], and Bin Cao[1]

[1] School of Computer Science and Engineering,
Hebei University of Technology, Tianjin, China
{lintao, caobin}@scse.hebut.edu.cn
[2] School of Control Science and Engineering,
Hebei University of Technology, Tianjin, China
1249226957@qq.com

Abstract. The integration of the cloud and mobile device enables users more convenient to access, retrieve the file, but due to the limitations of resources of mobile devices, how to shorten the search time and get more accurate target files, to avoid unnecessary consumption has become the focus of research. This paper presents an efficient searchable encryption scheme based on mobile cloud, the scheme combine with k nearest neighbor algorithm, design the initial trapdoor matching table (TMT), realizes the multi-keyword Boolean search, improve the query precision, shorten the searching time.

Keywords: Mobile cloud · Searchable encryption · k-nearest neighbor

1 Introduction

Since cloud computing can support dynamic services and provide economical storage and computing resources, data outsourcing, such as iCloud, is a new trend, which outsources a provider's data to cloud servers and accordingly provides a scalable and "always on" approach for data access. Also, providers don't need to worry about the burden of local data management and maintenance. However, outsourcing means that the providers lose the management of the data to the cloud, which may easily incur the leakage of sensitive personal information, such as personal photo, business files. To protect the security and privacy, the data and their indexes are usually encrypted before outsourcing to the server. When the user need to retrieve some certain documents, they first send keywords to the provider, then the provider generates encrypted keywords (trapdoors) and return them to the user. The user then send these trapdoors to the cloud. After receiving the trapdoors, the Cloud server uses the search algorithm to query a list of certain documents based on the encrypted indexes and generated trapdoors. Finally, the user receives the encrypted results and users the private key from the provider to decrypt documents. This architecture, as shown in Fig. 1, protects data security while entitles the providers to utilize the computation, storage power and network bandwidth of the Cloud for document query.

© Springer International Publishing Switzerland 2016
D.-S. Huang and K.-H. Jo (Eds.): ICIC 2016, Part II, LNCS 9772, pp. 659–669, 2016.
DOI: 10.1007/978-3-319-42294-7_59

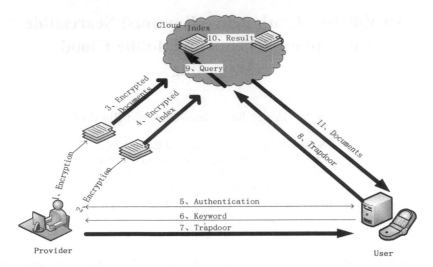

Fig. 1. Trapdoor encrypted search system over cloud

This study focuses on the search time inefficiency issues over the mobile cloud. We present an efficient Searchable Encrypted Data (ESED) scheme to tackle those problems. The presented search scheme leverages the secure k-nearest neighbor (KNN) technique and supports multi-keyword Boolean search and returns the results based on the relevant score.

2   Related Work

For data security and privacy, most of the previous encryption algorithms cannot directly apply to mobile cloud, because it is hard to achieve efficient network traffic and search time to tackle the important problems for mobile cloud. Agrawal et al. [7] proposed a one-to-one mapping order preserving encryption method but it would cause the leakage of sensitive information. Wang et al. [6] proposed a one-to-many mapping order preserving encryption method while it requires a complex computation process, and would cost the excessive resources in mobile cloud. Swaminathan et al. [8] employed an order-preserving encryption method to query data from encrypted cloud servers, which preserved security efficiently. However, this can't support Boolean query so that it only retrieves files in a coarse granularity. As network traffic and search time becoming important in the mobile cloud, a complicated algorithm is not suitable in mobile devices. So we choose the fast accumulated hash (FAH) [10–12]. In Boolean keyword searches, documents are searched by the presence and absence of keywords in a document. In other words, it returns "all-or-nothing," like [13–17]. Due to this effect, the Boolean query cannot efficiently denote the relevance between files and the keywords, and all files relevant to the keywords will be sent back to the users. In addition, this method is of high consumption, which is undesirable and reduces the users' quality of experience.

3 Traditional Searchable Encrypted System

As shown in Fig. 1, the traditional Searchable encrypted system on the cloud is composed of three different participants, Provider, Cloud, User, which are defined below.

The provider owns a list of documents and their indexes. He sends them to the cloud in order that user can contact the cloud for the query service. The Cloud is a commercial organization that provides computation, storage resources and network bandwidth in the form of virtual machines. The User is someone who submits keywords to search documents that contain these keywords. In our scenario, users would use mobile device such as smart phones to submit search requests.

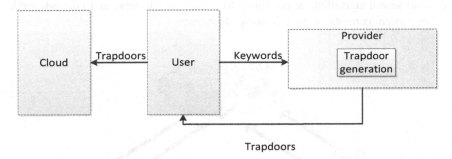

Fig. 2. Trapdoor generation in the traditional system

Figure 1 details the execution flow of a traditional encrypted search over the cloud, including three main flows: documents and indexes uploading, trapdoor generation process and document retrieval process. The weight of lines indicates the amount of data being transferred.

The search delay mainly composes the trapdoor generation time and document query time. Trapdoor generation time faces challenges in mobile wireless networks: high communication latency, poor connectivity and low network transmission rate [4].

According to Fig. 2, we could calculate the trapdoor generation time by Eq. (1).

$$T_{trr} = 2 * T_{net} + T_{gen} \tag{1}$$

where T_{trr} represents the total time delay, T_{net} is as the time delay of one round trip, T_{gen} is as the time delay of trapdoor generation.

Also according to our measurement, the trapdoor generation (steps 5 to 8 in Fig. 1) time accounts for around 60 % of the total search delay. On the other hand, document retrieval time depends on the search algorithm in the cloud. The Ranked Serial Search (RSS) algorithm is often used to retrieve documents in the cloud [4], which ranks the documents according to relevance scores. However, it can't support Boolean query, single keyword search usually returns coarse search results only. What mentioned above will cause serious resource consumption.

4 ESED Design

This section introduces the design of the ESED system and retrofitted trapdoor generation process in ESED. Compared the ESED system (Fig. 3) with traditional system (Fig. 1), the main difference is that (1) network traffic is reduced by a single round trip, (2) the search time is reduced by the TMT module; and (3) the computing burden for generating trapdoors is also offloaded by the TMT module, thus (4), it supports the Boolean query and provides the fine-grained results.

4.1 Architecture of the ESED System

Figure 3 shows the search flow in ESED system. The trapdoor generation process and the cloud search algorithm are retrofitted to reduce search delay and network traffic.
    The system consists of the following algorithms:

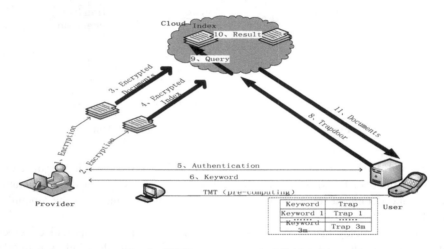

Fig. 3. ESED system over mobile cloud

Init: In the system, the provider sets a keyword dictionary W which contains m keywords. The provider randomly generates the secret key $K = (S, M_1, M_2)$, where S is a (m + 1)-dimensional binary vector, M_1 and M_2 are two (m + 1) × (m + 1) invertible matrices, m is the number of keywords in dictionary W. The provider sends (K, sk) to users, where sk is the symmetric key used to encrypt documents which would be outsourced to the cloud server.
    Build index: The data owner encrypts all the documents before outsourcing them to thecloud server using the symmetric key sk. For each encrypted document D_j, the provider generates an m-dimensional binary vector P, where P[i] indicates whether the encrypted document contains the keyword w_i in the keyword dictionary W or not, that is, P[i] = 1 is yes and P[i] = 0 is no. The vector P is then extended to a (m + 1)-dimensional vector P', and $P'[m+1] = 1$. The provider splits the vector P' to two

$(m + 1) \times (m + 1)$ dimensional vectors (p_a, p_b), the value of the (p_a, p_b) are shown as the Eq. (2).

$$\begin{cases} p_a[i] = \begin{cases} random, S[i] = 1 \\ P'[i], S[i] = 0 \end{cases} \\ p_b[i] = \begin{cases} random, S[i] = 1 \\ P'[i], S[i] = 0 \end{cases} \end{cases} \qquad (2)$$

If $p_a[i] = random \vee p_b[i] = random$, then $p_a[i] + p_b[i] = P'[i]$.

Then, the index of encrypted document C_j can be calculated as $I_j = (p_a M_1, p_b M_2)$. Finally, the provider sends all the encrypted documents and their associated encrypted index to the cloud.

Trapdoor Mapping Table Module. The trapdoor generation process utilizes a Trapdoor Mapping Table (TMT), which stores a large amount of frequently-used trapdoors calculated offline. The TMT module saves not only traffic but also search time. TMT module does not require the provider to compute trapdoors through expensive communication, while it only requires the user to look up trapdoors, avoiding re-computing trapdoors.

First, the provider chooses three completely identical keyword sets W_a, W_b, W_c corresponding to "or" "and" "no" operations. In the system, we denote the "or" "and" "no" by "$\vee$" "$\wedge$" "$\neg$". Thus, the Boolean query can be presented as $\wedge(\neg w_{c1} \wedge \neg w_{c2} \wedge \neg w_{c3} \wedge \ldots \neg w_{cm}) \wedge (w_{b1} \wedge w_{b2} \wedge w_{b3} \wedge \ldots w_{bm}) \wedge (\neg w_{c1} \wedge \neg w_{c2} \wedge \neg w_{c3} \wedge \ldots \neg w_{cm})$. We choose three sequences $\overrightarrow{a}, \overrightarrow{b}, \overrightarrow{c}$ for the "or" "and" "no" keywords. The value of

$\overrightarrow{a}, \overrightarrow{b}, \overrightarrow{c}$, where $\begin{cases} \overrightarrow{a} = \{a_1, a_2, a_3, \ldots\ldots a_m\} \\ \overrightarrow{b} = \{b_1, b_2, b_3, \ldots\ldots b_m\}, \\ \overrightarrow{c} = \{c_1, c_2, c_3, \ldots\ldots c_m\} \end{cases}$

$$\sum_{i=1}^{m} a_i < b_j (j = 1, 2, 3\ldots m)$$

$$\sum_{i=1}^{m} a_i + \sum_{j=1}^{m} b_j < c_k (k = 1, 2, 3\ldots m)$$

We set the initial trapdoors — trap based on the three sequences. Each trap is a $(1*3m)$-dimensional vectors, as shown in the Table 1.

(1) Trapdoor generation: When user queries the certain documents, he just need to look from the TMT module, and we utilize the Fast Accumulated Hash algorithm, then generate the final trapdoor and submit it to the cloud. The Trapdoor Generation Process (TGP) consists of the following algorithms:

The process:

1、Extract the keyword directory $\tilde{K} = \{K_1, K_2 \ldots \ldots K_l\}$

2、for K_n in $\tilde{K}$

if $(K_n \in W_a) \wedge (K_n \in W_b) \wedge (K_n \in W_c)$

    $Trapdoor_n \mathrel{+}= Trap_n$

else

    Hash it by H() and get its l-bit hash code ;

    Map by G(), which contains r bits

end if

3、duplicate removal algorithm $\operatorname{Re}move(Trapdoor_n)$

    $Trapdoor_n' = \operatorname{Re}move(Trapdoor_n)$

4、Generate trapdoor T_w

    (1) set a (1*3m)-dimensional vector   Trap, and each element are both set as 0 .

    (2)

      For $Trapdoor_k$ in $Trapdoor_n'$

$$\sum_{i=1}^{3m} Trap[i] \oplus Trapdoor_k[i]$$

. . . . (3);

  $\therefore Trap = [a_1, b_2, 0, \ldots \ldots - c_m]$;

    (3) Trap   are   splitted   into   three   (1*m) vectors, $Trap = [Trap_1, Trap_2, Trap_3]$.

$$Q = \sum_{i=1}^{m} Trap_1[i] \oplus Trap_2[i] \oplus Trap_3[i]$$

$$= [a_1, b_2, 0... - c_m] \quad \dots\dots\dots\dots\dots\dots\dots\dots\dots\dots\dots\dots\dots (4),$$

(4) In addition, we extends Q to an (m+1)-dimension vector $Q^{'}$ and $Q^{'}[m+1] = -s = -\sum_{i=1}^{m} b_i$. Then, $Q^{'}$ are splitted into two (1*m+1) matrices $[q_a, q_b]$, and the values of the $[q_a, q_b]$

$$q_a[i] = \begin{cases} random, S[i] = 0 \\ Q^{'}[i], S[i] = 1 \end{cases}$$

$$q_b[i] = \begin{cases} random, S[i] = 0 \\ Q^{'}[i], S[i] = 1 \end{cases}$$

are shown as Equation [5]

If $q_a[i] = random \vee q_b[i] = random$

$$q_a[i] + q_b[i] = Q^{'}[i] \quad \dots\dots\dots\dots\dots\dots\dots\dots\dots\dots\dots\dots\dots\dots\dots\dots\dots\dots\dots\dots\dots\dots\dots\dots\dots\dots\dots (5)$$

Then ,the final trapdoor can be calculated as $T_w = (M_1^{-1} q_a, M_2^{-1} q_b)$

5、Return the trapdoor T_w

Query: For each encrypted document C_j with the index I_j (j = 1, 2, ..., N), the cloud server calculates the relevance scores using the trapdoor $T \sim W$ as R_i,

$$R_i = I_j * T_w = (p_a M_1, p_b M_2) * \left(M_1^{-1} q_a, M_2^{-1} q_b \right)$$
$$= p_a q_a + p_b q_b \tag{6}$$
$$= [1, 0, 1, 0\dots\dots 1] * [a_1, b_2, 0\dots - c_m]$$

Table 1. Initial trapdoor

"or" keyword	Trap
W_{a1}	$Trap_{a1} = [a_1, 0 \ldots\ldots, 0]$
....	......
W_{am}	$Trap_{am} = [0, \ldots a_m \ldots 0]$
"and" keyword	Trap
W_{b1}	$Trap_{b1} = [0 \ldots b_1 \ldots 0]$
....	......
W_{bm}	$Trap_{bm} = [0, \ldots b_m \ldots 0]$
"no" keyword	Trap
W_{c1}	$Trap_{c1} = [0 \ldots -c_1 \ldots 0]$
....	......
W_{cm}	$Trap_{am} = [0 \ldots\ldots -c_1]$

If $R_i > 0$, the corresponding document can satisfy the "or," "and" and "no" operations of the mobile user, and by sorting the relevance scores, the cloud sends back the most relevant results to the user.

4.2 Evaluation

We analyze the performance of ESED in search time when generating trapdoors. Utilizing TMT module, ESED has only one network round trip used to search target documents, which is shown in Fig. 3. The total search delay of ESED when generating can be calculated by Eq. (7)

$$T_{srr} = 1 * T_{net} + T_{look} \tag{7}$$

where T_{srr} represents the total time delay, T_{net} is as the time delay of one round trip, T_{gen} is as the time delay of trapdoor lookup. Comparing Eq. (7) with Eq. (2), we find that $T_{srr} < T_{trr}$. This is because ESED is only required to look up, rather than encrypting trapdoors. As shown in Table 2, the encryption time accounts around 85 % for total delay time. Thus, the TMT module efficiently shorten search time.

Table 2. The encryption time in the traditional system

Trapdoor Generation Delay	One keyword	Two keywords	Three keywords
Encryption time	84.52 %	85.92 %	85.48 %
Other time	15.48 %	14.08 %	14.52 %

5 The Experiment

To further analyze the overall search process for a search request, we evaluated the detailed search time, as shown in Table 3. In this table, we see that two network round trips in Traditional text cost around 251.78 ms to build a trapdoor for one keyword,

Table 3. The comparison of search time

Search process	Traditional (ms)	ESED (ms)
Authentication	54.53	54.32
Transmitting a keyword (U to P)	28.57	N/A
Build a trapdoor	178.44	140.39
Transmitting a trapdoor (P to U)	44.77	N/A
Transmitting a trapdoor (U to C)	42.68	30.53
Searching documents in C	69.43	68.57
Returning documents	90.21	89.67
Total search time	508.63	383.48
(user submits one keyword)		
Search process	Traditional (ms)	ESED (ms)
Authentication	53.55	53.67
Transmitting three keywords (U to P)	28.74	N/A
Build a trapdoor	514.33	265.39
Transmitting a trapdoor (P to U)	50.64	N/A
Transmitting a trapdoor (U to C)	51.33	35.53
Searching documents in C	84.32	75.57
Returning documents	110.31	94.32
Total search time	893.22	524.48
(user submits three keywords)		

(U as user, P as provider, C as cloud)

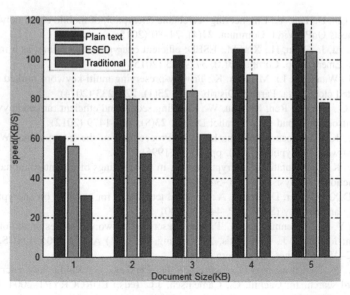

Fig. 4. The transmission speed comparison (Color figure online)

593.71 ms to build a trapdoor for three keywords, while single network round trip in ESED only spends 140.39 ms for one keyword, 265.39 ms for three keywords and the time reduced for building a trapdoor mainly benefits from the TMT module.

Assisted by TMT module and the KNN algorithm, the ESED costs less network traffic less than traditional system (TS). Figure 4 shows the comparison between ESED and TS.

We can see that the ESED's transmission speed is far more effective than the traditional's. In addition, the ESED's transmission speed is almost similar as that of Plain text. In a word, the ESED system outperforms the traditional system in terms of network traffic costs.

6 Conclusion

In this work, we proposed a novel encrypted search system ESED over the mobile cloud, which improves network traffic and search time efficiency compared with the traditional system. We started with a thorough analysis of the traditional encrypted search system and analyzed its bottlenecks in the mobile cloud: network traffic and search time inefficiency. Then we developed an efficient architecture of ESED which is suitable for the mobile cloud to address these issues, where we utilized the TMT module to cope with the inefficient search time issue and the KNN technique to support Boolean query. Finally our evaluation study experimentally demonstrates the performance advantages of ESED.

References

1. Li, H., Liu, D., Dai, Y.: Engineering searchable encryption of mobile cloud networks: when QoE meets QoP. Wirel. Commun. **22**(4), 74–80 (2015)
2. Ma, R., Li, J., Guan, H., Xia, M.: ESED: efficient encrypted data search as a mobile cloud service. Emerg. Top. Comput. **3**(3), 372–383 (2015)
3. Cao, N., Wang, C., Li, M., Ren, K.: Privacy-preserving multi-keyword ranked search over encrypted cloud data. Parallel Distrib. Syst. **25**(1), 222–233 (2014)
4. Wang, C., Cao, N., Ren, K., Lou, W.: Enabling secure and efficient ranked keyword search over outsourced cloud data. Parallel Distrib. **23**(8), 146–1479 (2012)
5. Nyberg, K.: Fast accumulated hashing. In: Proceedings of the International Workshop on Fast Software Encryption (FSE), pp. 83–87(1996)
6. Nyberg, K.: Commutativity in cryptography. In: Proceedings of the International Workshop on Functional Analysis, pp. 331–342(1996)
7. Song, D.X., Wagner, D., Perrig, A.: Practical techniques for searches on encrypted data. In: Security and Privacy (S&P), PP. 44–55(2000)
8. Chang, Y.-C., Mitzenmacher, M.: Privacy preserving keyword searches on remote encrypted data. In: Ioannidis, J., Keromytis, A.D., Yung, M. (eds.) ACNS 2005. LNCS, vol. 3531, pp. 442–455. Springer, Heidelberg (2005)
9. Boneh, D., Di Crescenzo, G., Ostrovsky, R., Persiano, G.: Public key encryption with keyword search. In: Cachin, C., Camenisch, J.L. (eds.) EUROCRYPT 2004. LNCS, vol. 3027, pp. 506–522. Springer, Heidelberg (2004)

10. Shen, Q.: Exploiting geodistributed clouds for e-Health monitoring system with minimum service delay and privacy preservation. Health Inform. **18**(2), 30–39 (2014)
11. Dong, Ma.: Multi cloud-based evacuation services for emergency management. Cloud Comput. **1**(4), 50–59 (2014)
12. Liang, H.: An SMDP-based service model for interdomain resource allocation in mobile cloud networks. Veh. Technol. **61**(5), 22–32 (2012)

Improvement of KMRNG Using n-Pendulum

Jae Jun Lee[(⊠)], Sungyoung Lee, and Taeseon Yoon

HAFS, Yongin, Korea
{chjjma, brianlsy}@naver.com, tsyoon@hafs.hs.kr

Abstract. The report mostly concentrates on 'Keyboard Mouse Random Number Generator (KMRNG)', which is made to overcome the limit of TRNG/PRNG and also including advantages of it being easy to be commercialized, by using Keyboard and Mouse inputs. Comparing SFMT (PRNG) and KMRNG with TestU01 showed us that even though KMRNG was proven suitable for random number generator, its statistical randomness was lower than that of SFMT. This also let us improve the KMRNG's statistical performance by applying the statistical characteristic of the multiple pendulum.

Keywords: KMRNG · PRNG · TRNG · RNG · Keyboard mouse random number generator · Pendulum · n-pendulum

1 Introduction

1.1 Problem

This research is for solving the main problem of KMRNG, which is its statistical performance being poor than SFMT. Even before refining KMRNG, it did its job as a random number generator but this research has its meaning on improvement of statistical performance.

1.2 Method of Study

At first, we planned how to make the program and then made it with computer program. Then we studied about randomness test, and found TestU01 instruction. After reading instruction, we compared and analyzed with SFMT (one kind of PRNG) and assured the suitability of KMRNG as RNG.

2 Theoretical Background

2.1 PRNG (Pseudo-Random Number Generator)

PRNG, also known as Deterministic random bit generator, is an algorithm for making a sequence similar to random sequence. Sequence made by PRNG is not really random, due to the fact that seeds determine sequences. However, PRNG is useful in that its generating speed is fast and the cycle of sequence is long.

© Springer International Publishing Switzerland 2016
D.-S. Huang and K.-H. Jo (Eds.): ICIC 2016, Part II, LNCS 9772, pp. 670–681, 2016.
DOI: 10.1007/978-3-319-42294-7_60

SFMT is a variant of Mersenne Twister introduced by Mutsuo Siato and Makoto Matsumoto. MT19937, which is mostly used, has a lot of advantages as PRNG. Its cycle is $2^{\wedge}19937 - 1$, which is huge. As we do not need that much of combination for the real use, huge cycle doesn't help much.

The name Mersenne Twister came from the fact that the cycle of number is Mersenne Prime number. Mersenne Twister is adapted in many programs due to its speed and quality, and MT19937 is mostly used. MT19937-64, which is 64 bit instead of 32 bit is also used much. SIMD-Oriented Fast Mersenne Twister, is twice faster than MT19937.

Despite its quality, Mersenne Twister is not secure for security. This means that if you know the range, period and finite numbers of previous sequence, you can predict the number.

Created numbers are uniformly distributed. We can't find any consistency even if we pair 623 numbers and put them is 623-dimension, which means connected numbers have low relationships. Due to dis fact, MT19937 is used in simulation. It passes probability tests such as Diehard. As it is programmed simply using bit calculation, it is really fast.

It is developed version of Mersenne Twister generator, have period starting from $2^{\wedge}602 - 1$ to $2^{\wedge}216091 - 1$, which enables us to choose period for the need of quantity of sequence.

2.2 TRNG (True Random Number Generator)

In computing, a hardware random number generator (TRNG, True Random Number Generator) is a device that generates random numbers from a physical process, rather than a computer program. Such devices are often based on microscopic phenomena that generate low-level, statistically random "noise" signals, such as thermal noise, the photoelectric effect, and other quantum phenomena. These processes are, in theory, completely unpredictable, and the theory's assertions of unpredictability are subject to experimental test. A hardware random number generator typically consists of a transducer to convert some aspect of the physical phenomena to an electrical signal, an amplifier and other electronic circuitry to increase the amplitude of the random fluctuations to a measurable level, and some type of analog to digital converter to convert the output into a digital number, often a simple binary digit 0 or 1. By repeatedly sampling the randomly varying signal, a series of random numbers is obtained.

2.3 Randomness Test (TestU01)

Randomness Test progresses the statistical test, which means it investigates the statistical distribution of the random number sequence. Statistically well distributed is one kind of random number sequence's characteristic since it doesn't focuses on particular number.

One randomness includes numerous tests and each tests deals with different kinds of problems. RNG which produces better statistically distributed number sequence, passes more complicated tests. We used TestU01 in our research.

Tests in TestU01 are Small Crush, Crush, Crush, Big Crush, Rabbit. We used Small Crush for analyzing and comparing random numbers made by KMRNG, and PRNG (SFMT in this case) statistically.

The criteria for well-passed is

1. p close to 0.5.
2. p not in the range of eps/eps1.

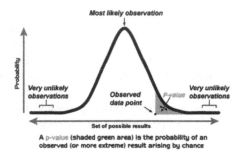

A p-value (shaded green area) is the probability of an observed (or more extreme) result arising by chance

Fig. 1. p-value for Test U01

2.4   Lagrangian Mechanics in n-Pendulum

Lagrangian mechanics is a theory Joseph Louis Lagrange presented in his dissertation 'Mécanique analytique' by renewing the classical mechanics. In Lagrangian mechanics, one can calculate Lagrangian and find the trace of an object by solving Euler-Lagrange equation with substitution of Lagrangian.

Newtonian mechanics focuses on the forces and mostly deals with vectors, but Lagrangian mechanics focuses on scalar such as kinetic energy and potential energy. Since scalar is manageable compared to vector, Lagrangian mechanics is more useful when analyzing complex motions (ex: analysis of multiple pendulum).

$$\frac{d}{dt}\frac{\partial L}{\partial q_\sigma} - \frac{\partial L}{\partial q_\sigma} = 0, \ \sigma = 1, 2, \cdots, 3N - k \tag{1}$$

q: generalized coordinates, σ: index indicating generalized coordinates
N: number of the particle, k: number of holonomic constrain, t: time
L: Lagrangian $(T - U)$, T: Kinetic Energy and U: Potential Energy.

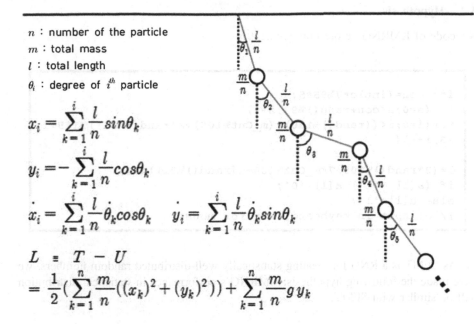

n : number of the particle

m : total mass

l : total length

θ_i : degree of ith particle

$$x_i = \sum_{k=1}^{i} \frac{l}{n} sin\theta_k$$

$$y_i = -\sum_{k=1}^{i} \frac{l}{n} cos\theta_k$$

$$\dot{x}_i = \sum_{k=1}^{i} \frac{l}{n} \dot{\theta}_k cos\theta_k \quad \dot{y}_i = \sum_{k=1}^{i} \frac{l}{n} \dot{\theta}_k sin\theta_k$$

$$L = T - U$$
$$= \frac{1}{2}(\sum_{k=1}^{n} \frac{m}{n}((\dot{x}_k)^2 + (\dot{y}_k)^2)) + \sum_{k=1}^{n} \frac{m}{n} g y_k$$

By solving L in Eq. 1 with Lagrangian Mechanics, we get n differential equation, and we used Mathematica to analyize the movement of the pendulum's end by changing n.

3   Body

3.1   Application

We had only choice, digit, to use the statistical distribution n-pendulum while saving the randomness. This was due to the fact that random number created by n-pendulum had numbers with relatively huge digits for 9957 numbers. (As the possibility of getting big number is much bigger than getting small number, we got only a few of numbers with smaller digits). However, it's distribution was well-distributed than KMRNG, so we brought the digit of n-pendulum random numbers to enhance KMRNG's statistical characteristic. As we used binary digit number for creating random number, we've made zero until.

2^k(k: the largest digit of random number for n-pendulum: 18-the number's n-pendulum >=14)th number from the back (=largest digit). By this, we were able to make KMRNG's statistical distribution distributed better than before.

Specifically, the maximum digit of random number created by n-pendulum is 18, and minimum is 14. Let's say the chosen random number's digit is i. We change 2^{18-i} numbers from the back to 0, from the largest part of binary digit number. In maximum, we change 16 numbers from the back to 0, which have no errors. In statistical distribution, numbers with 14 digits are little in statistical distribution, so the tendency of having big numbers much more than small numbers doesn't change.

3.2 Hypothesis

Keycode of KMRNG before improvement is

```
int con=(int)c*7%95*5;
for(s=0;s<con+rand()%97;s++)
for(i=0;i<((rand()%100)*(m_Cnt%100)+a[rand()%31])%119+1
13;i++){

l=(3*rand()%100+7*m_Cnt%100+a[rand()%31])%31;
if (a[l]=='1') a[l]='0';
else a[l]='1';
}//c:input in keyboard, x*y for mouse.
```

As SFMT is a RNG for creating statistically well-distributed random numbers, we have made the following hypothesis: KMRNG's performance in statistical distribution will be similar with SFMT.

3.3 Hypothesis 2

The following is code for the movement of n-pendulum and each graphs (Figs. 2, 3 and 4).

```
Needs["VariationalMethods`"];      n = Input("input n :
");      l ≡ 1; m ≡ 1; g ≡ 9.8;
icds = Table[{θᵢ[0] == π/2, θᵢ'[0] == 0}, {i, 1, n}];
L ≡ 1/2 * m/n * Σⁿⱼ₌₁((1/n * Σʲₖ₌₁((θ'ₖ[t]) * Cos[θₖ[t]]))² + (1/n * Σʲₖ₌₁((θ'ₖ[t]) *
Sin[θₖ[t]]))² + m/n * g * Σⁿⱼ₌₁(1/n Σʲₖ₌₁ Cos[θₖ[t]]) ;
deqns ≡ EulerEquations[L, Table[θᵢ[t], {i, 1, n}], t];

s ≡ NDSolve[{deqns, icds}, Table[θᵢ[t], {i, 1, n}],
{t,0,300}, Method → {"EquationSimplification" → "Residu-
al"}];
ParametricPlot[{1/n * (Σⁿₖ₌₁(Sin[θₖ[t]])), (-1)*l/n * (Σⁿₖ₌₁(Cos[θₖ[t]]))}/.s,
{t,0,300}, AxesLabel → {"x", "y"}, PlotRange → Full]
```

The movements of n-pendulum showed chaotic characteristic. However, when we compare each graphs by each n, we can figure out that the domain of pendulum gets

ParametricPlot(t,0,300) :

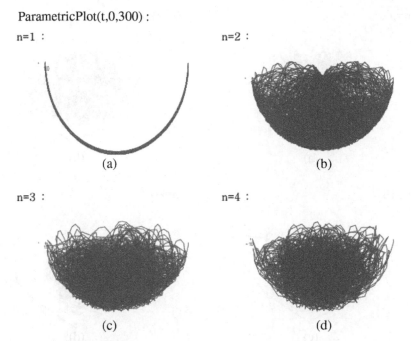

Fig. 2. (a)–(d) Trace of n-pendulum (n = 1, 2, 3, 4)

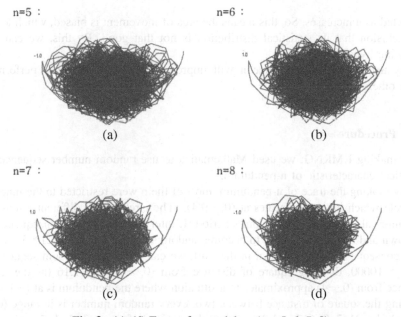

Fig. 3. (a)–(d) Trace of n-pendulum (n = 5, 6, 7, 8)

n=9 :   n=10 :

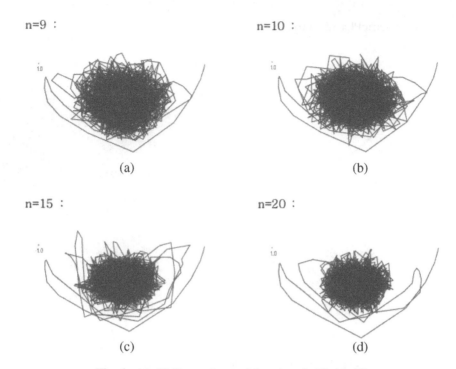

(a)   (b)

n=15 :   n=20 :

(c)   (d)

Fig. 4. (a)–(d) Trace of n-pendulum (n = 9, 10, 15, 20)

restricted as n increases. So, this means the area of movement is biased, which leads to a conclusion that its statistical distribution is not that good. By this, we can make following hypothesis:

By using n-pendulum, some n will improve KMRNG's statistical performance while others don't.

3.4   Procedure

After making KMRNG, we used Mathematica to use random number sequence with statistical characteristic of n-pendulum.

By looking the trace of n-pendulum, most of them were restricted to the inner side of circle which its center locates at (0, −0.4). (The circles have different radius). So, with interval of half of 1-pendulum's period (If interval is too small, consequent move is shown and this makes it hard to become random sequence), from (T/2) * 50 (If from 0, as consequent move is shown in the start, we can't take it as random sequence) to (T/2) * 10000, dividing square of distance from (0, −0.4) by 1.16 (as the longest distance from (0, −4) approximates to a situation where the pendulum is at $(\pm 1, 0)$, by dividing the square of distance between two, every random number is in range (0, 1)), multiply by 2147483647 and then delete the decimals for n = 2, 3, 4, 5, and created 9951 numbers each. The code is shown on the next page.

```
Needs["VariationalMethods'"];   n≡ Input[input n : ];
l≡ 1; m ≡ 1; g ≡ 9.8; T ≡ 2π√(l/g);
icds ≡ Table[{θᵢ[0] == π/2, θᵢ'[0] == 0}, {i, 1, n}];
L≡ 1/2 * m/n * Σⁿⱼ₌₁((1/n * Σʲₖ₌₁((θₖ'[t]) * Cos[θₖ[t]]))² + (1/n * Σʲₖ₌₁((θₖ'[t]) *
Sin[θₖ[t]]))² + m/n * g * Σⁿⱼ₌₁(1/n Σʲₖ₌₁ Cos[θₖ[t]]);
Deqns ≡ EulerEquations[L, Table[θᵢ[t], {i, 1, n}], t];
k≡10000;      #amount of data

s≡ NDSolve[{deqns, icds}, Table[θᵢ[t], {i, 1, n}],
{t,0,(T/2)*k}, Method → {"EquationSimplification"→
"Residual"}, StartingStepSize → T/2, MaxStepSize →
T/2, MaxSteps → ∞];
f≡ Flatten[Table[{((1/n*(Σⁿₖ₌₁(Sin[θₖ[t]])))² + (1/n*(Σⁿₖ₌₁(Cos[θₖ[t]]))+4*1/10)²)/1.16 *
2147483647} /.s, {t, (T/2)*50, (T/2)*k, T/2}]]
```

By using the way shown in 'Application' on KMRNG, we created new random sequence using n-pendulum.

The improved KMRNG's keycode is

```
int con=(int)c*7%95*5;
for(s=0;s<con+rand()%97;s++)
for(i=0;i<((rand()%100)*(m_Cnt%100)+a[rand()%31])%119+1
13;i++){
l=(3*rand()%100+7*m_Cnt%100+a[rand()%31])%31;
if (a[l]=='1') a[l]='0';
else a[l]='1';
for(k=0;k<imnum;k++)a[k]='0';
}//c:input in keyboard, x*y for mouse.
, while imnum is 2^(18-current number`s digit).
```

We compared the random number sequences of KMRNG (original), SFMT (period:), SFMT (period:), KMRNG (n = 2), KMRNG (n = 3), KMRNG (n = 4), KMRNG (n = 5). For our first hypothesis of how well the KMRNG is invented, we compared KMRNG (original), SFMT (period: $2^{602} - 1$), SFMT (period: $2^{216091} - 1$); and for our second hypothesis, we compared KMRNG (n = 2), KMRNG (n = 3), KMRNG (n = 4), KMRNG (n = 5) additionally to prove that the improvement of KMRNG was successful.

Experiment was done by using 'SmallCrush'. Tests that SmallCrush includes is as followed.

3. BirthdaySpacings
4. smultin_Multinomial
5. sknuth_Gap
6. sknuth_SimpPoker
7. sknuth_CouponCollector
8. sknuth_MaxOft
9. svaria_WeightDistrib
10. smarsa_MatrixRank
11. sstring_HammingIndep
12. swalk_RandomWalk1 (H, M, J, R, C)

Table 1. Results of Test U01 for each RNG

TEST	1	2	3	4	5	6	7	8	9	10	sum
KMRNG (orig)	eps	7.4e-3	0.13	0.69	0.36	eps, 1-eps1	0.30	0.37	0.66	eps	**3.19**
KMRNG (n=2)	eps	eps	0.02	0.80	0.83	eps, 1-eps1	0.95	0.66	0.30	eps	3.92
KMRNG (n=3)	eps	eps	0.38	0.65	0.60	eps, 1-eps1	0.50	0.87	0.47	eps	2.77
KMRNG (n=4)	eps	eps	0.56	0.74	0.88	eps, 1-eps1	0.36	0.32	0.06	eps	3.44
KMRNG (n=5)	eps	eps	0.71	0.03	0.85	eps, 1-eps1	0.32	0.58	0.33	eps	3.46
SFMT ($2^{602}-1$)	eps	0.70	0.12	0.14	0.45	eps, 0.56	0.58	0.54	0.05	0.484	2.508
SFMT ($2^{216091}-1$)	eps	0.91	0.43	0.63	0.46	eps, 0.39	0.54	0.74	0.86	0.342	2.397

3.5   Result

Range of eps is under 1.0e–300, and range of eps1 is under 1.0e–15.
Data in table is the value of p, and Sum is the sum of p in all experiments.
Data in test 10 is the average value of tests H, M, J, R, C

3.6   Analysis

Firstly, sum of KMRNG was 3.19, SFMT was 2.508, 2.397 each. Difference between
SFMT and KMRNG was about 0.6 and this means these two RNGs had an average
0.06 difference in one test (which is very small). Since SFMT was made to be statistically even, it showed us a little bit better result, but there was very small difference,
which fits our first hypothesis.

Secondly, KMRNG (n = 2, 3, 4, 5) had the sum of 3.92, 2.77, 3.44, 3.46 each. The
data indicates that the improvement of KMRNG was successful for only triple pendulum. Especially when n = 3, the sum was 2.77, so its statistical randomness was
better than that of KMRNG (original), which was 3.19. So we could say that we
improved the performance of KMRNG by applying multiple pendulum (Table 1).

Detailed analysis :

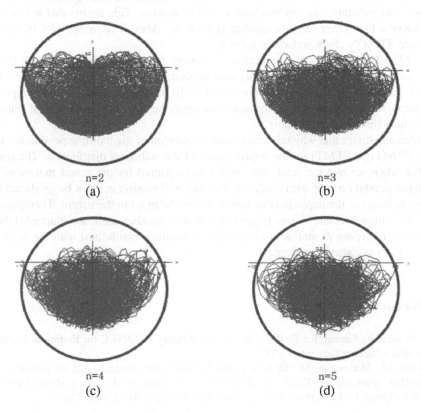

n=2
(a)

n=3
(b)

n=4
(c)

n=5
(d)

Fig. 5.   (a)–(d) Trace of n-pendulum (n = 2, 3, 4, 5) in the possible area

When we draw a circle of radius with its center (0, −0.4) in each graph, graph of n = 2 showed bilateral symmetry rather than a circle. In this point, the statistical distribution would have been judged weak. But for n = 3, the trace of the 3rd pendulum had more constant distance from (0, −0.4) compared to the double pendulum, and this feature would have made its statistical distribution better. But as we stated previously that the radius of the trace becomes limited when n is large, we can analyze that this feature limited the distribution, which let to worse statistical distribution (Fig. 5).

4 Conclusion

When we judge the performance of random number generator, we should consider the use. Hardware random number generator is specialized for security as it is unpredictable, and PRNG is considered to be best in simulation due to its uniform statistical distribution. As we consider the code and the result of Test U01, we can consider that KMRNG is more similar with Hardware random number generator than PRNG. However, as it is software, it has similarity with PRNG. This means that KMRNG has characteristics of both TRNG and PRNG. So, generated numbers are suitable for security, and KMRNG has an advantage in that it is easy to commercialize, as it is software.

In this research, we succeeded on making a random number generator, which a person can generate true random number when wanted. This means that it has more advantages than other random number generators. Also, it is meaningful in that we realized TRNG's characteristic in software.

KMRNG generates random number sequences much faster than Hardware RNGs. But compared to PRNG, the generation of random number sequence is slow and also PRNG has better statistical distribution than KMRNG. So by additional development, we improved the statistical performance by applying the chaotic motion of multiple pendulum (for triple pendulum).

One can doubt that why this study used the motion of the multiple pendulum rather than PRNG (ex: SFMT) for the improvement of the statistical distribution. The reason is that when we use the 'real' data which is measured by the actual motion of the multiple pendulum (not with mathematica), the real motion can not be predicted precisely because of the unpredictable factors (air resistance) in the nature. Therefore, we can use 2 unpredictable factors to generate random number sequence (unpredictability of keyboard typing & motion measurement of multiple pendulum) with the improvement of the statistical randomness.

References

1. L'Ecuyer, P., Simard, R.: TestU01: A Software Library in ANSI C for Empirical Testing of Random Number Generators (2007)
2. Saito, M., Matsumoto, M.: SIMD-oriented fast mersenne twister: a 128-bit pseudorandom number generator. In: Keller, A., Heinrich, S., Niederreiter, H. (eds.) Monte Carlo and Quasi-Monte Carlo Methods 2006, pp. 607–622. Springer, Heidelberg (2008)

3. Matsumoto, M., Nishimura, T.: Mersenne twister: a 623-dimensionally equidistributed uniform pseudo-random number generator. ACM trans. model. comput. simul. **8**(1), 3–30 (1998)
4. Blum, L., Blum, M., Shub, M.: A simple unpredictable pseudo-random number generator. SIAM J. Comput. **15**(2), 364–383 (1986)
5. Jones, R.M., Patel, K.N.: Examination of chaos in multiple pendulum systems through computer visualization in Java (2001)

XACML Policy Optimization Algorithm Based on Venn Diagram

Qiuru Lu, Jianping Chen$^{(\boxtimes)}$, Haiying Ma, and Weixu Chen

School of Computer Science and Technology, Nantong University,
Nantong 226019, Jiangsu, People's Republic of China
chen.jp@ntu.edu.cn

Abstract. This paper proposes an XACML (Extensible Access Control Markup Language) policy optimization algorithm to increase the efficiency of policy evaluation, which is based on the Venn graphic method of set theory. A three layer structure model for XACML is constructed. The policies and rules in the layers are mapped into sets and expressed with the Venn diagrams. According to the decision result of each layer and by setting the combining algorithm priority, the conflicts and the redundancies among access control policies and rules are detected and eliminated based on the intersection and union relations between sets. Experimental tests carried under the main evaluation engines show that the algorithm can decrease the evaluation time effectively and reduce the memory space occupancy as well.

Keywords: Access control · XACML · Policy evaluation · Venn diagram

1 Introduction

Access control is an important technology to protect network information security, which can effectively manage data resources in a network and have them used legally. As a standard language to describe secured access control policies, the Extensible Access Control Markup Language (XACML) [1] is widely used. In addition to describe access control policies, the XACML also has the policy executing mechanism to execute the decision of access control policies. When a user gives an access request to a resource, whether it is authorized is determined by evaluating the related access policies. With the rapid increases of users and resources, the number of the policies and rules for access control becomes larger and larger and the structure of the access control becomes more and more complex, which brings the problem of the conflicts and redundancies among access policies and rules. It is necessary to reduce the conflicts and redundancies and optimize access control policies to increase the efficiency of the policy evaluation.

Regarding to the optimization of access control policies, the commonly used XACML system such as Sun XACML [2] just provides a decision instance that is based on a file policy mode. It does not have the index structure to realize complex policy matching. A more efficient method is to establish an index structure to reduce the policy search space [3–7]. Kolovski et al. [8] used description logic (DL) to formalize XACML policies logically and then used it to verify and examine redundant

D.-S. Huang and K.-H. Jo (Eds.): ICIC 2016, Part II, LNCS 9772, pp. 682–690, 2016.
DOI: 10.1007/978-3-319-42294-7_61

rules. Fisler et al. [9] proposed a tool that can also be used to analyze redundant rules. Mourad et al.'s algorithm [10–12] described the original policy in semantics language and gave a method for judging defect, conflict and redundancy when evaluating rules. Wang and Feng [13] discussed the reasons of redundant rules and proposed a redundant decision theorem. References [14–16] made studies on rule weights and statistical measures for the efficiency of evaluation. The above research mostly focused on the aspects of the semantic description of the redundancies and conflicts. The problems of redundancy elimination and conflict removal have not been well solved. Aiming at this problem, we propose a method for detecting and removing the policy conflicts and redundancies. A three layer XACML structures are constructed mapped into sets and expressed using the Venn diagrams. The conflicts and redundancies are detected using the intuitive nature of the Venn diagram and removed based on the relationships of intersection and union between sets, which hence achieves the purpose of the policy optimization.

2 Related XACML Concepts

A XACML policy consists of a policy set and a combining algorithm. A policy set contains some policies or other policy sets. A policy consists of rules and a rule contains a target. Each policy has a target, which determines the relevance between policy and request. A rule set consists of some rules, which has three basic attributes (do not necessarily all exist): Subject, Resource (Re), Action (Ac) and Effect (E) suit for the rule.

XACML provides four combining algorithms, on the case of one request matches multiple rules, XACML uses evaluation combining algorithm to get the final evaluation result. The combining algorithm in XACML standards include: Permit-overrides which is that as long as a rule's evaluation is permit, the decision of final authorization is permit; Deny-overrides which is that as long as a rule's evaluation is deny, the decision of final response is deny; First-application, namely, the first-met relevant rule's evaluation result is the decision of final response; and Only-one-application, namely, only one evaluation result that adapts to the request rule is the decision of final response. When we use the above combining algorithm, XACML permits to build your own combining algorithm. Though XACML can restrain decisions through a variety of merging algorithm, it also can easily lead to complex conflict and redundancy.

The process of policy evaluation matches rules in turn based on the request of users. How to match them quickly determines the efficiency of policy evaluation, especially when it needs many times to judge. It is important to match rules suit for the request rapidly. So we define the priority of four typical algorithms for the following algorithm: Under the condition of permit-overrides, the highest priority is permit and the lowest priority is denied; on the contrary, under the condition of deny-overrides, the highest priority is denied and the lowest is permit. First-application is always prior to only-one-application.

For a policy sequence, the execution sequence of its priority is as follows:

$$Queue_{ps} = \{P_1, P_2, ..., P_n\}, prior(P_i) \geq prior(P_{i+1}), 1 \leq i \leq n-1.$$

For a rule sequence, the execution sequence of its priority is as follows:

$Queue_p = \{R_1, R_2, ..., R_n\}, prior(R_i) \geq prior(R_{i+1}), 1 \leq i \leq n-1$.

3 XACML Three Layer Structure Model

The XACML has a complex nested structure. The external part is policy set. The middle part is policy. And the internal part is rule. Each layer has its target which contains various basic attributes such as subject, resource and action. Each policy set or policy has combining algorithm and makes the final decision according to the relationship of policies.

In order to implement the algorithm, the nested XACML structure is represented as three layer model. The top is the policy set (PS) layer which contains one or more policies or policy sets. To simplify the algorithm, we assume that the subsets of one policy set are all policy. The middle is the policy (P) layer. There are one or more rules (R) in the policy layer. The bottom is the rule layer, which contains some basic elements such as subject, resource, action and so on. So there exist inclusion relations between the upper layer and the lower layer. The execution process begins at the bottom in bottom-up approach. The three layer structure model is shown in Fig. 1.

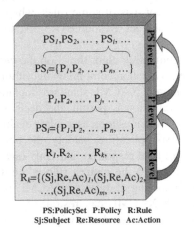

PS:PolicySet P:Policy R:Rule
Sj:Subject Re:Resource Ac:Action

Fig. 1. Three layer model of XACML

In the constructed three layer model of XACML, each layer contains a set of data that can be seen as a set from the abstract perspective. Conflicts and redundancies are similar to intersection, disjoint and union in the set. So we consider describing policy relationship with Venn diagrams for it can show the relationship between sets clearly. And then using the intersection and union relations to detect and eliminate conflict and redundancy.

4 Policy Optimization Algorithm Based on Venn Diagram

In the XACML three layer model, there exist rule combining algorithms between the rule layer and the policy layer and policy combining algorithms between the policy layer and policy set layer accordingly. This algorithm focuses on the discussion of the combining algorithms of permit-overrides and deny-overrides.

4.1 Conflict and Redundancy Detection and Elimination

Under the environment of XACML, the data conflicts or redundancies won't occur in two disjoint sets, so we only need to discuss the case of intersection sets. In this case, there are two types of conflicts and redundancies: properly inclusion and partial intersection, which are shown in Fig. 2 using the Venn diagram. E_i or E_j is a response (Effect) that is returned after the policy evaluations. The effect value is permit or deny.

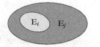   

Fig. 2(a). Properly inclusion   **Fig. 2(b).** Partial intersection

Fig. 2. Two cases of conflicts or redundancies

According to the set relations, we can find out how to judge conflict or redundancy in the two cases. In terms of conflict, E_i is different from E_j. Onc valuc is permit and the other value is deny. These two decisions are intersecting. In terms of redundancy, the values of E_i and E_j both are permit or deny, and there is an inclusion relation between these decisions. The detection and elimination methods of conflicts and redundancies in permit-overrides (solid lines with arrow) and deny-overrides (dotted lines with arrow) are described as follows.

Conflict detection and elimination is showed in Fig. 3. For the policies matched with the same request, if the decisions are different ($E_i \neq E_j$), then there is a conflict.

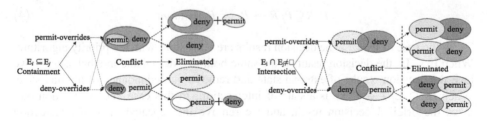

Fig. 3. Conflict detection and elimination

Conflict presented in the rule set layer is intersection among attributes of the rules:

$$(R_i.Sj \cap R_j.Sj \neq \phi \wedge R_i.Ac \cap R_j.Ac \neq \phi \wedge R_i.Re \cap R_j.Re \neq \phi) \rightarrow (R_i, R_j) = Conflict \quad (1)$$

Conflict presented in the policy set layer is the intersection among rules of the policies:

$$P_i.R \cap P_j.R \neq \phi \rightarrow (P_i, P_j) = Conflict \tag{2}$$

If the upper algorithm is permit-overrides, the value of permit have priority. Whereas the upper is deny-overrides, the effect value of deny receive precedence. And the value will override the intersection part. When $E_i \subseteq E_j$, if external state has high priority, we take the state of outer results. If internal state has high priority, we reserve the decision results of internal state, eliminating corresponding part of the external state. When $E_i \cap E_j \neq \phi$, there is a state of intersection. Taking the higher priority in the intersection as the decision results of conflicts.

Redundancy detection and removal is showed in Fig. 4. For the policies matched with the same request, if the decisions are the same ($E_i = E_j$), then there is a redundancy.

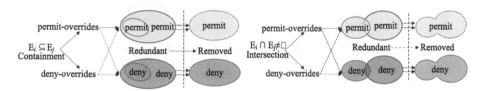

Fig. 4. Redundancy detection and removal

Redundancy presented in the rule set layer is the inclusion relation among attributes of the rules:

$$(R_i.Sj \subseteq R_j.Sj \wedge R_i.Ac \subseteq R_j.Ac \wedge R_i.Re \subseteq R_j.Re) \rightarrow R_i = Redundant \tag{3}$$

Redundancy presented in the policy set layer is the inclusion relation among rules of the policies:

$$P_i.R \subseteq P_j.R \rightarrow P_i = Redundant \tag{4}$$

In the case of $E_i = E_j$, the decision results are consistent with the priority algorithm. When $E_i \subseteq E_j$, the decision results is the same both in external and internal state, so we take the decision results of external state and remove the redundancy of internal state. When $E_i \cap E_j \neq \phi$, there is a state of intersection. In this case, the intersection part takes the original decision result, and we remove the repeated ones of intersection part. Thus, we achieve the goal of redundancy removal.

4.2   Policy Optimization Algorithm

As discussed above, the policy optimization between two layers follows the bottom-up order. This means that the optimization of policy set layer is performed after the policy layer. According to the method of conflict and redundancy detection mentioned above, we give the optimization algorithm based on the intersection and union operations on sets.

The first optimization is the policy layer. It has three inputs: the rule combining algorithm (P.RCA), the effects of the rule set R_i ($R_i.E$) and the effects of rule set R_j ($R_j.E$), and has one output: the decision results from the policy layer (P.E). The algorithm is described as follows.

```
if  R₁ ⊆ Rⱼ
    if P.RCA = permit-overrides
        if  R₁.E = Rⱼ.E return Rⱼ.E          //Redundant
        if  R₁.E = permit and Rⱼ.E = deny
            return R₁.E + ( Rⱼ.E - R₁.E)      //Conflict
        if  R₁.E = deny and Rⱼ.E = permit
            return Rⱼ.E
    else if  P.RCA = deny-overrides
        if  R₁.E = Rⱼ.E return Rⱼ.E          // Redundant
        if  R₁.E = permit and Rⱼ.E = deny
            return Rⱼ.E                       // Conflict
        if  R₁.E = deny and Rⱼ.E = permit
            return R₁.E + ( Rⱼ.E - R₁.E)
    else
        Queueₚ = { R₁.E, Rⱼ.E }, prior(R₁) ≥ prior(Rⱼ)
else if  R₁ Rⱼ ≠ ∅
    if P.RCA = permit-overrides
        if R₁.E = Rⱼ.E
            return R₁.E + ( Rⱼ.E - R₁.E   Rⱼ.E )  // Redundant
        if R₁.E = permit and Rⱼ.E = deny
            return R₁.E + ( Rⱼ.E - R₁.E   Rⱼ.E)  // Conflict
        if R₁.E = deny and Rⱼ.E =permit
            return Rⱼ.E + ( R₁.E - R₁.E   Rⱼ.E)
    else if  P.RCA = deny-overrides
        if R₁.E = Rⱼ.E
            return R₁.E + ( Rⱼ.E - R₁.E   Rⱼ.E) // Redundant
        if R₁.E = permit and Rⱼ.E = deny
            return Rⱼ.E + ( R₁.E - R₁.E   Rⱼ.E) // Conflict
        if R₁.E = deny and Rⱼ.E = permit
            return R₁.E + ( Rⱼ.E - R₁.E   Rⱼ.E)
    else
        Queueₚ = { R₁.E, Rⱼ.E }, prior(R₁) ≥ prior(Rⱼ)
else
    Queueₚ = { R₁.E,Rⱼ.E }, prior(R₁) ≥ prior(Rⱼ)
```

The second optimization is the policy set layer. It also has three inputs: the policy combining algorithm (PS.PCA), the effects of the policy set P_i (P_i.E) and the effects of policy set P_j (P_j.E), and has one output: the decision results of the policy set layer (PS.E). The detailed description of the algorithm is similar to the one of the policy set optimization described above and we do not give repeatedly.

5   Experiment Testing Results

Simulation experiments are carried out under the main evaluation engines to test the performance of the proposed optimization algorithm. The evaluation engines include Sun XACML, Enterprise XACML, XEngine and MLOBEE. The test cases used in the experiments are revised and extended on the basis of XACML official test package [17]. The experiment environment is the Inter(r) Pentium(r) 2.70 GHZ that has two core CPUs and 2 GB memories with the Windows 7 operation system and Java Runtime Environment 1.6.10.

5.1   Time Performance

The four evaluation engines mentioned above are used in the experiments to compare the differences in time performance. We calculate the average processing time to a request before and after the optimization. Three groups of test samples are used. The first group consists of 500 original policies, including 100 redundancy rules and 20 conflict rules; the second group consists of 1000 original policies, including 300 redundancy rules and 50 conflict rules; the third group consists of 2000 original policies, including 500 redundancy rules and 100 conflict rules. In the three groups of test samples, each policy contains five rules. According to the attributes in the policies, 1000 different requests are generated at random respectively. First, the requests are evaluated on the four engines. Then, the proposed algorithm is used to evaluate the requests again. We compare the average response times for the requests in the two cases. The results are shown in Fig. 5.

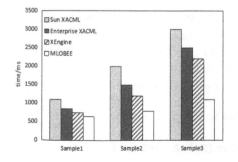
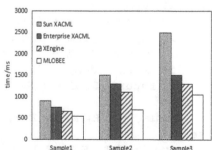

Fig. 5(a). Performance before optimization   **Fig. 5(b).** Performance after optimization

Fig. 5. Comparison of evaluation time

Compared Fig. 5(a) with (b), the evaluation time is reduced by 10 % ~ 30 % in the cases of three groups of samples. It can be seen that with the increase of the number of the rules, the time reduction becomes larger, especially in the case of Sun XACML. The results show that the proposed algorithm can effectively reduce the evaluation time of a request by eliminating the conflicts and redundancies.

5.2 Storage Performance

The proposed optimization algorithm not only reduces the evaluation time, but also reduces the memory space occupied by policies. In the Sun XACML, the memory spaces used before and after the optimization are compared. We gradually increase the number of rules and test the size of the memory space occupied. As shown in Fig. 6, when the number of rules reaches a certain scale (about 1000), the memory space occupied with the proposed algorithm begins to decrease. With the number of rules increases further, the effect of the space saving becomes more remarkable.

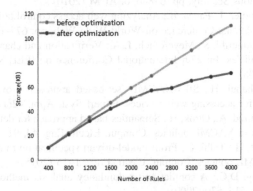

Fig. 6. Comparison of storage space

6   Conclusions

Through deep analysis of its hierarchical architecture, the original XACML is structured into a three layer model and expressed graphically and intuitively using the Venn diagrams in set theory. The conflicts and redundancies among the access control policies and rules are detected and eliminated according to the relationships of the intersection and union between sets. And a XACML policy optimization algorithm based on the Venn diagram is hence formed. By reducing the evaluation time and the memory space, the algorithm improves the efficiency of policy evaluation effectively. Next, we are going to investigate the problems of the privacy protection in the process of policy decisions.

Acknowledgments. This research work is financially supported by the National Natural Science Foundation of China (grant No. 61402244 and 61371111), the Nantong Municipal Application

Research Foundation of China (No. GY2015012), and the Funds of Natural Science Research (No. 15z06) and the Doctoral Start-up Scientific Research (No. 15B10) from Nantong University.

References

1. Extensible Access Control Markup Language (XACML) v3.0 (2012)
2. Sun XACML. http://sunxacml.sourceforge.net/
3. Enterprise XACML. http://code.google.com/p/enterprise-java-xacml/
4. Liu, A.X., Chen, F., Hwang, J.H.: Designing fast and scalable XACML policy evaluation engines. IEEE Trans. Comput. **60**(12), 1802–1817 (2011)
5. Wang, Y.Z., Feng, D.G., Zhang, L.W., Zhang, M.: XACML policy evaluation engine based on multi-level optimization technology. J. Softw. **22**, 323–338 (2011)
6. Niu, D.H., Ma, J.F., Ma, Z.: HPEngine: high performance XACML policy evaluation engine based on statistical analysis. J. Commun. **35**(8), 206–215 (2014)
7. Butler, B., Jennings, B., Botvich, D.: XACML policy performance evaluation using a flexible load testing framework. In: Proceedings of the 17th ACM Conference on Computer and Communications Security, pp. 648–650. ACM (2010)
8. Kolovski, V., Hendler, J., Parsia, B.: Analyzing web access control policies. In: Proceedings of the 16th International Conference on World Wide Web, pp. 677–686. ACM (2007)
9. Fisler, K., Krishnamurthi, S., Meyerovich, L.A.: Verification and change-impact analysis of access-control policies. In: 27th International Conference on IEEE Software Engineering, pp. 196–205 (2005)
10. Mourad, A., Jebbaoui, H.: SBA-XACML: set-based approach providing efficient policy decision process for accessing web services. Expert Syst. Appl. **42**(1), 165–178 (2015)
11. Jebbaoui, H., Mourad, A., Otrok, H.: Semantics-based approach for detecting flaws, conflicts and redundancies in XACML policies. Comput. Electr. Eng. **44**, 91–103 (2015)
12. Mourad, A., Tout, H., Talhi, C.: From model-driven specification to design-level set-based analysis of XACML policies. Comput. Electr. Eng., 1–15 (2015)
13. Wang, Y.Z., Feng, D.G.: A conflict and redundancy analysis method for XACML rules. J. Comput. **32**(3), 516–530 (2009)
14. Chen, W.H., Wang, N.N.: Research on XACML policy evaluation optimization technology. Appl. Res. Comput. **30**(3), 900–905 (2013)
15. Qi, Y., Chen, J., Li, Q.M.: XACML policy evaluation optimization method based on reordering. J. Nanjing Univ. Sci. Technol. **39**(2), 187–193 (2015)
16. Marouf, S., Shehab, M., Squicciarini, A.: Adaptive reordering and clustering-based framework for efficient XACML policy evaluation. IEEE Trans. Serv. Comput. **4**(4), 300–313 (2012)
17. XACML 2.0 Conformance Test. http://www.oasis-open.org/committees/download.php/14846/xacml2.0-ct-v.0.4.zip

Virtual Reality
and Human-Computer Interaction

Usability Evaluation of the Flight Simulator's Human-Computer Interaction

Yanbin Shi[1,2(✉)] and Dantong Ouyang[2]

[1] Aviation University of Air Force, Changchun 130022, China
shiyanbin_80@163.com
[2] Key Laboratory of Symbolic Computation and Knowledge Engineering
of Ministry of Education, Jilin University, Changchun 130012, China

Abstract. Attention is concentrated on the usability of the flight simulator's Human-Computer Interaction system, especially on the cockpit display systems. Based on the User-Centered Design tool, used the cloud model to evaluate the usability of the cockpit display system from learnability, efficiency, memorability, errors and satisfaction. Cloud model based on grey system theory and the normal grey whitenization weight function used in conversion between qualitative and quantitative uncertainty model.

Keywords: Usability · Flight simulator · User-Centered Design · Human-computer interaction · Multimodal

1 Introduction

A flight simulator is a device that artificially re-creates aircraft flight and the environment in which it flies, for pilot training, design, or other purposes [1]. It is used for a variety of reasons, including flight training (mainly of pilots), the design and development of the aircraft itself, and research into aircraft characteristics and control handling qualities.

User-Centered Design (UCD) is the process of designing a tool, such as flight simulator's or application's user interface, from the perspective of how it will be understood and used by a pilot. Rather than requiring pilots to adapt their attitudes and behaviors in order to learn and use simulator, simulator can be designed to support its intended pilots' existing beliefs, attitudes, and behaviors as they relate to the tasks that the simulator is being designed to support. Recent years, many organizations and researchers have studied the coordination design theory and methods for aircraft cockpit human-computer. Tang [2] and Zhang [3] considered the pilot's cognition characteristic and perception motion characteristic. Ai [4] researched the relationship of the main equipment's arrangement in the cockpit and the ergonomics. Wang [5] taking the ecological interface design (EID) theory as the basis, summarized the guidelines of designing the interface of the fighter cockpit.

D.-S. Huang and K.-H. Jo (Eds.): ICIC 2016, Part II, LNCS 9772, pp. 693–702, 2016.
DOI: 10.1007/978-3-319-42294-7_62

2 Usability

According to the international standard, ISO 9241-11 [6], the definition of usability is the extent to which a product can be used by specified users to achieve specified goals with effectiveness, efficiency and satisfaction in a specified context of use, which is affected by the users, their goals and the usage situation.

Usability is defined by 5 quality components [7]:

- Learnability: How easy is it for users to accomplish basic tasks the first time they encounter the design?
- Efficiency: Once users have learned the design, how quickly can they perform tasks?
- Memorability: When users return to the design after a period of not using it, how easily can they reestablish proficiency?
- Errors: How many errors do users make, how severe are these errors, and how easily can they recover from the errors?
- Satisfaction: How pleasant is it to use the design?

There are many methods for studying usability, but the most basic and useful is user testing, which has 3 components [8]:

- Get hold of some representative users.
- Ask the users to perform representative tasks with the design.
- Observe what the users do, where they succeed, and where they have difficulties with the user interface.

In order to evaluate the usability of the flight simulator's, many advanced and primary pilots respectively be selected to complete designated training missions. Recording what the pilots do and where they have difficulties with the cockpit.

3 Flight Simulator's Human-Computer Interaction

The cockpit display systems is the major human-computer interaction means, they provide a visual presentation of the information and data from the aircraft sensors and systems to the pilot to enable the pilot to operate the aircraft safely and carry out the mission [9]. They are thus vital to the operation of any aircraft as they provide the pilot, whether civil or military, with primary flight information, navigation information, engine data, airframe data and warning information. The military pilot has also a wide array of additional information to view, such as: infrared imaging sensors, radar, tactical mission data, weapon aiming, threat warnings.

The pilot is able to rapidly absorb and process substantial amounts of visual information but it is clear that the information must be displayed in a way which can be readily assimilated, and unnecessary information must be eliminated to ease the pilot's task in high work load situations. A number of developments have taken place to improve the pilot–display interaction and this is a continuing activity as new technology and components become available. Examples of these developments are Head up displays (HUD), Helmet mounted displays (HMD), Multi-function colour displays,

Digitally generated color moving map displays and Synthetic pictorial imagery, Displays management using intelligent knowledge based system (IKBS) technology, Improved understanding of human factors and involvement of human factors specialists from the initial cockpit design stage.

Equally important and complementary to the cockpit display systems in the human-computer interaction are the means provided for the pilot to control the operation of the avionic systems and to enter data [10]. Again, this is a field where continual development is taking place [11, 12]. Multi-function keyboards and multi-function touch panel displays are now widely used. Speech recognition technology has now reached sufficient maturity for 'direct voice input' control to be installed in the new generation of military aircraft. Audio warning systems are also now well established in both military and civil aircraft. The integration and management of all the display surfaces by audio/tactile inputs enables a very significant reduction in the pilot's workload to be achieved in the new generation of single seat fighter/strike aircraft. Other methods of data entry which are being evaluated include the use of eye trackers, sounds and body languages.

In this research, as the major HCI system, the power gauges, turn & slip, compass, altimeter, ASI, AI and VSI are considered.

4   Evaluation Index System

A usable system is created by getting hold of enough of the right usability information and using the information well. Usually people thinking that "measured values" are quantitative and that criteria application yields quantitative results, that it is important that the target value of a usability evaluation criterion can be either quantitative or qualitative. Traditional, there are generally three types of usability evaluation methods [13–16]: Testing, Inspection, and Inquiry. In Usability Testing approach, representative users work on typical tasks using the system (or the prototype) and the evaluators use the results to see how the user interface supports the users to do their tasks. In Usability Inspection approach, usability specialists – and sometimes software developers, users and other professionals – examine usability-related aspects of a user interface. In Usability Inquiry approach usability evaluators obtain information about users' likes, dislikes, needs, and understanding of the system by talking to them, observing them using the system in real work (not for the purpose of usability testing), or letting them answer questions verbally or in written form.

In order to evaluate the usability of the flight simulator, these three kinds of methods all be used in studying. The factors which influence the efficiency of the flight simulator's human-computer interaction mainly include: learnability, efficiency, memorability, errors and satisfaction. Here, based on the grey system theory, using cloud model [17–26] and the normal grey whitenization weight function to evaluate the usability of the flight simulator.

4.1   Index Quantization and Weight Value

The evaluation model of the flight simulator's human-computer interaction is multi-indexes issue, a prominent characteristic is that these indexes without unified a measuring standard, their fetching value ranges of initial data and tolerance units have nothing in common with each other, it is unable to compare each other. So, assess the question as to many indexes of a known index attribute value, before confirming the index weight and carrying on comprehensive evaluation, must have no dimension treatment (standardization or standardize to the attribute value of the index), its essence is to utilize certain mathematics to vary and turn the attribute value with different dimension, property into by comprehensive treatment "Quantization value", Generally vary every attribute value to the range of fetching value in unison. In assessing models, most index have clear number value, and it can utilize method of normalization their in unison for have dimension, distribute in [0, 1] number of the space to value. But some indexes can only be described in natural language by qualitative method, such as satisfaction, this kind index can utilize the cloud model to determine the natural to quantitative good conversion.

4.2   Normalization of the Index

The normalization method should carry on according to the request of index type, practical problem characteristic and policymaker, assess index can be divided into four kinds, they are benefit type, cost type, fixed type and interval type.

If labeled the subscripts set of all of the indexes $f_i(i = 1, 2, \cdots, m)$ as $\Phi = \{1, 2, \cdots, m\}$, x_{ij} is the original value, y_{ij} is the normalization value, the ways of the dimensionless are:

About the benefit type index valve: $y_{ij} = x_{ij}/\max x_{ij}$;

About the cost type index valve: $y_{ij} = \min x_{ij}/x_{ij}$;

About the fixed type index valve:

$$y_{ij} = \begin{cases} 1 & (x_{ij} - x_i^*) \\ 1 - \left\{ |x_{ij} - x_i^*| \Big/ \max_{1 \le j \le n} (|x_{ij} - x_i^*|) \right\} & (x_{ij} \ne x_i^*) \end{cases} ; \tag{1}$$

About the interval type index valve:

$$y_{ij} = \begin{cases} 1 - [(x_i^L - x_{ij})/\eta_i] & (x_{ij} < x_i^L) \\ 1 & (x_{ij} \in [x_i^L, x_i^U]) \\ 1 - [(x_{ij} - x_i^U)/\eta_i] & (x_{ij} > x_i^U) \end{cases} \tag{2}$$

Where, $i = 1, 2, \cdots, m$ (m is the number of the indexes), $j = 1, 2, \cdots, n$ (n is the number of the objects for evaluating); x_i^* is the optimal value of the fixed type index valve; closed interval $[x_i^L, x_i^U]$ is the optimal interval, and $\eta_i = \mathbf{max}(x_i^L - \mathbf{min} x_{ij}, \mathbf{max} x_{ij} - x_i^u)$.

4.3   Conversion Method of the Natural Language from Qualitative to Quantitative

The normal Cloud is the most basic tool to express the language value, and can be generated by the Cloud's digital character (*Ex, En, He*), and its mathematic expectation curve (MEC) is:

$$MEC_A(x) = \exp[-(x - Ex)^2/(2En^2)] \tag{3}$$

The generating algorithm of the normal Cloud is that:

(1) $x_i = G(Ex, En)e^{i\theta}$, generating a normal random number x_i, whose expected value is *Ex* and standard deviation is *En*;

(2) $E'_{ni} = G(En, He)$, generating a normal random number E_{ni}, whose expectation value is E_n and standard deviation is H_e;

(3) Calculate $\mu_i = \exp[-\frac{(x_i - E_x)^2}{2E'_{ni}}]$, (x_i, μ_i) is a cloud drop.

(4) Repeat the step 1 to step 4, until the Cloud drops were generated enough. Figure 1 gives out "well achievements" languages which were described by one-dimension Cloud.

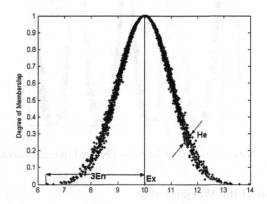

Fig. 1. Straight attitude cloud models and digital characteristics

Cloud is combined by many Cloud drops, and a Cloud drop may be footy, but the whole shape of the Cloud reflects the important characters of the quantity concept [6]. Cloud is used to compare to an uncertainty mapping between quantity and quality because this distribution looks like the Cloud in the sky, and it has a clear shape when far looking, but there is no borders in nearly. The digital character of Cloud can be expressed by *Ex*(Expected value), *En*(entropy) and *He*(Hyper entropy).

Expected value *Ex* is the center value of concept in the theory field, and it's the representative value of the qualitative concept.

Entropy *En* is the measuring of the fuzziness of qualitative concept, reflects the numerical range which can be accepted by this concept in the theory field, and embodies the uncertain margin of the qualitative concept. The bigger the entropy is, the bigger numerical range can be accepted by the concept, and the fuzzier are the concepts.

Hype Entropy *He* reflects the dispersion of the Cloud drops. The bigger the Hyper Entropy is, the bigger of its dispersion and the randomness of degree of membership, and this is the Cloud's thickness.

So, the fuzziness (the uncertainty of qualitative concept) and randomness (randomness of degree of membership) are completely combined together by three characteristic numbers of the Cloud models, which make up the mapping between quality and quantity as the foundation of knowledge expression.

According to the habit of natural languages of people, the performance of description things of determining the natural generally uses five graduation laws, i.e. "better", "good", "general", "difference" and "worse". Using the cloud model to express these five formulations, the desired value E_x can be expressed by 1, 0.75, 0.5, 0.25 and 0; but at the same time, these values are approximate, so they can be adjusted by the entropy, $En = 0.04$, by this way, the cloud drops can cover the [0, 1] interval; Hype Entropy $He = 0.005$, as showed by the Fig. 2.

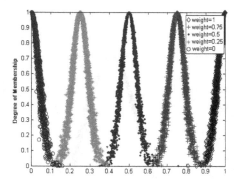

Fig. 2. Natural description language changed to quantitative cloud

4.4   Quantization of the Weight Value

The right value also can adopt cloud model, confirm right value adopt nine graduation law generally, the [0, 1] block turns nine graduations, i.e. 0.1, 0.2, 0.3, 0.4, 0.5, 0.6, 0.7, 0.8, 0.9, The important intensity shown is increased progressively sequentially, the importance can be adjusted through the determination of entropy value. Figure 3 shows the cloud of right value.

4.5   Arithmetic of Clouds

The arithmetic of the cloud is the algorithm changing the base cloud to the empty cloud. The so-called empty cloud means the digital characteristics of some already designated clouds will carry on a certain operation, cloud getting the new digital characteristic and

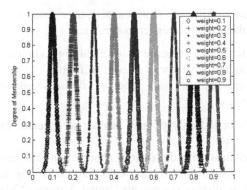

Fig. 3. Weight value cloud chart expresses by nine graduation law

constructing. These designated clouds used for calculating the empty cloud are called the base cloud. Because in exceeding the evaluation research of the efficiency of the flight simulator's Human-Computer Interaction system, only use the weighted sum algorithm, so only introduce addition and multiplication of the cloud here.

If there is a given cloud in fixed domain $C_1(E_{x1}, E_{n1}, H_{e1}), C_2(E_{x2}, E_{n2}, H_{e2})$, the calculated result of arithmetic operation for C_1 and C_2 is $C(E_x, E_n, H_e)$.

Its operation of multiplication can be expressed as:

$$E_x = E_{x1} \times E_{x2}; \tag{4}$$

$$E_n = |E_{x1}E_{x2}| \times \sqrt{\left(\frac{E_{n1}}{E_{x1}}\right)^2 + \left(\frac{E_{n2}}{E_{x2}}\right)^2}; \tag{5}$$

$$H_e = |E_{x1}E_{x2}| \times \sqrt{\left(\frac{H_{e1}}{E_{x1}}\right)^2 + \left(\frac{H_{e2}}{E_{x2}}\right)^2}; \tag{6}$$

Its operation of addition can be expressed as:

$$E_x = E_{x1} + E_{x2}; E_n \Sigma \sqrt{E_{n1}^2 + E_{n2}^2}; H_e = \sqrt{H_{e1}^2 + H_{e2}^2}$$

After get the positive-going, antidromic generate algorithm and the weighted sums algorithm, can use the cloud model theory to carry on the evaluation of efficiency for the Human-Computer Interaction system.

5   The Evaluation Example

In order to get the usability information, research chooses 10 pilots with different flying experience, aiming for a reasonable balance regarding age and rank, i.e. at least 40 % must be primary and at least 40 % must be senior rank pilots. About age, approximately one-third must be under 30 years old, between 30 and 40 and above 40 %.

Regarding operation skills, the representative user group will divide subjects into two groups, each balanced with respect to age, i.e.:

(1) Subjects who have little experience in specified flight simulator's human-computer interaction system but have strong interest in flying.
(2) Subjects who are familiar with specified flight simulator's human-computer interaction system.

So, the 10 pilots including 4 primary ranks and 4 senior ranks, the other 2 are middle rank pilots.

Through the meetings with discussion, observation and interaction with system three kinds methods to get the usability information. During the pilots completing interaction with cockpit display system, the Digital Airborne Video Recorder recorded the pilot's sight scan path in different training process or different mission.

Choose four kinds of flight simulator's human-computer interaction, about "availability" in Fig. 1, the index and affiliated subordinate index are calculated. The experts give the five indexes "learnability, efficiency, memorability, errors and satisfaction", respectively a different weight assignment "0.2", "0.2", "0.4", "0.2", and change them into digital characteristic of the cloud, likewise, the experts provides and appraises languages to the system among them, the digital characteristic changed into the cloud as shown in Table 1.

Table 1. The weight and evaluation cloud of the "availability"

Index	Weight	Weight cloud	Evaluation cloud
Learnability	0.2	(0.2, 0.04, 0.005)	(0.9, 0.04, 0.005)
Efficiency	0.2	(0.2, 0.02, 0.005)	(0.4, 0.02, 0.005)
Memorability	0.2	(0.2, 0.08, 0.005)	(0.3, 0.02, 0.005)
Errors	0.2	(0.2, 0.02, 0.005)	(0.6, 0.02, 0.005)
Satisfaction	0.2	(0.2, 0.02, 0.005)	(0.2, 0.01, 0.005)

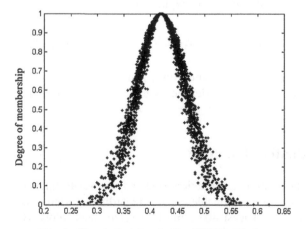

Fig. 4. Cloud chart for the "availability" index

According to number value described above and weighted sum operation method of the cloud, it can be obtained through calculating, the system is shown in Fig. 1, the digital characteristic of cloud of the "learnability" index is: (0.42, 0.0458, 0.0058), its cloud chart is shown as Fig. 4.

6 Conclusion

This paper analyzed the index system of Human-Computer Interaction system for usability evaluation, and representative evaluation model has been set up. The normalization method of the index and the conversion method of the natural language from qualitative to quantitative are described. Taking the flight simulator's human-computer interaction in four different periods to evaluate, which accords with the evaluation result. According to the model, sensitivity analysis can be carried on, by thus promote the usability of flight simulator's human-computer interaction system.

Acknowledgement. This work is supported by the Education Department of Jilin Province "13th Five-Year" scientific research planning project (2016-514).

References

1. Flight Simulator (2016). www.baidubaike.com
2. Tang, Z., Zhang, A., Bi, W.: Aircraft Cockpit Human-Computer Coordination Design Theory and Methods. Northwestern Polytechnical University Press, xi'an (2015)
3. Zhang, W., Ma, Z., Yu, J.: Human-Computer Integrative Design of the Civil Aircraft Cockpit. Northwestern Polytechnical University Press, xi'an (2015)
4. Ai, L.-Y.: Ergonomics-apply-research on cockpit's equipment arrangement. Aircr. Des. **32**, 78–80 (2012)
5. Wang, H., Bian, T., Xue, C.: Layout design of display interface for a new generation fighter. Electro-Mech. Eng. **27**, 57–61 (2011)
6. ISO 9241-11:1998: Ergonomic requirements for office work with visual display terminals (VDTs) – Part 11: Guidance on usability (2016). http://www.iso.org
7. Niels, O.B., Laila, D.: Multimodal Usability, p. 68. Springer, London (2009)
8. Collinson, R.P.G.: Introduction to Avionics Systems, 3rd edn, pp. 19–20. Springer, Heidelberg (2011)
9. Kong, Y., Qu, K.: Cockpit Engineering. Aviation industry press (2015)
10. Guo, Y., Gu, X., Shi, L.: Virtual Reality and Interaction Design. Wuhan University Press (2015)
11. Tan, H., Tan, Z., Jin, C.: Automative Human Machine Inerface Design. Publishing House of Electronics Industry (2015)
12. Zhao, C.: Usability Evaluation and Development for Mobile Phone Products Based on Simulator. Dalian Maritime University (2004)
13. Hua, Q.: Research on Usability Oriented Software Interaction Design of Smart Phone. Northwestern University (2011)
14. Whitten, A., Tygar, J.D.: Why Jonny can't encrypt: a usability evaluation of PGP 5.0. In: Proceedings of the Ninth USENIX Security Symposium, p. 14 (1999)

15. Usability Evaluation (2016). http://www.usabilityhome.com/
16. Wang, J., Xiao, W., Sheng, W.: An improved effectiveness evaluation method based on grey cloud model. Microcomput. Inf. **4**, 277–278 (2009)
17. Zhang, Y., Wu, L., Wei, G.: Improved ant colony algorithm based on membership cloud models. Comput. Eng. Appl. **27**, 11–14 (2009)
18. Yang, S.: An improved system cloud grey neural network model. In: Proceedings-2nd International Conference on Genetic and Evolutionary Computing, pp. 158–160 (2008)
19. Wang, G.: Rough set based uncertain knowledge expressing and processing. In: Kuznetsov, S.O., Ślęzak, D., Hepting, D.H., Mirkin, B.G. (eds.) RSFDGrC 2011. LNCS, vol. 6743, pp. 11–18. Springer, Heidelberg (2011)
20. Zhao, T., Hu, J.: Performance evaluation of parallel file system based on lustre and grey theory. In: Proceedings 9th International Conference on Grid and Cloud Computing, pp. 118–123 (2010)
21. Chen, Z., Yu, B., Hu, W., Luo, X., Qin, L.: Grey assessment and prediction of the urban heat island effect in city. J. Xi'an Jiaotong Univ. **38**, 985–988 (2004)
22. Chen, Y.: The comprehensive ranking evaluation of flood disaster based on grey-cloud whitening-weight function. In: Proceedings of 2011 International Conference on Electronic and Mechanical Engineering and Information Technology, vol. 4, pp. 1932–1934 (2011)
23. Zhang, G., Zhang, M., Zhong, H., Huang, X.: Research of transportation scientific and technological projects evaluation based on improved triangular whitenization weight function. 2010 International Conference on Logistics Systems and Intelligent Management, vol. 3, pp. 1500–1503 (2010)
24. Liu, J., Xiao, M., Li, B.: An evaluation model based on grey trigonometry whitenization weight function for logistic center. In: Proceedings of the 2nd International Conference on Transportation Engineering, pp. 345, 2055–2060 (2009)
25. Zeng, B., Liu, S.-F., Cui, J.: Prediction model for interval grey number with known whitenization weight function. Control Decis. **25**, 1815–1820 (2010)
26. Yang, X.-W., Wu, S.-L., Xu, C.-R., Jie, C., Shun-Peng, Z.: The comparison of two types of whitenization weight function. In: 2009 International Conference on Apperceiving Computing and Intelligence Analysis, pp. 69–72 (2009)

Healthcare Informatics Theory and Methods

Examining the Adoption and Use of Personally Controlled Electronic Health Record (PCEHR) System Among Australian Consumers: A Preliminary Study

Jun Xu$^{(\boxtimes)}$, Xiangzhu Gao, Golam Sorwar, and Nicky Antonius

Southern Cross University, Lismore, Australia
jun.xu@scu.edu.au

Abstract. This study looks at the adoption and use of Personally Controlled Electronic Health Record (PCEHR) system among consumers (individual users) in Australia. The specific aim of this study is to examine the current status of adoption and continued use of the PCEHR system among consumers in Australia. An online questionnaire survey was conducted, and 110 valid responses were received. The results of this study could contribute to the success of the ongoing roll-out of the PCEHR system in Australia and further studies in adoption and contused use of the system.

Keywords: Electronic health records · PCEHR system · Australia · Adoption · Use

1 Introduction

Electronic health records (EHRs) are electronic repositories of information regarding the health status of a subject of care, which can be managed electronically via Internet technologies and computing equipment, and they can include such health/medical information as: general information (e.g., geographic, demographic or habit information), physical or biological characteristics (e.g., height, weight, blood type), family health histories, doctor visits, medications, allergies & alerts, disabilities, chronic conditions, sexual health information, diagnoses, immunisations, problem lists, measurements, observations, pathology orders & results, imaging orders & results, Medicare/Veteran affairs data, care plans, test results, event summaries, discharge summaries, specialist Letters, referrals, consumer-entered notes, and personal data. In recent years, the notion that EHRs underpin all other e-health initiatives has been widely accepted [19, 20]. EHRs could contribute to enhance the health care quality, reduce medical errors, cut down healthcare costs, and empower consumers to better understand their health needs and make more informed decisions on their health [8, 20].

Jha et al. [12] argued that only those nationals willing to put in significant investments and take up the challenge of developing standards and interchanges will be able to succeed in their e-health record systems and be able to provide the benefits of e-health records to their people. Australia is one of such nationals and has invested

© Springer International Publishing Switzerland 2016
D.-S. Huang and K.-H. Jo (Eds.): ICIC 2016, Part II, LNCS 9772, pp. 705–718, 2016.
DOI: 10.1007/978-3-319-42294-7_63

significantly in e-health (i.e., over A\$ 2 billion e-health spending in 2012 alone [14]). Meanwhile Australian government has invested significantly in deploying the PCEHR system and its associated infrastructure (i.e., more than A\$1 billion since its introduction [13]), but the system has not been well received by consumers. And there is a lack of empirical studies specifically designed for examining the adoption and use of the PCEHR system. The research aims to close the gap via investigating the current status of PCEHR system adoption and use among consumers in Australia.

2 Background

Most of the medical information in Australia has been held in paper-based files or non-shared databases, and the medical information may be inconsistent between files, inaccurate because of lack of standards, incorrect because of manual operation, and is often not available in emergency situations [9, 18]. Some estimates of consequences arising from inaccurate and inconsistent information include: 18 % of medical errors in Australia occur from inadequate information; nearly 30 % of unplanned hospital admissions are associated with prescribing errors; and approximately 13 % of healthcare provider consultations suffer missing information [2, 3, 15]. E-health record initiatives like Australia's PCEHR system could be an effective solution to such problems, especially in emergency situations and special conditions [17]. The successful implementation of the PCEHR system could provide Australian consumers with better quality of health care and equitable access to healthcare [2, 3].

In addition, the successful implementation of the PCEHR system could bring in huge economic benefits to Australia. According to an estimate by international consulting firm Deloitte Consulting, the PCEHR system could generate approximately \$11.5 billion in net direct benefits over the period of 2010 to 2025, with \$9.5 billion to Australian governments and \$2 billion to the private sector including public hospitals, GPs, specialists, allied health clinics, private hospitals and private health insurance providers [9]. The current release of the PCEHR system in Australia is far from mature and suffers criticisms from major stakeholders. The system is facing various challenges, and users are not enthusiastic in registering with and using the system. Participation in the PCEHR system is essential for adoption and use of the system. At this stage, consumers have not been active in participation for different reasons and various concerns; and the interests of consumers have been properly represented in the development and implementation of the PCEHR system (e.g., even in the recent PCEHR review ordered by the Coalition Government, no review panel members are really representing consumers). Recently the opt-out model has been looked at by the Australian government [7] and trials of opt-out approach for the PCEHR system have been commenced in 2016 [5]. However even though the change to opt-out model could have more registrations for the PCEHR system, the long-term and sustainable success of the system requires far more things and efforts than initial registration for the system.

3 Research Design

Staff and students in an Australian university were invited to take part in the online questionnaire survey, which asks questions about the participants' adoption and use of the PCEHR system. Two rounds of e-mail invitations for participating in the research were sent to the two generic email lists of staff and students. Sending emails to these two email addresses needs the approval from the university and so we only did twice (an original one and a follow-up one in late October and early November 2015 respectively). The survey questions were developed from literature on innovation diffusion and technology acceptance (e.g., [6, 16]) and the authors' previous studies on PCERH system in Australia (e.g., [18, 19]). In addition, the researchers sent the online survey invitation to their networks. In the end, 110 valid questions were received. Frequency analysis via IBM SPSS software was adopted to examine the current status of adoption and use of the PCEHR system among Australian consumers.

4 Results of Online Survey

4.1 Demographic Information

Among 110 received responses, 63.6 % are from females and 35.5 % are from males. 28.2 % of the respondents are in the 40–49 age group, 24.5 % in the age group of 50–59, 22.7 % in the age group of 30–39, 9.1 % in the age group of 21–29, 6.4 % in the age group of 60–65, 2.7 % in the age group of 18–20, and 6.4 % in the other age groups (e.g., younger than 18 and older than 65 groups). 1.8 % of the respondents are Aboriginal Australian or Torres Strait Islander. 73.6 % of the respondents are from New South Wales, 20.9 % from Queensland, 3.7 % from Victoria, 0.9 % from both Tasmania and Australian Capital Territory. 77.2 % of the respondents have at least a Bachelor's degree, with 13.6 % having Graduate Certificate or Graduate Diploma, 12.7 % having a Master's degree, and 11.8 % having a Doctorate's degree. The distribution of the respondents by occupation is as follows: 20.7 % educational professionals, 11.7 % administrators, 10.6 % managers, 9.8 % technology and engineering professionals, 9 % students, 8.1 % health care professionals, 3.6 % technicians and trade workers, 3.6 % company directors and business owners (including self-employed), 4.5 % accounting, consulting and legal professionals, 2.7 % sales staff, 2.7 % community and personal services workers, 2.7 % tourism and hospitality staff, 1.8 % librarians, 0.9 % media professionals, 0.9 % public servant, and 0.9 % staying at home parent.

74.5 % of the respondents are native English speakers and 24.5 % able to communicate in English in both writing and speaking. The reported averages of weekly household income in 2014-2015 financial year are: 20 % more than $1,800, 18.2 % in the range of $1,501 to $1,800, 14.5 % in the range of $1,101 to $1,500, 12.7 % in the range of $901 to $1,100, 12.7 % in the range of $701 to $900, 10.9 % in the range of $401 to 700, and 9.1 % less than $401. The respondents report the following household compositions: 42.7 % more than one person, 26.4 % having child/children under the age of 14, 13.6 % only one person, 9.1 % having family member(s) older than 65 years,

6.4 % having family members with chronic conditions, mental health conditions or disabilities, 0.9 % having new born child/children, and 0.9 % living in an aged care home (nursing home).

4.2   E-health Experience

It can be seen from Table 1 the majority of the participants (52.7 %) don't have any e-health experience while others have varied e-health experiences, with top five experiences of (in the order) (1) using electronic systems for healthcare supportive services (e.g., online appointment scheduling, checking in via kiosks, claiming expense online), (2) using wearable health and fitness tracking technology, using health-related phone/tablets apps, (3) using health-related phone/tablets apps, (4) participating in online health programs (e.g., health and fitness through diet and exercise programs), and (5) using telemedicine (i.e., enabling doctors and nurses to see and diagnose patients remotely).

Table 1. Survey participants' e-health experience

Survey participants' e-health experience	Percentages of respondents
Don't have any e-health experience	52.7 %
Have experience of using electronic systems for healthcare supportive services (e.g., online appointment scheduling, checking in via kiosks, claiming expense online)	35.5 %
Have experience of using wearable health and fitness tracking technology	24.5 %
Have experience of using health-related phone/tablets apps	21.8 %
Have experience of participating in online health programs (e.g., health and fitness through diet and exercise programs)	15.5 %
Have experience of using telemedicine (i.e., enabling doctors and nurses to see and diagnose patients remotely)	10.9 %
Have experience of online treatments and assessments	8.2 %
Have gained e-health experience via registering with PCEHR system	0.9 %
Have experience of medicare online	0.9 %
Have e-health research experience	0.9 %
Have experience of working for the government's e-health projects	0.9 %

Australia has the one of the highest Internet penetration rates in the world (i.e., more than 93 % in 2015 [11], and Australians are very active in e-commerce (for example, according to Australian Bureau of Statistics [1], some top online activities of Australian consumers are (in the order): Banking, Social networking, Purchasing goods or services, and Entertainment). More education and empowerment of using the Internet for health-related purposes and activities should be encouraged and nurtured by healthcare providers and government agencies.

4.3   Involvement in the PCEHR System

From Table 2 it can be said that survey participants get to know the PCEHR system via different sources, such as from the media (including TV, website, radio, newspaper, magazine, or other media Channels), from other people (e.g., my friends, my family members, colleagues, neighbours or people I met/talked to), from healthcare providers, from promotion materials received from governments or/and their agencies, and even by taking part in this research. There are no dominating sources even though the media seems play a more important role than other sources (e.g., 29.1 % participants getting to know the PCEHR system via the media, comparing to 20.9 % via other people they know, 18.2 % via healthcare providers as well as 18.2 % from promotional materials). It is interesting to note that healthcare providers are not the most importance source of knowing the PCEHR system since it can be argued that consumers typically trust their doctors and other healthcare providers most when it comes to health-related matters. Many (if not most) doctors in Australia lack the confidence in the PCEHR system for concerns such as risk, liability, insurance, security and information currency. According to a survey of its members by Australian Doctor magazine (reported in Foo [10]), among 514 responded doctors about 58 % of them said they would never take part in the scheme and will not be promoting its use to patients, around 29 % said they would be taking part but have yet to write a health summary for the PCEHR system, and only 6 % have written health summaries for the PCEHR system. Logically convincing and enticing doctors and other healthcare providers to use and promote the PCEHR system is critical to the success of the PCEHR system.

Table 2. Knowing the PCEHR system

Sources/ways of knowing the PCEHR system	Percentages of respondents
From the media (including TV, website, radio, newspaper, magazine, or other media Channels)	29.1 %
From other people (e.g., my friends, my family members, colleagues, neighbours or people I met/talked to)	20.9 %
From healthcare providers	18.2 %
From promotional materials received from governments or/and their agencies	18.2 %
From this survey	7.2 %
From my previous employment related to e-health areas	1.8 %
By working it out myself	0.9 %
From using other government online services/systems	0.9 %

Table 3 presents the levels of participants' understanding of the PCEHR system and Table 4 indicates their involvement in the implementation and roll-out of the PCEHR system. The majority of the respondents (85.5 %) have no good understanding/knowledge of the PCEHR system, only 2.7 % reckon they have sufficient knowledge of the PCEHR system (see Table 3). Without good understanding of the system (e.g., benefits, functions, potential risks & associated risk management measures), people won't use the system since resisting the change and new things are human being's nature; they will ask questions such as "what is in it for me"?, "why should I change the current practices and make changes?". The governments and their agencies should be actively working on educating the consumers and promoting the realized benefits of the systems. When people see the realized benefits themselves, they will start using the system-There is a saying in Chinese, "Seeing is believing but not hearing".

Table 3. Level of understanding of the PCEHR system

Level of understanding of the PCEHR system	Percentages of respondents
I don't know much about the PCEHR system	37.3 %
I know the basics of the PCEHR system	26.4 %
I have no knowledge/understanding of the PCEHR system before doing this survey	21.8 %
I am well informed and I know most things I need to know	2.7 %

Table 4. Involvement in the implementation/rollout of the PCEHR system

Involvement	Percentages of respondents
Have not been involved in the implementation/rollout of the system at all	66.4 %
Have discussed the systems with other individuals	15.5 %
Have discussed the system with healthcare providers	10.9 %
Have attended the sessions on the system organized by governments or/and their agencies	5.5 %
Have been kept informed	2.7 %
Having experience of working with related e-health project or earlier versions of the PCEHR system	1.8 %
Have provided my feedback to governments or/and their agencies	1.8 %
My feedback has been sought	0.9 %
My feedback and concerns have been understood and considered	0.9 %
My recommendations have been accepted	0.9 %

Meanwhile, most of the respondents (66.4 %) have not been involved in the implementation/rollout of the system at all (see Table 4). Any system implementations, especially for large-scare and very complex systems like the PCEHR system, getting the users actively involved in the early stage of system development and implementation is a must for the success of the system. Past studies (e.g., Chen et al. [4]) have supported this notion. Users need to be ensured that they are an integral part of the system (and system development & implementation) and the system is designed for them to address their needs (i.e., managing their medical/health information), otherwise they won't feel the ownership of the system and won't contribute to and use the system, especially the use of the system is on the voluntary basis (opt-in approach, which is the current model of the PCEHR system).

It can be seen from Table 5 that most respondents (91.9 %) have trust in their GPs and other healthcare providers, and would not mind their access to their information in the PCEHR system even though there are differences in the extent and scope of access by GPs and other healthcare providers. It further enhances the notion mentioned earlier on that consumers typically trust their doctors and other healthcare providers most when it comes to health-related matters. The Australian government and its agencies need to ensure that GPs and other healthcare providers are supporters of the PCEHR system, otherwise the consumers (the patients/customers of the GPs and other healthcare providers) would very likely not register and use the PCEHR system as a result of the influence from their very much respected health-related experts (i.e., their GPs and other healthcare providers).

Table 5. PCERH system access by GPs and other healthcare providers

Controls of PCEHR system access by GPs and healthcare providers	Percentages of respondents
Beside my GP, I would allow other healthcare providers to access my record in a case-by-case manner	40 %
I don't know how to control the access, and would use the system's initial access setting	17.3 %
I would only allow my GP to access my full record	17.3 %
Beside my GP, I would allow other healthcare providers to access some of my record	15.5 %
Beside my GP, I would allow other healthcare providers to access my full record	12.7 %
I would consult others to make a decision	10.9 %
I would allow my GP to access some of my record	6.4 %
I myself would control the access to my record in the PCEHR system by my GP and other healthcare providers	2.7 %
Not feeling comfortable to storing such sensitive information in a government system	0.9 %

4.4   Adoption and Use of the PCEHR System

Only 11.8 % of the respondents (including for their dependants) have registered for the PCEHR system, 2.7 % have tried to register for the system but experienced difficulties, and 22.7 % stated that they will register for the PCEHR system in the future (see Table 6). Among those respondents who have registered for the PCEHR system (11.8 %), 85 % of those registered respondents have done their registration for the system themselves (see Table 7), and 77.1 % of them have registered online (see Table 8). It suggests that the earlier adopters of the PCEHR system are pro-active, willing to try new things like the PCEHR system and comfortable with doing things online.

Table 6. Registration for the PCEHR system

The current status of registration for the PCEHR system	Percentages of respondents
I don't know whether I will register for the system	39.1 %
I will register for the system	22.7 %
I have registered for the system or on behalf of my dependant	11.8 %
I will not register for the system	9.1 %
I have tried to register for the system but experienced difficulties	2.7 %

Table 7. System registration approaches

System registration approaches	Percentages of registered respondents	Percentages of all respondents
I registered myself	85 %	10 %
I registered with the assistance of the government departments or their agencies	7.6 %	0.9 %
I registered while I was in hospital	7.6 %	0.9 %

Table 8. The location of system registration

The location of system registration	Percentages of registered respondents	Percentages of all respondents
I registered online	77.1 %	9.1 %
I registered at my GP's practice/clinic	7.6 %	0.9 %
I registered at a mobile station/booth set up by the governments or their agencies	7.6 %	0.9 %
I registered at the place of a healthcare provider (other than my GP)	7.6 %	0.9 %

Rogers ([16], p. 261) classifies the population (e.g., all users of the PCEHR system) into five categories as per their personal innovativeness, which is the degree to which an individual or other unit of adoption is relatively earlier in adopting new ideas than other members of a "social system" (p. 261). The five categories consist of: the "innovators" (estimated 2.5 % of the population), the "early adopters" (estimated next 13.5 % of the population), the "early majority" (estimated next 34 % of the population), "late majority" (estimated next 34 % of the population), and the "laggards" (estimated next 16 % of the population). Personal innovativeness could be caused by such individual differences as: socioeconomic status, personality variables, and communication behaviour ([16], p. 261). It is interesting to note that the 11.8 % registered respondents and 2.7 % tried respondents (all together 14.5 %) are quite close to the Rogers' 16 % innovators and early adopters of an innovation.

The governments and their agencies need to find effective solutions to encourage wider adoption and more use of the PCEHR system. Even though the proposed change from opt-in to opt-out model would achieve more registrations for the system, the ultimate success (e.g., the actual and continuous use of the system by users) would not guaranteed until other factors such as looking after users' real needs, engaging the users early and letting them having the feeling of ownership of the system, effectively addressing privacy and security concerns, useful and user-friendly system, transparency and good governance of the management of the system, stable policies and sufficient resources, unified approach from different levels of government, being-patient with the system (i.e., taking long term view of the system), continuous education and promotion of the system's benefits (especially realized benefits) and best practices of using the system, and good management of system development and implementation (including project management, risk management, and change management), have been dealt with properly [18, 19].

Among those respondents who have registered for the PCEHR system, most of them (61 %) have been with the system for more than 12 months (see Table 9) and the majority of them (68.4 %) have not really used the system after the registration (either not at all or very little) (Table 10).

Table 9. System registration duration

The time length since system registration	Percentages of registered respondents	Percentages of all respondents
More than 12 months but less than 18 months	30.5 %	3.6 %
More than 24 months but less than 36 months	30.5 %	3.6 %
More than 6 months but less than 12 months	15.2 %	1.8 %
More than 1 month but less than 3 months	7.6 %	0.9 %
More than 18 months but less than 24 months	7.6 %	0.9 %
Less than 1 month	7.6 %	0.9 %

As previously argued, the ultimate success (e.g., the actual and continuous use of the system by users) would not guaranteed until other factors, such as looking after users' real needs, engaging the users early and letting them having the feeling of ownership of the system, effectively addressing privacy and security concerns, useful and user-friendly system, transparency and good governance of the management of the system, stable policies and sufficient resources, unified approach from different levels of government, being-patient with the system (i.e., taking long term view of the system), continuous education and promotion of the system's benefits (especially realized benefits) and best practices of using the system, and good management of system development and implementation (including project management, risk management, and change management), have been dealt with properly [18, 19]. Those views are not only supported by the past studies (e.g., Chen et al. [4]), but also by the participants' comments in the survey (See Table 11).

Table 10. The status of actual use of the PCEHR system

Actual use of the PCEHR system	Percentages of registered respondents	Percentages of all respondents
I have played around with the system, but have not really done anything yet	38 %	4.5 %
After registering with the system, I have not used the system at all	22.8 %	2.7 %
I have used the system to some extent (e.g., uploading some information into the system)	15.2 %	1.8 %
I have looked at and managed my records in the system along with my family	7.6 %	0.9 %
I have looked at and managed my records in the system without assistance from a health professional	7.6 %	0.9 %
The system seems to contain virtually none of my records and I don't know how to get them up and all showing.	7.6 %	0.9 %

Table 11. Participants' comments on the PCEHR Systems

Participants' comments on the PCEHR systems	Key factors/themes of participant's comments
• Inform us of the existence of the system. I consume the media but have never heard of this.... • ....This system has not been adequately publicised or promoted. It is very important to me to have continuity of medical records and I am very keen to use the system but I have been unaware of even what this system was named...	Poor Education and Promotion of the System
• Access to records is difficult.... • ...The elderly, those with disabilities, those with language issues, etc., will not be able to use the system..... • ...if software designed for the public is not easy to register and use it will fail.... • ....Even to register as a user of the federal government systems was atrociously tedious and difficult even for an experienced computer user like me...... After I'd registered with the system, virtually nothing appeared on it and I didn't know how (or did not try) to get all my information there because it all seemed too difficult... • The system is poorly designed to address fed/state inefficiency's thereby making it barely workable......	User Friendliness of the System

Table 11. (*Continued*)

Participants' comments on the PCEHR systems	Key factors/themes of participant's comments
• Privacy and Security are the largest concerns.... • As a national health database, I would be concerned about the systems online security and potential unauthorized access from international agencies/interests..... • Privacy and security would need to be absolutely assured though... • This PCEHR is another way the governments (Both federal and state) are attempting to erode our privacy and freedom of movement and speech..... • I would not want my health records shared with insurance agencies/companies simply so they could deny coverage for a perceived pre-existing condition.... • PCEHR must be secure and PRIVATE. No chance of hacking or information being lost, altered or stolen. Need assurances about the chances of identity theft especially with my medical records.... • Personal privacy laws need to be reviewed, created or changed to protect personal information..... • Personally I am very much in favour of the PCEHR but in discussion of this system with others have found that there is a distrust of the security of information...... I do not have a 'socially sensitive' illness (HIV/AIDS or Hep C etc.) and can well understand why people who do would not choose to have their medical information in this system. I hope that over the years the security of information and each individuals control over access to this information can be assured...... • Way better security than what is offered.....	Privacy & Security Concerns of the System
• Government spin will not be enough for me to adopt the system. Must be factual information provided based on evidence and performance alone....	Benefits/Usefulness and Realized Benefits/Usefulness of the System
• The data is owned by the user not the government..... The name suggests Personally Controlled but I still don't believe it, this will be a hard sell.... • There should be no option for owner of the profile to modify personal medical info, only control the access and be able to view everything on the file......	Ownership of the System
• The implementation of this system was appalling...... It's implementation has been a disgrace.....	Poor Implementation of the System

4.5   Future of the PCEHR System

Most of the respondents are not certain about the future of the PCEHR system, only 24.5 % believe the system will be successful in the future (see Table 12). Some top selected issues associated with the future of the PCEHR system include (in the order): the uncertainty in government continuity, the uncertainty in the development of information and communication technology (ICT) infrastructure (e.g., broadband, mobile devices) in Australia, the possibilities of having better technical solutions than such a system, the economic uncertainty affecting resources required for system development, and the consumers' concerns (e.g., privacy, security).

Table 12. The future of the implementation/roll out of the PCEHR system

Views of the future of the PCEHR system	Percentages of respondents
I am not sure because of the uncertainty in government continuity, which may cause changes in government budget, policies & regulations	40.9 %
I am not sure because of the uncertainty in the development of information and communication technology (ICT) infrastructure (e.g., broadband, mobile devices) in Australia	31.8 %
I am sure it will be successful in the future	24.5 %
I am not sure because there may be better technical solutions than such a system	20.9 %
I am not sure because of the economic uncertainty, which affects resources required for system development	8.2 %
I am not sure because of the consumers' concerns (e.g., privacy, security)	3.6 %
I am not sure because of the poor take-up by health care providers	0.9 %
I don't think the system will be successful for such reasons as extremely poor implementation/roll out and low take-up and usage of the system	0.9 %
I hope the system will be successful	0.9 %

In order to ensure the success of the PCEHR system, these issues and other factors/concerns discussed previously in the paper have to be effectively addressed.

5   Conclusions

This paper presents the results of a research studying the current status of adoption and use of the PCEHR system among Australian consumers. The outcomes of the research could provide assistance to the ongoing implementation of the PCEHR system in Australia in understanding the current status and challenges/issues of and in developing practical guidelines to the successful implementation of the PCEHR system. This research was a preliminary study and as a result of such considerations as time, cost, convenience, the research only collected data via the Internet-based survey. In the future, we will look at collecting data via such methods such as interviews, focus groups, expert panels, and observations. In addition, future research will look at consumers outside the university. Furthermore, subsequent studies could build on the findings of this research and explore the factors and concerns identified in this research further via developing and testing models of adoption and continued use of the PCEHR system.

References

1. Australian Bureau of Statistics (ABS): 8146.0 - Household Use of Information Technology, Australia, 2014–15, Released at 11:30 AM (CANBERRA TIME) 18 February 2016. http://www.abs.gov.au/ausstats/abs@.nsf/mf/8146.0. Accessed 2 Apr 2016
2. News, Australian Nursing Journal: National e-health rollout on. Aust. Nurs. J. **18**(10), 5 (2011)
3. Burmester, S.: Review the Progress of eHealth in Australia. Presentation by National E-Health Transition Authority, 20 March 2012
4. Chen, J.J., Su, W.C., Wang, P.W., Yen, H.C.: A CMMI-based approach for medical software project life cycle study. SpringerPlus **17**(2(1)), 266 (2013)
5. Coyne, A.: Australia's first opt-out e-health site to start trials this week, iTnews, 27 January 2016. http://www.itnews.com.au/news/australias-first-opt-out-ehealth-site-to-start-trials-this-week-414099. Accessed 22 Feb 2016
6. Davis, F.D.: Perceived usefulness, perceived ease of use, and user acceptance of information technology. MIS Q. **13**(3), 319–340 (1989)
7. Department of Health: Patients to get new myHealth Record: $485 m 'rescue' package to reboot Labor's e-health failures, Media Release, 10 May 2015. https://www.health.gov.au/internet/ministers/publishing.nsf/Content/health-mediarel-yr2015-ley050.htm. Accessed 18 Feb 2016
8. Department of Health and Aging: The eHealth Readiness of Australia's Allied Health Sector, Department of Health and Aging, May 2011
9. Department of Health and Aging: Personally Controlled Electronic Health Records: Stakeholder Briefing Guide, February 2012 Update, May 2012
10. Foo, F.: Most doctors reject e-health record system as white elephant, The Australian, 16 July 2013. http://www.theaustralian.com.au/business/technology/most-doctors-reject-e-health-record-system-as-white-elephant/story-fn4htb9o-1226679780209. Accessed 20 Mar 2016
11. Internetworldstats.com: Australia: Internet Usage Stats and Telecommunications Market Report (2016). http://www.internetworldstats.com/sp/au.htm. Accessed 13 Apr 2016
12. Jha, A.K., Doolan, D., Grandt, D., Scott, T., Bates, D.W.: The use of health information technology in seven nations. Int. J. Med. Inf. **77**, 848–854 (2008)
13. Kerlin, J., Heath, J.: E-health scheme to be revived after panel review. The Australian Financial Review, 24 May 2014. http://www.afr.com/business/health/pharmaceuticals/ehealth-scheme-to-be-revived-after-panel-review-20140523-iupi8. Accessed 18 Feb 2016
14. Lohman, T.: Australian e-health spending to top $2 billion in 2010, 15 April 2010
15. NETHA: Overview of the National eHealth Strategy and the Personally Controlled Electronic Health Record, May 2012
16. Rogers, E.M.: Diffusion of Innovations, 4th edn. The Free Press, New York (1995)

17. Townsend, R.: Doctors and patients uneasy about new e-health record system. The Conversation, 05 July 2012
18. Xu, J., Gao, X.Z., Sorwar, G., Croll, P.: Current Status, Challenges, and Outlook of E-Health Record Systems in Australia. In: 7th International Conference on Intelligence Systems and Knowledge Engineering, 15–17 December 2012, Beijing, China (2012)
19. Xu, J., Gao, X.J., Sorwar, G., Croll, P.: Implementation of E-Health Record Systems in Australia. Int. Technol. Manag. Rev. 3(2), 92–104 (2013)
20. Xu, J.: Managing Digital Enterprise: Ten Essential Topics. Atlantis Press, Paris (2014)

Artificial Bee Colony Algorithms

Improved Artificial Bee Colony Algorithm Based on Reinforcement Learning

Ping Ma and Hong-Li Zhang$^{(\boxtimes)}$

College of Electrical Engineering, Xinjiang University,
Urumqi 830047, Xinjiang, China
694073078@qq.com

Abstract. In order to overcome the basic artificial bee colony algorithm converges slowly and prematurely, the reinforcement learning is added into the artificial bee colony algorithm, in which several different updating strategies is mapped into an action used to update the nectar source location. According to the calculation of Q function value, each nectar source selects the optimal updating strategy to speed up the convergence rate. At the same time, the selection probability based on ranking is used instead of roulette wheel selection probability to keep population diversity and avoid premature convergence. Comparing with several different algorithms through the test functions and the parameter identification of Chaotic system. The results show that the proposed algorithm has higher accuracy and faster convergence rate, the feasibility and effectiveness of the algorithm is validated.

Keywords: Artificial bee colony · Reinforcement learning · Rank selection · Parameter identification

1 Introduction

The artificial bee colony algorithm (ABC) is a swarm intelligence algorithm optimization algorithm that was introduced by Karaboga [1, 2] in 2005, As similar with other intelligent algorithms, it is a global optimization algorithm by simulating intelligent behavior of creatures in nature and inspired by the foraging behavior of honey bees. The algorithm is as simple as particle swarm optimization (PSO) and differential evolution (DE) algorithms, uses only common control parameters such as colony size and maximum cycle number and therefore the algorithm have the advantages of simple in structure, control parameters and less robust. Since the development of artificial bee colony algorithm, it has been widely applied to different fields such as function optimization [3], sensor network [4], image processing [5, 6], artificial neural network [7], and so on.

In ABC algorithm, the search process was composed of three phases.: employed bees phase, onlookers phase and scouts phase. In the first two phases with the same solution search equation to find nectar source and scouts phase for a random initialization to the abandoned nectar source. We know that the performance of ABC depends on solution search equation substantially [8]. For this reason, people do a lot of research to search for how to improve solution search equation. Das et al. [9] proposed the distance factor to the solution search equation and which was capable of

© Springer International Publishing Switzerland 2016
D.-S. Huang and K.-H. Jo (Eds.): ICIC 2016, Part II, LNCS 9772, pp. 721–732, 2016.
DOI: 10.1007/978-3-319-42294-7_64

solving prematurity problems. Liu et al. [10] suggested to use local search strategy in the algorithm aimed at improve the convergence rate. Inspired by particle swarm optimization (PSO), Zhu and Kwong [11] introduced global optimal solution into solution search equation and which improved the exploitation and exploration capability of algorithm (GABC). Li and Yang [12] introduced the information entropy into the search process of onlookers bee and adapt it to realizes self-adaptive adjustment of ABC algorithm. Bi and Wang [13] also presented a new solution search equation that was based on the crossed factors. In the literature [14–16], Gao and Liu et al. proposed several different versions of improved solution search equation inspired by mutation strategy of DE. Zhou et al. [17] proposed a new scheme that employs the orthogonal experimental design to generate a new nectar source and therefore enhancing the search efficiency of ABC. In order to accelerate convergence of the method, Kiran et al. [18] proposed using a control parameter (modification rate-MR) for ABC and this approach is based on updating more design parameters than one.

In this paper, we suggested to applied a variety of update strategies to nectar source at the same time and according to the effect of multi-step update to choose one of the strategy. Based on this idea, in this work, the reinforcement learning was introduced to solution search equation and several different updating strategies is mapped into an action used to update the nectar source location and the selection probability based on ranking is used instead of roulette wheel selection probability to keep population diversity and avoid premature convergence. Through the test function comparing with several different algorithms. The results show that the proposed algorithm has higher accuracy and faster convergence rate, the feasibility and effectiveness of the algorithm is validated.

2   Artificial Bee Colony Algorithm

In ABC algorithm, the colony of honey bees make up of three groups: employed bees, onlookers and scouts. The employed bees have the responsibility to store their best nectar sources and sharing the information of these nectar sources to onlooker bees by dancing. According to this information, the onlooker bees tend to choose a good food sources. The nectar sources that has higher fitness will have a better chance to be selected than the one of lower fitness. An employed bee's nectar sources is regarded as low fitness by employed, and this employed bee will change to a scout bee to determine a new food source and replace with the abandoned one. In our model, the number of employed bees or the onlookers is equal to the number of nectar sources at the initial stages of the search.

At first, let $X_i = (x_{i1}, x_{i2}, \cdots, x_{iD})$ present the position of the nectar sources, where D is dimension size. Every employed bees produce a candidate nectar position V_i from the old one X_i in memory to update nectar sources, the position of the new nectar sources will be calculated from equation below.

$$V_{ij} = X_{ij} + \phi_{ij} \cdot (X_{ij} - X_{kj}) \tag{1}$$

In the Eq. (1), V_{ij} is a new nectar sources and modified from its previous food sources X_{ij}. $k \in (1, 2, \cdots, SN), j \in (1, 2, \cdots, D), k \neq i$ are randomly chosen indexes. ϕ_{ij} a random number between $[-1,1]$ and it is used to control the V_{ij} around X_{ij}, if the fitness value of V_{ij} is better than it's previous X_{ij} then update X_{ij} with V_{ij}, if not keep X_{ij} unchangeable.

In the next step, the onlooker bees make a decision to choose a good nectar source based on the fitness value P_i obtained from the employed bees, that can be calculated by an equation below.

$$P_i = \frac{F_i}{\sum_{i=1}^{SN} F_i} \tag{2}$$

In the third step, any food source's fitness value not changed at any limited time and the position will be abandoned and replaced by a new position that is chosen by scout bee randomly. The new position produced by scout bees can be calculated by the following expression (3).

$$X_{i,j} = X_{min}^{j} + rand() \cdot (X_{max}^{j} - X_{min}^{j}) \tag{3}$$

where X_{min}^{j} and X_{max}^{j} is the lower and upper bound of the nectar source position in dimension j.

Thus, we all know that there are three control parameters at ABC: nectar source's number (it is equal to the number of employed bees or onlooker bees), the value of limit, the maximum cycle time (max-cycle).

3 Improved Artificial Bee Colony Algorithm

3.1 Q-learning Algorithm

Q-learning is one of the major algorithm in reinforcement learning. It's regarded as a method to optimize solutions in Markov decision process problems. And it was used to search a strategy which maximizes the sum of the reinforcements in future.

The formula of Q learning's transition rule is:

$$Q(s, a) = R(s, a) + \gamma \max_{a'} Q(s', a') \tag{4}$$

where s and s' is the state of agent. a and a' represents an action of agent.

An agent observes the vector of state then chooses an action to get the immediate rewards (r(state, action)), then the process moves to next state. And apparently the Max {Q(next state, all actions) represents agent chooses the different action to get the maximum return rewards at next state. In order to ensures the convergence of the sum, the discount factor gamma is limited between 0 to 1($0 <= \gamma < 1$).

Q learning algorithm to get the optimal strategy through evaluate the optimal value of Q. We suppose n is the number of all actions that can be selected by state at any time, that is A = {a1, a2,..., an}. At the first step, agent choose the action a and after a few steps forward (We assume that the m steps), then we can get the rewards as follows:

$$Q(a) = R(a) + rQ(a^{(1)}) + r^2Q(a^{(2)}) + \cdots + r^mQ(a^{(m)}) \tag{5}$$

where $a, a^{(i)} \in A$ and $1 \leqq i \leqq m$.

3.2   Update Strategy of Food Sources

In each iteration of ABC algorithm, employed bees or onlooker bees update the positions of nectar sources by the solution search equation (Eq. 1).

The method of time-varying crossover factor was adopted to improve the global searching ability and convergence speed of ABC algorithm. Inspired by the literature [13], a new solution search equation is proposed to search the optimal solution as follows.

$$V_{ij} = F_1 \times X_{ij} + (-1 + 2 \times F_2) \cdot (X_{ij} - X_{kj}) \tag{6}$$

within this equation, F1, F2 is learning factor:

$$F_1 = 2 - e^{\frac{i}{\max-cycle}\cdot \ln 2}, \quad F_2 = e^{\frac{i}{\max-cycle}\cdot \ln 2} - 1$$

In this work, we introduce differential evolution to solution search equation based on differential evolution (DE) algorithms [19] and define it as follows:

$$V_{ij} = X_{ij} + C_1 \cdot (X_{ij} - X_{kj}) + C_2 \cdot (X_{ij} - X_{Pj}) \tag{7}$$

where C1, C2 is crossed factor, and C1 = C2 = 1. p is determined randomly and it should be different from k.

A large number of experiments show that those update strategy has their own advantages at different time in the iteration. Can be integrated application of these update strategy to make a better effect? For this purpose, we regard these different update strategy to an action and choose a best strategy to control the position of nectar sources in the search process of every time. In addition, we can considers a few steps forward in order to make the right decision.

Fortunately, we can use reinforcement learning to select the maximum agent with discount and get the optimal strategy by learning gradually. In ABC, we put several solution search equations (Eqs. 1, 6 and 7) as an action, so the solution search equations that employed bees or onlooker bees choose the optimal nectar source location can be regard as the choose of optimal action.

3.3   The Selection Probability Based on Ranking

In ABC algorithm, onlooker bee's choice of nectar source location depends on the fitness and in which high fitness value nectar source has a good chance to be selected.

But the poor fitness value nectar source may be exists useful information at the beginning of the iteration and its control the selection process and has great influence to the convergence. The selection probability based on ranking related with the order of the position of nectar source and there is no direct relationship with the objective function value. This method has simple operation and easily realizes advantages [10, 20]. In this paper, we follow the selection probability based on ranking to increase the probability of poor food source. Well, first, we have to sort the position of nectar source through evaluate the fitness of nectar source. The position of nectar source will be a higher rank when it is a high fitness value, we can calculate the probability of nectar source by the following expression:

$$P_k = \frac{1}{SN} + a(t) \cdot \frac{SN + 1 - 2k}{SN(SN + 1)}, \, k = 1, 2, \cdots, SN$$

$$a(t) = 0.2 + \frac{3t}{4max\_cycle}, \, t = 1, 2, \cdots, Max\_cycle$$

(8)

In the formula, a(t) is adaptive parameter. In order to maintain the diversity of population so a(t) is a small value in the early evolution stage of algorithm. With the increase of iteration, the lower individual varieties and the lower population diversity makes the algorithm trapping in local optimum. To avoid this situation, a(t)'s value should be larger.

3.4   Improved ABC Algorithm Based on Reinforcement Learning

The basic idea of the algorithm is: employed bees and onlooker bees find the optimal position of nectar source through solution search equation that was selected by Q learning, this method strengthen the exploitation ability of the algorithm, next, the selection probability based on ranking is a method to maintain the diversity of population. We are going to abbreviate this improved algorithm to QABC.

The detailed steps of this improved algorithm are as follows:

1: The employed bees in original nectar source location search for new nectar source based on the solution search equation that was obtained by Q learning. And sharing the information of these nectar sources to onlooker bees by dancing. The choice of the optimal solution search equation is as follows:

   a: For each employed bee or onlooker bee, using the given n action to generate n new nectar sources and set t = 0;

   b: Do while t <= m, The employed bee or onlooker bee at each new nectar source to generate n new nectar sources based on given n action and choose a nectar sources that has optimal fitness value. Then make t = t + 1;

   c: Evaluate the value of Q for each solution search equation and employed bees or onlooker bees choose the solution search equation that corresponds to optimal value of Q to update the position of nectar source.

2: The onlooker bees receive the information of nectar source and select the nectar source according to the given probability (evaluate by Eq. (7)).

3: The same as employed bees, onlooker bees search for new nectar source based on the solution search equation that was obtained by Q learning. (according to the step a, b, c).

4: Scouts random search nectar source by Eq. (3) from beginning to the end.

5: Determine whether meet the termination conditions, if so, end of the algorithm, on the other hand, continue to implement algorithm.

4 Experiments

We choose solution search Eqs. (1), (6) and (7) as a set of actions in Q learning, and we know that the searching process of bees have great randomness, so we define the discount factor $\gamma = 0$, and m = 0, that is, we only need get the immediate rewards and regardless of return rewards.

4.1 Benchmark Functions

In this experiments, in order to test the performance of the QABC on numerical functions optimization, we considered let it compare to basic ABC through several numerical benchmark functions [21] and given in Table 1. And D: Dimension, C: Characteristic, U: Unimodal, M: Multimodal, S: Separable, N: Non-Separable.

Table 1. Benchmark functions used in experiments 1.

No	D	Range	C	Function	Formulation
F1	30	[−5.12, 5.12]	US	Sphere	$f(x) = \sum_{i=1}^{n} x_i$
F2	30	[−2.048, 2.048]	UN	Rosenbrock	$f(x) = \sum_{i=1}^{n-1} [100(x_{i+1} - x_i^2)^2 + (x_i - 1)^2]$
F3	30	[−5.12, 5.12]	MS	Rastrigin	$f(x) = \sum_{i=1}^{n} [x_i^2 - 10\cos(2\pi x_i) + 10]$

In order to make a clear comparison, in all experiments, the population, limit value and the number of cycles are equal to each other. Population size is 30 and the limit value is detained as follow [9]:

$$\text{Limit} = SN * D \tag{9}$$

where SN is the number of nectar sources. The maximum cycle number is taken as 500 for functions.

The experimental results are given in Table 2 By comparing QABC with ABC, the QABC algorithm is effective to solve functions optimization problems.

Table 2. Statistical results of 30 runs obtained by two algorithm. The mean of the results.

Function	ABC	QABC
	Mean	Mean
Sphere	6.81E−007	**6.21E−e105**
Rosenbrock	1.92E−008	**1.85E−025**
Rastrigin	5.02E−008	**0.00E+00**

4.2   Parameter Identification of Chaotic System

4.2.1   Problem Formulation

For N-dimensional chaotic system [22],

$$\dot{X} = F(x, x_0, \theta) \tag{10}$$

where $x = (x_1, x_2, \cdots, x_n)^T \in R^n$ is the N-dimensional state vector of chaotic system, and x_0 represent the initial state, $\theta = (\theta_1, \theta_2, \cdots, \theta_n)^T$ is the parameter vector of original.

Assume that the structure of the original system is known, and the estimated system can be described as follows:

$$\tilde{X} = F(\tilde{x}, x_0, \tilde{\theta}) \tag{11}$$

where $\tilde{x} = (\tilde{x}_1, \tilde{x}_2, \cdots, \tilde{x}_n)^T \in R^n$ is the state vector and $\tilde{\theta} = (\tilde{\theta}_1, \tilde{\theta}_2, \cdots, \tilde{\theta}_n)^T$ is the parameter vector of estimated system.

Suppose X_k and Y_k present the state of the original systems and estimated systems at time k, and N is a total number of points and defined as 100. Then, the problem of parameters identification for chaotic system can be changed to a multi-dimensional optimization problem, it can be formulated as follows:

$$\min J - \min(\frac{1}{N} \sum_{k=1}^{N} \|X_k - Y_k\|^2) \tag{12}$$

The principle of parameter identification for chaotic system can be showed in Fig. 1.

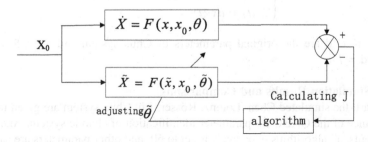

Fig. 1. The principle of parameter identification for chaotic system

4.2.2 Typical Chaotic Problem

In this part, we select five typical chaotic systems to simulation and comparisons, including Chen, Lorenz, Rösser, Lü, Chua system.

Chen system depicter as follows:

$$
\begin{cases}
\dot{x}_1(t) = a(x_2(t) - x_1(t)) \\
\dot{x}_2(t) = (c - a)x_1(t) - x_1(t)x_3(t) + cx_2(t) \\
\dot{x}_3(t) = x_1(t)x_2(t) - bx_3(t)
\end{cases}
\tag{13}
$$

where a, b, c are the original parameters of Chen system and a = 35, b = 3, c = 28.

Lorenz system depicter as follows:

$$
\begin{cases}
\dot{x}_1(t) = a(x_2(t) - x_1(t)) \\
\dot{x}_2(t) = bx_1(t) - x_1(t)x_3(t) - x_2(t) \\
\dot{x}_3(t) = x_1(t)x_2(t) - cx_3(t)
\end{cases}
\tag{14}
$$

where a, b, c are the original parameters of Lorenz system and a = 10, b = 28, c = 8/3.

Rösser system depicter as follows:

$$
\begin{cases}
\dot{x}_1(t) = -x_2(t) - x_3(t) \\
\dot{x}_2(t) = x_1(t) + ax_2(t) \\
\dot{x}_3(t) = b + x_1(t)x_3(t) - cx_3(t)
\end{cases}
\tag{15}
$$

where a, b, c are the original parameters of Rösser system and a = 0.2, b = 0.2, c = 5.7.

Lü system depicter as follows:

$$
\begin{cases}
\dot{x}_1(t) = a(x_2(t) - x_1(t)) \\
\dot{x}_2(t) = -x_1(t)x_3(t) + cx_2(t) \\
\dot{x}_3(t) = x_1(t)x_2(t) - bx_3(t)
\end{cases}
\tag{16}
$$

where a, b, c are the original parameters of Lü system and a = 30, b = 22.2, c = 8.8/3.

Chua system depicter as follows:

$$
\begin{cases}
\dot{x}_1(t) = ax_2(t) - bx_1(t) - cx_1(t)^3 \\
\dot{x}_2(t) = x_1(t) - x_2(t) - x_1(t)x_3(t) \\
\dot{x}_3(t) = dx_2(t)
\end{cases}
\tag{17}
$$

where a, b, c, d are the original parameters of Chua system and a = 5.5, b = 0.25, c = 0.1, d = 7.4.

4.2.3 Simulation Results and Comparisons

In this part, the simulated Chen, Lorenz, Rösser, Lü, Chua system are given to test the performance of the QABC for parameter identification of chaotic system. And at those experiments, all algorithms max_cycle is set to 60, and other parameters are as same as part 4.1. The searching ranges of the parameters are defined as follows: Chen system: $34 \leq a \leq 36, 2 \leq b \leq 4, 27 \leq c \leq 29$, Lorenz system: $9 \leq a \leq 10, 27 \leq b \leq 29, 2 \leq c \leq 4$, Rösser system: $0.01 \leq a \leq 0.5, 0.01 \leq b \leq 0.5, 4.5 \leq c \leq 6.5$, Lü system: $29 \leq a \leq 31$, $21 \leq b \leq 23, 2 \leq c \leq 4$, Chua system: $4.5 \leq a \leq 6.5, 0.01 \leq b \leq 0.5, 0.01 \leq c \leq 0.3$, $6.4 \leq d \leq 8.4$. In this experiment, all parameters of those five chaotic systems are unknown and we need to identified it by several typical algorithm, that is, the performance of QABC will be compared with ABC, GABC [10], MABC [16] The statistical results of 20 runs are given in Table 3 and including Jmean, Std and average estimation values.

Table 3. The statistical results of five typical chaotic systems obtained by four algorithms.

Chaotic system			Result	QABC	ABC	GABC	MABC
Chen system			Jmean	**0.0004**	0.0581	1.67E+003	2.97E+003
			Std	**0.0020**	0.0903	1.86E+003	5.87E+003
	a	35	$\hat{a}$	**35.0001**	35.0024	34.8640	35.0534
	b	3	$\hat{b}$	**3.0001**	2.9977	3.1903	2.8562
	c	28	$\hat{c}$	**28.0001**	27.9989	28.0678	28.1141
Lorenz system			Jmean	**1.29E−016**	0.0015	6.08E−012	3.15E−021
			Std	**1.89E−015**	0.0027	1.36E−011	5.34E−020
	a	10	$\hat{a}$	**10.0000**	10.0011	10.0000	10.0000
	b	28	$\hat{b}$	**28.0000**	27.9981	28.0000	28.0000
	c	8/3	$\hat{c}$	**2.6667**	2.6664	2.6667	2.6667
Rösser system			Jmean	**3.40E−015**	4.81E−008	1.80E−009	1.12E−010
			Std	**5.42E−015**	9.84E−008	4.02E−009	2.50E−010
	a	0.2	$\hat{a}$	**0.2000**	0.2000	0.2003	0.2000
	b	0.2	$\hat{b}$	**0.2000**	0.1997	0.1999	0.2000
	c	5.7	$\hat{c}$	**5.7000**	5.6913	5.6960	5.7003
Lü system			Jmean	**1.59E−013**	6.93E−007	8.43E−009	1.81E−005
			Std	**2.21E−013**	1.52E−006	1.65E−008	3.67E−005
	a	30	$\hat{a}$	**30.0000**	29.9920	30.0006	29.9085
	b	22.2	$\hat{b}$	**22.2000**	22.2035	22.1998	22.0592
	c	8.8/3	$\hat{c}$	**2.9333**	2.9334	2.9333	2.9336
Chua system			Jmean	**5.99E−010**	1.24E−006	1.13E−007	1.70E−006
			Std	**8.37E−010**	1.40E−006	2.21E−007	2.17E−006
	a	5.5	$\hat{a}$	**5.5001**	5.4825	5.4977	5.4782
	b	0.25	$\hat{b}$	**0.2500**	0.2468	0.2497	0.2460
	c	0.1	$\hat{c}$	**0.0998**	0.1296	0.1030	0.1542
	d	7.4	$\hat{d}$	**7.4000**	7.4004	7.4002	7.4013

We can clear know from Table 3 that the convergence precision of QABC is more better than other intelligent optimization algorithms in identifying parameters of five typical chaotic systems. For Chen system and Chua system, the ABC, GABC, QABC algorithms traps into local optima easily and has a low convergence accuracy. The results show that QABC are very close to the true values for compared with a big deviation of values from other algorithm. The estimated parameters of Chua system are shown in Fig. 2, from Fig. 2, we can see that only the estimated parameters by QABC can converge to the matching true values, other algorithms are not converges to a given value within a given cycle. It shows that QABC has great advantages to solve the problem of the parameter identification of chaotic system.

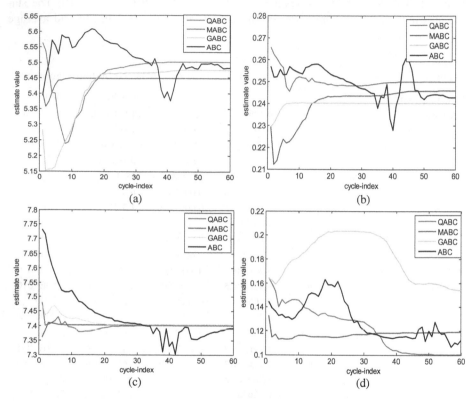

Fig. 2. Evolution of the estimated parameters values derived from QABC, MABC, GABC, ABC on the Chua system (a) The estimated parameter $\hat{a}$ of Chua system. (b) The estimated parameter $\hat{b}$ of Chua system. (c) The estimated parameter $\hat{c}$ of Chua system. (d) The estimated parameter $\hat{d}$ of Chua system. (Color figure online)

5 Conclusion

In this paper, a newly optimization algorithm-QABC is developed, through introducing the reinforcement learning into the solution search equation of ABC and proposing the selection probability based on ranking to onlooker bees phase. And by through the simulation tests of benchmark functions we know that this algorithm has higher accuracy and faster convergence rate. Then, QABC is applied to the parameters identification of five typical chaotic systems by translated them into multi-dimensional continuous optimization. Numerical simulation and comparison results show that the performance of QABC is more robust than other three intelligent optimization algorithm and it may be a viable way to deal with complex numerical optimization problem, such as, data mining, optimization of communication network and traveling salesman problem (TSP).

References

1. Karaboga, D.: An idea based on honey bee swarm for numerical optimization. Technical report-TR06, Engineering Faculty, Computer Engineering Department, Erciyes University (2005)
2. Karaboga, D., Basturk, B.: A powerful and efficient algorithm for numerical function optimization: artificial bee colony (ABC) algorithm. J. Global Optim. **39**(3), 459–471 (2007)
3. Liu, Y., Ma, L.: Bees algorithm for function optimization. J. Control Decis. **27**(6), 886–890 (2012)
4. Öztürk, C., Karaboğa, D., Görkemli, B.: Artificial bee colony algorithm for dynamic deployment of wireless sensor networks. Turk. J. Electr. Eng. Comput. Sci. **20**(2), 255–262 (2012)
5. Horng, M.H.: Multilevel thresholding selection based on the artificial bee colony algorithm for image segmentation. J. Expert Syst. Appl. **11**, 13785–13791 (2011)
6. Banharnsakun, A., Achalakul, T., Sirinaovakul, B.: The best-so-far selection in artificial bee colony algorithm. J. Appl. Soft Comput. **11**(2), 2888–2901 (2011)
7. Yeh, W.C., Hsieh, T.J.: Artificial bee colony algorithm-neural networks for S-system models of biochemical networks approximation. J. Neural Comput. Appl. **21**(2), 365–375 (2012)
8. Karaboga, D., Basturk, B.: On the performance of artificial bee colony (ABC) algorithm. J. Appl. Soft Comput. **8**(1), 687–697 (2008)
9. Das, S., Biswas, S., Kundu, S.: Synergizing fitness learning with proximity-based food source selection in artificial bee colony algorithm for numerical optimization. J. Appl. Soft Comput. **13**(12), 4676–4694 (2013)
10. Liu, S.Y., Zhang, P., Zhu, M.M.: Artificial bee colony algorithm based on local search. J. Control Decis. **29**(1), 123–128 (2014)
11. Zhu, G., Kwong, S.: Gbest-guided artificial bee colony algorithm for numerical function optimization. J. Appl. Math. Comput. **217**(7), 3166–3173 (2010)
12. Li, Y.C., Peng, Y.: Improved artificial bee colony algorithm based on information entropy. J. Control Decis. **30**(6), 1121–1125 (2015)
13. Bi, X.J., Wang, Y.J.: Artificial bee colony algorithm with fast convergence. J. Syst. Eng. Electr. **33**(12), 2755–2761 (2011)
14. Gao, W., Liu, S.: Improved artificial bee colony algorithm for global optimization. J. Inf. Process. Lett. **111**(17), 871–882 (2011)
15. Gao, W., Liu, S., Huang, L.: A global best artificial bee colony algorithm for global optimization. J. Comput. Appl. Math. **236**(11), 2741–2753 (2012)
16. Gao, W., Liu, S.: A modified artificial bee colony algorithm. J. Comput. Oper. Res. **39**(3), 687–697 (2012)
17. Zhou, X.Y., Wu, Z.J., Wang, M.W.: Artificial bee colony algorithm based on orthogonal experimental design. J. Softw. **26**(9), 2167–2190 (2015)
18. Kıran, M.S., Fındık, O.: A directed artificial bee colony algorithm. J. Appl. Soft Comput. **26**, 454–462 (2015)

19. Zhou, X.Y., Wu, Z.J., Wang, H., et al.: Elite opposition-based particle swarm optimization. J. Acta Electronica Sinica. **41**(8), 1647–1652 (2013)
20. Bao, L., Zeng, J.: Comparison and analysis of the selection mechanism in the artificial bee colony algorithm. In: Ninth International Conference on Hybrid Intelligent Systems, HIS 2009, vol. 1, pp. 411–416. IEEE (2009)
21. Karaboga, D., Akay, B.: A comparative study of artificial bee colony algorithm. J. Appl. Math. Comput. **214**(1), 108–132 (2009)
22. Jiang, Q., Wang, L., Hei, X.: Parameter identification of chaotic systems using artificial raindrop algorithm. J. Comput. Sci. **8**, 20–31 (2015)

Differential Evolution

Detect Method of Time Series' Abnormal
Value for Predictive Model

Yang Feng[✉]

School of Physics and Mechanical and Electrical Engineering, Hechi University,
Yizhou, Guangxi, China
oyfo@163.com

Abstract. Abnormal value that in the predictive model of Ad Hoc networks may affecting the whole system's working efficiency. We proposed a new detect method to dealing with this problem, constructed a forwarding model firstly, and then constructed a suitable model function through smoothing and modeling the time series. By using the mean shift model, we calculated the time series posterior probabilities and abnormal perturbation values, and then adjusted them, so as to weakening the influence of time series abnormal value. To verifying the efficiency of this forwarding method, we selected a time series with 300 observation points as the numerical example, statistics and analysis results indicate that it will be helpful to improve the efficiency of prediction models if we using this method.

Keywords: Posterior probabilities · Time series · Abnormal perturbation · Forwarding method · Ad Hoc networks

1 Introduction

It has significant research value to enhance the node's forwarding efficiency in Ad Hoc network, which is helpful to improve the QoS metrics and, by reducing unnecessary power consumption, to prolong the survival time of the network. Traditional mature forwarding models and methods, however, could not be directly applied to Ad Hoc Networks; because its forwarding table needs to be rebuilt frequently due to these innate characteristics as node failures, node mobility et al. In Ad Hoc network, therefore, the researches about construction method of predictive forwarding models have been a hot topic [1].

The present research focuses of construction of forwarding model are the following two: one is using social network knowledge, the other the contact time series prediction method [2–5]. With mature theoretical support, the method based on time series prediction has broader research basis. Its basic definition is as follows: contact time is the time from two nodes starting connecting to leaving the communication range of each other, and it has a characteristic of exponential distribution; the node's contact interval is the interval between two consecutive contacts, and it has power-law distribution. With contact time and contact interval, it is able to speculate the possibilities of any two nodes' encounter within a given time and the switching packet time, namely, the contact time

© Springer International Publishing Switzerland 2016
D.-S. Huang and K.-H. Jo (Eds.): ICIC 2016, Part II, LNCS 9772, pp. 735–742, 2016.
DOI: 10.1007/978-3-319-42294-7_65

after encounter, which can provide support for the construction of more efficient forwarding table [6].

But in actual operation, the time series observed may be abnormality on account of abnormal perturbations; hence the accuracy of the predicted series may be affected and even lead to erroneous predictions if a forwarding model is constructed without detecting and dealing with these abnormal values. Therefore, it is very important to formulate effective strategies to detecting the abnormal values of time series, which currently is a hot issue in academic research [7].

2   Related Works

Currently, there are three main methods to detect the time series' abnormal values; respectively they are recognition method based on deleting data points, test method on the generalized likelihood ratio (Score) [8, 9] and detect method on Bayesian statistical theory [10, 11]. The basic idea of first method is to treat the abnormal value as a missing one, and then replace it with an estimate value in the course of the actual prediction. The Score algorithm proposed by Fox is adding a simulation variable to each observed value, and then maximizing its likelihood function, and finally determining a measure basis by the significance of the statistics values. Subsequently, there are several scholars studying the improved method of the test statistic. The accuracy of this kind depends on the total amount of detection values; therefore, the scope is somewhat limited.

For the limitations that the above two methods are not treat priori information as appraised value, based on Bayesian statistical theory, some scholars have deduced the posterior distribution of the relevant parameters by using the abnormal values' priori probability that in the time series as a basis. Schervish and Tsay have studied the problem that abnormal values and models changed simultaneously, on the basis of solving the detection problem of AO and IO. However, these intuitive methods are difficult to achieve because of the large amount of calculations while calculating the posterior probability, so they are subject to certain limitations. For this problem, Justel introduced Gibbs sampling algorithm into the AR model, and achieved the detection of clustered abnormal points. However, if the priori distribution is not only random but also unstable, then the calculation precision would be decayed rapidly. Therefore, a smooth processing to original series is also the precursor step of constructing predictive forwarding model.

3   Predictive Forwarding Model Building

In order to provide the verification platform for the detection method of time series' abnormal values, we firstly using map to formal describing the topology of Ad Hoc network, then simplified the original fully connected map to a new one which can maintain connectivity subgraph, consider the optimization of overall degree of disturbance as principle. On the basis of analyze information of contact situation and move rule that between the nodes, we classified the nodes with similar contact probabilities in the whole topology diagram. Then, we build an effective forwarding model based on contact spacetime by combined the distribution of nodes' contact time, contact time interval and the variation of the node's spatial position. The specific process of model building is shown in Fig. 1.

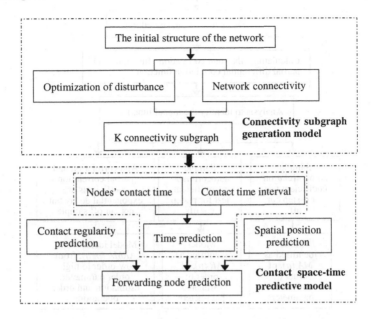

Fig. 1. The predictive forwarding model building schematic

4   Detection Method of Time Series

4.1   Time Series Modeling

Modeling of time series is the priority needs of implementing the following operations such as detection, statistical analysis and optimization for the prediction abnormal values of time series, in the predictive forwarding model. Its specific steps including identify the model' class, determine the model' loop order and estimate the model' parameters et al. In this paper, we use BJ (Box-Jenkins) method to modeling [12], the detailed modeling process is shown in Fig. 2.

Stability of the detected time series is an important prerequisite for the subsequent modeling. For a measured process, it can be regarded as a smooth series if its system parameters and surrounding runtime conditions does not change. Therefore, we firstly implement the smooth processing to time series by using the parametric test method [13] after inputting the prediction data, then carry on model identification and model order operations, the last step is model identification process, in this step, we create a mathematical model for the series according to its priori information and data overview. In this paper, the theoretical basis of model identification are based on the cut off property of $AR(p)$ series' partial correlation function $MA(q)$ series' autocorrelation function and the trailing of $ARMA(p, q)$ series' partial autocorrelation function.

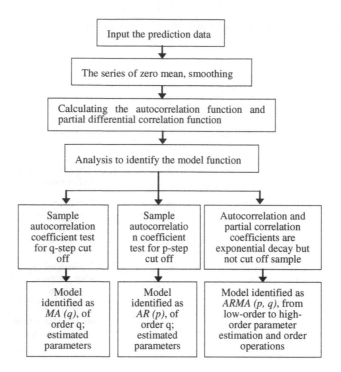

Fig. 2. Modeling method schematic

4.2   Time Series Abnormal Values Detection

After the Modeling of time series, we can detect its abnormal values. Without loss of generality, we can suppose $\{Z_i\}$ is a $AR(p)$ series, then the it meets

$$Z_i = \Phi_1 Z_{i-1} + \Phi_2 Z_{i-2} + \cdots + \Phi_p Z_{i-p} + \delta_i \tag{1}$$

Further describe it to matrix form $Y = X\Phi + \delta$, in which

$$Y = (Z_{p+1}, \cdots, Z_n)^z, \Phi = (\Phi_1, \cdots, \Phi_p)^z,$$
$$\delta = (\delta_{p+1}, \cdots, \delta_n)^z$$

$$\underset{(N-p)*p}{X} = \begin{bmatrix} Z_p & Z_{p-1} & \cdots & Z_1 \\ Z_{p+1} & Z_p & \cdots & Z_2 \\ \cdots & \cdots & \cdots & \cdots \\ Z_{N-1} & Z_{N-2} & \cdots & Z_{N-p} \end{bmatrix} \tag{2}$$

Suppose the first p detected values of this series are all normal values, that is $Z_1, Z_2, \cdots, Z_p$ are not abnormal values. In this condition, the abnormal value detection problem of the time series can be converted into abnormal value detection problem of linear regression model.

Suppose there is a dataset which meets linear regression models $(Y_i, X_i)^T$, $i = 1, \cdots,$ $N - p$, δ_i is the random error term of each data point and its standard distribution is $f(\delta_i)$, its alternate distribution is $g(\delta_i)$. Define $B_J = \{$subscript belong to J of K observations$\}$, event B_j happened equivalent to subscript belong to J of K observations δ_i obey alternative distribution $g(\delta_i)$, and the rest $N - p - k$ observation error Δ_i obey the standard distribution. Therefore, on the basis of relevant priori information, the key of diagnose abnormal points is calculate the posterior probability $p(B_j|Y)$ of event B_j. Using mean shift model [14, 15] to calculate the posterior probability, abnormal value discriminant standard is $p(B_j|Y)$ has a big value (dozens of times bigger than other values). To calculate the posterior probability, the following tags need to define: $g(\delta_i)$ and $f(\delta(J))$ respectively express the density function of δ_j and $\delta(J)$ when event B_J happening, $p(Y|B_j) = h(Y_{j\sim g}, Y(J) \sim f)$ express the density function of Y, $h(Y_{j\sim g}, Y(J) \sim f)$ express the density function of Y_j under the conditions that each component obey the standard distribution.

Considering the operability, we suppose the standard distribution and alternative distribution of δ_j are $N(0, \sigma^2)$ and $N(r, \sigma^2)(r \neq 0)$, there are known Abnormal value of k in the distribution, so, the priori probabilities are equal, that they are all abnormal values, and $N(0, \sigma^2)$ has normal prior information $\tau = \sigma^{-2}$ that is $(\phi, \tau) \sim NG(m, \sum^{-1}, a_0, a_1)$, Σ is $t \times t$ rank positive definite matrix, m is t-dimensional vector, a_0, a_1 are constants. If $X(J)$ is full rank matrix, that is, the posterior probability of event B_J can expressed as

$$p(B_J|Y) = \frac{|\Sigma + X(J)^T X(J)|^{-1/2}[a_1 + \frac{1}{2}RSS(J) + \frac{d}{2}]^{-(N-p-k+2a_0)/2}}{\sum_J \{|\Sigma + X(J)^T X(J)|^{-1/2}[a_1 + \frac{1}{2}RSS(J) + \frac{d}{2}]^{-(N-p-k+2a_0)/2}\}} \tag{3}$$

In   which $RSS(J) = [Y(J) - X(J)\hat{\phi}(J)]^T[Y(J) - X(J)\hat{\phi}(J)]$; $\hat{\phi}(J) =$ $(X(J)^T X(J))^{-1} X(J)^T Y(J)$, substitute it into formula (3), the corresponding d, u value can be obtained.

5   Numerical Example and Analysis

On the basis of the model described in Fig. 1, we first constructed the forwarding model, then modeling the detected time series by according to the steps that shown in Fig. 2. We selected a time series with 300 observation points as the numerical example, detecting the abnormal values of top 290 nodes by using the aforementioned algorithm, predictive and analysis the remaining 10 data.

Firstly, smooth processing was taken to the detected time series by using the difference method, the original time series and smoothed time series that in this numerical example are shown in Figs. 3 and 4.

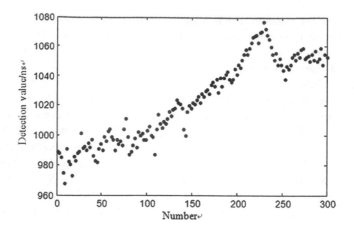

Fig. 3. Original time series

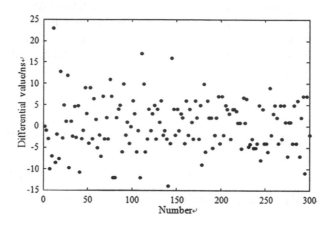

Fig. 4. Differential time series

Calculating the autocorrelation function and partial differential correlation function of the smooth handled differential time series, the ARMA model was approximatively Approached and the loop order is 2, estimated the posterior probability of event B_J by using the method described in Sect. 3. The results are shown in Table 1.

Table 1. B_J-s posterior probability table

Time series number	28	147	158	277	88	103	Other
$p(B_J|Y)$	0.8798	0.0135	0.0119	0.0077	0.0068	0.0052	≤0.0050

Table 1 listing six maximum when the K value is equal to 1, from the statistical results can be seen, the posterior probability of the twenty-eighth value as the abnormal value is much larger than the rest of BJ's posterior probability of the detection points as the abnormal value, and the maximum value is higher than the other for at least 65 times.

Therefore, we can determine that the value is abnormal value, calculated according to Eq. 3, can have abnormal disturbance value. Predicting time series was corrected after the introduction of abnormal disturbance value, the corrections are shown in Table 2.

Table 2. Comparison table of corrected prediction results

Time series number	291	292	293	294	295	296	297	298	299	300
Before correction	0.1066	0.0873	0.0890	0.0688	0.0651	0.1321	0.0843	0.0870	0.4164	0.0613
After correction	0.0869	0.0838	0.0847	0.0812	0.0786	0.0900	0.0848	0.0850	0.2279	0.0698

Through the data in Table 2 were analyzed, it can be seen, after the disturbance correction, 10 predictive points in nine are relatively stable distributed around reasonable value, only the 299th predictive value was the failure to get a better fix because large deviations, verified the feasibility of the method.

6   Summary and Outlook

In order to support the time prediction based prediction forwarding model for Ad Hoc network, we proposed a method to detecting and processing its time series abnormal values. According to the time series modeling, we firstly determined the specific applicable model function, then estimates the posterior probability based on the mean shift model. Through the above measures, we identified the series abnormal values and made the time series becoming smoothing. The prediction accuracy is also being enhanced by the correction operation for the predictive time series that based on the introduction of abnormal perturbations values. The effectiveness of this method has been proved by the count case analysis results. In the future work, we need to further improvement and optimization this method, especially need to improve the processing method for the predictive series abnormal values.

Acknowledgment. This work was supported by the University research projects of the education department of Guangxi, China (YB2014330).

References

1. Wang, B., Huang, C., Yang, W.: Adaptive opportunistic routing protocol based on forwarding-utility for delay tolerant networks. J. Commun. **31**(10), 36–47 (2010)
2. Cerf, V., Burleigh, S., Hooke, A., et al.: Delay-tolerant networking architecture. IETF RFC RFC4838 (2008)
3. Fan, X., Shan, Z., Zhang, B., et al.: State-of -the-Art of the architecture and techniques for delay-tolerant networks. Acta Electronica Sinica **36**(1), 161–170 (2008)
4. Xiong, Y.-P., Sun, L.-M., Niu, J.-W., et al.: Opportunistic networks. J. Softw. **20**(1), 124–137 (2009)
5. Zhang, L., Zhou, X.-W., Wang, J.-P., et al.: Routing protocols for delay and disruption tolerant networks. J. Softw. **21**(10), 2554–2572 (2010)

6. Conan, V., Leguay, J., Friedman, T., et al.: Fixed point opportunistic routing in delay tolerant networks. IEEE J. Sel. Areas Commun. **26**(5), 773–782 (2008)
7. Yong, L., Yurong, J., Depeng, J., et al.: Energy-efficient optimal opportunistic forwarding for delay-tolerant networks. IEEE Trans. Veh. Technol. **59**(9), 4500–4512 (2010)
8. Chen, C., Liu, L.M.: Joint estimation of model parameters and outlier effects in time series. J. Am. Stat. Assoc. **88**, 284–297 (1993)
9. Wu, L.S.Y., Hosking, J.R.M., Ravishanker, N.: Reallocation outliers in time series. J. Royal Stat. Soc. Ser. C **42**, 301–313 (1993)
10. McCulloch, R.E., Tsay, R.S.: Bayesian analysis of autoregressive time series via the Gibbs sampler. J. Time Ser. Anal. **15**, 235–250 (1994)
11. Justel, A., Pena, D., Tsay, R.S.: Detection of outlier patches in autoregressive time series. Working Paper, University of Chicago, Graduate School of Business (1998)
12. Yang, H., Zang, Y.: Comparisons of bias compensation methods and other identification approaches for Box-Jenkins models. Control Theor. Appl. **24**(02), 215–222 (2007)
13. Xiong, J., Li, Y., Yang, Z.: A new parametric test for time series. J. S. Chin. Normal Univ. (Natural Science Edition) **03**, 5–8 (2008)
14. Koch, K.R.: Einfuhrung in Die Bayes-Statistic. Springer, Berlin (2000)
15. Koch, K.R.: Parameter Estimation and Hypothesis Testing in Linear Models, 2nd edn. Springer, Berlin (1999)

Memetic Algorithms

Solving Bi-objective Unconstrained Binary Quadratic Programming Problem with Multi-objective Backbone Guided Search Algorithm

Li-Yuan Xue[1], Rong-Qiang Zeng[2,3(✉)], Yang Wang[4], and Ming-Sheng Shang[5]

[1] School of Electronic Engineering, University of Electronic Science and Technology of China, Chengdu 610054, Sichuan, People's Republic of China
xuely2013@gmail.com

[2] School of Mathematics, Southwest Jiaotong University, Chengdu 610031, Sichuan, People's Republic of China

[3] School of Computer Science and Engineering, University of Electronic Science and Technology of China, Chengdu 610054, Sichuan, People's Republic of China
zrq@home.swjtu.edu.cn

[4] School of Management, Northwestern Polytechnical University, Xi'an 710072, Shanxi, People's Republic of China
sparkle.wy@gmail.com

[5] Chongqing Institute of Green and Intelligent Technology, Chinese Academy of Sciences, Chongqing 400714, People's Republic of China
msshang@cigit.ac.cn

Abstract. This paper presents a multi-objective backbone guided search algorithm in order to optimize a bi-objective unconstrained binary quadratic programming problem. Our proposed algorithm consists of two main procedures which are hypervolume-based local search and backbone guided search. When the hypervolume-based local search procedure can not improve the Pareto approximation set any more, the backbone guided search procedure is applied for further improvements. Experimental results show that the proposed algorithm is very effective compared with the original multi-objective optimization algorithms.

Keywords: Multi-objective optimization · Backbone guided search · Local search · Hypervolume contribution · Unconstrained binary quadratic programming

1 Introduction

As a canonical NP-hard problem, the Unconstrained Binary Quadratic Programming (UBQP) problem is known for its ability to formulate a wide range of important problems [8, 14], which includes computer aided design [11], traffic management [7], machine scheduling [1], and so on. In [12], Liefooghe et al. have extended the conventional single-objective UBQP problem to the multi-objective case, where the multiple objectives are to be maximized simultaneously.

© Springer International Publishing Switzerland 2016
D.-S. Huang and K.-H. Jo (Eds.): ICIC 2016, Part II, LNCS 9772, pp. 745–753, 2016.
DOI: 10.1007/978-3-319-42294-7_66

In this work, we focus on solving bi-objective UBQP problem, which is formalized as follows [13]:

$$f_k = x'Q^k x = \sum_{i=1}^{n} \sum_{j=1}^{n} q_{ij}^k x_i x_j \tag{1}$$

where $Q^k = \left(q_{ij}^k \right)$ is an $n \times n$ matrix of constants and x is an n-vector of binary (zero-one) variables, i.e., $x_i \in \{0, 1\}$ $(i = 1, \cdots, n)$, $k = 1, 2$.

Motivated by the extensive range of applications of UBQP, a large number of heuristic and metaheuristic algorithms have been devised for tackling this problem in the literature [10]. The trajectory methods are very popular approaches, such as simulated annealing [2], tabu search [9], scatter search [3]. The population-based methods constitute another class of popular tools of finding high quality solutions for UBQP, including genetic algorithms [15], memetic algorithms [16], path relinking [17].

In this paper, we present a multi-objective backbone guided search algorithm to optimize the bi-objective UBQP problem. The proposed algorithm is an hybrid metaheuristic algorithm, which integrates the backbone guided search techniques for further improvements. Furthermore, this algorithm follows a general framework composed of two main procedures: hypervolume-based local search and backbone guided search. The local search procedure iteratively improves Pareto approximation set until it can not be improved any more. Then, the backbone guided search procedure is employed to further improve the quality of the Pareto approximation set. Experimental results show that the proposed algorithm is very competitive in comparison with the original multi-objective optimization algorithms.

The rest of this paper is organized as follows. In Sect. 2, we introduce some basic notations and definitions related to multi-objective optimization. Then, we briefly review the literature related to backbone guided search in Sect. 3. Afterwards, we present the ingredients of our Multi-Objective Backbone Guided Search Algorithm (MOBGSA), which includes a hypervolume-based local search procedure and backbone guided search procedure in Sect. 4. In Sect. 5, our experimental results are reported. Finally, the concluding remarks are given in Sect. 6.

2   Multi-objective Optimization

In this section, we introduce some useful notations and definitions of multi-objective optimization problems, which are taken from [5, 21]. Let X denote the search space of the optimization problem under consideration and Z the corresponding objective space. Without loss of generality, we assume that $Z = \Re^n$ and that all n objectives are to be maximized. Each $x \in X$ is assigned exactly one objective vector $z \in Z$ on the basis of a vector function $f : X \to Z$ with $z = f(x)$. The mapping f defines the evaluation of a solution $x \in X$, and often one is interested in those solutions that are Pareto optimal with respect to f. The relation $x_1 \succ x_2$ means that the solution x_1 is *preferable* to x_2. The dominance relation between two solutions x_1 and x_2 is usually defined as follows [5, 21]:

Definition 1 *(Domination). A decision vector x_1 is said to dominate another decision vector x_2 (written as $x_1 \succ x_2$), if $f_i(x_1) \geq f_i(x_2)$ for all $i \in \{1, \cdots, n\}$ and $f_j(x_1) > f_j(x_2)$ for at least one $j \in \{1, \cdots, n\}$.*

Definition 2 *(Non-Dominated Solution). $x \in S$ $(S \subset X)$ is said to be non-dominated if and only if there does not exist another solution $x' \in S$ such that x' dominates x.*

Definition 3 *(Pareto Optimal Solution). $x \in X$ is said to be Pareto optimal if and only if there does not exist another solution $x' \in X$ such that x' dominates x.*

Definition 4 *(Non-Dominated Set). S is said to be a non-dominated set if and only if any two solutions $x \in S$ and $y \in S$ which do not dominate each other.*

Definition 5 *(Pareto Optimal Set). S is said to be a Pareto optimal set if and only if S is composed of all the Pareto optimal solutions.*

In multi-objective optimization, there usually does not exist one optimal but a set of Pareto optimal solutions. The aim is to generate the Pareto optimal set, which keeps the best compromise among all the objectives. Nevertheless, in most cases, it is impossible to obtain the Pareto optimal set in a reasonable time. Thus, we are interested in finding a non-dominated set which is as close to the Pareto optimal set as possible, and the whole goal is often to identify a good Pareto approximation set.

3 Related Works

The backbone strategy originates from solving the well-known satisfiability problem (SAT) [20], which is also applied to other problems such as multi-dimensional knapsack problem [19] and single-objective UBQP problem [18]. In the following paragraphs, we present the studies related to the backbone-based heuristic algorithm and the resolution of multi-objective UBQP problem.

Wang et al. [18] propose a backbone guided tabu search algorithm for the UBQP problem, which is composed of two phases: a tabu search procedure to maximize the objective function as far as possible and a backbone guided search strategy to further improve the objective function. Experimental results show that the proposed algorithm is capable of finding the best known solutions on some large random instances.

Liefooghe et al. [12] first extend the single-objective UBQP to the multi-objective case, and propose a hybrid metaheuristic algorithm which combines an elitist evolutionary multi-objective optimization algorithm with an effective single-objective tabu search procedure based on the scalarizing function. The experimental analysis validates the effectiveness of the proposed algorithm on large instances with two and three objectives.

In order to solve the bi-objective UBQP problem more efficiently, Liefooghe et al. [13] design and analyze several local search algorithms, including two scalarizing approaches, a Pareto-based approach and a hybrid approach combining these two complementary search strategies. The computational results indicate the algorithms substantially improved the obtained Pareto approximation set on the large instances.

4   Multi-objective Backbone Guided Search Algorithm

The Multi-Objective Backbone Guided Search Algorithm (MOBGSA) is designed to deal with the bi-objective UBQP problem, which is composed of two main components: hypervolume-based local search and backbone guided search. The general scheme of our proposed algorithm is described in Algorithm 1, in which the main components are detailed in the following subsections.

Algorithm 1 Multi-Objective Backbone Guided Search Algorithm

Input: N (Population size)
Output: A: (Pareto approximation set)
Step 1 - Initialization: $P \leftarrow N$ randomly generated solutions
Step 2: $A \leftarrow \Phi$
Step 3 - Fitness Assignment: Assign a fitness value for each individual $x \in P$
Step 4:
while Running time is not reached do
  repeat
    Hypervolume-Based Local Search: $x \in P$
    1) $x^* \leftarrow$ one randomly chosen unexplored neighbors of x
    2) $P \leftarrow P \bigcup x^*$
    3) compute x^* fitness: $HC(x^*, P)$
    4) update all $z \in P$ fitness values
    5) $\omega \leftarrow$ worst individual in P
    6) $P \leftarrow P \backslash \{w\}$
    7) update all $z \in P$ fitness values
    8) if $w \neq x^*$, Progress $\leftarrow$ True
  until all neighbors are explored or Progress = True
    9) $A \leftarrow$ Non-dominated solutions of $A \bigcup P$
  Backbone Guided Search: $y \in A$
    for $i \leftarrow 1, \ldots, m$ $(m \leq n)$ do
      10) randomly choose the k_i^{th} variable of each individual $y \in A$
      11) $S_0 = \{y | y_{k_i} = 0\}$, $S_1 = \{y | y_{k_i} = 1\}$
      12) $S_0' = \{y' | y_{k_i} \rightarrow 1\}$, $S_1' = \{y' | y_{k_i} \rightarrow 0\}$
      13) compute a fitness value for each individual $y' \in S_0' \bigcup S_1'$
      14) $\nu \leftarrow$ best individual in $S_0' \bigcup S_1'$
      15) $A \leftarrow \nu$
    end for
end while
Step 5: Return A

In MOBGSA, all the individuals in an initial population are randomly generated, i.e., each variable of an individual is randomly assigned a value 0 and 1 (Step 1). Then, each individual in the population is optimized by the hypervolume-based local search procedure. Afterwards, we employ a backbone guided search procedure to further improve the whole population, which generates a high-quality Pareto approximation set at last.

4.1   Fitness Assignment

In single-objective optimization, a total order relation can be easily used to rank the solutions. However, such a natural total order relation does not exist in multi-objective optimization. Thus, we take the Hypervolume Contribution (HC) indicator defined in [4] to assign a fitness value to each solution in the population.

Based on the dominance relation, the HC indicator computes the contribution of each solution in the objective space. More precisely, the whole population is divided into two sets: dominated set and non-dominated set. The solution belonging to the dominated set is assigned a negative value, which is dominated by at least one another solution in the population. While the solution belonging to the non-dominated set is assigned a positive value, which dominates at least one another solution in the population. According to the fitness values, an order induced by the HC indicator can be used to rank the solutions in the selection process of multi-objective optimization.

4.2 Hypervolume-Based Local Search

After the fitness assignment for each solution, we apply the hypervolume-based local search procedure proposed in [4] to the initial population. In this procedure, a solution x^*, which is one of the unexplored neighbors of x in the population P, is assigned to a fitness value by the HC indicator. If x^* is dominated, the fitness values of all the solutions in P remain unchanged. If x^* is non-dominated, we need to update the fitness values of non-dominated neighbors of x^* in the objective space.

Then, the solution w with the worst fitness value is deleted from the population P. If w is dominated, the fitness values of the other solutions do not need to be updated. If w is non-dominated, the fitness values of the non-dominated neighbors of w need to be updated. The whole procedure will repeat until the termination criterion is satisfied.

4.3 Backbone Guided Search

The backbone strategy has been proved to be an effective approach by solving single-objective UBQP problem. As suggested in [18], we suppose some variables to be strongly determined if changing its assigned value in a high quality solution will cause the quality of that solution to deteriorate significantly. Essentially, we just borrow the terminology "backbone" from the SAT literature as a vehicle for naming our procedure.

In order to further improve the quality of the obtained Pareto approximation set by the hypervolume-based local search procedure, we use the backbone guided search procedure to achieve this goal. In this procedure, we first randomly choose the k_i^{th} variable of each solution y in the Pareto approximation set A. According to the value (0 or 1) of the chosen k_i^{th} variable of each solution $y \in A$, we divide all the solutions in A into two sets: $S_0 = \{y \mid y_{k_i} = 0\}$, $S_1 = \{y \mid y_{k_i} = 1\}$.

Then, we flip the value 0 of the k_i^{th} variable of each solution $y \in S_0$ to 1 and the value 1 of the k_i^{th} variable of each solution $y \in S_1$ to 0. With such move of flipping the variable, we obtain two new sets: $S_0' = \{y' \mid y_{k_i} \leftarrow 1\}$ and $S_1' = \{y' \mid y_{k_i} \leftarrow 0\}$. Actually, the move value Δ_i can be computed in linear time using the formula [14]:

$$\Delta_i = (1 - 2x_i)(q_{ii} + \sum_{j \in N, j \neq i, x_j = 1} q(i,j)) \tag{2}$$

Afterwards, we assign a fitness value to each solution y' both in S_0' and S_1' with the HC indicator. Moreover, we only select one solution ν with the best fitness value from two sets S_0' and S_1', and we insert this solution ν into the Pareto approximation set A for further improvements. The whole procedure will continue for at most n times, since the number of the variables of each solution is equal to n.

5   Experimental Results

In order to evaluate the efficiency of our proposed algorithm, we carry out experiments on 10 instances of bi-objective UBQP problem [12]. All the algorithms are coded in C ++ and compiled using Dev-C++ 5.0 compiler on a PC running Windows 7 with Core 2.50 GHz CPU and 4 GB RAM.

5.1   Parameter Settings

In this paper, all the test instances are generated with the tool provided in [12], and the dimension of these instances ranges from 1000 to 5000. Besides, the MOBGSA algorithm requires to set a few parameters, we mainly present two important ones: running time and population size. The exact information about the instances and the parameters is given in Table 1.

Table 1. Parameter settings used for bi-objective UBQP instances: instance dimension (n), population size (P) and running time (T).

	Dimension (n)	Population (P)	Time (T)
bubqp_1000_01	1000	10	100"
bubqp_1000_02	1000	10	100"
bubqp_2000_01	2000	20	200"
bubqp_2000_02	2000	20	200"
bubqp_3000_01	3000	30	300"
bubqp_3000_02	3000	30	300"
bubqp_4000_01	4000	40	400"
bubqp_4000_02	4000	40	400"
bubqp_5000_01	5000	50	500"
bubqp_5000_02	5000	50	500"

5.2   Performance Assessment Protocol

In this paper, we evaluate the effectiveness of multi-objective optimization algorithms using a test procedure that has been undertaken with the performance assessment package provided by Zitzler et al.[1]. The quality assessment protocol works as follows: we first create a set of 20 runs with different initial populations for each algorithm and each benchmark instance. Afterwards, we calculate the set PO^* in order to determine the quality of k different sets $A_0, \cdots, A_{k-1}$ of non-dominated solutions. Furthermore, we define a reference point $z = [w_1, w_2]$, where w_1 and w_2 represent the worst values for each objective function in $A_0 \cup \cdots \cup A_{k-1}$. Then, the evaluation of a set A_i of solutions can be determined by finding the hypervolume difference between A_i and PO^* [23]. This hypervolume difference has to be as close as possible to zero.

---

[1] http://www.tik.ee.ethz.ch/pisa/assessment.html.

5.3   Computational Results

In this subsection, we present the computational results obtained by the MOBGSA algorithm and make comparisons with two original multi-objective optimization algorithms, which are the fast elitist multi-objective genetic algorithm (NSGA-II) proposed in [6] and the strength Pareto evolutionary algorithm (SPEA2) proposed in [22].

The computational results are summarized in Table 2. Each line in this table contains a value **in grey**, which is the best result obtained on the considered instance. The values both **in italic** and **bold** mean that the corresponding algorithms are **not** statistically outperformed by the algorithm which obtains the best result (with a confidence level greater than 95 %).

Table 2. The computational results on bi-objective UBQP problem obtained by the algorithms: NSGA-II, SPEA2 and MOBSGA

Instance	Algorithm		
	NSGA-II	SPEA2	MOBSGA
bubqp_1000_01	0.576117	0.546847	0.048291
bubqp_1000_02	0.529058	0.508735	0.264665
bubqp_2000_01	0.583509	0.554699	0.188339
bubqp_2000_02	0.566623	0.552244	0.316849
bubqp_3000_01	0.573531	0.545882	0.149644
bubqp_3000_02	0.575149	0.546455	0.022204
bubqp_4000_01	0.555975	0.557967	0.124167
bubqp_4000_02	0.580607	0.578930	0.168716
bubqp_5000_01	0.506434	0.541808	0.136980
bubqp_5000_02	0.630504	0.636295	0.044326

From Table 2, we observe that all the best results are obtained by MOBSGA, which statistically outperforms the algorithms NSGA-II and SPEA2 on all the instances. Moreover, the most significant result is achieved on the instance bubqp_3000_02, where the average hypervolume difference value obtained by MOBSGA is much smaller (at least 10 times) than the values obtained by NSGA-II and SPEA2. For most instances, the average hypervolume difference values obtained by MOBSGA are 50 % of those values obtained by NSGA-II and SPEA2.

Compared with NSGA-II and SPEA2, we can see the distinct contribution of the backbone guided search techniques in MOBSGA. We suppose that, when the hypervolume-based local search can no longer improve the Pareto approximation set A, which means the algorithm is trapped in the local optima after a certain amount of running time. Then, the backbone guided search techniques provide a possible way to jump out of the local optima and broaden the search space. Furthermore, one solution selected from two generated sets S'_0 and S'_1 could continue updating the Pareto approximation set A. Therefore, MOBSGA has a better performance on all the instances.

6   Conclusions

In this paper, we have investigated the multi-objective backbone guided search algorithm for solving the bi-objective UBQP problem. The proposed algorithm is mainly composed of the hypervolume-based local search procedure and the backbone guided search procedure, in which the whole Pareto approximation set is evidently improved. We have carried out the experiments on 10 benchmark instances, and the computational results indicate that the MOBGSA algorithm is very competitive.

Acknowledgment. The work in this paper was supported by the Fundamental Research Funds for the Central Universities (Grant No. A0920502051408-25), supported by the Research Foundation for International Young Scientists of China (Grant No. 61450110443), supported by the Scientific Research Foundation for the Returned Overseas Chinese Scholars (Grant Nos. 2015S03007), supported by National Natural Science Foundation of China (Grant No. 61370150, 61433014 and 71501157) and supported by West Light Foundation of Chinese Academy of Science (Grant No: Y4C0011001). The authors would like to thank the anonymous referees for their valuable comments and suggestions.

References

1. Alidaee, B., Kochenberger, G.A., Ahmadian, A.: 0-1 quadratic programming approach for the optimal solution of two scheduling problems. Int. J. Syst. Sci. **25**, 401–408 (1994)
2. Alkhamis, T.M., Hasan, M., Ahmed, M.A.: Simulated annealing for the unconstrained binary quadratic pseudo-boolean function. Eur. J. Oper. Res. **108**, 641–652 (1998)
3. Amini, M., Alidaee, B., Kochenberger, G.: A scatter search approach to unconstrained quadratic binary programs. In: Cone, D., Dorigo, M., Glover, F. (eds.) New Methods in Optimization, pp. 317–330. McGraw-Hill, New York (1999)
4. Basseur, M., Zeng, R.-Q., Hao, J.-K.: Hypervolume-based multi-objective local search. Neural Comput. Appl. **21**(8), 1917–1929 (2012)
5. Coello, C.A., Lamont, G.B., Van Veldhuizen, D.A.: Evolutionary Algorithms for Solving Multi-objective Problems (Genetic and Evolutionary Computation). Springer, New York (2006)
6. Deb, K., Pratap, A., Agarwal, S., Meyarivan, T.: A fast elitist multi-objective genetic algorithm: NSGA-II. IEEE Trans. Evol. Comput. **6**, 182–197 (2000)
7. Gallo, G., Hammer, P., Simeone, B.: Quadratic knapsack problems. Math. Program. **12**, 132–149 (1980)
8. Garey, M.R., Johnson, D.S.: Computers and intractability: A guide to the theory of NP-completeness. Freeman, New York, USA (1978)
9. Glover, F., Kochenberger, G., Alidaee, B.: Adaptive memory tabu search for binary quadratic programs. Manage. Sci. **44**, 336–345 (1998)
10. Kochenberger, G., Hao, J.-K., Glover, F., Lewis, M., Lü, Z., Wang, H., Wang, Y.: The unconstrained binary quadratic programming problem: a survey. J. Comb. Optim. **28**, 58–81 (2014)
11. Krarup, J., Pruzan, A.: Computer aided layout design. Math. Program. Study **9**, 75–94 (1978)
12. Liefooghe, A., Verel, S., Hao, J.-K.: A hybrid metaheuristic for multiobjective unconstrained binary quadratic programming. Appl. Soft Comput. **16**, 10–19 (2014)

13. Liefooghe, A., Verel, S., Paquete, L., Hao, J.-K.: Experiments on local search for bi-objective unconstrained binary quadratic programming. In: Gaspar-Cunha, A., Henggeler Antunes, C., Coello, C.C. (eds.) EMO 2015. LNCS, vol. 9018, pp. 171–186. Springer, Heidelberg (2015)
14. Lü, Z., Glover, F., Hao, J.-K.: A hybrid metaheuristic approach to solving the UBQP problem. Eur. J. Oper. Res. **207**, 1254–1262 (2010)
15. Merz, P., Freisleben, B.: Genetic algorithms for binary quadratic programming. In: Proceedings of the 1st International Conference on Genetic and Evolutionary Computation Conference (GECCO 1999), Orlando, Florida, USA, pp. 417–424 (1999)
16. Merz, P., Katayama, K.: Memetic algorithms for the unconstrained binary quadratic programming problem. Biosystems **78**, 99–118 (2004)
17. Wang, Y., Lü, Z., Glover, F., Hao, J.-K.: Path relinking for unconstrained binary quadratic programming. Eur. J. Oper. Res. **223**, 595–604 (2012)
18. Wang, Y., Lü, Z.P., Glover, F., Hao, J.K.: Backbone guided tabu search for solving the UBQP problem. J. Heuristics **19**, 679–695 (2013)
19. Wilbaut, C., Salhi, S., Hanafi, S.: An iterative variable-based fixation heuristic for the 0-1 multidimensional knapsack problem. Eur. J. Oper. Res. **199**(2), 339–348 (2009)
20. Zhang, W.: Configuration landscape analysis and backbone guided local search. Part 1: satisfiability and maximum satisfiability. Artif. Intell. **158**, 1–26 (2004)
21. Zitzler, E., Künzli, S.: Indicator-based selection in multiobjective search. In: Yao, X., Burke, E.K., Lozano, J.A., Smith, J., Merelo-Guervós, J.J., Bullinaria, J.A., Rowe, J.E., Tiňo, P., Kabán, A., Schwefel, H.-P. (eds.) PPSN 2004. LNCS, vol. 3242, pp. 832–842. Springer, Heidelberg (2004)
22. Zitzler, E., Laumanns, M., Thiele, L.: SPEA2: improving the strength Pareto evolutionary algorithm for multiobjective optimization. TIK Report 103, Computer Engineering and Networks Laboratory (TIK), ETH Zurich, Zurich, Switzerland (2001)
23. Zitzler, E., Thiele, L.: Multiobjective evolutionary algorithms: a comparative case study and the strength pareto approach. Evol. Comput. **3**, 257–271 (1999)

Swarm Intelligence and Optimization

Discrete Chaotic Gravitational Search Algorithm for Unit Commitment Problem

Sheng Li[1], Tao Jiang[2], Huiqin Chen[3], Dongmei Shen[2], Yuki Todo[4],
and Shangce Gao[5(✉)]

[1] College of Computer Science and Technology,
Taizhou University, Taizhou, China
[2] College of Computer Science and Technology,
Donghua University, Shanghai, China
[3] Jiangsu Agri-animal Husbandry Vocational College, Taizhou, China
[4] School of Electrical and Computer Engineering,
Kanazawa University, Kanazawa, Japan
[5] Faculty of Engineering, University of Toyama, Toyama, Japan
gaosc@eng.u-toyama.ac.jp

Abstract. This paper presents a discrete chaotic gravitational search algorithm (DCGSA) to solve the unit commitment (UC) problem. Gravitational search algorithm (GSA) has been applied to a wide scope of global optimization problems. However, GSA still suffers from the inherent disadvantages of trapping in local minima and the slow convergence rates. The UC problem is a discrete optimization problem and the original GSA and chaos which belong in the realm of continuous space cannot be applied directly. Thus in this paper a data discretization method is implemented after the population initialization to make the improved algorithm available for coping with discrete variables. Two chaotic systems, including logistic map and piece wise linear chaotic map, are used to generate chaotic sequences and to perform local search. The simulation was carried out on small-scale UC problem with six-unit system and ten-unit system. Simulation results show lower fuel cost than other methods such as quadratic model, selective pruning method and iterative linear algorithm, confirming the potential and effectiveness of the proposed DCGSA for the UC problem.

Keywords: Chaotic search · Data discretization · Gravitational search · Global optimization · Unit commitment

1 Introduction

Unit commitment (UC) is the most significant optimization task in the operation scheduling of power systems [1]. The objective of UC problem refers to determine which units should be started up or be shut down over the scheduled horizon so as to minimize the total operating cost while satisfying a set of system constraints [2]. However, the solution of the UC problem is really a complex optimization problem with both discrete (unit commitment) and continuous (generation levels) variables. The optimal solution to the problem can be found by exhaustive enumeration of all feasible

© Springer International Publishing Switzerland 2016
D.-S. Huang and K.-H. Jo (Eds.): ICIC 2016, Part II, LNCS 9772, pp. 757–769, 2016.
DOI: 10.1007/978-3-319-42294-7_67

combinations of generating units [3]. But the computer execution time for this method is usually too immense for practical systems as the number of combinations of 0–1 variables grows in an exponential way [4].

Therefore, many research efforts have been focused on efficient, suboptimal UC algorithms which can be applied to realistic power systems. Among those methods, meta-heuristic techniques seem to be promising and evolving, and have come to be the most widely used tools for solving the UC problem [5, 6]. Gravitational search algorithm (GSA) is comparatively a novel optimization algorithm which has a flexible and well-balanced mechanism to enhance exploration and exploitation abilities. The advantages of robustness, adaptability and simplicity of GSA make it possible to be applied to a wide scope of optimization problems. However, few GSA algorithms have been applied to solve the unit commitment problem. GSA still suffers from the inherent disadvantages of trapping in local minima and the slow convergence rates that reduce the solution quality. So in our previous work [7], we put forward CGSA combining GSA with chaos mechanisms to improve the solution quality. The ergodicity of chaos has been viewed as an optimization mechanism to avoid falling into the local search process. We used CGSA to test continuous problems of six widely used benchmark optimization instances and it is verified that CGSA can directly improve the current solution found by GSA with a faster convergence speed and a higher probability of jumping out local optima.

In this paper, based on our previous works, an improved discrete chaotic gravitational search algorithm (DCGSA) is employed for solving the UC problem. Chaos is generated by two chaotic maps and is employed to do local search. Because the UC problem is a discrete optimization problem while chaos belongs in the realm of continuous space, data discretization is implemented in our work after the population initialization to make the proposed DCGSA available for dealing with discrete variables. The simulation was carried out on small-scale UC problem with six-unit system and ten-unit system. It shows that the total generation cost can be remarkably reduced while considering various constraints.

The rest of the paper is organized as follows: In Sect. 2, formulation of unit commitment is defined. The traditional GSA algorithm is discussed in Sect. 3. In Sect. 4, chaotic maps used in this paper are introduced. The proposed DCGSA algorithm is presented in Sect. 5. Section 6 gives the experimental results of DCGSA on small-scale UC problem with six-unit system and ten-unit system. Finally, some general remarks are presented to conclude the paper.

2 Formulation of Unit Commitment

2.1 Objective Function

The objective of the UC problem is to minimize the total production cost over the scheduled time horizon under various constraints. The objective function can be represented as [8]:

$$\min \sum_{i=1}^{N} \sum_{t=1}^{T} [C_i\left(P_i^t, I_i^t\right) + S_i^t \cdot I_i^t \cdot \left(1 - I_i^{t-1}\right)] \tag{1}$$

where N is number of generators, T is total scheduling period, P_i^t is the output power by unit i at t time interval, I_i^t is the on/off status of unit i at t time interval (on = 1 and off = 0), S_i^t is start-up cost of unit i at t time interval. $C_i(P_i^t, I_i^t)$ is the fuel cost function of the generator power output. Most frequently used cost function is in a quadratic form of

$$C_i\left(P_i^t\right) = a_i(P_i^t)^2 + b_i\left(P_i^t\right) + c_i \tag{2}$$

Where $a_i, b_i,$ and c_i are the cost function parameters of unit i.

The start-up cost depends on the down time of unit i. Here, the time-dependent cost is shown as [1]:

$$S_i = \begin{cases} S_{cold,i}, T_{i,t}^{off} > T_i^{off} + T_i^{cold} \\ S_{hot,i}, T_{i,t}^{off} \le T_i^{off} + T_i^{cold} \end{cases} \tag{3}$$

where S_i is the total start-up cost, $S_{cold,i}$ is cold start cost of unit i, $S_{hot,i}$ is hot start cost of unit i, T_i^{cold} is cold start time of unit i, T_i^{off} is minimum down time of unit i, $T_{i,t}^{off}$ is continuously off time of unit i up to time t.

2.2   Constraints

Generation power limits [9]	$I_i^t \cdot P_{i,min} \le P_{i,t} \le I_i^t \cdot P_{i,max}$	(4)
Ramp rate limits for unit generation changes	$\begin{cases} P_i^{t-1} - P_i^t \le P_i^{down} \\ P_i^t - P_i^{t-1} \le P_i^{up} \end{cases}$	(5)
Minimum down-time [1]	$\left(T_{i,t-1}^{off} - T_i^{off}\right)(I_i^t - I_i^{t-1}) \ge 0$	(6)
Minimum up-time	$\left(T_{i,t-1}^{on} - T_i^{on}\right)(I_i^{t-1} - I_i^t) \ge 0$	(7)
Power balance [6]	$\sum_{i=1}^{N} I_i^t \cdot P_i^t = P_D^t(t = 1, 2, \ldots, T)$	(8)
Spinning reserve [6]	$\sum_{i=1}^{N} I_i^t \cdot P_i^t = S_D^t(t = 1, 2, \ldots, T)$	(9)
Unit initial status [1]	The initial status at the start of the scheduling period must be taken into account.	

$P_{i,min}$ is the minimum generation limit of unit i. $P_{i,max}$ is the maximum generation limit of unit i.

P_i^{down} is the ramp-down limit of unit i, P_i^{up} is the ramp-up rate limit of unit i.

T_i^{off} is the minimum down time of the unit i, $T_{i,t-1}^{off}$ is continuously off time of unit i up to time $t-1$.

T_i^{on} is the minimum on time of the unit i. $T_{i,t-1}^{on}$ is continuously on time of unit i up to time $t-1$.

P_D^t is the power demand at the t-th hour. S_D^t is the spinning reserve at the t-th hour.

3   Gravitational Search Algorithm (GSA)

GSA is a global search strategy that can handle efficiently arbitrary optimization problems, in which agents are considered as objects and their performances are measured by masses. The heavier masses (which correspond to good solutions) move more slowly than the lighter ones, which guarantee the exploitation of the algorithm to find the optima around a good solution. Consider a system with N agents (objects), we define the position of the i-th agent by [8]:

$$X_i = \left(x_i^1, x_i^2 \ldots, x_i^d, \ldots, x_i^n\right) \qquad i = 1, 2, \ldots, N \tag{10}$$

where x_i^d is the position of the ith agent in the d-th dimension, n is the dimension of the search space.

At the t-th iteration, the gravitational force acting on the i-th object from the j-th object is represented as follows:

$$F_{ij}^d(t) = G(t) \frac{M_j(t) M_i(t)}{R_{ij}(t) + \varepsilon} \left(x_j^d(t) - x_i^d(t)\right) \tag{11}$$

where M_i and M_j are masses of agents. $G(t)$ is the gravitational constant at time t, ε is a very small constant and $R_{ij}(t)$ indicates the Euclidean distance between two agents i and j:

$$R_{ij}(t) = \left\| x_i(t), x_j(t) \right\|_2 \tag{12}$$

The gravitational constant $G(t)$ is given by:

$$G(t) = G_0 e^{-\alpha t / iter_{max}} \tag{13}$$

where G_0 is the initial value, α is a defined parameter, $iter_{max}$ is the maximum number of iterations.

The total force acting on the i-th agent is given by:

$$F_i^d(t) = \sum_{j=1, j \neq i}^{Kbest} rand_i F_{ij}^d(t) \tag{14}$$

where Kbest is the set of first K agents with better fitness (i.e. bigger mass). $rand_j$ is a random number in the interval $[0, 1]$. The acceleration $a_i^d(t)$ of the agent i at time t and in d-th dimension is given by:

$$a_i^d(t) = \frac{F_i^d(t)}{M_i(t)} \tag{15}$$

where $M_i(t)$ is calculated through the map of fitness defined as follows:

$$M_i(t) = \frac{m_i(t)}{\sum_{j=1}^{N} m_j(t)} \tag{16}$$

$$m_i = \frac{fit_i(t) - worst(t)}{best(t) - worst(t)} \tag{17}$$

where best(t) is the best fitness of all agents, worst(t) is the worst fitness of all agents, and $fit_i(t)$ represents the fitness of agent M_i by calculating the objective functions. The position and the velocity of the i-th agent at t-th iteration in the d-th dimension is calculated as follows:

$$v_i^d(t+1) = rand_i \times v_i^d(t) + a_i^d(t) \tag{18}$$

$$x_i^d(t+1) = x_i^d(t) + v_i^d(t+1) \tag{19}$$

where $rand_i$ is a uniform random variable generated in the interval [0, 1].

4 Chaotic Maps

Chaos is a kind of dynamic behavior of nonlinear systems and it has aroused much concern in different fields of sciences such as chaos control, pattern recognition and optimization theory. In this work, we use the Logistic Map and Piece Wise Linear Chaotic Map (PWLCM) to generate chaotic sequences and employ chaos to act as a local search approach.

4.1 Logistic Map

The logistic map was popularized in a seminal paper by the biologist May [10], in part as a discrete-time demographic model analogous to the logistic equation. Its mathematical expression is given by Eq. (20).

$$x_{k+1} = ax_k(1 - x_k) \quad k = 1, 2, \ldots, N \tag{20}$$

where x_k is the k-th chaotic number and k represents the iteration number, The parameter a is usually set to 4. The initial number $x_0 \in [0, 1]$ and $x_0 \notin \{0.0, 0.25, 0.5, 0.75, 1.0\}$. When the logistic map is combined with GSA, the hybrid algorithm is labeled as LDCGSA.

4.2 Piecewise Linear Chaotic Map

Piecewise linear chaotic map (PWLCM) has obtained much attention in chaos research recently for its simplicity in representation and good dynamical behavior. PWLCM has been known as ergodic and has uniform invariant density function on their definition intervals. The simplest PWLCM is defined in Eq. (21):

$$x_{k+1} = \begin{cases} x_k/px_k \in (0, p) \\ (1 - x_k)(1 - p)x_k \in [p, 1) \end{cases} \tag{21}$$

In the experiment, p is set to be 0.7. When PWLCM is combined with GSA, the hybrid algorithm is labeled as PDCGSA.

5 Discrete Chaotic Gravitational Search Algorithm (DCGSA)

5.1 Chaotic Local Search Algorithm

Optimization algorithms combined with chaos is easier to escape from local minimum points than traditional stochastic optimization algorithms. Chaotic local search can alleviate the blindness and randomness of the search process so that better solutions near the current optimal solutions can be reached more effectively. The general steps of CLS are given as following [8]:

Step 1. Set the parameters of a chaotic system and the number of chaotic search L
Step 2. According to the chaotic system, get a chaotic sequence whose length is N
Step 3. Choose the best individual v_c in the current population
Step 4. Record chaotic search initial counter as 0. **While(t < L),** Superimpose an item of the chaotic sequence on v_c in any dimension to form a new individual that is marked as v_n
Step 5. Calculation the fitness value of the new individual v_n and compute current velocity according to Eq. (18)
Step 6. **If** $f(v_c) > f(v_n)$, $v_c \leftarrow v_n$, **end-if**
Step 7. $T = t+1$, **end-while**

The search neighborhood of x_g is constructed in a hypercube whose center is x_g with a radius r.

5.2 Data Discretization

In the traditional GSA algorithm, population is initialized randomly in search space. But in our improved DCGSA algorithm, we use chaos, which belong in the realm of continuous space, to initialize the population and thus the values of obtained coordinates are continuous. However, the UC problem is a discrete problem which has variables of 0–1 values to represent on/off status of the units, so it is necessary to do data discretization after the population initialization to make DCGSA available for coping with discrete variables. Moreover, in DCGSA, we executed a chaotic local search to reach better solutions near the current optimal solutions. The coordinate values of the population will become further continuous after the chaotic local search, thus data discretization will be performed necessarily for the second time.

In our work, we discretize the coordinate values of the population into 0 and 1. If the value is bigger than 0.5, then it will go to 1. If the value is smaller than 0.5, then it will go to 0. To assign the unit status as 1 or 0, we use the sigmoid function which is given by [6]:

$$f(x_i) = \frac{1}{1 + \exp(-x_i)} \tag{22}$$

where x_i is the value of the i-th controlled variable obtained by using Eq. (19).

5.3   DCGSA Algorithm

Based on the GSA and chaotic local search, we propose an improved discrete chaotic gravitational search algorithm (DCGSA). In order to enable DCGSA to deal with the UC problem, we add data discretization into the algorithm. Different steps of UC based DCGSA is mentioned below:

Step 1:   Using Logistic Map and Piece Wise Linear Chaotic Map to initialize M agent position as Swarm(K) respectively which can be represented in a matrix as: $[\text{Swarm}(1) \ldots \text{Swarm}(i) \ldots \text{Swarm}(k)]^T$ and

$$\text{Swarm}(i) = [x_1^1 x_2^2 \ldots x_N^1 x_1^2 x_2^2 \ldots x_N^2 \ldots x_1^T x_2^T \ldots x_N^T] \tag{23}$$

where Swarm(i) is the position of the i-th agent. N is the number of unit, T is the time period usually defined to be 24. So the individual dimension becomes NT. x_1^T means the status of the first unit in T time period.

Step 2:   Simply discretize the initialized variables according to the discretization rule employed in our work. If $0.5 \leq x_1^j \leq 1, x_1^j = 1$. If $0 < x_1^j < 0.5, x_1^j = 0$.

Step 3:   Verify if the agents satisfy the constraints of unit commitment or not. If not, do modification of the variables based on Eqs. (18) and (19).

Step 4:   According to the objective function, calculate the fitness$_i$(t) value, t = 0, 1,2, … , T, i = 1, 2, … , M.

Step 5:   Based on the objective function value, modify the value of G(t), best(t), worst(t) and M_i(t) of each agent.

Step 6:   Update the acceleration, velocity and position of each agent using Eqs. (15), (18) and (19)

Step 7:   Find out the global best agent X_g, and build a super cube around X_g to implement chaotic local search approach.

Step 8:   Find out the global best agent after the chaotic local search process.

Step 9:   If the termination condition is satisfied, then finish the algorithm. Otherwise, repeat Step 2 to Step 8.

A penalty function method is used to deal with the ramp rate constraints. Thus the objective function in Eq. (1) will be extended as [6]:

$$F_T = F + \sum_{i=1}^{NP} \sum_{t=1}^{T} PF_i^t \tag{24}$$

Where F_T is the total production cost after penalty, NP is the number of generator. The value of T is defined to be 24 for there are 24 h a day. PF_i^t denotes the penalty function for the ramp rate constraint which is given by

$$PF_i^t = \begin{cases} K(P_i^t - P_i^{t-1} - P_i^{up})^2 P_i^t - P_i^{t-1} \geq P_i^{up} \\ K(P_i^{t-1} - P_i^t - P_i^{down})^2 P_i^{t-1} - P_i^t \geq P_i^{down} \\ 0 \; P_i^t - P_i^{t-1} \geq P_i^{up} \, and \, P_i^{t-1} - P_i^t \geq P_i^{down} \end{cases} \tag{25}$$

where K is the penalty factor.

6   Numerical Simulation

The simulation was carried out on small-scale UC problem with six-unit system and ten-unit system. The population size is set to be 40. The algorithm is carried out for 50 times and the maximum iteration number is 1000 in each run. The constants α and G_0 are set to 1.0E−100 and 100 respectively. The algorithm is developed in MATLAB R2010a on a personal PC.

6.1   Test of Six-Unit System

In this section, six-unit system has been tested in order to prove the applicability of the proposed DCGSA algorithm for solving the UC problem. The algorithm is tested in two sets of instances: one with ramp constraints and the other without ramp constraints. The parameters of six-unit system are given in Table 1 and Power demands for 24 h are given in Table 2. The DCGSA approach is applied to solve the UC problem considering all constraints such as generator constraints, reserve constraints, minimum up time and minimum down time. Scheduling of the generation obtained by DCGSA approach for six-unit system is given in Table 3, including the value neglecting ramp rate constraints and considering ramp rate constraints. As we can see in the Table 3, the unit output power is influenced both by rated power and the ramp rate constraints. Increase and decrease of the unit output must be limited within a certain range, otherwise the output unit needs to be redistributed. Thus there will be a significant difference in the decision making of unit commitment when considering ramp rate constraints. Here we will take the first hour as an example for analysis:

The output of unit 2 changed to 35 MW from 0 MW when neglecting ramp rate constraints and the change of unit power was affected only by the maximum and minimum power. However, the power of unit 2 could only increase 12 MW when

Table 1. The parameters of six-unit.

Parameters	Unit 1	Unit 2	Unit 3	Unit 4	Unit 5	Unit 6
a($/h)	0	0	0	0	0	0
b($/MWh)	2	1.7	1	3.25	3	3
C(MW2 h)	0.00375	0.0175	0.0625	0.00834	0.025	0.025
Hot start cost	80.00	60.00	50.00	25.00	30	40
Cold start cost	160.00	120	100	50	60	80
Cold start(h)	2	1	1	1	1	1
Pmax(MW)	200.00	80.00	50.00	35.00	30.00	30
Pmin(MW)	80.00	20.00	15.00	10.00	10.00	12
Min up-time(h)	5	3	1	2	1	1
Min down-time(h)	5	3	1	2	1	1
Initial unit status	5	−3	2	3	−2	2
Initial output	80.00	0.00	50.00	35.00	0.00	20

Table 2. Power demand for 24 h of six-unit

Hour	Load	Hour	Load	Hour	Load
1	166	9	192	17	246
2	196	10	161	18	241
3	229	11	147	19	236
4	267	12	160	20	225
5	283	13	170	21	204
6	272	14	185	22	182
7	246	15	208	23	161
8	213	16	232	24	131

considering the ramp rate constraints. Meanwhile, the output of unit 3 changed to 16 MW from 50 MW and the output of both unit 4 and unit 6 changed to 0 when neglecting ramp rate constraints.

While considering the ramp rate constraints of the unit, decrease and increase of its power must be limited within a certain range, and cannot be arbitrarily increased or decreased. In addition, we can also see that after the 10 h, unit 2 alone could meet the needs of system load when considering ramp rate constraints. The optimal generation scheduling of the output corresponding to the load change per hour would not exceed the limit of ramp rate constraints. Therefore, each time period after the 10 h, optimal generation scheduling of unit commitment regardless of ramp rate constraints is about the same as that considering ramp rate constraints.

The obtained best fuel cost is shown in Table 4, including the value neglecting ramp rate constraints and considering ramp rate constraints. Here in Table 4, we also compared the optimization results of DCGSA with that of mixed-integer linear programming (MILP) of the unit commitment problem. We can see that the value of cost obtained by DCGSA is smaller than that obtained by MILP. The possible reason for this result may be that we use a quadratic curve to deal with the burning cost while

Table 3. Scheduling of the generation without/with ramp rate constraints of six-unit.

Hour	Unit 1		Unit 2		Unit 3		Unit 4		Unit 5		Unit 6	
1	117	87	35	0	16	37	0	30	0	0	0	12
2	140	445	38	20	16	24	0	25	0	0	0	12
3	166	141	44	38	18	16	0	20	0	0	0	12
4	196	175	50	46	19	18	0	15	0	0	0	12
5	200	191	60	49	22	19	0	11	0	0	0	12
6	200	183	51	47	20	18	0	10	0	0	0	12
7	180	172	47	45	18	18	0	10	0	0	0	0
8	154	146	41	39	17	16	0	10	0	0	0	0
9	137	139	38	36	16	15	0	10	0	0	0	0
10	113	113	32	32	15	15	0	0	0	0	0	0
11	101	101	30	30	15	15	0	0	0	0	0	0
12	112	112	32	32	15	15	0	0	0	0	0	0
13	120	120	34	34	15	15	0	0	0	0	0	0
14	132	132	36	36	15	15	0	0	0	0	0	0
15	150	150	40	40	17	17	0	0	0	0	0	0
16	169	169	44	44	18	18	0	0	0	0	0	0
17	180	180	47	47	18	18	0	0	0	0	0	0
18	176	176	46	46	18	18	0	0	0	0	0	0
19	172	172	45	45	18	18	0	0	0	0	0	0
20	163	163	43	43	17	17	0	0	0	0	0	0
21	147	147	40	40	16	16	0	0	0	0	0	0
22	129	129	36	36	15	15	0	0	0	0	0	0
23	113	113	32	32	15	15	0	0	0	0	0	0
24	88	88	27	27	15	15	0	0	0	0	0	0

Table 4. Best fuel cost without/with ramp rate constraints of six-unit.

Method	Operational cost		Start-up cost		Total fuel cost	
DCGSA	12325.6	12450.6	155	155	12480.6	12505.6
MILP	–	12723.06	–	155	–	12898.06

MILP uses a linear process. So it can be concluded that DCGSA proposed in this paper is feasible for the UC problem while considering ramp rate constraints. In addition, according to the results of the above optimization we can clearly see that there is an increase of total unit operating costs when considering ramp rate constraints. This is because that the output unit is not only limited between maximum and minimum power, but also limited to the range of ramp rate constraints. Therefore, in the following section we will use DCGSA to solve the UC problem with 10-unit system without considering ramp rate constraints, to demonstrate the effectiveness and feasibility of proposed DCGSA.

6.2   Test of Ten-Unit System

To further validate the computational efficiency of DCGSA algorithm, we test the ten-unit system of UC problem neglecting ramp rate constraints. The fuel cost and generation scheduling of ten-unit system without ramp rate constraints obtained by DCGSA are shown in Table 5. It can be seen that unit 1 and unit 2 keep being on all over the time period showing that unit 1 and unit 2 play an important role in the process of power generation. Unless special circumstances, the two units are not allowed to be off.

Table 5. Scheduling of the generation of ten-unit system (DCGSA).

Hour	Start-up cost	Operating cost	Scheduling of the generation									
			1	2	3	4	5	6	7	8	9	10
1	0	13683.13	455	245	0	0	0	0	0	0	0	0
2	0	14554.5	455	295	0	0	0	0	0	0	0	0
3	900	16809.45	455	370	0	0	25	0	0	0	0	0
4	0	18597.67	455	455	0	0	40	0	0	0	0	0
5	560	20020.02	455	390	0	130	25	0	0	0	0	0
6	550	22387.04	455	360	130	130	25	0	0	0	0	0
7	0	23261.98	455	410	130	130	25	0	0	0	0	0
8	0	24150.34	455	455	130	130	30	0	0	0	0	0
9	860	27251.06	455	455	130	130	85	20	25	0	0	0
10	60	30075.86	455	455	130	130	162	33	25	0	10	0
11	60	31916.06	455	455	130	130	162	73	25	10	10	0
12	60	33890.16	455	455	130	130	162	80	25	43	10	10
13	0	30057.55	455	455	130	130	162	33	25	10	0	0
14	30	27321.53	455	455	130	130	100	20	0	0	10	0
15	0	24150.34	455	455	130	130	30	0	0	0	0	0
16	0	21513.66	455	310	130	130	25	0	0	0	0	0
17	0	20641.82	455	260	130	130	25	0	0	0	0	0
18	0	22387.04	455	360	130	130	25	0	0	0	0	0
19	0	24150.34	455	455	130	130	30	0	0	0	0	0
20	460	30057.55	455	455	130	130	162	33	25	10	0	0
21	0	27251.06	455	455	130	130	85	20	25	0	0	0
22	0	22735.52	455	455	0	0	145	20	25	0	0	0
23	0	17645.36	455	425	0	0	0	20	0	0	0	0
24	0	15427.42	455	345	0	0	0	0	0	0	0	0

In addition, according to Table 5, we can see that only unit 1 and 2 are turned on in the first 2 h. From the 3 h, there is one or several other units turned on. As system load increases, the power system will turn on the units which can satisfy the load demands and save cost simultaneously.

In addition, GSA is still a new optimization algorithm although it has been applied in few areas of power system. But it has few been used to solve the UC problem so far.

So to further demonstrate the effectiveness and feasibility of our new proved DCGSA, we compared the optimization results of the UC problem with quadratic model (QM), selective pruning method (SDPSP) and iterative linear algorithm (ILA), as shown in Table 6. It is proved that the algorithm proposed in this paper (DCGSA) has better global searching ability than other methods for solving the UC problem with 10-unit system and it also shows the better global searching ability when combining chaos with GSA.

Table 6. Fuel cost comparison among several methods

Method	Total fuel cost
QM	570396.4
SDPSP	564482
ILA	570396.4
DCGSA	560790

7 Conclusions

In this paper, a discrete chaotic gravitational search algorithm (DCGSA) is developed and implemented successfully to solve small-scale UC problem with six-unit system and ten-unit system. Data discretization is implemented after the population initialization to make the improved algorithm available for coping with discrete variables. Logistic map and piece wise linear chaotic map are used to generate chaotic sequences and to perform local search. It is observed from the simulation results that the fitness value obtained by DCGSA after satisfying all the constraints of all the systems, are better than other methods. It is also found that the proposed method DCGSA has superior features, including stable convergence characteristics and remittance of premature convergence. Moreover, DCGSA appears to be a robust and reliable optimization algorithm for solving the UC problem. Simulation results confirm the potential and effectiveness of the proposed DCGSA to generate better optimal solution for the UC problem.

Acknowledgment. This research was partially supported by the National Natural Science Foundation of China (Grant Nos. 61203325, 11572084, 11472061, and 61472284), the Shanghai Rising-Star Program (No. 14QA1400100) and JSPS KAKENHI Grant No. 15K00332 (Japan).

References

1. Yuan, X., Su, A., Nie, H., Yuan, Y., Wang, L.: Unit commitment problem using enhanced particle swarm optimization algorithm. Soft. Comput. **15**(1), 139–148 (2011)
2. Ting, T., Rao, M., Loo, C., Ngu, S.: Solving unit commitment problem using hybrid particle swarm optimization. J. Heuristics **9**(6), 507–520 (2003)

3. Dudek, G.: Unit commitment by genetic algorithm with specialized search operators. Electr. Power Syst. Res. **72**(3), 299–308 (2004)
4. Pappala, V., Erlich. I.: A new approach for solving the unit commitment problem by adaptive particle swarm optimization. In: IEEE Power and Energy Society General Meeting-Conversion and Delivery of Electrical Energy in the 21st Century, pp. 1–6 (2008)
5. Sun, L., Zhang, Y., Jiang, C.: A matrix real-coded genetic algorithm to the unit commitment problem. Electr. Power Syst. Res. **76**(9), 716–728 (2006)
6. Roy, P.: Solution of unit commitment problem using gravitational search algorithm. Int. J. Electr. Power Energy Syst. **53**, 85–94 (2013)
7. Shen, D., Jiang, T., Chen, W., Qian, S., Gao, S.: Improved chaotic gravitational search algorithms for global optimization. In: IEEE Congress on Evolutionary Computation (CEC), pp. 1220–1226 (2015)
8. Dang, C., Li, M.: A floating-point genetic algorithm for solving the unit commitment problem. Eur. J. Oper. Res. **181**(3), 1370–1395 (2007)
9. Padhy, N.: Unit commitment-a bibliographical survey. IEEE Trans. Power Syst. **19**(2), 1196–1205 (2004)
10. May, R.: Simple mathematical models with very complicated dynamics. Nature **261**(5560), 459–467 (1976)

A Discrete Biogeography-Based Optimization for Solving Tomato Planting Planning

Hong-li Zhang[✉] and Cong Wang

School of Electrical Engineering,
Xinjiang University, Urumqi 830047, China
641087385@qq.com

Abstract. The yield of tomato affects the processing ability of ketchup factory directly. To improve the imbalance supply of the materials during tomato sauce season, building the mathematical model of tomato planting planning, a discrete Biogeography-based Optimization is proposed for solving tomato planting planning model. Considering the tomato planting planning is a large-scale combinatorial optimization problem, tomato planting matrix can be compressed by sparse matrix compression method to achieve compression of the solution space. And a new kind discrete BBO with a new coding way was used for planting planning. A tomato plant provides data in Xinjiang as an example of simulation calculation, the results showed that tomato planting planning scheme calculated by the proposed algorithm can realize the balance supplement of tomato materials effectively.

Keywords: Balanced supply · Discrete biogeography-based optimization · Matrix compression method · Tomato planting planning

1 Introduction

The unique conditions of heat, water and soil in Xinjiang is conducive to the growth of higher quality tomatoes. The tomato sauce become the international market quality products for its characteristics with red pigment, the higher color difference value, the higher solid content and the lower mold. Tomato sauce production in Xinjiang accounted for nearly 90 % across the country. And the export of tomato sauce is close to 30 % of the total world trade. The tomato processing has become a characteristic "red" pillar industry in Xinjiang.

At present, the tomato industry in Xinjiang is facing a serious problem is that the tomato planting planning mainly depends on artificial experience to determine the planting area, planting variety and planting time. This will result a serious mismatch problem between tomatoes sold amount and factory processing capacity during the processing tomato season. The processing capacity of a tomato ketchup factory is fixed in one production period. And the yield of tomato appears an unbalanced situation that there is a lack of tomatoes in early and late yield and middle yield is excess. This leads material supply lacking, a large number of production lines idle and production costs rising in the early and late period of the tomato sauce processing enterprises. And in the medium period the material supplying exceeds demand, which is far more than the

© Springer International Publishing Switzerland 2016
D.-S. Huang and K.-H. Jo (Eds.): ICIC 2016, Part II, LNCS 9772, pp. 770–781, 2016.
DOI: 10.1007/978-3-319-42294-7_68

enterprise storage and processing capacity. With the continuous expansion of the scale of the growing scale of Xinjiang's tomato, waste problems caused by this extensive planning model have become more serious. To solve the problem of balanced supply of tomato raw materials has important application value.

In recent years, some scholars have tried to use swarm intelligence algorithm for solving crop planting planning problems. Sarker formulates a crop-planning problem as a multi-objective optimization model and solves two different versions of the problem using three different optimization approaches. And they compared the performance of their algorithm with other two algorithms and gained their algorithm delivers superior than others [1]. Adeyemo presents four strategies of a novel evolutionary algorithm, multi-objective differential evolution algorithm for solving the crop planting planning problem. The non-dominated solutions of maize, peanut, alfalfa and mountain walnut, were obtained by simulation and calculation tended to the Pareto front, which showed that the proposed algorithm was effective [2].

The Biogeography-based optimization [3] is a new kind of intelligent optimization algorithm as Genetic algorithm (GA) [4], Particle swarm optimization (PSO) [5, 6], which is proposed by Simon in 2008. Some performance of the algorithm is better than the traditional intelligent optimization algorithm. And it has been successfully applied to the practical problems such as economic load dispatch, traveling salesman problem and robot path planning. For examples, Aniruddha Bhattacharya presented a biogeography-based optimization to solve both convex and non-convex economic load dispatch problems of thermal plants [7]; Zheng used novel differential biogeography-based optimization algorithm to optimized the hybrid neuro-fuzzy network to solving online population classification in earthquakes [7]; Zhu and Duan presented chaotic biogeography-based optimization approach to receding horizon control for multiple UAVs formation flight [9].

In this paper, for a large number of tomato planting, a planning model is established. And the sparse matrix compression method is used to compress the planting matrix to make the transverse solution space compressed. In order to satisfy each constraint conditions and obtain the most reasonable planting plan, a discrete biogeography-based optimization is proposed. A tomato plant provides data in Xinjiang as an example of simulation calculation, the results showed that tomato planting planning scheme calculated by the proposed algorithm can realize the balance supplement of tomato materials effectively.

2 Mathematical Model of Tomato Planting Planning

2.1 Objective Function

In order to make the raw material waste lowest and production efficiency highest, the square sum of the error of the raw material and the total processing capacity must be the smallest.

$$\min f_1 = \sum_{j=\max(m_i)+1}^{\max(m_i)+T} \left(w_j - P_f \right) \tag{1}$$

Where, m_i is the mature period of ith kinds of tomato. T presents the tomato ketchup factory's production period. P_f is daily production capacity of all the tomato paste factory production line. According to the mature stage and the production stage, the planting time is $(j - m_i)$ and tomato i's cultivated area is $x_{i(j-m_i)}$ in $(j - m_i)$ day. w_j is the tomato yield in j day, which is related to $x_{i(j-m_i)}$ and the yield p_i of tomato i, as the formula (2):

$$w_j = \sum_{i=1}^{c} x_{i(j-m_i)} p_i \tag{2}$$

Where, c is the number of tomato varieties.

2.2 Constraint Conditions

In tomato planting planning problem, the tomato paste factory's planting area is limited. In order to reflect the actual situation of the tomato production, there are two following constraints.

(1) The land constraint of tomato paste factory
    According to the research of tomato planting data, it is shown that the planting area is composed by many independent plots. In this study, the plots are all numbered.
(2) The constraint of tomato yield in production
    Every variety of the tomato can be harvested every day during the production period, we defined as follows:

$$w_{i(j-m_i)} \geq 0 \tag{3}$$

(3) The constraints of lycopene and vitamin content in tomato paste factory
    For the production line requirements, there have:

$$t_{\mathrm{red}}(x) \geq R, \; t_{\mathrm{vit}}(x) \geq V \tag{4}$$

Where, R and V respectively represent lycopene and vitamin content of the factory production line requirements.

2.3 The Mathematical Model

The nonlinear mathematical model for the tomato planting planning is constructed as formula (5):

$$
\begin{cases}
\min f(X) \\
s.t \;\; t_{red}(X) \geq L \\
\quad\; t_{vit}(X) \geq V \\
\quad\; i = 1, 2, \cdots, c \\
\quad\; j = max(m_i) + 1, \cdots, T + max(m_i) \\
\quad\; X = [x_1 \cdots x_c]_{1*(T*c)}
\end{cases}
\tag{5}
$$

The nonlinear mathematical model of tomato planting planning is constructed by the penalty function method. And the constrained optimization problem can be transformed into an unconstrained optimization problem. We can get the following objective function after augmented.

$$
minF(X, M_k) = min\{f(X) + M_1 \times max(l(X) - L, 0) + M_2 \times max(v(X) - V, 0)\} \tag{6}
$$

Where, M_1 and M_2 are penalty factors.

3 The Discrete Biogeography-Based Optimization for Tomato Planning

3.1 The Biogeography-Based Optimization

In the BBO algorithm, each solution of the problem corresponds to the planting area of each tomato variety, which is called a suitability index variable (namely SIV). The measure function value of each solution corresponds to the objective function value of each planting scheme. The value of the objective function is called a fitness index (namely HSI). The details of the BBO are shown as follows [9, 10]. The two operators of BBO algorithm is summarized as follows:

(1) Migration operation of habitat

Inspired by the literature [11, 12], according to λ_i, the need move into habitat X_i is selected. After we determined X_i, according to X_i, the need move out habitat X_j is selected and randomly selected a habitat X_m. Then the SIV is selected from X_j and X_m to replace the corresponding SIV in X_i, is shown as:

$$
\begin{cases}
X_i(SIV) = \alpha X_i(SIV) + (1 - \alpha)X_j(SIV) & r_{rand} < \mu_i \\
X_i(SIV) = X_i(SIV) + r_{rand}(X_{best}(SIV) - X_m(SIV)) & else
\end{cases}
\tag{7}
$$

Where, α is the weighted factor. $r_{rand} \in (0, 1)$ is the random number.

(2) The mutation operation of habitat

According to the probability Ps of the number species s in the habitat, the characteristic variables of the habitat is mutated. The mutation rate of each habitat is calculated as formula (8):

$$M_s = M_{max} \cdot (1 - P_s)/P_{max} \tag{8}$$

Where, M_{max} is the maximum mutation rate. $P_{max} = \arg(\max(P_s))$, $s = 1, \cdots S_{max}$. P_s is the probability of the species number in the habitat. Then the need changed habitat is selected by M_s. The random selected SIV of the habitat will be replaced by one quantity, which is randomly generated by Gauss distribution [10].

3.2 A Discrete Biogeography-Based Optimization for Tomato Planting Planning

Tomato planting planning problem is a large-scale combinatorial optimization problem. In this paper, a programming method based on sparse matrix compression is proposed to reduce the scale of large-scale tomato planning problem solving space. At the same time, inspired by the interval crossover operation of genetic algorithm, a new discrete BBO algorithm is proposed, which is used to solve the problem of large-scale tomato planting planning problem.

Compression Storage Encoding of Tomato Planting Matrix. For the small scale tomato planning problem, this paper analyzes the 4 tomato varieties in the 240 place of the planting planning program, which means that the dimension of the problem is $D = 240$. The dimension in solving large-scale tomato planning problem is more than that. For solving this problem, we propose sparse matrix compression method for large scale tomato planning problem. We suppose the planting day T_1, m pieces of land and n classes of tomato varieties.

The specific operations are as follows:
The planting matrix before compression is expressed as follows:

T_1 *column:* *The ith column means the growing day i*

$$ZZ = \begin{bmatrix} x_{11} & x_{12} & \cdots & x_{1i} & \cdots & x_{1T_1} \\ x_{21} & x_{22} & & x_{2i} & & x_{2T_1} \\ \cdots & \cdots & & \vdots & & \cdots \\ x_{j1} & x_{j2} & & x_{ji} & & x_{jT_1} \\ x_{m1} & \cdots & \cdots & \vdots & \cdots & x_{jT_1} \end{bmatrix} \begin{matrix} 1th \\ 2nd \\ \cdots \\ jth \\ \cdots \end{matrix} \quad m \text{ row}: \text{ The jth land} \tag{9}$$

When we considered that cultivar k is planted in yield j on the i day, which is remembered $x_{ji} = k$. Other elements of the j line are zero. We say that the matrix in formula (9) is a sparse matrix. If we encoding as the last section, the dimension of the problem is $T_1 * m$. For example, if there are 105 planting days, 200 pieces of land and 8 tomato varieties, the dimension of the problem is $D = 105 * 300$. We will find that the solver runs slowly and the effect is not good for too large dimension. Through the compression of sparse matrix and holding non-null elements, the new planting matrix is obtained by using the storage method of the three tuples table, as:

$$ZZ' = \begin{bmatrix} 1 & i & x_{1i}, & 2 & i & x_{2i}, \cdots, & j & i & x_{ji}, \cdots, & m & i & x_{mi} \end{bmatrix}$$

$$\downarrow \qquad \downarrow \qquad \downarrow$$

land   planting   varieties

$$i = (1, 2, \dots, T_1)$$

number   time   number

(10)

At the same time, the number of the $1, 4, \ldots, 3 * (j - 1) + 1, ..$ dimension of the planting matrix ZZ' can be numbered 1 to m, which can be simplified the planting matrix and gain the new matrix ZZ'.

$$ZZ' = \begin{bmatrix} & 1th\ land & 2nd\ land & & jth\ land & \\ i & x_{1i}, & x_{2i} & , \cdots, & i & x_{ji}, & i & x_{mi} \end{bmatrix}$$

$$\downarrow \qquad \downarrow$$

planting   varieties

time   number

(11)

The dimension of the problem becomes $2 * m$, which is more simple than $T_1 * m$.

Implementation of Discrete BBO for Solving Tomato Programming

(1) The designing of position and speed update in discrete BBO

Inspired by the traditional crossover design by solving the TSP problem in the [13] literature, the migration operator of the discrete BBO using partially matched crossover method to update information as Genetic Algorithm (GA).

According to the λ_i and μ_j of BBO respectively to choose immigration habitat X_i and emigration habitat X_j. Then a non-repeated habitat X_k is randomly selected to result two crossing points. We suppose:

$$The\ selected\ cross\ regional$$

$$X_k = \begin{bmatrix} 2 & 4 & | & 1 & 8 & 3 & 7 & 5 & | & 5 & 1 & 6 \end{bmatrix}$$
$$X_i = \begin{bmatrix} 1 & 6 & | & 7 & 2 & 3 & 4 & 3 & | & 5 & 8 & 7 \end{bmatrix}$$
$$X_j = \begin{bmatrix} 3 & 6 & | & 2 & 7 & 5 & 4 & 8 & | & 1 & 2 & 1 \end{bmatrix}$$

(12)

After the crossing area of the habitat is obtained, the cross area of X_i is replaced by the crossing area of X_j. In order to increase the ability to jump out of the local optimal solution, the cross area of X_j is replaced by the crossing area of X_k. The final new habitat (the optimal solution) is shown as formula (13):

$$X_i = \begin{bmatrix} 1 & 6 & | & 2 & 7 & 5 & 4 & 8 & | & 5 & 8 & 7 \end{bmatrix}$$
$$X_j = \begin{bmatrix} 3 & 6 & | & 1 & 8 & 3 & 7 & 5 & | & 1 & 2 & 1 \end{bmatrix}$$

(13)

(2) The designing of mutation operator in discrete BBO

According to the mutation rate, the random discrete variables is replaced the selection variant vector, which is generated by the selection variant vector. In the discrete BBO, the solution of each habitat information is corresponding to the solution of the optimization problem, that is, the varieties matrix in tomato planning problem. Each fitness function value is a criterion for evaluating the quality of tomato planning matrix. The flow chart of discrete BBO for solving the problem of tomato planting planning is shown as Fig. 1.

4 Illustrative Example

4.1 Example Data

In order to verify the effectiveness of the discrete BBO and rationality of tomato planting planning model, 8 tomato varieties and 2 production lines of tomato paste factory is analyzed as an example.

In a Xinjiang tomato ketchup factory, total capacity of 2 production lines is 3 000 tons/day. The production cycle is 90 days. And it has 600 pieces of tomato planting area. The maturity period, tomato red pigment content and vitamin content of 8 kinds tomato yield can be seen in Refs. [14, 15]. The requirement on lycopene and vitamin content of 1 production line is 30 and 40 mg/100 g, and the 2 production line is 20 and 15 mg/100 g [15]. The example with 8 tomato varieties and 600 pieces of land was used to simulate.

And the tomato planting planning scheme of data obtained from 30 independent experiments and the results were shown in Table 1. In order to verify the effectiveness of discrete BBO, Fig. 2 given the average fitness curves of discrete BBO and real coding BBO. And Fig. 3 is tomato planting planning gantt chart, which is optimized by discrete BBO. Figure 4 is partial enlargement of tomato planting planning gantt chart. The number or color of the labeled data in the map is the number of varieties planted on the corresponding block number. The comparison between the forecast yield and the expected yield is shown in Fig. 5.

It can be seen from Fig. 2 that in solving tomato planting planning problem, the discrete BBO can obtain the optimal value after 500 iterations, and BBO can tend to the optimal value slowly after 1500 iterations. The convergence effect and convergence rate of the discrete BBO algorithm are better than the real coding BBO algorithm.

During the processing of tomato, the traditional planning method can't meet the requirements of the increasing demand for the balanced supply of raw materials. The Table 1 and Fig. 5 were shown that the discrete BBO algorithm is obviously able to achieve the balanced supply of tomato raw materials, and obtain a reasonable tomato

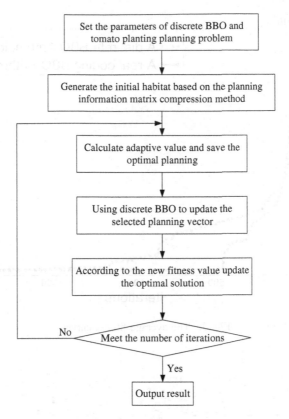

Fig. 1. The flow chart of discrete BBO algorithm for solving the problem of tomato planting planning

Table 1. Tomato planting planning data results

Discrete	Fitness value	Tomato planting area(mu)							
		1#tomato	2#tomato	3#tomato	4#tomato	5#tomato	6#tomato	7#tomato	8#tomato
The best value	5.42e+6	3784.6	6783.6	6783.6	5516.6	5606.6	7502.6	7137.6	3630.1
The mean value	6.15e+6	3784.6	6315.6	6222.6	6645.6	6070.6	6783.6	8152.1	2770.6
The worst value	6.54e+6	3393.6	6440.6	5606.6	6952.6	5606.6	6878.6	8217.1	3649.6

planting planning result than LS-SVM prediction method in Ref. [14], which is other method for solving prediction of Xinjiang tomato yield. At the same time, it is proved that the discrete BBO can better solve the large scale tomato planting planning problem than simplex method in Ref. [15], which is a other method our team proposed for tomato planting programming.

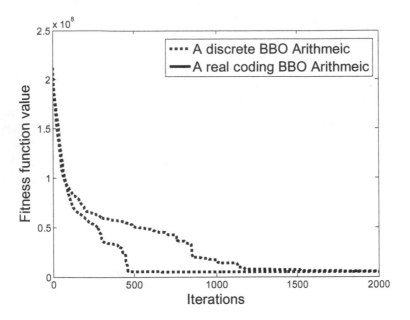

Fig. 2. The average fitness curves

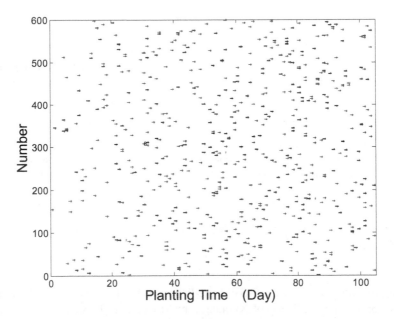

Fig. 3. Tomato planting planning gantt chart

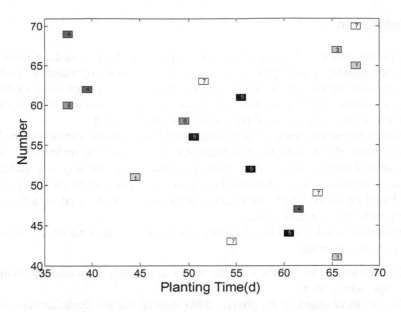

Fig. 4. Partial enlargement of tomato planting planning gantt chart

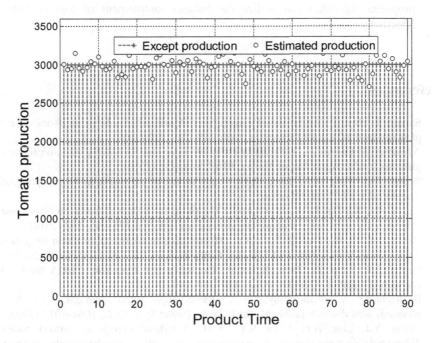

Fig. 5. The tomato production results obtained based on discrete BBO

5 Conclusion

This paper has established the large-scale tomato planting planning model. Considering the tomato planting planning is a large-scale combinatorial optimization problem, tomato planting matrix can be compressed by sparse matrix compression method to achieve compression of the solution space. And a new kind discrete BBO algorithm with a new coding way was used for solving planting planning.

Through the improvement of the position, speed and operator variation of the basic BBO algorithm, the discrete BBO is established to improve the performance of the optimization algorithm. And the dimension of the planning model is greatly reduced by using the compression matrix method. Finally, a large-scale tomato planting planning model based on discrete BBO algorithm is proposed and to be used on a factory in Xinjiang to verify its effectiveness.

Through the simulation of the planning problem of 8 tomato varieties, the following conclusions are obtained.

(1) The efficiency and accuracy of the model are greatly improved by using the compression matrix.
(2) The results obtained by the discrete BBO showed that the deviation between the forecast yield and expected yield are very small, which can achieve the balanced supply of period materials basically.
(3) The simulation results showed that tomato planning scheme calculated by the proposed algorithm can realize the balance supplement of tomato materials effectively.

References

1. Sarker, R., Ray, T.: An improved evolutionary algorithm for solving multi-objective crop planning models. Comput. Electron. Agric. **68**, 191–199 (2009)
2. Adeyemo, J., Otieno, F.: Differential evolution algorithm for solving multi-objective crop planning model. Agric. Water Manag. **97**(6), 848–856 (2010)
3. Wang, C.R., Wang, N.N., Duan, X.D., et al.: Survey of biogeography-based optimization. Comput. Sci. **37**(7), 34–38 (2010)
4. Xu, Y., Li, K., Hu, J., et al.: A genetic algorithm for task scheduling on heterogeneous computing systems using multiple priority queues. Inf. Sci. **270**, 255–287 (2014)
5. Ouyang, A., Li, K., Truong, T.K., et al.: Hybrid particle swarm optimization for parameter estimation of Muskingum model. Neural Comput. Appl. **25**(7–8), 1785–1799 (2015)
6. Ouyang, A., Tang, Z., Zhou, X., et al.: Parallel hybrid PSO with CUDA for LD heat conduction equation. Comput. Fluids **110**, 198–210 (2015)
7. Bhattacharya, A., Chattopadhyay, P.K.: Biogeography-based optimization for different economic load dispatch problems. IEEE Trans. Power Syst. **25**(2), 1064–1077 (2010)
8. Zheng, Y.J., Ling, H.F., Chen, S.Y., et al.: A hybrid neuro-fuzzy network based on differential biogeography-based optimization for online population classification in earthquakes. IEEE Trans. Fuzzy Syst. **23**(4), 1070–1083 (2015)

9. Zhu, W.R, Duan, H.B.: Chaotic biogeography-based optimization approach to receding horizon control for multiple UAVs formation flight. In: IFAC-Papers OnLine, vol. 48, no. 5, pp. 35–40 (2015)

10. Simon, D.: Biogeography-based optimization. IEEE Trans. Evol. Comput. **12**(6), 702–713 (2008)

11. Ma, H.P., Li, X., Lin, S.D.: Analysis of migration rate models for biogeography based optimization. J. South East Univ. (Natural Science Edition) **39**(1), 16–21 (2009)

12. Ma, H.P., Simon, D., Fei, M., et al.: Variations of biogeography-based optimization and Markov analysis. Inf. Sci. **220**, 492–506 (2013)

13. Mu, Y.C.: The application of genetic algorithm in the traveling salesperson problem. J. Tianjin Normal Univ., Tianjin (2004)

14. Gao, B.P., Jiang, B., Nan, X.Y.: The on-line prediction of tomato yield based on LS-SVM. Hubei Agric. Sci. **51**(5), 1025–1027 (2012)

15. Chen, F.F., Jiang, B.: Tomato planting programming using simplex method. Chin. Agric. Sci. Bull. **27**(25), 256–260 (2011)

A Multi-agent Approach for the Newsvendor Problem with Word-of-Mouth Marketing Strategies

Feng Li[(⊠)] and Ning Lin[(⊠)]

School of Business Administration, South China University of Technology,
Guangzhou, China
fenglee@scut.edu.cn, 1027781985@qq.com

Abstract. Word-of-mouth (WOM) marketing is increasingly playing an important role in consumers' purchase decision with the development of mobile Internet and various social media APP. We are particularly interested in such a problem as how to make decisions under effects of WOM campaigns? To answer this question, we develop a multi-agent model that emulates WOM or viral marketing process as spread of disease among people. Assume that each "infected" individual will purchase one unit of product. Then, the total "infected" people form the demand of the product, as an input of newsvendor problem. Besides finding the optimal order quantity of newsvendor problem, we also identify the most influential source node for kick off of the WOM marketing. The simulation results reveal that social network and WOM have a great influence on demand and profit of the firm. Even the source node has significant effect on output of WOM marketing. According to our simulation, the closeness centrality in social network analysis is the best measure to recognize the most influential source node, comparing to degree centrality, or betweenness centrality, etc. Finally, parameter analysis infers that profit of the firm will increase with higher the spreading probability or/and lower the resistant probability.

Keywords: Newsvendor problem · Word of mouth marketing · Multi-agent modeling · SIR model

1 Introduction

The past several years have witnessed an increasingly attention to viral marketing in business. Development of mobile Internet and smartphone APP facilitates information communications among potential and actual consumers. For example, actual customers always post product reviews on the B2C websites to comment products they have purchased. Meanwhile, potential consumers are more likely to follow these advices. Online social media APP make this kind of recommendation more aggressive. People are able to push their opinions to all their friends directly and simultaneously. On the other hand, audiences probably accept these opinions due to their relationship. It is a typical practice of word of mouth (WOM) marketing, which is one of the most efficient marketing strategies. According to [1], 91 % of respondents consult online reviews, blogs, and other user generated content (UGC) before purchasing a new product or service.

© Springer International Publishing Switzerland 2016
D.-S. Huang and K.-H. Jo (Eds.): ICIC 2016, Part II, LNCS 9772, pp. 782–792, 2016.
DOI: 10.1007/978-3-319-42294-7_69

WOM strategies are attractive because people are more likely to accept recommendations from their friends. Moreover, online users don't need to contact with their friends face to face, so the cost for WOM dramatically lower down to near zero. Thus, online WOM marketing becomes one of the most popular advertising strategies in practices. This raises a series of operation decisions, which are seldom targeted before by researchers. For instance, (1) how many products should be ordered with demand influenced by WOM marketing? (2) which user may be the best source node to startup WOM? (3) what kind of additional sales effort could be better ROI (return on investment)?

To address these problems, we develop a multi-agent model that captures dynamics of WOM propagating in social networks. Under the setting of classical single-period newsvendor model, we assume that demand is influenced by WOM practice, instead of an exogenous variable. Based on WOM results, we compute the optimal order quantity for this extended newsvendor problem. Moreover, we examine different WOM initial spreader selection tactics by means of social network analysis. Finally, key parameters are investigated to gain strategic implications for managers.

Our study reveals several important findings. First, effects of WOM campaigns in social networks may be highly uncertain. Thus, companies should carefully balance the risk and benefits. Fortunately, we also find that choosing a specific consumer as the initial information spreader can substantially increase effects of WOM strategies. Thirdly, profits of WOM campaigns rely on such critical factors as information spreading probability and resistant probability. These provide insights how to improve effects of WOM marketing.

The remainder of this paper is organized as follows: Sect. 2 briefly reviews the current literature and the contributions of this study. Section 3 presents our multi-agent model. Then, in Sects. 4 and 5 the numerical results and managerial insights are presented from multi-agent simulation and analysis on the given setting. Section 6 concludes the whole paper.

2 Literature Review

Most significant findings about WOM marketing are derived from empirical research. For examples, studies show motivations that consumers spread WOM autonomously include desire for social interactions, interest in the product and altruism, etc. [2, 3]. Besides personal factors, Wangenheim et al. shows that satisfied customers are more likely to spread WOM information [4]. Meanwhile, unsatisfied consumers would deliver a negative WOM with higher probability than those satisfied consumers [5]. Over- or under-expected experiences lead to high frequency of WOM sharing. On the other hand, different people have different opinions and attitudes towards WOM. For example, personality and expertise of information receivers are significantly correlated to acceptance of WOM [6, 7]. Interestingly, acting rather than as passively audiences, current consumers tend to search for product reviews on the Internet [8]. As a result, WOM plays a more important role in new product diffusion among consumers.

Social contagion models offer an alternative perspective. These models describe WOM spreading process as "viral contagion". Increasing research focused on these models because individual-level communication and contact are well presented, instead of aggregate data only. The following reasons account for this trend: (1) findings in complex system area increase the understanding of topological structure of social network, which is a fundament of WOM spreading dynamics; (2) online social networks and communities make it more available to trace and track process of information diffusion among users. For instance, based on famous Bass model, Goldenberg et al. introduced cellular automata method to model the underlying process of personal communications, which confirmed the so-called "strength of weak ties" theory [9]. That is the influence of weak ties (distant relationships) on the information dissemination is at least as strong as the influence of strong ties (closer relationships) [10].

Generally, researchers analyze WOM spreading for sales prediction. By simulating WOM spread in a small-word network, Garber et al. had predicted new product success in early stage [11]. Toubia et al. proposed to improve aggregate penetration forecasts using individual-level data on social interactions [12]. However, seldom research has combined sales data of WOM strategies with operation problems. Therefore, this study focuses on single-period newsvendor problem considering effects of WOM marketing.

3   The Extended Newsvendor Problem

3.1   The Mathematical Model for WOM Spreading

We refer to a well-accepted model, Susceptible-Infectious-Recovered (SIR) rumor spreading model [13], to describe dynamics process of WOM in a monopoly market. Assume that the potential consumer population is N. At each time step, each individual in the market belongs to one of the three compartments: (1) Ignorant: the number of individuals have not yet heard the WOM information and is susceptible to be informed; (2) Spreader: the number of individuals have accepted the WOM and purchased one unit of product, after that they prefer to re-transmit to their "Ignorant" neighbors; (3) Resistant: the compartment for individuals who had purchased the product, but no intent to transmit WOM to others.

The WOM spreading process evolves as follows. When a spreader meets an ignorant individual, the ignorant will turn into a new spreader with probability λ. However, a spreader becomes resistant with probability α if he encounters another spreader or resistant individual. Since that met spreader or resistant is an actual customer who had already persuaded by WOM to purchase the product, the spreader feels that the WOM information has lost "news value". So, he determines not to diffuse WOM further.

We denote the density of ignorant, spreader and resistant populations as the function of time by $i(t)$, $s(t)$ and $r(t)$. Hence, at any time step the normalization function is:

$$i(t) + s(t) + r(t) = 1 \tag{1}$$

Numerous empirical researches approve that most of the real-world online social networks present the "small-world" characteristic [14, 15]. Therefore, this study describes topological structure of the consumer social network by Watts-Strogatz (WS) small-world model [16]. In this social network, the links are undirected. It means node A is capable to contact with B and node B is also capable to contact A if there is a link between node A and B.

According to the mean-field theory, the differential equations for WOM information spreading in the consumer network are as follows:

$$\frac{di(t)}{dt} = -\lambda \langle k \rangle i(t)s(t) \tag{2}$$

$$\frac{ds(t)}{dt} = \lambda \langle k \rangle i(t)s(t) - \alpha \langle k \rangle s(t)[s(t) + r(t)] \tag{3}$$

$$\frac{dr(t)}{dt} = \alpha \langle k \rangle s(t)[s(t) + r(t)] \tag{4}$$

Initially (at time $t = 0$), there is only one spreader in the network, while others are ignorant.

$$i(0) = \frac{N-1}{N}, s(0) = \frac{1}{N}, r(0) = 0 \tag{5}$$

Consequently, the number of spreaders increases because the ignorant transit to spreaders, and then decreases because spreaders change to resistant state. The process will finally stop once all the spreaders became resistant. We defined density of resistant populations at the end of WOM spreading by R_{final}. It's easy to deduce that R_{final} will approach a well-defined limit with the parameters λ and α. In particular, R_{final} is equal to 0.7968 according to formula (7), when $\lambda = \alpha = 1.0$.

$$s(\infty) = 0, i(\infty) + r(\infty) = 1 \tag{6}$$

$$i(0) = \frac{N-1}{N} \approx 1, s(0) = \frac{1}{N} \approx 0$$

$$dr(t) = \frac{\alpha}{\lambda} di(t) - \frac{\alpha}{\lambda i(t)} di(t)$$

$$\Rightarrow \begin{cases} r(\infty) = 1 - e^{-\varepsilon r(\infty)} \\ i(\infty) = e^{\varepsilon(1-i(\infty))} \end{cases}, \varepsilon = 1 + \frac{\lambda}{\alpha} \tag{7}$$

3.2 The Agent-Based Mode for WOM Spreading

Generally, the mean-field model over-simplifies the dynamics process of WOM spreading in the small-world network. For example, it assumes that individuals in the network are homogeneous, and capable to contact others with equal probability. However, this is not the case for most real systems. Usually, individuals only contact with their friends, linked nodes in the social network. Considering this situation, we relax assumption that individual is capable to contact all population with same probability. Assume that each individual can only spread WOM to his linked neighbors of the small-world network. Meantime, the number of neighbors, degree in network terminology, is not uniformly distributed. The common distribution for degree of nodes is Poisson distribution.

According to SIR model and the assumptions mentioned above, we rebuild WOM process by multi-agent modeling approach. In the model, every consumer is represented as an intelligent agent. The evolution proceeds are as follows:

1. Initially, only one spreader agent of WOM is chosen according to the specific tactic. The left $N - 1$ agents stay in ignorant state.
2. At each time step, one of spreaders agents contacts one of linked neighbors randomly.
3. If the contacted neighbor is an ignorant agent, this neighbor becomes a new spreader agent with probability λ. Otherwise, the contacting spreader agent transfers into resistant state with probability α.
4. Steps 2 and 3 is repeated until there is no more spreaders agent in the market.

3.3   Newsvendor Model Based on WOM Spreading

The final result of WOM spreading in a small-world network is not a deterministic one as formula (7). Due to input parameters of random variables of λ and α, previous research and simulation data implied that the R_{final}, also the demand for newsvendor problem, is stochastic and bimodal distribution [17, 18]. Given uncertainty of the demand, companies need to decide the optimal order level Q^* and predict expected profits π^* of the product or service. That is a newsvendor problem. Try to find the optimal order level will be a time-consuming project. Therefore, this paper introduces a quadratic interpolation algorithm to search the solution. The pseudo-code of the algorithm is:

```
Order1 := max(Demand); Order2 := median(Demand);
Order3 := min(Demand); F1 := Profit(Order1);
F2 := Profit(Order2); F3 := Profit(Order3);
DO {
  H1 := Order2 - Order1; H2 := Order3 - Order2;
  Q1 := F2 - F1; Q2 := F3 - F2;
  a := (Q2 - Q1) / (H2 + H1);
  b := Q2 + a * H2;
  H3 := -0.5 * b / a;
  x := Order3 + H3; y := Profit(x);
  IF (x < Order2 && y > F2) Order3 := Order2, Order2 :=
    x, F3 := F2, F2 := y;
  IF (x > Order2 && y > F2) Order1 := Order2, Order2 :=
    x, F1 := F2, F2 := y;
  IF (x < Order2 && y <= F2) Order1 := x, F1 := y;
  IF (x > Order2 && y <= F2) Order3 := x, F3 := y;
}
LOOP UNTIL round(Order1) = round(Order2) = round(Order3);
OptimalOrderLevel = round(Order2);
OptimalExpectedProfit = Profit(OptimalOrderLevel);
Return (OptimalOrderLevel,OptimalExpectedProfit).
```

In the algorithm, functions *max()*, *min()*, and *mean()* are to find the maximum, minimum, and mean value of the Demand dataset, while *round()* is to calculate the closest integer of the input parameter. When the algorithm is finished, the optimal order level Q^* and predict expected profits π^* are given by *OptimalOrderLevel* and *OptimalExpectedProfit*, respectively.

4   Demonstration

4.1   Verification of Small-World Social Network

We generate a WS small-world network by Pajek 4.01, which is a free social network analysis software (http://vlado.fmf.uni-lj.si/pub/networks/pajek/). The major indexes of the network are listed in Table 1. The size of the social network is 5000, K (Number of Linked Neighbors on each Side of a Vertex) is 2, with rewiring probability $p = 0.3$. As is shown in Table 1, the average path length of generated WS small-world is slightly greater than E-R random network with the same size, while the average clustering coefficient is much larger. Hence, the generated consumer network is consistent with the small-world characteristic.

Table 1. Indexes of the consumer network.

Network	Average degree	Max degree	Min degree	Average path length	Diameter	Average clustering coefficient
WS small-world	4	10	0	7.435028	16	0.1820269
E-R random	4	–	–	6.143856	–	0.0008000

4.2   The Most Influential Source Node (Initial Spreader)

In order to identify the most influential source node to improve market demand, this study compares several different WOM initial spreader selection tactics by social network analysis method.

Generally, the position of an individual in the social network represents one's social capital. People situate in central sits have better access to information and more opportunities to spread information, thus have greater centrality. In social network analysis, centrality of vertices can be measured by several ways: (1) degree centrality ("Degree"); (2) betweenness centrality ("Betweenness"); (3) closeness centrality ("Closeness"); (4) eigenvector centrality ("Eigenvector"). Based on these measures, we choose the node with the maximum centrality as the initial spreader. If there are more than one nodes with the maximum centrality, we randomly choose one of the nodes. To evaluate WOM effects by these initial spreaders, simulations are repeated 1000 times. The parameters for SIR are $\lambda = \alpha = 1.0$.

As shown in Fig. 1, different policies have significantly different results (P-value << 0.05). The means with 95 % confident intervals are also presented in Fig. 1. Comparing with random source node ("Random"), initial spreaders with great

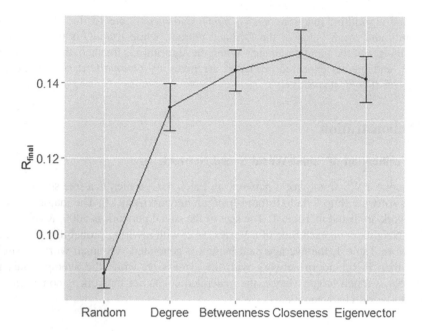

Fig. 1. Mean plot of WOM results with different initial spreader (95 % confident intervals)

centrality affect a wider range of consumer populations. Furthermore, it's counterintuitive that individuals linking with the most neighbors do not necessarily get highest R_{final}. Because degree centrality measures the local structure of a node, but ignores its position in the whole network. On the contrary, consumers who own great closeness centrality play the most important role in WOM spreading. Their released WOM messages are more likely to reach other individuals in the small-world network through "short cuts" rather than limit within clustering groups. The results explain the bridge function of weak ties in information diffusion.

Therefore, effects of WOM strategy can be improved, if companies are able to find out and encourage those consumers with max closeness centrality to spread WOM in social networks. To our knowledge, some companies do try to learn social relationships among potential customers and identifying those "opinion leader" via online surveys or data mining technology.

4.3 The Optimal Order Level

Assume that the firm has persuaded the consumer with greatest closeness centrality to diffuse WOM information. And, the unit sold profit of the product $p = 10$, unit unsold cost of the product $c = 5$. The expected profit π as a function of order level Q is shown in Fig. 2. The optimal order level Q^* and expected profit π^* can be computed by the quadratic interpolation algorithm.

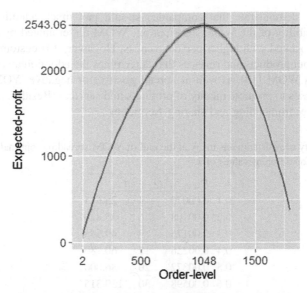

Fig. 2. Expected profit π as a function of order level Q

Repeating the whole procedures for 40 times, we get the final results under 95 % confidence level: the optimal order level $Q^* = 1039 \pm 23$ and expected profit $\pi^* = 2557.378 \pm 150.777$.

Meanwhile, if the firm predicts demand by mean-field approach, the "optimal" order level would be $Q = R_{final} * N = 3984$. In this case, expected profit $\pi = -10505.091$, which decreases from π^* by 510.78 %. Obviously, the loss caused by simplified aggregate level prediction of purchases is great.

In addition, if the firm misunderstands that the product demand follows a normal distribution. According to the value of critical point $p/(p + c) = 0.67$, the order level can be calculated as $Q = 887 \pm 14$, with expected profit $\pi = 2461.414$ decreasing by 3.75 %.

In fact, evaluation of WOM spreading effect exhibit more variance when relaxing the assumptions of mean-field model. The result is more consistent with real situation. Hence, companies should carefully balance the risk and benefits when making operation decision under WOM strategies.

5 Parameter Analysis

5.1 Spread Probability λ

Product demand and expected profits of WOM marketing depends to a great extent on the infected ability of WOM information among potential consumers. Table 2 shows density of resistant consumers at the end of WOM spreading, optimal order level and expected profit for different values of λ. It is seen that the effects of WOM and profits from the WOM campaign increase exponentially as spreading probability λ grows.

The results demonstrate that companies should pay much attention to improve spreading possibility of the WOM information. WOM disseminated by acquaintances is typically associated with higher persuasiveness. However, it is customers' perceived satisfaction about products and services that determines the adoption and willingness to re-transmission WOM [19], as well as spread positive or negative WOM [20]. These remind companies to promote quality of products and services. Besides, some incentive may increase recommending and sharing frequency.

Table 2. Density of resistant consumers at the end of WOM spreading, optimal order level and expected profit for different values of λ.

λ	R_{final}	Q^*	π^*
0.1	0.00070	4	25.615
0.2	0.00108	6	35.820
0.3	0.00175	9	49.545
0.4	0.00236	12	60.180
0.5	0.00372	20	86.000
0.6	0.00598	30	120.315
0.7	0.00992	46	177.250
0.8	0.02447	89	266.945
0.9	0.05484	279	577.965
1.0	0.12449	1039	2557.378

5.2   Resistant Probability α

Resistant mechanism of WOM spreading implies that consumers will stop spreading WOM about products and services, when realizing the information has lost its "news value". Table 3 presents density of resistant consumers at the end of WOM spreading, optimal order level and expected profit for different values of α. As shown, high probability of resistant leads to lower effectiveness of WOM spreading. Thus, order level and expected profit decrease as α grows.

Table 3. Density of resistant consumers at the end of WOM spreading, optimal order level and expected profit for different values of α.

α	R_{final}	Q^*	π^*
0.1	0.99818	4993	49889.735
0.2	0.98121	4924	48926.755
0.3	0.94085	4740	46758.225
0.4	0.86924	4438	42812.065
0.5	0.77729	4033	37855.210
0.6	0.65500	3551	30990.395
0.7	0.51165	2991	22837.995
0.8	0.39643	2433	16884.360
0.9	0.24031	1790	8069.570
1.0	0.12449	1039	2557.378

In order to reduce resistant probability of WOM, managers should take actions to improve the quality of WOM information. Although the content of WOM can not be directly controlled by sellers, it is feasible to provide customers with expertise and attractive information about products or services. Incorporated with these messages, consumer WOM may become more useful and interesting.

6 Conclusion

Popularity of mobile Internet and social media facilitates the personal communication about products and services. This study introduces a multi-agent based model that simulates WOM information spreading in consumer social network. We solve the newsvendor problem for the product demand generated by WOM. Moreover, we estimate different WOM initial spreader selection tactics by social network analysis approach. Finally, key parameters are investigated to gain implications for WOM strategies. The results reveal that demand created by WOM varies greatly. Managers should trade off expected profits and loss of WOM campaigns when make operation decisions. We find that consumers with great closeness centrality can increase the impacts of WOM spreading in the social network. Furthermore, to improve benefits from WOM, measures should be taken to enhance the spreading probability and reduce the resistant probability of spreading.

Acknowledgements. This research was supported by "Guangdong Natural Science Foundation (No. 2014A030313262)".

References

1. Cheung, C.M.K., Thadani, D.R.: The impact of electronic word-of-mouth communication: a literature analysis and integrative model. Decis. Support Syst. **54**, 461–470 (2012)
2. Godes, D., Mayzlin, D.: Using online conversations to study word-of-mouth communication. Mark. Sci. **23**, 545–560 (2004)
3. Hennig-Thurau, T., Gwinner, K.P., Walsh, G., Gremler, D.D.: Electronic word-of-mouth via consumer-opinion platforms: what motivates consumers to articulate themselves on the Internet? J. Interact. Mark. **18**, 38–52 (2004)
4. Wangenheim, F.V., Bayón, T.: The chain from customer satisfaction via word-of-mouth referrals to new customer acquisition. J. Acad. Mark. Sci. **35**, 233–249 (2007)
5. Blodgett, J.G., Granbois, D.H., Walters, R.G.: The effects of perceived justice on complainants' negative word-of-mouth behavior and repatronage intentions. J. Retail. **69**, 399–428 (1993)
6. Laroche, M., Pons, F., Mourali, M.: Individualistic orientation and consumer susceptibility to interpersonal influence. J. Serv. Mark. **19**, 164–173 (2005)
7. Voyer, P.A.: Word-of-mouth processes within a services purchase decision context. J. Serv. Res. **3**, 166–177 (2000)
8. Kulviwat, S., Guo, C., Engchanil, N.: Determinants of online information search: a critical review and assessment. Internet Res. **14**, 245–253 (2004)

9. Goldenberg, J., Libai, B., Muller, E.: Talk of the network: a complex systems look at the underlying process of word-of-mouth. Mark. Lett. **12**, 211–223 (2001)
10. Granovetter, M.S.: The strength of weak ties. Am. J. Sociol. **13**, 1360–1380 (1973)
11. Garber, T., Goldenberg, J., Libai, B., Muller, E.: From density to destiny: using spatial dimension of sales data for early prediction of new product success. Mark. Sci. **23**, 419–428 (2004)
12. Toubia, O., Goldenberg, J., Garcia, R.: Improving penetration forecasts using social interactions data. Manage. Sci. **60**, 3049–3066 (2014)
13. Yamir, M., Maziar, N., Pacheco, A.F.: Dynamics of rumor spreading in complex networks. Phys. Rev. E **69**, 279–307 (2003)
14. Albert, R., Barabasi, A.L.: Statistical mechanics of complex networks. Rev. Mod. Phys. **74**, 47–97 (2002)
15. Yan, Q., Wu, L., Zheng, L.: Social network based microblog user behavior analysis. Phys. A **392**, 1712–1723 (2013)
16. Watts, D.J., Strogatz, S.H.: Collective dynamics of 'small-world' networks. Nature **393**, 440–442 (1998)
17. Zanette, D.H.: Critical behavior of propagation on small-world networks. Phys. Rev. E **64**, 050901 (2001)
18. Zanette, D.H.: Dynamics of rumor propagation on small-world networks. Phys. Rev. E **65**, 041908 (2002)
19. Harrison-Walker, L.J.: The measurement of word-of-mouth communication and an investigation of service quality and customer commitment as potential antecedents. J. Serv. Res. **4**, 60–75 (2001)
20. Jones, M.A., Reynolds, K.E.: The role of retailer interest on shopping behavior. J. Retail. **82**, 115–126 (2006)

Economic Dispatch of Grids Based on Intelligent Coordination Between Electric Vehicle and Photovoltaic Power

Guangqing Bao[1], Weisheng Li[1(✉)], Dunwei Gong[2],
and Jiangwei Mao[3]

[1] College of Electrical Engineering and Information Engineering,
Lanzhou University of Technology, Lanzhou 730050, China
liweishenglws@163.com
[2] School of Information and Electrical Engineering,
China University of Mining and Technology,
Xuzhou 221008, China
[3] Faculty of Management and Economics,
Dalian University of Technology,
Dalian 116024, China

Abstract. There are growing interests in electric vehicles because of environment concern. When a large number of EVs are introduced a power system, there will be extensive impacts on the security and economy of power system operation. Given this background, a dispatch optimization model is built in this paper to mitigate the peak-to-valley ratio of equivalent loads and reduce the active power losses of the distributed grid, based on the driving characteristics of EVs and other demands in power system. Particle swarm optimization algorithm is adopted to solve this optimization problem. The proposed model and algorithm are applied to the IEEE 33-bus test system, and the simulation results indicate the feasibility and efficiency.

Keywords: Particle swarm optimization · Electric vehicles · Dispatch · Peak-to-valley ratio · Active power losses

1 Introduction

With the continued development of renewable energy generation technologies and increasing pressure to combat the global effects of greenhouse warming, electric vehicles (EVs) have received worldwide attention. Photovoltaic (PV) has obtained a wide range of applications with the development of renewable energy techniques. The penetration of PV is also increased in China with about 17.8 GW being installed in 2015, the greatest installation capacity in the world. Because of intermittent and limited predictable nature, PV raises a challenge in maintaining a safe and economic power grid [1]. Electric vehicles (EVs) have been becoming attractive for green transportation to cope with global energy crisis and environmental pollution. Previous statistics have

© Springer International Publishing Switzerland 2016
D.-S. Huang and K.-H. Jo (Eds.): ICIC 2016, Part II, LNCS 9772, pp. 793–804, 2016.
DOI: 10.1007/978-3-319-42294-7_70

shown that China has 120,000 EVs by 2014, with the growth rate of more than 70 % every year. It is expected that the occupancy rate of EVs in China will amount to about 25 % of the global market by 2020. The out-of-order charging of large-scale EVs will have a negative influence on the system losses, the operation costs, and the harmonic pollution to a safe and economic operation of the power grid. According to the survey in [2], a vast majority of EVs are stopped in 96 % of the time each day. By using the V2G technique [3], EVs are the loads of the power system in a state of charging, and are emergent power sources when discharging [4].

In recent years, integrating renewable energy generation with EVs' charging and discharging hasbeen becoming a research focus all over the world. In [5], in consideration of the wind power output distribution of the Jiuquan wind farm, a collaborative dispatching model with the objective to minimize the running costs of micro-grid system was developed. The feasibility and efficiency of accommodating wind power was validated through IEEE 9 node distribution system. By minimizing the output fluctuation of renewable generation and maximizing the income of EV users, a multi-objective optimization model was built in [6]. In this model, the grid-connectable EVs, the wind power generation systems, and the photovoltaic generation systems are simultaneously taken into account. According to the energy management of micro-grids with both EVs and the PV-energy storage, an energy management model was developed in [7] to reduce the operation costs, maintain the load balance, and enhance the service reliability. To analyze the possibility of dispatched EVs' charging to reduce the load fluctuation and increase its capability of accommodating the wind power, a multi-scale synergistic dispatch model was developed in [8]. PV-assisted charging stations for EVs have been becoming a typical integration of consuming renewable resources in [9]. To this end, a variety of countries have been building a great number of demonstration projects, and the converter and the control techniques related to EVs' charging have been verified. However, as the number of EVs increases, it is of necessity to consider reasonable charging/discharging optimization strategies. On one hand, taking the quantity of electricity stored in a battery, the charging/discharging power and the distributed power flow in an optimization strategy into consideration, mitigating the peak-to-valley ratio of equivalent loads and reducing the reserved capacity in adjusting the peak to making full use of renewable energy outputs are of necessity. On the other hand, according to PV generation and load forecasting, decreasing the active power losses of the distributed grid by controlling EVs' charging/discharging power and improving the economy and the security of the distributed network operation are also required. This issue belongs to a combinatorial optimization problem with constraints, and previous optimization algorithms have not solved it. Due to the simplicity of particle swarm optimization (PSO), PSO has been applied to solve various power system problems [10].

This paper proposes a optimal dispatch model to reduce the peak-to-valley ratio of equivalent loads and the active power losses of the distributed grid. To effectively solve the above model, an improved particle swarm optimization algorithm is proposed. The simulation results on the IEEE 33-bus test system show the effectiveness of the proposed model and method.

2 The Proposed Optimal Dispatch Model

EVs have such types of taxi, official cars, buses, and private cars. In this paper, only private cars are considered, and they interact with the power grid by optimizing the charging/discharging power since private cars often have long stopping time and enough quantity of electricity stored in a battery.

2.1 The Postulated Conditions of the Model

The postulated conditions of the model are as follows:

(1) An intelligent charging pile provides charging/discharging service for EVs that can be adjusted according to requirement.
(2) Only the optimization of the charging/discharging power and time are focused on.
(3) EVs are uniformly distributed among nodes.
(4) The communication between EVs and the distributed network is real-time.
(5) EVs connecting with the grid can be dispatched all the day.
(6) Equivalent loads and the EVs' charging and discharging power remain unchanged at each period.
(7) EVs' daily mileage obeys the lognormal distribution, and time in stopping the last travel obeys the normal distribution.

2.2 The Objective Functions

Mitigating the peak-to-valley ratio of equivalent loads and reducing the active power losses of the distributed grid are of considerable concern of the power system. On the premise of normally operating the system, optimizing the EVs' charging and discharging power is of importance to mitigate the peak-to-valley ratio, reduce the losses of the distributed grid, and improve the economy and the security of the distributed network operation.

The mean square deviation of the loads reflects the fluctuation of these loads, and the smaller the mean square deviation, the smaller the fluctuation. In this paper, the charging and discharging power of EVs aggregation under each node is regarded as the optimized variables. The optimal solution of the optimization problem should minimize the sum of the mean square deviations of equivalent daily loads, i.e.,

$$\min F_1 = \sum_{j=1}^{24} (P_{Lj} - P_{Sj} - P_{av,j} + \sum_{i=1}^{n} P_{ij})^2 \tag{1}$$

$$P_{av,j} = [\sum_{j=1}^{24} (P_{Lj} - P_{Sj})]/24 \tag{2}$$

where P_{Lj}, P_{Sj}, P_{avj}, and P_{ij} denote the active load of the system, the PV output, the average equivalent load, and the charging and discharging power of EVs of node i at the j_{th} time period, respectively.

The optimal solution of the optimization problem should minimize the active losses of the distributed system in a day, i.e.,

$$\min F_2 = \sum_{i=1}^{24} [\sum_{j=1}^{n} R_j(I_{aj}^2 + I_{bj}^2)]\Delta t \tag{3}$$

where R_j represents the resistance of branch j, $I_{aj} + jI_{bj}$ means the current of branch j, n is the total number of branches, Δt denotes the time interval.

The linear weighted method is adopted to convert the above two objectives into a single objective which is given as follows:

$$\min F = \gamma_1(F_1/F_{1\max}) + \gamma_2(F_2/F_{2\max}) \tag{4}$$

where $F_{1\max}$ denotes the sum of the mean square deviations of original equivalent loads, $F_{2\max}$ represents the original losses of the distributed system, γ_1, γ_2 refer to the weight coefficients of F_1 and F_2, respectively. The study in [11] has shown that the same value of these coefficients can achieve the comprehensive optimization, therefore, $\gamma_1 = \gamma_2 = 0.5$ are set in this paper.

2.3   The Constraints

The capacity of EVs' batteries should satisfy that

$$\Delta C_{imin} \leq \Delta C_{ij} \leq \Delta C_{imax} \tag{5}$$

where ΔC_{imin}, ΔC_{ij}, and ΔC_{imax} denote the minimal capacity of EVs' batteries, the capacity of EVs' batteries at period j, and the maximal capacity of EVs' batteries, respectively.

The charging/discharging power should satisfy that

$$P_{imin} \leq P_{ij} \leq P_{imax} \tag{6}$$

where P_{imin} represents the minimal power of charging and discharging EVs, P_{imax} means the maximal power of charging and discharging EVs.

The node voltage amplitude should satisfy that

$$U_{min} \leq U_i \leq U_{max} \tag{7}$$

where U_i is the voltage of node i, U_{min} and U_{max} are the minimal and the maximal voltages of each node, respectively.

The node power balance constraint is provided as

$$P_{Gi} + P_{Si} - P_{Li} - P_{EV,i} = U_i \sum_{j=1}^{n} U_j (G_{ij} \cos \delta_{ij} + B_{ij} \sin \delta_{ij}) \tag{8}$$

$$Q_{Gi} + Q_{Sj} - Q_{Li} - Q_{EV,i} = U_i \sum_{j=1}^{n} U_j (G_{ij} \sin \delta_{ij} - B_{ij} \cos \delta_{ij}) \tag{9}$$

where P_{Gi} and Q_{Gi} are the injected active and the reactive power from the external power grid, respectively, P_{Si} and Q_{Si} denote the active and the reactive power from the PV system, P_{Li} and Q_{Li} refer to the active and the reactive loads, $P_{EV,i}$ and $Q_{EV,i}$ denote the changing/discharging active and reactive power of EVs, U_i and U_j are the voltages of the first and the last nodes of each branch, G_{ij} and B_{ij} mean the real and the imaginary parts of the admittance of branch $i - j$, and δ_{ij} denotes the power factor angle of branch $i - j$.

3 The Optimization Algorithm

3.1 The Modified Particle Swarm Optimization Algorithm

The model formulated in the previous section is a multi-objective optimization model with constraints, and classical optimization algorithms, e.g., the simplex method, the polynomial algorithm, and the interior-point method, to say a few, have difficulties in solving it. PSO has been widely used in solving a great variety of nonlinear combinatorial optimization problems owning to its advantages of versatility, simple principle, and easy implementation.

The study in [12] has shown that PSO is apt to mature convergence when solving high- dimensional optimization problems. To overcome the drawback, a number of particles are selected with the probability of P_c in each iteration and put into a hybrid pool whose size is determined by a parameter, S_p. The offspring particles are generated by randomly blending two parent particles. In this way, offspring particles inherit the advantages of their parents', strength the capability in exploitation, and improve the search results. The position of an offspring particle is updated as follows:

$$\begin{cases} child_1(x) = \rho \cdot pt_1(x) + (1 - \rho) \cdot pt_2(x) \\ child_2(x) = (1 - \rho) \cdot pt_1(x) + \rho \cdot pt_2(x) \end{cases} \tag{10}$$

where $child(x)$ and $pt(x)$ are the positions of the offspring and the parent particles, respectively.

The speed of an offspring particle is provided as follows:

$$\begin{cases} child_1(v) = \dfrac{pt_1(v) + pt_2(v)}{|pt_1(v) + pt_2(v)|} |pt_1(x)| \\ child_2(v) = \dfrac{pt_1(v) + pt_2(v)}{|pt_1(v) + pt_2(v)|} |pt_2(x)| \end{cases} \tag{11}$$

where $child(v)$ and $pt(v)$ represent the speed of the offspring and the parent particles, respectively.

3.2   Algorithm Process

The process of tackling the optimization problem considered in this paper is provided as follows.

Step 1:   Set the values of parameters used in the proposed algorithm, and initialize a particle swarm.

Step 2:   Modify the position of a particle according to constraint condition.

Step 3:   Calculate the fitness of a particle, and store the individual and the global optimal position.

Step 4:   Update the position and velocity of a particle.

Step 5:   Select randomly some particle from swarm and produce the offspring particle.

Step 6:   Determine the position and velocity of a offspring particle and replace its parent particle.

Step 7:   Calculate the fitness of a particle, and update the individual and global optimal fitness.

Step 8:   Judge whether the condition of stopping the algorithm is met. If yes, output the optimal results; otherwise, go to Step 4.

4   A Case Study

The IEEE 33-bus benchmark system is employed to verify the proposed model and algorithm, whose initial topology is depicted in Fig. 1. Node 1 is connected with the infinite power grid, and regarded as the slack bus because its voltage amplitude is constant, and the PU value of its voltage amplitude is 1.05. The values of the other parameters are set as follows. The voltage reference value $U_B = 12.66$ kV, and the power reference value $S_B = 10$ MVA. For the peak load, the total active load is 3715 kW, and the total reactive load is 2300 kvar.

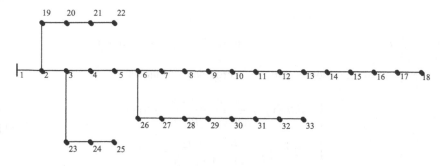

Fig. 1. IEEE 33-bus benchmark system

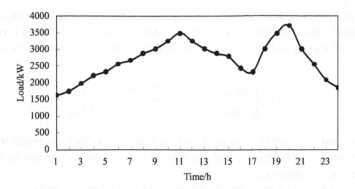

Fig. 2. The typical load curve of this case

The typical load data of this case come from [13], shown as Fig. 2.

4.1 The Parameter Settings

(1) The parameter settings and the assumptions of EVs

In [14], this region has 929 subscribers and approximate 4000 residents. In this paper, the number of EVs is assumed to be 200. The number of EVs under each node of the distributed network is equal. The EVs in running are Nissan Altra in this system. According to industry standards [15], the nominal capacity of a battery is 32.78 kkW·h, and the charging/discharging power is 6.5 kW. In addition, the energy consumption per kilometer is 0.142 kWh/km, and the power factor of an intelligent charging and discharging machine is 0.95. The penetration of EVs refers to the ratio of the number of scheduling EVs to the total number.

Table 1. The average values of intensity parameters at each time period

Time	Parameters (W/m²)	Time	Parameters (W/m²)
07:00	30	14:00	790
08:00	150	15:00	740
09:00	410	16:00	580
10:00	530	17:00	460
11:00	680	18:00	120
12:00	860	19:00	20
13:00	890		

(2) The parameter settings and the assumptions of PV generation systems

The PV array area of switching in each node is 150 m², and the photoelectric conversion efficiency is 15 %. In this paper, the reactive output of a PV system accounts for 10 % of the active output. The average values of intensity parameters at each time period are shown in Table 1.

(3) The parameter settings and the assumptions of the proposed PSO algorithm

To balance the capabilities between the global and the local search of an PSO algorithm, the inertia weight coefficient is nonlinear and dynamic, shown as follows:

$$\omega = \begin{cases} \omega_{\min} + \frac{(\omega_{\max} - \omega_{\min})(f - f_{\min})}{f_{avg} - f_{\min}}, & f \leq f_{avg} \\ \omega_{\max}, & f \leq f_{avg} \end{cases} \tag{12}$$

where $\omega_{\max}$ and $\omega_{\min}$ are the maximal and the minimal coefficients respectively, f denotes the fitness of a particle, f_{avg} and $f_{\min}$ refer to the average and the minimal fitness of all the particles, respectively.

To make a particle have a larger self- learning ability and a smaller social learning ability in the initial optimization stage, and a smaller self-learning ability and a larger social learning ability in the later period, the following asynchronously changing learning factors are adopted

$$\begin{cases} c1 = c1_0 + \dfrac{c1_1 - c1_0}{t_{\max}} t \\ c2 = c2_0 + \dfrac{c2_1 - c2_0}{t_{\max}} t \end{cases} \tag{13}$$

where $c1_0$ and $c2_0$ are the initial values of $c1$ and $c2$, respectively, $c1_1$ and $c2_1$ mean the final values of $c1$ and $c2$ respectively, $t_{\max}$ denotes the largest number of iterations, and t is the number of iterations.

The swarm size is set to 40, and the value of ρ is randomly chosen in the range of [0, 1]. The range of a particle's velocity is [−1, 1], and the other parameters are set to $P_c = 0.9$, $S_p = 0.2$, $\omega_{\max} = 0.8$, $\omega_{\min} = 0.4$, $F_{1\max} = 6.2856 \text{ MW}^2$, and $F_{2\max} = 2.449 \text{ kW·h}$. In addition, the values of the learning factors are chosen as follows: $c1_0 = 2.5$, $c2_0 = 0.5$, $c1_1 = 0.5$, $c2_1 = 2.5$.

4.2 The Simulation Results

The curves of the equivalent loads with 50 % and 100 % penetration are given in Fig. 3.

Figure 3 demonstrates that the load fluctuation is large, and the random charging load of EVs is increased, deteriorating the load curve of the system. Under the condition of the optimization, EVs charge from the power grid at the period of the valley loads, and discharge to the power grid at the period of the perk loads so as to effectively reduce the peak-to-valley ratio. As the penetration of EVs increases, the effect of EVs to mitigate the peak-to-valley ratio of equivalent loads will be obvious, reducing the reserved capacity in adjusting the peak and enhancing the economy and the stability of the system.

The optimized charging and discharging power curve of a typical node with 100 % penetration is depicted in Fig. 4. It can be observed from this figure that EVs frequently charge at 1:00–07:00, 15:00–17:00 and 23:00–24:00, and discharge at 18:00–22:00,

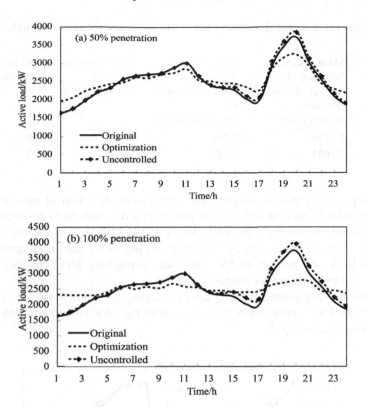

Fig. 3. The curves of the equivalent loads with 50 % and 100 % penetration

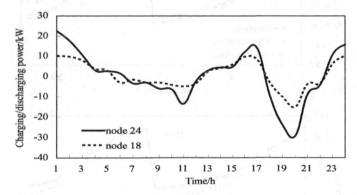

Fig. 4. The optimized charging and discharging power curve of a typical node with 100 % penetration

suggesting a important role of shifting peak loads on the lag, and avoiding additional charging loads to the power grid at rush hours.

The Table 2 reports that optimized charging and discharging power of EVs can reduce mean square deviation of equivalent loads. Comparing optimized charging and

Table 2. The mean square deviation of equivalent loads and the active network loss of the system

Penetration	Mean square deviation of equivalent loads of randomly charging/MW^2	Mean square deviation of equivalent loads via optimization/MW^2	Active losses of randomly charging/MW·h	Active losses via optimization/MW·h
50 %	6.7724	2.3047	2.495	2.438
100 %	7.1633	0.4476	2.553	2.452

discharging with randomly charging, the mean square deviation of equivalent loads reduces by 65.9 %, and the active losses reduce 57 kW·h with 50 % penetration. With 100 % penetration, it reduces the active losses of 101 kWh, saving the electric energy of 36865 kW·h and approximately ¥18800 according to the current electrovalence of ¥0.51/kW·h. As the number of EVs increases, dispatching EVs' charging and discharging will produce greater economic benefits.

Further more, optimized charging and discharging can improve the level of the bus voltage. With 100 % penetration, the amplitude of the node voltage between 1:00 and 20:00 is depicted in Fig. 5.

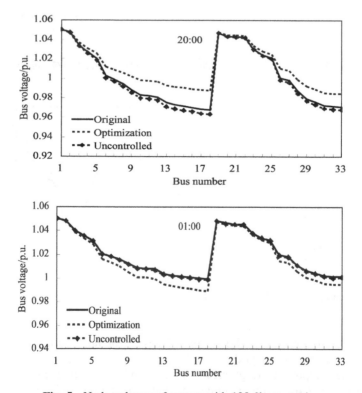

Fig. 5. Node voltages of system with 100 % penetration

Table 3. The comparison between algorithms

Solving algorithm	Standard particle swarm algorithm	Modified particle swarm algorithm
Average fitness	0.7742	0.7631

At 01:00, the minimal voltage amplitude of randomly charging is higher than that of optimized charging and discharging, since the power grid dispatches EVs to charge at the moment, introducing larger charging loads and dropping the node voltage. But the PU value of the minimal voltage amplitude is 0.9892. During peak hour 20:00, the EVs discharge power to the system under the condition of optimization. The minimal amplitude of the bus voltage is 0.9876, where as that of randomly charging is 0.9634. The minimal amplitude of the node voltage increases by 2.51 %, and dispatching EVs' charging and discharging can improve the voltage amplitude of the terminal node and improve the voltage level of the system.

Table 3 lists the comparison between standard PSO algorithm and modified PSO counterpart. From this table, the modified PSO algorithm has better performances.

5 Conclusion

As the capacity of PV generation systems and penetration of EVs increase, new challenges will be raised to guarantee a safe and economic power grid. In this paper, the photovoltaic generation systems and EVs with the capacity of V2G operation are integrated in the distributed grid. A multi-objective optimization model is built to regulate charging and discharging of EVs, with the purpose of reducing the line losses and balancing the load distribution. The test results on the IEEE 33 node system verify that the proposed model and algorithm can flat the curve of equivalent loads, reduce the reserved capacity in adjusting the peak, optimize the active power losses and provide the voltage support for the system.

Acknowledgment. The work presented in this paper is result of the research project National Natural Science Foundation of China (51267011), partly financed by Ministry of Human Resources and Social Security of the People's Republic of China (1202ZBB136).

References

1. Infield, D., Li, F.: Integrating micro-generation into distribution systems: a review of recent research. In: 2008 IEEE Power and Energy Society General Meeting-Conversion and Delivery of Electrical Energy in the 21st Century, pp. 1–4. IEEE Press, Pittsburgh (2008)
2. Kempton, W., Letendre, S.: Electric vehicles as a new power sources for electric utilities. Transp. Res. Part D Transp. Environ. **2**(3), 157–175 (1997)
3. Kempton, W., Tomic, J.: Vehicle-to-grid power implementation: from stabilizing the grid to supporting large-scale renewable energy. J. Power Sour. **144**(1), 280–294 (2005)
4. Ipakchi, A., Albuyeh, A.: Grid of the future. IEEE Power Energ. Mag. **7**(2), 52–62 (2010)

5. Bao, G.Q., Xu, X.: Economic dispatch of micro-grids based on coordination between electric vehicle and wind power. Acta Energiae Solaris Sinica **36**(9), 2300–2306 (2015)
6. Zhang, Z.S., Wen, L.Y., Li, G., Zhang, W.: Multi-objective coordinated scheduling of electric vehicles and renewable generation based on improved chemical reaction optimization algorithm. Power Syst. Technol. **38**(3), 634–637 (2014)
7. Su, S., Jiang, X.C., Wang, W., Jiang, J.C., Agelidis, V.G., Geng, Q.: Optimal energy management for microgrids considering electric vehicles and photovoltaic-energy storage. Autom. Electr. Power Syst. **39**(9), 164–171 (2015)
8. Yu, D.Y., Song, S.G., Zhang, B., Han, X.S.: Synergistic dispatch of PEVs charging and wind power in Chinese regional power grids. Autom. Electr. Power Syst. **35**(14), 24–29 (2011)
9. Lu, X.Y., Liu, N., Chen, Z., Zhang, J.H., Xiao, X.N.: Multi-objective optimal scheduling for PV-assisted charging station of electric vehicles. Trans. Chin. Electrotech. Soc. **29**(8), 46–56 (2014)
10. Del, V.Y., Venayagamoorthy, G.K., Mohagheghi, S.: Particle swarm optimization: Basic concepts, variants and applications in power system. IEEE Tans. Evol. Comput. **12**(2), 171–195 (2008)
11. Dawes, R.W., Corrigan, B.: Linear models in decision making. Psychol. Bull. **181**, 95–106 (1974)
12. Higashi, N., Iba, H.: Particle swarm optimization with gaussian mutation. In: 2013 IEEE Symposium on Swarm Intelligence, pp. 72–79. IEEE Press, Singapore (2013)
13. Yang, X.J., Bai, X.Q., Li, P.J., Wei, H.: Charging optimization of massive electric vehicles in distribution network. Electr. Power Autom. Equip. **35**(6), 31–36 (2015)
14. Zhan, K.Q., Song, Y.H., Hu, Z.C., Xu, Z.W., Jia, L.: Coordination of electric vehicle charging to minimize active power losses. Proc. CSEE **32**(31), 11–18 (2012)
15. Piao, T., Jeevarajan, J., Bragg, B., Zhang, J.: Performance evaluation of lithium ion cell. In: The Battery Conference on Applications and Advances, pp. 101–105. IEEE Press, Long Beach (1999)

Study on Tracking and Detecting Weak Multi-target Based on KF-GMPHDA in Multi-radar Networking

Hai-Long Ding[✉], Wen-Bo Zhao, and Luo-Zheng Zhang

Army Officer Academy of PLA, Hefei 230031, Anhui, China
656797226@qq.com

Abstract. Gaussian mixture probability hypothesis density algorithm (GMPHDA), which is suitable for tracking weak signal to noise ratio (WSNR) multi-target, has rigorous theoretical foundation. The states and number of WSNR multi-target are tracked accurately by GMPHDA application in multi-radar networking, forming KF-GMPHDA. A suite of algorithm about KF-GMPHDA in multi-radar networking is proposed, improving track and detect algorithm in multi-radar networking. Simulation results show that all WSNR multi-target are tracked in multi-radar networking, which gets target tracks corresponding one to one with real targets by the proposed KF-GMPHDA. And then these guarantee higher-up to make full use of track information to acquire real targets states and judge battlefield.

Keywords: KF-GMPHDA · WSNR · Multi-radar networking

1 Introduction

For detecting and tracking echo signal of targets which are flooded in strong false alarm and clutter (Targets in false alarm and clutter area such as stealth aircraft and UAV, which are focused detected by multi-radar networking system [1, 2], are called RNWT in this paper), there are some related research methods such as Hough transform method [3], dynamic programming method [4] and particle filter method [5]. Hough transform method is used for image straight line detection originally. It can be used to track uniform linear motion RNWT. But it is difficult to track uniform linear motion RNWT. Dynamic planning method, based on optimization principle, whose optimization objective function is target measurement range accumulation or likelihood ratio, combining more measurement data (image) to search the possible target trajectory. Dynamic planning method is suitable for detecting and tracking image target or radar target which is based on the signal level detection. Particle filter method, based on bayesian filtering model, uses set of random sampling particles with weights to approximate posterior probability distribution of target motion state. Due to the particle degeneration phenomenon and high load calculation demand, the use of particle filter method is seriously hindered. Probability hypothesis density filter algorithm (PHDA) [6, 8], emerging in recent years, based on random finite set point estimation theory [9] and Bayesian filtering model [10], sets motion state set of multi-targets and measurements of sensors as a stochastic finite set using Bayesian single target tracking method for reference.

© Springer International Publishing Switzerland 2016
D.-S. Huang and K.-H. Jo (Eds.): ICIC 2016, Part II, LNCS 9772, pp. 805–812, 2016.
DOI: 10.1007/978-3-319-42294-7_71

PHDA realizes joint estimation of number and state of targets based on first moment of random finite set. Without of traditional data processing such as track origination, points association, filter estimation and track management, PHDA avoids complex data correlation, which help to solve problem of detecting and tracking RNWT.

2 Probability Hypothesis Density Filter Algorithm

Essential of PHDA is describing the posterior distribution by passing the first moment. Let first moment of multi-target posterior probability distribution as probability hypothesis density (PHD) $D_k(x|Z^k)$ at k moment, the integral of $D_k(x|Z^k)$ in the region S is target number, where D_k means hypothetic probability density of posterior state X_k of each target. D_k contains target state information (a moment) and probability information of target. PHDFA approximates posterior distribution of multi-target through density D_k.

PHD is first moment of random set, and the integral of itself is also target number. PHD can be understood as the expectations of the Dirac delta function. Suppose that δ_x is a special Dirac delta function, the expectation of δ_x is more bigger (even to 1) when independent variable x is closer from true state, more smaller (even to 0) when farther from true state. The expectation of the hypothetic δ_x is PHD D_x. D_x is differential of trust function. It means that D_x expresses minimum probability that x is believed to exist. That is,

$$D_S(\{X\}) = \frac{\delta\beta_S}{\delta X}(X) = \frac{\delta G_S}{\delta X} \tag{1}$$

PHD can be further understood as probability distribution that is not normalized [44]. PHD also can be intuitive understood as: PHD $D_s(X)$ expresses probability (density) that n targets have state $X = \{x_1, x_2 \ldots \ldots x_n\}$ in the region S. $D_s(x)$ means target density at the expectation of x location. The peak point of $D_s(x)$ is state estimation of target. As shown in Fig. 1:

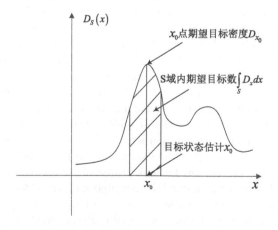

Fig. 1. Probability hypothesis density

PHDFA ignore information of moment which higher than one moment. PHDFA should need to have three conditions at the same time:

Condition 1: Each target evolved independently of each other, cause measurements independently of each other;

Condition 2: False alarm follows Poisson distribution and is independent of measurements caused by targets.

Condition 3: Prediction distribution $f_{k+1|k}(X|Z^k)$ of multi-target set follows Poisson distribution. (If there is no derived target and target state and state noise don't follow Poisson distribution, the hypothesis is easy to meet).

Conditions 1 and 2, are also common in other filter. Condition 3 is unique. PHDFA can iterate independently and avoid the complicated calculation caused by measurements association under three conditions above. The iteration equation of PHDFA is:

$$D_{k+1|k}(X)= \int p_{s,k+1}(\zeta)f_{k+1|k}(X|\zeta)D_k(\zeta)d\zeta + \int \beta_{k+1|k}(X|\zeta)D_k(\zeta)d\zeta + \gamma_{k+1}(X) \quad (2)$$

$$D_{k+1}(X)=\left[1 - p_{D,k+1}(X)\right]D_{k+1|k}(X)$$
$$+ \sum_{Z \in Z_{k+1}} \frac{p_{D,k+1}(X)L_{k+1}(Z|X)D_{k+1|k}(X)}{\kappa_{k+1}(Z) + \int p_{D,k+1}(X)L_{k+1}(Z|X)D_{k+1|k}(X)} \quad (3)$$

Formula (2) is recursive equation of PHD. $D_{k+1|k}(X)$ is predicted PHD. Formula (3) is updating recursive equation of PHD. $D_{k+1|k}(X)$ is updated PHD. X is stochastic finite set of targets state. Z is stochastic finite set of targets measurements. $p_{s,k+1}(\zeta)$ is probability that target with state ζ can survive. $f_{k+1|k}(X|\zeta)$ is probability density of state transition. $\beta_{k+1|k}(X|\zeta)$ is PHD that target with state ζ derive other target at k + 1 moment. $\gamma_{k+1}(X)$ is birth PHD at k + 1 moment. $p_{D,k+1}(X)$ is probability that target X is detected. $L_{k+1}(Z|X)$ is likelihood function. $\kappa_{k+1}(Z)$ is density of clutter false-alarm stochastic finite set.

3   KF-GMPHDA in Multi-radar Networking

Mainly based on PHD, PHDFA is applied to realize the Bayesian recursion when used in multi-radar networking to detect and track target, including predict recursion and PHD recursion. PHD mainly includes survive, spawn, birth and clutter density. Survive density express statistical properties that the target at this moment continue to exist at next moment. Spawn density express statistical properties that the target at this moment spawn new target at next moment, such as fighters firing missiles. Birth density express statistical properties that the new target birth.

This no analytical solution for mostly PHDFA [11]. But Gaussian mixture probability hypothesis density algorithm (GMPHDA) [12–14] is one of PHDFA which is mature and has analytical solution. The Gaussian elements of GMPHDA are predicted and updated through KFA. We use GMPHDA to detect and track RNWT. KFA of GMPHDA is replaced by VOKFA [15], getting Gaussian Mixture probability hypothesis density detect and track algorithm (called KF-GMPHDA).

For the problem that stochastic finite set of RNWT is unknown and complicated, based on condition of linear Gaussian environment, KF-GMPHDA is used to detect and track RNWT. With the help of KF-GMPHDA, we describes probability distribution with first moment (PHD), and interprets probability density recursion of Bayesian filter model as PHD recursion. KF-GMPHDA uses mixture of a limited number of Gaussian functions to realize integral of Bayesian probability density, solving the problem of detecting and tracking RNWT with time-varying number and state in multi-radar networking system.

KF-GMPHDA, combining Gaussian elements predict and update methods of VOKFA has two Bayesian filter processes and three auxiliary processes. Where Bayesian filter process contain predict and update process, auxiliary process contain cut merger process, number of targets estimation and state extraction processes.

Step 1 prediction process

Prediction process includes newborn prediction, spawn prediction and survive prediction processes. Newborn prediction is based on priori information such as airport location information. Spawn prediction is based on tactical usage information of part of the survival targets. Survive prediction means that targets continue to exist based on survive probability and transfer density. Prediction of newborn, spawn and survive targets is form of Gaussian mixture. As follows:

$$
\begin{cases}
D_{\gamma,k+1|k}(x) = \sum_{i=1}^{J_{\gamma,k+1}} \omega_{\gamma,k+1}^i N\left(x; F_{\gamma,k}^i m_{\gamma,k}^i, P_{\gamma,k+1}^i\right) \\
D_{\beta,k+1|k}\left(x|m_k^j\right) = \sum_{j=1}^{J_k} \sum_{l=1}^{J_{\beta,k+1}} \omega_k^j \omega_{\beta,k+1}^l N\left(x; F_{\beta,k}^i m_{\beta,k}^i, P_{\beta,k+1|k}^{j,l}\right) \\
D_{S,k+1|k}(x) = p_{S,k+1} \sum_{j=1}^{J_k} \omega_k^j N\left(x; F_k m_k^j, P_{S,k+1|k}^j\right)
\end{cases}
\tag{4}
$$

Step 2 update process

Update process includes two cases, target is undetected and detected. As follows:

$$
\begin{cases}
D_{1-D,k+1}(x; z) = (1 - p_D)D_{k+1|k}(x) \\
D_{D,k+1}(x; z) = \sum_{j=1}^{J_{k+1|k}} \omega_{k+1}^j(z) N\left(x; m_{k+1}^j, P_{k+1}^j\right)
\end{cases}
\tag{5}
$$

Mixing the two cases based on detection probability.

$$
D_{k+1}(x) = (1 - p_D)D_{k+1|k}(x) + \sum_{z \in Z_{k+1}} D_{D,k+1}(x; z)
\tag{5}
$$

Step 3 process of trimming and merging Gaussian elements

Process of trimming and merging Gaussian elements means trimming Gaussian elements which are below a certain threshold and merging Gaussian elements which are close enough. Condition of merging is as follows:

$$\left(m_{k+1}^i - m_{k+1}^j\right)^T \left(P_{k+1}^i\right)^{-1} \left(m_{k+1}^i - m_{k+1}^j\right) \leq U \tag{6}$$

Merging way is as follows:

$$\tilde{m}_{k+1} = \frac{1}{\tilde{\omega}_{k+1}} \sum_{i=1}^l \omega_{k+1}^i m_{k+1}^i \tag{7}$$

Where l is the number of Gaussian elements over weight threshold.

The trimmed and merged Gaussian elements are Gaussian elements that can survive at next moment.

Step 4 target state and number estimation

Finally extracting states of multi-target according to the peaks of posterior probability density distribution, the number of the peaks is estimated number of targets, and the state corresponding to the peak is estimated state $\widehat{X}_{k+1}$.

4 Simulation and Conclusion Analysis

In order to verify the correctness of using KF-GMPHDA in multi-radar networking in this paper, the method of MATLAB software simulation is used to test and verify. The simulation hardware environment is: the Pentium (R) Dual-Core CPU, HP E5200, 2.5 GHz CPU Frequency, 2.00 G of Memory. Simulation scenario is as follows: two networking radar, radar 1 and 2, the distance accuracy of radar 1 is 130 m, azimuth accuracy is $0.3°$ and itching angle accuracy is $0.2°$, configuration location is $[118° \ 29°$ 120 m]. The distance accuracy of radar 2 is 90 m, azimuth accuracy is $0.4°$ and itching angle accuracy is $0.1°$, configuration location is $[117° \ 31° \ 20 \ m]^T$. The configuration location of the fusion is $[117° \ 30° \ 170 \ m]^T$. In order to verify the simplicity of calculation, we assume that two networking radars scans every second, the fusion center receives detection data on time every second, and the whole detection time is 100 s, the detected targets is in strong clutter environment, and the number of targets is unknown, the detected target is with flight level of 6000 m in the whole courses, with two-dimensional detection region $[- \ 20000, 20000] \times [- \ 20000, 20000]$ and survival probability $P_S = 0.99$. The survival targets subject to linear Gaussian distribution, the state transition matrix and the process noise variance are:

$$F_k = \begin{bmatrix} 1 & 0 & \Delta t & 0 \\ 0 & 1 & 0 & \Delta t \\ 0 & 0 & 1 & 0 \\ 0 & 0 & 0 & 1 \end{bmatrix}$$

Where time interval Δt is one second and standard deviation of process noise σ is 5 m/s^2. Birth and spawn targets accord to Gaussian distribution as the form of Gaussian mixture. Where two positions of birth targets are $m_\lambda^1 = [-5000, 15000, 0, 0]^T$ and $m_\lambda^2 = [-1700, 3000, 0, 0]^T$. Target 1 and 2 do linear uniform motion with velocity $V_1 = [150 - 300]^T$ and $V_2 = [320 - 130]^T$. Covariance matrix R of measurement noise refer to literature [16]. The probability P_D that target is detected is 0.98. Clutter obey the Poisson distribution. Its probability hypothesis density is as follows:

$$K_k(z) = \lambda_c V u(z)$$

The region of survival target is $V = 16 \times 10^8$ m^2. Unit clutter of each square is $\lambda_c = (50/V)$ m^2. (50 clutters are produced in observation area every time). u(z) is density of uniform distribution in survival target region.

At 46th second, target 1 spawn target 3, which do linear uniform motion with velocity $V_3 = [-200 -250]^T$. Combining threshold U = 4 in the process of trimming and combining Gaussian elements. The maximum number allowed to anticipate calculating is J = 100. Now we compare five kinds of situations with each other, decomposing every residual covariance S with under rules in this paper. At the beginning of the track, we let target 1 and 2 as birth targets. After tracking, error between track points and target real points is measured with optimal subpattern assignment (OSPA). Upper limit of distance in this simulation is 200 m. In order to overcome the random factors, we repeat simulation for 100 times, getting statistical result of 100 times repeated simulation. The simulation results are as follow figures.

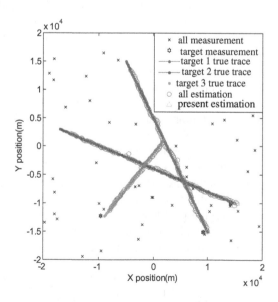

Fig. 2. Whole scene of simulation

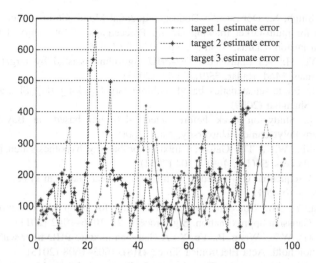

Fig. 3. Error of track points and true points

Figure 2 is whole scene of simulation, where the red 'o' type points are KF-GMPHDA target tracking points. The red and blue '.' type points are rule trace of two survival targets. Black ' × ' type points are measurement points of clutter in current time. '☆' type points are measurement points of three targets in current time. 'Δ' type points are estimation points of KF-GMPHDA in current time. Figure 3 show the distance error between track points and true points. Red '.' type dotted line, black '*' type dotted line and blue '.' type solid line respectively indicate distance error of target 1, 2 and 3 in Fig. 3, where mean square of target 1, 2 and 3 track error is 65.6 m, 75.8 m and 131.0 m respectively. The track traces can approximate to true target trace, achieving the goal of judging the battlefield situation by track trace.

References

1. Xiong, J.L., Xu, H., Han, Z.Z., He, Q., Feng, J.P.: A study on intermittent target tracking technology in fire-control radar network. Mod. Radar **33**(8), 13–16 (2011)
2. Chaomou, Y., Jianjiang, D., Jinjian, L.V.: Resource control function model for radar networking based on modalization. Syst. Eng. Electron. **35**(9), 1979–1982 (2013)
3. Wu, Z.M., Zhang, L., Liu, H., Tian, C.: Centralized 3D track initialization using random hough transformation. Acta Electronica Sinica **41**(5), 840–847 (2013)
4. Daikun, Z., Shouyong, W., Jun, Y.: A multi-frame association dynamic programming track before detect algorithm based on second order Markov target state model. J. Electron. Inf. Technol. **34**(4), 885–890 (2014)
5. Yun, S., Guohong, W., Shuncheng, T.: A TBD algorithm for maneuvering stealthy target based on auxiliary particle filtering. Electron. Opt. Control **20**(7), 28–31 (2013)
6. Stein, M.C., Winter, C.L.: An additive theory of Bayesian evidence accural. Report LA-UR-93-3336, Los Alamos National Laboratories (1993)

7. Vo, B.N., Singh, S., Doucet, A.: Doucet, sequential Monte Carlo implementation of the PHD filter for multi-target tracking [J]. In: Proceedings of International Conference on Information Fusion, Cairns, Australia, pp. 792–799 (2003)

8. Wu, W., Ye, H.: An improved SMC-PHD algorithm assisted by target initiation and continuity rules. Mod. Radar. **35**(9), 24–29 (2003)

9. Zhang, H.J.: Finite-set statistics based multiple target tacking [D]. Shanghai Jiao Tong University, Shanghai (2009)

10. Zhang, H.J.: Study on track before detect algorithm based on Baysian filter [D]. Northwestern Polytechnical University, Xian (2006)

11. Vo, B.N., Ma, W.K.: The Gaussian mixture probability hypothesis density filter. IEEE Trans. Sig. Process. **54**(11), 4091–4104 (2004)

12. Liu, Z.X.: A sequential GM-based PHD filter for a linear Gaussian system. Sci. China Inf. Sci. **56**(10), 1–10 (2013)

13. Li, W., Jia, Y.: The Gaussian mixture PHD filter for jump Markov models based on best-fitting Gaussian approximation. Sig. Process. **91**, 1036–1042 (2011)

14. Liu, Z.X., Xie, W.X., Wang, P., Yu, Y.: A Gaussian mixture PHD filter with the capability of information hold. Acta Electronica Sinica **41**(8), 1603–1608 (2013)

15. Wenbo, Z., Ding Hailong, Q., Chenghua, M.J.: Study on virtual-observation Kalman filter algorithm of multi-radar networking. J. Artillery Acad. **27**(4), 851–858 (2015)

16. Wenbo, Z., Jiyan, D.: Study on statistical properties of radar network noise in inertial coordinate system. J. Artillery Acad. **126**(5), 91–95 (2010)

Study on Important Parameters of Tracking and Detecting RNWT Based on GMPHDA in Radar Networking

Hai-Long Ding[✉], Wen-Bo Zhao, and Guo-Chun Zhu

Army Officer Academy of PLA, Hefei 230031, Anhui, China
656797226@qq.com

Abstract. Gaussian mixture probability hypothesis density filter (PHDF), which is suitable for tracking RNWT (Targets in false alarm and clutter area such as stealth aircraft and UAV), has rigorous mathematical foundation. But the distribution covariance P and truncation threshold T of Gaussian elements in PHDF have not got reasonable calculation rules as yet, which bring bad influence to PHDF. Because the residual covariance S will be inverse calculated when it is involved in the gain calculation. If S is non-positive definite, it would lead to divergence calculation. To determine the P and T calculation rules we do probability statistics derivation. To solve the S calculation problem we do Cholesk and QR decomposition. The simulation compares demonstrate that PHDF in radar networking, using the proposed calculation rules of P, T and S, can precisely track RNWT multi-target, containing exist, birth and spawn targets, bring no extra calculation burden.

Keywords: Radar networking · Gaussian mixture probability hypothesis density filter · Parameter

1 Introduction

Mainly based on PHD (Probability hypothesis density), PHDFA [1, 2] (PHD filter algorithm) is applied to realize the Bayesian recursion when used in radar networking to detect and track target, including predict recursion and PHD recursion.

This no analytical solution for mostly PHDFA. But GMPHDA (Gaussian mixture probability hypothesis density algorithm) [3–5] is one of PHDFA which is mature and has analytical solution. The Gaussian elements of GMPHDA are predicted and updated through KFA. We use GMPHDA to detect and track RNWT (Targets in false alarm and clutter area such as stealth aircraft and UAV, which are focused detected by radar networking system, are called RNWT in this paper) [6]. With the help of GMPHDA, application of engineering practice of PHDFA is realized. But there is no reasonable value for trim threshold T, covariance P_λ, P_β of spawn target distribution yet. They are

The National Nature Science Fund Project 61273001, Anhui Province Nature Science Fund Project 11040606M130.

© Springer International Publishing Switzerland 2016
D.-S. Huang and K.-H. Jo (Eds.): ICIC 2016, Part II, LNCS 9772, pp. 813–824, 2016.
DOI: 10.1007/978-3-319-42294-7_72

always valued based on experience. Useful information is easy to be lost if T is too high, which result in detect and track failure. Excessive Gaussian elements, including a large number of clutter Gaussian elements, will be calculated in the next merging calculation, which increases calculation account and influences the merging calculation precision. We need to calculate inverse matrix S^{-1} of residual covariance during calculating the filter gain K. And non-positive definite matrix S can cause computational divergence. Because T is used to compare with PHD of Gaussian elements, specific value of T is decided by the PHD of Gaussian elements, which according to Gaussian distribution. And PHD is influenced by covariance P of distribution. In this sense, we should first confirm value of covariance P of each Gaussian element distribution, to determine the value of T. The reasonable value rule of P and T is derived through mathematical analysis in this paper. The reasonable calculating rule of residual covariance S is derived through Cholesky and QR decomposition in this paper, solving the problem of divergence during calculating gain. We prove the rationality of P, T value rule and S calculating rule through experiment that RNWT is successful detected and tracked in radar networking, achieving well results.

2 Determine Value Rule of Trim Threshold T

Covariance P_λ, P_β of birth and spawn target distribution is P_λ, P_β. Let σ be their standard deviation, then Gaussian element m obey Gaussian distribution $N(x; m; \sigma^2)$. The probability density is:

$$f = \frac{1}{\sqrt{2\pi}\sigma} e^{\frac{-(x-m)^2}{2\sigma^2}} \tag{1}$$

If PHD of Gaussian elements meet follow formula during trimming Gaussian elements, the Gaussian elements should be trimmed and don't participate in next calculation.

$$D = \omega N\left(x; m, \sigma^2\right) < T \tag{2}$$

We must choose specific value of threshold T to make sure that most of effective Gaussian elements can participate in next calculation. But T cant too low, or else clutter elements with small weight will be introduced to next emerging calculation, which influences precision and increases calculating account. And T cant too high, or else only small number of Gaussian elements can participate in calculation, lost useful information. Translate probability function correspond with probability density f of formula (1) as follow [7]:

$$\frac{1}{\sqrt{2\pi}\sigma} e^{\frac{-(x-m)^2}{2\sigma^2}} \Rightarrow \Phi\left(\frac{x-m}{\sigma}\right) \tag{3}$$

Value of T must ensure that 99.7 % of Gaussian elements can participate in next calculation. So:

$$\Phi\left(\frac{x-m}{\sigma}\right) = 99.7\%$$ (4)

We can get that value range of state x of Gaussian elements is $x \in (m - 3\sigma, m + 3\sigma)$ through checking normal distribution function table. As follow Fig. 1:

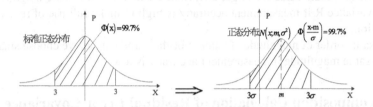

Fig. 1. Range of state value

Bring value of state x into formula (2), so we can ensure determine value rule of trim threshold T as follow:

$$T \approx \omega \frac{1}{\sqrt{2\pi}\sigma} e^{\frac{-9\sigma^2}{2\sigma^2}} \approx \frac{4.4\omega \times 10^{-3}}{\sigma}$$ (5)

3   Determine Value Rule of Variance of Target Distribution

Covariance P_λ, P_β, and P_S of birth, spawn and survival target distribution describe divergence of Gaussian elements near the average, in order to more accurately describe state distribution of real target. Covariance P_S of survival target distribution is associated with process noise and depends on state motion equation.

Predicted covariance of birth, spawn and survival target is:

$$\begin{cases} P^j_{\gamma,k+1|k} = P^i_{\gamma,k+1} \\ P^{j,l}_{\beta,k+1|k} = P^l_{\beta,k} + F^l_{\beta,k} P^j_{\beta,k}\left(F^l_{\beta,k}\right)^T \\ P^j_{S,k+1|k} = Q_k + F_{k-1}P^j_k\left(F_k\right)^T \end{cases}$$ (6)

So predicted covariance contains covariance information (distribution information).

In the process of calculating filter gain, residual covariance which participate in calculate state update gain is:

$$S^j_{k+1} = H_{k+1}P^j_{k+1|k}H^T_{k+1} + R_{radar}$$ (7)

Where predicted covariance $P^j_{k+1/k}$ of Gaussian element is decided by formula (6) and covariance P_λ of target distribution. $P^j_{k+1/k}$ and P_λ have the same magnitude. Based on formula (7), to make full use of measurement and prediction information of target Gaussian distribution, the magnitudes of $P^j_{k+1/k}$ and P_λ are the same after ensuring covariance R of measurement noise. So covariance P_λ and P_β of birth and spawn target distribution should be consistent with covariance R of measurement noise. As result, we should choose small value for target covariance which has the same magnitude with noise covariance R if measurement accuracy is high to make full use of residual error information.

The same order of magnitude of value. For the same reason, we choose small value with the same magnitude if measurement accuracy is low.

4   Decomposition Calculation of Residual Error Covariance

Residual covariance S may lead to computational divergence and filter failure if it is non-positive definite matrix during calculating gain K. To solve this problem, we can do Cholesky and QR decomposition. Because result of Cholesky decomposition is $chol(S) = R$ if $S = R^T*R$. Again we do QR decomposition, getting upper triangle matrix which can be inversed. So S can be approximated as $S = qr(chol(S))^T*qr(chol(S))$.

Let

$$CholS = chol\left(R_k + H_k P^j_{k|k-1} H^T_k\right) \tag{8}$$

Then do QR decomposition for formula (8)

$$QrCholS = qr(CholS) \tag{9}$$

Converting covariance of residual error to formula below
So we can calculate gain below

$$K^j_k = P^j_{k|k-1} H^T_k (QrCholS)^{-1}(QrCholS)^{-T} \tag{10}$$

QrCholS, got from decomposition, is positive definite upper triangular matrix. So we solve the problem of divergence during calculating gain K.

5   Simulation and Conclusion Analysis

In order to verify the correctness of this paper, the method of MATLAB software simulation is used to test and verify. The simulation hardware environment is: the Pentium (R) Dual - Core CPU, HP E5200, 2.5 GHz CPU Frequency, 2.00G of Memory. Simulation scenario is as follows: two networking radar, radar 1 and 2, the distance accuracy of radar 1 is 130 m, azimuth accuracy is 0.3° and itching angle accuracy is 0.2°, configuration location is [118° 29° 120 m]. The distance accuracy of radar 2 is 90 m, azimuth accuracy is 0.4° and itching angle accuracy is 0.1°, configuration location is [117° 31° 20 m]^T. The configuration location of the fusion is [117° 30° 170 m]^T. In order to verify the simplicity of calculation, we assume that two networking radars scans every second, the fusion center receives detection data on time every second, and the whole detection time is 100 s, the detected targets is in strong clutter environment, and the number of targets is unknown, the detected target is with flight level of 6000 meters in the whole courses, with two-dimensional detection region $[-10^5, 10^5] \times [-10^5, 10^5]$ and survival probability $P_S = 0.99$. The survival targets subject to linear gaussian distribution, the state transition matrix and the process noise variance are:

$$F_k = \begin{bmatrix} 1 & 0 & \Delta t & 0 \\ 0 & 1 & 0 & \Delta t \\ 0 & 0 & 1 & 0 \\ 0 & 0 & 0 & 1 \end{bmatrix}, F_k = \sigma^2 \begin{bmatrix} \Delta t^4/4 & 0 & \Delta t^3/2 & 0 \\ 0 & \Delta t^4/4 & 0 & \Delta t^3/2 \\ \Delta t^3/2 & 0 & \Delta t^2 & 0 \\ 0 & \Delta t^3/2 & 0 & \Delta t^2 \end{bmatrix}$$

Where time interval Δt is one second and standard deviation of process noise σ is 5 m/s^2. Birth and spawn targets accord to Gaussian distribution. The form of Gaussian mixture is:

$$\gamma_{k+1}(x) = 0.1N\left(x; m_\gamma^1, P_\gamma\right) + 0.1N\left(x; m_\gamma^2, P_\gamma\right)$$

$$\beta_{k+1|k}(x|\zeta) = 0.05N\left(x; \zeta, Q_\beta\right)$$

Where two positions of birth targets are $m_\lambda^1 = [2500, 2500, 0, 0]^T$ and $m_\lambda^1 = [-2500, -2500, 0, 0]^T$. Target 1 and 2 do linear uniform motion with velocity $V_1 = [25\ -115]^T$ and $V_2 = [115\ -25]^T$. Covariance matrix R of measurement noise refer to literature [8]. The probability P_D that target is detected is 0.98. Clutter obey the Poisson distribution. Its probability hypothesis density is as follows:

$$K_k(z) = \lambda_c V u(z)$$

The region of survival target is $V = 4 \times 10^8$ m^2. Unit clutter of each square is $\lambda_c = (50/V)$ m^2. (50 clutters are produced in observation area every time). u(z) is density of uniform distribution in survival target region.

At 26th second, target 1 spawn target 3, which do linear uniform motion with velocity $V_3 = [-100\ -20]^T$. Combining threshold U = 4 in the process of trimming and combining Gaussian elements. The maximum number allowed to anticipate calculating is J = 100. Now we compare five kinds of situations with each other, decomposing every residual covariance S with under rules in this paper. At the beginning of the track, we let target 1 and 2 as birth targets. After tracking, error between track points and target real points is measured with optimal subpattern assignment (OSPA), which is calculated according to reference [9, 10]. Upper limit of distance in this simulation is 200 m. In order to overcome the random factors, we repeat simulation for 100 times, getting statistical result of 100 times repeated simulation.

(1) Trim threshold T is valued according to experience $T = 10^{-8}$. The value of distribution covariance of birth and spawn target is:

$$P_\gamma = diag([10^5, 10^5, 25, 25]^T)$$

$$P_\beta = diag([10^5, 10^5, 400, 400]^T)$$

In Fig. 2, the red 'o' type points are KF- GMPHDA target tracking points. The red and blue '.' type points are rule trace of two survival targets. Black '×' type points are measurement points of clutter in current time (at the 100th s). '☆' type points are measurement points of three targets in current time. 'Δ' type points are estimation points of GMPHDA in current time. Red 'o' type points are track points of GMPHDA in tracking process. Blue 'o' type points are measurement points of targets. Black '×' type points are clutter of the whole process. Upper subgraph is location map of all points in X direction. Down subgraph is location map of all points in Y direction. Figure 4 show OSPA distance of track and true points. Mean square root of OSPA distance error is 30 m. As we can see in Figs. 2 and 3, we can track all targets by using GMPHDA. As we can see in Fig. 4, because too few Gaussian elements are trimmed, the number of Gaussian elements participate in filter increase, containing amount of clutter elements, which increases calculation burden and influences computational accuracy. The whole calculation time is 185.61 s. Mean square root of OSPA distance error is 30 m.

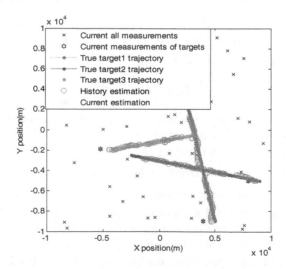

Fig. 2. GMPHDA track, true points and measurement points (Color figure online)

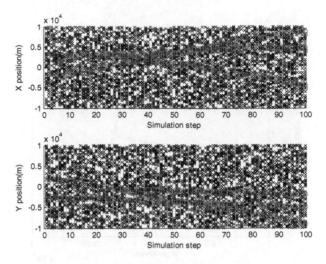

Fig. 3. GMPHDA track and measurement points in X and Y direction (Color figure online)

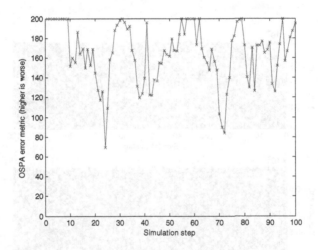

Fig. 4. OSPA distance of track points and true points (OSPA root-mean-square error is 30 m)

(2) Trim threshold is valued based on experience: $T = 10^{-6}$. Covariance of birth and spawn targets distribution:

$$P_\gamma = diag([10^5, 10^5, 25, 25]^T)$$

$$P_\beta = diag([10^5, 10^5, 400, 400]^T)$$

As we can see in Figs. 5 and 6, we can't track survival and spawn targets by using GMPHDA, which lead to tracking failure.

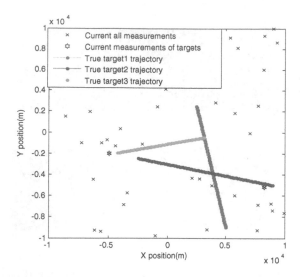

Fig. 5. GMPHDA track, true points and measurement points (Color figure online)

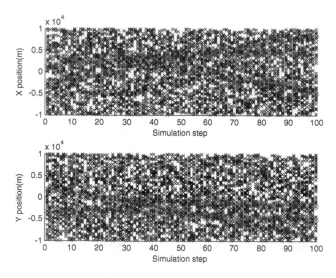

Fig. 6. GMPHDA track and measurement points in X and Y direction (Color figure online)

(3) Trim threshold is valued based on experience: $T = 10^{-7}$. Covariance of birth and spawn targets distribution:

$$P_\gamma = diag([10^6, 10^6, 25, 25]^T)$$

$$P_\beta = diag([10^6, 10^6, 400, 400]^T)$$

Tracking failed in this situation. Tracking effect is as Figs. 5 and 6.

(4) Trim threshold is valued based on experience: $T = 10^{-7}$. Covariance of birth and spawn targets distribution:

$$P_\gamma = diag([10^4, 10^4, 25, 25]^T)$$

$$P_\beta = diag([10^4, 10^4, 400, 400]^T)$$

As we can see in Figs. 7 and 8, if covariance of targets distribution decreases, we can track survival targets after a certain amount of information accumulation, can't track spawn targets.

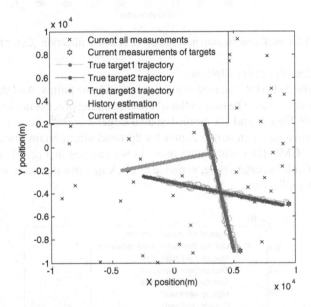

Fig. 7. GMPHDA track, true points and measurement points (Color figure online)

(5) According to value rule of this paper, value of covariance of birth and spawn targets distribution should be consistent with covariance R of virtual measurement noise. Trim threshold T is calculated according to follow formula:

$$P_\gamma = diag([10^5, 10^5, 25, 25]^T)$$

$$P_\beta = diag([10^5, 10^5, 400, 400]^T)$$

$$T = 10^{-7}$$

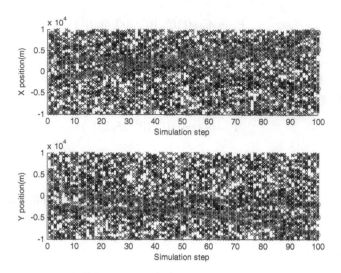

Fig. 8. GMPHDA track and measurement points in X and Y direction (Color figure online)

Tracking effect figure is as follow:

Figure 11 reflects OSPA distance of track points and true points. And the OSPA root-mean-square error is 25 m. The time of the whole calculating process in Matlab is 89.18 s. As we can see in Figs. 9 and 10, target 1 spawn target 3 at 26th second. We can track survival and spawn target in strong clutter background after certain information accumulation, using GMPHDA with value rule. As we can see in Fig. 11, track error of GMPHDA is small. In certain time, variance of tracking error decrease with time, suitable for project application.

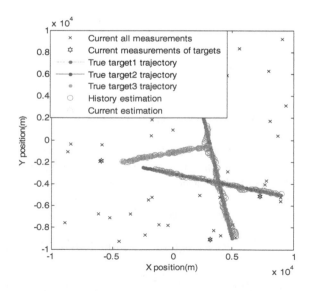

Fig. 9. GMPHDA track, true points and measurement points (Color figure online)

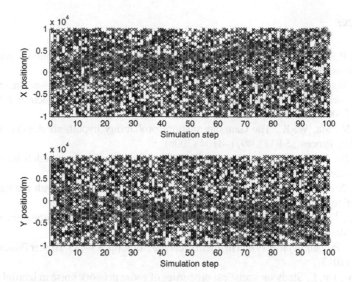

Fig. 10. GMPHDA track and measurement points in X and Y direction (Color figure online)

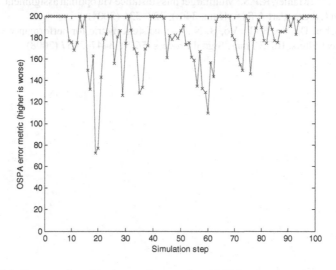

Fig. 11. OSPA distance of track points and true points (OSPA root-mean-square error is 25 m)

Comparison of five kinds of simulation show that we can precisely track birth, spawn and survival targets during tracking RNWT using GMPHDA with value rule of variance P, trim threshold T and covariance S of residual error in this paper. So value rule of GMPHDA parameter in this paper is scientific and rational.

References

1. Mahler, R.P.S., Martin, L.: Multitarget Bayes filtering via first-order multitarget moments. IEEE Trans. Aerosp. Electron. Syst. **39**(4), 1152–1178 (2003)
2. Tan, S., Wang, G., Na, W., et al.: A probability hypothesis density filter and data association based algorithm for multitarget tracking with pulse doppler radar. J. Electron. Inf. Technol. **35**(11), 2700–2706 (2013)
3. Vo, B.-N., Ma, W.-K.: The Gaussian mixture probability hypothesis density filter. IEEE Trans. Sig. Process. **54**(11), 4091–4104 (2006)
4. Li, W., Jia, Y.: The Gaussian mixture PHD filter for jump Markov models based on best-fitting Gaussian approximation. Sig. Process. **91**, 1036–1042 (2011)
5. Liu, Z., Xie, W., Wang, P., et al.: A Gaussian mixture PHD filter with the capability of information hold. Acta Electronica Sinica. **41**(8), 1603–1608 (2013)
6. Zhan, L., Tang, Z., Zhu, Z.: An overview on track-before-detect algorithms for radar weak targets. Mod. Radar **35**(4), 45–52, 57 (2013)
7. Long, Y.: Probability theory and mathematical statistics, pp. 33–113. Higher Education Press, Beijing (2013)
8. Zhao, W., Du, J.: Study on statistical properties of radar network noise in inertial coordinate system. J. Artill. Acad. **126**(5), 91–95 (2013)
9. Hoffman, J.R., Mahler, R.P.S.: Multitarget miss distance via optimal assignment. IEEE Trans. Syst. Man Cybern. Part A Syst. Hum. **34**(3), 327–336 (2004)
10. Schuhmacher, D., Vo, B.-T., Vo, B.-N.: A consistent metric for performance evaluation of multi-object filters. IEEE Trans. Signal Process. **56**(8), 3447–3457 (2008)

Tracking Number Time-Varying Nonlinear Targets Based on SQUF-GMPHDA in Radar Networking

Hai-Long Ding[✉], Wen-Bo Zhao[✉], and Luo-Zheng Zhang[✉]

Army Officer Academy of PLA, Hefei 230031, Anhui, China
656797226@qq.com

Abstract. For the problem that detecting and tracking targets in false alarm and clutter area (RNWT) is under constraints that motion model and measurement model are both nonlinear, firstly square root unscented filter algorithm (SQUFA) is used in nonlinear tracking in multi-radar networking. Then SQUFA is introduced to GMPHDA to form square root unscented filter Gaussian mixture probability hypothesis density filter algorithm (SQUF-GMPHDA), where newborn targets, spawn targets and existing targets are independently sampled, predicted and updated based on SQUFA. So the problem of high-precisely tracking RNWT under nonlinear condition is solved.

Keywords: SQUFA · SQUF-GMPHDA · RNWT

1 Introduction

Gaussian mixture probability hypothesis density algorithm (GMPHDA) [1, 2], which realize analytical solutions of probability hypothesis density algorithm (PHDA) [3, 4] through the form of Kalman Gaussian mixture, can solve the problem of detect and track RNWT (Targets in false alarm and clutter area such as stealth aircraft and UAV, which are focused detected by radar networking system, are called RNWT in this paper) [5, 6]. However, when the target motion model or measuring model is nonlinear, it will be difficult to track RNWT by using GMPHDA. Because of the good performance of unscented filter algorithm (UFA) [7, 8] about tracking nonlinear target, we introduce square root unscented filter algorithm (SQUFA) to PHDA. And similar to GMPHDA, we achieve its analytical solution through mixed Gaussian elements, sampling, forecast and update birth, spawn and survival targets separately. Next we trim and merge the updated Gaussian elements. Finally, we extract target state, getting square root unscented filter Gaussian mixture probability hypothesis density filter algorithm (SQUF-GMPHDA) in multi-radar networking. Simulation about detecting and tracking nonlinear number/state time-varying RNWT using SQUF-GMPHDA in multi-radar networking prove its well performance.

The National Nature Science Fund Project 61273001, Anhui Province Nature Science Fund Project 11040606M130.

D.-S. Huang and K.-H. Jo (Eds.): ICIC 2016, Part II, LNCS 9772, pp. 825–837, 2016.
DOI: 10.1007/978-3-319-42294-7_73

2 Square Root Unscented Filter Algorithm

Unscented filter algorithm is based on UT transform, approximating the probability distribution of the nonlinear state/observation equation through particle collector (Sigma point set), which is deterministically chosen based on the principle of statistical convergence. UFA is formed through a combination of UT and linear minimum variance estimation algorithm. SQUFA means updating Cholesky decomposition factor $S^{ZZ}_{k+1/k}$, $S^{XX}_{k+1/k}$ of predicted state variance $P^{XX}_{k+1/k}$ and predicted observation variance $P^{ZZ}_{k+1/k}$ through QR decomposition and Cholesky factor, using their Cholesky decomposition factor to replace variance of estimation error to participate in interating to avoid divergence caused by Cholesky decomposing P_{k+1}.

3 SQUF-GMPHDA

SQUF-GMPHDA means introducing SQUFA to GMPHDA in multi-radar networking, getting analytical solutions of PHDA through form of SQUF Gaussian mixture to realizing project application. Because UFA is suitable for nonlinear system, we can well detect and track RNWT with nonlinear state and measure model. Using SQUF-GMPHDA to track linear RNWT in multi-radar contains several steps. Firstly, we predict and update Gaussian elements of birth, spawn and survival targets. Secondly, we trim and merge updated Gaussian elements. Finally, we extract state and judge whether there is birth and spawn target according to lasted measurement. Its implementation steps are as follows.

Step 1: Get measurement, precision and location information of netted radar and location information of fusion center at current moment. Location scope of birth target is known. Set ratio corrected parameters λ, α and κ.

Step 2: Predict birth target in detecting area of multi-radar networking. Sample Gaussian elements of birth target in detect area according to ratio correction. Set each Gaussian element as sample mean to get σ point. J means number of Gaussian element. L means number of σ point of every Gaussian element. N means state dimension of tracked target. The distribution variance $P^j_{\gamma,k+1}$ of birth target is known (So we can calculate decomposition factor $S^j_{\gamma,k+1}$ of $P^j_{\gamma,k+1}$).

Do ratio correction

$$\begin{cases} W^m_0 = \dfrac{\lambda}{n+\lambda} \\[2mm] W^c_0 = \dfrac{\lambda}{n+\lambda} - \alpha^2 + \beta + 1 \\[2mm] W^m_l = W^c_l = \dfrac{1}{2(n+\lambda)} \quad l = 1, \cdots, 2n \\[2mm] \lambda = \alpha^2(n+\kappa) - n \end{cases} \qquad (1)$$

Sampled σ point set of Gaussian element j of birth target is

$$
\begin{cases}
\chi^{j,0}_{\gamma,k+1} = m^{j}_{\gamma,k+1} \\
\chi^{j,l}_{\gamma,k+1} = m^{j}_{\gamma,k+1} + \left(S^{j}_{\gamma,k+1}\sqrt{(n+\lambda)} \right)_{l}, l = 1, \cdots, n \\
\chi^{j,l}_{\gamma,k+1} = m^{j}_{\gamma,k+1} - \left(S^{j}_{\gamma,k+1}\sqrt{(n+\lambda)} \right)_{l-n}, l = n+1, \cdots, 2n
\end{cases}
\tag{2}
$$

Calculate statistical properties of σ point set of predicted state of birth target.

$$
\begin{cases}
\chi^{j,l}_{\gamma,k+1|k} = \chi^{j,l}_{\gamma,k+1} \\
m^{j}_{\gamma,k+1/k} = \sum_{l=0}^{2n} W^{m}_{l} \chi^{j,l}_{\gamma,k+1|k}
\end{cases}
\tag{3}
$$

Let

$$
\begin{cases}
e_{0}^{m^{j}_{\gamma,k+1/k}} = \sqrt{|W^{c}_{0}|} \left(\chi^{j,0}_{\gamma,k+1|k} - m^{j}_{\gamma,k+1/k} \right) \\
e_{l}^{m^{j}_{\gamma,k+1/k}} = \sqrt{|W^{c}_{l}|} \left(\chi^{j,l}_{\gamma,k+1|k} - m^{j}_{\gamma,k+1/k} \right) l = 1, 2, \ldots, 2n \\
E^{T}_{\gamma} = \left[e_{1}^{m^{j}_{\gamma,k+1/k}} \quad e_{2}^{m^{j}_{\gamma,k+1/k}} \quad \ldots\ldots \quad e_{2n}^{m^{j}_{\gamma,k+1/k}} \quad \sqrt{Q_{k+1}} \right]^{T}
\end{cases}
\tag{4}
$$

Calculate the Cholesky decomposition factor $S^{XX,j}_{\gamma,k+1/k}$ of predicted state variance $P^{XX,j}_{\gamma,k+1/k}$.

$$
\begin{cases}
\Re^{X}_{\gamma} = \begin{bmatrix} A^{X}_{\gamma} \\ 0 \end{bmatrix} = qr\left(E^{T}_{\gamma} \right) \\
S^{XX,j}_{\gamma,k+1/k} = cholupdate\left(A^{X}_{\gamma}, e_{0}^{m^{j}_{\gamma,k+1/k}}, '+' \right) W^{c}_{0} \geq 0 \\
S^{XX,j}_{\gamma,k+1/k} = cholupdate\left(A^{X}_{\gamma}, e_{0}^{m^{j}_{\gamma,k+1/k}}, '-' \right) W^{c}_{0} < 0
\end{cases}
\tag{5}
$$

Calculate statistical properties of σ point set of predicted measurement of birth target. R_{radar}, covariance of measurement noise, is precision of netted radar. R_{radar} is chose according to radar which produces measurement just at that moment.

$$\begin{cases} \Im_{\gamma,k+1/k}^{j,l} = h\left(\chi_{\gamma,k+1}^{j,l}\right) + V_{k+1}^{j} \\ z_{\gamma,k+1/k}^{j} = \sum_{l=0}^{2n} W_{l}^{m}\Im_{\gamma,k+1|k}^{j,l} \\ P_{\gamma,k+1/k}^{zz,j} = \sum_{l=0}^{2n} W_{l}^{c}\left\{\left(\chi_{k+1/k}^{j,l} - m_{\gamma,k+1/k}^{j}\right)\left(\Im_{\gamma,k+1/k}^{j,l} - z_{\gamma,k+1/k}^{j}\right)^{T}\right\} \end{cases} \tag{6}$$

Calculate the Cholesky decomposition factor $S_{\gamma,k+1/k}^{ZZ,j}$ of predicted measure variance $P_{\gamma,k+1/k}^{ZZ,j}$.

$$\begin{cases} \text{(when } W_{0}^{c} \geq 0)S_{\gamma,k+1/k}^{ZZ,j} = \\ cholupdate\left(A_{\gamma}^{Z}, \sqrt{|W_{0}^{c}|}\left(\Im_{\gamma,k+1/k}^{j,0} - z_{\gamma,k+1/k}^{j}\right), '+'\right) \\ \text{(when } W_{0}^{c} < 0)S_{\gamma,k+1/k}^{ZZ,j} = \\ cholupdate\left(A_{\gamma}^{Z}, \sqrt{|W_{0}^{c}|}\left(\Im_{\gamma,k+1/k}^{j,0} - z_{\gamma,k+1/k}^{j}\right), '-'\right) \end{cases} \tag{7}$$

Calculate weight, gain of birth target Gaussian element and Cholesky decomposition factor S_{k+1} of error variance used in iterative computation. After replacing $\sqrt{(n+\lambda)P_{k+1}}$ with $\sqrt{(n+\lambda)S_{k+1}^{T}}$, we can do the calculation above before getting measurement the present moment.

$$\begin{cases} \omega_{\gamma,k+1/k}^{j} = \omega_{\gamma,k+1}^{j} \\ K_{\gamma,k+1}^{j} = \left(P_{\gamma,k+1/k}^{xz,j}\Big/S_{\gamma,k+1/k}^{ZZ,j}\right)\Big/\left(S_{\gamma,k+1/k}^{ZZ,j}\right)^{T} \\ S_{\gamma,k+1}^{j} = Seqcholupdate(S_{\gamma,k+1/k}^{XX,j}, K_{k}\left(S_{\gamma,k+1/k}^{ZZ,j}\right)^{T}, '-') \end{cases} \tag{8}$$

Step 3: Predict spawn and survival target in detected area of multi-radar networking. Sample σ points of Gaussian element h. One Gaussian element correspond one σ point set.

$$\begin{cases} \chi_{k}^{h,0} = m_{k}^{h} \\ \chi_{k}^{h,l} = m_{k}^{h} + \left(S_{k}^{h}\sqrt{(n+\lambda)}\right)_{l}, l = 1, \cdots, n \\ \chi_{k}^{h,l} = m_{k}^{h} - \left(S_{k}^{h}\sqrt{(n+\lambda)}\right)_{l-n}, l = n+1, \cdots, 2n \end{cases} \tag{9}$$

Calculate statistical feature of σ point set of predicted state of survival target.

$$\begin{cases} \chi_{S,k+1|k}^{h,l} = F_{S,k}\chi_k^{h,l} \\ m_{S,k+1/k}^h = \sum_{l=0}^{2n} W_l^m \chi_{S,k+1|k}^{h,l} \end{cases} \quad (10)$$

Let

$$\begin{cases} e_0^{m_{S,k+1/k}^h} = \sqrt{|W_0^c|}\left(\chi_{S,k+1|k}^{h,0} - m_{S,k+1/k}^h\right) \\ e_l^{m_{S,k+1/k}^h} = \sqrt{|W_l^c|}\left(\chi_{S,k+1|k}^{h,l} - m_{S,k+1/k}^h\right) l = 1,2,\ldots,2n \\ E_S^T = \left[e_1^{m_{S,k+1/k}^h} \; e_2^{m_{S,k+1/k}^h} \; \ldots\ldots e_{2n}^{m_{S,k+1/k}^h} \; \sqrt{Q_{k+1}}\right]^T \end{cases} \quad (11)$$

Calculate the Cholesky decomposition factor $S_{S,k+1/k}^{XX,h}$ of variance $P_{S,k+1/k}^{XX,h}$ of predicted state.

$$\begin{cases} \Re_S^X = \begin{bmatrix} A_S^X \\ 0 \end{bmatrix} = qr(E_S^T) \\ S_{S,k+1/k}^{XX,h} = cholupdate\left(A_S^X, e_0^{m_{S,k+1/k}^h}, '+'\right) \; W_0^c \geq 0 \\ S_{S,k+1/k}^{XX,h} = cholupdate\left(A_S^X, e_0^{m_{S,k+1/k}^h}, '-'\right) \; W_0^c < 0 \end{cases} \quad (12)$$

Calculate statistical feature of σ point set of predicted measurement of survival target.

$$\begin{cases} \Im_{S,k+1/k}^{h,l} = h\left(\chi_{S,k+1/k}^{h,l}\right) + V_{k+1}^h \\ \zeta_{S,k+1/k}^h = \sum_{l=0}^{2n} W_l^m \Im_{S,k+1|k}^{h,l} \\ P_{S,k+1/k}^{(xz),h} = \sum_{l=0}^{2n} W_l^c\left\{\left(\chi_{S,k+1/k}^{h,l} - m_{S,k+1/k}^h\right)\left(\Im_{S,k+1/k}^{h,l} - \zeta_{S,k+1/k}^h\right)^T\right\} \end{cases} \quad (13)$$

Calculate the Cholesky decomposition factor $S_{S,k+1/k}^{ZZ,h}$ of variance $P_{S,k+1/k}^{ZZ,h}$ of predicted state. Let

$$
\begin{cases}
e_0^{zz} = \sqrt{|W_0^c|}\left(\Im_{S,k+1/k}^{h,l} - z_{S,k+1/k}^h\right) \\
e_l^{z_{S,k+1/k}^h} = \sqrt{|W_l^c|}\left(\Im_{S,k+1/k}^{h,l} - z_{S,k+1/k}^h\right) l = 1,2,\ldots,2n
\end{cases}
\tag{14}
$$

Calculate the Cholesky decomposition factor $S_{S,k+1/k}^{ZZ,h}$ of $P_{S,k+1/k}^{ZZ,h}$ through Cholupdate.

$$
\begin{cases}
\Re_S^Z = \begin{bmatrix} A_S^Z \\ 0 \end{bmatrix} = \\
qr\left(\left[e_1^{z_{S,k+1/k}^h} \quad e_2^{z_{S,k+1/k}^h} \quad \cdots\cdots \quad e_{2n}^{z_{S,k+1/k}^h} \quad \sqrt{R_{radar}} \right]^T\right) \\
cholupdate\left(A_S^Z, \sqrt{|W_0^c|}\left(\Im_{S,k+1/k}^{h,0} - z_{S,k+1/k}^h\right), '+'\right) \quad W_0^c \geq 0 \\
S_{S,k+1/k}^{ZZ,h} = \\
cholupdate\left(A_S^Z, \sqrt{|W_0^c|}\left(\Im_{S,k+1/k}^{h,0} - z_{S,k+1/k}^h\right), '-'\right) \quad W_0^c < 0
\end{cases}
\tag{15}
$$

Calculate weight of Gaussian element of survival target, gain and Cholesky decomposition factor $S_{S,k+1}^h$ of error variance used for sampling. After replacing $\sqrt{(n+\lambda)P_{S,k+1}}$ with $\sqrt{(n+\lambda)}\left(S_{S,k+1}^h\right)^T$, we can do the calculation above before getting measurement the present moment.

$$
\begin{cases}
\omega_{S,k+1/k}^j = p_S\omega_{S,k}^j \\
K_{S,k+1}^h = \left(P_{S,k+1/k}^{xz,h} / S_{S,k+1/k}^{ZZ,h}\right) / \left(S_{S,k+1/k}^{ZZ,h}\right)^T \\
S_{S,k+1}^h = Seqcholupdate(S_{S,k+1/k}^{XX,h}, K_k\left(S_{S,k+1/k}^{ZZ,h}\right)^T, '-')
\end{cases}
\tag{16}
$$

Calculate statistical feature of σ point set of predicted state and measurement of spawn target, gain and weight according to the same principle. $\omega_{\beta(Z),k+1}^j$, calculated according to measurement at the present moment, is weight of moment.

if t >= spawn-time

$$
\begin{cases}
\chi_{\beta,k+1|k}^{h,l} = F_{\beta,k}\chi_k^{h,l} \\
m_{\beta,k+1/k}^h = \sum_{l=0}^{2n} W_l^m \chi_{\beta,k+1|k}^{h,l}
\end{cases}
\tag{17}
$$

$$
\begin{cases}
e_0^{m_{\beta,k+1/k}^h} = \sqrt{W_0^c}\left(\chi_{\beta,k+1|k}^{h,0} - m_{\beta,k+1/k}^h\right) \\[2mm]
e_l^{m_{\beta,k+1/k}^h} = \sqrt{W_l^c}\left(\chi_{\beta,k+1|k}^{h,l} - m_{\beta,k+1/k}^h\right) \\[2mm]
l = 1,2,\ldots 2n \\[2mm]
E_\beta^T = \left[e_1^{m_{\beta,k+1/k}^h}\, e_2^{m_{\beta,k+1/k}^h}\ldots\ldots e_{2n}^{m_{\beta,k+1/k}^h}\, \sqrt{Q_k}\right]^T
\end{cases} \tag{18}
$$

Calculate the Cholesky decomposition factor $S_{\beta,k+1/k}^{XX,h}$ of variance $P_{\beta,k+1/k}^{XX,h}$ of predicted state.

$$
\begin{cases}
\Re_\beta^X = \begin{bmatrix} A_\beta^X \\ 0 \end{bmatrix} = qr\left(E_\beta^T\right) \\[3mm]
S_{\beta,k+1/k}^{XX,h} = cholupdate\left(A^X, e_0^{m_{\beta,k+1/k}^h}, '+'\right) \quad W_0^c \geq 0 \\[3mm]
S_{\beta,k+1/k}^{XX,h} = cholupdate\left(A^X, e_0^{m_{\beta,k+1/k}^h}, '-'\right) \quad W_0^c < 0
\end{cases} \tag{19}
$$

Calculate statistical feature of σ point set of predicted state of spawn target.

$$
\begin{cases}
\Im_{\beta,k+1/k}^{h,l} = h\left(\chi_{\beta,k+1/k}^{h,l}\right) + V_{k+1}^j \\[3mm]
z_{\beta,k+1/k}^h = \sum_{l=0}^{2n} W_l^m \Im_{\beta,k+1|k}^{h,l} \\[3mm]
P_{\beta,k+1/k}^{(xz),h} = \sum_{l=0}^{2n} W_l^c\left\{\left(\chi_{\beta,k+1/k}^{h,l} - m_{\beta,k+1/k}^h\right)\left(\Im_{\beta,k+1/k}^{h,l} - z_{\beta,k+1/k}^h\right)^T\right\}
\end{cases} \tag{20}
$$

Calculate the Cholesky decomposition factor $S_{\beta,k+1/k}^{ZZ,h}$ of variance $P_{\beta,k+1/k}^{ZZ,h}$ of predicted measurement.

$$
\begin{cases}
e_0^{zz} = \sqrt{|W_0^c|}\left(\Im_{\beta,k+1/k}^{h,l} - z_{\beta,k+1/k}^h\right) \\[3mm]
e_l^{z_{\beta,k+1/k}^h} = \sqrt{|W_l^c|}\left(\Im_{\beta,k+1/k}^{h,l} - z_{\beta,k+1/k}^h\right) \\[3mm]
l = 1,2,\ldots 2n
\end{cases} \tag{21}
$$

Calculate the Cholesky decomposition factor $S_{\beta,k+1/k}^{ZZ,h}$ of variance $P_{\beta,k+1/k}^{ZZ,h}$ through Cholupdate.

$$
\begin{cases}
\Re_\beta^Z = \begin{bmatrix} A_\beta^Z \\ 0 \end{bmatrix} = qr\left(\left[e_1^{z_{\beta,k+1/k}^h} e_2^{z_{\beta,k+1/k}^h} \cdots e_{2n}^{z_{\beta,k+1/k}^h} \sqrt{R_{radar}} \right]^T\right) \\[2mm]
S_{\beta,k+1/k}^{ZZ,h} = cholupdate \\
\quad \left(A_\beta^Z, \sqrt{|W_0^c|}\left(\Im_{\beta,k+1/k}^{h,0} - z_{\beta,k+1/k}^h \right), '+' \right) \quad W_0^c \ge 0 \\[2mm]
S_{\beta,k+1/k}^{ZZ,h} = cholupdate \\
\quad \left(A_\beta^Z, \sqrt{|W_0^c|}\left(\Im_{\beta,k+1/k}^{h,0} - z_{\beta,k+1/k}^h \right), '-' \right) \quad W_0^c < 0
\end{cases}
\tag{22}
$$

Calculate weight of Gaussian element of spawn target, gain and Cholesky decomposition factor $S_{\beta,k+1}^h$ of error variance used for sampling.

$$
\begin{cases}
\omega_{\beta,k+1/k}^j = \omega_{S,k}^j \omega_{\beta(Z),k+1}^j \\[2mm]
K_{\beta,k+1}^h = \left(P_{\beta,k+1/k}^{xz,h} \big/ S_{\beta,k+1/k}^{ZZ,h} \right) \big/ \left(S_{\beta,k+1/k}^{ZZ,h} \right)^T \\[2mm]
S_{\beta,k+1}^h = Seqcholupdate(S_{\beta,k+1/k}^{XX,h}, K_k \left(S_{\beta,k+1/k}^{ZZ,h} \right)^T, '-')
\end{cases}
\tag{23}
$$

end

Step 4: After getting measurement present moment, update statistical feature of Gaussian element of birth, survival and spawn target.

for each z_{k+1}:

Update birth target

$$
\begin{cases}
m_{\gamma,k+1}^j = m_{\gamma,k+1/k}^j + K_{\gamma,k+1}^j\left(z_{k+1} - z_{\gamma,k+1/k}^j \right) \\[2mm]
\omega_{\gamma,k+1}^j = p_{D,k+1}\omega_{\gamma,k+1/k}^j \\[2mm]
N\left(z_{k+1}; z_{\gamma,k+1/k}^j, \left(S_{\gamma,k+1/k}^{(zz),j} \right)^T S_{\gamma,k+1/k}^{(zz),j} \right)
\end{cases}
\tag{24}
$$

Update spawn target

$$
\begin{cases}
m_{\beta,k+1}^h = m_{\beta,k+1/k}^h + K_{\beta,k+1}^h\left(z_{k+1} - z_{\beta,k+1/k}^h \right) \\[2mm]
\omega_{\beta,k+1}^j = p_{D,k+1}\omega_{\beta,k+1/k}^j \\[2mm]
N\left(z_{k+1}; z_{\beta,k+1/k}^j, \left(S_{\beta,k+1/k}^{(zz),j} \right)^T S_{\beta,k+1/k}^{(zz),j} \right)
\end{cases}
\tag{25}
$$

Update survival target

$$
\begin{cases}
m_{S,k+1}^h = m_{S,k+1/k}^h + K_{S,k+1}^h \left(z_{k+1} - z_{S,k+1/k}^h \right) \\
\omega_{S,k+1}^j = p_{D,k+1} \omega_{S,k+1/k}^j \\
N \left(z_{k+1}; z_{S,k+1/k}^j, \left(S_{S,k+1/k}^{(zz)j} \right)^T S_{S,k+1/k}^{(zz)j} \right)
\end{cases}
\tag{26}
$$

Update weight of each element

$$
\begin{cases}
\omega_{\gamma,k+1}^j = \dfrac{\omega_{\gamma,k+1}^j}{\kappa + \sum\limits_{i=1}^{J} \omega_{\gamma,k+1}^i + \sum\limits_{i=1}^{J} \omega_{\beta,k+1}^i + \sum\limits_{i=1}^{J} \omega_{S,k+1}^i} \\[3mm]
\omega_{\beta,k+1}^j = \dfrac{\omega_{\beta,k+1}^j}{\kappa + \sum\limits_{i=1}^{J} \omega_{\gamma,k+1}^i + \sum\limits_{i=1}^{J} \omega_{\beta,k+1}^i + \sum\limits_{i=1}^{J} \omega_{S,k+1}^i} \\[3mm]
\omega_{S,k+1}^j = \dfrac{\omega_{S,k+1}^j}{\kappa + \sum\limits_{i=1}^{J} \omega_{\gamma,k+1}^i + \sum\limits_{i=1}^{J} \omega_{\beta,k+1}^i + \sum\limits_{i=1}^{J} \omega_{S,k+1}^i}
\end{cases}
\tag{27}
$$

Step 5: After trim and merge updated Gaussian elements, we get peak of probability hypothesis density. Next we extract state combining weight. ω_{k+1}^j is weight of all Gaussian elements. T is threshold of weight. U is threshold of merging elements. W is threshold of extracting state.

If $\omega_{k+1}^j < T$ $(j = 1, 2, \cdots J_{k+1})$
          $D_{k+1}^i = \phi$

elseif $\omega_{k+1}^j > T$ $(j = 1, 2, \cdots J_{k+1})$

   if $\left(m_{k+1}^i - m_{k+1}^j \right)^T \left(P_{k+1}^i \right)^{-1} \left(m_{k+1}^i - m_{k+1}^j \right) \leq U$

          $\widetilde{\omega}_{k+1} = \omega_{k+1}^i + \omega_{k+1}^j$

          $\widetilde{m}_{k+1} = \dfrac{\omega_{k+1}^i}{\widetilde{\omega}_{k+1}} m_{k+1}^i + \dfrac{\omega_{k+1}^j}{\widetilde{\omega}_{k+1}} m_{k+1}^j$

      end

   end

   if $\widetilde{\omega}_{k+1} > W$

   $\widehat{X}_{k+1} = \widetilde{m}_{k+1}$

   end

Step 6: Calculate PHD and judge the existence of birth and spawn target according to the latest measurements from netted radars. One measurement only can correspond one birth or spawn target.

For each z_{k+1}:

$$m_{\gamma,k+1}^i = m_{\beta,k+1}^i = f^{-1}(z_{k+1})$$

$$D_{\gamma,k+1}(x) = N(x; m_{\gamma,k+1}^i, P_{\gamma,k+1}^i)$$

$$D_{\beta,k+1}(x) = \omega_{\beta,k+1}^i N(x; m_{\beta,k+1}^i, P_{\beta,k+1}^i)$$

If $D_{\gamma,k+1}(x) > D_{\beta,k+1}(x)$

$$D_{\beta,k+1}(x) = \phi, \ m_{\beta,k+1}^i = \phi$$

elseif $D_{\gamma,k+1}(x) < D_{\beta,k+1}(x)$

$$D_{\beta,k+1}(x) = \phi, \ m_{\beta,k+1}^i = \phi$$

end

4   Simulation and Conclusion Analysis

To verify effectiveness of detecting and tracking RNWT with number/state time-varying using SQUF-GMPHDA in radar networking. We use the method of MATLAB software simulation to test and verify. The simulation hardware environment is: the Pentium (R) Dual - Core CPU, HP E5200, 2.5 GHz CPU Frequency, 2.00 G of Memory. Simulation scenario is as follows: two networking radar, radar 1 and 2, the distance accuracy of radar 1 is 100 m, azimuth accuracy is 0.2° and itching angle accuracy is 0.3°, configuration location is [118° 29° 120 m]. The distance accuracy of radar 2 is 80 m, azimuth accuracy is 0.3° and itching angle accuracy is 0.2°, configuration location is $[117° \ 29° \ 150 \ m]^T$. The configuration location of the fusion is $[117° \ 31° \ 30 \ m]^T$. In order to verify the simplicity of calculation, we assume that two networking radars scans every second, the fusion center receives detection data on time every second, and the whole detection time is 100 s, the detected targets is in strong clutter environment, and the number of targets is unknown, the detected target is with flight level of 5000 meters in the whole courses, with two-dimensional detection region $[-20000, 20000] \times [-20000, 20000]$ and survival probability $P_S = 0.99$. The detected targets subject to linear Gaussian distribution, the state transition matrix:

$$F_k = \begin{bmatrix} 1 & 0 & \sin(w\Delta t)/w & -(1 - \cos(w\Delta t))/w \\ 0 & 1 & (1 - \cos(w\Delta t))/w & \sin(w\Delta t)/w \\ 0 & 0 & \cos(w\Delta t) & -\sin(w\Delta t) \\ 0 & 0 & \sin(w\Delta t) & \cos(w\Delta t) \end{bmatrix}$$

Where time interval Δt is one second and anticlockwise angular velocity is $\omega = 0.03$ rad/s. Standard deviation of process noise.

σ is 5 m/s². State equation is nonlinear. Birth and spawn targets accord to Gaussian distribution as the form of Gaussian mixture. Where two positions of birth targets are

$m_\lambda^1 = [-9000, 1000, 0, 0]^T$ and $m_\lambda^2 = [-1700, -5000, 0, 0]^T$ at the begin of detection. At 35th second, birth target may appear at range of $m_\lambda^3 = [-10000, -1000, 0, 0]^T$.

Target 1 and 2 do linear uniform motion with velocity $V_1 = [230 \ -240]^T$ and $V_2 = [320 \ -50]^T$. At 35th second, target 3 do uniform anticlockwise angular velocity with initial velocity $V_3 = [-200 \ -250]^T$.

Covariance matrix R of measurement noise refer to literature [9]. The probability P_D that target is detected is 0.98. Clutter obey the Poisson distribution. Its probability hypothesis density is as follows:

$$K_k(z) = \lambda_c V u(z)$$

The region of survival target is $V = 16 \times 10^8$ m^2. Unit clutter of each square is $\lambda_c = (60/V)$ m^2. (60 clutters are produced in observation area every time). u(z) is density of uniform distribution in survival target region. Combining threshold U = 4 in the process of trimming and combining Gaussian elements. The maximum number allowed to anticipate calculating is J = 100. Parameters of UF are set as $\sigma = 0.1$, $\kappa = 0$, $\beta = 2$. After tracking, error between track points and target real points is measured with optimal subpattern assignment (OSPA) [10, 11]. In order to overcome the random factors, we repeat simulation for 20 times, getting statistical result of 20 times repeated simulation. The simulation results are as follow figures.

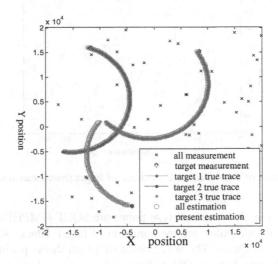

Fig. 1. Tracking trace, true trace and measurement by radar networking SQUF-GMPHDA (Color figure online)

Figure 1 is whole scene of simulation, where the red 'o' type points are SQUF-GMPHDA target tracking points in whole journey. The red and blue '.' type points are rule trace of two survival targets 1 and 2. Black '×' type points are measurement points of clutter in current time. '☆' type points are measurement points of three targets in current time. '△' type points are estimation points of SQUF-GMPHDA

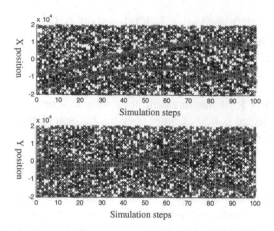

Fig. 2. Tracking trace and measurement by SQUF-GMPHDA in X/Y direction (Color figure online)

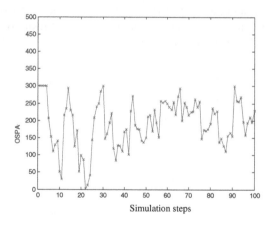

Fig. 3. OSPA distance of track trace and true trace of target (root-mean-square error is 58 m)

in current time. In Fig. 2, the red 'o' type points are SQUF-GMPHDA target tracking points. The blue 'o' type points are measurement of target. Black '×' type points are measurement points of clutter. The above/down subgraph shows position of each point in X/Y direction. Figure 3 shows OSPA distance between tracking and true points. As you can see in Figs. 1 and 2, we can successfully track number time-varying nonlinear target in strong clutter background, containing target 3 birthing at 35th second. Generally the precision of tracking single target by Kalman filter algorithm in engineering practice is about 80 m. But Fig. 3 shows that precision of tracking multi-target by SQUF-GMPHDA in radar networking is 58 m. This tracking error is small. And in a certain time, the tracking error variance (root-mean-square error of OSPA) decrease with time. So the SQUF-GMPHDA is suitable for engineering application.

References

1. Vo, B.-N., Ma, W.-K.: The Gaussian mixture probability hypothesis density filter. IEEE Trans. Sig. Process. **54**(11), 4091–4104 (2006)
2. Liu, Z.X., Xie, W.X., Wang, P., et al.: A Gaussian mixture PHD filter with the capability of information hold. Acta Electronica Sinica **41**(8), 1603–1608 (2013)
3. Mahler, R.P.S., Martin, L.: Multitarget Bayes filtering via first-order multitarget moments. IEEE Trans. Aerosp. Electron. Syst. **39**(4), 1152–1178 (2003)
4. Shun-cheng, T., Guo-hong, W., Na, W., et al.: A probability hypothesis density filter and data association based algorithm for multitarget tracking with pulse doppler radar. J. Electron. Inf. Technol. **35**(11), 2700–2706 (2013)
5. Lixiao, Z., Ziyue, T., Zhenbo, Z.: An overview on track-before-detect algorithms for radar weak targets. Mod. Radar **35**(4), 45–52, 57 (2013)
6. Hailong, D., Wenbo, Z.: Survey on detect and track algorithm for radar weak signal to noise ratio targets based on plot. J. Microwaves **30**(Supplement), 627–635 (2014)
7. Julier, S., Uhlmann, J., Durrant-Whyte, H.F.: A new method for the nonlinear transformation of means and covariances in filters and estimators. IEEE Trans. Autom. Control **45**(3), 477–482 (2004)
8. Yan, Z., Wei, F.: Attitude estimation algorithm of four rotor unmanned aircraft based on unscented Kalman filter. J. Test Meas. Technol. **28**(3), 194–198 (2014)
9. Wenbo, Z., Jiyan, D.: Study on statistical properties of radar network noise in inertial coordinate system. J. Artillery Acad. **126**(5), 91–95 (2010)
10. Hoffman, J.R., Mahler, R.P.S.: Multitarget miss distance via optimal assignment. IEEE Trans. Syst. Man. Cybern. Part A Syst. Hum. **34**(3), 327–336 (2004)
11. Schuhmacher, D., Vo, B.-T., Vo, B.-N.: A consistent metric for performance evaluation of multi-object filters. IEEE Trans. Sig. Process. **56**(8), 3447–3457 (2008)

Study on Tracking Strong Maneuvering Targets Based on IMM-GMPHDA

Hai-Long Ding$^{(\boxtimes)}$, Wen-Bo Zhao$^{(\boxtimes)}$, and Luo-Zheng Zhang$^{(\boxtimes)}$

Army Officer Academy of PLA, Hefei 230031, Anhui, China
656797226@qq.com

Abstract. Gaussian mixture probability hypothesis density filter algorithm (GMPHDA), which is effective method for tracking unknown number of multi-target in strong clutter environment, has solid theoretical basis. But it is hard to track target by GMPHDA when the targets maneuver. To model maneuvering target, we introduce interacting multi-model (IMM) in GMPHDA by modeling maneuvering model of survival target and fusing probability hypothesis density of each model filter based on latest model probability, getting IMM-GMPHDA. The simulation results show that we can real-time track strong maneuvering and supersonic multi-target with IMM-GMPHDA, whose tracking precision can reach 70 m in multi-radar networking system, which meets the project requirement.

Keywords: GMPHDA · Interacting multi-model · Strong maneuvering target · Multi-radar networking

1 Introduction

States of targets vary with time during intelligence radars tracking airspace moving targets. Number of airspace targets vary with time when new target birth and old target disappear in radar detecting area. We can receive lot of clutter information exceptecho information in the process of detecting target with netted radar. Measurement information of target mix in large amount of clutter information. Although multi-radar networking [1, 2] realize complementation and multiplication in performance and function of netted multi-radar, improving stability, reliability and confidence level of detecting airspace target, true measurement information of target also mix in large amount of clutter information (strong clutter environment) [3]. Detected airspace multi-target is always of strong maneuver model with motion model change over time, not simply following single motion model. Random number of target, strong clutter environment and maneuver model make tracking airspace target with radar more complicated.

About the problem that number and state of target vary with time, the traditional processing method is to association measurement with true target. This method need

---

The National Nature Science Fund Project 61273001, Anhui Province Nature Science Fund Project 11040606M130.

D.-S. Huang and K.-H. Jo (Eds.): ICIC 2016, Part II, LNCS 9772, pp. 838–849, 2016.
DOI: 10.1007/978-3-319-42294-7_74

large amount of calculation and is prone to cause calculation explosion. Probability hypothesis density filter algorithm (PHDA) [4–6] is based on random finite set and Bayesian filtering theory. Drawing lessons from Bayesian method of tracking single target, PHDA respectively treats states and measurement of multi-target as finite set, avoiding complex data correlation. PHDA, positive researched by relevant scholars, is a promising method. GMPHDA [7–9], realize engineering practice application of PHDA, is an analytical solution form for PHDA. GMPHDA is suitable for tracking single target, not suitable for tracking strong maneuvering target. Interacting multiple model algorithm (IMMA) [10, 11] is effective subprime multiple model algorithm, also the most effective method for tracking strong maneuvering target. We combine IMMA and GMPHDA as IMM-GMPHDA. Because birth and spawn target don't maneuvering. IMM-GMPHDA only solves maneuvering problem of survival target. IMM-GMPHDA builds parallel filter for survival target in each filter. In the meantime combining with updated model probability, IMM-GMPHDA get probability hypothesis density (PHD) of survival target, solving maneuvering problem of multi-target.

2 Interacting Multiple Model Kalman Filter

Basic train of thoughts of IMM-GMPHDA in multi-radar networking is to select a limited number of model combination to describe maneuvering situation. Based on Kalman filter algorithm (KFA), each model of the model combination independently estimates airspace target. State and covariance of airspace target are got by method of weighting model probability.

Every step of iteration needs restart initialization of filter, which means combining estimation situation at former moment, inherent state transfer situation and extent of participation to recalculate initial state and covariance estimation.

$$\begin{cases} \hat{X}_k^{0j} = \sum_{i=1}^{r} \hat{X}_k^i \mu_k^{i/j} \\ P_k^{0j} = \sum_{i=1}^{r} \mu_k^{i/j} \left\{ P_k^i + \left[\hat{X}_k^i - \hat{X}_k^{0j} \right] \left[\hat{X}_k^i - \hat{X}_k^{0j} \right]^T \right\} \end{cases} \tag{1}$$

Where mixed probability of model and normalization constant are

$$\begin{cases} \mu_k^{i/j} = P\{M_k = i/M_k = j, Z^k\} = \frac{1}{\bar{c}_j} p_{i/j} \mu_k^i \\ \bar{c}_j = \sum_{i=1}^{n} p_{i/j} \mu_k^i \end{cases} \tag{2}$$

Calculate predicted state and covariance

$$\begin{cases} \hat{X}_{k+1/k}^j = \Phi_{k+1/k} \hat{X}_k^{0j} \\ P_{k+1/k}^j = \Phi_{k+1/k} P_k^{0j} \Phi_{k+1/k}^T + \Gamma_{k+1/k} Q_k \Gamma_{k+1/k}^T \end{cases} \tag{3}$$

State updating

$$\hat{X}_{k+1}^j = \hat{X}_{k+1/k}^j + K_{k+1}(Z_{k+1} - H_{k+1}\hat{X}_{k+1/k}^j) \tag{4}$$

Calculate likelihood function of each model filter

$$\Lambda_{k+1}^j = \frac{1}{\sqrt{(2\pi)^m |S_{k+1}^j|}} \exp\left\{-\frac{1}{2} v_j^T (S_{k+1}^j)^{-1} v_j\right\} \tag{5}$$

Where residual error and covariance of residual error is

$$\begin{cases} v_{k+1}^j = Z_{k+1} - H_{k+1}\hat{X}_{k+1/k}^j \\ S_{k+1}^j = H_{k+1}P_{k+1/k}^j H_{k+1}^T + R_{k+1} \end{cases} \tag{6}$$

Calculate probability of each model

$$\begin{aligned} \mu_{k+1}^j &= P\{Z_{k+1}/M_{k+1} = j, Z^k\}P\{M_{k+1} = j, Z^{k+1}\} \\ &= \frac{1}{c}\Lambda_{k+1}^j \sum_{i=1}^{n} p_{ij}\mu_k^i = \Lambda_{k+1}^j \bar{c}_j/c \end{aligned} \tag{7}$$

Where

$$c = \sum_{j=1}^{r} \Lambda_{k+1}^j \bar{c}_j \tag{8}$$

Weight state estimation and its covariance according to model probability.

$$\begin{cases} \hat{X}_{k+1} = \sum_{j=1}^{n} \hat{X}_{k+1}^j \mu_{k+1}^j \\ P_{k+1} = \sum_{j=1}^{n} \mu_{k+1}^j \left\{P_{k+1}^j + \left[\hat{X}_{k+1}^j - \hat{X}_{k+1}\right]\left[\hat{X}_{k+1}^j - \hat{X}_{k+1}\right]^T\right\} \end{cases} \tag{9}$$

3 IMM-GMPHDA

Target in detect area maneuvering means maneuvering at current moment relative to former moment. So there is no maneuvering about birth and spawn target. We only deal with maneuvering problem of survival target through IMM. Predicted value of survival target is calculated according to each motion model. Next the predicted Gaussian components are updated. During updating process, Gaussian component of each model is multiplied by the updated probability of each model to fused as one mix-model component. Weight is fused by the same way. Fusion of multi-model don't influence number of Gaussian elements. Each model filter parallel filtering. After getting measurement, we fuse results of parallel filter. Probability of Gaussian element model is calculated according to likelihood function of Gaussian element. Gaussian element and probability of Gaussian element model is one-to-one corresponding. After getting model probability, we fuse each model probability based on weight of each model. Element probabilities belong to the same model are fused as one model probability.

After getting predicted value. We calculate estimation value of survival target in two different cases. One case is that there is no detected target(no corresponding target measurements). In this case, predicted value is used as current estimation value (updated value). The other case is that target is detected. In this case, we update predicted value according to measure information in each parallel model filter, getting estimation value of each model, next fusing estimation value according to model probability.

Algorithm steps of IMM-GMPHDA are

Step 1. Transform measurement of radar from polar coordinate system of radar to state coordinate system of radar (fusion center rectangular coordinate system) as state of birth target. There is no spawn and survival target at initial moment. It is set to an empty set:

$$m_{\gamma,0}^i = f^{-1}(z_0), \ m_{\beta,0}^i = \phi, \ m_{S,0}^i = \phi \tag{10}$$

Step 2. Calculate predicted probability hypothesis density (PHD) of birth, spawn and survival target: For k = 0, 1, 2, 3–endtime

$$
\begin{cases}
D_{\gamma,k+1|k}(x) = \sum_{i=1}^{J_{\gamma,k+1}} \omega_{\gamma,k+1}^i N\left(x; F_{\gamma,k}^i m_{\gamma,k}^i, P_{\gamma,k+1}^i\right) \\[2mm]
D_{\beta,k+1|k}\left(x|m_k^j\right) = \sum_{j=1}^{J_k} \sum_{l=1}^{J_{\beta,k+1}} \omega_k^j \omega_{\beta,k+1}^l N\left(x; F_{\beta,k}^i m_{\beta,k}^i, P_{\beta,k+1|k}^{j,l}\right) \\[2mm]
P_{\beta,k+1|k}^{j,l} = P_{\beta,k}^l + F_{\beta,k}^l P_{\beta,k}^j \left(F_{\beta,k}^l\right)^T \\[2mm]
D_{S,k+1|k}^J(x) = p_{S,k+1} \sum_{j=1}^{J_k} \omega_k^j N\left(x; m_{S,k+1|k}^{j,J}, P_{S,k+1|k}^{j,J}\right)
\end{cases}
\tag{11}
$$

Where predicted state and covariance of survival target are (J means model J, j means element j of survival target):

$$
\begin{cases}
m_{S,k+1|k}^{j,J} = F_k^J m_{S,k}^j (j = 1, 2, \ldots \ldots, J_k) \\[2mm]
P_{S,k+1|k}^{j,J} = Q_k^J + F_k^J P_{S,k}^j \left(F_k^J\right)^T
\end{cases}
\tag{12}
$$

Where predicted state belongs to each model at current moment is calculated according to state and covariance of survival target at previous moment. Don't separately calculate it in each filter. Because estimated state and covariance from previous model filter are fused according to model probability. Next the fused state and covariance are together trimmed and merged. After trimming and merging, each model filter cannot be separated

Step 3. Update survival target

If not detected, let previous moment probability as current model probability.

$$
\begin{cases}
D_{1-D,k+1}(x) = (1 - p_D)D_{k+1|k}(x) \\
\mu_{1-D,k+1}^J = \mu_k^J
\end{cases}
\tag{13}
$$

Update state value of current moment survival target

$$
\begin{cases}
K_{S,k+1}^{j,J} = P_{S,k+1|k}^{j,J}H_{k+1}^T\left(H_{k+1}P_{S,k+1|k}^{j,J}H_{k+1}^T + R_{k+1}\right)^{-1} \\
m_{S,k+1}^{j,J}(z) = m_{S,k+1|k}^{j,J} + K_{S,k+1}^{j,J}\left(z - H_{k+1}m_{S,k+1|k}^{j,J}\right)
\end{cases}
\tag{14}
$$

Calculate likelihood function of survival target

$$
\begin{cases}
P_{S,k+1}^{j,J} = \left[I - K_{S,k+1}^{j,J}H_k\right]P_{S,k+1|k}^{j,J} \\
v_{S,k+1}^{j,J} = z - H_{k+1}m_{S,k+1|k}^{j,J} \\
S_{S,k+1}^{j,J} = H_{k+1}P_{S,k+1|k}^{j,J}H_{k+1}^T + R_{k+1} \\
\Lambda_{S,k+1}^{j,J} = \dfrac{1}{\sqrt{(2\pi)^m\left|S_{S,k+1}^{j,J}\right|}}\exp\left\{-\dfrac{1}{2}\left(v_{S,k+1}^{j,J}\right)^T\left(S_{S,k+1}^{j,J}\right)^{-1}v_{S,k+1}^{j,J}\right\}
\end{cases}
\tag{15}
$$

Update model probability of survival target

$$
\begin{cases}
c_S^J = \displaystyle\sum_{J=1}^{r}\Lambda_{S,k+1}^{j,J}\bar{c}_S^J \\
\mu_{k+1}^{j,J} = \dfrac{1}{c_S^J}\Lambda_{S,k+1}^{j,J}\displaystyle\sum_{I=1}^{r}p_{IJ}\mu_k^I = \Lambda_{S,k+1}^{j,J}\bar{c}_S^J/c_S^J
\end{cases}
\tag{16}
$$

Fuse each model filter according to updated model probability: estimated state and weight of survival target

$$
\begin{cases}
m_{S,k+1}^{j}(z) = \sum_{J=1}^{r} \mu_{k+1}^{J} m_{S,k+1}^{j,J}(z) \\
P_{S,k+1} = \sum_{j=1}^{n} \mu_{k+1}^{J} \left\{ P_{S,k+1}^{j,J} + \right. \\
\left. \left[m_{S,k+1}^{j,J}(z) - m_{S,k+1}^{j}(z) \right] \left[m_{S,k+1}^{j,J}(z) - m_{S,k+1}^{j}(z) \right]^{T} \right\} \\
\omega_{S,k+1}^{j,J}(z) = \dfrac{p_D \omega_{S,k+1|k}^{j,J} q_{S,k+1}^{j,J}(z)}{\kappa_{S,k+1}(z) + p_D \sum_{l=1}^{J_{k+1|k}} \omega_{S,k+1|k}^{l,J} q_{S,k}^{l,J}(z)} \\
\omega_{S,k+1}^{j}(z) = \sum_{J=1}^{r} \mu_{k+1}^{j,J} \omega_{S,k+1}^{j,J}(z)
\end{cases} \tag{17}
$$

Step 4. Update the current moment estimation state of spawn and birth target

$$
\begin{cases}
K_{\beta,\gamma,k+1}^{j} = P_{\beta,\gamma,k+1|k}^{j} H_{k+1}^{T} \left(H_{k+1} P_{\beta,\gamma,k+1|k}^{j} H_{k+1}^{T} + R_{k+1} \right)^{-1} \\
m_{\beta,\gamma,k+1}^{j}(z) = m_{\beta,\gamma,k+1|k}^{j} + K_{k+1}^{j} \left(z - H_{k+1} m_{\beta,\gamma,k+1|k}^{j} \right) \\
P_{\beta,\gamma,k+1}^{j} = \left[I - K_{k+1}^{j} H_{k} \right] P_{\beta,\gamma,k+1|k}^{j}
\end{cases} \tag{18}
$$

Step 5. Calculate current Gaussian mixture elements of birth, spawn and survival targets.

$$
\begin{cases}
m_{k+1}^{j} = [m_{\beta,\gamma,k+1}^{j}(z), m_{S,k+1}^{j}(z)] \\
D_{D,k+1}(x;z) = \sum_{j=1}^{J_{k+1|k}} \omega_{k+1}^{j}(z) N\left(x; m_{k+1}^{j}, P_{k+1}^{j}\right) \\
D_{k+1}(x) = (1 - p_D) D_{k+1|k}(x) + \sum_{z \in Z_{k+1}} D_{D,k+1}(x;z)
\end{cases} \tag{19}
$$

Step 6. Trim and merge current Gaussian mixture elements if $D_{k+1}^i < T$ $(i = 1, 2, \cdots J_{k+1})$

$D_{k+1}^i = \phi$

else if $D_{k+1}^i > T$ $(i = 1, 2, \cdots J_{k+1})$

if $\left(m_{k+1}^i - m_{k+1}^j\right)^T \left(P_{k+1}^i\right)^{-1} \left(m_{k+1}^i - m_{k+1}^j\right) \leq U$

$$
\begin{cases}
\tilde{\omega}_{k+1} = \omega_{k+1}^i + \omega_{k+1}^j \\[2mm]
\tilde{\mu}_{k+1} = \dfrac{\omega_{k+1}^i}{\tilde{\omega}_{k+1}} \mu_{k+1}^i + \dfrac{\omega_{k+1}^j}{\tilde{\omega}_{k+1}} \mu_{k+1}^j \\[2mm]
\tilde{m}_{k+1} = \dfrac{\omega_{k+1}^i}{\tilde{\omega}_{k+1}} m_{k+1}^i + \dfrac{\omega_{k+1}^j}{\tilde{\omega}_{k+1}} m_{k+1}^j
\end{cases}
\tag{20}
$$

end
end

Step 7. Extract number and state of target

if $\tilde{\omega}_{k+1} > W$

$$
\begin{cases}
\hat{X}_{k+1} = \tilde{m}_{k+1} \\[2mm]
\hat{\mu}_{k+1} = \tilde{\mu}_{k+1}
\end{cases}
\tag{21}
$$

end

Step 8. Judge PHD of birth and spawn target according to current measurement. Let this PHD as next moment PHD of birth and spawn.

if   k>=2

$$
\begin{cases}
m^i_{\gamma,k+1} = m^i_{\beta,k+1} = f^{-1}(z_{k+1}) \\
D_{\gamma,k+1}(x) = \sum_{i=1}^{J_{\gamma,k+1}} \omega^i_{\gamma,k+1} N\left(x; m^i_{\gamma,k+1}, P^i_{\gamma,k+1}\right) \\
D_{\beta,k+1}(x) = \sum_{i=1}^{J_{\beta,k+1}} \omega^i_{\beta,k+1} N\left(x; m^i_{\beta,k+1}, P^i_{\beta,k+1}\right)
\end{cases}
$$

(22)

if $D_{\gamma,k+1}(x) > D_{\beta,k+1}(x)$

$$D_{\beta,k+1}(x) = \phi, \quad m^i_{\beta,k+1} = \phi$$

elseif $D_{\gamma,k+1}(x) < D_{\beta,k+1}(x)$

$$D_{\gamma,k+1}(x) = \phi, \quad m^i_{\gamma,k+1} = \phi$$

end

end

Before estimated result of each single model filter of survival target be fused according model probability. The result only reflect estimation situation of single model filter. So we cannot fuse the result with birth and spawn target. Before be trimmed and merged, estimated result of each single model filter must be first fused according to model probability. Secondly, trim and merge the fused estimated result and $D_{\gamma,k+1}(x) = \phi$, $m^i_{\gamma,k+1} = \phi$ of birth and spawn targets. After trimming and merging, Gaussian elements cannot be separated according to single model filter, which results that state $\hat{X}^i_k$ and P^i_k covariance of survival target used in origination of single model filter next moment cannot be got, only can be replaced by $m^j_{S,k}$ and $P^j_{S,k}$ (results of trimming and merging).

4   Simulation and Conclusion Analysis

To verify effectiveness and practicability of IMM-GMPHDA proposed in this paper. We use the method of MATLAB software simulation to test and verify. The simulation hardware environment is: the Pentium (R) Dual - Core CPU, HP E5200, 2.5 GHz CPU Frequency, 2.00 G of Memory. Simulation scenario is as follows: two networking radar, radar 1 and 2, the scan period T is 2 s, the distance accuracy of radar 1 is 130 m, azimuth accuracy is 0.3° and itching angle accuracy is 0.2°, configuration location is [118° 29° 120 m]. The distance accuracy of radar 2 is 90 m, azimuth accuracy is 0.2° and itching angle accuracy is 0.1°, configuration location is $[117° \ 31° \ 20 \ m]^T$. The configuration location of the fusion is $[117° \ 30° \ 170 \ m]^T$. In order to verify the simplicity of calculation, we assume that two networking radars scans every second, the fusion center receives detection data on time every second, and the whole detection time is 100 s, the detected targets is in strong clutter environment, and the number of targets is unknown, the detected target is with flight level of 6000 m in the whole courses, with two-dimensional detection region [−20000, 20000] × [−20000, 20000] and survival probability P_S = 0.99. The survival targets subject to linear Gaussian distribution. The detected targets have supersonic speed and strong maneuvering, which following model of uniform linear motion (CV model), turning clockwise model (ACT model) and turn counterclockwise model (CCT model). In the first 30 s, target 1 and 2 fly according to CV model with speed respectively $[180–300]^T$ m/s and $[320–130]^T$ m/s. Between 30th and 60th second, they fly according to ACT model with the same linear speed and angular velocity0.07 rad/s. Between 60th and 100th second, they fly according to CCT model with the same linear speed and angular velocity −0.07 rad/s. Target 3, flying according to CV model with speed $[−220–280]^T$ m/s, spawn from target 1 at 46th second. Markov state transition probability between models is:

$$P_{ij} = \begin{bmatrix} 0.95 + 0.05/3 & 0.05/3 & 0.05/3 \\ 0.05/3 & 0.95 + 0.05/3 & 0.05/3 \\ 0.05/3 & 0.05/3 & 0.95 + 0.05/3 \end{bmatrix}$$

Initial model probability is:

$$\mu_0 = [0.95 + 0.05/3 \quad 0.05/3 \quad 0.05/3]^T$$

Standard deviation of the three targets is:

$$\sigma = 15\,\text{m/s}$$

Birth and spawn targets all obey the Poisson distribution with Gaussian mixture forms as follows:

$$\gamma_k(x) = N\left(x; m_\gamma^1, P_\gamma\right) + N\left(x; m_\gamma^2, P_\gamma\right)$$

$$\beta_{k|k-1}(x|\zeta) = N(x; \zeta, Q_\beta)$$

Where the area birth may spear is $m_\gamma^1 = [−5000, 15000, 0, 0]^T$ or $m_\gamma^2 = [−5000, 15000, 0, 0]^T$. Covariance matrix Rof virtual noise is according to literature [12]. The probability that target can be detected is P_D = 0.98. Clutter obeys Poisson distribution with PHD as follow:

$$K_k(z) = \lambda_c Vu(z)$$

Where unit clutter of each square is $\lambda_c = 16 \times 10^{-8}$. Acreage of target survival area is $V = 4 \times 10^{-8}$. u(z) is uniform distribution density of clutter in target survival area. Allowed top speed of algorithm is 600 m/s. Threshold of merging in process of trimming and merging Gaussian elements is U = 5. The top number of Gaussian elements allowed to participate in calculation is J = 150. Probability threshold of extracting state is 0.5. The allowed max distance of OSPA is 500 m. In order to overcome the random factors, we repeat simulation for 20 times, getting statistical result of 20 times repeated simulation. The simulation results are as follow figures.

Figure 1 is whole scene of simulation, where the red 'o' type points are IMM-GMPHDA target tracking points in whole journey. The red and blue '.' type points are rule trace of two survival strong maneuvering targets 1 and 2. Green '.' type points are rule trace of spawn targets. Black '×' type points are measurement points in current time(including clutter). '☆' type points are measurement points of three targets

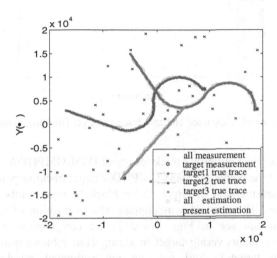

Fig. 1. Tracking trace, true trace and measurement by IMM-GMPHDA (Color figure online)

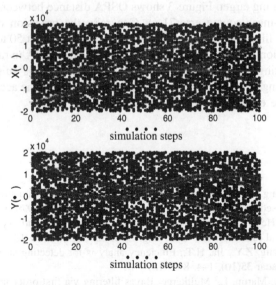

Fig. 2. Tracking trace and measurement by IMM-GMPHDA in X/Y direction (Color figure online)

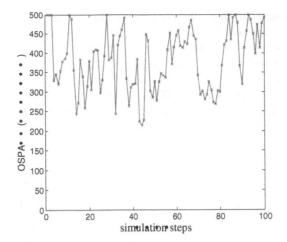

Fig. 3. OSPA distance of track trace and true trace of target (root-mean-square error is 58 m)

in current time. 'Δ' type points are track results of IMM-GMPHDA in current time. In Fig. 2, the red 'o' type points are IMM-GMPHDA target tracking points. The blue '∇' type points are measurement of target. Other black '.' type points are measurement points of all clutter. The above/down subgraph shows position of each point in X/Y direction. As you can see in Figs. 1 and 2, we can successfully track number time-varying strong maneuvering target in strong clutter background (including survival and spawn targets). And we can get estimated number of target by IMM-GMPHDA. Of course, we can get target estimation only after points information accumulation at initial time. Because of using IMM, we can track maneuvering target in process of tracking target. Figure 3 shows OSPA distance between tracking and true points with root-mean-square error71 m. Generally the precision of tracking single target by Kalman filter algorithm in engineering practice is about 50 to 100 m. Figure 3 shows that precision of tracking multi-target by IMM-GMPHDA in radar networking is 71 m. This tracking error is small. So the IMM-GMPHDA is suitable for tracking strong maneuvering target in engineering application, meeting the accuracy requirment. (root-mean-square error is estimated accuracy index).

References

1. Ye, C.M., Ding, J.J., Zhang, W., et al.: Resource control function model for radar networking based on modalization. Syst. Eng. Electron. **35**(9), 1979–1982 (2013)
2. Zhu, H.W., He, Y.: Joint estimation of target height and systematic error for two-dimensional radar network. Syst. Eng. Electron. **35**(9), 1861–1866 (2013)
3. Ren, A.Z., Zhang, Z.Y., Jia, H.T.: Efficiency analysis for detecting stealth target by netted radar. Mod. Radar **35**(10), 1–4, 8 (2013)
4. Mahler, R.P.S., Martin, L.: Multitarget Bayes filtering via first-order multitarget moments. IEEE Trans. Aerosp. Electron. Syst. **39**(4), 1152–1178 (2003)

5. Yang, J.L., Ji, H.B.: Gauss-Hermite particle PHD filter for bearings-only multi-target tracking. Syst. Eng. Electron. **35**(3), 457–462 (2013)
6. Lian, F., Han, C., Liu, W., Chen, H.: Joint spatial registration and multi-target tracking using an extended probability hypothesis density filter. IET Radar Sonar Navig. **5**(4), 441–448 (2011)
7. Vo, B.-N., Ma, W.-K.: The Gaussian mixture probability hypothesis density filter. IEEE Trans. Sign. Process. **54**(11), 4091–4104 (2006)
8. ZongXiang, L.: A sequential GM-based PHD filter for a linear Gaussian system. Sci. Chin. (Information Science) **56**(10), 1–10 (2013)
9. Wenling, L., Jia, Y.: The Gaussian mixture PHD filter for jump Markov models based on best-fitting Gaussian approximation. Sign. Process. **91**, 1036–1042 (2011)
10. Blom, H.A., Bar-Shalom, Y.: The interacting multiple model algorithm for systems with Markovian switching coefficients. IEEE Trans. Autom. Control **33**(8), 780–783 (1988)
11. Foo, P.H., Ng, G.W.: Combining the interacting multiple model method with particle filters for manoeuvring target tracking. IET Radar Sonar Navig. **5**(3), 234–255 (2011)
12. Zhao, W.B., Du, J.Y.: Study on statistical properties of radar network noise in inertial coordinate system. J. Artillery Acad. **126**(5), 91–95 (2010)

Network Topology Management Optimization of Wireless Sensor Network (WSN)

Chun Kit Ng[1], Chun Ho Wu[1(✉)], W.H. Ip[1], J. Zhang[2], G.T.S. Ho[1], and C.Y. Chan[1]

[1] Department of Industrial and Systems Engineering, The Hong Kong Polytechnic University,
Hung Hom, Kowloon, Hong Kong
{felix.ng.ise,jack.wu}@connect.polyu.hk,
{wh.ip,george.ho,cy.chan}@polyu.edu.hk
[2] School of Computer Science and Engineering, South China University of Technology,
Guangzhou 510006, People's Republic of China
junzhanghk@gmail.com

Abstract. Network topology management is one of the critical concerns when designing a Wireless Sensor Network (WSN). In this research, four basic factors including the total production cost, sensing coverage, network connectivity and fault tolerance are considered. A mathematical model is proposed to optimize four optimization metrics corresponding to the four design factors. This approach attaches a weighting coefficient to each optimization metric to adjust their importance in the optimization model. To solve the proposed model, an Ant Colony Optimization (ACO) based metaheuristics method, called *MAX–MIN* Ant System (*MMAS*) is used. In the experiment, Greedy algorithm (Greedy) and Genetic Algorithm (GA) are also adopted to solve the proposed model. The results indicate that *MMAS* shows a satisfactory performance on solving the proposed model, which there is an improvement on the number of sensor nodes comparing to the result of Greedy, and a better fitness value than the result of GA.

Keywords: WSNs · Sensing coverage · Network connectivity · Fault tolerance · ACO · MMAS

1 Introduction

In recent decades, Wireless Sensor Network (WSN) has raised significant attentions among researchers and practitioners. Plenty of applications from different areas successively emerge. It shows a high potential to be a highlighted technology in the next five to ten years [1]. Similar to many new technologies, the WSN has a relatively high threshold to be adopted. One of the most critical concerns is the energy efficiency which highly relates to the network life time. As the WSN is usually deployed in outdoor or large areas such as forests, deserts and highways, battery is almost the only choice for the power supply due to its portability and relatively low cost compared to the wire power supply system. Because of the limited power source (i.e. battery), communication protocol, clustering method and duty cycle management should be carefully designed. Another critical concern is the sensor network topology management. To effectively

© Springer International Publishing Switzerland 2016
D.-S. Huang and K.-H. Jo (Eds.): ICIC 2016, Part II, LNCS 9772, pp. 850–859, 2016.
DOI: 10.1007/978-3-319-42294-7_75

manage the network topology of a WSN, a series of factors should be considered. The first factor to be concerned is the number of sensor nodes. Although the cost of a sensor node is relatively low, the amount of total production cost of all the sensor nodes can be unexpectedly high, especially for large scale WSNs. Therefore, optimizing the number of sensor nodes is an essential objective to be fulfilled. At the same time, a WSN should maintain a certain level of basic Quality of Service (QoS) in terms of sensing coverage, network connectivity and fault tolerance. Sensing coverage and network connectivity are the two factors always being considered together when designing a WSN since the coverage ensures the data collection from the region of interest while the connectivity ensures the path of data reporting from sensor nodes to the sink [2, 3]. If one of them is ignored, the WSN cannot be considered as a satisfactory solution. Fault tolerance refers to the ability of a WSN that can tolerate a number of node failures and remain functional. This factor is vital since node failures can lead to network function disorder, network partition and failure and finally terminate the WSN much earlier than expected. In the ordinary course of events, deployment of a number of redundant sensor nodes to maintain the network connectivity is a common practice to achieve fault tolerance. However, this approach may increase the cost significantly. Therefore, how to determine the appropriate number and position of sensor nodes, which are able to balance the tradeoffs between the production cost and the required level of QoS, becomes an essential topic to be considered when designing a robust, user or application oriented WSN.

In this paper, a mathematical model is proposed to optimize the network topology of a WSN, in which well balance the total production cost, sensing coverage, network connectivity and fault tolerance. The model focuses on dealing with homogenous WSN and will be solved by using an Ant Colony Optimization (ACO) based metaheuristics method, called *MAX–MIN* Ant System (*MMAS*), which is proposed by [4].

2   Related Works

As mentioned before, keeping well balance of the tradeoffs between the cost and basic factors of QoS is critical for a WSN system. Thus, researchers utilized different approaches to deal with the tradeoffs. By considering the cost and sensing coverage, [5] utilized Integer Linear Programming (ILP) and alternative divide-and-conquer approaches to obtain the minimum cost of sensor nodes for a complete covered region. [6] proposed a coverage control scheme based on a multi-objective genetic algorithm (GA) to select the minimum number of sensor nodes from a WSN that can preserve full coverage to be activated, and put other sensor nodes into sleep mode. For the same purpose, [7] proposed a framework that takes dynamically periodic reconstruction strategies to select a subset of nodes for communication and sensing tasks only when the residual energy of network drops to a threshold. Considering the factor of fault tolerance, both [8, 9] proposed algorithms to add additional nodes into an existing WSN to improve the k-connectivity, which [8]'s algorithm results a k-connected or partially k-connected network while [9]'s algorithm results a k-connected network, for any desired k.

When the optimization is taking more design factors into account, the multi-objective optimization approach is introduced. [10] proposed an algorithm for the optimal design of WSNs in precision agriculture based on genetic algorithm (GA). In this approach, energy efficiency, sensing coverage, network connectivity and an application specific design factor (i.e. deployment uniformity) are the objectives needed to be optimized while considering grid deployment and different node operating modes. Likewise, [11] proposed an algorithm called flexible algorithm for sensor placement (FLEX) to optimize the placement of sensor nodes by using multi-objective GA. This algorithm aims to maximize coverage, desired k-connectivity and minimize energy consumption simultaneously.

To summarize, adequate works on optimizing sensing coverage and network connectivity can be found in the literature. Some researchers might also take fault tolerance into account for the network topology optimization. However, only a few literatures considered the cost and the three basic QoS factors together in homogeneous WSN. This stimulates the research work described in this paper, to propose a mathematical model for balancing the tradeoffs between the total production cost, sensing coverage, network connectivity and fault tolerance.

3   Problem Formulation

The aim of this research is to propose an approximately optimized network topology for homogeneous WSN in a 2-D region of interest such that the sensing coverage and network connectivity are maximized, while a certain level of fault tolerance is maintained. Simultaneously, a minimum production cost is achieved. Before describing the proposed model, some assumptions are considered, which are listed below:

1. The deployed wireless sensor nodes are homogeneous, which all have the same sensing and communication range.
2. The sensing model is defined as Boolean sensing disc model:

$$c_{xy}(s_i) = \begin{cases} 1, & \text{if } d(s_i, P) < r \\ 0, & \text{otherwise} \end{cases} \tag{1}$$

Where $c_{xy}(s_i)$ is the Boolean sensing function. s_i is a sensor node, P is a point in the region of interest, r is the sensing range of a sensor node, $d(s_i, P)$ is the Euclidean distance between the sensor node s_i and the point P.

3. The communication model is also defined as Boolean communication disc model:

$$c(s_i, s_j) = \begin{cases} 1, & \text{if } d(s_i, s_j) < R \\ 0, & \text{otherwise} \end{cases} \tag{2}$$

Where $c(s_i, s_j)$ is the Boolean communication function. s_i and s_j are two sensor nodes, R is the communication range of a sensor node, $d(s_i, s_j)$ is the Euclidean distance between the sensor node s_i and s_j.

4. The optimization model works under an ideal network, which is no data loss during communication.

Based on the assumptions, the preliminary settings of the proposed optimization model can be defined. The nodes deployment method is proposed to be grid deployment, since the position of each node can be simply determined and the node density may require less than other deployment strategies like uniformly randomly distribution strategy and Poisson point process distribution strategy [12].

In the proposed optimization model, four basic factors which dominate the design and the performance of a WSN are addressed. They are total production cost, sensing coverage, network connectivity and fault tolerance. The energy efficiency may not be considered in the proposed model since this factor can be optimized via energy efficient communication protocol, better clustering method or duty cycle management after the WSN is formed. The objectives of the optimization model are to maximize the sensing coverage, the network connectivity and the fault tolerance with a desire level, and simultaneously minimize the total production cost. In order to combine these four optimization metrics into one single objective, weighted sum approach is applied. This approach attaches a weighting coefficient to each optimization metric to adjust their importance in the optimization model. Thus, the objective function of the proposed optimization model will be formed as:

$$f = \alpha_1 \frac{p_{cov}}{p_{tot}} + \alpha_2 \frac{n_{conn}}{n_{dep}} + \alpha_3 \frac{n_{ft}}{n_{dep}} + \alpha_4 \frac{1}{c \cdot n_{dep}} \tag{3}$$

Where α_x is the weighting coefficient and $\alpha_x \geq 0$. Each part attached with the weighting coefficient represents one optimization metric. The detail will be described in the following:

3.1 Coverage Ratio

The coverage ratio can be represented as the number of grid points covered by the largest connected cover of the WSN (p_{cov}) divided by the total number of intersection points in the grid (p_{tot}). The total number of points in the grid can be obtained by multiplying the number of row lines in the grid (m) and the number of column lines in the grid (n).

$$p_{cov}/p_{tot}$$
$$p_{tot} = m \cdot n \tag{4}$$

3.2 Connectivity Ratio

The connectivity ratio can be represented as the number of sensor nodes in the largest connected cover (n_{conn}) divided by the total number of wireless sensor nodes deployed in the grid (n_{dep}).

$$n_{conn}/n_{dep} \tag{5}$$

The same as the previous optimization metric, the largest connected cover should be obtained first before calculating these two ratios. Breadth-first search (BFS) and depth-first search (DFS) are two graph algorithms for searching in a graph, which can be applied to find out the largest connected cover. In this case, the DFS is selected because it is more efficient in search connected components. When searching starts at a source, the algorithm tries to search as deep as possible along a branch path. Once the end of the branch path is reached, the algorithm backtracks to another connect branch path to perform the search. This process continues until all the connected nodes have been visited.

3.3 Fault Tolerance Ratio

The fault tolerance ratio can be represented as the number of sensor nodes with desired degree of fault tolerance (n_{ft}) divided by the total number of wireless sensor nodes deployed in the grid (n_{dep}).

$$n_{ft}/n_{dep} \tag{6}$$

The degree of fault tolerance is measured by the number of vertex-disjoint (node-disjoint) paths between a sensor node and the sink. To find out the number of vertex-disjoint paths, the problem should be firstly transformed into max-flow min-cut problem based on Menger's theorem [13]. Then, the maximum flow or minimum cut between a source node to a sink in a graph can be found by using Ford-Fulkerson algorithm according to max-flow min-cut theorem [14]. The Ford-Fulkerson algorithm is used to find out the maximum flow of a graph by iteratively increasing the value of the flow. At the beginning, all the flows of edges in the graph are set to 0. The flows value will be increased along an augmenting path found in an associated residual network at every iteration. This process will be repeated until no augmenting paths exist in the residual network. The term "residual network" refers to a sub-network that consists of edges with capacities and an augmenting path is a simple path from the source node to the sink in the residual network.

3.4 Cost

The cost can be represented as the unit cost of a sensor node (c) multiplied by the total number of wireless sensor nodes deployed in the grid (n_{dep}).

$$c \cdot n_{dep} \tag{7}$$

As the objective of this metric is to minimize the cost, therefore this metric is reciprocal in the objective function as shown below.

$$\frac{1}{c \cdot n_{dep}} \tag{8}$$

In addition, the total number of wireless sensor nodes deployed in the grid (n_{dep}) can be obtained through the equation shown below.

$$n_{dep} = \sum_{i=1}^{m} \sum_{j=1}^{n} p_{ij} \tag{9}$$

Where

$$p_{ij} = \begin{cases} 1, & \text{if } a \text{ sensor node is deployed} \\ 0, & \text{otherwise} \end{cases}$$

In the proposed optimization model, the four optimization metrics including the coverage ratio, the connectivity ratio, the fault tolerance ratio and the cost factor are addressed. These four metrics will be connected by using weighted sum method. In general, the weighting coefficients are set with equal importance for each metric and they can also be determined based on the experience of specialists according to different applications.

4 Approach

Ant Colony Optimization (ACO) is a metaheuristics optimization algorithm developed based on the food searching behavior of ants. ACO is commonly used in solving NP-hard combinatorial optimization problems such as Traveling Salesman Problem (TSP) and Quadratic Assignment Problem (QAP). ACO algorithms make use of artificial ants to construct candidate solutions for an optimization problem every iteration, and each solution will be associated with a pheromone trail. The candidate solutions will then be evaluated according to their fitness value. Based on the result, the pheromone trail with better fitness value will be deposited more pheromone, and the pheromone evaporates every iteration. Following this mechanism, an approximately optimized solution will be obtained when the ACO algorithm is converged. In this experiment, *MAX-MIN* Ant System (*MMAS*) is adopted to solve the proposed optimization model. *MMAS* is a modified ACO algorithm, which better utilizes the search results of every iteration, and more effectively avoids premature convergence. Among most of ACO based algorithms, *MMAS* shows the best performance for many different combinatorial optimization problems especially for the TSP and QAP [4].

When applying *MMAS* to solve the proposed model, the area for WSN deployment is represented as a grid $G = (V, E)$ with V being the set of vertexes and E being the set of edges of the grid. The base station is placed at the lower left corner of the grid. To start constructing the solution, an ant moved from its initial position, which is a random node around the base station. The path selection of the ant is based on the probability assigned to each path, which is calculated by:

$$P_{i,j} = \frac{\tau_{i,j}{}^{\alpha} \cdot \eta_{i,j}{}^{\beta}}{\sum \tau_{i,j}{}^{\alpha} \cdot \eta_{i,j}{}^{\beta}} \tag{10}$$

Where $\tau_{i,j}$ is the amount of pheromone between vertex i and j, $\eta_{i,j}$ is the local sensor coverage heuristic information, and α and β are the parameters to determine the degree of affection of the pheromone. $\eta_{i,j}$ can be calculated following the equation proposed in [15], which is:

$$\eta_{i,j} = S_{i,j} \cdot l_{i,j}\left(1 - b_{i,j}\right) \tag{11}$$

Where $S_{i,j}$ is the available area in the grid covered by the new sensor node at the position where an ant will move to. $l_{i,j}$ is a binary coefficient, which its value is either 1 or 0, and is determined based on whether there is a connection existing between the edge $E(i, j)$ and the constructed path. $b_{i,j}$ is used to checked the availability for the new sensor node placement. If the position is already occupied, $b_{i,j} = 1$, otherwise, $b_{i,j} = 0$.

In every iteration of the *MMAS* algorithm, every ant constructs a path in the grid for a solution. All the solutions are then examined using the proposed objective function. After the solution examination, the pheromone trails are updated. The evaporation process is first performed on the trails. In the second step, only the ant with the best iteration solution is allowed to deposit pheromone on its constructed path. Particularly, the pheromone update process is represented as:

$$\tau_{i,j} = \rho \cdot \tau_{i-1,j-1} + \Delta\tau_{i,j} \tag{12}$$

$$\Delta\tau_{i,j} = 1/C(V) \tag{13}$$

Where ρ denotes the pheromone residual factor and $C(V)$ represents the fitness value of either the iteration best solution or the global best solution. The limits of pheromone are set as $[\tau_{min}, \tau_{max}]$ with the relationship given by

$$\tau_{min} = 0.087 \cdot \tau_{max} \tag{14}$$

$$\tau_{max} = 1/[(1 - r) \cdot C(V)] \tag{15}$$

To avoid ants trapping in local optimal solutions, roulette wheel selection method is used for an ant to select the next node to construct its path when the probabilities assigned to all the available points are calculated. This approach allows the point with lower probability to be chosen which is significant for global search of ACO.

When ants start searching, there are three conditions for stopping the path searching process: (1) no more available points for search; (2) the rest of available points have been covered; (3) it is impossible to reach other available points. At the end of each iteration, the best path with highest fitness value is chosen to update the pheromone of all the points, and the new pheromone information will be passed to the next iteration.

5 Experiment

In this section, the experiment for the proposed model and the results are described. The results obtained using *MMAS* are compared with the results obtained using Greedy algorithm (Greedy) and Genetic Algorithm (GA). Initially, the region for WSN deployment is set as a 20 by 20 grid with 1 unit cell size. The base station is placed at the lower left corner. All sensors are homogeneous, which the sensing range r of a sensor node is equal to the cell size of the grid and the communication range R is twice the sensing range, which is $R = 2r$. In the result of Greedy shown in Fig. 1, the required number of sensor nodes is 200 and the fitness value is 3.4875. For the GA approach, the algorithm will be run for 100 generations with the crossover rate 0.75 and the mutation rate 0.1. The population size is set to 300. Figure 2 shows the sensor deployment result, which the number of used sensor nodes is 192 and the fitness value is 3.4473. For MMAS, the parameter settings are listed as follow: $\alpha = 1$, $\beta = 4$, $\rho = 0.5$ and the number of ants placed in the grid is 5. As Fig. 3 shown, the number of used sensor nodes is 147 and the fitness value is 3.4475.

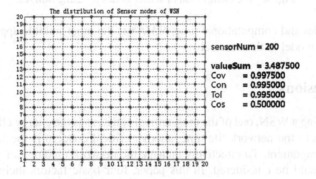

Fig. 1. The distribution of sensor nodes by using Greedy algorithm.

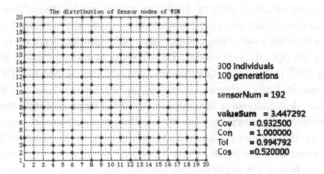

Fig. 2. The distribution of sensor nodes by using GA.

The result of data analysis indicated that the result of Greedy included the best fitness value, but required the largest number of sensor nodes. In contrast, the result of GA required 4 % less number of sensor nodes, but had lower fitness value than the result of Greedy. The result of MMAS showed 26.5 % improvement on the number of sensor nodes comparing to the result of Greedy, and shows a better fitness value than the result of GA. For the computational time, the Greedy and MMAS approaches showed a better performance than the GA approach. To conclude, considering the fitness value, number

858   C.K. Ng et al.

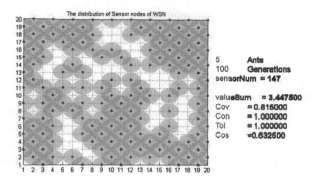

Fig. 3. The distribution of sensor nodes by using *MM*AS.

of sensor nodes and computational time, *MM*AS is the most suitable approach to solve the proposed model in this research.

6   Conclusions

When designing a WSN, one of the most critical concerns is the energy efficiency which highly relates to the network life time, another critical concern is the sensor network topology management. To effectively manage the network topology of WSN, a series of factors should be considered. In this paper, four basic factors including the total production cost, sensing coverage, network connectivity and fault tolerance are considered. Based on these four factors, a mathematical model is proposed to optimize four corresponding optimization metrics. In the proposed model, these four optimization metrics are combined into one single objective by using the weighted sum approach. This approach attaches a weighting coefficient to each optimization metric to adjust their importance in the optimization model. To solve the proposed model, an Ant Colony Optimization (ACO) based metaheuristics method, called *MAX–MIN* Ant System (*MM*AS) is used. In the experiment, Greedy algorithm (Greedy) and Genetic Algorithm (GA) were also adopted to solve the proposed model. The results indicated that *MM*AS showed a satisfactory performance on solving the proposed model, which there is an improvement on the number of sensor nodes comparing to the result of Greedy, and a better fitness value than the result of GA. However, the fitness value obtained through *MM*AS still has rooms for improvement and the proposed model is mainly focus on homogenous WSN, which heterogeneous WSN may not be considered. Therefore, our future work will focus on enhancing the proposed model to cover heterogeneous WSN and exploring other metaheuristics and heuristics approaches for the modified model to obtain better results. It is believed that this research work can contribute to various areas such as real-time locating system [16] and pervasive health care services [17].

Acknowledgments. We would like to acknowledge the support received from the Innovation and Technology Commission (ITC) of the Government of the HKSAR of the People's Republic of China (PRC) for this project (ITS/161/13FX). Our gratitude is also extended to the Research

Committee and the Department of ISE of The Hong Kong Polytechnic University for their support on this project (G-UB97) and financial assistance to the involved research student (RT3C).

References

1. Gubbi, J., Buyya, R., Marusic, S., Palaniswami, M.: Internet of Things (IoT): a vision, architectural elements, and future directions. Future Gener. Comput. Syst. **29**, 1645–1660 (2013)
2. Ammari, H.M.: Connected k-coverage in two-dimensional deployment fields. In: Ammari, H.M. (ed.) Challenges and Opportunities of Connected k-Covered Wireless Sensor Networks, pp. 73–109. Springer, Heidelberg (2009)
3. Lin, Y., Zhang, J., Chung, H.S.H., Ip, W.H., Li, Y., Shi, Y.H.: An ant colony optimization approach for maximizing the lifetime of heterogeneous wireless sensor networks. IEEE Trans. Syst. Man Cybern. Part C Appl. Rev. **42**(3), 408–420 (2012)
4. Stutzle, T., Hoos, H.: MAX-MIN Ant System. Future Gener. Comput. Syst. **16**, 889–914 (2000)
5. Chakrabarty, K., Iyengar, S., Qi, H., Cho, E.: Grid coverage for surveillance and target location in distributed sensor networks. IEEE Trans. Comput. **51**(12), 1448–1453 (2002)
6. Jia, J., Chen, J., Chang, G., Tan, Z.: Energy efficient coverage control in wireless sensor networks based on multi-objective genetic algorithm. Comput. Math. Appl. **57**(11–12), 1756–1766 (2009)
7. Zeng, Y., Sreenan, C., Xiong, N., Yang, L., Park, J.: Connectivity and coverage maintenance in wireless sensor networks. J. Supercomput. **52**, 23–46 (2010)
8. Pu, J., Xiong, Z., Lu, X.: Fault-tolerant deployment with k-connectivity and partial k-connectivity in sensor networks. Wireless Commun. Mob. Comput. **9**(7), 909–919 (2009)
9. Bredin, J., Demaine, E., Hajiaghayi, M., Rus, D.: Deploying sensor networks with guaranteed fault tolerance. IEEE/ACM Trans. Netw. **18**(1), 216–228 (2010)
10. Ferentinos, K.P., Tsiligiridis, T.A.: Adaptive design optimization of wireless sensor networks using genetic algorithms. Comput. Netw. **51**(4), 1031–1051 (2007)
11. Chaudhry, S.B., Hung, V.C., Guha, R.K., Stanley, K.O.: Pareto-based evolutionary computational approach for wireless sensor placement. Eng. Appl. Artif. Intell. **24**(3), 409–425 (2011)
12. Zhang, H., Hou, J.: Is deterministic deployment worse than random deployment for wireless sensor networks? In: Proceedings of 25th IEEE International Conference on Computer Communications, INFOCOM 2006, pp. 1–13. IEEE Press, Barcelona (2006)
13. Oellermann, O.R.: Menger's theorem. In: Topics in Structural Graph Theory (Encyclopedia of Mathematics and its Applications), pp. 13–39 (2013)
14. Even, S.: Graph Algorithms, 2nd edn. Cambridge University Press, New York (2011)
15. Fidanova, S., Marinov, P.: Optimal wireless sensor network coverage with Ant Colony Optimization. In: Proceedings of ICSI 2011: International Conference on Swarm Intelligence, Cergy, France, pp. 1–7 (2011)
16. Zhai, C., Zou, Z., Zhou, Q., Mao, J., Chen, Q., Tenhunen, H., Zheng, L., Xu, L.: A 2.4-GHz ISM RF and UWB hybrid RFID real-time locating system for industrial enterprise Internet of Things. Enterp. Inf. Syst. (2016). http://dx.doi.org/10.1080/17517575.2016.1152401
17. Pang, Z., Zheng, L., Tian, J., Kao-Walter, S., Dubrova, E., Chen, Q.: Design of a terminal solution for integration of in-home health care devices and services towards the Internet-of-Things. Enterp. Inf. Syst. **9**(1), 86–116 (2015)

Soft Computing

Tourism Network Comments Sentiment Analysis and Early Warning System Based on Ontology

Yanxia Yang(✉) and Xiaoli Lin

City College, Wuhan University of Science and Technology,
Hubei 430083, China
{yxy_job, aneya}@163.com

Abstract. Mine Tourism information and opinion, intelligent analysis user emotion, to improve tourism products and services, is the key to the success of tourism e-commerce. This paper embarks from the tourism network review information, researches how to build the microblog emotional vocabulary ontology and how to classify emotion based on Naive Bayes classification algorithm, implements a tourism network comments sentiment analysis and early warning system based on ontology. It not only save a large amount of manpower and material resources, but also have a certain reference value to establish reasonable tourism policy.

Keywords: Sentiment analysis · Early warning · Naive Bayes classifier · Ontology

1 Introduction

Under the background of modern information age, information dissemination is rocketing. The production of a piece of network review cannot be ignored. Due to the strong timeliness of network reviews, if the demands on suggestions are not responded in due course, negative impact on corporate image would be made. In the field of tourism, the guiding effect of network consensus is attached great importance in particular. Text sentiment analysis is mainly carried out on the judgment of emotional polarity, which is to judge whether the emotion expressed by a comment is positive, negative or neutral [1]. Sentiment analysis aims to excavate users' viewpoint and emotional polarity from the text, and make the machine understand the emotional tendency of the text by means of supervised learning or unsupervised learning. With the sentiment analysis, it is possible for automatic analysis and early warning of the network. In this situation, the system is designed, which is used to capture the network comments of the tourism industry and analyze the emotional tendency of the text and push appropriate early warning information to users [2].

© Springer International Publishing Switzerland 2016
D.-S. Huang and K.-H. Jo (Eds.): ICIC 2016, Part II, LNCS 9772, pp. 863–870, 2016.
DOI: 10.1007/978-3-319-42294-7_76

2 Related Theory and Technology

2.1 Ontology Concept and Construction Method

The application of ontology is to construct domain model. For example, in the knowledge engineering, an ontology provides a glossary of terms concepts and relation, through which modeling can be done to a domain. In the semantic Web, ontology has a very important position, which is the basis of solving the Web information sharing and exchange in the semantic level.

The ontology is divided into two spheres: the degree of detail and domain dependence. The degree of detail is a relative and relatively vague concept, which refers to the extent of describing and depicting modeling objects. The high level of detail is called the reference ontology while the low level of detail is called the share ontology. According to domain dependence, it can be divided into four categories, which are top-level ontology, domain ontology, task ontology and application ontology. The top-level ontology describes the most common concepts and relation between concepts, such as space, time, event, behavior, etc., which is irrelevant to concrete applications; and other types of ontologies are special cases of this kind. Domain ontology describes relation between concept and concept in some specific domain (such as medicine, geography, etc.). The task ontology describes relation between concept in a specific task or behavior and concept. The application ontology describes the relation between concept that is dependent on specific domains or asks and concept.

The construction of emotional vocabulary ontology is to fully express the semantic information contained in emotional vocabulary, such as the emotional tendency of vocabulary, similarity between vocabularies, progression and adversative relation between words, and so on., which is convenient for organization and sharing of emotional vocabulary so as to offer effective analysis evidence to analysis of the tendency of microblog topics.

2.2 The Construction of Microblog Emotional Vocabulary Ontology

Chinese microblog, as a product of the Internet, has a diversity of text information form and the words used to express the tendency are in constant change. Therefore, the core ontology is not required to be completed in one time. So, at this stage, it's only necessary to collect relatively important core concepts and relations that can express people's standpoints, and to establish basic emotional word ontology. In this paper, core words that sentiment analysis words centralize are extracted from HowNet, which is used as information source of constructing ontology.

(1) Collection and analysis of emotional vocabulary knowledge

The vocabulary of basic emotional words ontology is selected from labeled sentiment analysis wording set in HowNet. Categories (in Chinese) and the number of vocabulary contained by this wording set as well as the number of emotional vocabulary selected in ontology is shown in Table 1.

Table 1. HowNet the number of the emotional analysis words and ontology language words

Vocabulary types	Quantity of vocabulary	Number of ontology language
Negative words	3116	216
Negative emotional words	1254	275
Positive words	3730	595
Positive emotional words	836	227
Advocate words	38	21

Negative words and adverbs of degree, as well as logical connectives that indicate transition and progression, will have an effect on the tendency of the subjective sentence. Therefore, this paper also set up a set of negative words, degree adverbs and conjunctions. According to wording set that HowNet released of negative words, degree adverbs and conjunctions, 18 negative adverbs, 188 degree adverbs and 40 conjunctions are included, which are specifically shown in Table 2 [3].

Table 2. Artial negative words, adverbs of degree and conjunction sets

Part of speech	Type	Number
Negative words	Negative	18
Adverbs of degree	Severe	110
	Moderate	37
	Light	41
Conjunction	Transitional conjunction	21
	Progressive conjunction	19

(2) Formalized representation of emotional vocabulary ontology

After construction of emotional vocabulary ontology is completed, this paper uses OWL to depict language's formalized description to emotional vocabulary ontology, which is to use defined meta-ontology of OWL language to formally depict the extracted concept and relations on attribute. The most important thing is to depict classes, subclasses, attributes and their respective characteristics. This paper uses protege tools to build ontology, and the finished ontology is saved in the OWL file format whose suffix is OWL.

Expressions of concept categories have two basic concepts in protege: whole concept and relational concept. The editing interface of whole concept is shown in Fig. 1.

Interface diagram that uses protege tools to build important categories and its properties of emotional vocabulary ontology:

Network emotional words needs long-playing attention and collection and there is no available emotional dictionary nowadays, therefore collecting, labeling and adding vocabulary with emotional color into emotional vocabulary ontology is a necessary complement through social networks, blogs, BBS, reviews, microblog (Fig. 2).

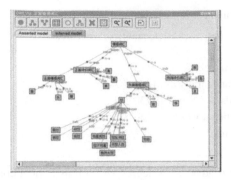

Fig. 1. The editing interface of whole concept

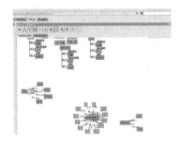

Fig. 2. Interface diagram of protege construction category

2.3 The Composition of Corpus

This paper mainly classifies comment information on tourism network, and uses octopus collector to crawl comment information on tourism network.

3 System Implementation

3.1 System Framework

See Fig. 3.

3.2 The Realization of Each Module

(1) Crawling data of tourism comments

Collection of tourism comment information refers to extract the corresponding comments from tourist website information, which is the basis of the review emotion text analysis. The collection of comment information is obtained by using web crawlers.

(2) Text preprocessing

Text preprocessing is the key step in the whole system, because machine cannot automatically judge category attributes of the whole text and only human beings can

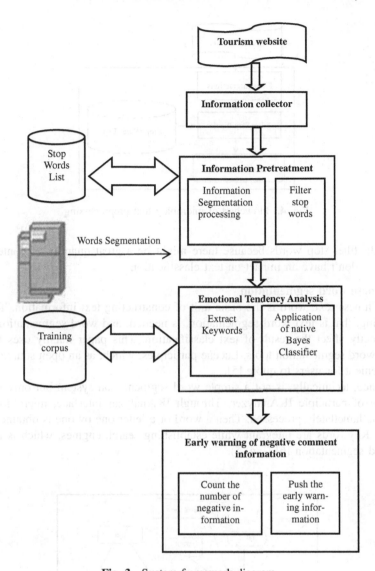

Fig. 3. System framework diagram

turn Chinese text data into data that machines can recognize, in which way data will be better handled [4]. The implementation process is shown in Fig. 4:

Step 1: Preparation. Prepare microblog data that has been crawled for subsequent text preprocessing.

Step 2: Because there are a great deal of crawled comment information without practical significance, in order to avoid affecting the text classification result, it is necessary to process text and filter needless information which is unnecessary to be handled and is directly filtered out.

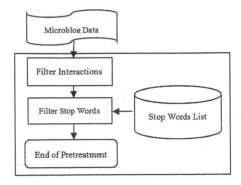

Fig. 4. Flowchart of microblog text preprocessing

Step 3: filter stop words because there might be a great quantity of contents that don't have an impact on text classification.

(3) Chinese word segmentation

In Chinese text, word is the basic unit of constructing text information. Text pre-processing, that is text word segmentation, is needed, and word segmentation results will directly affect the result of text classification. This paper mainly uses the very mature word segmentation tools, Lucene participles, which, as an open source project, is convenient for users to quote [5].

Luence, technically, is not a simple word segmentation system but only provides interface of participle IKAnalyzer. Through IKAnalyzer interface, microblog information is lamellately processed. Then a word or a letter one by one is obtained and is labeled. Keywords are integrant while establishing search engines, which is acquired via word segmentation [6] (Fig. 5).

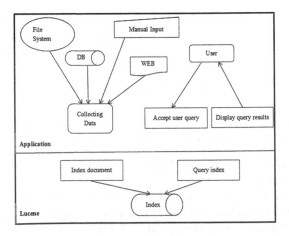

Fig. 5. Luence word segmentation flowchart

(4) Text classification implementation

This paper uses Bayesian algorithm to classify text and to interpret the article key words' belonging to a certain class of probability. Then, by comparing different types of probability, text of the maximum probability is classified as the class directly [7–11]. The concrete implementation process of naive Bayesian algorithm text categorization [7–11] is shown in Fig. 6:

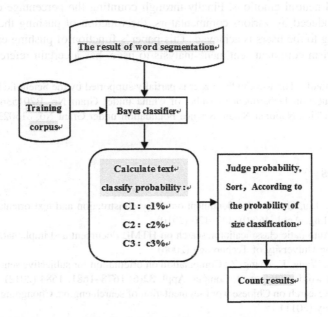

Fig. 6. Bayes classifier flowchart

Bayesian algorithm implementation process includes the following steps:

Step 1: return to the Lucene text information processed by word segmentation system, including word segmentation results and keyword extraction results.

Step 2: for the backward results, calculate probability value of the keywords in training repertoire through Bayesian algorithm and find out the value with the highest probability via ranking.

Step 3: for results of sequencing, according to certain classification rules that texts of the highest probability will be classified as one type, final classification results are gained.

(5) Early warning of negative comment information

Count the number of negative information, calculate the percentage of negative information on the total information, if the proportion reaches the threshold (user-defined), then push the appropriate early warning information to users.

4   Conclusion

Carrying out sentiment analysis aiming at tourist information comments, firstly needs to analyze Chinese emotional words ontology and understand the construction of ontology. Then web crawler collects tourism website comments information, calculate, filter stop words through the Naive Bayes algorithm and complete counting emotional tendency of test text by means of Lucene participle classification, including positive, negative and neutral emotions. Finally through counting the percentage of emotional tendency produced by various commentators, the function of pushing the appropriate early warning to the users is achieved. This paper's function of pushing early warning towards tourism comment sentiment analysis to users has a certain reference value.

Acknowledgment. The work in this paper is partially supported by the Scientific Research Plan Project of Education Department of Hubei of China under Grant No. B2015360 and in part supported by China National Nature Science Foundation under Grant No. 61502356.

References

1. Wang, X., Liu, Q., Tao, X.: Sentiment ontology construction and text orientation analysis. Comput. Eng. Appl. **46**(30), 117–120 (2010)
2. Liu, D.: Automatic classification research on HTML document and implantation of the toll. Chongqing University of Technology (2006)
3. Wang, X., Wang, J., Zhang, Z.: Computation on orientation for subjective sentence based on sentiment words ontology. J. Comput. Appl. **32**(6),1678–1681, 1684 (2012)
4. Ren, L.: Research on Chinese word segmentation of search engine. Chongqing University of Technology (2011)
5. Guan, R., Lu, B.: A novel phrase_extraction method for Chinese text. J. Mech. Electr. Eng. **27**(9), 123–126 (2010)
6. Zheng, J., Lu, J.: Study of an improved keywords distillation method. Comput. Eng. **31**(18), 194–196 (2005)
7. Guan, R.: Research of keywords extraction algorithm for Chinese text based on gene expression programming. Hangzhou Dianzi University (2009)
8. Zou, Y., Chen, X., Wang, W.: Research on focused crawler based on Bayes classifier. Appl. Res. Comput. **9**(26), 3418–3420, 3439 (2009)
9. http://www.core.org.cn/NR/rdonlyres/Civil-and-Environmental-Engineering/1-017Computing-and-Data-Analysis-for-Environmental-ApplicationsFall2003/62A96B91-D71B-4734-80E8-E57 63316BFA9/0/class03_6.pdf
10. Li, Y.: Chinese web page classification research overview. Mod. Comput. **15**, 3–7 (2012)
11. Verma, A., Fiasché, M., Cuzzola, M., Iacopino, P., Morabito, F.C., Kasabov, N.: Ontology based personalized modeling for type 2 diabetes risk analysis: an integrated approach. In: Leung, C.S., Lee, M., Chan, J.H. (eds.) ICONIP 2009, Part II. LNCS, vol. 5864, pp. 360–366. Springer, Heidelberg (2009)

Protein Structure
and Function Prediction

Prediction of Lysine Acetylation Sites
Based on Neural Network

Wenzheng Bao[1], Zhichao Jiang[1], Kyungsook Han[2], and De-Shuang Huang[1,3(✉)]

[1] Institute of Machine Learning and Systems Biology,
College of Electronics and Information Engineering, Tongji University, Shanghai 201804, China
baowz55555@126.com, 1015553840@qq.com
[2] Department of Computer Science and Engineering, Inha University, Incheon, South Korea
khan@inha.ac.kr
[3] Institute of Machine Learning and Systems Biology,
School of Electronics and Information Engineering, Tongji University, Caoan Road 4800,
Shanghai 201804, China
dshuang@tongji.edu.cn

Abstract. Lysine acetylation is a crucial type of protein post-translational modification, which is involved in many important cellular processes and serious diseases. In practice, identification of protein acetylated sites through traditional experiment methods is time-consuming and laborious. Computational methods are not suitable to identify a large number of acetylated sites quickly. Therefore, machine learning methods are still very valuable to accelerate lysine acetylated site finding. In this study, many biological characteristics of acetylated sites have been investigated, such as the amino acid sequence around the acetylated sites, the physicochemical property of the amino acids and the transition probability of adjacent amino acids. A special structure neural network, which is named flexible neural tree (FNT), was then utilized to integrate such information for generating a novel lysine acetylation prediction system named LA+FNT. When compared with existing methods, our proposed method overwhelms most of state-of-the-art methods. Such method has the ability to integrating different biological features to predict lysine acetylation with high accuracy.

Keywords: Lysine acetylation · Flexible neural tree · Post-translational modification

1 Introduction

Protein post-translational modification (*PTM*) has emerged as a major contributor to variation, localization and control of proteins. It has been suggested that the incongruity between the complexity of vertebrate organisms and the size of their encoded genomes is compensated by the large number of PTMs available [1]. Of the hundreds of PTMs identified, the most intensively studied is protein phosphorylation, in which protein kinases (PK) attach phosphate moieties to *Ser*, *Thr*, or *Tyr* residues. Protein phosphorylation appears to play an essential role in many diverse and critical cellular processes, and inhibitors of their function have emerged as important modes of therapy for malignant diseases.

© Springer International Publishing Switzerland 2016
D.-S. Huang and K.-H. Jo (Eds.): ICIC 2016, Part II, LNCS 9772, pp. 873–879, 2016.
DOI: 10.1007/978-3-319-42294-7_77

Lysine acetylation is a dynamic and reversible post-translational modification (PTM) that is highly conserved in prokaryotes and eukaryotes. Such critical protein function process neuralizes the positive charge on an amino acid and regulates DNA binding, protein-protein interaction, and protein stability [1]. Lysine acetylation conjugates on $N\alpha$-amino group of lysine residues and is regulated by a highly balanced enzyme system containing lysine acetyltransferases (KATs, also known as histone acetyltransferases HATs) and histone deacetylase (HDACs). Furthermore, acetylation also occurs on α-amino groups of $N\alpha$-terminal residues (N-lysine acetylation) [2–5]. In addition, lysine acetylation is involved in diverse biological consequences including transcriptional activity, cell survival, and subcellular localization [6]. Most importantly, it has been reported that aberrant lysine acetylation is linked to many pathological diseases, such as cancer, neurodegenerative diseases, and metabolic diseases [7]. Following the identification of nuclear KATs, a number of nonhistone proteins have been identified as substrates, including DNA-binding proteins (transcription factors), nonnuclear proteins, and shuttle proteins from the nucleus to cytoplasm [8–11]. To date, over 2000 human acetylated proteins and 4000 lysine sites have been identified by traditional experiments and large-scale mass spectrometry-based proteomic analysis [12–14]; nevertheless, determining the acetylation substrate specificity of KATs remains a challenge. Moreover, KATs responsible for nonhistone proteins are still unclear.

In this paper, a novel classification model of two-classification framework has been proposed for lysine acetylation sites of protein. And then flexible neural tree (FNT), a structural alternative neural network, is employed as the classifiers in our tree structural classification model. Several properties of amino acid residues have been treated as the classification features in such model. The interaction between amino acid side residues and predicted lysine acetylation sites, which have been treated as the center amino acid residues in the protein sequences, has been introduced in this work.

2   Methods and Materials

2.1   Data Collection

In order to make the positive dataset make the same size with the negative ones, the 600 protein sequences have been selected in this research. The experimentally validated lysine acetylation sites are extracted from a database for post-translational modification (PTM) called scbit (http://www.scbit.org/iPTM/) [15] database. In scbit database, 300 acetylated lysine (K) sites from 300 protein sequences are retrieved. At the same time, the 300 non-acetylated lysine peptides have been retrieved from other 300 protein sequences. Several selected protein sequence information information will give a further description.

2.2   Feature Selection

A great deal of protein sequence features have been proposed, such as amino acid compositions, dipeptide compositions, pseudo amino acid compositions [16]. Nevertheless, such features can hardly represent the lysine acetylation process in the field of post translational modification. In our model, one main type of features was utilized to predict lysine acetylation: amino acid physicochemical property (AAPP).

Amino acid physicochemical property is the most important features for protein biochemical reactions. Amino Acid index (AAindex) [17] is a database of numerical indices representing various physicochemical and biochemical properties of amino acids. There are 541 amino acid indices in current release of the database (version 9.1), and 10 of these indices contain descriptions like 'NA'. In order to unify the input format, we replaced the 'NA' character with number 0. For a peptide, its value of a physicochemical property was calculated through followed Eq. (1):

$$P_s = \frac{1}{l\_seq} \sum_{j=1}^{l\_seq} p_j \qquad (1)$$

Where, l_seq was length of the peptide; p_j was index value of the jth residue. The features of this model selection showed in the [18].

When it comes to the issue of amino acid sites' prediction, the length of l_seq play a critical role in the field. Therefore, in this research, the lysine acetylation sites in the amino acid sequences may more likely to influencing by the adjacent amino acid residues. For such reason, the Energy Interaction Attenuation Curve (EIAC), which will have the ability to describe the interaction between the center sites and the amino acid residues, has been proposed in this research. In this curve, the assumption of attenuation relations could be computed by the linear relationship. The picture on the EIAC has show in Eq. (2) and Fig. 1.

$$f(d)_{AAI} = \begin{cases} 10 - d & 10 > d > 1 \\ 0 & d = 0 \\ d & -10 < d < -1 \end{cases} \qquad (2)$$

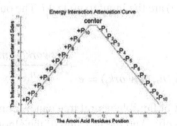

Fig. 1. Energy interaction attenuation curve of lysine acetylation sites

3   Machine Learning Methods

Supervised learning is an important part in the field of machine learning. A supervised learner analyzes the training data and produces a model which can be used for predicting the class of test samples (or unknown examples). In the following some supervised algorithms are discussed that we have used in our proposed method.

3.1 Flexible Neural Tree

Flexible neural tree, which is an alternative tree structural neural network, put forward by Chen [18, 20]. The model has the ability to create the tree structure network. The tree network will achieve the different input in different node of the tree structure.

Firstly, the using instruction set for generating the FNT is showed in (3).

$$Instructor\_Set = \{+_2, +_3, \cdots, +_n, x_1, x_2, \cdots, x_n\} \tag{3}$$

In Eq. (6), the parameter $+N$ denote the non-leaf nodes instructor and taking an argument. The variable $x_1, x_2, \cdots, x_n$ are the leaf nodes instructor and taking no argument each. On the other hand, the output of each non-leaf node may be treated as a single neuron model. For this reason the non-leaf nodes, which called $+a$, are also called a-inputs neuron operator model. In the creation process of FNT, if a non-terminal instruction, $+a (a = 2, 3, 4, \cdots, N)$ is selected, a real values are randomly generated and used for representing the connection strength between the node $+a$ and its children. In addition, these two parameters m_i and n_i are randomly created as flexible activation function parameters and attach them to node $+a$. So in our research, the used flexible activation function is described as:

$$f(m_i, n_i, x) = e^{-(\frac{x - m_i}{n_i})^2} \tag{4}$$

Then, the output of a FNN can be calculated in a recursive way. For any non-terminal node $+i$, the total excitation is calculated as:

$$network_i = \sum_{j=1}^{i} \omega_j \times y_j \tag{5}$$

Where $y_j (j = 1, 2, \cdots, i)$ are the input to node $+i$. The output of the node $+i$ are then calculated by (6):

$$out_i = f(m_i, n_i, network_i) = e^{-(\frac{network_i - m_i}{n_i})^2} \tag{6}$$

Thus, the overall output of flexible neural tree can be computed from left to right by depth-first method, recursively. And the last step of the algorithm is the objective function. During this step, the fitness function used for *PSO* should be given by mean square error (*MSE*):

$$Fitness\ (i) = \frac{1}{N\_train} \sum_{j=1}^{N\_train} (y_{AO}^j - y_{MO}^j) \tag{7}$$

Where, N_train is the total number of training samples, y_j actual_output and y_j model_output are the actual and model outputs of jth sample. Fitness (i) denotes the fitness value of i-th individual.

3.2 Classification Model

In this research, we have introduced FNT as the basic classification model. Considering the speciality of this prediction issue, the lysine acetylation sites prediction seem to a classical binary classification problems. Therefore, the positive samples could easily get the label 1. At the same time, the negative samples will get the label 0 in this classification framework. Not insignificant, however, is the 7 groups of features hard to get the desired results. So the above mentioned classifier with feature groups will treat as the weak classifiers. Employing an integrated strategy, 7 groups of features run different in two classification model, respectively. And then, the results of each predicted sites will be decided by voting.

4 Results

The above mentioned features of lysine acetylation sites have been trained by the FNT model in the classification framework, which was proposed in this paper. At the same time, the performances and capabilities will be compared with other existing methods. In order to further assess performance of our model, comparison in the independent dataset was carried out for LA+FNT and other existing methods. Currently, many acetylation prediction software has been developed, but some of them had broken links so they could not be tested in our study. Actually, EnsemblePail, PHOSIDA, PLMLA and PSKAcePred were included in the comparison. The comparison results were shown in Table 1. In terms of sensitivity and specificity, LA+FNT achieved 61.33 % and 75.40 %, which suggested that LA+FNT had a relatively balanced performance in positive and negative datasets. In contrast, there was a great divergence between sensitivity and specificity in PHOSIDA, PLMLA and PSKAcePred. In terms of accuracy, the value of LA+FNT was 68.37 %, which overwhelmed all other methods. By compared with state-of-art methods, it was worth pointing out that LA+FNT had a fairly good capability to predict lysine acetylation. Comparison results showed in Table 1.

Table 1. Comparison among LA + FNT and existing methods

MethodS	Sn (%)	Sp (%)	Acc (%)
EnsemblePail	49.33	62.67	56.00
PHOSIDA	42.33	92.33	67.33
PLMLA	78.92	44.29	61.64
PSKAcePred	72.24	49.66	60.97
LA + ANN	61.33	75.40	68.37
LA + FNT	79.21	86.74	70.57

5 Discussions and Conclusions

From the above research, seven groups of amino acid residues features have been selected. So, the assumption of Energy Interaction Attenuation (EIA) may be regarded as the special relationship between side animo acids residues and the center predicted animo acids residues. The Lysine acetylation sites seem to be a classical two-classification issue. Nevertheless, several challenges still have to be resolved in such field. Firstly, the lysine acetylation in the PTM process needs to be explained or described in the field of bioinformatics more clearly and detailed. Secondly, the relationship between the side animo acids residues and the center predicted animo acids residues may seem to be other kind of function relationship. So, in the next step of work, the relationship of the above mentioned residues will try to discover. It was not ignored that the seven features play different importance role in this classification framework. So, in the next step work, the seven features should be integrated by the classification model. At the same time, the other features seem to be integrated in such model. Finally, the special structure neural network will find out the fit structure in such similar biological field data and features.

Acknowledgements. This work was supported by the grants of the National Science Foundation of China, Nos. 61133010, 61520106006, 31571364, 61532008, 61572364, 61373105, 61303111, 61411140249, 61402334, 61472282, 61472280, 61472173, 61572447, and 61373098, China Postdoctoral Science Foundation Grant, Nos. 2014M561513 and 2015M580352.

References

1. Armengaud, J.: Proteogenomics and systems biology: quest for the ultimate missing parts. Expert Rev. Proteomics **14**, 2360–2675 (2014)
2. Filippakopoulos, P., Knapp, S.: Targeting bromodomains: epigenetic readers of lysine acetylation. Nat. Rev. Drug Discov. **13**(5), 337–356 (2014)
3. Scholz, C., Weinert, B., Wagner, S.: Acetylation site specificities of lysine deacetylase inhibitors in human cells. Nat. Biotechnol. **33**(4), 415–423 (2015)
4. Zhu, L., Ping, D.-S., Huang, D.S.: A two-stage geometric method for pruning unreliable links in protein-protein networks. IEEE Trans. Nanobiosci. **14**(5), 528–534 (2015)
5. Zhu, L., Guo, W., Deng, S.-P., Huang, D.S.: ChIP-PIT: enhancing the analysis of ChIP-Seq data using convex-relaxed pair-wise interaction tensor decomposition. IEEE/ACM Trans. Comput. Biol. Bioinform. **13**(1), 55–63 (2016)
6. Aram, R.Z., Charkari, N.M.: A two-layer classification framework for protein fold recognition. J. Theor. Biol. **365**, 32–39 (2015)
7. Kouranov, A., et al.: The RCSB PDB information portal for structural genomics. Nucleic Acids Res. **34**(Suppl 1), 302–305 (2006)
8. Deng, S.-P., Zhu, L., Huang, D.S.: Mining the bladder cancer-associated genes by an integrated strategy for the construction and analysis of differential co-expression networks. BMC Genom. **16**(Suppl 3), S4 (2015)
9. Yang, X., Seto, E.: Lysine acetylation: codified crosstalk with other posttranslational modifications. Mol. Cell **31**(4), 449–461 (2008)

10. Zhao, D., Zou, S., Liu, Y.: Lysine-5 acetylation negatively regulates lactate dehydrogenase A and is decreased in pancreatic cancer. Cancer Cell **23**(4), 464–476 (2013)
11. Huang, D.S., Zhang, L., Han, K., Deng, S., Yang, K., Zhang, H.: Prediction of protein-protein interactions based on protein-protein correlation using least squares regression. Curr. Protein Pept. Sci. **15**(6), 553–560 (2014)
12. Wu, X., Oh, M., Schwarz, E.: Lysine acetylation is a widespread protein modification for diverse proteins in arabidopsis. Plant Physiol. **155**(4), 1769–1778 (2011)
13. Sadoul, K., Wang, J., Diagouraga, B.: The tale of protein lysine acetylation in the cytoplasm. BioMed Res. Int. (2010)
14. Huang, D.S., Yu, H.-J.: Normalized feature vectors: a novel alignment-free sequence comparison method based on the numbers of adjacent amino acids. IEEE/ACM Trans. Comput. Biol. Bioinform. **10**(2), 457–467 (2013)
15. Li, Z.R., Lin, H.H., Han, L.Y., Jiang, L., Chen, X., Chen, Y.Z.: PROFEAT: a web server for computing structural and physicochemical features of proteins and peptides from amino acid sequence. Nucleic Acids Res. **34**, W32–W37 (2006)
16. Rao, H.B., Zhu, F., Yang, G.B., Li, Z.R., Chen, Y.Z.: Update of PROFEAT: a web server for computing structural and physicochemical features of proteins and peptides from amino acid sequence. Nucleic Acids Res. **39**, W385–W390 (2011)
17. Deng, S.-P., Huang, D.S.: SFAPS: an R package for structure/function analysis of protein sequences based on informational spectrum method. Methods **69**(3), 207–212 (2014)
18. Bao, W., Chen, Y., Wang, D.: Prediction of protein structure classes with flexible neural tree. Bio-Med. Mater. Eng. **24**, 3797–3806 (2014)
19. Huang, D.S.: Systematic Theory of Neural Networks for Pattern Recognition. Publishing House of Electronic Industry, China (1996). (in Chinese)
20. Yang, B., Chen, Y.H., Jiang, M.Y.: Reverse engineering of gene regulatory networks using flexible neural tree models. Neurocomputing **99**, 458–466 (2013)
21. Zhu, L., You, Z.-H., Huang, D.S., Wang, B.: t-LSE: a novel robust geometric approach for modeling protein-protein interaction networks. PLoS ONE **8**(4), e58368 (2013). doi:10.1371/journal.pone.0058368
22. Huang, D.S., Zhang, L., Han, K., Deng, S., Yang, K., Zhang, H.: Prediction of protein-protein interactions based on protein-protein correlation using least squares regression. Curr. Protein Pept. **15**(6), 553–560 (2014)
23. Zheng, C.-H., Zhang, L., Ng, V.T.-Y., Shiu, S.C.-K., Huang, D.S.: Molecular pattern discovery based on penalized matrix decomposition. IEEE/ACM Trans. Comput. Biol. Bioinform. **8**(6), 1592–1603 (2011)

A Parallel Multiple K-Means Clustering and Application on Detect Near Native Model

Hongjie Wu[1]([⊠]), Chuang Wu[1], Chen cheng[1], Longfei Song[1],
and Min Jiang[2]

[1] School of Electronic and Information Engineering,
Suzhou University of Science and Technology, Suzhou 215009, China
Hongjie.wu@qq.com
[2] The First Affiliated Hospital of Soochow University, Suzhou 215006, China

Abstract. Protein structure clustering is an important and essential step in protein 3D structure prediction. However, two issues limited current methods. But the large-scale candidate models in the decoy and undistinguished metric limit current methods to identify the near-native models. In this paper we proposed a novel method based on parallel multiple K-means cluster algorithms to identify the near-native structures. Parallel is introduced to reduce the memory and time consumption and multiple K-means to fusion different metrics of protein 3D similarity. Tested on 56 proteins, MK-means can well identify 33 (58.9 %) proteins which are better or the same to SPICKER selected and 10 of the 33 proteins is the same results to the SPICKER. It indicates the performance of MK-means is similar to the top protein clustered tools SPICKER.

Keywords: 3D structural cluster · K-means · Protein prediction · Root mean square deviation

1 Introduction

Protein structure clustering is an important step in protein 3D structure prediction, function and interaction prediction [1–3]. In protein structure prediction methodology, the task of protein structure clustering is identifying the best near-native models from large-scale decoys, generated by the free modeling or template modeling, based on the 3D structure similar to the clustering algorithm [4–6]. Current RMSD and TM-score are two common metrics for the 3D structure similarity of the candidates. The following refinement step will be carried out on the cluster results.

Zhang et al. [7] have developed SPICKER, a simple and efficient strategy to identify near-native folds by clustering, in which clustering is performed in a one-step procedure using a shrunken but representative set of decoy conformations and the pairwise RMSD cutoff is determined by self-adjusting iteration. After benchmarking on

This paper is supported by grants no. 61540058, 61202290 under the National Natural Science Foundation of China (http://www.nsfc.gov.cn) and grants no. BK20131154 under Natural Science Foundation of Jiangsu Province.

© Springer International Publishing Switzerland 2016
D.-S. Huang and K.-H. Jo (Eds.): ICIC 2016, Part II, LNCS 9772, pp. 880–887, 2016.
DOI: 10.1007/978-3-319-42294-7_78

set of 1489 nonhomologous proteins that represent all protein structures in the PDB >= 200 residues. Clusco [8] is fast and easy-to-use software for high-throughput comparison of protein models with different similarity measures (cRMSD, dRMSD, GDT TS, TM-Score, MaxSub, Contact Map Overlap) and clustering of the comparison results with standard methods: K-means Clustering or Hierarchical Agglomerative Clustering. The application was highly optimized and written in C/C++, including the code for parallel execution on CPU and GPU, which resulted in a significant speedup over similar clustering and scoring computation programs. Francois et al. [9] proposed a fast method that works even on large decoy sets. This method is implemented in software called Durandal. Durandal has been shown to be consistently faster than other software performing fast exact clustering. In some cases, Duran-dal can even outperform the speed of an approximate method. Durandal uses the triangular inequality to accelerate exact clustering, without compromising the distance function.

However, two issues limited current methods. Firstly, since the decoys contain large-scale, ranged from 10 k–100 k, candidate models. How to handle the memory exceeding and the time-consumed task becomes an urgent issue [10–12]. For example a decoy set, which contains 50 k candidate models, required $5*10^4 * 5*10^4 *4 = 25*10^{10}$ (232 Mega)bytes memory to store the distance matrix if each float data type is 4 bytes. The other is how to precisely metric the similarity between two proteins. RMSD and TM-score [13] are the two wide-used metrics for proteins [14]. But it is difficult for us to distinguish which is the best metric for 3D similarity between proteins, since their different emphasis in local and global region.

So, in this paper we proposed a new method based on parallel multiple K-means cluster algorithm to identify the near-native structures. Parallel is introduced to reduce the memory and time consumption and multiple K-means to fusion different metrics of protein 3D similarity. In the paper, firstly we introduced the metrics of 3D structural similarity and the datasets. Then, the mechanism of MK-Means was introduced. At last the results were compared and discussed.

2   Materials and Methods

2.1   Data Set of Benchmark

We use I-TASSER Decoy Set-II as the benchmark which is widely used for evaluating the performance of protein decoys clustered algorithm [15, 16]. I-TASSER Set-II contains the whole-set atomic structure decoys of 56 non-homologous small proteins. The backbone structures were generated by the I-TASSER ab initio modeling; the side-chain atoms were added using Pulchra [17] Table 1.

Table 1. Summary of the data set

Data set	Proteins	Avg length	Avg number of decoy
I-TASSER Set-II	56	80.875	439.1964

2.2   3D Distance Metrics of Two Proteins

(1) **Root Mean Square Deviation**

The similarity between two models is usually assessed by the root mean square deviation (RMSD) between equivalent atoms in the model and native structures after the optimal superimposition [18–20]. RMSD is defined as:

$$RMSD = \sqrt{\frac{1}{N}\sum_1^N \left(x_i^2 - x_j^2\right) + \left(y_i^2 - y_j^2\right) + \left(z_i^2 - z_j^2\right)} \tag{1}$$

Where, N is the number of corresponds atoms between two protein i and j whose coordinates are and respectively.

(2) **Template Modeling Score**

Due to RMSD alone is not sufficient for globally estimating the similarity between the two proteins because the alignment coverage can be very different from approaches. Obviously, a template with a 2Å RMSD to native having 50 % alignment coverage is not necessarily better for structure modeling than one with an RMSD of 3Å but having 80 % alignment coverage. While the template aligned regions are better in the former because fewer residues are aligned, the resulting full-length model might be of poorer quality. Template Modeling Score (TM-score) function is a variation on the Levitt–Gerstein (LG) score [21, 22], which was first used for sequence independent structure alignments. TM-score is defined as:

$$TM\text{-}score = Max\left[\frac{1}{L_n}\sum_i^{L_a}\frac{1}{1 + \left(\frac{d_i}{d_0}\right)^2}\right] \tag{2}$$

where L_n is the length of the native structure, L_a is the length of the aligned residues to the template structure, di is the distance between the ith pair of aligned residues and d_0 is a scale to normalize the match difference. 'Max' denotes the maximum value after optimal spatial superposition.

2.3   Multiple K-Means Cluster Algorithm

Classical single K-means cluster algorithm includes mean step and updating centroid step. After the randomly initiating the k centroids from the whole decoys (1st line, Algorithm 1), the mean step calculates the distance from each decoy to the each centroid and groups the decoy the nearest centroid.

The updating centroid step selects the new centroid from the kth cluster. In our multiple parallel K-means algorithm, the mean step builds the main thread as the monitor thread, which forks N new threads to perform K-means cluster independently (4th–6th lines, Algorithm 1) after a local data division (3rd line, Algorithm 1). Then an energy-based weights are employed to merge the decoys from different cluster into a whole decoy set (7th line, Algorithm 1). After that, in the updating step, the monitor thread forks K thread again. Each thread responses to update one cluster centroid (9th–11th lines, Algorithm 1). Then we repeat to divide the V until the difference between less than the threshold (9th–12th lines, Algorithm 1) which indicates the algorithm is convergence or not. The algorithm is demonstrated as below. OpenMP [23–25] is used in MK-means to support multiple threading.

Algorithm 1: Multiple K-means cluster algorithm, MK-means(V,E,K,T)

Input: Distance matrix V for N decoys, E is the energy score set for the decoys, K is the number of clusters; k is the index of clusters; T is the number of threads; t is the index of threads.

Output: Cluster set: $C_1..C_k$, k is the index of the cluster.

1: Initialize(V, K);

2: while do

3: V^TDivide(V,E,T);

4: for $t = 1$ to T do

5:    K-means(V^t,K);

6: end for

7: VMerge($,E$,T,K);

8: V^KDivide(V,E,K);

9: for k = 1 to K do

10: Update(V^K,k);

11: end for

12: ;

13: end while

Output: $C_1..C_k$

---

3   Experiments and Discussion

We compared the results with the wide-used protein cluster tools, SPICKER, on I-TASSER Set-II. The results demonstrated on Table 2.

Existing clustering approaches have approximately the same performance on the SPICKER set. However, Table 2 shows that the models at the centers of the clusters identified by MK-means are superior to those identified by SPICKER. MK-means can

Table 2. Comparison between SPICKER and MK-means on 56 protein decoys

PDB	Len[a]	Size[b]	Best[c]	SPICKER[d]	MK-means[e]
1abv_	103	526	4.90	8.222	*5.048*
1af7_	72	527	3.04	4.181	4.383
1ah9_	63	510	2.22	6.262	*5.598*
1aoy_	65	529	2.69	4.233	4.233*
1b4bA	71	460	4.23	8.983	*6.54*
1b72A	49	534	2.50	4.381	*2.888*
1bm8_	99	329	6.70	6.864	13.806
1bq9A	53	573	3.96	7.847	*6.722*
1cewI	108	452	3.25	3.857	3.858
1cqkA	101	284	1.45	1.683	1.683*
1csp_	67	315	2.04	2.441	2.441*
1cy5A	92	273	1.30	1.396	1.586
1dcjA_	73	525	9.42	12.67	*12.576*
1di2A_	69	374	1.49	2.201	*2.114*
1dtjA_	74	285	1.77	2.146	3.544
1egxA	115	352	2.08	2.593	2.593*
1fadA	92	514	2.95	3.616	3.616*

PDB	Len[a]	Size[b]	Best[c]	SPICKER[d]	MK-means[e]
1mkyA3	81	285	3.78	5.502	*5.501*
1mla_2	70	335	2.36	3.23	*3.229*
1mm8A	84	545	5.32	6.808	7.456
1n0uA4	69	301	3.42	4.934	4.934*
1ne3A	56	566	3.49	6.522	7.381
1no5A	93	426	6.26	10.938	*10.715*
1npsA	88	469	1.97	2.21	11.373
1o2fB_	77	510	4.17	11.021	*8.84*
1of9A	77	507	3.16	3.94	*3.606*
1ogwA_	72	520	1.12	1.315	*1.262*
1orgA	118	442	2.18	2.745	2.745*
1pgx_	59	562	2.80	5.802	*3.263*
1r69_	61	291	1.39	2.068	2.068*
1sfp_	111	308	4.82	5.309	5.309*
1shfA	59	536	1.42	3.338	*2.928*
1sro_	71	515	2.60	3.859	4.755
1ten_	87	294	1.55	1.848	2.01

(Continued)

Table 2. (Continued)

	a	b	c	d	e		a	b	c	d	e
1fo5A	85	340	3.63	3.832	3.959	1tfi	47	339	2.46	4.694	***4.61***
1g1cA	98	307	2.26	2.672	2.772	1thx_	108	302	1.72	2.256	**2.255**
1gjxA	77	525	5.22	7.123	7.793	1tif_	59	542	6.57	7.75	9.66
1gnuA	117	553	4.12	10.563	_9.168_	1tig_	88	565	3.38	9.45	10.279
1gpt_	47	469	3.20	5.141	7.847	1vcc_	76	551	4.84	6.536	6.843
1gyvA	117	337	3.03	3.458	3.51	256bA	106	506	2.74	3.716	***3.409***
1hbkA	89	300	2.83	3.463	3.463*	2a0b_	118	282	2.19	2.42	***2.419***
1itpA	68	526	4.43	8.074	11.593	2cr7A	60	540	2.56	3.487	***3.062***
1jnuA	104	269	2.48	2.951	3.204	2f3nA	65	485	1.65	2.35	4.506
1kjs_	74	548	4.68	8.453	8.722	2pcy_	99	435	3.91	4.556	4.971
1kviA	68	550	1.65	2.291	_2.026_	2reb_2	60	550	4.38	7.07	_6.23_

a The length of the protein sequence
b The number of the models in the decoy
c The best(minimum) RMSD of the models in the decoy
d The RMSD of centroid model in the largest cluster selected by SPICKER
e The RMSD of centroid model in the largest cluster selected by MK-means(**Bold** and underline indicates better than SPICKER)
* indicates the same result between MK-means and SPICKER.

well identify 23(41 %) out of 56 proteins which are better near native than SPICKER selected.

To evaluate the overall performance of specified cluster methods, Z-score is employed to calculate the root-mean-square deviation between minimal RMSD and the RMSD of first cluster centroid as the Eq. (3). For MK-means, the Z-score is 2.70 which increase 50 % from 1.80 of SPICKER.

$$Z - score = \frac{\sum_1^N \sqrt{RMSD^2_{Cluster} - RMSD^2_{Min}}}{N} \tag{3}$$

4 Conclusions

This paper explored the ability of parallel multiple K-means (MK-means) clustered model in identifying near native models from the protein decoy. Tested on 56 proteins. The results show that MK-means can well identify 33(58.9 %) proteins which are better or the same to SPICKER selected and 10 of the 33 proteins is the same results to the SPICKER.

But, at the same time, we also found that MK-means did not achieve the results as we original expected. For further study, there is still some improvement can be taken basing on MK-means. For example, more valuable weights need to be explored. The weights here only contain information about the final energy of the proteins, while actually the weights should reflect the current model's affection to others. So the protein structural clustering is still a challenge and important task for protein 3D structural prediction.

Acknowledgments. This paper is supported by grants no. 61540058, 61202290 under the National Natural Science Foundation of China (http://www.nsfc.gov.cn) and grants no. BK20131154 under Natural Science Foundation of Jiangsu Province. The funders had no role in study design, data collection and analysis, decision to publish, or preparation of the paper. Chuang Wu and Longfei Song wrote the codes, paper and implemented the experiments, Hongjie Wu designed the algorithm, experiments and wrote the paper, Min Jiang prepared the datasets.

References

1. Huang, D.S., Zhang, L., Han, K., et al.: Prediction of protein-protein interactions based on protein-protein correlation using least squares regression. Curr. Protein Pept. Sci. **15**(6), 553–560 (2014)
2. Wu, H., Lü, Q., Quan, L., et al.: patGPCR: a multitemplate approach for improving 3D structure prediction of transmembrane helices of G-protein-coupled receptors. Comput. Math. Methods Med. **2013**(1), 151–164 (2013)
3. Yang, J., Yan, R., Roy, A., et al.: The I-TASSER suite: protein structure and function prediction. Nat. Methods **12**(1), 7–8 (2014)

4. You, Z.H., Lei, Y.K., Zhu, L., et al.: Prediction of protein-protein interactions from amino acid sequences with ensemble extreme learning machines and principal component analysis. BMC Bioinformatics 14(8), 69–75 (2013)
5. Ravinder, A., Bray, J.K., Goddard, W.A.: Bihelix: towards de novo structure prediction of an ensemble of G-protein coupled receptor conformations. Proteins Struct. Funct. Bioinformatics 80(2), 505–518 (2012)
6. Roy, A., Xu, D., Poisson, J., et al.: A protocol for computer-based protein structure and function prediction. J. Visualized Exp. 57(57), e3259–e3259 (2012)
7. Zhang, Y., Skolnick, J.: SPICKER: a clustering approach to identify near-native protein folds. J. Comput. Chem. 25(25), 865–871 (2004)
8. Jamroz, M., Kolinski, A.: ClusCo: Clustering and comparison of protein models. BMC Bioinformatics 14(1), 898 (2013)
9. Francois, B., Rojan, S., Yong, Z., et al.: Durandal: fast exact clustering of protein decoys. J. Comput. Chem. 33(4), 471–474 (2012)
10. Zhu, L., Huang, D.S.: A Rayleigh-Ritz style method for large-scale discriminant analysis. Pattern Recogn. 47(4), 1698–1708 (2014)
11. Tim, H., Mikael, B., Wouter, B., et al.: Fast large-scale clustering of protein structures using Gauss integrals. Bioinformatics 28(4), 510–515 (2012)
12. Zhang, J., Xu, D.: Fast Algorithm for Clustering a Large Number of Protein Structural Decoys. In: Proceedings of the 2011 IEEE International Conference on Bioinformatics and Biomedicine, pp. 30–36. IEEE Computer Society (2011)
13. Zhang, Y., Skolnick, J.: Scoring function for automated assessment of protein structure template quality. Proteins Struct. Funct. Bioinformatics 68(4), 702–710 (2007)
14. Shatabda, S., Newton, M.A., Rashid, M.A., et al.: How good are simplified models for protein structure prediction? Adv. Bioinformatics 2014, 867179 (2014)
15. Zhou, J., Wishart, D.S.: An improved method to detect correct protein folds using partial clustering. BMC Bioinformatics 14(1), 101 (2013)
16. Tan, C.W., Jones, D.T.: Using neural networks and evolutionary information in decoy discrimination for protein tertiary structure prediction. BMC Bioinformatics 9(4), 1–23 (2008)
17. Wu, S., Skolnick, J., Zhang, Y.: Ab initio modeling of small proteins by iterative TASSER simulations. BMC Biol. 5, 17 (2007)
18. Kabsch, W.: A solution for the best rotation to relate two sets of vectors. Acta Cryst A32, 922–923 (1976)
19. Kabsch, W.: A discussion of the solution for the best rotation to relate two sets of vectors. Acta Cryst A34, 827–828 (1978)
20. Dehzangi, A., Paliwal, K., Lyons, J., et al.: Proposing a highly accurate protein structural class predictor using segmentation-based features. BMC Genom. 15(Suppl 1), 133–139 (2014)
21. Levitt, M., Gerstein, M.: A unified statistical framework for sequence comparison and structure comparison. Proc. Nat. Acad. Sci. U.S.A. 95(11), 5913–5920 (1998)
22. Zhang, J., Xu, D.: Fast algorithm for population-based protein structural model analysis. Proteomics 13(2), 221–229 (2013)
23. Dagum, L., Menon, R.: OpenMP: an industry-standard API for shared-memory programming. IEEE Comput. Sci. Eng. 5(1), 46–55 (1998)
24. Jain, Anil K.: Data Clustering: 50 Years Beyond K-means. Pattern Recogn. Lett. 31(8), 651–666 (2015)
25. Huang, T., Lu, D.T., Li, X., et al.: GPU-based SNESIM implementation for multiple-point statistical simulation. Comput. Geosci. 54(4), 75–87 (2013)

Computational Analysis of Similar Protein-DNA Complexes from Different Organisms to Understand Organism Specific Recognition

R. Nagarajan and M. Michael Gromiha[✉]

Department of Biotechnology, Bhupat and Jyoti Metha School of Biosciences,
Indian Institute of Technology Madras, Chennai 600036, Tamilnadu, India
gromiha@iitm.ac.in

Abstract. Protein-DNA interactions play vital roles in many cellular processes. It is not clear whether the recognition of same protein-DNA complexes in different organisms is similar or different. In this work, we have analyzed the similarities and variations in interactions and interacting patterns in a set of 41 similar protein-DNA complexes from different organisms based on several features such as propensity of binding site residues, preference of binding segments, preferred amino acid-nucleotide pairs, etc. Based on the analysis, we showed the variations in interactions and interacting patterns of similar protein-DNA complexes from different organisms, which possibly lead to difference in recognition mechanism.

Keywords: Protein-DNA interaction · Binding site residues · Recognition mechanism · Propensity · Binding specificity

1 Introduction

Protein-DNA interactions play crucial roles in many cellular processes including DNA replication, transcription, gene regulation, DNA repair and packaging. Identifying DNA-binding residues and determining the specificity of protein-DNA interactions will help in understanding the specific recognition mechanism of protein–DNA complexes. The recognition mechanisms with its functional significance were investigated using several experiments and computational methods [1–5]. The structures of protein-DNA complexes obtained from three dimensional structure determinations [6] have been effectively utilized to reveal important factors in the study of protein-DNA recognition. The binding site residues have been identified using several approaches including atomic distance based criteria, solvent accessibility and energy based approach [7–9].

Investigating protein-DNA complexes using the properties of amino acids, secondary structures, conservation, conformational changes and non-covalent interactions has revealed the importance of hydrogen bonds, electrostatic, hydrophobic, van der Waals interactions and weak interactions [1–5, 10–20] in protein-DNA recognition. In addition, the contributions of energetic terms along with the combination of inter- and intra-molecular interactions have been used to understand the recognition mechanism [18].

© Springer International Publishing Switzerland 2016
D.-S. Huang and K.-H. Jo (Eds.): ICIC 2016, Part II, LNCS 9772, pp. 888–894, 2016.
DOI: 10.1007/978-3-319-42294-7_79

Further, statistical potentials derived from the contacts between protein and DNA has been utilized successfully to predict the specificity of complexes [21, 22]. On the other hand, several computational methods have been developed to predict DNA binding site residues based on protein structural and/or sequence features such as side chain pK_a, hydrophobicity index, molecular mass, position specific scoring matrices [PSSM], evolutionary conservation, secondary structure and solvent accessibility [23, 24].

In our earlier work, we have reported the interacting patterns of same protein-RNA complexes vary among different organisms [25]. In this work, we have analyzed the binding site residues of same protein-DNA complexes from different organisms in terms of binding propensity, preference of neighboring residues, binding segments, amino acid-nucleotide pair preference, etc. From the analysis, we have found that positively charged residues are highly preferred over negatively charged and most aliphatic residues. Ser, Tyr and Asn are highly preferred in thermophiles whereas Trp and Thr are preferred in bacteria. Binding sites/segments with single residues are more common; segments of 5 residues or more are not preferred, particularly in fungi. Further analysis showed that the preference of neighboring residues and amino acid-nucleotide pairs are unique in different organisms.

2 Materials and Methods

2.1 Dataset Preparation

We have collected the available protein-DNA complex structures from PDB [6] and constructed 41 sets of protein-DNA complexes belonging to different organisms with the following criteria: [i] the structures must be available for at least two organisms, [ii] the protein should contain a minimum of 30 residues, [iii] the DNA should have at least 5 nucleotides and [iv] the sequence identity should be more than 30 % in each set. As of our knowledge, this is the first dataset of similar protein-DNA complexes from different organisms in the literature. Further, we have classified the protein-DNA complexes into 5 major groups based on the source organism: *Homo sapiens, Mus-musculus*, fungi, bacteria and thermophiles.

2.2 Identification of Binding Site Residues [Nucleotides]

We have identified the binding site residues [nucleotides] for the considered protein-DNA complexes by following distance based approach. Using this approach, we have calculated the distance between two heavy atoms from protein and DNA and if the distance is less than 3.5Å, then the respective residues [nucleotides] are identified as binding site residues [nucleotides].

2.3 Binding Propensity

We have calculated the binding propensity of amino acids and nucleotides using the following approach [9, 26]:

$$P_{bind}[i] = f_b[i] * 100/f_t[i]$$

Where, i denotes each of the 20 amino acids and 4 nucleotides, f_b is the frequency of occurrence of amino acid residues [nucleotides] in binding sites and f_t is the frequency of occurrence of amino acid residues [nucleotides] in the protein [DNA] as a whole.

In addition, we identified the conserved sites in each protein chain using the Consurf server [27] and computed the binding propensity of amino acids in the conserved sites to analyze the variations with the binding propensity of whole protein.

2.4   Preference of Binding Segments

We have analyzed the binding sites in terms of binding segments, which is based on the number of consecutive binding site residues [nucleotides]. For example, a 5-residue binding segment denotes a stretch of five consecutive binding site residues. We have considered the binding segments with sizes one, two, three, four, five, six and more than six residues.

3   Results and Discussion

3.1   Binding Site Residues in Protein-DNA Complexes

The analysis on binding site residues and nucleotides showed their variations based on the organism in similar protein-DNA complexes. For example, in Methyl-CpG-binding domain protein 4, we observed 12.9 % and 17.8 % of binding site residues in the protein from *H. sapiens* and *M. musculus*, respectively. In the case of DNA, 37.5 % and 40.9 % of binding nucleotides are observed from *H. sapiens* and *M. musculus*, respectively. Similar trends are observed in other protein-DNA and protein-RNA complexes [25].

3.2   Binding Propensity of Amino Acid Residues in Different Organisms

The binding propensity of all 20 amino acids showed that Ser, Thr, Asn, Tyr as well as positively charged residues are highly preferred in most organisms with propensity greater than 1 whereas other residues including aliphatic residues are less preferred with a propensity of less than 1. Further, Trp is highly preferred only in bacteria with a propensity value of more than 2 and Ser and Asn are preferred in thermophiles. In addition, Arg is highly preferred in *M. musculus* with the highest propensity of 3.8 and Tyr, Asn, Lys and Arg are preferred in *H. sapiens*. The observed differences in the preferences indicate the variations in the interactions among the organisms.

3.3   Binding Propensity of Nucleotides in Different Organisms

The binding propensity of each nucleotide in different organisms revealed that guanine is highly preferred in *H. sapiens*, *M. musculus* and bacteria whereas thymine is preferred in Fungi. In addition, it is noticed that the preference of all the nucleotides are high in thermophiles, which clearly shows the larger number of binding nucleotides than in other organisms.

3.4   Binding Propensity of Amino Acid Residues in Conserved Regions

We have further retrieved the conserved sites for all the protein-DNA complexes using the Consurf server [27] and have analyzed the binding propensity of amino acids at conserved regions [Fig. 1]. We observed similar trends in the binding preference of amino acids in both conserved regions and the whole protein. In addition, we observed changes in the propensities of some residues at conserved sites compared to the whole protein. The preference of His in fungi, Glu in bacteria and fungi and Arg in thermophiles are increased while that of Tyr in Fungi and Phe and Gly in *H. sapiens* are decreased compared to the binding propensities in the whole protein. These results clearly indicate that the binding preference of residues also varies at conserved regions.

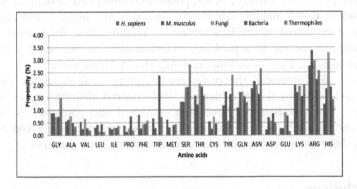

Fig. 1. Binding propensity of amino acid residues in conserved regions (Color figure online)

3.5   Binding Segments in Protein-DNA Complexes

We have identified and analyzed the preference of binding segments in proteins. Similar to reports in protein-RNA complexes, single-residue segments are highly preferred in all organisms followed by segments with two residues. In addition, we observed that the binding segments of 5 residues or more are not preferred in Fungi completely.

Similar to proteins, we have computed the preference of binding segments in DNA and the results are shown in Fig. 2. Unlike proteins, the preferences are vary among the organisms. Single nucleotide segments in *H. sapiens* and fungi, segment with 3 nucleotides in *M. musculus*, segments with 5 nucleotides in bacteria and segments with

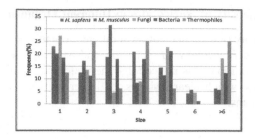

Fig. 2. Preference of binding segments of DNA in different organisms (Color figure online)

2, 4 and 6 nucleotides in thermophiles are highly preferred binding segments in each organism. Segments with 6 nucleotides are commonly less preferred in all the organisms. These results reveal the variations in the preference of binding segments in different organisms.

3.6   Preference of Neighboring Residues and Amino Acid-Nucleotide Pairs

We have computed the preference of 400 possible combinations of both categories [XB and BX, where B is a binding site residue and X is any residue] and analyzed the preference of neighbouring residues around the binding sites. We noticed that some residue pairs are specific to a particular organism such as Ser-Ser in thermophiles, Asn-Lys in bacteria, Thr-Thr in fungi and Asp-His in *M. musculus*.

We have also evaluated the preference of amino acid-nucleotide pairs at the interface and normalized the preference with overall distribution to obtain the propensities. These propensities have been converted into potentials to identify the preferred and avoided pairs. We noticed that these preferred and avoided pairs are specific to different organisms and the detailed analysis is on progress.

4   Conclusions

We have investigated the similarities and differences in the interactions and interacting patterns of similar protein-DNA complexes from different organisms based on various features such as binding propensity of amino acids and nucleotides, propensity in the conserved sites, preference of binding segments, neighboring residues and amino acid-nucleotide pairs to understand the organism specific protein-DNA recognition. The results showed that the preference of amino acids, nucleotides, neighboring residues and amino acid-nucleotide pairs vary in different organism. Further, these results can be utilized for organism specific prediction of DNA binding site residues and specificity of protein-DNA complexes. As this is the first report and dataset in the literature about protein-DNA complexes from different organism, it will helpful to further investigate and understand the organism specific recognition of protein-DNA complexes.

Acknowledgements. RN and MMG thank and Indian Institute of Technology Madras for computational facilities.

References

1. Cherstvy, A.G., Kolomeisky, A.B., Kornyshev, A.A.: Protein DNA interactions: reaching and recognizing the targets. J. Phys. Chem. B **112**, 4741–4750 (2008)
2. Fuxreiter, M., Simon, I., Bondos, S.: Dynamic protein-DNA recognition: beyond what can be seen. Trends Biochem. Sci. **36**(8), 15–23 (2011)
3. Gromiha, M.M., Fukui, K.: Scoring function based approach for locating binding sites and understanding the recognition mechanism of protein-DNA complexes. J. Chem. Inf. Model. **51**(3), 7–9 (2011)
4. Nadassy, K., Wodak, S.J., Janin, J.: Structural features of protein-nucleic acid recognition sites. Biochemistry **38**, 1999–2017 (1999)
5. Paillard, G., Lavery, R.: Analyzing protein-DNA recognition mechanisms. Structure **12**(1), 11–22 (2004)
6. Rose, P.W., Prlić, A., Bi, C., Bluhm, W.F., Christie, C.H., Dutta, S., Green, R.K., Goodsell, D.S., Westbrook, J.D., Woo, J., Young, J., Zardecki, C., Berman, H.M., Bourne, P.E., Burley, S.K.: The RCSB Protein Data Bank: views of structural biology for basic and applied research and education. Nucleic Acids Res. **43**, 45–56 (2015)
7. Ahmad, S., Gromiha, M.M., Sarai, A.: Analysis and prediction of DNA-binding proteins and their binding residues based on composition, sequence and structural information. Bioinformatics **20**, 477–486 (2004)
8. Tjong, H., Zhou, H.-X.: DISPLAR: an accurate method for predicting DNA-binding sites on protein surfaces. Nucleic Acid Res. **35**, 1465–1477 (2007)
9. Gromiha, M., Selvaraj, S., Jayaram, B., Fukui, K.: Identification and analysis of binding site residues in protein complexes: energy based approach. In: Huang, D.-S., Zhao, Z., Bevilacqua, V., Figueroa, J.C. (eds.) ICIC 2010. LNCS, vol. 6215, pp. 626–633. Springer, Heidelberg (2010)
10. Sarai, A., Kono, H.: Protein-DNA recognition patterns and predictions. Annu. Rev. Biophys. Biomol. Struct. **34**, 379–398 (2005)
11. Hogan, M.E., Austin, R.H.: Importance of DNA stiffness in protein-DNA binding specificity. Nature **329**, 263–266 (1987)
12. Gromiha, M.M., Munteanu, M.G., Simon, I., Pongor, S.: The role of DNA bending in Cro protein-DNA interactions. Biophys. Chem. **69**, 153–160 (1987)
13. Olson, W.K., Gorin, A.A., Lu, X.J., Hock, L.M., Zhurkin, V.B.: DNA sequence-dependent deformability deduced from protein-DNA crystal complexes. Proc. Natl. Acad. Sci. USA **95**, 11163–11168 (1998)
14. Gromiha, M.M.: Influence of DNA stiffness in protein-DNA recognition. J. Biotechnol. **117**, 137–145 (2005)
15. Mandel-Gutfreund, Y., Margalit, H., Jernigan, R.L., Zhurkin, V.B.: A role for CH···O interactions in protein-DNA recognition. J. Mol. Biol. **277**, 1129–1140 (1998)
16. Jones, S., van Heyningen, P., Berman, H.M., Thornton, J.M.: Protein-DNA interactions: a structural analysis. J. Mol. Biol. **287**, 877–896 (1999)
17. Jayaram, B., McConnell, K., Dixit, S.B., Das, A., Beveridge, D.L.: Free-energy component analysis of 40 protein-DNA complexes: a consensus view on the thermodynamics of binding at the molecular level. J. Comput. Chem. **23**, 1–14 (2002)

18. Gromiha, M.M., Siebers, J.G., Selvaraj, S., Kono, H., Sarai, A.: Intermolecular and intramolecular readout mechanisms in protein-DNA recognition. J. Mol. Biol. **2004**(337), 285–294 (2004)

19. Prabakaran, P., Siebers, J.G., Ahmad, S., Gromiha, M.M., Singarayan, M.G., Sarai, A.: Classification of protein-DNA complexes based on structural descriptors. Structure **14**, 1355–1367 (2006)

20. Gromiha, M.M., Santhosh, C., Suwa, W.: Influence of cation-pi interactions in protein-DNA complexes. Polymer **45**, 633–639 (2004)

21. Kono, H., Sarai, A.: Structure-based prediction of DNA target sites by regulatory proteins. Proteins **35**, 114–131 (1999)

22. Donald, J.E., Chen, W.W., Shakhnovich, E.I.: Energetics of protein-DNA interactions. Nucleic Acids Res. **35**, 1039–1047 (2007)

23. Gromiha, M.M., Nagarajan, R.: Computational approaches for predicting the binding sites and understanding the recognition mechanism of protein-DNA complexes. Adv. Protein Chem. Struct. Biol. **91**, 65–99 (2013)

24. Nagarajan, R., Ahmad, S., Gromiha, M.M.: Novel approach for selecting the best predictor for identifying the binding sites in DNA binding proteins. Nucleic Acids Res. **41**, 7606–7614 (2013)

25. Nagarajan, R., Chothani, S.P., Ramakrishnan, C., Sekijima, M., Gromiha, M.M.: Structure based approach for understanding organism specific recognition of protein-RNA complexes. Biol. Direct **10**, 8 (2015)

26. Gromiha, M.M., Yokota, K., Fukui, K.: Energy based approach for understanding the recognition mechanism in protein-protein complexes. Mol. BioSyst. **5**, 1779–1786 (2009)

27. Glaser, F., Pupko, T., Paz, I., Bell, R.E., Bechor, D., Martz, E., Ben-Tal, N.: ConSurf: identification of functional regions in proteins by surface mapping of phylogenetic information. Bioinformatics **19**, 163–164 (2003)

Advances in Swarm Intelligence:
Algorithms and Applications

Adaptive Structure-Redesigned-Based Bacterial Foraging Optimization

L.J. Tan[1], W.J. Yi[2(✉)], C. Yang[2(✉)], and Y.Y. Feng[2]

[1] Department of Business Management,
Shenzhen Institute of Information Technology,
Shenzhen 518172, China
[2] College of Management, Shenzhen University, Shenzhen 518060, China
yiwenjie1006@gmail.com, yangc@szu.edu.cn

Abstract. This paper proposes an adaptive structure-redesigned-based bacterial foraging optimization called ASRBFO. In this improved algorithm, the chemotaxis step of SRBFO is adaptively adjusted based on the bacterial searching status. The personal current and best positions of bacteria as well as the mean of all bacterial positions are taken and used to calculate the chemotaxis step during the searching process. The goal of the study is to improve the convergence efficiency and the accuracy of SRBFO. To demonstrate the performance, six different benchmark functions are chosen to the experiment, and other three SRBFOs are used to compare with the proposed algorithm. The results show that ASRBFO outperforms other SRBFOs.

Keywords: Adaptive adjustment · Structure-redesigned-based bacterial foraging optimization (SRBFO) · ASRBFO

1 Introduction

Bacterial Foraging Optimization Algorithm (BFO), originally proposed by Passino [1] in 2002, which is inspired by the foraging behaviors of the E. coli bacteria, including chemotaxis, reproduction, elimination-dispersal, has became a member of the field of intelligent computing. BFO has already been used to solve plenty of engineering problems, such as economic dispatch [2], nurse scheduling [3], image registration [4], vehicle routing problem [5], and portfolio optimization [6], etc.

After the introduction of chemotaxis step into the algorithm, Passino [1] observed a phenomenon, that is, at the beginning of an algorithm, the larger the step is, the more facilitative the comprehensive search is. While at the end, the smaller the step is, the more facilitative the intensive search is. In recent years, many scholars have begun to study the step change strategy for the purpose of improving the algorithm's performance. Niu et al. [7] put forward a linear decreasing strategy called BFO-LDC. Xu et al. [8] presented a new chemotactic mechanism, which is inspired by the time-varying acceleration coefficients in hierarchical PSO, and thus, proposed a novel adaptive BFO called ABSA, and so on.

© Springer International Publishing Switzerland 2016
D.-S. Huang and K.-H. Jo (Eds.): ICIC 2016, Part II, LNCS 9772, pp. 897–907, 2016.
DOI: 10.1007/978-3-319-42294-7_80

In the previous researches of chemotaxis step adjustment strategy, most of them are linearly or nonlinearly decreasing based on the relationship between the current iteration algebra and the maximum iteration, which do not consider the effect of the variation of the bacterial fitness. In this paper, we propose a new adjustment strategy which is made up of three steps. The first step is to make chemotaxis step change in a given range by using nonlinear strategy based on the current iteration and the maximum number of iterations. Then we calculate two parameters defined in this paper respectively. Furthermore, the two parameters are introduced in the form of proportional or inverse ratio.

The organizational structure of the paper is as follows: Sect. 1 presents brief introduction of BFO. Sections 2 and 3 provide simple principle introduction of BFO as well as SRBFO. In Sect. 4, a detailed introduction about ASRBFO which is the algorithm proposed by this paper, is given. Section 5 discusses the experimental setting, benchmark function selection and the result analysis. Finally, Sect. 6 concludes the results of the work.

2 Bacterial Foraging Optimization

Bacterial Foraging Optimization Algorithm is a novel heuristic algorithm that searches for the optimal solution randomly, which is inspired by social behaviors of bacteria. This method first proposed by Passino [1] has raised wide attention of scholars recently.

BFO mainly simulates three important bacterial behaviors in the process of foraging, including chemotaxis, reproduction, elimination-dispersal, and then solves the optimization problem. Bacterial foraging process imitates the bacterial cell's movement that through tumbling and swimming by means of flagella. It tends to move to the region which is full of nutrient and avoid the harmful environment. After intaking proper nutrition, bacteria begins to replicate reproduction for maintaining the stability of the population. But during the process of chemotaxis and copying process, the environment may be damaged so that the bacterial will die or migrate to another region. Through such evolution, the survival bacterial health quality can be guaranteed.

BFO contains three levels of circulation. From the inside to the outside are chemotaxis, reproduction, extinction and dispersal. The loop structure is nested, which performs high memory consumption and low efficiency.

3 Structure-Redesigned-Based Bacterial Foraging Optimization

Original BFO algorithm has inherent defects, such as high memory consumption and slow execution. In order to improve the convergence efficiency and lower the computational complexity, Niu et al. [9] proposed the Structure-Redesigned-Based Bacterial Foraging Optimization Algorithm (SRBFO) by using the single-loop structure to redesign the BFO's chemotactic-reproduce-disperse structure. Through this way, it's greatly simplified and the operation efficiency is improved. SRBFO is successfully

Table 1. The optimization process of SRBFO

Step 1	Initialize the population, and record the current fitness J, let J $_{last}$ = J
Step 2	Chemotaxis cycle: $l = l+1$, execute chemotactic operation, and update J
Step 3	If J < J $_{last}$, continue to move along the current direction, otherwise, turn to the next step
Step 4	If mod (l, Fre) ==0, execute reproductive operation, otherwise, turn to the next step
Step 5	If mod(l, Fed) ==0, execute disperse operation, otherwise, turn to the next step
Step 6	If l < Nc, turn to Step 2, otherwise, finish the algorithm and output the result

applied in solving the investment portfolio problem and others. Optimization process of SRBFO is listed in Table 1.

4 Adaptive SRBFO

In the previous study of step adjustment strategy, most of them are linearly or non-linearly decreasing based on the relationship between the current iteration algebra ad the maximum iteration, in which the effect of the variation of the bacterial fitness on the chemotaxis step isn't taken into account.

In this paper, a new adjustment strategy is proposed. The main ideas are as follows:

(i) Based on the current iteration and the maximum number of iterations, we make chemotaxis step change with a given range by using nonlinear strategy. With the increase in the number of iterations, the step shows a decreasing trend. As is shown below:

$$A = \left[(c_{max} - c_{min} - k_1 - k_2) * \cos\left(\frac{iter}{iter_{max}} * \frac{\pi}{2}\right) + c_{min} + k_2 \right] \quad (1)$$

$$A \in [c_{min} + k_2, c_{max} - k_1]$$

(ii) After determining the initial value of the current iteration (i.e. A), the step begins to adjust adaptively based on the aggregation degree of the bacteria. The aggregation degree parameter is shown below:

$$B = \frac{\pi}{2} * \arctan \frac{J}{J_{avg}} \quad (2)$$

$$B \in [0, 1]$$

with the increase of B, the value of the chemotaxis step will increase, that is, B is proportional to the step size.

(iii) After considering the influence of the degree of aggregation, the effect of particle adjustment on the step size is further considered. The adjustment parameter is shown below:

$$C = \frac{\min(J, J_t)}{\max(J, J_t)}$$

(3)

$$C \in [0, 1]$$

The value of C is smaller, fitness adjusted speed is faster, indicating that bacteria are globally searching in large areas, and therefore the bacteria need larger the chemotactic step. The value of C is bigger, fitness adjustment speed is slower, indicating that bacteria are locally searching in the small areas, and therefore the bacteria need smaller the chemotactic step. That is, the chemotactic step is inversely proportional to the value of C.

(iv) Based on the points mentioned above, this paper proposes an adaptive adjustment strategy as follows:

Table 2. Pseudo code of the ASRBFO algorithm

While $iter<=iter_{max}$

-------------%Chemotaxis---

-------------%Swim---

   m=0;

   while m<Ns

      if $J_t(1,st)>J(1,st)$

      J_{avg}=mean($J(1,st)$);

   C=(((C_{max}-k_1)-($iter/iter_{max}$)*(C_{max}-C_{min}-k_1-k_2))+..

k_1*(2/pi)*atan($J(1,st)/J_{avg}$)-k_2*($J(1,st)$/ $J_t(1,st)$))*ones(Dim,SS);

      $J_t(1,st)=J(1,st)$;

------ %move follow the profitable direction--------------------

---

$$C(j) = \left[(c_{max} - c_{min} - k_1 - k_2) * \cos(\frac{iter}{iter_{max}} * \frac{\pi}{2}) + c_{min} + k_2 \right]$$
$$+ k_1 * \frac{\pi}{2} * \arctan\frac{J}{J_{avg}} - k_2 * \frac{\min(J, J_t)}{\max(J, J_t)}$$

(4)

Among the formula, c_{max}. represents the initial step size as well as the maximum step size and c_{min} represents the termination step size and the minimum step size. $.k_1$ and k_2 are two positive constants and both of their value are determined by experiment. *iter* is the current iteration number, and $iter_{max}$ represents the maximum of the iteration. Besides, J and J_{avg} represent the current bacterial fitness and the mean value of all the bacterial fitness. The last parameter J_t is the individual optimal fitness value before moving, which is the temporary variable for comparison.

The Pseudo Code of ASRBFO Algorithm is given in Table 2.

5   Experiments and Results

5.1   Benchmark Functions and Experiment Setting

To verify the efficiency of the proposed algorithm, three well-known unimodal benchmark functions and another three multimodal benchmark functions have been chosen. The minimum values of all the six functions are the same and are equal to zero. Table 3 shows the different search ranges of the chosen functions whereas Table 4 represents the parameters of four different SRBFOs. There is no difference in most parameters of the entire algorithm, but, it should be noticed that the chemotactic step of SRBFO is a fixed value named c_{sz}, which is equal to the c_{min} of the other three algorithms. In addition, c_{max} is set to be 0.3, and k_1, k_2 of the ASRBFO are set to be 2, 0.4 respectively. The change laws of the chemotactic step in four different SRBFOs are shown in Table 5.

Table 3. The search ranges of the benchmark function

Name	Function	Range
Sphere	$f(x) = \sum_{i=1}^{n} x_i^2$	[−100, 100]
Rosenbrock	$f(x) = \sum_{i=1}^{n} 100 * (x_{i+1} - x_i^2) + (1 - x_i^2)^2$	[−100, 100]
Schwefel221	$f(x) =_{i=1}^{n} \max\{|x_i|\}$	[−100, 100]

(Continued)

Table 3. (*Continued*)

Name	Function	Range
Ackley	$f(x) = -20 \exp\left(-0.2 * \sqrt{\frac{1}{30}\sum_{i=1}^{n} x_i^2}\right)$ $- \exp\left(\frac{1}{30} * \sum_{i=1}^{n} \cos 2\pi x_i\right) + 20 + e$	$[-32, 32]$
Griewank	$f(x) = \frac{1}{4000}\sum_{i=1}^{n} x_i^2 - \prod_{i=1}^{n}\cos\left(\frac{x_i}{\sqrt{i}}\right) + 1$	$[-600, 600]$
Weierstrass	$f(x) = \sum_{i=1}^{n}\left(\sum_{k=0}^{20}\left[0.5^k\cos\left(2\pi * 3^k(x_i + 0.5)\right)\right]\right)$ $-n\sum_{k=0}^{20}\left[\left(0.5^k\cos\left(3^k * \pi\right)\right)\right]$	$[-0.5, 0.5]$

Table 4. The parameter of setting of four different SRBFOs

Algorithm	S	N_S	P_{ed}	F_{re}	MaxFEs	c_{sz}/c_{min}	c_{max}	MaxFEs
SRBFO	100	10	0.25	55000	110000	0.15		300000
SRBFOLDC/ SRBFONDC	100	10	0.25	55000	110000	0.15	0.3	300000
ASRBFO	100	10	0.25	55000	110000	0.15	0.3	300000

Table 5. The chemotactic step of four different SRBFOs

Algorithm	Chemotactic step
SRBFO	Fixed valve: c_{sz}
SRBFOLDC	Linear decreasing: $C(j) = c_{max} - \frac{iter}{iter_{max}} * (c_{max} - c_{min})$
SRBFONDC	Nonlinear decreasing: $C(j) = (c_{max} - c_{min}) * \frac{2}{\pi} * ar\cos\left(\frac{iter}{iter_{max}}\right) + c_{min}$
ASRBFO	Adaptive adjustment: $C(j) = \left[(c_{max} - c_{min} - k_1 - k_2) * \cos\left(\frac{iter}{iter_{max}} * \frac{\pi}{2}\right) + c_{min} + k_2\right]$ $+ k_1 * \frac{\pi}{2} * \arctan\frac{J}{J_{avg}} - k_2 * \frac{\min(J,J_t)}{\max(J,J_t)}$

5.2   Results and Analyses

Table 6 and Fig. 1 present the results on six different benchmark functions in 10-D whereas Table 7 and Fig. 2 present the results on the same benchmark functions in 30-D. Tables 6 and 7 show the convergence characteristics in terms of the min value, the max value, the mean value, and standard deviation of the results for every benchmark function. The optimum values obtained are in bold in Tables 6 and 7.

Table 6. Numerical results of six benchmark functions for 10-D

Algorithm	Sphere	Rosenbrock	Schwefel	Ackley	Griewank	Weierstrass
SRBFO	0.6783	136.6659	9.59e−01	3.83e+00	2.91e+01	**7.23e+00**
	0.7797	671.8181	3.76e+00	6.49e+00	6.55e+01	8.34e+00
	0.7363	444.8181	2.01e+00	4.78e+00	4.16e+01	7.95e+00
	0.0522	276.6518	1.53e+00	1.49e+00	2.07e+01	6.24e−01
SRBFOLDC	0.2733	59.4166	5.25e−01	2.65e+00	3.71e+01	7.41e+00
	0.2966	70.3905	8.37e−01	3.78e+00	4.90e+01	8.49e+00
	0.2827	63.8916	7.08e−01	3.13e+00	4.33e+01	**7.84e+00**
	0.0123	5.7601	1.63e−01	5.82e−01	5.96e+00	5.75e−01
SRBFONDC	0.3519	32.9517	6.21e−01	3.35e+00	1.45e+01	7.95e+00
	0.5111	92.7531	7.90e−01	3.60e+00	4.71e+01	8.63e+00
	0.4066	64.3259	7.08e−01	3.49e+00	3.27e+01	8.33e+00
	0.0906	30.0090	8.43e−02	**1.30e−01**	1.66e+01	3.48e−01
ASRBFO	**0.0399**	**25.7351**	**4.07e−01**	**1.73e+00**	**1.96e+00**	7.74e+00
	0.1190	**30.3252**	**5.75e−01**	**2.45e+00**	**4.43e+00**	**8.24e+00**
	0.0931	**28.2867**	**4.76e−01**	**2.11e+00**	**2.99e+00**	7.96e+00
	0.0466	**2.4274**	**8.81e−02**	3.64e−01	**1.28e+00**	**2.57e−01**

For Sphere Function, as is illustrated in Figs. 1 and 2, ASRBFO has a better ability to search the minimum fitness within 300000 generations. As is shown in Tables 6 and 7, comparing with the other three algorithms, ASRBFO gets the best value in the minimum, maximum and mean although it does not get the minimum standard deviation. For Rosenbrock Function, ASRBFO gets the best value of all the four parameters whether it is in the 10 dimension or in the 30 dimension. For the third unimodal functions, ASRBFO still performs very well as the other two.

For Ackley Function, ASRBFO can find the minimum fitness with greatly faster speed in 10-D, but it suffers from premature convergence the same as SRBFO and SRBFONDC in 30-D. For Griewank Function, as is shown above, the algorithm that this paper proposed obtains the best value of four parameters of algorithm, including the minimum value, the maximum value, the mean value as well as the standard deviation in whichever dimension. Although ASRBFO gets good results both in Ackley and Griewank, it still falls into premature convergence the same as the other algorithms when it comes to the Weierstrass Function.

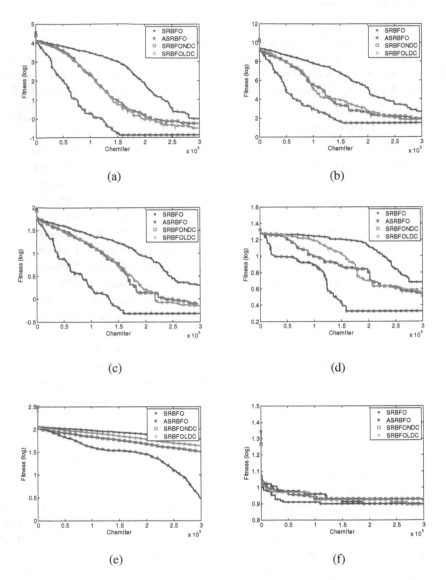

Fig. 1. Convergence characteristics on six benchmark functions in 10-D (a) Sphere Function (b) Rosenbrock Function (c) Schwefel221 Function (d) Ackley Function (e) Griewank Function (f) Weierstrass Function (Color figure online)

Table 7. Numerical results of six benchmark functions for 30-D

Algorithm	Sphere	Rosenbrock	Schwefel	Ackley	Griewank	Weierstrass
SRBFO	6.37e+03	2.70e+08	4.08e+01	1.93e+01	3.47e+02	**3.75e+01**
	7.90e+03	5.61e+08	4.46e+01	2.01e+01	4.90e+02	3.96e+01
	7.04e+03	4.08e+08	4.30e+01	1.97e+01	4.11e+02	**3.87e+01**
	7.87e+02	1.46e+08	2.01e+00	**3.77e−01**	7.25e+01	1.07e+00
SRBFOLDC	3.94e+02	2.81e+06	3.21e+01	**1.73e+01**	3.30e+02	3.85e+01
	7.35e+02	7.00e+06	3.59e+01	2.00e+01	3.79e+02	3.95e+01
	5.76e+02	5.22e+06	3.35e+01	**1.90e+01**	3.57e+02	3.89e+01
	1.71e+02	2.16e+06	2.10e+00	1.50e+00	2.48e+01	**5.44e−01**
SRBFONDC	3.15e+02	**3.10e+05**	2.69e+01	1.96e+01	3.66e+02	3.88e+01
	3.86e+02	1.24e+06	2.93e+01	2.02e+01	4.16e+02	3.99e+01
	3.45e+02	8.94e+05	2.77e+01	2.00e+01	3.93e+02	3.93e+01
	3.68e+01	5.09e+05	1.37e+00	3.01e−01	2.55e+01	5.69e−01
ASRBFO	**1.69e+02**	4.75e+05	**2.14e+01**	1.96e+01	**1.55e+02**	3.81e+01
	2.51e+02	**5.71e+05**	**2.31e+01**	**1.99e+01**	**1.81e+02**	**3.94e+01**
	2.04e+02	**5.21e+05**	**2.22e+01**	1.98e+01	**1.72e+02**	3.88e+01
	4.25e+01	**4.77e+04**	**8.56e−01**	1.46e−01	**1.47e+01**	6.74e−01

Based on the experimental results, ASRBFO performs better than other algorithms in most benchmark functions whether it is unimodal or multimodal both in 10-D and 30-D. It can achieve better convergence efficiency and the accuracy while avoiding premature convergence. But for Weierstrass Function, the algorithm failed to do that.

6 Conclusions and Further Work

In this paper, an adaptive structure-redesigned-based bacterial foraging optimization called ASRBFO is presented. Compared with the other algorithms, the effect of the variation of the bacterial fitness is considered.

In considering the effect of the variation of the fitness, bacteria can adjust step size more flexibly. As a result, ASRBFO can avoid getting struck at the local optima and converge to a better fitness at a faster speed. In order to validate the proposed algorithm, three unimodal functions and three multimodal functions are chosen to the experiment both in 10-D and 30-D. As is shown in the experimental results, ASRBFO can achieve the convergence efficiency and the accuracy while avoid premature convergence. In a word, ASRBFO has better convergence efficiency compared with other algorithms.

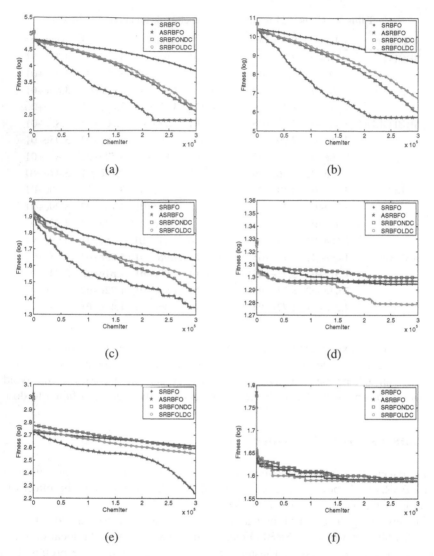

Fig. 2. Convergence characteristics on six benchmark functions in 30-D (a) Sphere Function (b) Rosenbrock Function (c) Schwefel221 Function (d) Ackley Function (e) Griewank Function (f) Weierstrass Function (Color figure online)

Acknowledgment. This work is partially supported by The National Natural Science Foundation of China (Grants nos. 71571120, 71271140, 71461027, 71471158) and the Natural Science Foundation of Guangdong Province (Grant no. 2016A030310074).

References

1. Passino, K.M.: Biomimicry of bacterial foraging for distributed optimization and control. IEEE Control Syst. **22**(3), 52–67 (2002)
2. Hassan, E.E., Zakaria, Z., Rahman, T.K.: Improved Adaptive Tumbling Bacterial Foraging Optimization (ATBFO) for emission constrained economic dispatch problem. Lect. Notes Eng. Comput. Sci. **2198**(1), 975–978 (2012)
3. Niu, B., Wang, C., Liu, J., Gan, J., Yuan, L.: Improved bacterial foraging optimization algorithm with information communication mechanism for nurse scheduling. In: Huang, D.-S., Jo, K.-H., Hussain, A. (eds.) ICIC 2015. LNCS, vol. 9226, pp. 701–707. Springer, Heidelberg (2015)
4. Bermejo, E., Valsecchi, A., Damas, S., Cordon, O.: Bacterial foraging optimization for intensity-based medical image registration. In: Evolutionary Computation. IEEE (2015)
5. Tan, L., Lin, F., Wang, H.: Adaptive comprehensive learning bacterial foraging optimization and its application on vehicle routing problem with time windows. Neurocomputing **151**, 1208–1215 (2015)
6. Niu, B., Fan, Y., Xiao, H., Xue, B.: Bacterial foraging based approaches to portfolio optimization with liquidity risk. Neurocomputing **98**(18), 90–100 (2012)
7. Niu, B., Fan, Y., Zhao, P., Xue, B., Li, L., Chai, Y: A novel bacterial foraging optimizer with linear decreasing chemotaxis step. In: 2010 2nd International Workshop on Intelligent Systems and Applications (ISA), pp. 1–4. IEEE (2010)
8. Xu, X., Liu, Y.H., Wang, A.M., Wang, G.: A new adaptive bacterial swarm algorithm. In: International Conference on Natural Computation, pp. 991–995. IEEE (2012)
9. Niu, B., Bi, Y., Xie, T.: Structure-redesign-based bacterial foraging optimization for portfolio selection. In: Huang, D.-S., Han, K., Gromiha, M. (eds.) ICIC 2014. LNCS, vol. 8590, pp. 424–430. Springer, Heidelberg (2014)

Artificial Bee Colony Optimization for Yard Truck Scheduling and Storage Allocation Problem

Fangfang Zhang, Li Li[✉], Jing Liu, and Xianghua Chu

College of Management, Shenzhen University, Shenzhen, China
lli318@163.com, x.chu@szu.edu.cn

Abstract. The yard truck scheduling (YTS) and the storage allocation problem (SAP) are two significant sub-issues in container terminal operations. This paper takes them as a whole optimization problem (YTS-SAP) and analyzes the factor of different travel speeds of trucks based on different loads. The goal is to minimize the total time cost of the summation of the delay of requests and the travel time of yard trucks. Due to the simplicity and easy implementation of artificial bee colony (ABC), the algorithm is applied to address the issue. Computational experiment is employed to examine and analyze the problem solutions and the performance of ABC algorithm. Particle Swarm Optimization (PSO) and Genetic Algorithm (GA) are chosen as contrastive algorithms. From the results of the computational experiment, it is found that ABC algorithm can achieve better solution for the YTS-ASP problem.

Keywords: Yard truck scheduling · Storage allocation problem · Artificial bee colony · Travel speed

1 Introduction

Statistics from the Chinese industry information network shows that the port transportation has been developing rapidly in the past decade [1]. As seen in Fig. 1, the throughput of the nationwide port shows an increasing trend year by year. It has grown from 117.67 million tons in 2013 to 124.52 million tons in 2014. The container throughput has grown from 1.9 twenty-foot equivalent unit (TEU) in 2013 to 2.02 TEU in 2014, with an increase of 6.3 % over the previous year. The throughput of coastal port and inland port are 1.7 TEU and 0.2 TEU in 2013, respectively. For such a speedy development, it is necessary to increase the terminal throughput. Unfortunately, in many countries, there is an important pressure from the political and business sectors. It is obvious that port handling capacity becomes more and more important for terminal executives to contend with that of other terminals [2].

More and more researchers have been attracted to do research on the container terminal operation problems [3–5]. As two significant sub-problems in terminal operations, Yard truck scheduling (YTS) refers to the scheduling of yard trucks to convey the containers between the seaside and the yard side, while the (SAP) relates to the optimizing of the storage space. It has shown its remarkable research value in both science and project [3, 5].

D.-S. Huang and K.-H. Jo (Eds.): ICIC 2016, Part II, LNCS 9772, pp. 908–917, 2016.
DOI: 10.1007/978-3-319-42294-7_81

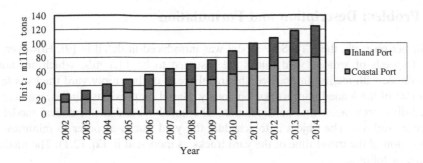

Fig. 1. The growth of the national port throughput from 2002 to 2014. (Color figure online)

Many researchers devote themselves to optimizing the YTS and SAP problems. Nishimura [6] et al. proposed single truck and multi-trucks scheduling models based on the number of work trucks, respectively. Taking the various processing times and ready times into consideration, Ng [7] et al. researched the issue of truck scheduling. For storage allocation problems, Kim [8] et al. considered how to distribute storage space for import containers. Lee [9] et al. considered both the charging and loading operations and developed a new heuristic method to find the scheduling scheme.

It is noticed that the YTS problems are highly correlated to the SAP problems. Bish [10] et al. was the first researcher that took these two issues into consideration as a whole. Lee [11] et al. also raised an integrated model for yard truck scheduling and storage allocation for import containers. Niu [12, 13] et al. used the particle swarm optimization algorithm and bacterial colony optimization algorithm to optimize the integrated YTS and SAP.

In 2005, a new optimization algorithm based on the intelligent foraging behavior of honey bee swarm was introduced by Karaboga, called ABC algorithm [14]. Since then, the ABC algorithm has been successfully applied in solving combinatorial optimization problems, such as job-shop scheduling problem [15], image segmentation [16], uninhabited combat air vehicle path planning [17], training neural networks [18], TSP problems [19], long-term prediction and feature selection. In this paper, we employ ABC algorithm to solve the yard truck scheduling and storage allocation problem.

In general, there are many container transportation related problems, like berth allocation problem, quay crane scheduling problem, yard truck scheduling problem, yard crane scheduling problem and storage allocation problem. This paper commits to solving the combination of yard truck scheduling and storage allocation problem using ABC.

The remaining sections of this paper are organized as follows. Section 2 provides the problem description and formulation. ABC algorithm is presented for problem solutions in Sect. 3. Some computational experiments are illustrated in Sect. 4. Finally, Sect. 5 provides concluding remarks of this paper.

2 Problem Description and Formulation

In former research, the YTS-SAP model was introduced in detail in [9]. However, the travel speeds of various yard trucks are assumed to be 11.1 m/s, which are not in accordance with reality. In practice, the travel speed of the empty yard trucks is faster than that of the loaded yard trucks, obviously. Based on previous studies, we take the speed-difference factor into consideration. The notations and mathematical model are given as follows. The object is to schedule the yard trucks in order to minimize the summation of the travel time of the yard trucks as modelled in Eq. (2.1). The model is shown as follows.

$$Minimize : Z = \alpha_1 \sum_{i \in J} d_i + \alpha_2 \left(\sum_{i \in J} t_i + \sum_{i,j \in J} s_{ij} y_{ij} \right) \tag{2.1}$$

$$\sum_{i \in J^-} x_{ik} = 1 \quad \forall k \in K \tag{2.2}$$

$$\sum_{k \in K} x_{ik} = 1 \quad \forall i \in J^- \tag{2.3}$$

$$\sum_{i \in J''} y_{ij} = 1 \quad \forall i \in J' \tag{2.4}$$

$$\sum_{i \in J'} y_{ij} = 1 \quad \forall j \in J'' \tag{2.5}$$

$$w_i \geq a_i \quad \forall i \in J' \cup J'' \tag{2.6}$$

$$d_i \geq w_i + t_i - b_i \quad \forall i \in J' \cup J'' \tag{2.7}$$

$$w_j + M(1 - y_{ij}) \geq w_i + t_i + S_{ij} \quad \forall i \in J' \ \text{ and } \ \forall j \in J'' \tag{2.8}$$

$$t_i = \tau_{o_i, e_i} \quad \forall \in J^+ \tag{2.9}$$

$$t_i = \sum_{k \in K} \tau_{o_i, \varsigma_k} x_{ik} \quad \forall i \in J^- \tag{2.10}$$

$$s_{ij} = \tau_{e_i, o_j} \quad \forall i \in J^+ \ \text{ and } \ j \in J \tag{2.11}$$

$$s_{ij} = \sum_{k \in K} \tau_{o_i, \varsigma_i} x_{ik} \quad \forall i \in J^- \text{ and } \forall j \in J \tag{2.12}$$

$$\tau_{o_i, e_i} = dis_{o_i, e_i} / V_{load} \quad \forall i \in J^+ \tag{2.13}$$

$$\tau_{o_i, \varsigma_k} = dis_{o_i, \varsigma_i} / V_{load} \quad \forall i \in J^- \ \text{ and } \ k \in SL \tag{2.14}$$

$$\tau_{e_i,o_j} = dis_{e_i,o_j}/V_{empty} \quad \forall i \in J^+ \ \text{ and } \ j \in J \tag{2.15}$$

$$\tau_{\varsigma_k,o_i} = dis_{\varsigma_k,o_i}/V_{empty} \quad \forall j \in J \ \text{ and } \ k \in SL \tag{2.16}$$

$$x_{ik}, y_{ij} \in \{0,1\}, \forall i \in J', \forall j \in J'' \ \text{ and } \ \forall k \in K \tag{2.17}$$

$$w_i \in R \ \ \forall i \in J' \cup J'' \tag{2.18}$$

$$t_i \in R \ \ \forall i \in J \tag{2.19}$$

$$S_{ij} \in R \ \ \forall i \in J \ \text{ and } \ \forall j \in J \tag{2.20}$$

$$d_i \geq 0 \ \ \forall i \in J' \cup J'' \tag{2.21}$$

The movement of a container from its origin to destination is defined as a job, denoted by i and j. Two types of requests are considered in this paper, loading requests and discharging requests. Let J^+ and J^- donate the set of loading requests and the set of discharging requests, respectively. V_{load} and V_{empty} donates the travel speed of loaded trucks and empty trucks, respectively. A soft time window $[a_i, b_i)$ for each job is given as a constant.

Constraints (2.2) insure that each unloading container can be located in a storage location. Constraints (2.3) mean each storage location can save one and only one unloading container. Constraints (2.4) and (2.5) guarantee each route is a one-to-one assignment. Constraints (2.6) imply that the requests can only be performed after the earliest given time. Constraints (2.7) compute the delay time of requests. Constraints (2.8) limit the relationship of the start time of two neighboring requests in the same route. Constraints (2.9) and (2.10) represent the processing time of job i. Constraints (2.11) and (2.12) obtain the setup time from the destination of job i to the origin of job j. Constraints (2.13)–(2.16) limit the speed of the trucks based on different load. Constraints (2.17)–(2.21) define the range for w_i, t_i, s_{ij} and d_i.

3 Artificial Bee Colony Optimization

In ABC algorithm, three types of bees are included, which are employed bees, onlookers and scouts. The process of looking for food sources in ABC is equal to the process of finding the optimum solution. Specifically, the activities for each role of bees are demonstrated as follows.

- Employed bees
  This type of bees go to the food source found by themselves previously. The mount of employed bees and food sources are often set to be equal.
- Onlooker bees
  Onlookers are placed on the food sources by using a probability based selection process. With the growing amount of the nectar, the probability value for the food source is preferred by onlooker's increases.

- Scout bees

  Scouts, which are mainly concerned with finding any kind of food source, are used to doing random search. The average number of scouts is 5 %–20 % of bees.

The Pseudo-code of the ABC algorithm is presented in Fig. 2.

```
ABC Algorithm
Initialize operation;
WHILE ((Iter < MaxCycle))
//Stage 1: Employed Bees
FOR(i = 1:(FoodNumber))
  Form a new food source;
  Calculate the fitness of the new food source;
  Greedy selection;
  END FOR
  Calculate the probability p;
//Stage 2:Onlooker Bees
FOR(i = 1:(FoodNumber))
  Parameter P is set randomly;
  Onlooker bees find food sources depending on P;
  Form a new food source;
  Evaluate the fitness of the new food  source;
  Greedy selection;
END FOR
//Stage 3:Scout Bees
IF(any employed bee turns to scout bee)
  Parameter p is set randomly;
  The scout bees find food sources depending on p
  ;
END IF
Record the best solution;
Iter = Iter + 1;
END WHILE
```

Fig. 2. Pseudo-code of the ABC algorithm

ABC was initially used to solve continuous problems, while the YTS-SAP problem is a typical discrete combinatorial optimization problem. It is necessary to make modifications according to YTS-SAP problem by encoding and decoding of the individuals. To transfer continuous variables to discrete scheduling ones effectively, the random key representation and the smallest position value (SPV) rule are employed in this study. The application of ABC to YTS-SAP is given in the following section.

4   Computational Experiment

In this part, we examine the problem proposed in Sect. 2 by illustrating a set of computational experiments. In the computation test, we assume that 20 containers should be operated. The travel speed of empty yard trucks is assumed to be 11.1 m/s and the weight parameters $\alpha 1$ and $\alpha 2$ are 0.6, 0.4, respectively [9]. But the travel speed of loaded yard trucks is assumed to be 5.5 m/s. Each instance is run for 500 iterations and 20 replications for each instance to collect the results for statistical analysis.

A set of computational experiments are conducted to examine the problems we proposed. To show the optimization performance and effectiveness of YTS-SAP problem, three representative instances are selected, as shown in Table 1. We can see that the solution is feasible, given arbitrary subset of jobs, there is always a feasible assignment of jobs to yard trucks. It is obvious that the routes of trucks are one-to-one assignment.

Table 1. The scheduling result of three test instances

Case	Job	Truck	Scheduling	Value
1	10	1	Route: $6 \rightarrow 9 \rightarrow 10 \rightarrow 3 \rightarrow 7 \rightarrow 2 \rightarrow 8 \rightarrow 5 \rightarrow 1 \rightarrow 4$	368.2
2	8	2	Route1: $6 \rightarrow 3 \rightarrow 5$ Route2: $7 \rightarrow 2 \rightarrow 8 \rightarrow 1 \rightarrow 4$	243.5
3	7	3	Route1: $1 \rightarrow 4$ Route2: $6 \rightarrow 3 \rightarrow 5$ Route3: $7 \rightarrow 2$	224.1

To verify the optimization performance of ABC algorithm on YTS-SAP problem, five representative instances with different scales are selected as test problems and the PSO, GA algorithms are chosen as comparable algorithms, as shown in Table 2. The amount of job and truck are denoted by n, m, respectively. From the mean values of the objectives, we can observe that the ABC algorithm performs better than the PSO and GA algorithms. This is not meant to be exhaustive, but rather indicative of the different scale instances related to the problem analyzed in the current paper.

We take the instance (n = 25, m = 7) as an example. The convergence progress of the mean fitness values is shown in Fig. 3. It is obvious that the ABC algorithm performs better than PSO and GA algorithms.

With the same amount of travel jobs and the same kind of discharging and loading requests, Table 3 shows the change of the objective values, according to the number of trucks. In term of the objective values, they are changed from 379.3 (m = 1) to 308.7 (m = 2), with the reduction of 18.6 %. Similarly, they are changed from 308.7 (m = 2) to 291.2 (m = 3), with the reduction of 5.7 %.

It is consistent with our experience that increasing the number of trucks can reduce the cost of the total time. But with the increasing of the number of trucks, the reduction of the objective becomes more and more slow. It may be because of the delay of setup time caused by truck congestion.

Table 2. Performance of three algorithms on five test instances

Case	(n, m)	Algorithm	Mean
1	(10, 3)	ABC	291
		PSO	314
		GA	335
2	(15, 5)	ABC	424
		PSO	442
		GA	471
3	(25, 7)	ABC	1834
		PSO	1839
		GA	1842
4	(30, 9)	ABC	1967
		PSO	2197
		GA	2201
5	(35, 10)	ABC	2055
		PSO	2364
		GA	2016

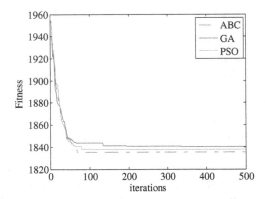

Fig. 3. Objective values at iterations. (Color figure online)

Figure 4 shows the objective values based on the different number of routes. The total time is closely related to the number of yard trucks distributed to work at the specified jobs. Namely, the cost time greatly depends on the ratio of the number of containers to the number of yard trucks. In this experiment, the reduction of the cost time will become less if the number of yard trucks is more than 3.

Table 3. The objective with different number of trucks

Case	Discharging	Loading	Job	Truck	Value
1	5	5	10	1	379.3
2	5	5	10	2	308.7
3	5	5	10	3	291.2
4	5	5	10	4	278.7
5	5	5	10	5	268.3
6	5	5	10	6	261.7
7	5	5	10	7	255.2
8	5	5	10	8	251.2
9	5	5	10	9	248.3
10	5	5	10	10	247

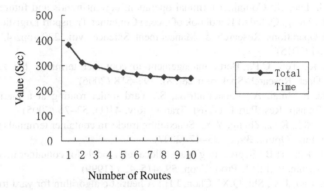

Fig. 4. Effect of number of routes

5   Conclusions and Future Work

Yard truck scheduling and storage allocation problem are two intractable issues in container terminal operation. This paper analyzes the yard truck scheduling and storage allocation problem as a whole. The objective is to minimize the total time cost of the summation of the delay of requests and the travel time of yard trucks.

Aiming to improve the efficiency of the container terminal operations, artificial bee colony algorithm is employed to address YTS-SAP due to its success in many other engineering problems.

To verify the optimization performance of ABC algorithm, we compare it with PSO and GA algorithms. Experimental results on five different scale instances show that ABC algorithm performs better than PSO and GA algorithms. Also we focus on how the number of trucks impacts the global objective. In future, other efficient algorithms need to probe into solving YTS-SAP problem by taking some other objectives into consideration, such as the uncertain factors, the cost of trucks and so on.

Acknowledgments. This work is partially supported by the National Natural Science Foundation of China (Grants nos. 71571120, 71271140, 71461027, 71471158 and 71501132), the Natural Science Foundation of Guangdong Province (Grant nos. 1614050000376, 2016A030310067, 2016A030310074 and 2015A030313556) and the Innovation and Development Fund Project of Shenzhen University (Grant no. 16XSCX04).

References

1. The Chinese Industry Information Network. http://www.chyxx.com/ (2015)
2. Stahlbock, R., VoB, S.: Operations research at container terminals: a literature update. OR Spectr. **30**, 1–52 (2008)
3. Zhu, M., Fan, X., Cheng, H., He, Q.: Modeling and simulation of automated container terminal operation. J. Comput. **5**(6), 951–957 (2010)
4. Kim, K.H., Lee, H.: Container terminal operation: current trends and future challenge. In: Lee, C.-Y., Meng, Q. (eds.) Handbook of Ocean Container Transport Logistics. International Series in Operations Research & Management Science, vol. 220, pp. 43–73. Springer, Switzerland (2015)
5. Moorthy, R., Teo, C.P.: Berth management in container terminal: the template design problem. Oper. Research-Spektrum **28**(28), 495–518 (2006)
6. Nishimura, E., Imai, A., Papadimitriou, S.: Yard trailer routing at a maritime container terminal. Transp. Res. Part E Logist. Transp. Rev. **41**(1), 53–76 (2005)
7. Ng, W.C., Mak, K.L., Zhang, Y.X.: Scheduling trucks in container terminals using a genetic algorithm. Eng. Optim. **39**(1), 33–47 (2007)
8. Kim, K.H., Kim, H.B.: Segregating space allocation models for container inventories in port container terminals. Int. J. Prod. Econ. **59**, 415–423 (1999)
9. Lee, D.H., Cao, J.X., Shi, Q.X., Chen, J.H.: A heuristic algorithm for yard truck scheduling and storage allocation problems. Transp. Res. Part E Logist. Transp. Rev. **45**(5), 810–820 (2009)
10. Bish, E.K., Leong, T.Y., Li, C.L., Ng, J.W.C., Simchi-Levi, D.: Analysis of a new vehicle scheduling and location problem. Naval Res. Logist. **48**, 363–385 (2001)
11. Lee, D.H., Cao, J.X., Shi, Q.: Integrated model for truck scheduling and storage allocation problem at container terminals. In: The Proceedings of the TRB Meeting, pp. 211–216 (2008)
12. Niu, B., Xie, T., Bi, Y., Liu, J.: Bacterial colony optimization for integrated yard truck scheduling and storage allocation problem. In: Huang, D.-S., Han, K., Gromiha, M. (eds.) ICIC 2014. L:NCS, vol. 8590, pp. 431–437. Springer, Switzerland (2014)
13. Niu, B., Xie, T., Duan, Q.Q., Tan, L.J.: Particle swarm optimization for integrated yard truck scheduling and storage allocation problem. In: IEEE Congress on Evolutionary Computation, pp. 634–639 (2014)
14. Karaboga, D.: An idea based on honey bee swarm for numerical optimization. Technical report TR06, Erciyes University, Engineering Faculty, Computer Engineering Department (2005)
15. Zhou, G., Wang, L., Xu, Y., Wang, S.: An effective artificial bee colony algorithm for multi-objective flexible job-shop scheduling problem. In: Huang, D.-S., Gan, Y., Gupta, P., Michael Gromiha, M. (eds.) ICIC 2011. LNCS, vol. 6839, pp. 1–8. Springer, Heidelberg (2012)
16. Horng, M.H.: Multilevel thresholding selection based on the artificial bee colony algorithm for image segmentation. Expert Syst. Appl. **38**(11), 13785–13791 (2011)

17. Xu, C., Duan, H., Liu, F.: Chaotic artificial bee colony approach to uninhabited combat air vehicle (UCAV) path planning. Aerosp. Sci. Technol. **14**(8), 535–541 (2010)
18. Bullinaria, J.A., AlYahya, K.: Artificial bee colony training of neural networks. In: Terrazas, G., Otero, F.E.B., Masegosa, A.D. (eds.) Nature Inspired Cooperative Strategies for Optimization (NICSO 2013). Studies in Computational Intelligence, vol. 512, pp. 191–201. Springer, Switzerland (2014)
19. Li, L., Cheng, Y., Tan, L.J., Niu, B.: A Discrete Artificial Bee Colony Algorithm for TSP Problem, pp. 566–573. Springer, Heidelberg (2012)

A Cooperative Structure-Redesigned-Based Bacterial Foraging Optimization with Guided and Stochastic Movements

Ben Niu[(✉)], Jing Liu, Fangfang Zhang, and Wenjie Yi

College of Management, Shenzhen University, Shenzhen 518060, China
drniuben@gmail.com

Abstract. The nested loop adopted in the original bacterial foraging optimization (BFO) is quite time-consuming and is the main reason for the complex computational process. Thus, in our previous work, an improved BFO with structure redesigned mechanism (SRBFO) is used to address this problem. Since the bacterial chemotaxis with stochastic direction in the original BFO has an adverse effect on the convergence rate, this paper proposes a new cooperative chemotactic movement strategy. In this cooperative strategy, some bacteria are selected to move toward a guided direction based on a predefined probability, while the other bacteria still swim to a stochastic direction as exhibited in the original BFO. By this strategy, all the bacteria alternatively use the guided movement and the stochastic movement to cooperatively balance global search and local search. The proposed improved algorithm is called Cooperative SRBFO (CSRBFO). A comparison of the CSRBFO with other BFOs has been made to demonstrate the superiority of the proposed algorithm.

Keywords: Cooperative bacterial foraging optimization · Movement strategy · Structure-redesigned

1 Introduction

Inspired from natural phenomena or biological characteristics, lots of bio-inspired optimization algorithms have been developed in recent decades such as the Particle Swarm Optimization (PSO) [1], the Ant Colony Optimization (ACO) [2] and the Bacterial Foraging Optimization (BFO) [3]. Among these heuristic algorithms, the BFO is one of the most well-known nature-inspired algorithms which was presented by Passino [3] in 2002 by getting inspiration from the social foraging behavior of *E.coli* bacteria.

After its acceptance, BFO has attracted increasing attention of scholars and has been applied to diverse real life problems like vehicle routing problem [4], voltage collapse detection [5] and so on. Besides, the enhancement of BFO algorithm is also an important research field on BFO. Lately, more real world problems have been investigated by using new variations of BFO. For example, Awadallah [6] used five versions of BFO to extract parameters of photovoltaic modules from nameplate data. Yi et al. [7] proposed a process fault detection approach: extended bacterial foraging optimization

© Springer International Publishing Switzerland 2016
D.-S. Huang and K.-H. Jo (Eds.): ICIC 2016, Part II, LNCS 9772, pp. 918–927, 2016.
DOI: 10.1007/978-3-319-42294-7_82

to optimize relative transformation matrix. The results proved the better accuracy of the proposed approach. Mohammadi et al. [8] developed an adaption scheme to enhance BFO's performance and used it to address optimizing the planning of passive power filters and distributed generations. We also have done numerous investigations on the improved BFOs. The adaptive chemotaxis step [9], the information communication mechanism [10, 11] and the redesigned algorithm structure [12] were proposed to incorporate into BFO to enhance the search capability. The primary experimental results proved the superiority of proposed algorithms.

In this paper, a new movement strategy with alterative guided and stochastic chemotactic movements is proposed to further keep the balance of exploration and exploitation. The movement of bacteria in the standard BFO is randomly determined, easily resulting in the low convergence speed. The combination of exploration using the stochastic movement method and exploitation using the guided movement method will facilitate bacteria to cooperatively find the global optima faster and more accurately. Besides, the aforementioned adaptive chemotaxis step and the redesigned algorithm structure are also used in our proposed algorithm. The proposed algorithm is named Cooperative SRBFO (CSRBFO).

The rest of the paper is organized as follows: the proposed cooperative structure-redesigned-Based bacterial foraging optimization is presented in Sect. 2. In Sect. 3, the original BFO and the proposed BFO have been applied to a series of benchmarks. Optimization results and analyses are also provided in Sect. 3, followed by conclusions in Sect. 4.

2　Cooperative Structure-Redesigned-Based BFO

2.1　A New Movement Strategy

The standard BFO consists of four primary steps, including chemotaxis, swarming, reproduction, and elimination & dispersal. These steps simulate the movement of an *E.coli* bacterium, the cell-to-cell signaling in the swarm and the death caused by sudden changes of the local environment.

In the standard BFO, the movements of bacteria in chemotaxis step are randomly decided. The lack of guided search may cause the problem of inefficient convergence rate. Moving toward to a stochastic direction will have some advantages, e.g. conserving high diversity, escaping from local minima. However, it may bring out some disadvantages, e.g. low convergence rate, low search accuracy. Inspired from the literature [13], we propose a new chemotactic movement strategy to address this issue.

In addition to moving toward a stochastic direction (stochastic movement) for exploration, bacteria in the new movement strategy may swim to a guided direction for exploitation, including the global best (*gbest* movement) or a determined target (target movement). The target is determined randomly and could be any bacterium with different fitness value such as the global best or even the worst bacterium. In other words, bacteria in CSRBFO can learn information from any other bacteria to decide the direction of movement instead of decide it stochastically. At each iteration, there are three movement methods to choose for each bacterium to update the location and each bacterium chooses its own method according to the random index S:

$$S = round(rand \times SS) \tag{1}$$

where SS is the swarm size of bacteria and $rand$ is a uniform distributed random number between 0 to 1. Thus the index S ranges from 0 to SS. If the index S of the ith bacterium belongs to the range from 0 to T_{lo} which is the lower bound of target range, the bacterium moves toward a stochastic direction Δ that is the same as it was in standard BFO. Moving toward the global best will occur when S ranges from T_{up} which is the upper bound of target range to SS. The determined target is the moving direction when S belongs to the target range from T_{lo} to T_{up}.

On the other hand, as the value of the chemotactic step size is static, it's difficult to keep the balance of global search and local search. Inspired from this, we proposed an adaptive chemotaxis step for global optimization in [9]. In the beginning of evolution process, the step size is set to a maximum value C_{max}. When the number of iteration increases, the value of step size will be smaller until it reaches the minimum value C_{min}. The updating equations for each bacterium's direction and location are given as follows:

$$D(i) = \begin{cases} \theta^{GBest}(j,k,l) - \theta^i(j,k,l) & \text{if } S \in (T_{up}, SS) \\ \theta^S(j,k,l) - \theta^i(j,k,l) & \text{if } S \in (T_{lo}, T_{up}) \\ \Delta(i) \Big/ \sqrt{\Delta^T(i)\Delta(i)} & \text{if } S \in (0, T_{lo}) \end{cases} \tag{2}$$

$$C(i) = C_{min} + (MaxFEs - ChemIter)/MaxFEs \times (C_{max} - C_{min}) \tag{3}$$

$$\theta^i(j+1,k,l) = \theta^i(j,k,l) + C(i)D(i) \tag{4}$$

where $\theta^i(j,k,l)$ means the position of the ith bacterium at the jth chemotactic, k^{th} reproductive and l^{th} elimination and dispersal step. $C(i)$ is the chemotactic step size and $D(i)$ is the moving direction for the ith bacterium. Δ represents a vector in the stochastic direction. *ChemIter* is the iteration counter while *MaxFEs* is the maximum number of iteration counters.

T_{lo} and T_{up} are calculated as round($\alpha_1 \times SS$) and round($\alpha_2 \times SS$) where α_1 and α_2 are used to control the ranges of stochastic movement, *gbest* movement and target movement. If users set α_1 to 0.1 and α_2 to 0.8, it means that 10 % of bacterial swarm can be randomly selected as uniform bacteria that use stochastic movement strategy and 20 % of bacteria can be chosen as bacteria moving toward the global best bacterium. The remaining 70 % of bacterial population can move toward the target bacteria. Specifically, the smaller the value of α_1, less exploration using stochastic movement and more exploitation using guided movement are considered. And the bigger the value of α_2, more bacteria are guided to the location of global best and less bacteria are guided to the target bacteria which means more exploitation considering best bacterium only and less exploitation considering any bacterium with different fitness value.

2.2   Redesign of Algorithm Structure

In the standard BFO, the chemotactic loop is nested inside the reproduction loop which is nested inside the elimination & dispersal loop once more. The complex structure of nested loop in standard BFO causes some problems, e.g., the standard BFO usually requests lots of time and memory consumption when handling problems with high dimension space. In order to address this issue, we proposed a Structure-Redesigned-Based Bacterial Foraging Optimization (SRBFO) in [12] for lower computational complexity and less memory consumption.

Table 1. The pseudo-code of BFO	Table 2. The pseudo-code of SRBFOs
Initialize parameters and the location of bacteria	Initialize parameters and the location of bacteria
For ($l = 1: N_{ed}$):	While $ChemIter \leq MaxFEs$
For ($k = 1: N_{re}$):	Tumble;
For ($j = 1: N_c$): chemotactic loop	Swimming;
Tumble;	Update position (using equations (2)- (4) in CSRBFO) and fitness;
Swimming;	$ChemIter = ChemIter + 1$;
Update position and fitness	If Mod ($ChemIter, Fre$)=0
End	Reproduction step;
reproduction loop	End
End	If Mod ($ChemIter, Fed$=0)
elimination&dispersal loop	Elimination&dispersal event;
End	End
	End

As the reproduction and elimination & dispersal steps take place after the chemotactic behavior is finished, iteration counters can be used to take the place of the nested loop. When a number of iterations have been completed, the reproduction and elimination & dispersal event can be triggered by the iteration counter $ChemIter$. For instance, when the total number of bacteria $SS = 10$, the number of chemotactic steps

$N_c = 10$, the number of reproduction steps $N_{re} = 4$ and the number of elimination & dispersal events $N_{ed} = 3$, the maximum number of iteration counters in total *MaxFEs* can be calculated as:

$$MaxFEs = SS \times N_c \times N_{re} \times N_{ed} \qquad (5)$$

to get the value of *MaxFEs* 1200. The counter for N_{re} can be replaced by an event every 240 iterations (*Fre*), and the counter for N_{ed} also can be replaced by an event every 300 iterations (*Fed*). The pseudo-codes of BFO and SRBFO/CSRBFO are presented in Tables 1 and 2.

3   Experiments and Analyses

3.1   Benchmark Functions and Experimental Parameters

To demonstrate the superiority of the proposed algorithm CSRBFO, CSRBFO is used to optimize eight well-known benchmark functions that are shown in Table 3, including Sphere, Quartic, SumPowers, Rosenbrock, Ackley, Griewank, Rastrigin and Schwefel2.22. These selected benchmark functions are classified as two categories: unimodal and multimodal functions. The unimodal functions (Sphere, Quartic, SumPowers, Rosenbrock) have a single optimal solution. On the contrary, the multimodal functions (Ackley, Griewank, Rastrigin and Schwefel2.22) have two or more local optima, resulting in the high possibility of being trapped into local optima and the difficulty in searching global optima.

As the proposed algorithm is a variant of the standard BFO, we compared the proposed algorithm with the standard BFO [3] and the structure-redesigned-based BFO (SRBFO) [12]. The three competitive algorithms are set to have similar experimental settings. For example, for each benchmark functions these algorithms are conducted independently for approximately 15 times with the search space dimension 30. The swimming length N_s is 10 and the probability of elimination & dispersal P_{ed} is set to 0.2. More parameters setting of BFO, SRBFO and SCRBFO are presented as follows:

- In BFO, SRBFO and CSRBFO, MaxFEs = 80000,
- In BFO, $N_c = 100$; $N_{re} = 4$; $N_{ed} = 2$; $C = 0.15$,
- In SRBFO and CSRBFO, $F_{re} = 16000$; $F_{ed} = 26667$,
- IN CSRBFO, $C_{max} = 0.15$; $C_{min} = 0.01$; $\alpha_1 = 0.2$; $\alpha_2 = 0.8$

3.2   Experimental Results and Analyses

MATLAB R2011a is used for coding these optimization algorithms. BFO, SRBFO and SRBFO are executed to get optimal numerical results for four unimodal functions and four multimodal functions. The mean and the standard deviation of optimal solutions obtained in 20 run times are presented in Table 4. The computational time used by the three algorithms is also given in Table 4. The best of all the numerical values obtained by the three competitive algorithms are emphasized by using a bold type. Figures 1, 2, 3, 4, 5, 6, 7 and 8 are convergence graphs of the average values obtained by all the competitive algorithms for the eight test functions.

From these Table and Figs we can conclude the following findings:

- CSRBFO algorithm gets the optimal solutions among the three competitive algorithms for all the 8 algorithms. It means that CSRBFO outperforms other two algorithms in terms of the search accuracy no matter the categories of the functions.
- CSRBFO algorithm acquires far better standard deviation values than BFO and SRBFO in these all cases. It means that the search stability of CSRBFO is usually better as compared to BFO and SRBFO.
- In terms of the computational time, CSRBFO uses the least time on SumPowers and Griewank and SRBFO have the best performance when dealing with the other 6 functions. Both the performance of CSRBFO and SRBFO is much better than that of BFO, which means the redesign of algorithm structure is effective in reducing computational time.
- Although SRBFO successfully lower the computational complexity and consumed time, SRBFO cannot always get better optima than BFO can get. This problem can be successfully solved by CSRBFO which maintains the low computational complexity of SRBFO but possesses a much better search accuracy.

Table 3. The details of benchmark functions

Function name	Formula	Search range
Sphere	$f(x) = \sum_{i=1}^{n} x_i^2$	$[-100,100]^n$
Quartic	$f(x) = \sum_{i=1}^{n} ix_i^4$	$[-1.28,1.28]^n$
SumPowers	$f(x) = \sum_{i=1}^{n} |x_i|^{i+1}$	$[-10,10]^n$
Rosenbrock	$f(x) = \sum_{i=1}^{n} 100 \times (x_{i+1} - x_i^2)^2 + (1 - x_i)^2$	$[-100,100]^n$
Ackley	$f(x) = -20\exp(-0.2 \times \sqrt{\frac{1}{30}\sum_{i=1}^{n} x_i^2}) - \exp(\frac{1}{30}\sum_{i=1}^{n} \cos 2\pi x_i) + 20 + e$	$[-32.32]^n$
Griewank	$f(x) = \frac{1}{4000}\sum_{i=1}^{n} x_i^2 - \prod_{i=1}^{n} \cos(\frac{x_i}{\sqrt{i}}) + 1$	$[-600,600]^n$
Rastrigrin	$f(x) = \sum_{i=1}^{n} (x_i^2 - 10\cos(2\pi x_i) + 10)^2$	$[-5.12,5.12]^n$
Schwefel2.22	$f(x) = \sum_{i=1}^{n} |x_i| + \prod_{i=1}^{n} |x_i|$	$[-10,10]^n$

Table 4. Optimal results of benchmark functions obtained by algorithms

Algorithm		Sphere	Quartic
BFO	mean	2.161233e+004	1.830501e+006
	SD	4.121535e+003	8.000643e+005
	time	1.688580e+002	2.740903e+002
SRBFO	mean	3.455068e+002	1.687083e+005
	SD	1.554593e+002	7.721484e+004
	time	**9.940292e+000**	**1.140782e+002**
CSRBFO	mean	**5.00E−04**	**1.51E−05**
	SD	**1.20E−04**	**7.83E−06**
	time	2.707242e+001	1.253987e+002
		SumPowers	Rosenbrock
BFO	mean	1.536610e+002	9.362519e+006
	SD	2.064995e+002	4.326234e+006
	time	1.659585e+002	1.791519e+002
SRBFO	SD	2.064995e+002	4.326234e+006
	time	1.659585e+002	1.791519e+002
	mean	1.776366e+006	9.857264e+003
CSRBFO	mean	**5.61E−06**	**3.903938e+001**
	SD	**4.34E−06**	**2.387991e+001**
	time	**1.141269e+002**	3.990767e+001
		Ackley	Griewank
BFO	mean	1.964264e+001	4.756842e+002
	SD	1.54E−01	7.031533e+001
	time	9.490752e+001	2.864881e+002
SRBFO	mean	1.977602e+001	3.895342e+002
	SD	2.25E−01	4.157214e+001
	time	**2.030747e+001**	1.187595e+002
CSRBFO	mean	**7.648295e+000**	**2.340888e+000**
	SD	**1.201768e+000**	**1.079698e+000**
	time	4.284316e+001	**1.163687e+002**
		Rastrigrin	Schwefel2.22
BFO	mean	2.243968e+002	8.458650e+001
	SD	1.033303e+001	1.040385e+001
	time	8.805903e+001	9.168354e+001
SRBFO	mean	3.563115e+002	9.530744e+001
	SD	2.002577e+001	8.057249e+000
	time	**1.474507e+001**	**1.555536e+001**
CSRBFO	mean	**1.134836e+001**	**2.879101e+000**
	SD	**4.094528e+000**	**1.209057e+000**
	time	3.074022e+001	3.427830e+001

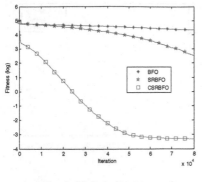

Fig. 1. Sphere function

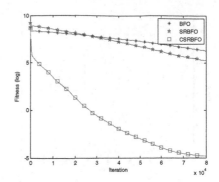

Fig. 2. Quartic function

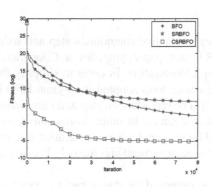

Fig. 3. SumPower function

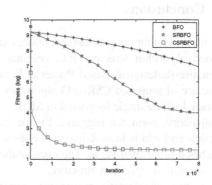

Fig. 4. Rosenbrock function

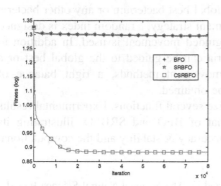

Fig. 5. Ackley function

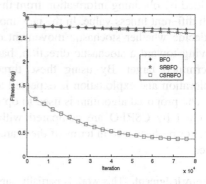

Fig. 6. Griewank function

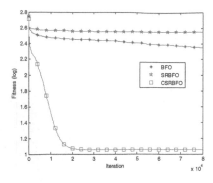 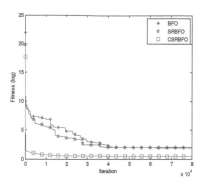

Fig. 7. Rastrigin function   **Fig. 8.** Schwefel2.22 function

4   Conclusions

By incorporating cooperative movement strategy, adaptive chemotaxis step and redesigned algorithm structure into original BFO, this paper proposes a Cooperative Structure-Redesigned-Based Bacterial Foraging Optimization. In order to simplify the structure of proposed CSRBFO algorithm, the nested loop adopted in original BFO is replaced with a single loop used in SRBFO [12]. The reproduction step and elimination & dispersal event are triggered by the iteration counter. In other words, the reproduction and elimination &dispersal events will be triggered when the iteration counter reaches a predefined value. The computational time is reduced significantly because of the simplified algorithm structure.

A new chemotactic movement strategy is proposed in this paper to speed up convergence rate and enhance search accuracy. Instead of the stochastic chemotactic movement in aforementioned BFO, bacterial moving directions in CSRBFO are decided by obtaining information from the global best bacterium or any other bacteria with different fitness value. In the new movement strategy, a random index is designed to decide whether stochastic movement or guided movement is used. In addition to moving toward a stochastic direction, bacteria can be guided to the global best or a determined target. By using these three movement methods, a right balance of exploration and exploitation is expected to be obtained.

The proposed algorithm is used to optimize several functions. Experimental results obtained by CSRFO are compared with that of BFO and SRBFO, illustrating its superior performance in terms of the search accuracy & stability and the computational time.

Acknowledgment. This work is partially supported by The National Natural Science Foundation of China (Grants nos. 71571120, 71271140, 71461027, 71471158) and the Natural Science Foundation of Guangdong Province (Grant no. 2016A030310074).

References

1. Kennedy, J., Eberhart, R.C.: Particle swarm optimization. In: IEEE International Conference on Neural Networks, pp. 1942–1948, Piscataway (1995)
2. Dorigo, M., Gambardella, L.M.: Ant colony system: a cooperative learning approach to the traveling salesman problem. IEEE Trans. Evol. Comput. 1(1), 53–66 (1997)
3. Passino, K.M.: Biomimicry of bacterial foraging for distributed optimization and control. IEEE Control Mech. Mag. 22(3), 52–67 (2002)
4. Niu, B., Wang, H., Tan, L.-J., Li, L., Wang, J.-W.: Vehicle routing problem with time windows based on adaptive bacterial foraging optimization. In: Huang, D.-S., Ma, J., Jo, K.-H., Gromiha, M. (eds.) ICIC 2012. LNCS, vol. 7390, pp. 672–679. Springer, Heidelberg (2012)
5. Talwar, S., Gray, M., Morsi, W.G.: Application of directed bacterial foraging optimization for voltage collapse detection. In: IEEE on Electrical Power and Energy Conference, EPEC 2015, pp. 402–407, London (2015)
6. Awadallah, M.A.: Variations of the bacterial foraging algorithm for the extraction of PV module parameters from nameplate data. Energy Convers. Manag. 113, 312–320 (2016)
7. Yi, J., Yi, J., Huang, D., Fu, S., He, H., Li, T.: Optimized relative transformation matrix using bacterial foraging algorithm for process fault detection. IEEE Trans. Industr. Electron. 63(4), 2595–2605 (2016)
8. Mohammadi, M., Rozbahani, A.M., Montazeri, M.: Multi criteria simultaneous planning of passive filters and distributed generation simultaneously in distribution system considering nonlinear loads with adaptive bacterial foraging optimization approach. Int. J. Electr. Power Energy Syst. 79, 253–262 (2016)
9. Niu, B., Wang, H., Tan, L.J., Li, L.: Improved BFO with adaptive chemotaxis step for global optimization. In: The 7th International Conference on Computational Intelligence and Security, CIS 2011, pp. 76–80, Sanya (2011)
10. Gu, Q., Yin, K., Niu, B., Xing, K., Tan, L., Li, L.: BFO with information communicational system based on different topologies structure. In: Huang, D.-S., Jo, K.-H., Zhou, Y.-Q., Han, K. (eds.) ICIC 2013. LNCS, vol. 7996, pp. 633–640. Springer, Heidelberg (2013)
11. Niu, B., Liu, J., Bi, Y., Xie, T., Tan, L.J.: Improved bacterial foraging optimization algorithm with information communication mechanism. In: The 10th International Conference on Computational Intelligence and Security, CIS 2014, pp. 47–51, Kunming (2014)
12. Niu, B., Bi, Y., Xie, T.: Structure-redesign-based bacterial foraging optimization for portfolio selection. In: 10th International Conference on Intelligent Computing Theories and Applications, ICIC 2014, pp. 431–437, Taiyuan (2014)
13. Ngo, T.T., Sadollah, A., Kim, J.H.: A cooperative particle swarm optimizer with stochastic movements for computationally expensive numerical optimization problems. J. Comput. Sci. 13, 68–82 (2016)

Author Index

Printed in the United States
By Bookmasters

Printed in the United States
By Bookmasters